Electrochemical Engineering

Electrochemical Engineering

Thomas F. Fuller
Georgia Institute of Technology, US

John N. Harb
Brigham Young University, US

This edition first published 2018
© 2018 John Wiley & Sons, Inc

All rights reserved. No part of this publication may be reproduced, stored in a retrieval system, or transmitted, in any form or by any means, electronic, mechanical, photocopying, recording or otherwise, except as permitted by law. Advice on how to obtain permission to reuse material from this title is available at http://www.wiley.com/go/permissions.

The right of Thomas F Fuller and John N Harb to be identified as the author(s) of this work has been asserted in accordance with law.

Registered Office
John Wiley & Sons, Inc., 111 River Street, Hoboken, NJ 07030, USA

Editorial Office
111 River Street, Hoboken, NJ 07030, USA

For details of our global editorial offices, customer services, and more information about Wiley products visit us at www.wiley.com.

Wiley also publishes its books in a variety of electronic formats and by print-on-demand. Some content that appears in standard print versions of this book may not be available in other formats.

Limit of Liability/Disclaimer of Warranty
In view of ongoing research, equipment modifications, changes in governmental regulations, and the constant flow of information relating to the use of experimental reagents, equipment, and devices, the reader is urged to review and evaluate the information provided in the package insert or instructions for each chemical, piece of equipment, reagent, or device for, among other things, any changes in the instructions or indication of usage and for added warnings and precautions. While the publisher and authors have used their best efforts in preparing this work, they make no representations or warranties with respect to the accuracy or completeness of the contents of this work and specifically disclaim all warranties, including without limitation any implied warranties of merchantability or fitness for a particular purpose. No warranty may be created or extended by sales representatives, written sales materials or promotional statements for this work. The fact that an organization, website, or product is referred to in this work as a citation and/or potential source of further information does not mean that the publisher and authors endorse the information or services the organization, website, or product may provide or recommendations it may make. This work is sold with the understanding that the publisher is not engaged in rendering professional services. The advice and strategies contained herein may not be suitable for your situation. You should consult with a specialist where appropriate. Further, readers should be aware that websites listed in this work may have changed or disappeared between when this work was written and when it is read. Neither the publisher nor authors shall be liable for any loss of profit or any other commercial damages, including but not limited to special, incidental, consequential, or other damages.

Library of Congress Cataloging-in-Publication Data

Names: Fuller, Thomas Francis, author. | Harb, John Naim, 1959- author.
Title: Electrochemical engineering / Thomas F. Fuller & John N. Harb.
Description: First edition. | Hoboken, NJ, USA : Wiley, 2018. | Includes
  index. | Identifiers: LCCN 2017044994 (print) | LCCN 2017047266 (ebook) |
ISBN 9781119446583 (pdf) | ISBN 9781119446590 (epub) | ISBN 9781119004257
  (hardback)
Subjects: LCSH: Electrochemistry, Industrial. | BISAC: SCIENCE / Chemistry /
  Physical & Theoretical.
Classification: LCC TP255 (ebook) | LCC TP255 .F85 2018 (print) | DDC
  660/.297–dc23
LC record available at https://lccn.loc.gov/2017044994

Cover image: Courtesy of Thomas F. Fuller; © crisserbug / Getty Images
Cover design by Wiley

Set in 10.25/12 pt TimesLTStd-Roman by Thomson Digital, Noida, India

V10007454_010919

Contents

Preface ix
List of Symbols xi
About the Companion Website xv

1. Introduction and Basic Principles 1
Charles W. Tobias

   1.1 Electrochemical Cells 1
   1.2 Characteristics of Electrochemical Reactions 2
   1.3 Importance of Electrochemical Systems 4
   1.4 Scientific Units, Constants, Conventions 5
   1.5 Faraday's Law 6
   1.6 Faradaic Efficiency 8
   1.7 Current Density 9
   1.8 Potential and Ohm's Law 9
   1.9 Electrochemical Systems: Example 10
        Closure 13
        Further Reading 13
        Problems 13

2. Cell Potential and Thermodynamics 15
Wendell Mitchell Latimer

   2.1 Electrochemical Reactions 15
   2.2 Cell Potential 15
   2.3 Expression for Cell Potential 17
   2.4 Standard Potentials 18
   2.5 Effect of Temperature on Standard Potential 21
   2.6 Simplified Activity Correction 22
   2.7 Use of the Cell Potential 24
   2.8 Equilibrium Constants 25
   2.9 Pourbaix Diagrams 25
   2.10 Cells with a Liquid Junction 27
   2.11 Reference Electrodes 27
   2.12 Equilibrium at Electrode Interface 30
   2.13 Potential in Solution Due to Charge: Debye–Hückel Theory 31
   2.14 Activities and Activity Coefficients 33
   2.15 Estimation of Activity Coefficients 35
        Closure 36
        Further Reading 36
        Problems 36

3. Electrochemical Kinetics 41
Alexander Naumovich Frumkin

   3.1 Double Layer 41
   3.2 Impact of Potential on Reaction Rate 42
   3.3 Use of the Butler–Volmer Kinetic Expression 46
   3.4 Reaction Fundamentals 49
   3.5 Simplified Forms of the Butler–Volmer Equation 50
   3.6 Direct Fitting of the Butler–Volmer Equation 52
   3.7 The Influence of Mass Transfer on the Reaction Rate 54
   3.8 Use of Kinetic Expressions in Full Cells 55
   3.9 Current Efficiency 58
        Closure 58
        Further Reading 59
        Problems 59

4. Transport 63
Carl Wagner

   4.1 Fick's Law 63
   4.2 Nernst–Planck Equation 63
   4.3 Conservation of Material 65
   4.4 Transference Numbers, Mobilities, and Migration 71

vi   Contents

    4.5 Convective Mass Transfer   75
    4.6 Concentration Overpotential   79
    4.7 Current Distribution   81
    4.8 Membrane Transport   86
         Closure   87
         Further Reading   88
         Problems   88

5. Electrode Structures and Configurations   93

John Newman

    5.1 Mathematical Description of Porous Electrodes   94
    5.2 Characterization of Porous Electrodes   96
    5.3 Impact of Porous Electrode on Transport   97
    5.4 Current Distributions in Porous Electrodes   98
    5.5 The Gas–Liquid Interface in Porous Electrodes   102
    5.6 Three-Phase Electrodes   103
    5.7 Electrodes with Flow   105
         Closure   108
         Further Reading   108
         Problems   108

6. Electroanalytical Techniques and Analysis of Electrochemical Systems   113

Jaroslav Heyrovský

    6.1 Electrochemical Cells, Instrumentation, and Some Practical Issues   113
    6.2 Overview   115
    6.3 Step Change in Potential or Current for a Semi-Infinite Planar Electrode in a Stagnant Electrolyte   116
    6.4 Electrode Kinetics and Double-Layer Charging   118
    6.5 Cyclic Voltammetry   122
    6.6 Stripping Analyses   127
    6.7 Electrochemical Impedance   129
    6.8 Rotating Disk Electrodes   136
    6.9 iR Compensation   139
    6.10 Microelectrodes   141
         Closure   145
         Further Reading   145
         Problems   145

7. Battery Fundamentals   151

John B. Goodenough

    7.1 Components of a Cell   151
    7.2 Classification of Batteries and Cell Chemistries   152
    7.3 Theoretical Capacity and State of Charge   156
    7.4 Cell Characteristics and Electrochemical Performance   158
    7.5 Ragone Plots   163
    7.6 Heat Generation   164
    7.7 Efficiency of Secondary Cells   166
    7.8 Charge Retention and Self-Discharge   167
    7.9 Capacity Fade in Secondary Cells   168
         Closure   169
         Further Reading   169
         Problems   169

8. Battery Applications: Cell and Battery Pack Design   175

Esther Sans Takeuchi

    8.1 Introduction to Battery Design   175
    8.2 Battery Layout Using a Specific Cell Design   176
    8.3 Scaling of Cells to Adjust Capacity   178
    8.4 Electrode and Cell Design to Achieve Rate Capability   181
    8.5 Cell Construction   183
    8.6 Charging of Batteries   184
    8.7 Use of Resistance to Characterize Battery Peformance   185
    8.8 Battery Management   186
    8.9 Thermal Management Systems   188
    8.10 Mechanical Considerations   190
         Closure   191
         Further Reading   191
         Problems   191

9. Fuel-Cell Fundamentals   195

Supramaniam Srinivasan

    9.1 Introduction   195
    9.2 Types of Fuel Cells   197
    9.3 Current–Voltage Characteristics and Polarizations   198
    9.4 Effect of Operating Conditions and Maximum Power   202
    9.5 Electrode Structure   205
    9.6 Proton-Exchange Membrane (PEM) Fuel Cells   206
    9.7 Solid Oxide Fuel Cells   211
         Closure   215
         Further Reading   215
         Problems   216

Contents   vii

10. Fuel-Cell Stack and System Design   223
Francis Thomas Bacon

- 10.1 Introduction and Overview of Systems Analysis   223
- 10.2 Basic Stack Design Concepts   226
- 10.3 Cell Stack Configurations   228
- 10.4 Basic Construction and Components   229
- 10.5 Utilization of Oxidant and Fuel   231
- 10.6 Flow-Field Design   235
- 10.7 Water and Thermal Management   238
- 10.8 Structural–Mechanical Considerations   241
- 10.9 Case Study   245
  - Closure   247
  - Further Reading   247
  - Problems   247

11. Electrochemical Double-Layer Capacitors   251
Brian Evans Conway

- 11.1 Capacitor Introduction   251
- 11.2 Electrical Double-Layer Capacitance   253
- 11.3 Current–Voltage Relationship for Capacitors   259
- 11.4 Porous EDLC Electrodes   261
- 11.5 Impedance Analysis of EDLCs   263
- 11.6 Full Cell EDLC Analysis   266
- 11.7 Power and Energy Capabilities   267
- 11.8 Cell Design, Practical Operation, and Electrochemical Capacitor Performance   269
- 11.9 Pseudo-Capacitance   271
  - Closure   273
  - Further Reading   273
  - Problems   273

12. Energy Storage and Conversion for Hybrid and Electrical Vehicles   277
Ferdinand Porsche

- 12.1 Why Electric and Hybrid-Electric Systems?   277
- 12.2 Driving Schedules and Power Demand in Vehicles   279
- 12.3 Regenerative Braking   281
- 12.4 Battery Electrical Vehicle   282
- 12.5 Hybrid Vehicle Architectures   284
- 12.6 Start–Stop Hybrid   285
- 12.7 Batteries for Full-Hybrid Electric Vehicles   287
- 12.8 Fuel-Cell Hybrid Systems for Vehicles   291
  - Closure   293
  - Further Reading   294
  - Problems   294
  - Appendix: Primer on Vehicle Dynamics   295

13. Electrodeposition   299
Richard C. Alkire

- 13.1 Overview   299
- 13.2 Faraday's Law and Deposit Thickness   300
- 13.3 Electrodeposition Fundamentals   300
- 13.4 Formation of Stable Nuclei   303
- 13.5 Nucleation Rates   305
- 13.6 Growth of Nuclei   308
- 13.7 Deposit Morphology   310
- 13.8 Additives   311
- 13.9 Impact of Current Distribution   312
- 13.10 Impact of Side Reactions   314
- 13.11 Resistive Substrates   316
  - Closure   319
  - Further Reading   319
  - Problems   319

14. Industrial Electrolysis, Electrochemical Reactors, and Redox-Flow Batteries   323
Fumio Hine

- 14.1 Overview of Industrial Electrolysis   323
- 14.2 Performance Measures   324
- 14.3 Voltage Losses and the Polarization Curve   328
- 14.4 Design of Electrochemical Reactors for Industrial Applications   331
- 14.5 Examples of Industrial Electrolytic Processes   337
- 14.6 Thermal Management and Cell Operation   341
- 14.7 Electrolytic Processes for a Sustainable Future   343
- 14.8 Redox-Flow Batteries   348
  - Closure   350
  - Further Reading   350
  - Problems   350

15. Semiconductor Electrodes and Photoelectrochemical Cells   355
Heinz Gerischer

- 15.1 Semiconductor Basics   355
- 15.2 Energy Scales   358
- 15.3 Semiconductor–Electrolyte Interface   360
- 15.4 Current Flow in the Dark   363
- 15.5 Light Absorption   366

15.6 Photoelectrochemical Effects   368
15.7 Open-Circuit Voltage for Illuminated Electrodes   369
15.8 Photo-Electrochemical Cells   370
Closure   375
Further Reading   375
Problems   375

16. Corrosion   379

Ulick Richardson Evans

16.1 Corrosion Fundamentals   379
16.2 Thermodynamics of Corrosion Systems   380
16.3 Corrosion Rate for Uniform Corrosion   383
16.4 Localized Corrosion   390
16.5 Corrosion Protection   394

Closure   399
Further Reading   399
Problems   399

Appendix A: Electrochemical Reactions and Standard Potentials   403

Appendix B: Fundamental Constants   404

Appendix C: Thermodynamic Data   405

Appendix D: Mechanics of Materials   408

Index   413

Preface

Given the existing books on electrochemistry and electrochemical engineering, what do we hope to accomplish with this text? The answer is twofold. First, it is our goal to provide enough of the fundamentals to enable students to approach interesting problems, but not so much that new students are overwhelmed. In doing this, we have attempted to write with the student in mind, and choices regarding content, examples, and applications have been made with the aim of helping students understand and experience the important and exciting discipline of electrochemical engineering. The text has a large number of worked out examples, includes homework problems for each chapter, and uses consistent nomenclature throughout in order to facilitate student learning. Our second objective is to integrate key applications of electrochemical engineering with the relevant fundamentals in a single book. We have found applications to be of great interest to students. Importantly, applications provide students with the opportunity to practice in a meaningful way the fundamentals that they have learned. We have made a deliberate effort to tie the investigation of the applications directly back to the fundamental principles.

The book is intended for senior-level engineering students and entering graduate students in engineering and the sciences. No knowledge of electrochemistry is assumed, but an undergraduate training in thermodynamics, transport, and freshman chemistry is needed. A basic knowledge of differential equations is also recommended. Historically, electrochemical engineering is closely associated with chemical engineering and chemistry. However, the lines between disciplines are becoming increasingly blurred, and today we find mechanical engineers, material science engineers, and physicists engaged in electrochemical engineering. We have attempted to broaden our view of electrochemical engineering both to appeal to a larger audience and to expand the exposure of students to new areas and opportunities.

The text is divided into two parts. The first six chapters cover essential topics in a familiar order: an introduction, thermodynamics, kinetics, and mass transport. Also included in these fundamentals is a chapter on electrode structures and one on electroanalytical techniques. The remaining chapters cover applications of electrochemical engineering. It is assumed that the student will have mastered the fundamental chapters before starting on the applications. The application-focused chapters comprise batteries, fuel cells, double-layer capacitors, energy storage systems for vehicles, electrodeposition, industrial electrolysis, semiconductor electrodes, and corrosion. For the most part, the instructor can pick and choose which application chapters to cover; and these applications can be presented in any order. Although not exhaustive, the applications selected represent important applications of electrochemical engineering.

Each chapter begins with a short biography of one of the founders in electrochemistry and electrochemical engineering from the twentieth century. Even though not essential to the technical material presented, writing these short sketches was often personal and deeply satisfying. In general, our focus was on the accomplishments and impact of these remarkable individuals. The selection of whom to profile was also personal and, without question, subjective. Other than space limitations, there can be no justification for omitting pioneers such as Alan Bard and Norbert Ibl.

The authors have used this material to teach introductory electrochemical engineering courses at Georgia Institute of Technology and Brigham Young University. Many of the chapters were drastically revised after the first teaching. Again, our objective was to create a textbook suitable for classroom teaching. Many of the problems presented at the end of the chapters have been honed through homework assignments and on examinations. In fact, these problems and our lectures guided what content to include in the text—we sought to eliminate from the book content not addressed in class or not suitable for practice in the form of problems. Our experience has been that in a one semester course only a couple of the application chapters can be completed in addition to the first six fundamental chapters. We have purposely included references only sparingly in the text itself. While important references for each topic are found at the end of each chapter, we made no attempt to meticulously ascribe credit—our thinking was a highly annotated text is not important to beginning students and actually presents an additional barrier to learning.

The journey of writing this book has been immensely rewarding and not surprisingly a bit longer than anticipated. We have so many colleagues, students, friends, and family members to thank. Their support and encouragement was beyond heartening.

2017

THOMAS F. FULLER
JOHN N. HARB

List of Symbols

a	specific interfacial area, m^{-1}	h	thickness, gap in electrolysis cells, m
a_i	activity of species i, unitless	h	Planck's constant, J-s
A	area, m^2	H	Enthalpy, J mol^{-1}
A_c	cell or electrode area, m^2	$i, \mathbf{i}$	current density, A m^{-2}
Ar	Archimedes number, unitless	I	current, A
b	Tafel slope, V per decade	i_o	exchange current density, A·m^{-2}
B	Debye constant for solvent, (kg/mol)^{½} m^{-1}	I''	photon flux, s^{-1}·m^{-2}
c	total concentration or salt concentration, mol m^{-3}	$\mathbf{J}_i$	molar flux of species i, mol m^{-2} s^{-1}
c	speed of light, 2.99792×10^8 m s^{-1}	k_a, k_c	anodic and cathodic reaction rate constants
c_i	concentration of species i, mol m^{-3}	k_c	mass transfer coefficient, m s^{-1}
C	capacitance, F	k	thermal conductivity, W·m^{-1}·K^{-1}
C_d	differential capacitance, F	K	equilibrium constant, unitless
C_{DL}	double layer capacitance per unit area, F m^{-2}	L	characteristic length, m
C_p	battery capacity offset, unitless	L_p	diffusion length, m
d	distance, m	m	mass, kg
d_h	hydraulic diameter, m	m	molality, mol kg^{-1}
D_i	molecular diffusivity of species i, m^2·s^{-1}	$\dot{m}$	mass flow, kg s^{-1}
E	energy for battery or double layer capacitor, J or W·h	M_i	molecular weight, g·mol^{-1} or kg·mol^{-1}
E	elastic modulus, Pa	n	number of electrons transferred
E_a	activation energy, J/mol	n	number density of electrons, m^{-3}
E_g	energy gap, eV	n_i	number of moles of species i, mol
f	frequency, Hz	n_i	number density of electrons in intrinsic semiconductor, m^{-3}
f_i	fugacity of species i, Pa	N_{AV}	Avogadro's number
F	Faraday's constant, 96485 C per equivalent	N_A	number density of acceptor atoms, m^{-3}
F_x	component of force, N	N_D	number density of donor atoms, m^{-3}
g	acceleration of gravity, m s^{-2}	$\mathbf{N}_i$	molar flux of species i relative to stationary coordinates, mol m^{-2} s^{-1}
G	Gibbs energy, J mol^{-1}	p	pressure, Pa
G	conductance, Ω^{-1}	p	number density of holes, m^{-3}
Gr	Grashof number, unitless	P	power, W

xii   List of Symbols

P	perimeter, m	ε	porosity, unitless
q	fundamental charge, 1.602×10^{-19} C	ε_o	permittivity of free space, 8.854×10^{-12} C V^{-1} m^{-1}
q	charge per unit surface area, C·m^{-2}	ε	permittivity, C V^{-1}m^{-1} or F m^{-1}
$\dot{q}$	rate of heat generation, W	ε_r	relative permittivity, unitless
Q	charge, C or A·h	ϵ	strain, unitless
r	radius, m	η_c	current efficiency
R	universal gas constant, 8.314 J mol^{-1}K^{-1}	η_{coul}	coulombic efficiency, unitless
R_Ω	ohmic resistance, Ω or Ω-m^2	η_f	faradaic efficiency, unitless
Re	Reynolds number, unitless	η_S	surface overpotential, V
$\mathcal{R}_i$	homogeneous reaction rate, mol m^{-3}s^{-1}	κ	electrical conductivity, S m^{-1}
s_i	stoichiometric coefficient, unitless	λ	Debye length, or wavelength, m
S	electrical conductance, S	λ_i	absolute activity
S	entropy, J mol^{-1}K^{-1}	λ_i	equivalent conductance, S m^{-2}equivilent^{-1}
Sh	Sherwood number, unitless	Λ	conductance, S m^{-1}
Sc	Schmidt number, unitless	μ	viscosity, Pa-s
t	time, s	μ_i	chemical potential, J mol^{-1}
t_i	transference number of species i, unitless	ν	kinematic viscosity, m^2 s^{-1}
T	Temperature, K, or °C	ν	scan rate, V s^{-1}
U	equilibrium potential, V	ν_i	stoichiometric coefficient, unitless
u_i	mobility of ion, m^2 mol J^{-1}s^{-1}	ξ	electro-osmotic drag coefficient, unitless
v	velocity, m s^{-1}	ρ	mass density, kg m^{-3}
V	electric potential or voltage, V	ρ_e	charge density, C m^{-3}
v_s	superficial velocity, m s^{-1}	σ	electrical conductivity of solid, S m^{-1}; or surface energy, J m^{-2}
$\dot{\mathbb{V}}$	volumetric flowrate m^3 s^{-1}		
$\mathbb{V}$	volume, m^3	τ	characteristic time, s
W	work, J	τ	tortuosity, dimensionless
W	width of depletion layer, m	ϕ	potential, V
w	molar flowrate, mol s^{-1}	Φ	quantum yield
Wa	Wagner number, unitless	φ	gas holdup, osmotic coefficient, unitless
x_i	mole fraction of species i in condensed phase, unitless	ω	angular frequency, rad s^{-1}
y_i	mole fraction of species i in gas phase, unitless		
z_i	charge number		

Greek Symbols

α	Debye constant for solvent, (kg/mol)$^{1/2}$
α	relative permeability, unitless
α	absorption coefficient, m^{-1}
α_a, α_c	transfer coefficient for anodic and cathodic reactions, unitless
β	symmetry factor, unitless
γ_i	activity coefficient, unitless

Subscripts and Superscripts

a	anode
avg	average value
b	bubble
c	charge, cathode, or cutoff frequency
cb	conduction band
ct	charge transfer
d	discharge
e	electrolyte
eff	effective quantity
f	formation

fb	flat band	*Rx*	reaction
h	electrical portion of hybrid system	*s*	solution, separator, or surface
i	species	*sc*	space charge
int	interface	*vb*	valence band
lim	limiting value	∞	bulk value far from electrode
M	metal	*o*	initial value
o	value at standard or reference conditions	θ	equilibrium potential at reference conditions
ocv	open circuit voltage	*	ideal gas
ph	photon		

About the Companion Website

This book is accompanied by a companion website:

www.wiley.com/go/fuller/electrochemicalengineering

The website features:

- Powerpoint slides for figures from the text
- Complete solution manual to end of chapter problems
- Excel files for data sets used in student problems and chapter illustrations

Charles W. Tobias

Charles W. Tobias was born on November 2, 1920 in Budapest, Hungary and died in 1996 in Orinda, California. He describes his family as an "engineering family" and realized that he would be an engineer from the age of 4 or 5. Music was a lifelong passion, but engineering was his career. Tobias was inspired to select chemical engineering by his high school chemistry teacher. He attended the University of Technical Sciences in Budapest, where he obtained his Ph.D. in 1946. He was part of the Hungarian diaspora, fleeing Hungary in 1947 as the Communists were taking over the government. He traveled to Berkeley to pursue postdoctoral studies at the University of California where his brother was already working for Ernest O. Lawrence at the University's radiation laboratory. W.M. Latimer, a pioneer in the thermodynamics of electrolytes and Tobias's new boss, directed him to "... build up engineering electrochemistry here, and it should be the best in the world." By all measures, he succeeded. He began as an instructor in the newly formed Department of Chemical Engineering and then became an Assistant Professor in 1950 and Professor in 1960. He also chaired the department from 1967 until 1972.

Charles Tobias is considered by many to be the father of modern electrochemical engineering. He added discipline and rigor, putting many of the principles on a sound theoretical footing. He studied fluid flow, electric potential fields, thermodynamic and materials properties, and mass transfer, linking his findings into a new discipline. Tobias was instrumental in moving the field from largely empirical approaches to a quantitative science that addressed scale-up and optimization. Professor Tobias is also credited with founding the electrochemical research program at Lawrence Berkeley National Laboratory, which for decades was the most influential electrochemistry group. He is reported to have started the first "electrochemical engineering" course. Professor Tobias joined with Paul Delahay in 1961 to edit a new monograph series entitled *Advances in Electrochemistry and Electrochemical Engineering*.

Professor Tobias had a long and productive involvement in the Electrochemical Society. He won several awards, including the Society's highest honor, the Acheson award. He served as President of the Electrochemical Society from 1970 to 1971 as well as an editor for the *Journal of the Electrochemical Society* for many years. As a result, he had a large influence on the structure and operation of the Society. Tobias brought the same philosophy of increasing the scientific rigor and quantitative engineering to the Society. He was also President of the International Society of Electrochemistry (1977–1978). In 2003, the Tobias Award of the Electrochemical Society was established to recognize outstanding scientific and/or engineering work in fundamental or applied electrochemistry by a young investigator.

Although Tobias had an immeasurable impact on the field of electrochemical engineering and was elected a member of the National Academy of Engineering, he was not particularly aggressive in publishing. One of his most significant findings was the use of propylene and ethylene carbonates for lithium batteries, but it was never published beyond a laboratory report. Nonetheless, Dr. Tobias believed that knowledge should always be shared—only ignorance needs secrecy. Much of the information here was derived from the chemical heritage foundation's oral history, which delivers an excellent portrait of his life.

http://www.chemheritage.org/discover/collections/oral-histories/details/tobias-charles-w.aspx

Image Source: Courtesy of UC Berkeley College of Chemistry.

Chapter 1

Introduction and Basic Principles

Welcome to this introductory text about electrochemical engineering. If you are like most people, you probably have little idea of what "electrochemical engineering" is and why it is important. That's okay. This book is to help you answer those questions and to prepare you to work in this exciting area. Electrochemistry is a branch of chemistry that studies the chemical changes that occur due to the flow of electrical current or, conversely, the production of electricity from chemical changes. As engineers, we are interested in the application of scientific knowledge, mathematical principles, and economic analyses to the design, manufacture, and maintenance of products that benefit society. It's not surprising, then, that electrochemical engineering has strong historical ties to the profession of chemical engineering.

Almost invariably, application of the principles of electrochemistry to develop products or devices requires consideration of an *electrochemical system*. An engineering system consists of multiple components that work together in a concerted manner. In addition to electrochemistry, other topics such as heat transfer, structural analysis, and materials science are critical to the design and operation of electrochemical systems. Today, we find engineers and scientists of all disciplines engaged in electrochemistry and collaborating in the design of electrochemical systems.

As with any topic, electrochemical engineering includes a new set of terminology that must be learned. There are also several conventions of practical importance. Hence, this chapter will introduce you to the key vocabulary of the discipline, as well as to some of the central aspects of electrochemical systems. In order to do this, we begin by looking at an *electrochemical cell*.

1.1 ELECTROCHEMICAL CELLS

Electrochemical cells, such as the cell illustrated in Figure 1.1, lie at the heart of electrochemical systems. A typical electrochemical cell consists of two electrodes: an *anode* where oxidation occurs and a *cathode* where reduction takes place. Electrons move through an external circuit via an electronic conductor that connects the anode and cathode. The liquid solution that is between the two electrodes is the *electrolyte*. The electrolyte does not conduct electrons and does not contain any free electrons. It does, however, contain a mixture of negatively charged ions (*anions*) and positively charged ions (*cations*). These ions are free to move, which allows them to carry current in the electrolyte.

The reactions take place at the electrode surface and are called *heterogeneous electron-transfer reactions*. For example, the electrodeposition of copper in the cell shown in Figure 1.1 can be written as

$$Cu^{2+}_{(aq)} + 2e^- \rightarrow Cu_{(s)} \quad (1.1)$$

Copper ions in solution accept two electrons from the metal and form solid copper. The reaction is described as heterogeneous because it takes place at the electrode surface rather than in the bulk solution; remember, there are no free electrons in the solution. Importantly, then, we see that electrochemical reactions are surface reactions. The metal that accepts or supplies electrons is the *electrode*. As written, copper ions gain electrons and therefore are *reduced* to form copper metal. When reduction occurs, the electrode is called the cathode.

In the same cell, the reaction that takes place on the other electrode is

$$Zn_{(s)} \rightarrow Zn^{2+}_{(aq)} + 2e^- \quad (1.2)$$

The zinc metal is oxidized, giving up two electrons and forming zinc ions in solution. When oxidation occurs at the surface, the electrode is called the anode.

Both the copper and zinc reactions written above (Equations 1.1 and 1.2) are *half-cell reactions*, meaning that they describe a reaction that takes place at one of the electrodes. Half-cell reactions always have electrons as either a reactant or a product. Also, charge is always balanced in half-cell reactions. Charge balance means

---

Electrochemical Engineering, First Edition. Thomas F. Fuller and John N. Harb.
© 2018 Thomas F. Fuller and John N. Harb. Published 2018 by John Wiley & Sons, Inc.
Companion Website: www.wiley.com/go/fuller/electrochemicalengineering

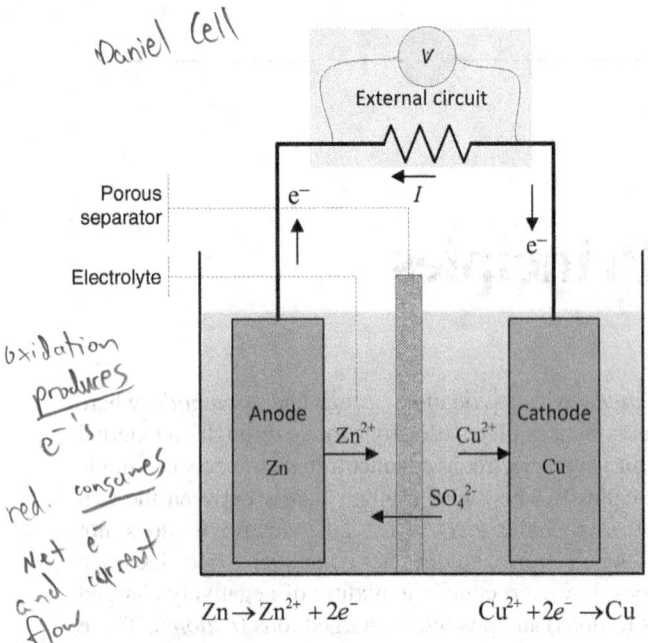
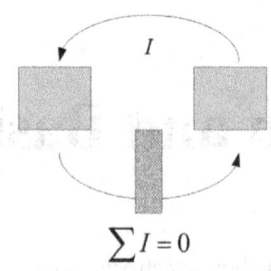
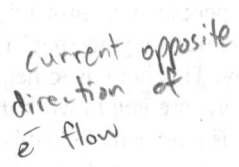

Figure 1.1 A Daniell cell is an example of an electrochemical cell. During steady operation, a constant current flows throughout the cell. For any given volume, the current entering and leaving must sum to zero since charge is conserved.

that the net charge on one side of the equation must equal the net charge on the other side of the equation. For example, in the above equation, the net charge on the Zn (left side of the equation) is zero. This value is equal to the net charge for the other side, which is $(+2) + 2(-1) = 0$. Charge balance is no surprise since electrochemical reactions simply add or remove electrons.

The cell illustrated in Figure 1.1 is called a Daniell cell and is named after John Frederic Daniell, who invented the cell in 1836. In this cell, the electrodes themselves (solid Zn and Cu) participate in the electrochemical reactions as either a reactant or a product. This is not always the case, as we shall see later. The electrons for the reduction of Cu^{2+} come from the other electrode where the oxidation of Zn takes place. The electrons that are liberated from zinc oxidation at the anode travel through an external circuit to participate in the reduction reaction at the cathode. The movement of electrons corresponds to an electrical current flowing through the external circuit. The anode and cathode reactions can be combined to give an overall cell reaction, also called the *full-cell reaction*. Note that there are no electrons in the overall reaction, but charge is still balanced.

$$Zn_{(s)} + Cu^{2+}_{(aq)} \rightarrow Zn^{2+}_{(aq)} + Cu_{(s)} \quad (1.3)$$

In other words, the exact number of electrons released from oxidation is used for reduction. Similarly, for each copper ion that is reduced, one zinc ion is formed. Because there is a production of charge at the anode and a consumption of charge at the cathode, there is a net movement of charge in solution from the anode to the cathode. Thus, there is also a current flowing in the solution, but the net charge in the solution is unchanged.

The flow of current in an electrochemical cell is shown in Figure 1.1. Note that, by definition, current in the external circuit is opposite (cathode to anode) to the direction that electrons flow. As illustrated in the figure, current flows counterclockwise in a continuous circuit that includes the solution and the external circuit. The electrochemical reactions are the means by which current flows from the electrode to the solution (anode) and from the solution to the electrode (cathode) as part of this circuit. These reactions will not take place if the circuit is broken. This circuit is most commonly "broken" by disconnecting the external connection between the electrodes. In the absence of this external connection, the cell is at *open circuit*, and there is no net reaction at either electrode. Finally, we note that the name "electrochemical" refers to the fact that the *chemical* changes are connected to the flow of *electric* current. Some of the additional aspects of electrochemical reactions are discussed in the next section.

1.2 CHARACTERISTICS OF ELECTROCHEMICAL REACTIONS

Electrochemical reactions are reactions where electrons are transferred through a conductor from the species being oxidized to that being reduced. Most of the unique and important properties of electrochemical reactions are the result of the way that these electrons are transferred. These characteristics include the following:

Separation of the oxidation (anodic) and reduction (cathodic) reactions. Because electrons are transferred through a conductor, the oxidation and reduction can take place on two different electrodes. This situation is the most common type of configuration for an electrochemical cell, as already illustrated.

Use of electrons to perform work. In cells like the Daniell cell introduced in Section 1.1, electrons move spontaneously from the anode to the cathode when the circuit is closed (completed), because the energy of electrons in the anode is higher than the energy of electrons in the cathode. Some of that extra electron energy can be used to perform work by passing the electrons (current) through a load (e.g., an electronic device). Perhaps the most familiar application of this concept is in the battery that you use to power your phone or computer, where you are using electron energy to send messages to your friends or to complete your homework.

Direct measurement of reaction rates. Typically, it is difficult to measure directly the rates of chemical reactions. However, since electron transfer in electrochemical reactions takes place through a conductor, the reaction rate is easy to measure—you just measure the electric current passing in the wire between the electrodes. That current is directly related to the reaction rate at each of the two electrodes. We will explore this later in the chapter when we discuss Faraday's law.

Control of the direction and rate of reaction. Because electrons participate in both the oxidation and reduction reactions as either a product or a reactant, we can change the rate of reaction by changing the difference in potential between the electrodes. The electric potential is a measure of the electron energy. By adjusting the potential, we can speed up the reactions, slow them down, or even make them go in the reverse direction.

This textbook will help you to understand and to use the powerful properties of electrochemical reactions to design useful products. In the application chapters of the book, you will have the opportunity to see what others have done with electrochemical systems to create devices and processes that benefit humankind. In preparation, the next section of this chapter provides a brief overview of the applications that will be considered in order to illustrate the importance of electrochemical systems. However, before doing that, let's consider some additional aspects of electrochemical reactions.

In the Zn/Cu cell considered previously, the electrodes participated directly as either a reactant (Zn) or a product (Cu). In many instances, the electrodes do not participate beyond supplying or removing electrons. In these instances, the electrodes are called *inert*. Each of the electrochemical reactions considered in the Daniell cell involved one solid species and one aqueous species. This is by no means a requirement or even typical. Reactants and products in electrochemical reactions can be solids, liquids, dissolved species, or gases. The evolution of hydrogen gas, for instance,

$$2H^+_{(aq)} + 2e^- \rightarrow H_{2(g)} \tag{1.4}$$

is a reduction (cathodic) reaction that occurs at an inert electrode. Here, the reactant is dissolved in solution and the product is a gas. The oxidation of iron(II) to iron(III) provides another example:

$$Fe^{2+}_{(aq)} \rightarrow Fe^{3+}_{(aq)} + e^- \tag{1.5}$$

In this case, both the reactant and product are dissolved species.

In looking at the reactions for the Daniell cell, we note that copper ions gained two electrons, zinc lost two electrons to form zinc ions, hydrogen ions gained one electron each to produce hydrogen gas, and a ferrous ion lost an electron to form a ferric ion. You may be wondering how we know how many electrons will be transferred and the direction in which the reaction will go. The direction depends on potential as we will see in the next two chapters. However, the oxidation and reduction products are determined by the electronic structure of the participating species. You undoubtedly learned about oxidation states in your beginning chemistry class, and those same principles apply here. Let's consider a couple of simple examples just to illustrate the concept. Sodium is an alkali metal and has a single electron in its outer electron shell. Whenever possible, sodium tends to give up that electron, leaving it with a +1 charge and a full outer shell. Thus, the stable state for a *sodium ion* is Na$^+$. Similar logic can be used for the other alkali metals. In the second column of the Periodic Table you have alkaline-earth metals, such as magnesium or calcium. These elements have two electrons in their outer shell and, therefore, yield ions with a +2 charge such as Mg^{2+} or Ca^{2+}, using rationale similar to that used for the alkali metals. Moving to the other side of the Periodic Table, we find chlorine that has seven electrons in its outer shell. Chlorine likes to pick up an extra electron in order to complete that shell. Therefore, it forms Cl$^-$. Finally, the electronic structure of some elements (e.g., most transition metals) allows them to form ions with different oxidation states; for example, iron can form Fe^{2+} or Fe^{3+}.

Fortunately, you don't need to memorize the Periodic Table or to look up the electronic structure of an element every time to determine the relevant oxidation state or half-cell reaction. Tables of known electrochemical reactions are available, and a short list appears in Appendix A of this book. These tables not only show the reactions, but also give the standard potential for each reaction. This standard potential is extremely important as we will see in the next chapter. Note that, by convention, the reactions in Appendix A are written as reduction reactions, but we can treat them as being reversible.

> **ILLUSTRATION 1.1**
>
> Use the table in Appendix A to find the two half-cell reactions and combine them to determine the full-cell reaction.
>
> **a.** *Corrosion of iron in an oxygen-saturated acid solution.* On one electrode, iron metal is oxidized to form Fe^{2+}; on the other electrode, oxygen is reduced. Entry 18 shows the reduction of iron:
>
> $$Fe^{2+} + 2e^- \rightarrow Fe \quad -0.440\,V$$
>
> We reverse the direction to get the oxidation of iron:
>
> $$Fe \rightarrow Fe^{2+} + 2e^-$$
>
> The reduction of oxygen in acid is given by entry 4:
>
> $$O_2 + 4H^+ + 4e^- \rightarrow 2H_2O$$
>
> These two reactions are then combined. The iron oxidation reaction is first multiplied by two so that the electrons cancel out when the equations are added together:
>
> $$O_2 + 4H^+ + 2Fe \rightarrow 2H_2O + 2Fe^{2+}$$
>
> **b.** *Deposition of silver from an alkaline solution.*
>
> $$Ag^+ + e^- \rightarrow Ag \quad \text{entry 6}$$
> $$4OH^- \rightarrow O_2 + 2H_2O + 4e^- \quad \text{entry 9}$$
>
> The silver reduction reaction is multiplied by four and added to the oxidation reaction to provide the full-cell reaction:
>
> $$4Ag^+ + 4OH^- \rightarrow O_2 + 2H_2O + 4Ag$$

1.3 IMPORTANCE OF ELECTROCHEMICAL SYSTEMS

Electrochemical systems are not only essential for our society but are also common in everyday life. Imagine a world without batteries to power your personal electronic devices. How would travel change without low-cost aluminum that is essential for aircraft? What if corrosion of the steel in bridges, the hulls of ships, superstructures of buildings, and pipelines couldn't be controlled? These examples, electrochemical energy storage, industrial electrolysis, and corrosion, are some of the major applications of electrochemical engineering. Following development of the fundamental principles of electrochemical engineering, these and other applications are covered in separate chapters.

We all know that the storage of energy using batteries (Chapters 7 and 8) is an essential feature for countless technologies. The size of the global battery market is more than 50 billion dollars annually. Applications of batteries are widespread in society, ranging from basic consumer electronics, to automobiles, implanted medical devices, and space travel. Energy storage is needed where portability is required, emergency power is desired, or simply when electrical demand and supply are not matched. Batteries store energy chemically and convert that energy to electricity as needed. For many applications, the conversion between chemical and electrical energy must be extremely efficient and, in some cases, nearly reversible in order for batteries to be practical. Electrochemical double-layer capacitors (Chapter 11) can also be used to store energy.

The purpose of industrial electrolysis is to use electrical energy to convert raw materials into desired products. This transformation of raw materials takes place in an electrochemical reactor. Industrial electrolytic processes (Chapter 14) consume about of 6% of the U.S. electricity supply. Although a number of metals and other materials can be produced electrochemically, the main products of electrolysis are aluminum, chlorine, and sodium hydroxide (caustic soda). Low-cost electricity and electrochemical routes for production transformed aluminum from an expensive and rarely used metal to a commodity material that is essential for society. Chlorine and sodium hydroxide, produced simultaneously on a huge scale by electrolysis, are required for the production of a variety of materials, including plastics, paper, soaps, solvents, and a large number of other chemicals.

The many desirable properties of metals have led to their widespread use in industry and, frankly, in nearly every aspect of our lives. Corrosion (Chapter 16) is the unwanted attack of metals by their environment. The manufacture of metals is sometimes done electrochemically as described above for the production of aluminum. We can think of corrosion as the reverse process, by which metals are converted back into their oxide form. This cycle is illustrated in Figure 1.2. The process of corrosion is electrochemical in nature. The global economic cost of corrosion has been estimated to be several trillion dollars annually. Therefore, it is important that engineers understand the conditions under which corrosion is likely to occur. They should also be able to measure, predict, and mitigate the negative impacts of corrosion. Avoidance of corrosion is incorporated into the appliances found in our homes. Additionally, corrosion prevention is essential for maintaining the integrity of the bridges over which we travel, and an integral part of the design of natural gas and oil pipelines that supply our energy. These are just a few of the areas where understanding corrosion is important.

Electrodeposition is the electrochemical deposition of a metal onto a surface (Chapter 13). Electrodeposition is

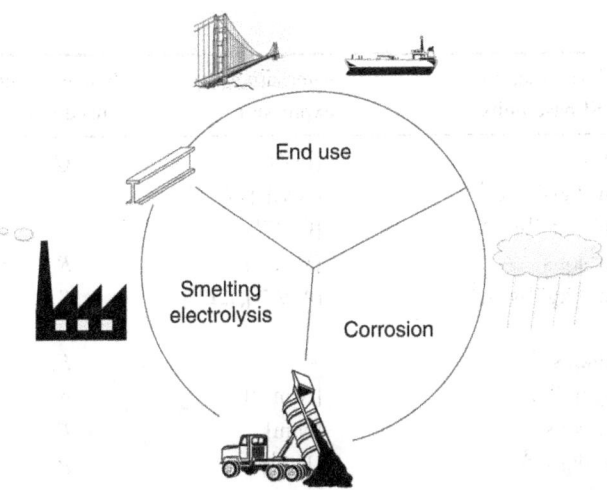

Figure 1.2 Life cycle of metals. Many of the processes are electrochemical.

used for decoration, to provide protective coatings, and is essential in the fabrication of interconnects for large-scale integrated circuits. Most metal fasteners (screws, nails, clips, and bolts to name but a few) are coated for appearance, to improve their corrosion resistance, or to reduce friction. Many of these fasteners use electrochemical processes to provide these coatings.

Roughly 10% of the world's population is afflicted with diabetes. Since treatment requires regular monitoring of blood glucose levels, the measurement of glucose concentration is one of the most frequent tests conducted. Did you know that most of the billions of these tests are done electrochemically? These sensors work by measuring the current generated from electrochemical oxidation of glucose. Modern automobiles use oxygen sensors to allow efficient operation of the fuel injection system. These sensors are also electrochemical devices.

We hope that this section has helped you gain an increased appreciation of the importance of electrochemical systems. In addition to the applications noted above, there are other promising technologies in development. These include fuel cells (Chapters 9 and 10), which convert clean fuels into energy at efficiencies much higher than that of a combustion engine. Fuel cells have been essential for manned space flight, but have yet to achieve widespread commercial success. Semiconductor electrochemistry (Chapter 15) offers opportunities for photoelectrochemical processes to convert sunlight into electricity or to produce useful chemicals through artificial photosynthesis. The list of developing applications extends well beyond these two examples, and electrochemistry will undoubtedly continue to provide new products and processes that will improve our quality of life and benefit humankind.

Before we can examine applications in detail, some fundamental principles must be mastered. These fundamental principles provide the foundation for all of the applications mentioned above. We begin by introducing some basic terminology, as well as reviewing scientific units and the conventions adopted in this text.

1.4 SCIENTIFIC UNITS, CONSTANTS, CONVENTIONS

There are seven base units within the International System of Units (SI system), as shown in Table 1.1 for quick reference. Except for luminous intensity, all of the quantities are used extensively in this text. Among the base units is electric current, I, which is critical for the examination of electrochemical systems. Electric current is measured in amperes [A].

Table 1.2 presents derived units that are used widely in this text and are associated with electrochemical engineering. For each quantity, the name and the SI symbol for the quantity is presented. Our general practice will be to place units in brackets, for example, [Pa·s]. When working problems, dimensional consistency is critical; it will also help you to avoid calculation errors. Therefore, for convenience, the table provides the dimensions of these derived quantities written in terms of both the base units and in an alternative form convenient for calculations. For instance, the units for capacitance $[m^{-2} \cdot kg^{-1} \cdot s^4 \cdot A^2]$ can be written more compactly in terms of other derived units $[C \cdot V^{-1}]$, which will usually be more intuitive and easier to remember when checking for dimensional consistency. Finally, the last column provides the variable symbol used in this text to represent the physical quantity. Where possible, we have used the most common nomenclature to represent these quantities. A detailed list of the nomenclature is provided at the beginning of the book. It is important to note that sometimes the nomenclature for a variable may be the same as the unit. To minimize confusion, remember that variables and fundamental physical constants are in *italics*, whereas the units will not be italicized. Thus, V

Table 1.1 The Seven SI Base Units; All Except Candela Are Used Frequently in This Book

Quantity	Name of unit	Symbol for unit	Nomenclature used in text
Length	meter	[m]	L
Mass	kilogram	[kg]	m
Mole	mole	[mol]	n_i
Time	second	[s]	t
Temperature	kelvin	[K]	T
Current	ampere	[A]	I
Luminous intensity	candela	[cd]	–

Table 1.2 SI Derived Units That Are Important for This Text

Quantity	Name of unit	Symbol for unit	Expressed in SI base units	Alternative expression	Nomenclature used in text
Electric charge	coulomb	[C]	A·s	[F·V]	Q
Potential	volt	[V]	$m^2 \cdot kg \cdot s^{-3} \cdot A^{-1}$	[A·Ω], [J·C^{-1}]	V, ϕ
Capacitance	farad	[F]	$m^{-2} \cdot kg^{-1} \cdot s^4 \cdot A^2$	[C·V^{-1}]	C
Resistance	ohm	[Ω]	$m^2 \cdot kg \cdot s^{-3} \cdot A^{-2}$	[V·A^{-1}]	R_Ω
Electrical conductance	siemens	[S]	$m^{-2} \cdot kg^{-1} \cdot s^3 \cdot A^2$	[A·V^{-1}], [Ω^{-1}]	S
Force	newton	[N]	$m \cdot kg \cdot s^{-2}$	–	F_x
Pressure	pascal	[Pa]	$kg \cdot m^{-1} \cdot s^{-2}$	[N·m^{-2}]	p
Energy	joule	[J]	$m^2 \cdot kg \cdot s^{-2}$	[N·m]	E
Power	watt	[W]	$m^2 \cdot kg \cdot s^{-3}$	[J·s^{-1}]	P
Frequency	hertz	[Hz]	s^{-1}	–	f

represents the potential of an electrochemical cell, whose units are volts [V]. Similarly, A is area [m^2], whereas [A] is the unit for current in amperes. Most often we will be dealing with scalar quantities. However, occasionally we will encounter vector quantities, ones that have both magnitude and direction. These quantities will be written in bold face. For example, velocity, a vector quantity, is written with the symbol **v**.

The coulomb [C] is an SI-derived unit of electrical charge—simply the amount of charge that passes with a current of 1 ampere for 1 second. The elementary charge, denoted with q, is a fundamental physical constant. The value of q is 1.602177×10^{-19} C and equal to the charge of a single proton. Additional essential fundamental constants are provided in Appendix B.

1.5 FARADAY'S LAW

In this section, we examine Faraday's law, which is the relationship between the amount of current that flows through the external circuit and the amount of material that is either consumed or produced in a half-cell reaction. To explore how this works, let's return once again to the zinc reaction:

$$Zn_{(s)} \rightarrow Zn^{2+}_{(aq)} + 2e^- \quad (1.6)$$

From the half-cell equation, it is apparent that two electrons are produced for every zinc atom that reacts. Typically, it is more convenient to work in terms of moles rather than atoms. Therefore, two moles of electrons are produced for every mole of zinc atoms that is oxidized. As you can see, we can easily relate the moles of electrons to the moles reacted or produced for any species in a given half-cell reaction.

The next step is to relate the moles of electrons to the current that we measure through the external circuit. To do this, it is customary to introduce a new unit of charge—an *equivalent*. An equivalent is defined as a mole of charge (either positive charge or negative charge, it does not matter). The number of equivalents of a compound is simply the amount of the substance (in moles) multiplied by the absolute value of its charge, z_i. For instance, one mole of Na$^+$ is 1 equivalent, whereas one mole of Ca^{2+} would be 2 equivalents. Because the sign of the charge does not matter, *one mole of electrons is equal to 1 equivalent of charge*.

The external current is expressed in amperes [A or C·s^{-1}]. Therefore, we need a relationship between coulombs [C] and equivalents. That relationship is called *Faraday's constant*, F, which has a value of 96,485 C/equivalent. Faraday's constant can also be expressed in terms of two other constants: the fundamental unit of charge, q, and Avogadro's number, N_{AV},

$$F = qN_{AV}. \quad (1.7)$$

This expression is another way of stating that an equivalent is a mole of charge, since q is the amount of charge on a proton in coulombs. We also need a relationship between the current, I [A], and the total charge passed in coulombs. That relationship is

$$Q = \int I\,dt, \quad (1.8)$$

where Q is the charge in [C], I in [A], and t in [s]. In situations where the current is constant, Equation 1.8 simplifies to

$$Q = It. \quad (1.9)$$

We now have the pieces that we need to write the relationship between the current in the external circuit and the amount of a substance that is either reacted or produced. Let's apply what we have learned to the zinc electrode.

ILLUSTRATION 1.2

For the oxidation of zinc, a current of 12 A passes for 2 hours. How much zinc reacts? Provide the answer in terms of mass and the moles of zinc. The reaction is

$$Zn_{(s)} \rightarrow Zn^{2+}_{(aq)} + 2e^-$$

First, the total charge passed is determined from Equation 1.9.

$$Q = It = \frac{12[C]}{[s]} \times \frac{2[h]}{} \frac{3600\,[s]}{[h]} = 86,400\,[C].$$

The number of moles of zinc is proportional to the charge passed. From the stoichiometry of the oxidation reaction, for each mole of zinc two moles of electrons are released. Thus,

$$n_{Zn} = \frac{Q}{2F} = \frac{(86400\,[C])}{(2\,[equiv/mol])(96485\,[C/equiv])} = 0.448\,\text{mol Zn}$$

The mass of zinc consumed is simply the moles times the molecular weight:

$$m_{Zn} = M_{Zn} n_{Zn} = (65.38)(0.448) = 29.3\,\text{g Zn}.$$

Let's now generalize what we have done to this point. Any half-cell reaction can be expressed in the following form:

$$\sum_i s_i A_i = ne^-, \qquad (1.10)$$

where s_i is the stoichiometric coefficient for species i, which can be either positive or negative, A_i is the symbol for the specific species (e.g., Zn^{2+}), and n is the number of electrons involved in the reaction. The mass of species i that is either consumed or produced is equal to

$$m_i = \frac{M_i Q}{\left(\frac{n}{|s_i|}\right) F} = \frac{M_i \left[\frac{g}{mol}\right] Q[C]}{\frac{n}{|s_i|} \left[\frac{equiv}{mol}\right] F \left[\frac{C}{equiv}\right]}, \qquad (1.11a)$$

where M_i is the molecular weight of species i. This equation is known as *Faraday's law*. We have been careful to include the stoichiometric coefficient with Equation 1.11a. However, most often Faraday's law is written as

$$m_i = \frac{M_i Q}{nF}. \qquad (1.11b)$$

We will also use this form. When using Equation 1.11b, you need to recognize that n represents the number of electrons per species i. Look carefully at the examples that follow.

Although Equation 1.11 is useful, the most important aspect of this section is the process that we used to develop the equation. We started with charge, Q, in standard units of [C]. This charge is directly related to the current that we measure in [A] as a function of time. We then converted that charge to moles of electrons (equivalents). The moles of electrons were then related through the half-cell reaction to the moles of species i that were either reacted or produced. Finally, we used the molecular weight to convert moles to the desired mass of species i.

A similar procedure can be used to calculate quantities such as the total moles reacted, the molar rate of reaction, or the molar flux due to reaction at the surface. The use of Faraday's law and the more general procedure used to derive it are demonstrated in Illustration 1.3.

ILLUSTRATION 1.3

For a lead–acid battery, calculate the rate at which Pb reacts and H^+ is consumed for a cell operating at 20 [A]. Express the rate of reaction in terms of a molar reaction rate.

The overall reaction for discharge is

$$Pb + PbO_2 + 2H_2SO_4 \rightarrow 2PbSO_4 + 2H_2O$$

The reaction at the positive electrode is

$$PbO_2 + SO_4^{2-} + 4H^+ + 2e^- \rightarrow PbSO_4 + 2H_2O, \quad \text{entry 2}$$

at the negative electrode

$$Pb + SO_4^{2-} \rightarrow PbSO_4 + 2e^-, \quad \text{entry 17}$$

The molar reaction rate of Pb (lead) consumption is obtained from Equation 1.11, Q is replaced with I to convert to a rate, and the molecular weight is not needed:

$$\text{molar reaction rate of Pb} = \frac{I}{2F} = \frac{20}{2F}$$
$$= 1.04 \times 10^{-4}\,\text{mol s}^{-1}.$$

The molar reaction rate of H^+ is also obtained from Equation 1.11 in a similar fashion. However, the n value is different. In this case, we have 2 equivalents of charge (the 2 electrons in the half-cell reaction) for every 4 moles of H^+. Therefore, n for H^+ is 2/4 or 1/2 [equiv mol^{-1} H^+].

$$\text{molar reaction rate of } H^+ = \frac{I}{\frac{1}{2}F} = \frac{(2)20}{F}$$
$$= 4.15 \times 10^{-4}\,\text{mol s}^{-1}.$$

Note that sulfate ions are produced at both electrodes. Based on the stoichiometry for the overall reaction, we expect 2 moles of sulfate ion to be consumed for each mole of Pb. Or, we can say that one mole of electrons react for each mole of sulfate. Formally, this is written as

$$\text{molar reaction rate of } SO_4^{2-} = \frac{I}{nF} = \frac{20}{F}$$
$$= 2.07 \times 10^{-4} \text{ mol s}^{-1}.$$

Finally, we note that the reactants and products may be charged or neutral. They can also take several forms, such as a species dissolved in the electrolyte, a solid that deposits on the electrode, or a gas. Illustration 1.3 involved a solid reaction product, $PbSO_{4(s)}$. Now let's consider the oxidation of chloride ions in solution to form chlorine gas:

$$2Cl^- \rightarrow Cl_2 + 2e^- \quad (1.12)$$

Two chloride ions from solution each give up one electron and they combine at the metal surface to form the gas.

ILLUSTRATION 1.4

A current of 20 [A] passes for 10 minutes. How many grams of chlorine gas are produced assuming Equation 1.12 represents the electrode reaction? For ideal gas behavior, what volume of gas does this represent at atmospheric pressure (100 kPa) and at 25 °C?

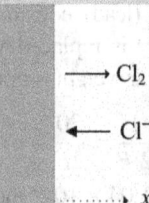

Use Equation 1.11, $n = 2$, $s_{Cl_2} = 1$, $M_{Cl_2} = 70.906$ g·mol^{-1}:

$$m_i = \frac{M_i Q}{2F} = \frac{70.906(20)(10)(60)}{2(96485)} = 4.41 \text{ g}.$$

The mass of Cl_2 is converted to moles (4.41/70.906), and the ideal gas law ($p\mathbb{V} = n_{Cl_2}RT$) is used to determine the volume:

$$\mathbb{V} = \frac{(m_i/M_i)RT}{p} = \frac{4.41/70.906(8.314)(298)}{100,000}$$
$$= 0.0015 \text{ m}^3.$$

1.6 FARADAIC EFFICIENCY

The electrochemical reactions that we have been discussing are called *faradaic reactions* since they involve electron transfer that is directly related to the consumption of reactants and the formation of products as described by Faraday's law. To this point, we have only considered one reaction at each electrode. However, it is possible, and in fact common, for multiple reactions to take place at a single electrode in a real system. Frequently, one of the reactions is the desired reaction, and the other reactions are referred to as *side reactions*. In such cases, not all of the current goes into the desired reaction. We can define a *faradaic efficiency* to characterize the fraction of the total current that drives the desired reaction.

$$\text{faradaic efficiency} =$$
$$\eta_f = \frac{\text{amount of desired material produced}}{\text{amount that could be produced with the coulombs supplied}}$$
(1.13)

Use of this efficiency is demonstrated in Illustration 1.5.

ILLUSTRATION 1.5

Nickel is electrodeposited from a bath of $NiSO_4$. A current of 1.0 A is passed for 3 hours and 3.15 g of Ni are deposited. What is the faradaic efficiency of this deposition?

The reaction is $Ni^{2+} + 2e^- \rightarrow Ni$.

First, use Faraday's law to determine how much Ni could possibly be deposited.

$$m_i = \frac{M_i Q}{nF} = \frac{M_i It}{nF}$$
$$= \frac{58.7 [\text{g mol}^{-1}]}{2 [\text{equiv/mol}]} \frac{(1.0 [\text{C s}^{-1}])(3[\text{h}])(3600[\text{s/h}])}{96485 [\text{C/equiv}]}$$
$$= 3.285 \text{ g Ni},$$

$$\eta_f = \frac{3.15}{3.285} = 0.96.$$

One of the common side reactions for aqueous metal deposition is water hydrolysis to produce hydrogen gas.

1.7 CURRENT DENSITY

To this point in the chapter, we have considered the total current, I. However, as you know from your experience, some electrochemical devices are small like a hearing aid battery. Others are significantly larger, such as the battery used to start your car. You would not expect the total current from these devices to be similar. Because electrochemical reactions take place on surfaces, we frequently normalize the total current by the surface area in order for us to better understand and characterize the system. *Current density* is defined as the current divided by the area of the electrode and will be used extensively throughout this book. Similarly, the molar rate of reaction is often expressed as a flux, [mol·m^{-2}·s^{-1}]. These molar fluxes are used extensively in studying mass transfer. Let's illustrate these concepts with an example.

ILLUSTRATION 1.6

a. The electrode from Illustration 1.4 has an area of 80 cm^2, what is the current density?

$$i = \frac{I}{A} = \frac{20\,[A]}{0.008\,[m^2]} = 2500\,[A\cdot m^{-2}].$$

From Faraday's law we can relate the charge to the amount of chloride ions that are consumed. Similar to current density, we define the flux of Cl$^-$ to the electrode surface. Namely,

$$N_{Cl^-} = \frac{i}{nF} = \frac{2500\,[C/s-m^2]}{(1[equiv/mol])\,96485\,[C/equiv]}$$
$$= 0.026\,[mol\cdot m^{-2}s^{-1}].$$

b. For the lead–acid battery described in Illustration 1.3, convert the molar reaction rates to molar fluxes given that the electrode area is 0.04 m^2. The previous results are divided by the electrode area.

$$\text{molar flux of Pb} = \frac{\text{rate of reaction}}{\text{electrode area}}$$
$$= \frac{1.04 \times 10^{-4}}{0.04}$$
$$= 0.0026 \text{ mol m}^{-2}\text{ s}^{-1}.$$

This quantity represents the rate of dissolution of Pb from the negative electrode into the electrolyte.

Next, consider the flux of sulfate ion (see Illustration 1.3). Because sulfate is consumed at both electrodes, we must calculate the sulfate flux for each electrode.

For the negative electrode,

$$\text{molar flux of SO}_4^{2-} = \frac{\frac{I}{A}}{nF} = \frac{\frac{20}{0.04}}{2F}$$
$$= 0.0026 \text{ mol m}^{-2}\text{s}^{-1}.$$

For the positive electrode,

$$\text{molar flux of SO}_4^{2-} = \frac{\frac{I}{A}}{nF} = \frac{\frac{20}{0.04}}{2F}$$
$$= 0.0026 \text{ mol m}^{-2}\text{s}^{-1}.$$

Thus, we have 0.0026 mol m^{-2} s^{-1} of sulfate ions moving from the electrolyte toward the negative electrode and an equal flux moving toward the positive electrode.

Strictly speaking, molar flux, and therefore current density, are vector quantities—ones that have both a magnitude and direction. The current density, i, is defined as

$$i = \frac{I}{A} = F \sum z_i N_i, \quad (1.14)$$

where N_i is the molar flux [mol·m^{-2}·s^{-1}] and z_i is the electrical charge of species i. This equation indicates that for current to flow, there needs to be a net movement of charged species. We will only occasionally need to treat these quantities as having more than one directional component; therefore, our practice will be to not make any special effort to identify these quantities as vectors unless needed. However, even for one-dimensional current flow, we need to pay attention to the direction, which of course depends on how our coordinate system is defined. Referring back to Illustration 1.4, the flux of chloride ions is from right to left, which is a negative flux (in the negative x direction). The charge of the chloride ion, z_i, is also negative. Therefore, current flow is positive in that example.

1.8 POTENTIAL AND OHM'S LAW

Another important quantity in the study of electrochemical systems is the potential. We can define an electrostatic potential, ϕ, in terms of the work required to move a unit (positive) charge from infinity to a specific position in the metal or in solution. This work can also be thought of in terms of energy. The unit for potential is a volt [V], which

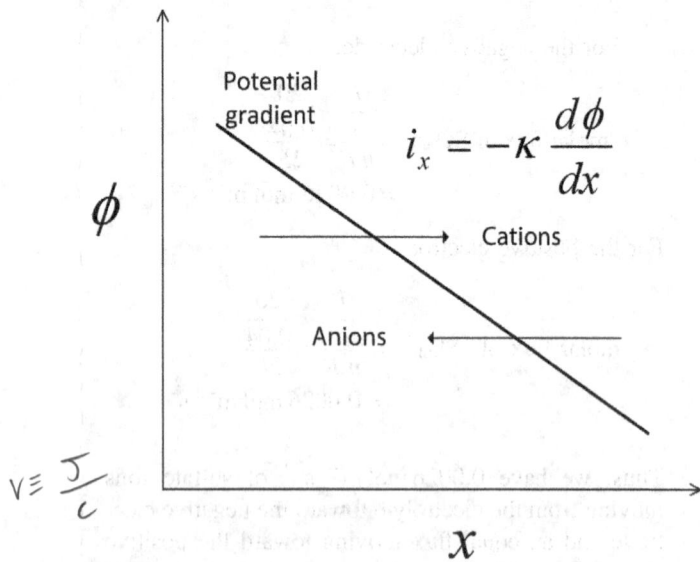

Figure 1.3 Gradient in potential and flow of current. Current flow is from left to right.

is also [J·C^{-1}]. Since we are frequently interested in the potential of the electrodes, and since these are generally metals where electrons are the carriers of current, we can think of potential as a measure of the energy of the electron in the metal. Thus, the cell potential measured with a voltmeter in Figure 1.1 represents the difference in electron energy in the two electrodes.

As you may have guessed, the flow of current in metals and in electrolyte solutions is critical to the understanding and analysis of electrochemical systems. A potential difference in a metal or in an electrolyte leads to the flow of current. The difference is often expressed in terms of a potential gradient, which is the rate of change in potential with distance. By convention, electric current is defined as flowing from high potential to low potential, which is the direction that a positive charge would move due to the difference in the potential. This situation is illustrated in Figure 1.3. Current in a metal (e.g., wire) is due to electron flow. Since electrons are negatively charged, electrons travel from low potential to high potential. *Ohm's law* relates the current and potential. For one-dimensional conduction (current flow),

$$i_x = -\kappa \frac{d\phi}{dx}, \quad (1.15)$$

where κ is the electrical conductivity of the electrolyte. The electrical conductivity is a property that characterizes how well the material moves electrical charge; it has units of [S·m^{-1}]. If the current density is constant, Equation 1.15 can be integrated to give

$$\Delta\phi = \frac{iL}{\kappa}. \quad (1.16)$$

This equation applies to one-dimensional current flow in an electrolyte of constant composition. A similar expression also applies to metals in general.

ILLUSTRATION 1.7

A fuel cell used for an automotive application operates at a current of 350 A. If the electrode area of each cell is 0.04 m^2, what is the current density? If the distance between the electrodes is 0.07 mm, and the conductivity of the electrolyte is 10 S·m^{-1}, what is the potential drop (difference in volts) across the electrolyte?

The current density, i, is simply the current divided by the electrode area

$$i = \frac{I}{A} = \frac{350}{0.04} = 8750 \text{ A} \cdot \text{m}^{-2}.$$

Ohm's law is used to calculate the drop in potential across the electrolyte:

$$\Delta\phi = \frac{iL}{\kappa} = \frac{8750(0.00007)}{10} = 0.06125 \text{V} = 61 \text{ mV}.$$

Compare these results with the potential drop in 10 cm long copper wire with a diameter of 1 cm. The conductivity of copper is much higher: 5.85×10^7 S·m^{-1}. The same current flows through the wire, but the cross-sectional area ($A_w = 7.9 \times 10^{-5}$ m^{-2}) is smaller

$$\Delta\phi = \frac{iL}{\kappa} = \frac{\frac{I}{A_w}L_w}{\kappa_w} = \frac{\frac{350}{7.9 \times 10^{-5}}(0.1)}{58,500,000} = 7.6 \text{ mV}.$$

1.9 ELECTROCHEMICAL SYSTEMS: EXAMPLE

We end this chapter with an example of an industrial electrolytic process—chlorine production by the *chlor-alkali* route. Our objectives in this section are twofold. First, to review principles and terminology within the context of an important electrochemical system. Second, to introduce the concept of the *current–voltage relationship*, also called the *I–V* curve. The connection between current and voltage is essential to any analysis of electrochemical systems. Finding and understanding this *I–V* relationship is a persistent theme of the book.

A simple representation of the electrochemical system for the chlor-alkali process is shown in Figure 1.4. The overall reaction can be written as

$$2H_2O + 2Cl^- \rightarrow Cl_2 + H_2 + 2OH^- \quad (1.17)$$

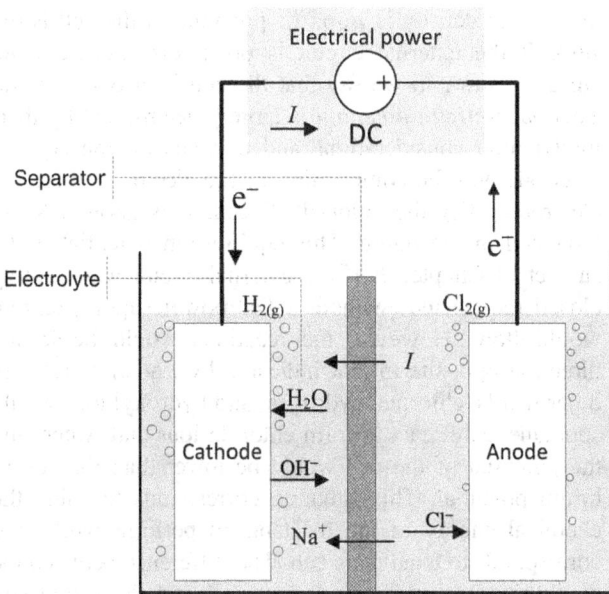

Figure 1.4 Electrochemical cell for the chlor-alkali process.

In order to operate without interruption, reactants are supplied and products removed continuously. The reactant stream for the process is a purified brine of sodium chloride (NaCl), essentially a concentrated solution of salt water. The solution flows at a steady rate into the electrochemical cell, also called the electrochemical reactor. In addition to chlorine gas, sodium hydroxide and hydrogen are produced simultaneously in the process. Despite the appearance from the above equation, this reaction does not occur homogeneously. Instead, two heterogeneous electron-transfer reactions take place on separate metal surfaces:

$$2Cl^- \rightarrow Cl_2 + 2e^-$$

$$2H_2O + 2e^- \rightarrow H_2 + 2OH^-$$

The two metals on which the reactions occur are the electrodes, which are inert since they do not participate in the reactions. The electron-transfer reactions are faradaic processes and are governed by Faraday's law. Oxidation takes place at the first electrode of our example, where two chloride ions lose electrons to form chlorine gas. Therefore, this electrode is the anode. The electrons enter the metal and travel through an external circuit. At the other electrode, the cathode, water is reduced to form hydrogen gas. The electrons consumed at the cathode are supplied by the external circuit, and the number of electrons produced at the anode is the same as the number consumed at the cathode. The electrolyte that is between the two electrodes contains ions of Na^+, Cl^-, and OH^-. In this instance, there is a membrane separator that serves two functions: It keeps the two gaseous products apart and it allows sodium ions to move from the anode compartment to the cathode compartment, but excludes transport of anions. It is this movement of sodium ions that carries the current in the electrolyte. Note that there are three phases: solid (electrodes), liquid (electrolyte), and gas (reaction products). Furthermore, not all species are present in every phase. For instance, free electrons are found only in the metals, ions are confined to the electrolyte, and hydrogen gas is limited to the cathode. This heterogeneity is one of the defining characteristics of electrochemical systems—Almost all of the action occurs at interfaces between the electrode and the electrolyte.

As you are probably beginning to appreciate, electrochemical systems are inherently complex. We've already seen that both electrodes are necessary and must be considered together. Many things happen at once: electron transfer, adsorption and desorption, transport of reactants and products, current flow in the electrolyte, and surface-mediated combination. In order to understand and analyze electrochemical systems, we'll need to simultaneously apply our knowledge of thermodynamics, kinetics, and transport. These fundamental topics will be considered systematically in the chapters that follow.

Electrochemical cells are sometimes broken into two categories: *galvanic* and *electrolytic*. The two types have more similarities than differences, and are governed by the same engineering principles. The main distinction between them is that energy or electrical work is an output of a galvanic cell (e.g., a fuel cell or a battery discharging), whereas electrical energy is an input to electrolytic cells. We see in Figure 1.4 that a direct current (DC) power supply adds electrical work in the chlor-alkali process; therefore, this process is electrolytic.

An important feature of electrochemical systems is that current is an expression of the rate of reaction. In our example, the electrical current through the cell is directly related to the formation and consumption rates of hydrogen, chlorine, and sodium chloride through Faraday's law. We can control the rate of reaction by controlling current. How much work is associated with this current flow? To answer this question, we need to know the potential of the cell. The potential is an expression of energy for charged species (electrons and ions), and provides the driving force needed for electrochemical reactions and for the flow of current in both metals and electrolytes.

We cannot simultaneously control the potential and current of an electrochemical system. If the potential applied to our electrolysis cell is increased, then the rate of hydrogen/chlorine production will also be increased. Alternatively, we might establish the rate of production by setting the current, but we would need to accept the resulting potential. These are the two basic modes of operation for electrochemical cells: *potentiostatic* (constant potential) or *galvanostatic* (constant current).

As mentioned above, a key objective in the analysis of any electrochemical system is to find the relationship between current and potential. The relationship can be complex, but with careful examination of the thermodynamics, kinetics, and mass transport, it can be readily understood. In the first chapters of this book, we will develop the fundamental principles needed. These will systematically be applied to a number of electrochemical systems, and we will return repeatedly to the relationship between the current and voltage as characterized by the *I–V* curve. Much of this textbook will be dedicated to the development and application of the relationships that describe the current (the reaction rate) as a function of potential (the driving force). You already know one such relationship, the resistor. Using an equivalent form of Ohm's law (Equation 1.15), we can relate the current and potential for a resistor. If the current density is constant, Equation 1.16 can be written in terms of resistance to give

$$\Delta \phi = \frac{iL}{\kappa} = iR_\Omega. \quad (1.18)$$

This equation shows a linear relationship between potential and current density. However, the current–voltage relationship is generally more complex for an electrochemical system, and we will need to use thermodynamics (Chapter 2), electrode kinetics (Chapter 3), and mass transfer (Chapter 4) to describe it. Here, we simply want to introduce some of the basic features of such a curve.

An example of a current–voltage curve is shown in Figure 1.5 for steady-state operation. First, note that when the current density is zero, the potential of the cell is *not* zero. If the external circuit is open (disconnected), no current flows and we say that the cell is at open circuit. This *open-circuit potential* is largely determined by thermodynamic considerations and depends on the type of electrode and the composition of the electrolyte near the electrodes. For the chlor-alkali cell, it is about 2 V, as indicated in the figure. This equilibrium potential is the subject of Chapter 2. If the external circuit were simply closed (connected) without a DC power supply, current would flow. However, the reactions would be in the direction opposite to that indicated by Equation 1.17. In other words, chlorine, hydrogen, and hydroxyl ions would spontaneously react to form chloride ions and water, and the potential of the cell would be lower than the equilibrium potential. This situation corresponds to using the chemical energy of the reactants to perform work, and corresponds to a galvanic cell. The difference between the equilibrium or open-circuit potential and the actual cell voltage is called *polarization* (see Figure 1.5).

Of course, this is not how a chlor-alkali cell is intended to operate. We want to produce chlorine gas and caustic soda. By the addition of work, we can drive the reaction in the opposite direction, the direction indicated in Equation 1.17. To do this, the cell potential must be greater than the equilibrium potential. Because we are adding work to the cell, we refer to the cell as an electrolytic cell. As shown in Figure 1.4, a voltage source is added to the chlor-alkali cell in order to provide the electrical work needed to drive the reaction in the desired direction. The difference between the cell potential and the equilibrium potential is a measure of how much work or energy must be added to drive the reaction at a particular rate.

Returning to Figure 1.5, which shows the *I–V* curve for the complete electrochemical cell, we have seen that an electrochemical cell has a nonzero potential at open circuit, where the current is zero. This potential is described by thermodynamics and represents the difference in the energy of electrons in the two electrodes. Operation of the cell in a galvanic mode takes place at cell potentials below the equilibrium potential, and corresponds to the conversion of chemical energy into electrical energy. In contrast, operation of the cell in an electrolytic mode occurs at potentials greater than the equilibrium potential, and requires that energy be added to the cell. In both cases, the difference between the equilibrium potential and the cell potential during current flow is called the polarization. As shown in the figure, the relationship between current and cell potential is, for the most part, nonlinear. The details will be covered in subsequent chapters. Also note that the curve is not symmetric. Although the basic shape of the curve under galvanic operation is similar to that of the electrolytic curve, the details are different.

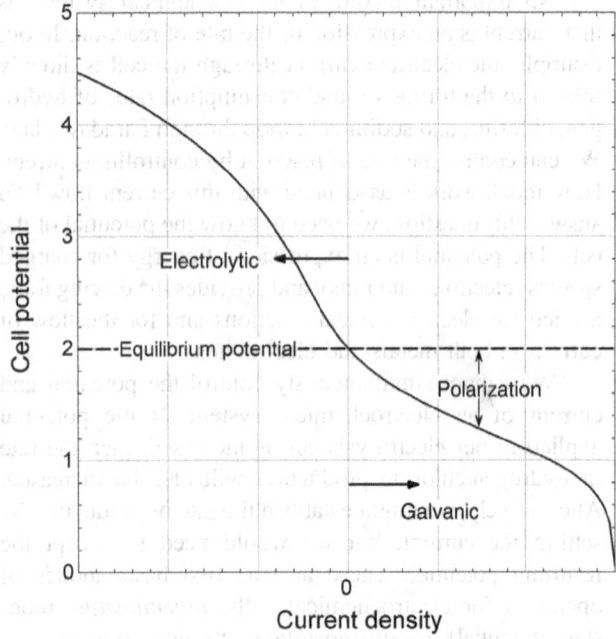

Figure 1.5 Representative relationship between current and potential at steady state. The dividing line between galvanic and electrolytic operation is at a current density of zero.

CLOSURE

In this chapter we have introduced the electrochemical cell and electrochemical systems. Common terminology for electrochemical systems has been presented, and will be used throughout the book. One of the most fundamental concepts, Faraday's law, was discussed. This law relates electrical charge to the amount of reactants consumed or products produced in an electrochemical reaction, and is essential for the analysis of electrochemical systems.

FURTHER READING

LeFrou, C., Fabry, P., and Poignet, C. (2012) *Electrochemistry: The Basics with Examples*, Springer Berlin.

Pletcher, D. (2009) *A First Course in Electrode Processes*, Royal Society of Chemistry, Cambridge.

West, A.C. (2013) *Electrochemistry and Electrochemical Engineering: An Introduction*. CreateSpace Independent Publishing Platform.

PROBLEMS

1.1 The original "International Ampere" was defined electrochemically as the current required to deposit 1.118 mg of silver per second from a solution of silver nitrate. Using this definition, how does the international ampere compare to the SI version? (Note that the SI version is based on the Ampere force law).

1.2 Molybdenum is deposited from a molten salt. 12.85 g are deposited in 1 hour using a current of 7 A. How many electrons are consumed per mole of Mo reacted? Given that Mo is present in the molten salt as an ion, what is its oxidation state in the molten salt?

1.3 How much hydrogen is needed to operate a 50 kW hydrogen/oxygen fuel cell for 3 hours? The potential of the cell is 0.7 V (remember that Power = IV). The reaction at the anode is

$$H_2 \rightarrow 2H^+ + 2e^-$$

1.4 Calculate the daily aluminum production of a 150,000 [A] aluminum cell that operates at a faradaic efficiency of 89%. The cell reaction is

$$2Al_2O_3 + 3C \rightarrow 4Al + 3CO_2$$

1.5 The annual production of Cl_2 is about 45 million tons per year. Assume that a typical plant is operational 90% of the year. The operating voltage of a cell is 3.4 V (considerably higher than the equilibrium voltage).

(a) Write down the half-cell reaction for the oxidation chloride to form chlorine.

(b) Determine the total current worldwide needed to generate the global supply of Cl_2.

(c) Calculate the electrical power needed to produce the global supply of chlorine using electrolysis.

1.6 A 25 A current is passed through a molten aluminum chloride melt. What are the likely reactions at the two electrodes? How long must this current be in place to deposit 50 g of Al. During this same time, what is the volume of gas that is evolved at STP (standard temperature and pressure: 273 K, 100 kPa).

1.7 An industrial aluminum cell is operating at 4.2 V with a current of 200 kA. The faradaic efficiency is 95 %, which means that 95% of the current goes to Al production. What is the rate of production of Al in kg per day? The starting material is Al_2O_3. As an aside, a Boeing 747 aircraft is made from over 60,000 kg of Al.

1.8 Continuous sheet copper is made by electrodeposition from a solution containing $CuSO_4$ onto a rotating drum of lead. For the conditions given below, what should be the rotation speed of the drum (revolution per hour)?

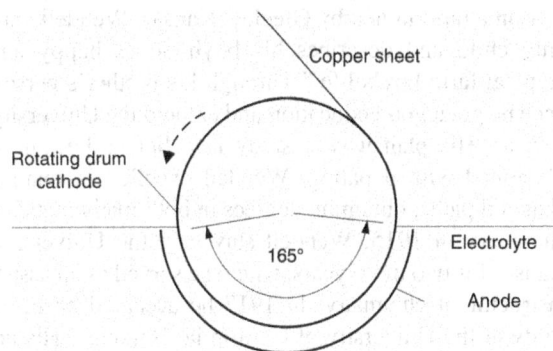

Cathode current density = 1750 [A·m^{-2}]

Current efficiency = 95%

Desired "thickness" (mass per area) = 1.22 [kg·m^{-2}]

Angle of cathode immersion = 165°

1.9 How many grams of lithium are in a 1320 [mAh] cell phone battery? Note that [mAh] is a unit of charge.

1.10 A plate of steel has lost 50 g to corrosion over the past year. Determine the current that would be associated with this rate of corrosion?

1.11 Corrosion of stainless steel in concrete is an important engineering problem. Because of the long times associated with the low rates of natural corrosion, accelerated testing is often used. One form of accelerated testing is the application of a potential to the metal of interest. However, under the accelerated conditions, the current efficiency may be low. In competition with the oxidation of iron to Fe(II), oxygen can be evolved at high pH:

$$2OH^- \rightarrow \frac{1}{2}O_2 + 2H_2O + 2e^-$$

A current of 1.4 mA is passed for 100 hours and the mass loss is 0.11 g. What is the faradaic efficiency for the iron oxidation reaction? How many moles of oxygen are evolved?

1.12 Consider a nickel–zinc battery operating at a current density of 4500 A·m^{-2}. The space between the electrodes is filled with an alkaline electrolyte. If the conductivity of the electrolyte is 60 S·m^{-1} and the distance between the two electrodes is 2 mm, what is the potential drop across the cell due to ohmic losses?

WENDELL MITCHELL LATIMER

Wendell Mitchell Latimer was born on April 22, 1893, in Garnett, Kansas. When Wendell was three, his father, a banker, moved the family to Kansas City to take advantage of the opportunities afforded by a larger city. The tragic death of his father from typhoid 5 years later put a financial strain on the family and forced Wendell and his mother to live with his grandfather on a farm in nearby Greeley, Kansas. Wendell was an only child and describes his boyhood as happy and a "typical farm boy's life." Through his mother's perseverance he got a good education and entered the University of Kansas. His plan was to study law, but he became disillusioned with debating. Wendell excelled in math and changed paths, obtaining degrees in both mathematics and chemistry in 1915. Wendell stayed at the University of Kansas for two more years, where he served as an assistant instructor of chemistry. In 1917 he accepted an offer to study at the University of California, Berkeley. His advisor, George E. Gibson, mentored a generation of leaders in physical chemistry and low-temperature calorimetry, including two Nobel Prize winners. Under Gibson's direction, Latimer completed his Ph.D. in 1919.

Wendell Latimer stayed on at Berkeley as a lecturer in chemistry. Latimer and a postdoctoral researcher named W.H. Rodebush were the first to publish findings on a special bonding of a proton between two strongly electronegative elements. Although the phrase was not used in their seminal 1920 publication, the phenomenon is now known as *hydrogen bonding*.

Much of Latimer's early research focused on low-temperature calorimetry and the determination of heat capacities and absolute entropies. To facilitate this research, he was the first in the United States to build and operate an experimental device to liquefy hydrogen. Latimer is largely responsible for developing the concept of single-ion entropies. He coauthored a couple of books on general and inorganic chemistry, but his legacy is *The Oxidation States of the Elements and Their Potentials in Aqueous Solution*. First published 1938, this invaluable monograph provided detailed thermodynamic data for elements in aqueous solutions and was reprinted in 1952. Although these data are best known for their use in the analysis of electrochemical systems, Latimer's interests appear to be more in elucidating redox chemistry. For example, he developed what is known as the *Latimer diagram*, which provides a convenient and compact way of viewing the complex reactions of iron or manganese.

One of Latimer's students, Kenneth Pitzer (Ph.D., 1937), went on to become an influential physical chemist in his own right, continuing the study of thermodynamics. During World War II, Latimer played a key role in the study of radiochemistry. From 1942 to 1946, he was director of a Manhattan Engineering District contract to study the chemistry of plutonium. Latimer was a member of the Electrochemical Society and was elected to the National Academy of Sciences, where he served as chairman of the chemistry section from 1947 to 1950. Wendell Latimer died on July 6, 1955.

Image Source: Reproduced with permission of G. PAUL BISHOP - PORTRAITS.

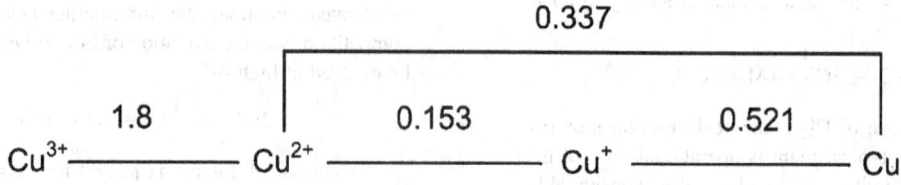

Latimer Diagram for copper.

Chapter 2

Cell Potential and Thermodynamics

The objectives of this chapter are to introduce the potential of an electrochemical cell and to develop a relationship between the cell potential and its chemical environment at equilibrium. The thermodynamics of electrochemical systems is no different than that of other systems—a dynamic equilibrium exists where an infinitesimal change in any driving force will force the system to shift reversibly to establish a new balance. One characteristic of electrochemical systems is that charged species and electron-transfer reactions are included. Hence, a new property, the electrical state, is important in describing equilibrium in electrochemical systems.

For an electrochemical system to be in equilibrium, there can be no net current flow. The condition of a cell where the external current is zero is termed *open circuit*, and the associated potential is the *open-circuit potential*. The equilibrium potential that we are interested in here is even more restrictive than the open-circuit potential. Not only is the external current zero, but a more general dynamic equilibrium exists in the cell. The equilibrium or thermodynamic potential is a critical characteristic that affects the design and operation of electrochemical devices. In this chapter, we use thermodynamics to calculate the cell potential as a function of the chemical components that make up the cell. As you will remember, an electrochemical cell is used to convert between chemical energy and electrical energy in order to (i) produce energy from stored chemicals, or (ii) use energy to produce chemical changes. The distinction pivots around the thermodynamic potential of the cell.

2.1 ELECTROCHEMICAL REACTIONS

An electrochemical reaction is a reaction where the transfer of electrons from a species being oxidized to a species undergoing reduction takes place through an electronic conductor. Typically, that conductor is a metal. Because the electron transfer takes place through a conductor rather than directly between the reacting species, we can separate the two electron-transfer reactions and use the flow of electrons (current) between them to do work. The oxidation or anodic reaction takes place at the anode, and the reduction reaction or cathodic reaction takes place at the cathode. An example of an anodic reaction is

$$Fe^{2+} \rightarrow Fe^{3+} + e^-$$

where 1 mol of electrons is produced for every mole of Fe (II) that is oxidized. An example of a cathodic reaction is

$$H^+ + e^- \rightarrow \frac{1}{2}H_2$$

Both an oxidation reaction and a reduction reaction are required to make an electrochemical cell. Therefore, an anodic or a cathodic reaction is referred to as a half-cell reaction as described in Chapter 1. The full-cell reaction is obtained by adding two half-cell reactions. For example, the two reactions above can be added to yield the following full-cell reaction:

$$Fe^{2+} + H^+ \rightarrow Fe^{3+} + \frac{1}{2}H_2$$

Note that there are no net electrons in the full-cell reaction, and that the charge on both sides of the reaction must balance. Identifying and understanding these half-cell reactions is essential to the analysis of electrochemical systems. Common reactions are tabulated in Appendix A.

2.2 CELL POTENTIAL

At open circuit, no current flows between the electrodes. Furthermore, for our thermodynamic analysis, each electrode half-cell reaction is at equilibrium (no net anodic or cathodic reaction occurs at either electrode). Under these conditions, it is appropriate to write the half-cell reactions

---

Electrochemical Engineering, First Edition. Thomas F. Fuller and John N. Harb.
© 2018 Thomas F. Fuller and John N. Harb. Published 2018 by John Wiley & Sons, Inc.
Companion Website: www.wiley.com/go/fuller/electrochemicalengineering

as reversible reactions. Since electrons are participants in each of the two reactions, the energy of the electrons in each conductor is determined by the reaction equilibrium at that electrode, and the electrical potential is different for each electrode. The difference in energy of the electrons at the two electrodes can be easily measured with a voltmeter as a voltage or potential difference. That potential difference at equilibrium is the principal topic of this chapter. This potential is also called the thermodynamic potential, U, since the half-cell reactions are at equilibrium.

Before proceeding with the analysis of electrochemical systems, it is productive to consider a simple, chemical-reacting system. A common exercise is to determine whether a reaction will occur spontaneously.

$$aA + bB \rightarrow cC + dD$$

The criterion used is the sign of the change in Gibbs energy for the reaction as written, namely,

$$\Delta G_{Rx} = \Delta G_f^{products} - \Delta G_f^{reactants},$$

where ΔG_f is the Gibbs energy of formation of the compound. If the change in Gibbs energy of the reaction, ΔG_{Rx}, is less than zero, then the reaction is said to proceed spontaneously. The first thing to note is that the sign of the change in Gibbs energy for the reaction depends on how the reaction is written, more specifically what is considered a product and what is treated as a reactant. Thus, it is more precise to say that the reaction will occur spontaneously as written. If the reaction is spontaneous, then we can obtain work from the reaction. If it is not spontaneous as written, then we would need to add work to force the reaction to go in that direction.

You are undoubtedly familiar with the reaction of hydrogen and oxygen, which combust to form water.

$$2H_2 + O_2 \rightarrow 2H_2O$$

Using values from Appendix C for the Gibbs energy of formation for the reactants and products at their *standard states*, a large negative ΔG_{Rx} is calculated. Thus, we conclude that the reaction as written will occur spontaneously, which is consistent with our expectations. Now of course, pure hydrogen and oxygen present together at near atmospheric pressure doesn't sound like the true thermodynamic equilibrium we seek. In fact, if these three species were together in equilibrium, we would find almost all water and only trace amounts of oxygen and hydrogen gas in our system. At this equilibrium state, the reactants and products are far from their standard states, and the change in Gibbs energy would be zero.

A hallmark of equilibrium of electrochemical systems is that there are two electrodes and at least one of the species is absent from each electrode. Furthermore, as noted earlier, electrons are involved in the half-cell reactions and, while the circuit is open, no electrons can be transferred between the two electrodes. Therefore, the half-cell reaction at each electrode will reach a dynamic equilibrium involving electrons where the rates of the forward and reverse reactions at that electrode are equal and there is no net current. The energy of the electrons for each of the half-cell reactions at equilibrium is different; this difference in electron energy, characterized by the potential difference, is what drives the full-cell reaction when the circuit is closed.

The same overall reaction of hydrogen and oxygen described above can occur electrochemically with two half-cell reactions. For instance,

$$4e^- + 4H^+ + O_2 \leftrightarrow 2H_2O$$

and

$$2H_2 \leftrightarrow 4H^+ + 4e^-$$

Notably, hydrogen is absent from the oxygen electrode, and oxygen is missing from the hydrogen electrode. At open circuit (i.e., no external current flow, and allowing sufficient time for equilibrium to be established), we can use thermodynamics to develop a relationship between the equilibrium cell potential, U, and the thermodynamic state of the reactants and products. In contrast to the combustion situation, it is possible for equilibrium to exist with significant amounts of hydrogen, oxygen, and water all present in the same electrochemical cell. This is because the hydrogen and oxygen are separated, with the hydrogen reaction in equilibrium at one electrode and the oxygen reaction in equilibrium at the other. In fact, we can now vary the amount of hydrogen, oxygen, and water independent of each other. However, the equilibrium potential of the cell, U, changes, and it is precisely this change in potential that is of interest here.

The laws of thermodynamics for a closed reversible system at constant temperature and pressure tell us that the maximum work that can be done by the system on the surroundings is equal to the change in the Gibbs energy per mol, ΔG [J·mol^{-1}]. In the electrochemical system of interest, that work is electrical work. In other words,

$$\Delta G = -W_{el}, \quad (2.1)$$

where work done by the system on the surroundings is defined as positive.

The cell potential, U, has units of joules per coulomb (J·C^{-1}, which is also V) and represents the work per unit charge required to move charge from one electrode to the other. The amount of charge transferred in the full reaction is equal to nF, where n is the number of moles of electrons in the reaction *as written*, and F is Faraday's constant, the number of coulombs per mole of electrons. The electrical work is therefore

$$W_{el} = nFU. \quad (2.2)$$

Combining Equations (2.1) and (2.2) yields

$$\Delta G = -nFU. \quad (2.3)$$

Equation 2.3 applies to a full cell consisting of an anodic and cathodic reaction, and relates the change in Gibbs energy for the chemical species to the difference in the electron energy at the two electrodes. Generally, we treat the potential of a full cell, U, as the potential difference between the more positive electrode and the more negative electrode. When this is done, U is positive and ΔG is negative for a full electrochemical cell at equilibrium. As described previously, a negative change in Gibbs energy implies a spontaneous reaction. This is in fact correct; when the circuit is closed for any cell at equilibrium that has a nonzero potential, current flows spontaneously. In order to ascertain the reactions that are occurring and the direction of the current, we simply choose the direction of the reaction such that the change in Gibbs energy is negative. In our example with hydrogen and oxygen, writing the reaction as

$$2H_2 + O_2 \rightarrow 2H_2O$$

resulted in a negative change in Gibbs energy, indicating that at one electrode hydrogen is oxidized (anode), and at the other electrode oxygen is reduced (cathode) to produce water when the circuit is closed. We now turn our attention to calculating the equilibrium potential of cells before the circuit is completed; i.e., at open circuit.

2.3 EXPRESSION FOR CELL POTENTIAL

Note that Equation 2.3 relates the Gibbs energy to the cell potential, U, which is the difference between the potential of the two electrodes. From a practical standpoint, that potential will be positive if you measure it one way, and negative if you switch the wires and measure it again. The fact that the potential can be measured either way illustrates the need to establish a convention so that the connection between the half-cell reactions, the cell potential, and the direction of the full-cell reaction is clear. To do this, we use a diagram similar to that illustrated in Figure 2.1 for a Daniell cell consisting of a zinc electrode in a zinc sulfate solution on one side of a semipermeable separator, and a Cu electrode in a copper sulfate solution on the other. The physical device (left side) is represented as the cell (right side) for thermodynamic analysis.

The vertical lines in the cell diagram indicate phase boundaries. In this example, we have four phase boundaries and five phases. On the left is zinc metal. The zinc electrode is in contact with an aqueous solution of zinc sulfate, which does not conduct electrons but allows for ions to move between the two metal electrodes. For simplicity, here we assume that the zinc sulfate solution is separated from the copper sulfate solution by a semipermeable membrane. This membrane allows for the transport of sulfate ions, but excludes all other species. Recall that at least one species is absent from each electrode. At the interface between the Zn and the solution of zinc sulfate, copper ions are absent. On the other side, zinc ions are absent. On the right is the copper electrode, which is more positive at standard conditions as we'll see shortly. The two electrochemical half-cell reactions are

$$Zn \leftrightarrow Zn^{2+} + 2e^- \quad (2.4)$$

and

$$Cu \leftrightarrow Cu^{2+} + 2e^- \quad (2.5)$$

The bidirectional arrow is used to emphasize the fact that each half-cell reaction is at equilibrium. By convention, the equilibrium potential of the cell, U, is equal to the potential of the electrode on the right minus the potential of the

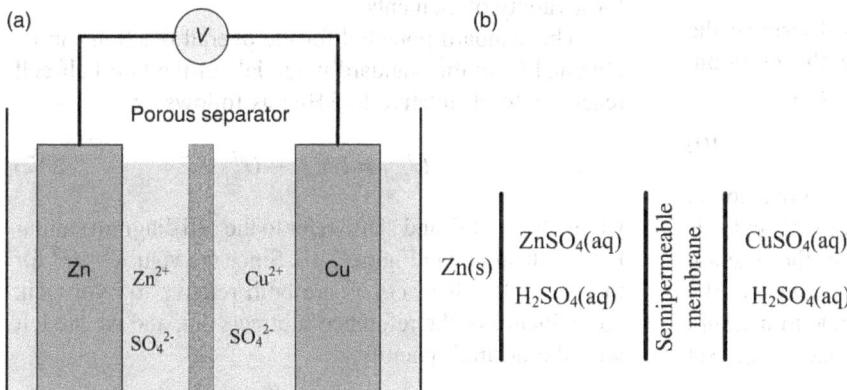

Figure 2.1 Daniel cell. (a) The physical representation. (b) Representation of the cell for thermodynamic analysis.

electrode on the left, or

$$U = \phi_{right} - \phi_{left}. \quad (2.6)$$

Since we generally define U so that it is positive, we put the more positive electrode on the right and the negative electrode on the left as shown in Figure 2.1. At this point, it is instructive to consider briefly what happens when the circuit is closed and a small amount of current is allowed to flow. Electrons naturally flow from low to high potential because of their negative charge. Therefore, if U is positive, the electrons would flow from the electrode on the left to the electrode on the right when the circuit is closed. Applying this knowledge to the cell considered above, we see that oxidation is occurring at the left electrode; that is, the left electrode is giving up electrons. Conversely, reduction is occurring at the right electrode:

$$Zn \rightarrow Zn^{2+} + 2e^- \text{ (anodic reaction on left electrode)} \quad (2.7)$$

and

$$Cu^{2+} + 2e^- \rightarrow Cu \text{ (cathodic reaction on right electrode)} \quad (2.8)$$

Thus, the spontaneous path is for zinc to be oxidized at the electrode on the left, and copper to be reduced at the other electrode.

$$Zn + Cu^{2+} \rightarrow Zn^{2+} + Cu \quad (2.9)$$

In other words, the direction of the spontaneous full-cell reaction that correctly corresponds to the cell potential is obtained by writing the reaction on the right as the cathodic reaction and that on the left as the anodic reaction. We now have a well-defined connection between the half-cell reactions, the overall or full reaction for the cell, and the equilibrium cell potential. In addition, Equation 2.3 provides a relationship between the Gibbs energy of the reaction and the equilibrium potential of the cell.

How can we use this information to determine the equilibrium potential of the cell? To answer this question, we next write the following expression for ΔG:

$$\Delta G = \Delta G° + RT \ln \prod a_i^{s_i}, \quad (2.10)$$

where $\Delta G°$ is the Gibbs energy change for the reaction at standard conditions, a_i is the activity of species i, and s_i is the stoichiometric coefficient of species i in the reaction (positive for products and negative for reactants) at the conditions of the cell. Because it is infeasible to measure and list the Gibbs energy for every reaction under every set of conditions, we tabulate the Gibbs energy at an arbitrary standard condition (25 °C, 1 bar) for species in a reference state (more on this later), and then correct this value for the particular conditions of interest. Combining Equation 2.10 with Equation 2.3 yields

$$U = U_{cell}^{\theta} - \frac{RT}{nF} \ln \prod a_i^{s_i}, \quad (2.11)$$

where U_{cell}^{θ} is the standard potential for the full cell ($\Delta G° = -nFU^{\theta}$) at 25 °C and is defined at the same standard conditions as $\Delta G°$. Also, similar to $\Delta G°$, U_{cell}^{θ} is not an absolute quantity, but a difference or relative value. One way to determine the standard potential for the full cell is to calculate it from the standard change in Gibbs energy for the reaction. However, it is often easier to use tabulated values of the standard potential for the reactions of interest as described in the next section.

2.4 STANDARD POTENTIALS

For convenience, values of standard potentials have been determined and tabulated for a large number of electrochemical reactions, a subset of which is found in Appendix A. Although listed as half-cell reactions, the potentials represent the difference between the potential of the reaction of interest and a reference reaction. The universal reference is the standard hydrogen electrode (SHE). Hence, the standard potential for the hydrogen reaction is defined to be zero. Tables of standard potentials are typically written as cathodic (reduction) reactions. Be mindful that for our thermodynamic analysis these reactions are assumed to be in equilibrium. Reactions whose standard potentials are positive relative to hydrogen naturally act as cathodes when coupled with hydrogen. In contrast, reactions with negative potentials are more anodic than hydrogen. A table such as that found in Appendix A is also a good place to start when you are unsure of how a particular compound might react. For example, the number of electrons transferred and the products produced can be readily determined from the table for a variety of elements.

The standard potential for the overall reaction can be obtained from the standard potentials of the two half-cell reactions (each relative to SHE) as follows:

$$U_{cell}^{\theta} = U_{right}^{\theta} - U_{left}^{\theta}, \quad (2.12)$$

where the "right" and "left" refer to the cell diagram similar to that illustrated in Figure 2.1b. Since the values of U^{θ} for the two half-cell reactions are both relative to hydrogen, the influence of the reference subtracts out, and we are left with the desired quantity.

ILLUSTRATION 2.1

Let's determine the standard potential for the Daniell cell shown in Figure 2.1. We see that the standard potential of the Cu is above that of the Zn reaction in Appendix A. Thus, the Cu electrode is shown on the right in the analysis diagram. The full-cell reaction is

$$Zn + Cu^{2+} \rightarrow Zn^{2+} + Cu,$$

where $n = 2$. The standard potential is

$$U^{\theta}_{cell} = U^{\theta}_{right} - U^{\theta}_{left} = U^{\theta}_{Cu/Cu^{2+}} - U^{\theta}_{Zn/Zn^{2+}}$$
$$= 0.337 - (-0.763) = 1.100 \text{ V}.$$

Let's consider a second example where the negative reaction is still the Zn reaction, but the positive reaction is replaced by the oxygen reaction on the right side: $O_2 + 4H^+ + 4e^- \leftrightarrow 2H_2O$.

The full-cell reaction is $O_2 + 4H^+ + 2Zn \rightarrow 2Zn^{2+} + 2H_2O$, where $n = 4$.

$$U^{\theta}_{cell} = U^{\theta}_{right} - U^{\theta}_{left} = U^{\theta}_{O_2} - U^{\theta}_{Zn/Zn^{2+}}$$
$$= 1.229 - (-0.763) = 1.992 \text{ V}.$$

Note that in the second example shown in the illustration it was necessary to multiply the stoichiometry of the negative reaction by two in order to get the electrons to balance and yield the correct overall reaction. However, we did *not* multiply the standard potential by the same factor. This practice may seem counterintuitive, but is correct. The standard potential represents the energy per charge, which does not change when you multiply the equation by the factor needed to make the electrons balance. The Gibbs energy changes because more electrons are involved, but the value per charge remains the same. This invariability will persist as long as the ratio of the number of electrons per mole(s) of reacting species remains the same. In other words, it doesn't matter if you have 2 electrons per mole of zinc, 4 electrons per 2 mol of zinc, or 5000 electrons per 2500 mol of zinc; the energy per unit charge is the same. That ratio always remains constant for the situation that we have considered involving a balanced anodic and cathodic reaction. Reworking the problem above with use of $\Delta G°$ helps illustrate the point:

Oxygen reaction: $O_2 + 4H^+ + 4e^- \rightarrow 2H_2O$;

$$\Delta G° = -4FU^{\theta}_{O_2}$$

Zinc reaction: $Zn^{2+} + 2e^- \rightarrow Zn$; $\Delta G° = -2FU^{\theta}_{Zn}$

Zinc reaction (doubled): $2Zn^{2+} + 4e^- \rightarrow 2Zn$;

$$\Delta G° = -4FU^{\theta}_{Zn}$$

Full cell: $O_2 + 4H^+ + 2Zn \rightarrow 2Zn^{2+} + 2H_2O$;

$$\Delta G° = -4FU^{\theta}_{O_2} - \left(-4FU^{\theta}_{Zn}\right)$$
$$\Delta G°_{cell} = -4F\left(U^{\theta}_{O_2} - U^{\theta}_{Zn}\right)$$
$$U^{\theta}_{cell} = \frac{\Delta G°_{cell}}{-4F} = U^{\theta}_{O_2} - U^{\theta}_{Zn}$$

As you can see, the total number of electrons for the anodic and cathodic reactions is the same, which results in U^{θ}_{cell} equal to the difference between that of the two half-cell values (Equation 2.12).

In the above example, it appears that we used Equation 2.3 to describe $\Delta G°$ of the half-cell reactions. However, as apparent from its derivation, Equation 2.3 only applies to full-cell reactions where the electrons are balanced; this equation relates the Gibbs energy change of the chemical species to the electrical work resulting from the difference in the electron energies that correspond to each of the half-cell reactions at equilibrium. Implicit in the use of Equation 2.3 for a "half-cell" is the fact that the other half is the standard hydrogen electrode, whose Gibbs energy is, by definition, equal to zero. The hydrogen portion simply subtracts out when we calculate $\Delta G°_{cell}$ as shown above. Subtracting off a balanced full-cell reaction does indeed reverse the sign of its $\Delta G°$ as shown in the example. In contrast to the above, calculation of the Gibbs energy change *for the half-cell alone* must involve the chemical potential of the electrons as demonstrated in Section 2.12.

Half-Cell Potentials Not in Table

What do we do if the reaction that we need is not in the table of standard potentials? One possibility is to combine existing half-cell reactions to create the desired half-cell reaction. In doing so, we always go through the Gibbs energy route (just demonstrated above) in order to avoid errors. The creation of a new half-cell reaction will of necessity not balance electrons since we want electrons in the final expression.

To illustrate:

Desired reaction: $Cr^{3+} + e^- \rightarrow Cr^{2+}$

Reaction 1 from Appendix A:

$$Cr^{2+} + 2e^- \rightarrow Cr; \quad \Delta G° = -2FU^{\theta}$$

Reaction 2 from Appendix A:

$$Cr^{3+} + 3e^- \rightarrow Cr; \quad \Delta G° = -3FU^{\theta}$$

To get the desired reaction, we subtract Reaction 1 from Reaction 2:

$$Cr^{3+} + e^- \rightarrow Cr^{2+};$$

$$\Delta G^{\circ}_{Cr^{3+}/Cr^{2+}} = -3FU^{\theta}_{Cr^{3+}/Cr} - \left(-2FU^{\theta}_{Cr^{2+}/Cr}\right)$$

$$= -FU^{\theta}_{Cr^{3+}/Cr^{2+}}$$

$$U^{\theta}_{Cr^{3+}/Cr^{2+}} = \frac{\Delta G^{\circ}}{-F} = 3U^{\theta}_{Cr^{3+}/Cr} - 2U^{\theta}_{Cr^{2+}/Cr}$$

$$U^{\theta}_{Cr^{3+}/Cr^{2+}} = 3(-0.74) - 2(-0.91) = -0.40 \text{ V}.$$

As you can see, the standard potential for the new half-cell reaction is not just the difference between half-cell potentials of the reactions that were combined. The reason for this adjustment is that the Gibbs energies above actually correspond to full-cell values with hydrogen on the left as the anode. In cases where we have electrons left over, the hydrogen reaction does not cancel out as it does when the electrons balance, and we are left with the difference between the half-cell of interest and the hydrogen electrode, consistent with other standard potentials. Consequently, it is important to use Gibbs energies when combining half-cell reactions to form a new half-cell reaction.

Standard Potential from Thermodynamic Data

Standard potentials for the cell can also be calculated directly from Gibbs energy data using Equation 2.3. The standard Gibbs energy of formation is tabulated for compounds in different states (gas, liquid, aqueous, and solid) assuming they are formed from the elements taken at 25 °C, 1 bar pressure. For a cell reaction,

$$\Delta G^{\circ}_{Rx} = \sum_i s_i \Delta G^{\circ}_{f,i}, \qquad (2.13)$$

where s_i are the stoichiometric coefficients for the reactants and products. By convention, s_i is positive for products and negative for reactants. As we have seen above, reactants and products of electrochemical reactions usually include ions. Because of the definition of the standard state for ions in electrolyte systems, which includes only interactions between the solvent and the ion of interest, the standard state Gibbs energy does not depend on the counterion(s) in the system. Rather, the impact of the counterions is included in the activity term. Therefore, tabulated values of ΔG°_f *for the ions* involved in the reaction should be used where available (see Appendix C), and are frequently tabulated for aqueous systems. When single-ion values are not available, the difference in the Gibbs energy of formation for ionic species can be calculated from ΔG°_f for two neutral aqueous (aq) species with the same counterion.

For the dissociation of a neutral species,

$$M_{\nu_+} X_{\nu_-} \rightarrow \nu_+ M^{z+} + \nu_- X^{z-};$$

and from Equation 2.13

$$\Delta G^{\circ}_{f,M_{\nu_+}X_{\nu_-}} = \nu_+ \Delta G^{\circ}_{f,M^{z+}} + \nu_- \Delta G^{\circ}_{f,X^{z-}}. \qquad (2.14)$$

Therefore, the counterion portion will cancel out when the aqueous values are subtracted (see Illustration 2.2). Finally, a combination of ΔH°_f and S° data can be used to estimate ΔG°_{Rx}. The following illustration demonstrates how to determine U^{θ}_{cell} for each of the situations mentioned above.

ILLUSTRATION 2.2

The balanced reaction for the oxidation of Fe by oxygen is

$$2Fe + O_2 + 4H^+ \rightarrow 2Fe^{2+} + 2H_2O$$

where four electrons are transferred in the reaction. We have already used half-cell potentials to calculate the standard potential. We will now calculate the potential from Gibbs energy data (ions and neutral species), and from enthalpy and entropy data. Note that the calculation below is for the reaction as written, specifically for 1 mol of oxygen.

1. Gibbs energy data using ionic species:
   $\Delta G^{\circ}_{f,Fe} = 0$
   $\Delta G^{\circ}_{f,O_2} = 0$
   $\Delta G^{\circ}_{f,H^+} = 0$
   $\Delta G^{\circ}_{f,Fe^{2+}} = -78.9 \text{ kJ mol}^{-1}$
   $\Delta G^{\circ}_{f,H_2O} = -237.2 \text{ kJ mol}^{-1}$ for liquid water
   From Equation 2.13,
   $\Delta G^{\circ}_{Rx} = 2 \text{ mol}\left(-78.9 \frac{kJ}{mol}\right) + 2 \text{ mol}\left(-237.2 \frac{kJ}{mol}\right) = -632 \text{ kJ}$
   Therefore,
   $U^{\theta} = \frac{-\Delta G^{\circ}_{Rx}}{4F} = \frac{-(-632.2 kJ)\left(1000 \frac{J}{kJ}\right)}{(4 \text{equiv})\left(96485 \frac{C}{equiv}\right)} = 1.64 \text{ J C}^{-1} = 1.64 \text{ V}$

2. Gibbs energy data using neutral aqueous species:
   $\Delta G^{\circ}_{f,Fe} = 0,$
   $\Delta G^{\circ}_{f,O_2} = 0,$
   $\Delta G^{\circ}_{f,HCl} = -131.26 \text{ kJ mol}^{-1}$ (aq),
   $\Delta G^{\circ}_{f,FeCl_2} = -341.373 \text{ kJ mol}^{-1}$ (aq),
   $\Delta G^{\circ}_{f,H_2O} = -237.2 \text{ kJ mol}^{-1},$
   $\Delta G^{\circ}_{Rx} = 2 \text{ mol}\left(-341.373 \frac{kJ}{mol}\right) +$
   $\quad 2 \text{ mol}\left(-237.2 \frac{kJ}{mol}\right) - 4 \text{ mol}\left(-131.26 \frac{kJ}{mol}\right)$
   $\quad = -157.7 \text{ kJ} + 2 \text{ mol}\left(-237.2 \frac{kJ}{mol}\right)$
   $\quad = -632 \text{ kJ},$

which is the same value as above. Note also that the difference between the iron chloride and the hydrogen chloride terms yielded the same value as obtained from the Gibbs energies of formation for the ionic species since the chloride contribution cancels out.

3. Enthalpy and entropy data:

$\Delta H^\circ_{f,Fe} = 0$,
$\Delta H^\circ_{f,O_2} = 0$,
$\Delta H^\circ_{f,H^+} = 0 \text{ kJ mol}^{-1}$ (aq),
$\Delta H^\circ_{f,Fe^{2+}} = -89.119 \text{ kJ mol}^{-1}$ (aq)
$\Delta H^\circ_{f,H_2O} = 285.830 \text{ kJ mol}^{-1}$,
$S^\circ_{Fe} = 27.28 \text{ J K}^{-1} \text{ mol}^{-1}$,
$S^\circ_{O_2} = 205.03 \text{ J K}^{-1} \text{ mol}^{-1}$,
$S^\circ_{H^+} = 0 \text{ J mol}^{-1}$ (aq)
$S^\circ_{Fe^{2+}} = -137.654 \text{ J K}^{-1} \text{ mol}^{-1}$ (aq)
$S^\circ_{H_2O} = 69.91 \text{ J K}^{-1} \text{ mol}^{-1}$,

$\Delta H^\circ_{Rx} = 2 \text{ mol}(-89.119 \tfrac{kJ}{mol}) + 2 \text{ mol}(-285.830 \tfrac{kJ}{mol}) = -749.898 \text{ kJ}$,

$\Delta S^\circ_{Rx} = 2 \text{ mol}(-137.654 \tfrac{J}{K\,mol})$
$\quad + 2 \text{ mol}(69.91 \tfrac{J}{K\,mol})$
$\quad - 1 \text{ mol}(205.03 \tfrac{J}{K\,mol})$
$\quad - 2 \text{ mol}(27.28 \tfrac{J}{K\,mol}) = -395.08 \tfrac{J}{K}$,

$\Delta G^\circ_{Rx} = \Delta H^\circ_{Rx} - T\Delta S^\circ_{Rx} = -749.898 \text{ kJ}$
$\quad - 298 \text{ K}\left(-395.08 \tfrac{J}{K}\right)\left(\tfrac{kJ}{1000J}\right)$
$\quad = -632 \text{ kJ}$,

which again is the same as above. Note that the entropy of an elemental species is typically not zero.

2.5 EFFECT OF TEMPERATURE ON STANDARD POTENTIAL

The cell potential in the relationships shown previously was determined at 25 °C, where the standard values have been tabulated. However, it is often desirable to calculate the cell voltage at a different temperature. To do so, we need the standard potential at that temperature. This section describes how to correct the standard potential to the temperature of interest. Beginning with one of the fundamental relations from thermodynamics,

$$\left(\frac{\partial G}{\partial T}\right)_p = -S, \qquad (2.15)$$

we find that

$$nF\left(\frac{\partial U}{\partial T}\right)_p = \Delta S. \qquad (2.16a)$$

If we further assume that ΔS does not change significantly over the temperature range of interest, it follows that

$$\left(\frac{\partial U^o}{\partial T}\right)_p = \frac{\Delta S^o}{nF} \qquad (2.17)$$

and

$$U^o|_T \approx U^\theta|_{25°C} + [T(°C) - 25]\left(\frac{\partial U^o}{\partial T}\right)_p. \qquad (2.18)$$

Tabulated values of $\partial U^o/\partial T$ are available in the literature. If the entropy change varies significantly with temperature over the range of interest, then integration of the temperature-dependent entropy term may be required as follows:

$$\int dU^o = \frac{1}{nF}\int \Delta S^o dT. \qquad (2.19)$$

Once the standard potential is known at the new temperature, activity corrections can be made as usual. This process is illustrated in the example at the end of the next section.

The following alternative to Equation 2.16 is equally rigorous:

$$\left(\frac{\partial U/T}{\partial T}\right)_p = \frac{\Delta H}{nFT^2}. \qquad (2.16b)$$

This approach has an advantage when the temperature range is sufficiently large that the assumption of constant ΔS or ΔH is not valid. If constant pressure heat capacity data (C_p) are available, then the change in enthalpy can be calculated as a function of temperature,

$$\frac{\partial \Delta H}{\partial T} = \Delta C_p. \qquad (2.20)$$

See Problem 2.29 to explore further this method.

ILLUSTRATION 2.3

A chlor-alkali cell is used in industry to produce chlorine, hydrogen, and sodium hydroxide. The cell is divided into two sides, the anodic side and the cathodic side. The anodic reaction is the oxidation of chloride ion, which takes place under slightly acidic conditions at a pH of 4. The cathodic reaction is the evolution of hydrogen under basic conditions (pH~14). The salt solution on the anodic side is 5 M NaCl. Please determine the standard potential for the cell at 65 °C.

22 Electrochemical Engineering

Solution:

We begin by writing the equations for the anodic and cathodic reactions; that is, the half-cell reactions. As usual, we put the positive electrode on the right side (see Figure 2.1b).

Positive: $Cl_2 + 2e^- \leftrightarrow 2Cl^-$; $U^\theta = 1.360$ V (25 °C)

Negative: $2H_2O + 2e^- \leftrightarrow H_2 + 2OH^-$;
$$U^\theta = -0.828 \text{ V} (25 °C)$$

Full cell: $Cl_2(g) + H_2(g) + 2OH^- \rightarrow 2H_2O + 2Cl^-$

$$U^\theta_{cell}(25 °C) = U^\theta_{Cl_2} - U^\theta_{H_2}$$
$$= 1.360 - (-0.828) = 2.188 \text{ V}.$$

Note that if the circuit were closed for this cell, chlorine, hydrogen, and hydroxyl would spontaneously be consumed to generate water and chloride. This is opposite the desired direction. Therefore, in practice, the potential of the cell is raised above the equilibrium potential, and work is added to the system in order to force the reaction in the reverse direction in order to produce hydrogen and chlorine.

To correct the standard potential for temperature:

$$\Delta S^o = (2 \text{ mol } Cl^-)(56.48 \text{ J K}^{-1}\text{mol}^{-1})$$
$$+ (2 \text{ mol } H_2O)(69.91 \text{ J K}^{-1}\text{mol}^{-1})$$
$$- (1 \text{ mol } Cl_2)(223.1 \text{ J K}^{-1}\text{mol}^{-1})$$
$$- (1 \text{ mol } H_2)(130.7 \text{ J K}^{-1}\text{mol}^{-1})$$
$$- (2 \text{ mol } OH^-)(-10.75 \text{ J K}^{-1}\text{mol}^{-1})$$
$$= -79.52 \text{ J K}^{-1}$$

$$U^o_{cell}(65 °C) = U^\theta_{cell}(25 °C)$$
$$+ (65 °C - 25 °C)\frac{\Delta S^o}{nF}$$
$$= 2.188 + (40 \text{ K})(-79.52 \text{ J K}^{-1})/$$
$$2 \text{ equiv}/96485 \text{ [C/equiv]}$$
$$= 2.188 - 0.0165 = 2.172 \text{ V}.$$

2.6 SIMPLIFIED ACTIVITY CORRECTION

Now that we have a way of obtaining U^θ for use in Equation 2.11, we must add the activity correction to get the desired expression for U. A discussion of activities and standard states for electrolytes is provided later in Section 2.14. Activity is a dimensionless quantity that depends on the standard state for each species. As a first approximation, we simply use the following:

Ionic species: $a_i = \dfrac{c_i}{c_i^o}$. (2.21a)

Solid species (pure): $a_i = 1$. (2.21b)

Solvent (e.g., H_2O): $a_i = 1$. (2.21c)

Gas-phase species (e.g., H_2 or O_2): $a_i = \dfrac{p_i}{p^o}$. (2.21d)

The standard state for gaseous species is an ideal gas at 1 bar, and for ions in solution it is an ideal 1 molal solution. As a first approximation, we will use concentration as a proxy for molality. With these assumptions, Equation 2.11 becomes

$$U = U^\theta_{cell} - \frac{RT}{nF} \ln \left(\prod_{\substack{ionic \\ species}} \left(\frac{c_i}{c_i^o}\right)^{s_i} \prod_{\substack{gas \\ species}} \left(\frac{p_i}{p^o}\right)^{s_i} \right), \quad (2.22)$$

where s_i is positive for products and negative for reactants. This equation is similar to the classical *Nernst equation*. The assumptions implicit therein are those most frequently used to approximate cell potential. Note that because of our assumption of unit activity for any solid reactants and the solvent, these species do not appear in Equation 2.22. The following illustration demonstrates the use of this equation and the process developed above to calculate the cell potential.

ILLUSTRATION 2.4

For the chlor-alkali cell described in Illustration 2.3, please determine the equilibrium potential at 25 °C and 2 bar (each side is at this pressure).

As usual, we write the half-cell reactions with the positive electrode on the right side (see Figure 2.1b).

Positive: $Cl_2 + 2e^- \leftrightarrow 2Cl^-$; $U^\theta = 1.360$ V (25 °C)

Negative: $2H_2O + 2e^- \leftrightarrow H_2 + 2OH^-$;
$$U^\theta = -0.828 \text{ V } (25 °C)$$

Full cell: $Cl_2(g) + H_2(g) + 2OH^- \rightarrow 2H_2O + 2Cl^-$

$$U^\theta_{cell}(25 °C) = U^\theta_{Cl_2} - U^\theta_{H_2} = 1.360 - (-0.828)$$
$$= 2.188 \text{ V}$$

> The full-cell reaction is written in the spontaneous direction, which gives a negative change in Gibbs energy and a positive potential. The number of electrons in the balanced reaction is 2, with products on the right. The OH⁻ concentration of interest is that on the alkali side. This is important since the hydroxide concentration is different on the two sides of the cell. pH = 14; $c_{OH^-} \approx 1$ M
>
> $$U_{cell}(25\,°C) = U^o_{cell}(25\,°C) - \frac{RT}{2F} \ln\left(\frac{c^2_{Cl^-}}{c^2_{OH^-}}\frac{p_{Cl_2}p_{H_2}}{(p^o)^2}\right)$$
>
> $$= 2.188\,\text{V} - \frac{(8.314)(273.15+25)}{(2)(96485)} \ln\left(\frac{(5)^2}{(1)^2(2)(2)}\right)$$
>
> $$= 2.164\,\text{V}.$$

Our approach to calculating the cell potential in the above illustration, consistent with the discussion in the chapter to this point, has been the following:

1. Write down the two half-cell reactions of interest and the standard potential for each reaction.
2. Put the reaction with the most positive standard potential on right side of the electrochemical cell as the cathode. The other reaction is the anodic reaction.
3. Write the full-cell reaction that appropriately combines the cathodic and anodic reactions. Make sure that the electrons are balanced.
4. Take the difference between the two standard half-cell potentials to get the standard potential for the full cell (cathode potential − anode potential).
5. Use the simplified activity corrections to correct the cell potential with the products on the top and the reactants on the bottom. Use the full-cell reaction to get the stoichiometric coefficient for each species and the total number of electrons for the reaction.

In using this approach, it is necessary to keep track of the location of the reactant and product species in order to make the required activity corrections, since those corrections, of necessity, use the environment adjacent to the electrode of interest. For example, in Illustration 2.4, the pH on the anodic side of the cell is approximately 4, and that on the cathodic side is about 14. Therefore, we have two different pH values that might be used to make the activity correction. In the example, we correctly selected a pH of 14 since the cathodic reaction depends on pH, whereas the anodic reaction does not. In other words, the activity corrections require us to use the concentration(s) that applies locally to each half cell. An alternative approach to finding the cell potential that many students find easier to use is to first determine the potential of each half cell relative to the SHE, *including activity corrections*, and then take the difference between the two half-cell potentials to get the full-cell potential. This approach is analogous to that used to calculate the standard potential, but has been expanded to include the needed activity corrections. The potential of any reaction versus SHE can be found from the following equation:

$$U - U^\theta_{SHE} = U^\theta - U^\theta_{SHE}$$
$$- \frac{RT}{nF} \ln\left(\prod_{\substack{ionic\\species}} \left(\frac{c_i}{c_i^o}\right)^{s_i} \prod_{\substack{gas\\species}} \left(\frac{p_i}{p^o}\right)^{s_i}\right), \quad (2.23)$$

where all of the activity corrections belong to the half-cell reaction of interest since there are no activity corrections, by definition, for the SHE. Note also that $U^\theta_{SHE} = 0$, and that the half-cell reaction was assumed to be expressed as

$$\sum_i s_i A_i = n e^-. \quad (1.10)$$

For half-cell reactions that are of the simple form

$$A_{ox} + ne^- = A_{red}, \quad (2.24)$$

U can be written as

$$U - U^\theta_{SHE} = U^\theta - \frac{RT}{nF}\ln\left(\frac{c_{red}}{c_{ox}}\right) = U^\theta + \frac{RT}{nF}\ln\left(\frac{c_{ox}}{c_{red}}\right), \quad (2.25)$$

where we have kept the standard potential for hydrogen in the equation to emphasize the fact that this potential is relative to SHE. This equation assumes that both the oxidized and reduced species are ions in solution, where the oxidized form of A is on the same side as the electrons in Equation 2.24. The reference concentrations were not explicitly included for the purpose of simplification since they are equal to unity. The following illustration demonstrates the alternative approach to calculating the full-cell potential.

ILLUSTRATION 2.5

For the chlor-alkali cell described in Illustration 2.3, please determine the equilibrium potential at 25 °C and 2 bar (each side is at this pressure).

As usual, we write the half-cell reactions with the positive electrode on the right side (see Figure 2.1b).

However, we will now correct each electrode separately for activity effects.

Positive: $Cl_2 + 2e^- \leftrightarrow 2Cl^-$; $U^\theta = 1.360$ V (25°C)

Correcting this electrode for composition gives the *half-cell potential relative to SHE*:

$$U_{Cl_2/Cl^-} = U^\theta_{Cl_2/Cl^-} - \frac{RT}{2F} \ln\left(\frac{c^2_{Cl^-}}{\frac{p_{Cl_2}}{p^o}}\right)$$

$$= 1.360\text{ V} - \frac{(8.314)(273.15+25)}{(2)(96485)} \ln\left(\frac{5^2}{2}\right)$$

$$= 1.3276\text{ V}.$$

Now repeat this same procedure for the negative electrode.

Negative: $2H_2O + 2e^- \leftrightarrow H_2 + 2OH^-$;
$U^\theta = -0.828$ V (25°C)

$$U_{H_2/OH^-} = U^\theta_{H_2/OH^-} - \frac{RT}{2F} \ln\left(\frac{c^2_{OH^-} p_{H_2}}{p^o}\right)$$

$$= -0.828\text{ V} - \frac{(8.314)(298.15)}{(2)(96485)} \ln\left(\frac{1^2(2)}{1}\right)$$

$$= -0.8369\text{ V}$$

$$U_{cell} = 1.3276 - (-0.8369) = 2.164\text{ V}$$

This value is the same answer as above. However, there was no need to write and balance the full-cell reaction, and there was no confusion regarding which concentration to use (that adjacent to the half-cell of interest). As a reminder, the spontaneous direction is where the more positive electrode is the cathode.

In these illustrations, the activity correction was small relative to the equilibrium voltage determined under standard conditions. While this is often the case, there are situations where the corrections make a significant difference and change the conclusions that would otherwise be made.

2.7 USE OF THE CELL POTENTIAL

Now that we have a way of determining the equilibrium potential, let's look at how we can use it. First, the convention we have adopted tells us that when the circuit is closed, a spontaneous reaction will occur. More specifically, a cathodic or reduction reaction takes place on the positive electrode, and an anodic or oxidation reaction occurs on the negative electrode. Electrochemical cells in which the reactions take place spontaneously to produce electrical current are called *galvanic cells*. A battery that is discharging is an example of such a cell. The maximum work that can be extracted from such a cell is equal to nFU (see Equation 2.3). However, to achieve this maximum, the cell would need to be discharged at an infinitely slow rate to avoid irreversible losses associated with the generation and movement of current in the cell. We will learn more about current and irreversible losses later in this book. In practice, as a consequence of irreversible processes, the operating voltage of a galvanic cell during discharge is less than the equilibrium voltage. But, as you can see, the equilibrium or thermodynamic potential serves as an essential reference point for the analysis of electrochemical cells.

In contrast to galvanic cells, an electrochemical cell in which energy must be added in order to drive the reactions in the desired direction is called an *electrolytic cell*. The magnitude of the applied potential, which is the operating voltage of the cell, must be greater than the equilibrium potential since energy must be added to overcome the equilibrium voltage in order to generate and move current through the cell. The chlor-alkali cell in Illustration 2.4 is an example of an electrolytic cell, where the full-cell reaction corresponding to the positive cell potential was opposite the desired direction. Thus, a potential greater than 2.164 V would need to be applied to such a cell in order to produce the desired products.

ILLUSTRATION 2.6

A commonly used battery is the lead–acid battery that you have in your car. The reactions for this battery are as follows:

Negative: $Pb(s) + SO_4^{2-} \leftrightarrow PbSO_4(s) + 2e^-$;
$U^\theta = -0.356$ V

Positive: $PbO_2(s) + SO_4^{2-} + 4H^+ + 2e^- \leftrightarrow PbSO_4(s) + 2H_2O$; $U^\theta = 1.685$ V

The electrolyte in the battery is about 5 M sulfuric acid. What is the thermodynamic potential of the battery at 25 °C?

The standard cell potential is 1.685 V − (−0.356 V) = 2.041 V.

The complete cell reaction is $PbO_2(s) + Pb(s) + 2SO_4^{2-} + 4H^+ \rightarrow 2PbSO_4(s) + 2H_2O$.

The equilibrium potential (assuming full dissociation of the acid and ion activities equal to the concentration ratios) is

$$U = 2.041\text{ V} - \frac{(8.314)(298)}{2(96485)} \ln\left(\frac{1}{5^2[(2)(5)]^4}\right) = 2.20\text{ V}.$$

> You may be wondering how your car has a 12 V battery and we calculated a cell voltage of just over 2 V. Your car battery contains six battery cells in series. Finally, full dissociation and unit activity coefficients are not good assumptions for the concentrated acid in these batteries, and the actual open-circuit voltage is about 2.13 V at room temperature.

2.8 EQUILIBRIUM CONSTANTS

The equilibrium constant for a chemical reaction is defined as

$$K \equiv \prod_i a_i^{s_i}. \quad (2.26)$$

Making use of Equation 2.11, the equilibrium constant can be related to the standard potential for the cell,

$$U = U_{cell}^\theta - \frac{RT}{nF} \ln \prod a_i^{s_i} = U_{cell}^\theta - \frac{RT}{nF} \ln K. \quad (2.27)$$

Also, since U for a reaction at equilibrium is equal to zero (as is ΔG),

$$\ln K = \frac{nF}{RT} U_{cell}^\theta \quad (2.28a)$$

and

$$K = \exp\left(\frac{nF}{RT} U_{cell}^\theta\right). \quad (2.28b)$$

You are probably familiar with the solubility product for a solute dissolving in water. Now we want to see how to use standard cell potential data to calculate it. To do this, let's determine the solubility product for AgCl. The relevant reaction is the following equilibrium reaction:

$$AgCl_{(s)} \leftrightarrow Ag^+ + Cl^-$$

This reaction can be viewed as the sum of the following two half-cell reactions:

$$AgCl_{(s)} + e^- \leftrightarrow Ag + Cl^-; \quad U^\theta = 0.222 \text{ V(cathode)}$$

and

$$Ag \leftrightarrow Ag^+ + e^-; \quad U^\theta = 0.7991 \text{ V(anode)}$$

Therefore, at 25 °C, $U_{cell}^\theta = U_{cathode}^\theta - U_{anode}^\theta = 0.222 - 0.7991 = -0.5771$ V. From Equations 2.23 and 2.28b,

$$K = \prod a_i^{s_i} = \frac{a_{Ag^+} a_{Cl^-}}{a_{AgCl_{(s)}}} = a_{Ag^+} a_{Cl^-} = \exp\left(\frac{F}{RT} U_{cell}^\theta\right)$$
$$= 1.76 \times 10^{-10}.$$

Thus, the equilibrium constant can be readily calculated for any reaction that can be written as the sum of two or more known half-cell reactions. Note that U_{cell}^θ must be known at the temperature of interest.

The ratio of activities and hence concentrations (if the activity coefficients are known) can also be determined for a half-cell at equilibrium if the cell potential at open circuit, U, is known and the other electrode reaction is fully specified. In this case, the full cell is not at equilibrium.

2.9 POURBAIX DIAGRAMS

A knowledge of the equilibrium potential for reactions involving a specified set of elements allows us to determine the species that are thermodynamically stable under a particular set of conditions. A common way to present such data in aqueous media is with a Pourbaix diagram, which has been particularly useful for studying corrosion. For example, the Pourbaix diagram for Zn at 25 °C is shown in Figure 2.2. This diagram presents a regional map of stable species as a function of the potential (versus SHE) and the pH. The construction of this diagram is outlined in the paragraphs that follow.

First, each diagram contains two reference lines (dashed in the figure) that represent the reactions for hydrogen and oxygen. For convenience, we reference reactions and equations to the corresponding lines on the Pourbaix diagram:

(a) $2H^+ + 2e^- \leftrightarrow H_2$

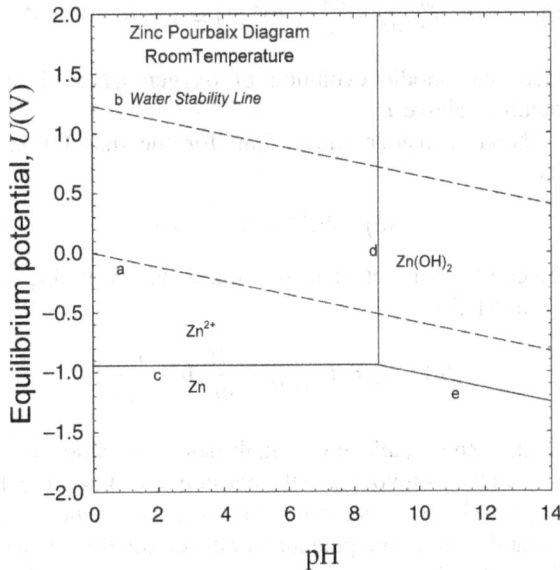

Figure 2.2 Simplified Pourbaix diagram for Zn.

and

$$\text{(b)} \quad \frac{1}{2}O_2 + 2H^+ + 2e^- \leftrightarrow H_2O$$

We can express the potential of these two reactions using the methodology described above. Reaction (a) relative to a standard hydrogen electrode is

$$\text{(a)} \quad U_a = U^\theta_{a/SHE} - \frac{RT}{nF}\ln\frac{\frac{p_{H_2}}{p^o}}{\left(\frac{c_{H^+}}{c^o}\right)^2} = -\frac{RT}{nF}\ln\frac{\frac{p_{H_2}}{p^o}}{\left(\frac{c_{H^+}}{c^o}\right)^2},$$

since $U^\theta_{a/SHE}$ is zero. If the hydrogen pressure is unchanged (at standard pressure), the cell potential varies only with the proton concentration. Assuming a reference state of ions as 1 M, this variation is most commonly expressed as a function of the pH as follows:

$$\text{(a)} \quad U_a = -\frac{RT}{2F}\ln\frac{1}{\left(\frac{c_{H^+}}{c^o}\right)^2} = \frac{RT}{F}\ln\left(\frac{c_{H^+}}{c^o}\right)$$

$$= -\frac{RT}{F}2.303\,\text{pH} = -0.0592\,\text{pH}$$

at 25 °C. Note that the switch from concentration to pH required us to change to a base 10 logarithm and then apply the simplified definition of pH ($-\log_{10} c$, where c is in mole per liter). U_a is represented as a line on the Pourbaix diagram as shown in Figure 2.2. At potentials greater than U_a, the anodic reaction is favored, and H^+ is the stable species. Conversely, the cathodic reaction is favored below U_a, and the stable species is H_2. The line, of course, represents the equilibrium potential.

Similarly, the potential for reaction (b) is

$$\text{(b)} \quad U_b = U^\theta_{b/SHE} - \frac{RT}{F}2.303\,\text{pH} = 1.229 - 0.0592\,\text{pH},$$

where the anodic evolution of oxygen takes place at potentials above U_b.

Next, consider equilibrium for the dissolution of zinc:

$$\text{(c)} \quad Zn^{2+} + 2e^- \leftrightarrow Zn$$

The equilibrium potential for the dissolution of zinc relative to SHE is

$$\text{(c)} \quad U_c = U^\theta_{c/SHE} - \frac{RT}{nF}\ln\frac{c^o}{c_{Zn^{2+}}}.$$

Metallic Zn is stable at potentials below U_c. Since neither H^+ nor OH^- is involved in the reaction, it makes sense that the potential does not depend on pH. Also, whereas the standard equilibrium potential for the dissolution of zinc is $U^\theta_c = -0.763$ V (Appendix A), this is not the value plotted on the graph. As seen by equation (c), if the concentration of the zinc ion changes, the equilibrium potential shifts. A series of lines could be plotted for the dissolution of Zn corresponding to different concentrations of ions in solution. By convention, a concentration of 10^{-6} M is usually assumed. At this concentration, the potential shifts negatively to -0.94 V, which is the value plotted in Figure 2.2.

As you may begin to realize, the Pourbaix diagram and associated calculations depend on the selection of species. This choice is not always clear. Up to 16 different species have been used in just the Pourbaix diagram for zinc. We will limit ourselves to just a few of these in order to get an idea of how these diagrams are constructed and how they can be used.

Consider a different type of reaction:

$$\text{(d)} \quad Zn^{2+} + 2OH^- \leftrightarrow Zn(OH)_2$$

How might this reaction be represented on the diagram? Since this is not an electron-transfer reaction, the reaction is not a function of potential and is therefore represented as a vertical line on the diagram. For our purposes, we use this equation to define the stability boundary between Zn^{2+} and $Zn(OH)_2$. Specifically, we are looking for the pH where $Zn(OH)_2$ is in equilibrium with 10^{-6} M Zn^{2+}. Using the methods described in the previous section, we find

$$K_{sp} = \frac{a_{Zn^{2+}}a^2_{OH^-}}{a_{Zn(OH)_2}} = 3 \times 10^{-17}.$$

With the concentration of zinc ions specified, the pH can be determined. The calculated pH is 8.74 (see Figure 2.2). Note that the choice of zinc concentration is arbitrary and represents the concentration above which Zn^{2+} is considered to be the stable species. The specified value of 10^{-6} M is frequently used for the analysis of corrosion systems.

One more reaction is considered:

$$\text{(e)} \quad Zn(OH)_2 + 2H^+ + 2e^- \leftrightarrow Zn + 2H_2O$$

This equilibrium is described by

$$U_e = -0.425 - \frac{RT}{F}2.303\,\text{pH},$$

and the line is labeled e on the diagram.

Now is a good time to review the meaning of these lines. These Pourbaix diagrams indicate the regions of stability of different phases and ionic species in equilibrium with the solid phases. Referring to line c, if the potential is below -0.94 V, Zn is stable. When the potential rises above -0.94 V, dissolution of Zn occurs. At higher potentials, when the pH of the solution is increased above 8.74, zinc hydroxide precipitates. Finally, at pH values above 8.74 and potential values higher than those indicated by line e, zinc can react directly with water to form zinc hydroxide.

2.10 CELLS WITH A LIQUID JUNCTION

The Daniell cell that was examined previously presupposed a selectively permeable membrane separator. In practice, simple porous media are often used. In this case, bulk mixing is avoided, but ions are able to move between electrodes. In fact, we often have electrochemical cells where the two electrodes are in solutions of different concentration and/or composition. The thermodynamic analysis described above accounts for the effect of the local solution composition on the equilibrium potential. However, there is also a small potential difference at open circuit associated with the junction between the two liquids of different composition. This potential difference is sometimes called the *liquid junction potential*, and is the topic of this section.

The liquid junction of interest is the region of varying composition between the two different electrolyte solutions. In practice, it is often a porous membrane that inhibits mixing of the two solutions, although several other physical possibilities exist such as use of a capillary tube to form a stable liquid junction. Ions must be able to move through the junction in order for current to flow in the electrochemical cell since ions are the current carriers. However, under the open-circuit conditions discussed in this chapter, the current is zero. Under such conditions, diffusion can still take place, but there can be no net transfer of charge. This situation violates two tenets of our thermodynamic analysis. The first is that the system is at true equilibrium—with concentration gradients and transport this condition is not strictly met. The second assumption is that at least one species is absent from each electrode. This condition too can no longer be guaranteed.

With this brief background, we can now describe the origin of the liquid junction potential. For illustration purposes, we can consider the situation where we have different concentrations of the same 1 : 1 binary electrolyte on opposite sides of a porous membrane as shown in Figure 2.3. At open circuit, there will be a diffusion driving force for ions to move from the high concentration side to the low concentration side. The problem is that the cation and anion typically have different diffusion coefficients. The ion with the largest diffusivity will initially move faster than the counterion. This difference in velocity causes a slight imbalance of charge, which results in a potential difference across the junction. The associated potential field in the junction serves to slow down the faster ion and speed up the slower ion so they move at the same rate. This potential difference is the liquid junction potential.

As you may have noticed from the above description, the liquid junction potential is the result of transport and is not thermodynamic in origin. The magnitude of this potential ranges from less than a millivolt to a few tens of millivolts (e.g., 20–30 mV), so that it is not a large correction. Hence, for the most part we will ignore it. Several methods exist for estimating the liquid junction potential. In general, the potential of a cell with a liquid junction can only be calculated with detailed knowledge of the concentration profile across the junction region. However, simplified methods such as the Henderson equation provide a common and straightforward way to estimate this potential. Awareness of potential errors from the junction potential may be important in selecting a reference electrode and in correcting measurements. Please refer to references in the "Further Reading" section for additional information.

There is one other aspect that is worth noting. Since the liquid junction potential is the result of the different transport rates of the anions and cations, the magnitude of this potential can be minimized by choosing anions and cations with similar diffusivities. Consequently, KCl is frequently used to minimize the junction potential. Finally, while we have described the situation for a binary electrolyte, the same physics and principles apply to multicomponent junctions.

2.11 REFERENCE ELECTRODES

The potential scale in Appendix A is based on the SHE. This scale is arbitrary, and by convention is taken to be zero at standard state as mentioned previously. For experimental work it is generally desirable to have a reference electrode in the system. The purpose of the reference electrode is to provide a known, stable potential against which other potentials can be measured. In principle, no current is passed through the reference electrode; therefore, it remains at its equilibrium potential—a potential that is known and well defined. In practice, a very, very small current is passed through the reference electrode to allow measurement of the potential; however, this current is not sufficient to move the electrode from its equilibrium potential. The desired characteristics for a reference electrode include the following:

- Reversible reactions
- Stable and well-defined potential

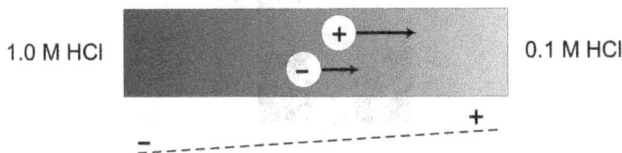

Figure 2.3 Schematic illustration of a liquid junction where the arrows are proportional to the diffusivity of the ion.

- Ion(s) that participates in the reference electrode reaction is present in the solution
- No liquid junctions that cause an offset in potential

The purpose of this section is to provide several examples of reference electrodes and to demonstrate calculation of the potential for some of these electrodes. The book by Ives and Janz should be consulted for more details.

Hydrogen Electrode

The hydrogen electrode (Figure 2.4) can be used in aqueous solutions over a wide range of pH values. It consists of a metal, such as platinum, on which hydrogen reacts rapidly and reversibly. The electrode is immersed in an aqueous solution, and hydrogen gas is bubbled around it. The hydrogen ion concentration (pH) is known in the aqueous solution that surrounds the electrode. The pressure above the solution is a combination of the hydrogen gas pressure and the water pressure. For example, the vapor pressure of water is 5 kPa at room temperature, so that a total pressure of 100 kPa corresponds to a hydrogen pressure of 95 kPa. Hydrogen is vented through a trap, which prevents air from diffusing into the cell. The reference electrode is connected to the point of interest through a capillary. The hydrogen electrode is appropriate for most aqueous solutions, but is not practical for many situations where hydrogen gas must be avoided. It is more difficult to use in unbuffered neutral solutions because of the challenge of maintaining a constant solution composition under such conditions. The hydrogen reactions under acidic and basic conditions are

$$H_2 \leftrightarrow 2H^+ + 2e^- \quad (0\ V)$$

and

$$H_2 + 2OH^- \leftrightarrow 2H_2O + 2e^- \quad (-0.828\ V)$$

The traditional hydrogen reference electrode is a *normal hydrogen electrode* (NHE), which has a hydrogen ion concentration in solution of one molar and a hydrogen gas pressure of 1 bar, and operates at a temperature of 25 °C. The NHE potential is very close to the SHE potential, but differs slightly due to non-ideal effects that are not present in the theoretical SHE.

Calomel Electrode

Another common reference electrode is the calomel electrode (Figure 2.5). Calomel refers to mercury(I) chloride, a sparingly soluble salt. This electrode is based on the reaction between Hg and Hg_2Cl_2:

$$2Hg_{(\ell)} + 2Cl^- \leftrightarrow Hg_2Cl_2 + 2e^- \quad (0.2676\ V)$$

For a saturated calomel electrode (SCE), the solution is often kept saturated by the addition of crystals of KCl to maintain a constant concentration of Cl^-. Note that the standard potential given above corresponds to an ideal 1 molal solution, rather than the Cl^- concentration of a saturated solution. For the saturated solution, the potential is about 0.242 V (SHE). The electrode consisting of the Hg (ℓ), Hg_2Cl_2(s) and saturated KCl is connected to the electrolyte solution of interest through a porous frit; this porous frit is equivalent to a salt bridge or junction region. These electrodes will generally have a liquid junction (see Section 2.10), although the correction in potential is not accounted for in most measurements.

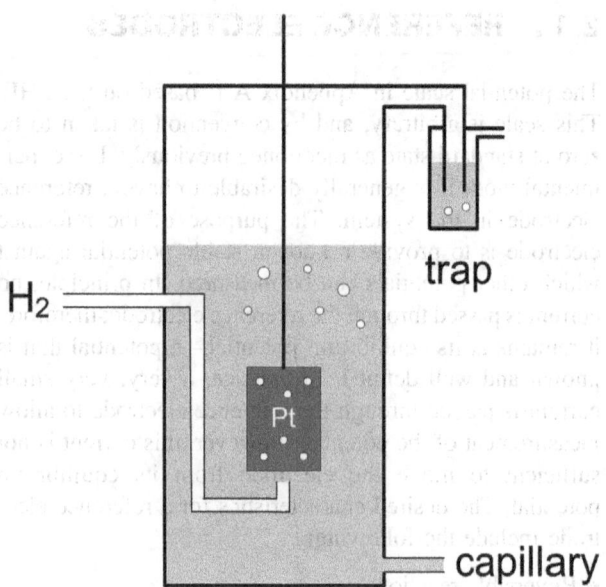

Figure 2.4 Hydrogen reference electrode.

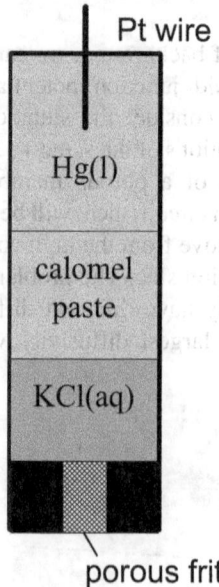

Figure 2.5 Calomel electrode.

Calomel electrodes are best suited to electrolytes that contain Cl⁻ and, conversely, should not be used in situations where low levels of chloride contamination are not acceptable. Additional mercury salt electrodes such as Hg/Hg_2SO_4 (second in popularity to the calomel electrode) and Hg/HgO (alkaline solutions) are available for use with other electrolytes.

Silver–Silver Chloride

Another popular reference electrode is the Ag/AgCl electrode, which is based on the following reaction:

$$AgCl_{(s)} + e^- = Ag_{(s)} + Cl^- \quad (0.2225 \text{ V})$$

The Ag/AgCl electrode could consist of a simple silver wire upon which a silver chloride layer has been formed. Alternatively, a base metal such as Pt can be used for the deposition of both silver and the silver halide. These electrodes are small and compact and can be used in any orientation. They can be inserted directly into the electrolyte solution of interest with no significant contamination. Ag/AgCl electrodes can be formed by either electrolytic or thermal methods. Bromide and iodide electrodes can likewise be formed. The thermodynamic properties of these electrodes do depend slightly upon the method of preparation. Commercial Ag/AgCl electrodes have a controlled solution concentration, which further increases their stability and reproducibility.

A related reference electrode is the silver sulfate electrode, which is suitable for use in a lead–acid battery since it shares the sulfate ion with the acid.

$$Ag_2SO_{4(s)} + 2e^- = 2Ag_{(s)} + SO_4^{2-}$$

ILLUSTRATION 2.7

Electrochemical experiments are frequently performed with a three-electrode experimental cell consisting of the anode, cathode and a reference electrode. Current flow, if present, occurs only between the anode and cathode; no current flows through the reference electrode. For our purposes here, let's consider a laboratory-scale cell for the electrodeposition of copper from an acidic sulfate solution. The composition of the electrolyte is 0.25M $CuSO_4$ and 1.8M H_2SO_4. Assume that the cell does not have a separator, and that the electrolyte is saturated with oxygen. The pressure is 1 bar. The cell reactions are as follows:

$$Cu^{2+} + 2e^- \leftrightarrow Cu$$

and

$$O_2 + 4H^+ + 4e^- \rightleftharpoons 2H_2O$$

A suitable reference electrode for this cell is the Hg/Hg_2SO_4 electrode, whose half-cell reaction is

$$Hg_2SO_4 + 2e^- \leftrightarrow 2Hg + SO_4^{2-}, \quad U^\theta = 0.6125 \text{ V}$$

The actual reference electrode solution is saturated K_2SO_4 (solubility ~120 g·L⁻¹). Please determine the equilibrium potential for both the oxygen and copper electrodes versus the reference electrode. What is the equilibrium potential for the anode versus the cathode? Note: there is a liquid junction between the saturated K_2SO_4 solution of the reference electrode and the acidic sulfate electrolyte of the cell, which you may ignore.

Solution:

First let's find the potential between the oxygen electrode and the reference electrode. To do this, we will first find the potential of the oxygen electrode and the reference electrode relative to SHE. We then subtract the reference electrode potential from that of the oxygen electrode to get the desired value.

Oxygen versus SHE:

$$U_{O_2/SHE} = U^\theta_{O_2} - \frac{RT}{nF} \ln \left(\frac{1}{\left(\frac{c_{H^+}}{1M}\right)^4 \frac{p_{O_2}}{1 \text{ bar}}} \right)$$

$$= 1.229 - \frac{(8.314)(298.15)}{4(96485)} \ln \left(\frac{1}{(3.6)^4 (0.21)} \right)$$

$$= 1.252 \text{ V}.$$

Reference versus SHE:

$$U_{Hg_2SO_4/SHE} = U^\theta_{Hg/Hg_2SO_4} - \frac{RT}{nF} \ln \left(\frac{c_{SO_4^{2-}}}{1M} \right)$$

$$= 0.6125 - \frac{(8.314)(298.15)}{2(96485)} \ln(0.689)$$

$$= 0.6173 \text{ V},$$

where $c_{SO_4^{2-}}$ is the saturation value for the reference electrode. The water activity was assumed to be one. Using these values,

$$U_{O_2/ref} = U_{O_2/SHE} - U_{Hg_2SO_4/SHE} = 0.635 \text{ V}$$

Similarly, for the copper electrode versus the reference electrode:

Copper versus SHE:

$$U_{Cu/SHE} = U^\theta_{Cu/Cu^{2+}} - \frac{RT}{nF}\ln\left(\frac{1}{c_{Cu^{2+}}}\right)$$

$$= 0.337 - \frac{(8.314)(298.15)}{2(96485)}\ln\left(\frac{1}{(0.25)}\right)$$

$$= 0.3192 \text{ V}.$$

Since we already have the potential of the reference electrode versus SHE, we can calculate

$$U_{Cu/ref} = U_{Cu/SHE} - U_{Hg_2SO_4/SHE} = -0.298 \text{ V}.$$

The potential between the anode and cathode can be determined from these values:

$$U_{O_2/Cu} = U_{O_2/ref} - U_{Cu/ref} = 0.635 - (-0.298)$$
$$= 0.933 \text{ V},$$

which can be compared with the value calculated directly (with O_2 on the right):

$$U_{O_2/Cu} = U_{O_2/SHE} - U_{Cu/SHE} = 1.252 - 0.3192$$
$$= 0.933 \text{ V}$$

Therefore, the reference electrode essentially splits the cell potential between the other electrodes into two pieces. This separation is a very valuable concept that will be used later.

2.12 EQUILIBRIUM AT ELECTRODE INTERFACE

In this chapter, we have shown how to calculate the thermodynamic potential and the relationship between this equilibrium value and the change in the Gibbs energy for the full-cell reaction. We have also referred to this potential as an equilibrium potential. Clearly, however, the full-cell reaction is not in equilibrium, which would mean that $\Delta G_{Rx} = 0$. Why, then, do we refer to this potential as the equilibrium voltage? What exactly is in equilibrium?

When there is no flow through an external circuit, then the half-cell reaction at each electrode approaches equilibrium. At equilibrium, the energy of the electrons in the metal is characteristic of the reaction, and related to the standard potential—more on this in a minute. First, what does it mean to approach equilibrium, and what exactly is in equilibrium?

An example in the form of a thought experiment may be helpful at this point. Let's imagine that we have a copper metal electrode in an acidic sulfate electrolyte. If the Gibbs energy for dissolution of the metal is negative, then some of the copper atoms will give up their electrons, leave the copper metal lattice, and move into solution as cations. This leaves excess charge in the copper metal, resulting in an increase in the electron energy. The addition of cations to the solution increases the copper ion activity. Since the half-cell reaction of interest is

$$Cu^{2+} + 2e^- \leftrightarrow Cu,$$

both the increase in electron energy and the increase in copper ion activity increase the rate of the reduction reaction relative to that of the oxidation reaction. The process continues until equilibrium is reached and the rates of the forward and reverse reactions are equal. While the above is a simplification of the actual physics, it provides a conceptual context for understanding the equilibrium of half-cell reactions. Thus, the half-cell voltage really does represent the equilibrium point for the reaction of interest on the hydrogen scale.

The Gibbs energy change, ΔG_{Rx}, is equal to zero for a half-cell reaction *at equilibrium*, which is approached under open-circuit conditions. At constant T and p, the Gibbs energy change for the reaction can be expressed in terms of the electrochemical potentials of the species that participate in the reaction:

$$\Delta G_{Rx} = \sum_i s_i \mu_i = 0, \quad (2.29)$$

where s_i is the stoichiometric coefficient for species i as we have defined and used it previously in this book. The electrochemical potential, μ_i, is defined as

$$\mu_i = \mu_i^{chem} + \mu_i^{electrical} = \mu_i^0 + RT\ln a_i + z_i F\phi, \quad (2.30)$$

which includes a chemical portion (μ_i^{chem}) and an electrical portion ($z_i F\phi$). The chemical portion is the chemical potential about which you may have learned in a course on chemical thermodynamics. The electrical portion represents the work required to bring a charge from infinity to a location inside the solution, and is known as the Galvani potential or inner potential. It is equal to zero for uncharged species. It is also zero for charged species in a bulk solution with no electric field or without surfaces where charges align to create fields. Thus, the columbic interactions of ions in a neutral bulk solution are included in the chemical term. In contrast, the impact of the electric field on ions in solution due to, for example, charge on the electrodes, is included in the electrical portion of the electrochemical potential.

Application of Equations 2.29 and 2.30 to the copper reaction yields, at equilibrium:

$$\sum_i s_i \mu_i = 0 = \mu_{Cu(M)} - \mu_{Cu^{2+}(aq)} - 2\mu_{e^-(M)}, \quad (2.31a)$$

where

$$\mu_{Cu^{+2}(aq)} = \mu^0_{Cu^{2+}(aq)} + RT \ln a_{Cu^{2+}(aq)} + 2F\phi_s, \quad (2.31b)$$

$$\mu_{Cu(M)} = \mu^0_{Cu(M)} + RT \ln a_{Cu(M)} = \mu^0_{Cu(M)}, \quad (2.31c)$$

$$\mu_{e^-(M)} = \mu^0_{e^-(M)} + RT \ln a_{e^-(M)} - F\phi_M = \mu^0_{e^-(M)} - F\phi_M, \quad (2.31d)$$

since the copper metal and the electrons are assumed to be in their standard state. Substituting Equations 2.31b–2.31d into Equation 2.31a and rearranging yields

$$\Delta\phi = \phi_M - \phi_s = \frac{\mu^0_{Cu^{2+}(aq)} + 2\mu^0_{e^-(M)} - \mu^0_{Cu(M)}}{2F}$$
$$+ \frac{RT}{2F} \ln a_{Cu^{2+}(aq)}, \quad (2.32)$$

$$= \Delta\phi^0 + \frac{RT}{2F} \ln a_{Cu^{2+}(aq)}, \quad (2.33)$$

where $\Delta\phi^0$ is the potential difference between the metal and the solution under standard conditions. The quantity $\Delta\phi = \phi_M - \phi_s$ *is the potential difference across the interface at equilibrium*. This interfacial potential plays an important role in electrochemical systems. For example, a potential difference across the interface that is greater than the value at equilibrium will cause the reaction to take place in the anodic direction. Conversely, the cathodic reaction will take place if the potential across the interface is less than the equilibrium value. We will examine this again in Chapter 3 when we discuss reaction kinetics. The purpose of this section was to illustrate the role of the potential in establishing interfacial equilibrium for single electrochemical reactions. This, of course, is unique to electrochemical reactions.

Unfortunately, $\Delta\phi = \phi_M - \phi_s$ is not accessible experimentally, since there is no way to measure the solution potential without introducing another interface, and therefore another interfacial potential drop, into the system. Consequently, how do we measure and quantitatively characterize the equilibrium potential for a half-cell reaction? We have already seen the results of the answer to this question. Since we can measure the potential between two electrodes, and we know that reversible half-cell reactions are at equilibrium at open circuit, we simply define a half-cell as our reference point and measure all other potentials relative to that reference. As long as the reference half-cell reaction remains at equilibrium, the process of defining a reference electrode is equivalent to adding a constant to the potential across the interface of the electrode of interest. The universal reference is the SHE, whose potential has been defined as zero. The hydrogen reaction is often highly reversible and reproducible. The potential of an electrode relative to a hydrogen electrode is measurable. Also, by making appropriate corrections, the potential of an electrode measured in a practical system relative to any other electrode at equilibrium can be quantitatively related to the potential of that electrode versus SHE. Thus, we have a well-defined way of determining the equilibrium potential of a half-cell reaction.

2.13 POTENTIAL IN SOLUTION DUE TO CHARGE: DEBYE–HÜCKEL THEORY

The previous section described a potential difference between a surface and the adjacent solution. This potential difference is due to unbalanced charge in solution as a result of charge on a surface or on an ion. Both of these situations (surface and ion) have been described in the literature using very similar approaches, and the key parameter that results is the same in both cases. Here we present the Debye–Hückel solution for the potential field surrounding a single charged central ion as shown in Figure 2.6. Our goal in presenting this material is to help you understand the basic physics of the problem and to introduce the *Debye length*. Later in this chapter we will use the solution developed here to provide a first approximation to the activity coefficient of an ion.

As shown in Figure 2.6, we shall evaluate the potential field surrounding a positively charged ion in solution. Ions near this central ion are affected by its charge, with negative ions drawn toward the central ion and positive ions pushed away. Ions in solution also experience random thermal motion, which tends to make the concentration more uniform. These two counteracting effects, the potential field, which tends to

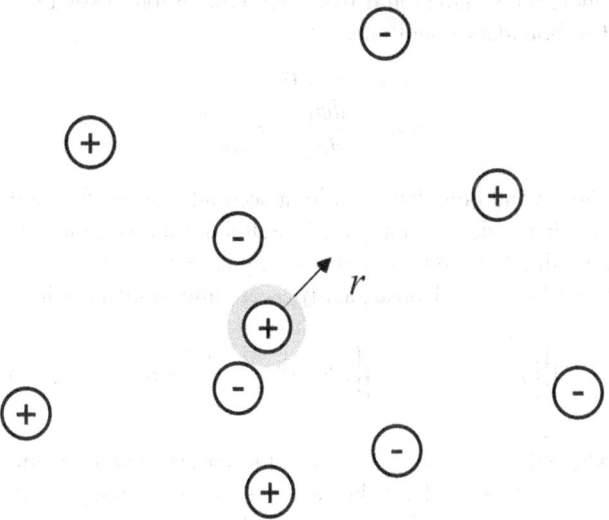

Figure 2.6 Central ion surrounded by charges.

separate ions, and random thermal motion, which tends to make concentrations more uniform, can be expressed quantitatively by the Boltzmann factor:

$$\frac{c_i}{c_i^\infty} = \exp\left(\frac{-z_i F\phi}{RT}\right). \qquad (2.34)$$

The quantity $z_i F\phi$ represents the work required to move a mole of ions to a different energy state as characterized by the local potential. This expression correctly yields a concentration of negative ions near the central ion that is greater than the bulk concentration, and a concentration of positive ions near the central ion that is less than the bulk. Also, the expression is quite sensitive to the potential. For example, a potential difference of only 10 mV leads to an increase in concentration of the negative ions of almost 50%. Note that the treatment presented in this section ignores the discrete nature of the ions and treats the concentration distributions as continuous.

Far away from the central ion, the concentration is that of the bulk; here the potential, ϕ, is arbitrarily set to zero. The potential distribution is given by Poisson's equation:

$$\nabla^2 \phi = \frac{-\rho_e}{\varepsilon} = \frac{-F}{\varepsilon}\sum_i z_i c_i, \qquad (2.35)$$

where ε is the permittivity with units [C·V^{-1}·m^{-1}]. Assuming spherical symmetry and substituting the Boltzmann distribution above (Equation 2.34) for the concentrations, Equation 2.35 becomes

$$\frac{1}{r^2}\frac{d}{dr}\left(r^2\frac{d\phi}{dr}\right) = \frac{-\rho_e}{\varepsilon} = \frac{-F}{\varepsilon}\sum_i z_i c_i^\infty \exp\left(\frac{-z_i F\phi}{RT}\right),$$

$$(2.36)$$

where the summation is over each type of ion in solution. The boundary conditions are

$$\begin{aligned} r \to \infty & \quad \phi = 0 \\ r = a & \quad \left.\frac{d\phi}{dr}\right|_{r=a} = \frac{-z_c q}{4\pi\varepsilon a^2}. \end{aligned}$$

The second boundary condition accounts for the fact that the charge density integrated throughout the volume surrounding the central ion must be equal to the charge of the central ion (z_c). Consequently, according to Gauss's law,

$$\iiint_V \rho_e dV = z_c q = \iint_S \varepsilon E \cdot dS = -\varepsilon \frac{d\phi}{dr} 4\pi a^2, \qquad (2.37)$$

where the potential gradient and the surface area of the ion are constants and can be removed from the integral. In order to simplify Equation 2.36, the exponential term is approximated by the first two terms of a Maclaurin series,

which is accurate for small values of the term in the exponent.

$$\exp\left(\frac{-z_i F\phi}{RT}\right) \approx 1 - \frac{z_i F\phi}{RT}.$$

Substituting this expression back into the original differential equation (Equation 2.36) results in

$$\frac{1}{r^2}\frac{d}{dr}\left(r^2\frac{d\phi}{dr}\right) = \frac{\phi}{\lambda^2}. \qquad (2.38)$$

Here, the Debye length, λ, has been introduced. This parameter is critical in the study of electrochemical systems, and is defined as

$$\lambda \equiv \sqrt{\frac{\varepsilon RT}{F^2 \sum_i z_i^2 c_i^\infty}}. \qquad (2.39)$$

The Debye length is important in describing the potential distribution; more specifically, it is the characteristic length over which the charge density in solution varies as a result of the central ion. The solution of (2.38) is

$$\phi = \frac{z_c q}{4\pi\varepsilon r}\frac{\exp\left[\frac{(a-r)}{\lambda}\right]}{1 + \frac{a}{\lambda}}, \qquad (2.40)$$

which represents the variation of the solution potential with position beginning at the surface of the central ion and moving outward. The Debye length characterizes the distance over which the potential changes as shown in Figure 2.7, where the potential in the less concentrated

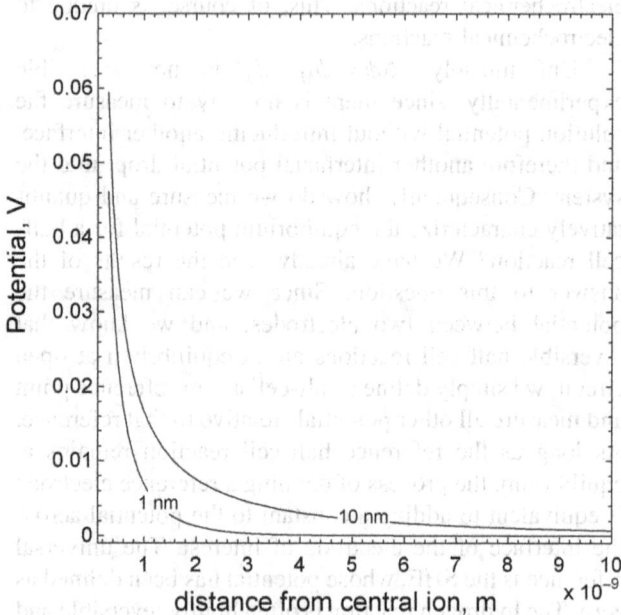

Figure 2.7 Variation of potential with distance near a central charged ion. The Debye length is a parameter.

solution takes significantly longer to decay. The potential is the result of the field from the central ion and the shielding effects of other ions in solution. The Debye length decreases with increasing concentration (see Equation 2.39), as more ions are available to shield the central ion. Typical λ values in aqueous solutions are on the order of 1 nm. While a nanometer may seem quite small, this characteristic distance is large relative to other types of interaction forces between molecules. The variation of the charge density in solution as a result of charge on an electrode surface also scales with the Debye length as discussed later in Chapter 3.

2.14 ACTIVITIES AND ACTIVITY COEFFICIENTS

The accuracy of our expression for cell potential can be increased by including the full activity corrections, rather than the approximations used earlier. The complexity of the calculations, however, increases significantly. The treatment below assumes that the reader has been exposed to the concepts of activity and fugacity. If these are new to you, you may want to learn more about them from a book on physical chemistry or chemical equilibrium (see Further Reading section at the end of the chapter). The fugacity is related to the chemical potential and is used as a surrogate for the chemical potential in phase-equilibrium calculations. The activity of a species is defined as the ratio of the fugacity of species i to the fugacity of pure i at the standard state:

$$a_i = \frac{\hat{f}_i}{f_i^o}. \tag{2.41}$$

Note that the activity is dimensionless.

For our analysis of electrochemical systems, we need the activity of solid, gas, solvent, and solute species. For solid species, the standard state is typically taken as the pure species. In this text, we will consider only pure solid species. Therefore, the fugacity is equal to the pure component fugacity, f_i, and

$$a_{i,pure\ solid} = 1, \tag{2.42}$$

since the solid is in its standard state.

For the gas-phase species, the standard state fugacity is an ideal gas at a pressure of 1 bar. The fugacity of species i in a mixture is defined as

$$\hat{f}_i = \hat{\varphi}_i y_i p, \tag{2.43}$$

where $\hat{\varphi}_i$ is the fugacity coefficient of species i in a mixture, y_i is the mole fraction of the component in the gas phase, and p is the total pressure. For the purposes of this book, we assume that $\hat{\varphi}_i = 1$, which is equivalent to assuming ideal gas behavior. With this assumption, the activity of the gas is

$$a_{i,gas} = \frac{p_i}{p^o} = \frac{y_i p}{p^o}. \tag{2.44}$$

Remember, a_i is dimensionless, and the pressure units in the numerator cancel out with those of the standard state fugacity, p^o. However, since the numerical value of f_i^o is 1 bar, sometimes it is left off the standard state when writing the activity. Do so with great care.

Electrochemical systems include an electrolyte, which is a material in which current flows due to the movement of ions. A common liquid electrolyte consists of a solvent (e.g., water) into which one or more salts are dissolved to provide the ionic species. For electrolyte solutions, molality (m_i = moles solute i per kg solvent) is the most commonly used form of expressing the composition when dealing with nonideal solutions and activities, and hence will be used here. Molality is convenient from an experimental perspective because it depends only upon the masses of the components in the electrolyte solution, and does not require a separate determination of density. The temperature dependence of the density may also introduce error when dealing with concentration rather than molality. The relationship between molality and concentration is

$$m_i = \frac{c_i}{\rho - \sum_{j \neq 0}^{c_j} M_j} = \frac{c_i}{c_0 M_0}, \tag{2.45}$$

where the subscript "0" refers to the solvent and M is the molecular weight. The summation in the denominator is simply the total mass of solute species per volume of solution.

We first consider the activity of the solvent. Since the concentration of the solvent is usually much higher than that of the dissolved solute species, the standard state fugacity is that of the pure solvent at the same pressure and phase of the system (e.g., liquid water).

$$a_{solvent} = \frac{\hat{f}_{solvent}}{f_{solvent}^o} \approx \frac{p_{solvent}}{p_{solvent}^o}. \tag{2.46}$$

In most cases, the vapor pressures are sufficiently low that the fugacity can be approximated by the pressure, as shown. An osmotic coefficient is usually used to express the activity of the solvent as follows:

$$\ln a_{solvent} = -\varphi \frac{M_{solvent}}{1000} \sum_i m_i, \tag{2.47}$$

where the summation is over ionic species and does not include the solvent. $M_{solvent}$ is the molecular weight of the solvent and φ is the osmotic coefficient. For a single salt (binary electrolyte), Equation 2.47 becomes

$$\ln a_{solvent} = -\nu m \varphi \frac{M_{solvent}}{1000}, \tag{2.48}$$

where m is the molality of the salt and ν is defined by Equation 2.54. When the ion concentration is zero, the right sides of Equations 2.47 and 2.48 both go to zero, and the solvent activity is equal to unity as expected. For our purposes in this text, we will assume unit activity for the solvent, unless otherwise specified.

We now turn our attention to the activity of the solute, which is the activity correction that is most frequently of concern in electrolyte solutions. The activity of a single ion is defined as

$$a_i = \frac{\gamma_i m_i}{f_i^o} = \frac{\gamma_i m_i}{\text{ideal 1 molal solution}} = \gamma_i m_i, \quad (2.49)$$

where γ_i is the single-ion activity coefficient (unitless). Again, we have left off the standard state in the final expression, since it has a value of one. Remember, however, that the activity is dimensionless.

Where does the standard state come from, and why is it used? What is meant by an ideal solution in this context? A thought experiment is useful for answering these questions. Think about an ion in solution. That ion will interact with the solvent (e.g., water molecules) and with other ions of all types in the solution. However, as the ion concentration in the electrolyte approaches zero, only the ion–solvent interactions are important. Under such conditions, the fugacity of the ion is equal to m_i. In other words, the fugacity of the ion depends only on the amount present. This is because the ions interact only with the solvent, and the nature of the interactions does not change with molality as long as ion–ion interactions remain insignificant (i.e., as long as changing interactions do not contribute to the fugacity). We define an ideal solution as a solution in which only ion–solvent interactions are important. Such behavior is approximated in real systems as the concentration approaches small values, and is seen as a linear asymptote in a plot of the activity as a function of composition.

The standard state defined in Equation 2.49 represents the fugacity of an ideal solution where only ion–solvent interactions are important at a concentration of one molal. This state is clearly hypothetical because ion-ion interactions are important in real electrolytes at this concentration. However, it is a convenient reference state and is widely used. The activity coefficient is used to account for deviations from this ideal state due to ion–ion interactions, including complex formation in solution. From the above, it follows that

$$\gamma_i \to 1 \text{ as } m_i \to 0. \quad (2.50)$$

There is, however, a practical problem with the single-ion activity and activity coefficient. Electrolyte solutions are electrically neutral and solutions containing just a single ion don't exist. Therefore, single-ion activities cannot be measured, although they can be approximated analytically under some conditions (see Section 2.15).

To address this issue, we define a measureable activity that is related to the single-ion activities defined above. A single salt containing ν_+ positive ions and ν_- negative ions dissociates as follows to form a binary electrolyte:

$$M_{\nu_+} X_{\nu_-} \to \nu_+ M^{z+} + \nu_- X^{z-}. \quad (2.51)$$

The activity of the salt in solution is

$$a_{M_{\nu_+} X_{\nu_-}} = (a_{M^{z+}})^{\nu_+} (a_{X^{z-}})^{\nu_-} = a_+^{\nu_+} a_-^{\nu_-} \quad (2.52)$$

where the $+$ and $-$ subscripts are introduced for convenience to represent the positive and negative ions of the salt, respectively. We now define a mean ionic activity in terms of the single-ion activities:

$$a_\pm^\nu \equiv a_+^{\nu_+} a_-^{\nu_-}, \quad (2.53)$$

where

$$\nu = \nu_+ + \nu_-. \quad (2.54)$$

Similarly, we define the mean ionic activity coefficient in terms of the single-ion activity coefficients

$$\gamma_\pm^\nu \equiv \gamma_+^{\nu_+} \gamma_-^{\nu_-}. \quad (2.55)$$

Assuming complete dissociation, the molality of the neutral salt in solution (moles of salt per kg solvent) is related to the molality of the individual ions according to

$$m = \frac{m_+}{\nu_+} = \frac{m_-}{\nu_-}. \quad (2.56)$$

Combining Equations 2.52–2.56 with the definition of the single-ion activity coefficient (Equation 2.49) yields the following for the salt in solution:

$$a_{M_{\nu_+} X_{\nu_-}} = a_\pm^\nu = (\gamma_+ m_+)^{\nu_+} (\gamma_- m_-)^{\nu_-} = (\gamma_\pm m)^\nu (\nu_+^{\nu_+} \nu_-^{\nu_-}). \quad (2.57)$$

Consistent with the above, the following limits are reached at low concentrations:

$$\gamma_\pm \to 1 \text{ as } m^\nu \to 0. \quad (2.58)$$

Equations 2.51–2.57 apply to a single salt in solution or binary electrolyte and permit activity corrections without requiring single-ion activity coefficients. Several systems of practical importance, such as Li-ion batteries and lead–acid batteries, have either binary electrolytes or electrolytes that can be approximated as binary. The mean ionic activity coefficient, $\gamma_\pm$, has been measured for a number of binary solutions, and the results have been correlated and can be found in the literature.

The activity relationships above provide a practical, measurable way to include activity corrections for binary electrolytes. Measurements for binary systems are typically fit to models in order to provide the needed activity coefficients in an accessible, usable way. Under some conditions, prediction of activity coefficients is possible. These issues are discussed briefly in the section that follows.

2.15 ESTIMATION OF ACTIVITY COEFFICIENTS

In dilute solutions, long-range electrical interactions between ions dominate, and the activity coefficient can be estimated from a determination of the electrical contribution alone. The chemical potential is equal to the reversible work of transferring 1 mol of the species to a large volume of the solution at constant temperature and pressure. The nonideal portion for dilute electrolytes is the electrical portion, which is the work required to charge 1 mol of ions in a solution where all of the other ions are already charged. This work can be calculated directly from the Debye–Hückel potential distribution that we determined earlier in Section 2.13. The result is

$$\mu_{i,electrical} = -\frac{z_i^2 Fq}{8\pi\varepsilon\lambda}\frac{1}{1+a/\lambda}, \quad (2.59)$$

where the variables were defined in Section 2.13. Since $\mu_{i,electrical} = RT\ln\gamma_i$, it follows directly that

$$\ln\gamma_i = \frac{-z_i^2 Fq}{8\pi\varepsilon RT\lambda}\frac{1}{1+a/\lambda} = \frac{-z_i^2 \alpha\sqrt{I}}{1+Ba\sqrt{I}}. \quad (2.60)$$

The expression on the right arises after introducing the following definitions:

The molal ionic strength $\quad I \equiv \frac{1}{2}\sum_i z_i^2 m_i, \quad (2.61)$

Solvent constant $\quad \alpha \equiv \frac{F^2 q\sqrt{2}}{8\pi(\varepsilon RT)^{1.5}}\sqrt{\rho_o}. \quad (2.62)$

Solvent constant $\quad B \equiv \frac{F}{\sqrt{\varepsilon RT/2}}\sqrt{\rho_o}. \quad (2.63)$

Finally, we can express the mean activity coefficient as

$$\ln\gamma_\pm = \frac{z_+ z_- \alpha\sqrt{I}}{1+Ba\sqrt{I}}, \quad (2.64)$$

where a is the mean ionic radius of the two hydrated ions. This equation utilizes the relationship shown in Equation 2.55 and takes advantage of the fact that $z_+\nu_+ = -z_-\nu_-$. Ionic strength is the key solution property, which depends on the amount of solute. The only other two parameters are the solvent density and the permittivity. Figure 2.8 shows the activity coefficient as a function of ionic strength for two salts in water. Thus, we can see that

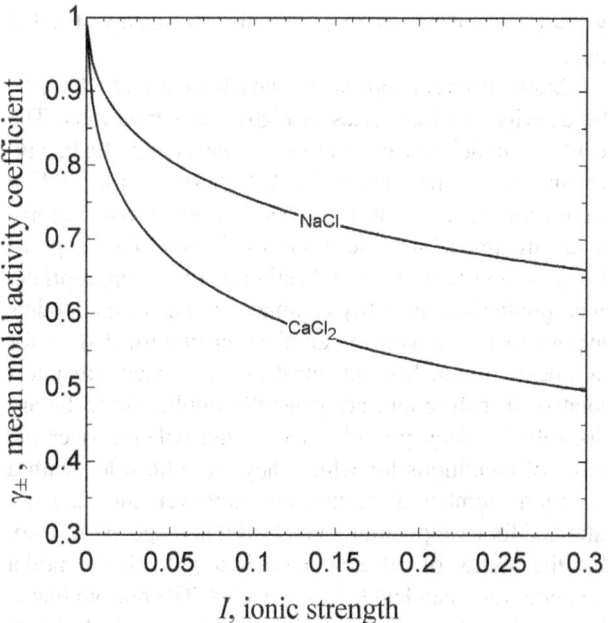

Figure 2.8 Activity coefficients from Debye–Hückel theory.

the activity coefficient is always less than one, and for dilute solutions, the logarithm of the activity coefficient is proportional to the square root of ionic strength. Also note that at the same ionic strength the 1 : 2 electrolyte has a lower activity coefficient (i.e., is more nonideal). As the ionic strength approaches zero, the limiting form of Equation 2.60, known as Debye–Hückel limiting law, is

$$\ln\gamma_\pm = z_+ z_- \alpha\sqrt{I}. \quad (2.65)$$

In spite of the approximations made in the derivation, the Debye–Hückel model represents the activity coefficients reasonably well for dilute solutions. In fact, the limiting law is usually within 10% of measured values for concentrations less than 0.1 M. Equation 2.64 further increases accuracy. The size parameter is usually treated as a fitting parameter, although it approximates the expected sizes of ions with typical values from 0.3 to 0.5 nm.

Beyond Debye–Hückel Theory

Debye–Hückel theory considers only long-range interactions between ions, and does not treat ion–solvent interactions or short-range interactions between ions. Therefore, it does not perform well for solutions at concentrations where these additional effects are important. Its shortcomings are evident from the measured behavior for electrolytes at finite concentrations. For example, some electrolytes show large positive deviations from ideality

($\gamma_\pm \gg 1$), which cannot be described by the Debye–Hückel model.

Many different approaches have been used to describe the activity of electrolytes at higher concentrations. The resulting models correctly represent the asymptotic behavior shown in the Debye–Hückel model while adding framework to account for other important interactions. A full treatment of these models is beyond the scope of this text; however, a generalization or two is appropriate. First, predictive capability is limited, and all of the models depend to one degree or another on empirical data for parameter fitting. Second, most of the models are interpolative in nature and are generally applicable to binary electrolytes. They provide precise interpolation over the range of conditions for which they were fit, often with a minimum number of parameters; however, they are not intended for extrapolation outside of this range. Note, also, that the ability of different models to effectively model temperature-dependent behavior varies. The bottom line is that data have been measured for many binary electrolytes, and a number of different types of models have been used to successfully fit those data over a broad range of concentrations.

The situation is different for multicomponent solutions where the mean ionic activity coefficient, $\gamma_\pm$, is not appropriate, and the activity of a given ion is influenced by all of the different types of ions in solution. In this situation, it is necessary to approximate single-ion activity coefficients for the multicomponent system of interest. To do this, Newman describes a method to approximate the activity coefficient of ions in a moderately dilute multicomponent solution using parameters from the corresponding binary systems. Under certain conditions, Meissner's corresponding states model can be extended to predict multicomponent systems from binary system parameters for a broad range of concentrations. Local composition models for electrolytes are extensions of similar types of models for nonelectrolytes and have built-in generalization to multicomponent solutions. The most used of these models is perhaps the model of Chen and Evans, which is included as part of a popular process simulation program where it is used to handle electrolyte thermodynamics. This model has been used successfully for many different systems over a broad range of concentrations.

ILLUSTRATION 2.8

Calculate the activity coefficient for aqueous 0.1 molal solutions of NaCl and CaCl$_2$ using the Debye–Hückel approximation. Use a permittivity of 6.9331×10^{-10} F m^{-1}. The two solvent constants for water at room temperature are

$$\alpha = \frac{F^2 q \sqrt{2}}{8\pi(\varepsilon RT)^{1.5}} \sqrt{\rho_o}$$

$$= \frac{F^2 1.602 \times 10^{-19} \sqrt{2}}{8\pi(6.9331 \times 10^{-10} R(298))^{1.5}} \sqrt{997.10}$$

$$= 1.176 \left(\frac{\text{kg}}{\text{mol}}\right)^{1/2},$$

$$B = \frac{F}{\sqrt{\varepsilon RT/2}} \sqrt{\rho_o} = 3.387 \times 10^9 \left(\frac{\text{kg}}{\text{mol}}\right)^{1/2} \text{m}^{-1}.$$

The ionic strength is given by Equation 2.61. For a 0.1 molal solution, $I_{\text{NaCl}} = 0.1$ $I_{\text{CaCl}_2} = 0.3$

The values for Ba are reported as 0.9992 and 1.536 for NaCl and CaCl$_2$, respectively. From Equation 2.64 the mean activity coefficients are $\gamma_{\text{NaCl}} = 0.754$ and $\gamma_{\text{CaCl}_2} = 0.497$.

CLOSURE

The focus of this chapter has been on the potential of electrochemical cells at equilibrium. Specifically, we have shown how to relate this potential to the environmental conditions at the electrode: temperature, pressure, and composition. Principles of classical thermodynamics have provided the framework for these calculations. The reference electrode has also been introduced. The potential measured with a reference electrode will be a critical variable for subsequent chapters.

FURTHER READING

Bard, A.J., Parsons, R., and Jordan, J. (Eds.) (1985) *Standard Potentials in Aqueous Solutions*, Marcel Dekker, New York.

Ives, D.J. and Janz, G.J. (Eds.) (1961) *Reference Electrodes Theory and Practice*, Academic Press, New York.

S. A. Newman (Ed.) (1980) *Thermodynamics of Aqueous Systems with Industrial Applications*, ACS Symposium Series, 133, American Chemical Society.

Newman, J. and Thomas-Alyea, K.E. (2004) *Electrochemical Systems*, John Wiley & Sons, Inc., Hoboken, N. J.

Pourbaix, M. (1974) *Atlas of Electrochemical Equilibria in Aqueous Solutions*, NACE, Houston.

Tester, J.W. and Modell, M. (1996) Models of electrolyte solutions, in *Thermodynamics and Its Applications*, Prentice Hall.

PROBLEMS

2.1 Write the associated electrochemical reactions and calculate the standard potential, U^θ from ΔG° for the following cells:

(a) Chlor-alkali process to produce hydrogen and chlorine from a brine of NaCl (aqueous salt solution). Use the hydrogen reaction for an alkaline solution.

(b) Acetic acid/oxygen fuel cell with acidic electrolyte, where the acetic acid reacts to form liquid water and carbon dioxide. The reaction at the negative electrode is

$$2H_2O + CH_3COOH \rightarrow 2CO_2 + 8H^+ + 8e^-$$

2.2 Does the redox reaction as written below proceed spontaneously at 25 °C and standard conditions?

$$2Ag^+ + H_2 \rightarrow 2Ag_s + 2H^+$$

2.3 What is the standard half-cell potential for the oxidation of methane under acidic conditions? The reaction for methane is as follows:

$$CH_4(g) + 2H_2O(\ell) \rightarrow CO_2 + 8H^+ + 8e^-$$

Which element is oxidized and how does its oxidation state change?

2.4 What is the standard cell potential for a methane/oxygen fuel cell? The oxidation of methane produces CO_2 as shown in Problem 2.3, but here assume the product water is a gas, rather than a liquid.

2.5 Let's consider the oxidation of methane in a fuel cell that utilizes an oxygen conductor (O^{2-}) rather than a proton conductor as the electrolyte.

(a) At which electrode (oxygen or methane) is O^{2-} produced and at which is it consumed?

(b) In which direction does O^{2-} move through the electrolyte? Why?

(c) Propose two electrochemical half-cell reactions.

(d) Does U^θ change for this fuel cell relative to a fuel cell that utilizes a proton conductor? Why or why not?

2.6 Determine the equilibrium potential of the cell shown below.

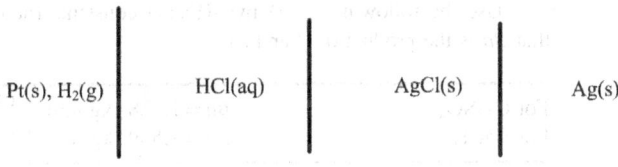

2.7 Consider the electrochemical reactions shown below. Mercury(I) chloride, also known as *calomel*, is a solid used in reference electrodes. The two reactions are

$$Zn \leftrightarrow Zn^{2+} + 2e^-$$
$$Hg_2Cl_2 + 2e^- \leftrightarrow 2Cl^- + 2Hg$$

(a) What is the overall chemical reaction?

(b) Develop an expression for U, the equilibrium potential of the cell.

(c) Write down an expression for the standard potential of the cell in terms of the standard Gibbs energies of formation.

(d) Use standard half-cell potentials from the table to determine the standard Gibbs energy of formation for aqueous $ZnCl_2$. Why is this value different than the value for solid $ZnCl_2$?

(e) What is the standard Gibbs energy of formation for Hg_2Cl_2?

2.8 The lithium air cell offers the possibility of a very high-energy battery. At the negative electrode,

$$Li \leftrightarrow Li^+ + e^-$$

At the positive electrodes, the following reactions are postulated:

$$2Li^+ + O_2 + 2e^- \leftrightarrow Li_2O_2$$
$$2Li^+ + \frac{1}{2}O_2 + 2e^- \leftrightarrow Li_2O$$

Estimate the standard potential for each of the two possible reactions at the positive electrode paired with a lithium anode.

2.9 Develop an expression for the equilibrium potential for the cell below. The first reaction is the negative electrode of the Edison cell (battery).

$$Fe + 2OH^- \rightarrow Fe(OH)_2 + 2e^-$$
$$O_2 + 4e^- + 2H_2O \rightarrow 4OH^-$$

The Gibbs energy of formation for $Fe(OH)_2$ is $-486.6\,kJ \cdot mol^{-1}$.

2.10 Develop an expression for the equilibrium potential of a hydrogen–oxygen fuel cell operating under acidic conditions. The two electrochemical reactions are

$$H_2 \rightarrow 2H^+ + 2e^- \quad \text{and} \quad O_2 + 4H^+ + 4e^- \rightarrow 2H_2O$$

Use the data in the Appendix C for standard Gibbs energy of formation. Compare with the value calculated from standard electrode potentials to identify whether the standard state for water in the table of Appendix A is liquid or gas.

2.11 Write the associated electrochemical reactions, and calculate the standard potential, U^θ, from $\Delta G°$ for the following cells:

(a) Propane fuel cell with solid oxygen conductor electrolyte.

(b) Electrolysis of aluminum, where aluminum is produced from Al_2O_3 and carbon. Note that carbon is oxidized at the anode.

2.12 Calculate the equilibrium potential for peroxide formation in an acid fuel cell

$$O_2 + 2H^+ + 2e^- \rightarrow H_2O_2(aq)$$

2.13 Use the half-cell reactions for the reduction of cupric ion (Cu^{2+}) to copper metal and cuprous ion (Cu^+) to copper metal to calculate the standard potential for the reduction of cupric ion to cuprous ion. Check your answer against the value given in Appendix A.

2.14 Consider the electrochemical cell below.

Pt(s), H$_2$(g) | HCl(aq) | Pt(s), Cl$_2$(g)

The two reactions are

$$H_2 \leftrightarrow 2H^+ + 2e^-$$

$$Cl_2 + 2e^- \leftrightarrow 2Cl^-$$

Find an expression for U. If the pressure of hydrogen is 250 kPa and that of chlorine is 150 kPa, what is the numerical value of U at 25 °C in 1 molal HCl? Include the simplified activity corrections (you may neglect activity coefficients).

2.15 Create a Pourbaix diagram for Pb. Treat the following reactions:

$$Pb = Pb^{2+} + 2e^-$$
$$Pb^{2+} + H_2O = PbO + 2H^+$$
$$Pb + H_2O = PbO + 2H^+ + 2e^-$$
$$Pb^{2+} + 2H_2O = PbO_2 + 4H^+ + 2e^-$$
$$3PbO + H_2O = Pb_3O_4 + 2H^+ + 2e^-$$

2.16 Create a Pourbaix diagram for Pt. Focus on the low pH range ($-2 \leq pH \leq 1$), and consider the following reactions:

$$Pt = Pt^{2+} + 2e^- \quad (1.188\ V)$$
$$Pt^{2+} + H_2O = PtO + 2H^+$$
$$Pt + H_2O = PtO + 2H^+ + 2e^- \quad (0.980\ V)$$
$$PtO + H_2O = PtO_2 + 2H^+ + 2e^- \quad (1.045\ V)$$
$$Pt^{2+} + 2H_2O = PtO_2 + 4H^+ + 2e^- \quad (0.837\ V)$$

2.17 Create a Pourbaix diagram for Fe. Treat the following reactions:

$$Fe = Fe^{2+} + 2e^- \quad (-0.440\ V)$$
$$Fe^{2+} = Fe^{3+} + e^- \quad (0.771\ V)$$
$$3Fe + 4H_2O = Fe_3O_4 + 8H^+ + 8e^- \quad (-0.085\ V)$$
$$3Fe^{2+} + 4H_2O = Fe_3O_4 + 8H^+ + 2e^- \quad (0.980\ V)$$
$$2Fe^{2+} + 3H_2O = Fe_2O_3 + 6H^+ + 2e^- \quad (0.728\ V)$$
$$2Fe^{3+} + 3H_2O = Fe_2O_3 + 6H^+$$
$$2Fe_3O_4 + H_2O = 3Fe_2O_3 + 2H^+ + 2e^- \quad (0.221\ V)$$
$$3HFeO_2^- = Fe_3O_4 + H_2O + OH^- + 2e^- \quad (-1.819\ V)$$
$$Fe + 2H_2O = HFeO_2^- + 3H^+ + 2e^- \quad (0.493\ V)$$

2.18 Use the information in Appendix A to determine the dissociation constant for water, K_w.

$$H_2O \leftrightarrow H^+ + OH^-$$

2.19 Determine the solubility product K_{sp} for PbSO$_4$.

2.20 Estimate the equilibrium constant for the disproportionation of copper.

$$Cu + Cu^{2+} = 2Cu^+$$

2.21 Explain what a liquid junction is and why the potential of cells with liquid junctions cannot be determined from thermodynamics alone.

2.22 LiPF$_6$ is a common salt used in lithium batteries. For lithium hexafluorophosphate, LiPF$_6$, dissolved in propylene carbonate, calculate the Debye length and the activity coefficient (using Debye–Hückel theory) of LiPF$_6$ at 30 °C. The concentration is 0.1 M. The dielectric constant, $\varepsilon_r = \varepsilon/\varepsilon_o$, for the solvent is 64 and the density 1.205 g·cm^{-3}. The density of the 0.1 M electrolyte is 1.286 g·cm^{-3}.

2.23 Consider the electrochemical cell below. Iron corrodes to form Fe^{2+}. Develop an expression for U, and determine the value of the standard potential.

Pt(s), H$_2$(g) | HCl(aq) | Fe | Pt(s)

2.24 Find the expression for the equilibrium potential of the cell at 25 °C.

Ag(s) | AgCl(s) | ZnCl$_2$(aq) | Zn(s)

The two electrochemical reactions are as follows:

$$Ag + Cl^- = AgCl_{(s)} + e^- \quad U^\theta = 0.222\ V$$
$$Zn = Zn^{2+} + 2e^- \quad U^\theta = -0.763\ V$$

2.25 Calculate the activity coefficient for (a) 0.05 m Cs$_2$SO$_4$ and (b) 0.05 m BaCl$_2$ in water at 25 °C.

Use the following for Debye–Hückel constants (note that Ba is the product of B and a).

For Cs$_2$SO$_4$	$Ba = 1.328$ (kg·mol^{-1})$^{1/2}$
For BaCl$_2$	$Ba = 1.559$ (kg·mol^{-1})$^{1/2}$

2.26 Use the Debye–Hückel theory to calculate the activity coefficient for 0.1 m MgCl$_2$ in water at 25 °C. Use the following ionic radii for the two ions: Mg^{2+} = 8 Å, Cl$^-$ = 3 Å.

2.27 The following activity coefficient data have been measured for NaCl solutions as a function of molality (radius for Na$^+$ = 4 Å, Cl$^-$ = 3 Å). The temperature is 25 °C. Please do the following:

(a) Use the Debye–Hückel theory to calculate the activity coefficients.

(b) Use the Debye–Hückel limiting law to calculate the activity coefficients.

(c) Compare the results of (a) and (b) with the experimental data and comment on the applicability of the Debye–Hückel equations.

m	$\gamma_\pm$
0.001	0.965
0.005	0.927
0.01	0.902
0.05	0.821
0.1	0.778
1	0.657

2.28 Before concerns about mercury became widespread, the calomel electrode was commonly used. Crystals of KCl are added to produce a saturated solution. What advantage does a saturated solution provide? The saturated calomel electrode has an equilibrium potential of 0.242 V, which is lower than the standard potential of 0.2676. Can this 25 mV difference be determined from thermodynamics? Why or why not? The solubility of KCl in water at 25 °C is 360 g KCl per 100 g water.

2.29 A solid oxide fuel cell operates at 1000 °C. The overall reaction is

$$0.5 O_2 + H_2 \leftrightarrow H_2O$$

(a) Calculate the standard potential at 25 °C assuming that reactants and products are gases.

(b) Calculate the standard potential at 1000 °C using Equation 2.18.

(c) Using the correlation for heat capacity as a function of temperature shown below, calculate the standard potential at 1000 °C. Comment on the assumption used in part (b) that $\Delta S°$ is constant.

$$C_p = A + BT + CT^{-2}$$

	A [J·mol^{-1}·K^{-1}]	$10^3 B$ [J·mol^{-1}·K^{-2}]	$10^{-5} C$ [J·K·mol^{-1}]
H_2O	30.54	10.29	0
O_2	29.96	4.184	−16.7
H_2	27.28	3.26	0.50

2.30 Alloys of LiSn are possible electrodes for batteries. There are many phases possible, but we want to focus on the reaction

$$3 LiSn + 4 Li^+ + 4 e^- = Li_7 Sn_3$$

The standard potential of this reaction at 25 °C is 0.530 V (versus reference Li electrode). If the enthalpy of the reaction

$$3 LiSn + 4 Li = Li_7 Sn_3$$

is -226 kJ·mol^{-1} Li$_7$Sn$_3$, estimate the standard potential at 400 °C.

2.31 Find the equilibrium constant, K_{eq}, for Pt dissolution reaction at 25 °C.

$$PtO + 2 H^+ = Pt^{2+} + H_2O$$

The following thermodynamic data are provided, $\Delta G°_{f, Pt^{2+}} = 229.248$ kJ mol^{-1}, and

$$PtO + 2 H^+ + 2 e^- = Pt + H_2O, \quad U^\theta = 0.980 \text{ V}$$

2.32

(a) Write the overall reaction and determine the standard potential for the lead–acid battery.

(b) Develop an expression for the equilibrium potential, U, for the lead–acid battery as a function of electrolyte composition. The final expression should include the molality and activity coefficient.

(c) The potentials of the two electrodes relative to a Hg-Hg$_2$SO$_4$ reference electrode at 25 °C has been measured. What is the likely reaction at the reference electrode? If the standard Gibbs energy of formation of Hg$_2$SO$_4$ is -625.8 kJ·mol^{-1} and the standard Gibbs energy of formation of SO$_4^{2-}$ is -744.62 kJ·mol^{-1}, determine the standard potential of the positive (lead oxide electrode) relative to the reference electrode. How does this value compare to the measured value of 0.96 V? Explain the possible cause of any difference.

(d) Given that the potential difference between the positive electrode and the reference electrode is 1.14 V, estimate the activity coefficient for sulfuric acid ($\gamma_\pm$) at 6 m. For this evaluation, you may assume that the activity of water is one.

(e) Does the potential of the lead electrode (Pb) relative to the reference electrode depend on the molality of the sulfuric acid? Explain your answer.

2.33 Rework Illustration 2.7 (reference electrode example) with a Ag$_2$SO$_4$ reference electrode rather than a Hg$_2$SO$_4$ reference electrode. The standard potential for this reference electrode reaction (below) is 0.654 V.

$$Ag_2SO_4 + 2 e^- = 2 Ag_{(s)} + SO_4^{2-}$$

ALEXANDER NAUMOVICH FRUMKIN

Alexander Naumovich Frumkin was born on October 24, 1895 in the city of Kishinev in what is now Moldova. As a young child, he and his family moved to nearby Odessa in the Ukraine. He left in 1912 to study at Strasbourg and then at the University of Bern. Frumkin published his first two manuscripts at the age of 19. He then returned to Odessa and obtained a degree from Novorossia University (now Odessa University) in 1915. He subsequently worked as a lab assistant at the University and published his first paper in electrochemistry in 1917. The topic was the movement of a mercury drop under the influence of current, an area that would occupy much of his attention over a long career. The Russian revolution temporarily eliminated the doctoral degree, and as a result Frumkin never received his Ph.D. Nonetheless, his thesis, "Electro-Capillary Phenomena and Electrode Potential," was completed in 1919 and contained the underpinnings of many of his subsequent research thrusts.

Frumkin moved to Moscow in 1920 and took a position at what is now the Karpov Institute of Physical Chemistry. In 1930 he was elected Professor at Lomonosov Moscow State University, and in 1932 he became a full member of the Academy of Sciences of the USSR. Following World War II, decades of tension between the Soviet Union and the West spurred efforts in science and technology. Frumkin founded the Institute of Electrochemistry of the Academy of the USSR, and directed it from 1958 until his death. Under his guidance, a generation of electrochemists was trained there; because Frumkin and the Institute remained relatively open to the West and to working with all scientists, the influence of these mentees was greater than those from other organizations in the Soviet Union. In 1983 the Institute was renamed A.N. Frumkin Institute of Physical Chemistry and Electrochemistry of the Russian Academy of Sciences. Frumkin also founded the *Russian Journal of Electrochemistry* (*Elektrokhimiya*) in 1965, which covers all aspects of modern electrochemistry.

One of Frumkin's most important contributions was the realization that the structure of the double layer affects the kinetics of electrode processes. Roger Parsons, a principal figure in electrochemistry in his own right, noted in 1969 that "It is to him (Frumkin) that we owe a broad picture of the effect of the double layer changes on the rate of electrode reactions. . . ." At the time of Frumkin's early work, the tools to investigate the double layer were quite limited. One of the most fruitful approaches related the surface energy of a mercury drop to the electrical potential and composition of the electrolyte; this phenomenon is known as electrocapillarity. Frumkin made many contributions in this area, and the theory is now well developed. At this point its importance is mostly historical—validating the models of the electrical double layer. This understanding helped to make the connections between the structure of the double layer and a description of electrode kinetics. Another important contribution made by Frumkin was to show the importance of removing impurities from the electrode and electrolyte—often still a vexing issue today. Frumkin developed the concept of potential of zero charge (first introduced in his thesis) and identified the PZC as an important property in describing electrochemical processes of metals. Frumkin, along with Dolin and Ershler, introduced the term "exchange-current density," which we will see plays an important role in describing the kinetics of electron transfer reactions. He made many other contributions, including collaborating with Levich in the development of the rotating ring-disk electrode, describing what is now known as the Frumkin effect for changes in reactant concentration at the electrode surface, extending the Langmuir isotherm, and showing the link between the transfer coefficient in electrochemical reaction and the Brønsted coefficient used in homogeneous catalysis, to name but a few.

In 1959 Frumkin received the Olin Palladium Award of the Electrochemical Society. This recognition is awarded for distinguished contributions to the field of electrochemical or corrosion science. His award address was entitled "The Double Layer in Electrochemistry." Frumkin was President of the International Society of Electrochemistry during 1965–1966; this organization established the Frumkin memorial medal, which is awarded biennially to recognize the contributions of individuals to fundamental electrochemistry. For more than 50 years, he led the field and was active until the end. Alexander Frumkin died on May 26, 1976 in Tula, Russia.

Image Source: Courtesy of Prof. Evgeny Antipov, Lomonosov Moscow State University.

Chapter 3

Electrochemical Kinetics

One of the real advantages of electrochemical reactions is that it is possible to control the reaction rate and even the direction of the reaction by changing the potential. This control is possible because electron transfer in electrochemical reactions takes place through a conductor, and the potential of a conductor can easily be changed or controlled. In this chapter we learn about reaction rates and how to describe those rates quantitatively. Before we do so, however, we need to look carefully at the interface between a metal electrode and the adjacent electrolyte solution. That interface is critical to understanding and manipulating electrode reactions. In the previous chapter, we determined the potential of an electrode at equilibrium, U, in the absence of current. Now we want to find the potential when current is flowing. The difference between the potential with current flowing and the equilibrium potential is called the overpotential.

3.1 DOUBLE LAYER

A metal electrode in an electrolyte solution is typically charged, even at equilibrium. Excess charge is present on the outer surface of the electrode, adjacent to the electrode–electrolyte interface. The amount of charge in the metal electrode is a measure of the energy of the electrons in the metal and can be changed by using a power supply to force electrons into or out of the metal. Excess electrons result in higher energy of electron because electrons tend to repel each other.

How is the electrolyte solution influenced by this excess charge? As you might expect, charged species in the electrolyte solution align themselves at the interface to balance the excess charge on the metal. The result is the formation of a *double layer*, as shown in Figure 3.1. Also shown in the figure is the diffusion layer; only a portion of this layer is shown because it is so much larger than the double layer and too large for a drawing of this scale (see numbers below). The solution side of the double layer is a region of nonuniform charge and, as it takes enormous energy to separate charge, it is a region where the electric field is very large. The potential difference between the metal and the solution just outside the double layer is critical to our discussion of electrode kinetics. At the open-circuit conditions we discussed in Chapter 2, where the overall current is zero, this potential difference adjusts itself through the transfer of electrons (i.e., reaction) so that the rates of the forward and reverse reactions are equal. In other words, equilibrium is achieved across the interface. The potential difference between the electrode and the solution at equilibrium provides a reference point for our discussion of kinetics later in this chapter.

The actual structure of the double layer is more complex than the simple arrangement presented above. A more complete view, along with the common nomenclature used to describe it, is shown in Figure 3.2. Note the importance of the solvent (water in this instance) molecules. Water molecules, which are polar, surround the ions in solution to form a hydration layer or solvent sheath. They also tend to align at the interface because of their dipole nature and the charge on the surface. The inner Helmholtz plane (IHP) is the position of the centers of ions or molecules that adsorb directly onto the electrode surface. They may experience van der Waals interactions with the surface in addition to coulombic interactions.

To adsorb directly on the surface, an ion must at least partially shed its waters of hydration. Negatively charged ions (anions) hydrate less strongly than cations and are more frequently found in the inner Helmholtz plane. They can even adsorb to a negatively charged surface. As a general rule, the weaker the solvation of an anion, the more strongly it tends to undergo specific (direct) absorption to the surface. The outer Helmholtz plane (OHP) is the plane of closest approach for solvated ions, where the solvent sheath prevents specific adsorption. Its position is defined by the radius, including the waters of hydration, of the largest solvated ions (~0.2 nm). The last portion of the

---

Electrochemical Engineering, First Edition. Thomas F. Fuller and John N. Harb.
© 2018 Thomas F. Fuller and John N. Harb. Published 2018 by John Wiley & Sons, Inc.
Companion Website: www.wiley.com/go/fuller/electrochemicalengineering

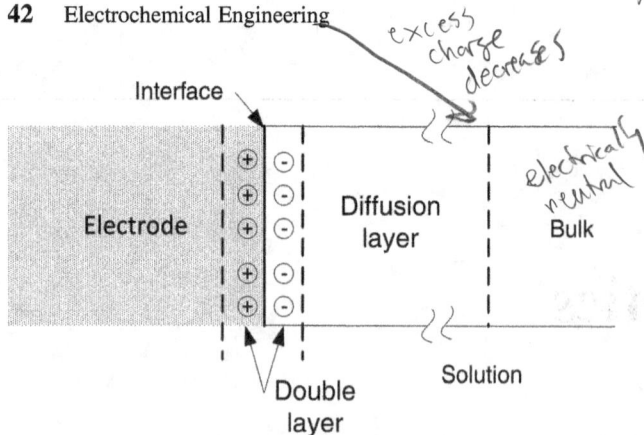

Figure 3.1 Simplified description of electrical double layer. *Source*: Adapted with permission from L. Faulkner, *J. Chem. Ed.*, **60**, 262 (1983).

double layer is called the diffuse double layer. This layer is characterized by a balance of Brownian motion and Coulombic forces. This balance is analogous to the one introduced in Chapter 2 in the development of the Debye–Hückel theory for activity coefficients. Both positive and negative charges are present, and the amount of excess charge decreases gradually from the OHP to the solution side of the diffuse double layer where the solution is electrically neutral. The thickness of the diffuse double layer is characterized by the Debye length:

$$\lambda = \sqrt{\frac{\varepsilon RT}{F^2 \sum z_i^2 c_{i,bulk}}} \quad (3.1)$$

and is a function of concentration. The diffuse part of the double layer is important at low concentrations and has a thickness of about 10 nm at an ionic strength of 1 mM. The diffuse part of the double layer is typically not as important at concentrations used in practical electrochemical systems, but the structure of the double layer is critical to the understanding of charge transfer. In contrast to the thin diffuse double layer, the diffusion layer is on the order of 1 mm; thus, the diffusion layer is typically orders of magnitude larger than the thickness of the double layer. The potential drop from the potential of the metal (ϕ_m) to the potential of the solution just outside the double layer (ϕ_s) is also shown in Figure 3.2. For a potential difference of 1 V, the electric field is

$$\approx \frac{\Delta\phi}{\lambda} = \frac{1}{1 \times 10^{-8}} = 10^8 \ [\text{V m}^{-1}],$$

which is quite high. There is also a capacitance associated with the separation of charge across the double layer. In fact, reference is often made to electrochemical capacitors or double-layer capacitors. The capacitance, C, is simply

$$C \equiv \frac{Q}{\Delta\phi} \quad (3.2)$$

where Q is the charge on the electrode. By adjusting the potential, this charge can be changed. Here an ideally polarizable electrode is envisaged, which is an electrode where an arbitrary potential can be applied without causing a faradaic reaction to occur. Under these conditions, there is no faradaic reaction associated with the electrical current, only simple movement of the charges. The capacitance is discussed in more detail in Chapter 11, along with devices that take advantage of this effect. In this chapter, our focus will be on faradaic reactions, which are reactions that involve the transfer of electrons across the interface in order to change the oxidation state of at least one of the participating species. The half-cell reactions that we wrote in Chapters 1 and 2 are examples of faradaic reactions. Faraday's law applies to faradaic reactions and relates the current to the amount of a species reacted or formed.

3.2 IMPACT OF POTENTIAL ON REACTION RATE

Electrochemical reactions take place at the surface of the electrode as electrons are transferred to and from ions or neutral species in solution. The rate of reaction is influenced by the potential drop across the double layer. At open circuit (no external current), this potential drop reaches an equilibrium value where the forward and reverse reactions are equal. One of the real advantages of electrochemical reactions is that the rate of reaction can be controlled by changing the potential of the metal, which changes the potential drop across the double layer and, therefore, the rate of reaction across the interface. By changing the potential of the metal, we can control the rate and even the direction of the reaction. The main purpose of this chapter is to develop a relationship between the rate, expressed as a current density, and the potential, and to illustrate how this relationship can be used.

The reference point for our discussion of reaction rate is the equilibrium potential where the net rate of reaction is zero. What happens when we increase the potential or, in other words, make the potential more positive than the equilibrium potential? Increasing the potential lowers the energy of electrons in the electrode (remember, electrons are negatively charged) and makes it easier for molecules or atoms to lose electrons to the electrode. Thus, values of the potential that are positive relative to the equilibrium potential promote the anodic reaction (oxidation). As you may have guessed, the opposite is true for potentials lower than the equilibrium potential. Lowering the potential increases the energy of electrons in the electrode and promotes the transfer of electrons from the electrode to, for example, a species in solution (reduction). Therefore, the cathodic reaction is favored as the potential is lowered relative to the equilibrium potential. The situation is summarized in Figure 3.3 using the reaction of ferric (Fe^{3+}) and ferrous (Fe^{2+}) ions as an example.

Chapter 3 Electrochemical Kinetics 43

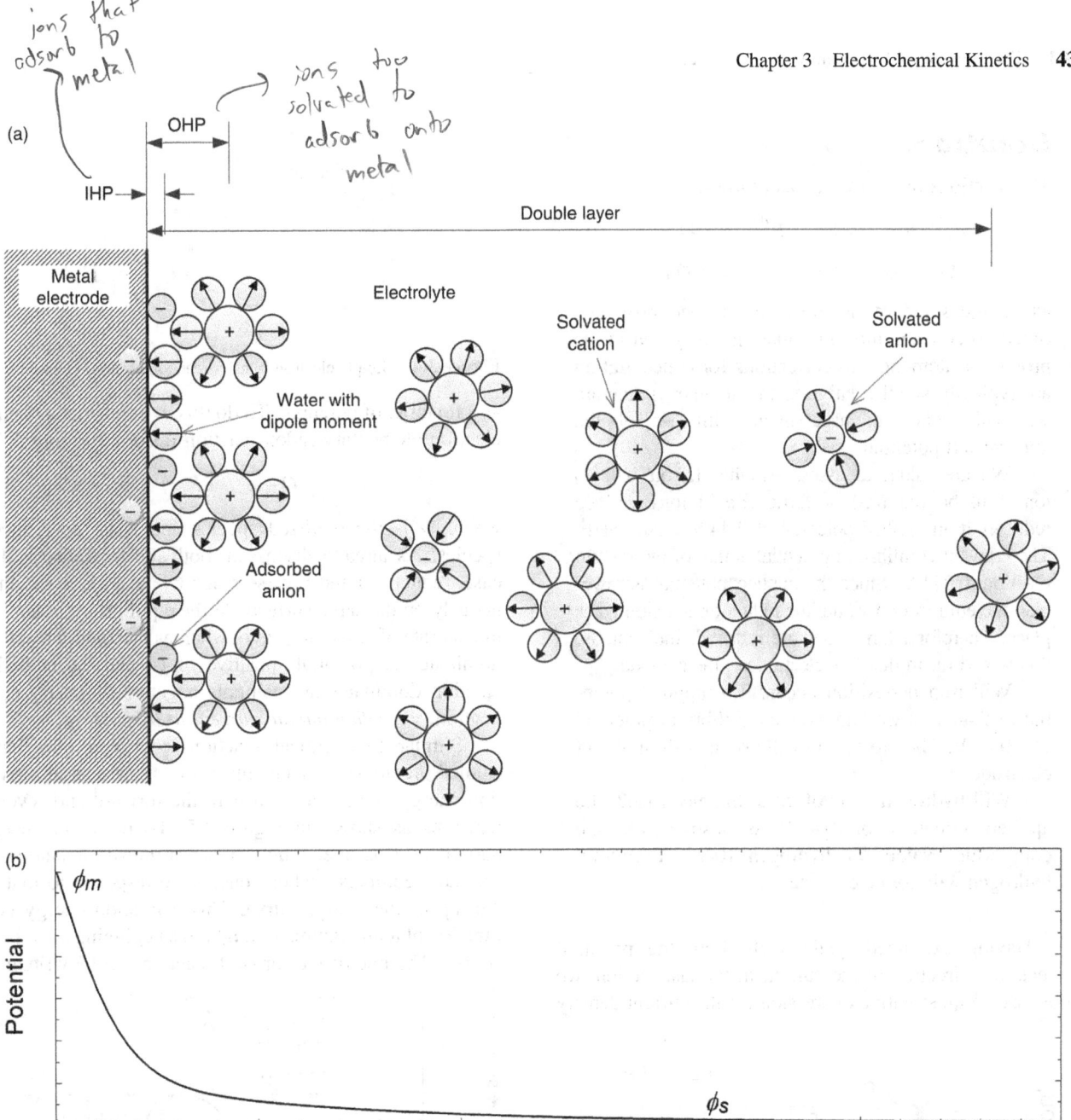

Figure 3.2 (a) Detailed structure of electrical double layer. (b) Sketch of potential across the double layer.

There is one slight problem. The potential difference ($\phi_m - \phi_s$) between the metal and the solution just outside the double layer (see Figure 3.2) cannot be measured directly. The reason for this is easy to understand. Potentials are measured between two electrodes. However, if we put another electrode into solution to measure ϕ_s, it will also form its own double layer with its own potential drop. Therefore, what you end up measuring is a combination of the potential drop associated with the two electrodes. Fortunately, this problem is easy to overcome, as mentioned in Chapter 2. We simply express all potentials relative to a reference electrode, which defines our potential scale. Thus, what you learned about reference electrodes and potential in the previous chapter provides the needed reference point for our discussion of kinetics.

ILLUSTRATION 3.1

You have been given a solution containing Fe^{3+} and Fe^{2+} ions at a pH of 3. Two inert Pt electrodes are placed into the beaker and one of the electrodes is set at a potential of 0.44 V versus SHE. Will ferrous ions be oxidized or will ferric ions be reduced at that electrode? Is there any chance of depositing metallic iron on the electrode? Your lab partner claims that at 0.44 V you will also evolve hydrogen. Is she correct?

44 Electrochemical Engineering

Solution:

The reactions of interest are as follows:

$$Fe^{2+} + 2e^- = Fe; \quad U^\theta = -0.44 \text{ V}$$

$$Fe^{3+} + e^- = Fe^{2+}; \quad U^\theta = 0.771 \text{ V}$$

where the standard potentials have been provided. Since no concentration information was given in the problem statement, and corrections for concentration are typically small relative to the standard potential, we will make our assessment with use of the uncorrected potentials.

We are asked to assess whether ferrous (Fe^{2+}) ions will be oxidized or ferric (Fe^{3+}) ions will be reduced at an applied potential of 0.44 V versus SHE. The relevant equilibrium potential is that of the second reaction, 0.771 V. Since the applied potential is lower than the equilibrium value, the reduction reaction takes place. Therefore, ferric ions are reduced. Incidentally, this would mean that this electrode is the cathode.

Will iron deposition occur? The applied potential of 0.44 V is well above the equilibrium potential of −0.44 V. Therefore, iron will not deposit on the Pt electrode.

Will hydrogen be evolved at the electrode? The applied potential of 0.44 V is also above the equilibrium value for hydrogen (0 V). Therefore, hydrogen will not be evolved.

Having examined qualitatively how the potential affects the direction of the current in the last section, we now seek a quantitative expression for the current density

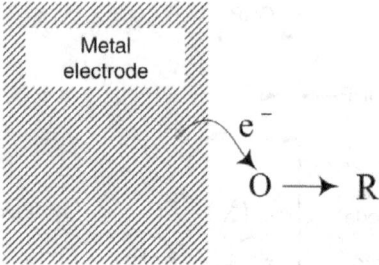

Figure 3.4 Simple electron transfer reaction at metal electrode.

as a function of potential. To do this, we consider a single-electron elementary redox reaction (also see Figure 3.4):

$$O + e^- \leftrightarrow R, \tag{3.3}$$

where O is the oxidized species and R is the reduced species. As already discussed, both the forward (in this case, reduction) and reverse reactions take place simultaneously on the same surface. At the equilibrium potential, the net rate of reaction is zero, whereas the anodic reaction dominates at potentials positive of U, and the cathodic reaction dominates at potentials negative of U. By convention, *we define anodic current to be positive*.

Similar to chemical reactions that you may have studied previously in a chemistry class, there is an activation energy associated with both the forward and reverse reactions as shown in Figure 3.5. To react, an energy barrier must be overcome to reach a transition state (i.e., activated complex), whose energy corresponds to that at the top of the energy curve. This activation energy is a function of temperature, as taught in a beginning chemistry course. The rate of the forward reaction can be written in

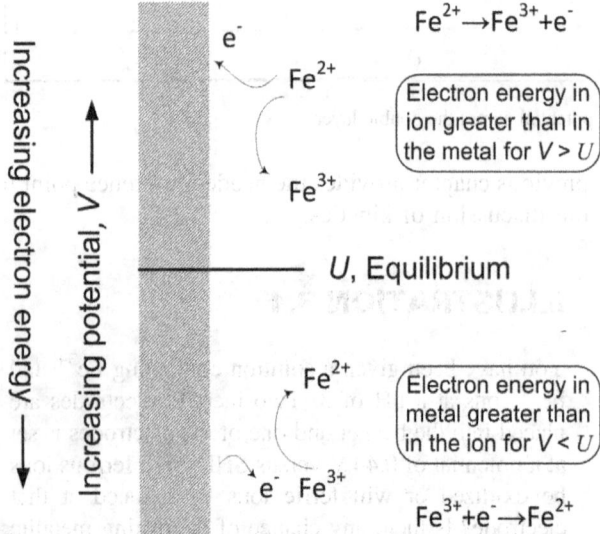

Figure 3.3 The relationship between potential, electron energy, and the direction of a faradaic reaction.

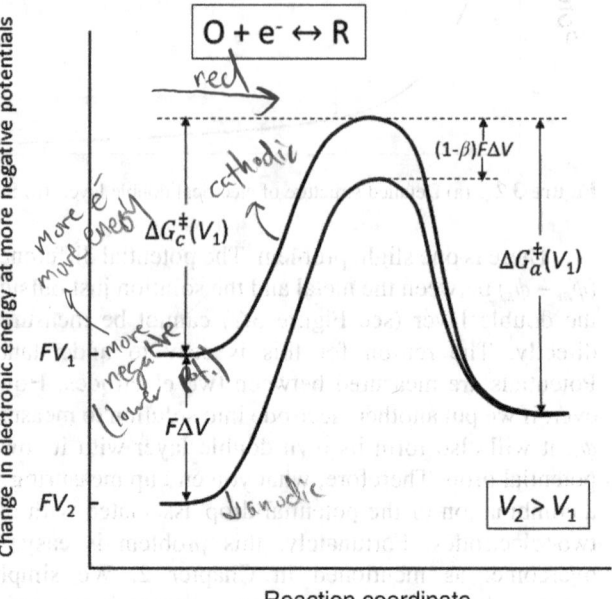

Figure 3.5 Change in energy associated with reaction at an electrode surface (subscript a = anodic and c = cathodic).

terms of the reactant concentration and a rate constant, k_f, that is a function of $\Delta G_f^\ddagger$, the Gibbs energy difference between the initial reactants and the transition state. Assuming a first-order elementary reaction (see Equation 3.3),

$$r = k_f c_O = k_f^o c_O \exp\left(\frac{-\Delta G_f^\ddagger}{RT}\right), \quad (3.4)$$

where c_O is the concentration of the oxidized species. A decrease in $\Delta G_f^\ddagger$ corresponds to a lower activation energy and will cause the rate of the forward reaction to increase. A similar expression can be written for the reverse reaction. The same type of model applies to electrochemical reactions, with one important difference.

We can modify the activation energy of electrochemical reactions by changing the electrode potential, as shown in Figure 3.5 for a single-electron reaction. Since we are only interested in energy differences, we have arbitrarily set the energy of the reduced state to be the same for the two curves. The upper curve is at a lower (more negative) potential since a decrease in potential corresponds to an increase in the energy of the electrons. Therefore, the energy is higher for V_1 (the lower potential), and $\Delta V = V_2 - V_1 > 0$.

The reduction reaction represented by Equation 3.3 takes place as one moves along the reaction coordinate from the left to the right of the diagram. At an electrode potential of V_1, the reduction reaction is favored as the energy of the reduced state is lower than that of the oxidized state. The activation barrier for the reduction is also lower than that for oxidation (the reverse reaction). However, the situation changes when the potential of the electrode is changed from V_1 to V_2 so that the oxidation reaction becomes favored by the increase in potential (see Figure 3.5).

The potential change from V_1 to V_2 affects the energy of an electron in the oxidized state relative to that in the reduced state by $F\Delta V$. The activation energies of both the reduction (cathodic) and oxidation (anodic) reactions are also changed. We define β as the fraction of the total energy change that impacts the activation energy for the cathodic reaction and $(1-\beta)$ as the fraction that impacts the anodic activation energy. Therefore, the shift in potential from V_1 to V_2 changes the activation energies as follows for the single-electron reaction under consideration:

For the cathodic reaction:
$$\Delta G_c^\ddagger(V_2) = \Delta G_c^\ddagger(V_1) + F\Delta V - (1-\beta)F\Delta V$$
$$= \Delta G_c^\ddagger(V_1) + \beta F\Delta V, \quad (3.5)$$

and for the anodic reaction:
$$\Delta G_a^\ddagger(V_2) = \Delta G_a^\ddagger(V_1) - (1-\beta)F\Delta V, \quad (3.6)$$

where c represents cathodic, a represents anodic, and ΔV is positive since $V_2 > V_1$. We can now use Equation 3.5 to write an expression for the cathodic reaction at potential V_2:

$$r_c(V_2) = k_c^o c_O \exp\left(-\frac{\Delta G_c^\ddagger(V_1) + \beta F(V_2 - V_1)}{RT}\right). \quad (3.7)$$

Since the precise values of V_2 and V_1 are arbitrary, we can define a reference potential so that V_1 is equal to zero. The activation energy at the reference potential ($\Delta G_c^\ddagger(V_1)$) can then be treated as a constant and be incorporated into the rate constant $k_c = k_c^o \exp\left[-(\Delta G_c^\ddagger(0)/RT)\right]$. We can also drop the subscript on V_2. This convention leaves us with the following expression for the cathodic reaction:

$$r_c(V) = k_c c_O \exp\left(-\frac{\beta FV}{RT}\right). \quad (3.8)$$

We can now use this equation to write an expression for the cathodic current density. Note that the direction of cathodic current is from the solution to the electrode (electrons move in the direction opposite to the current or, in this case, from the electrode to the species in solution that is reduced). *Cathodic current is negative by convention.* Consistent with this convention and the transfer of a single electron,

$$\frac{i_c}{F} = -r_c(V) = -k_c c_O \exp\left(-\frac{\beta FV}{RT}\right). \quad (3.9)$$

Following a similar procedure, we obtain the following for the anode:

$$\frac{i_a}{F} = r_a(V) = k_a c_R \exp\left(\frac{(1-\beta)FV}{RT}\right). \quad (3.10)$$

At the equilibrium potential, U, the net current is zero and the magnitude of the anodic current is equal to that of the cathodic current. This value of the current density at equilibrium is defined as the exchange-current density, i_o (remember that the net current is zero). Therefore,

$$\frac{i_o}{F} \equiv k_a c_R \exp\left(\frac{(1-\beta)FU}{RT}\right) = k_c c_O \exp\left(-\frac{\beta FU}{RT}\right). \quad (3.11)$$

If we multiply and divide Equation 3.10 by the anodic expression for i_o, we obtain

$$\frac{i_a}{F} = \left[k_a c_R \exp\left(\frac{(1-\beta)FU}{RT}\right)\right] \frac{\left[k_a c_R \exp\left(\frac{(1-\beta)FV}{RT}\right)\right]}{\left[k_a c_R \exp\left(\frac{(1-\beta)FU}{RT}\right)\right]}, \quad (3.12)$$

which simplifies to

$$i_a = i_o \left[\exp\left(\frac{(1-\beta)F(V-U)}{RT}\right)\right]. \quad (3.13)$$

The above procedure yields a similar expression for the cathodic current density:

$$i_c = -i_o \left[\exp\left(-\frac{\beta F(V-U)}{RT}\right)\right]. \quad (3.14)$$

We now define the surface overpotential, η_s, which is the driving force for the reaction:

$$\eta_s \equiv V - U, \quad (3.15a)$$

where V and U are the potential and equilibrium potential relative to the same reference electrode located just outside the double layer. As such, these potentials are measurable and well defined. η_s is also equal to the difference between the voltage drop across the double layer at the potential of interest (not measurable) and the voltage drop across the double layer at equilibrium (also not measurable). The choice of reference electrode is arbitrary since both V and U are referred to the same reference. The surface overpotential is quite important and will be used repeatedly throughout the course.

Although the definition of the surface overpotential in Equation 3.15a is adequate for many problems, it is often necessary to fix further the value of the potential when analyzing more complex systems. To facilitate this, it is useful to further expand our expression for η_s as follows:

$$\eta_s = V - U = \left[(\phi_{metal} - \phi_{solution}) - U\right] = \left[(\phi_1 - \phi_2) - U\right], \quad (3.15b)$$

where ϕ_1 is the potential of the electrode (e.g., metal) and ϕ_2 is the potential measured by a specific reference electrode located in the solution just outside the double layer. U is the equilibrium potential defined against that same reference electrode. Be careful since the value of U will change with the reference electrode chosen.

We can now write the expression for the net current density by combining Equations 3.13, 3.14, and 3.15a:

$$i = i_a + i_c = i_o \left[\exp\left(\frac{(1-\beta)F\eta_s}{RT}\right) - \exp\left(-\frac{\beta F \eta_s}{RT}\right) \right]. \quad (3.16)$$

This equation is known as the *Butler–Volmer* (BV) *equation*, shown here for an elementary reaction involving the transfer of a single electron. The value of β for this type of reaction is typically ~0.5. A derivation similar to that provided above can be performed for any elementary reaction, and procedures have been developed to handle sets of elementary reactions.

The form of Equation 3.16 is useful for describing a wide variety of elementary and nonelementary reactions and is used broadly in electrochemical engineering. It can be generalized by defining anodic and cathodic transfer coefficients α_a and α_c to yield

$$i = i_o \left[\exp\left(\frac{\alpha_a F \eta_s}{RT}\right) - \exp\left(-\frac{\alpha_c F \eta_s}{RT}\right) \right], \quad (3.17)$$

which is a frequently used form of the BV equation. Note that $\alpha_a + \alpha_c = n$ is generally true, where n is the number of electrons transferred in the reaction of interest.

3.3 USE OF THE BUTLER–VOLMER KINETIC EXPRESSION

The purpose of this section is to help you understand and learn how to use Equation 3.17, the Butler–Volmer equation. The BV equation is a relationship between the current density (i) and the charge transfer or surface overpotential (η_s). It contains three parameters: the exchange-current density (i_o), the anodic charge transfer coefficient (α_a), and the cathodic charge transfer coefficient (α_c). It is frequently used to represent experimental data by fitting these three parameters to the data.

The charge transfer overpotential (η_s) is defined by Equation 3.15, where V represents the difference between the potential of the electrode (ϕ_1, often a metal) and the potential of a reference electrode located just outside the double layer (ϕ_2). U is the difference between the potential of the electrode and the reference electrode at equilibrium. As already mentioned, V and U must be relative to the same reference electrode, although the precise choice of reference electrode is not critical as long as it is appropriate for the system. Note that $U = 0$ if the reference electrode is the same as the electrode of interest. For example, U would be zero for the combination of a zinc electrode and a zinc reference electrode (Why?). The methods that you learned in Chapter 2 can be used to calculate U, which is a function of the concentration at the surface.

The current density (i) is the current per area, with typical units of mA·cm^{-2} or A·m^{-2}. In theory, it is based on the actual area over which the reaction is taking place. The area, however, can be an issue in practice. For example, if the measured value of i_o is based on the superficial or apparent area in a system where the microscopic area is significantly greater than the superficial area, i_o will not transfer to another system with a different microstructure (see Figure 3.6). The surface roughness illustrated in Figure 3.6b results in a larger surface area for electrochemical reaction.

A surface roughness factor, R_a, can be defined as the ratio of the true surface area and the apparent surface

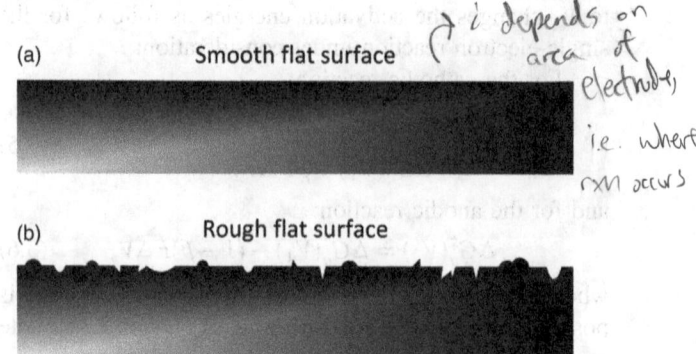

Figure 3.6 The area of the electrode with a rough surface is much larger than that of the one with the smooth surface.

Table 3.1 i_o for Different Reactions

Reaction	i_o [A·m^{-2}]
$O_2 + 4H^+ + 4e^- = 2H_2O$ on Pt	4×10^{-9}
$NiOOH + H_2O + e^- \rightarrow Ni(OH)_2 + OH^-$	6.1×10^{-1}
$H_2 = 2H^+ + 2e^-$ on Pt in 1 N HCl	10
$Fe^{3+} + e^- = Fe^{2+}$	20
$Zn + 2OH^- \rightarrow Zn(OH)_2 + 2e^-$	600
Ferri/ferrocyanide, 0.001 M	230

area, a ratio that can be quite large. Consequently, care should be taken to accurately account for the area for which the current density is defined. We will consider this topic in some detail in Chapter 5, and in other places in the text.

The exchange-current density, i_o, represents the rate per area of the forward and reverse reactions at equilibrium, where the total current is zero (see Equation 3.16). The exchange-current density for different reactions can vary over many orders of magnitude, as shown in Table 3.1 for some commonly encountered reactions. The value of i_o for a given reaction can also vary widely as a function of the electrode surface. Table 3.2 provides values for the hydrogen reaction on different metals. Reactions with higher exchange-current densities occur more readily and result in higher current densities for a given overpotential. In contrast, reactions with very low exchange-current densities are slow or sluggish. The units on i_o are the same as those on i since the exponential terms must be unitless. The exchange-current density is also based on the same area as i. The exchange-current density includes the traditional chemical rate constant and concentration terms and is therefore a function of both temperature and concentration.

The temperature dependence is exponential and can be represented by an Arrhenius expression. Values of i_o at different temperatures can be used to determine the activation energy needed to describe the temperature dependence. The dependence is described with

$$i_o(T) = A \exp\left\{\frac{-E_a}{RT}\right\}. \quad (3.18)$$

Here A is a constant and E_a is the activation energy in J·mol^{-1} for the reaction.

The concentration dependence can be determined explicitly for the elementary reaction considered in the previous section as follows. First, the forward and reverse reactions in Equation 3.11 can be solved together for U to yield:

$$U = \frac{RT}{F} \ln \frac{k_c c_O}{k_a c_R}, \quad (3.19)$$

which is a form of the Nernst equation. This expression can be substituted back into either the forward or reverse rate expressions in Equation 3.11 to give

$$\frac{i_o}{F} = k_a^\beta k_c^{1-\beta} c_R^\beta c_O^{1-\beta}. \quad (3.20a)$$

Equation 3.20a applies to the single-electron elementary reaction that we have considered. In general, i_o varies with the surface concentrations of reacting species raised to a power. The concentration exponents can be determined from a mechanistic understanding of the reaction (see Equation 3.3) or by measuring i_o at different concentrations. Once the exponents are known, it is common practice to report i_o values as

$$i_o = i_{o,ref} \left(\frac{c_1}{c_{1,ref}}\right)^{\gamma_1} \left(\frac{c_2}{c_{2,ref}}\right)^{\gamma_2}, \quad (3.20b)$$

where the subscript *ref* refers to a reference concentration.

Finally, α_a and α_c are charge transfer coefficients and, as defined in Equation 3.17, have units of equiv mol^{-1}. These coefficients are positive and bounded (typically, $\alpha_a + \alpha_c = n$, the number of electrons transferred in the reaction of interest). Note that the terms $\alpha F\eta/RT$ must be unitless, and the units for R and T should be chosen accordingly. Information regarding the reaction mechanism can be inferred from experimentally measured values of α. For example, $\alpha_a = \alpha_c = 0.5$ indicates an elementary single-electron reaction.

ILLUSTRATION 3.2

The reaction

$$[Fe(CN)_6]^{3-} + e^- \leftrightarrow [Fe(CN)_6]^{4-}$$

is sometimes used in analytical electrochemistry. This reaction can be assumed to be an elementary reaction. The exchange-current density for this reaction on a platinum electrode is 229 A·m^{-2} and

Table 3.2 i_o for Hydrogen Reaction in 1 m H_2SO_4 at 25 °C, values provide order of magnitude estimates

Metal	i_o [A·m^{-2}]
Pb, Hg	10^{-8}
Zn	10^{-7}
Sn, Al, Be	10^{-6}
Ni, Ag, Cu, Cd	10^{-3}
Fe, Au, Mo	10^{-2}
W, Co, Ta	10^{-1}
Rh, Ir	2.5
Pd, Pt	10

was measured at room temperature (25 °C) where both the reactant and product concentrations were 0.001 M. Experiments are being performed in a neutral 0.1 M KCl solution with a ferricyanide concentration of 0.02 M and a ferrocyanide concentration of 0.015 M.

a. Calculate the current that you would expect to measure from a 0.5 cm × 0.75 cm piece of platinum foil at a potential of 0.10 V versus a saturated calomel reference electrode (SCE). You may use simple activity corrections (neglect activity coefficients); liquid junction potentials may also be neglected. Please comment on the applicability of these assumptions.

b. A potentiostat is used to hold the potential of the Pt working electrode at the desired value versus the SCE reference electrode. What reaction do you think is being driven at the counter electrode?

Solution:

a. First, we need the equilibrium potential for the reaction. A search online reveals that the standard potential for the reaction is 0.36 V. We will use Equation 2.25 to correct this value for concentration:

$$U = U^\theta - \frac{RT}{nF} \ln\left(\frac{c_{[Fe(CN)_6]^{4-}}}{c_{[Fe(CN)_6]^{3-}}}\right)$$

$$= 0.36\,V - \frac{\left(\frac{8.314\,J}{mol \cdot K}\right)(298K)}{\left(\frac{1\,equiv}{mol}\right)\left(\frac{96{,}485\,C}{equiv}\right)} \ln\left(\frac{0.015\,M}{0.02\,M}\right)$$

$$= 0.3674\,V.$$

This potential is relative to a SHE. We have assumed that the bulk concentrations given apply at the surface of the electrode. This assumption will not be good at high rates of reaction. Activity coefficient corrections should not be excessively large because of the relatively low concentration of ions. Also, the corrections to the numerator and denominator would be similar and tend to cancel. The liquid junction potential should be small since K^+ and Cl^- have similar transport properties (remember that this correction is due to a potential created by unbalanced transport of the ions). Even so, we have undoubtedly included more significant figures than is warranted and will want to consider this in evaluating our final answer.

In order to get the overpotential for the BV equation, the equilibrium potential and the electrode potential need to be relative to the same reference. 0.10 V SCE = 0.342 V SHE (you should verify this for yourself).

The overpotential η_s is $0.342 - 0.3674 = -0.0254$ V. The negative overpotential indicates that the cathodic reaction will dominate.

We also need the exchange-current density at the concentrations of interest. From Equation 3.20, we get

$$i_o = 229\,A \cdot m^{-2} \left(\frac{0.015}{0.001}\right)^{0.5} \left(\frac{0.02}{0.001}\right)^{0.5}$$

$$= 3966\,A \cdot m^{-2}.$$

Note that the 0.5 values for the exponents are a result of the fact that we have a one-electron elementary reaction (see Equation 3.19). We can now plug the values into the BV equation to calculate the current density.

$$i = 3966\,A \cdot m^{-2} \left[\exp\left(\frac{0.5F(-0.0254\,V)}{RT}\right)\right.$$

$$\left. - \exp\left(-\frac{0.5F(-0.0254\,V)}{RT}\right)\right] = -4080\,A \cdot m^{-2}.$$

The current density is negative since it is cathodic. To get the current, we must multiply by the area of the electrode. We assume that both sides of the electrode are active to give a total area of 0.75 cm². The resulting current from the electrode is therefore −0.306 A.

b. In order to set the potential of the Pt electrode at the given value, the potentiostat will drive an oxidation reaction at the opposite electrode. The easiest reaction to drive will be the oxidation of the ferrocyanide. If that reaction is not sufficiently rapid to provide the needed oxidation current, additional current will likely be provided by the oxidation of water to form O_2, a reaction that takes place frequently in aqueous electrochemical cells.

In summary, the Butler–Volmer equation was derived for a single-electron elementary reaction and then generalized. From a practical perspective, the BV equation is a three-parameter equation that can be fit to experimental data from a wide variety of electrochemical reactions.

3.4 REACTION FUNDAMENTALS

We will find the Butler–Volmer formulation of kinetics extraordinarily useful for the study of electrochemical systems. Nonetheless, it is largely a phenomenological or empirical equation with three parameters, i_o and two transfer coefficients, α_a and α_c. We saw vast variations in exchange-current densities based on the reactions and the electrode surface. What causes one reaction to be facile and others to be sluggish? The answer to these questions must be rooted in the chemical environment of the reactants and products as well as the structure of the interface between the electrode and the electrolyte. These fundamental studies have been pursued extensively, and in fact two Nobel Prizes for Chemistry have been awarded in the field of electron transfer reactions in chemical systems (Henry Taube, 1983 and Rudy Marcus, 1992). Although the theory is still far from complete, it is instructive to consider a few aspects and results from these works.

First, we should like to distinguish between two types of electron transfer reactions: *inner sphere* and *outer sphere*. For outer sphere reactions, the reactants and products do not interact with the electrode surface. Rather than being adsorbed onto the surface, they may be a solvent layer away from the surface. You might expect that for these types of reactions, the nature of the electrode surface is less important. In contrast, inner sphere reactions have strong interactions with the electrode surface and often involve specific adsorption. The oxygen and hydrogen reactions are two examples of inner sphere reactions. As we saw from Table 3.2, the electrode material plays a critical role in determining the exchange-current density. As a specific example, this dependence may be explained in part by understanding the mechanism of the hydrogen reaction in more detail. We can view the metal surface as a collection of possible sites for adsorption of hydrogen. The hydrogen reaction proceeds through two steps, referred to as the Volmer–Tafel mechanism. Dissociative adsorption of molecular hydrogen onto two metal sites, S, takes place first, and is followed by the electron transfer reaction.

$$H_2 + 2S = 2HS \quad \text{(Tafel)}$$
$$HS = H^+ + e^- \quad \text{(Volmer)}$$

The adsorption step is relatively slow and thus rate determining. If hydrogen adsorbs either too weakly or too strongly, the reaction is inhibited. First principles calculations of the strength of adsorption on different metals predict the trends in exchange-current density with different materials.

Reaction theory is most developed for the so-called outer sphere reactions. These reactions are more tractable because the interactions with the electrode surface are absent. Let's compare a reaction already encountered in this chapter and treated with the phenomenological Butler–Volmer equation to an isotope exchange reaction:

$$[Fe(CN)_6]^{3-} + e^- = [Fe(CN)_6]^{4-} \quad (U^\theta = 0.361 \text{ V})$$

$$Fe^{2+} + Fe^{*3+} = Fe^{3+} + Fe^{*2+}$$

The first is the ferri/ferrocyanide reaction and the second is the reaction between Fe(II) and Fe(III). The asterisk refers to an isotope of iron, and this reaction is an isotope exchange reaction. In both cases, no chemical bonds are broken or formed. In contrast to what might be intuitive, the ferri/ferrocyanide reaction is relatively fast compared to the isotope exchange reaction. This result can be understood in terms of the reorganization energy, λ_{re} (not to be confused with the Debye length, which also uses λ). There are two components of this reorganizational energy representing solvation and vibrational energy terms. When, because of electron transfer, the charge of the species changes, its chemical environment must adjust to the new equilibrium. The theory predicts that the change in Gibbs energy of the activated complex is proportional to this reorganizational energy:

$$\lambda_{re} \propto \frac{\Delta G^\ddagger}{RT}. \quad (3.21)$$

Thus, the larger the reorganization energy, the smaller the rate constant. For the bulkier ferricyanide complex, the change in charge has a relatively smaller effect and the reorganization needed is much smaller, and therefore the reaction is faster.

A second key result from the theory has to do with the transfer coefficient:

$$\alpha = \frac{1}{2} + \frac{F(V-U)}{2\lambda_{re}}. \quad (3.22)$$

Here we note that generally the term involving the reorganizational energy λ_{re} is small, and therefore a transfer coefficient of 0.5 for an elementary single-electron transfer reaction has some theoretical basis. Furthermore, we see that there is no dependence of the transfer coefficient on temperature, but a small dependence on potential. It is also important to note that Equation 3.22 is for outer sphere reactions. For inner sphere reactions, the interactions with the electrode must be considered, and one can neither expect the transfer coefficient to be 0.5 nor to be independent of temperature and potential.

3.5 SIMPLIFIED FORMS OF THE BUTLER–VOLMER EQUATION

Tafel Approximation

The BV equation has two exponential terms, one that represents the anodic current ($i > 0$, $\eta_s > 0$) and the other that represents the cathodic current ($i < 0$, $\eta_s < 0$). What happens to the relative magnitude of the two terms as η_s becomes more positive? What about as η_s becomes more negative?

When η_s is large and positive, the anodic term of the BV equation dominates and the cathodic term does not contribute significantly to the current. Basically, the activation energy changes as a function of the potential until that corresponding to the cathodic reaction becomes much larger than that for the anodic reaction. Under such conditions,

$$i \approx i_o \exp\left(\frac{\alpha_a F}{RT} \eta_s\right). \quad (3.23)$$

This simplified expression is called the Tafel equation. It can be readily solved for the overpotential as a function of the current:

$$\eta_s = \frac{RT}{\alpha_a F} \ln i - \frac{RT}{\alpha_a F} \ln i_o. \quad (3.24)$$

Equation 3.24 is the equation of a line for η_s versus $\ln i$. Both α_a and i_o can be determined from experimental current–voltage data by fitting a line to the data. One of the advantages of the Tafel equation is that there is one less parameter to fit (two rather than three). In the past, when most analysis was done graphically, it was common practice to use the base$_{10}$ logarithm in analysis rather than the natural logarithm. We also use base$_{10}$ logarithms in log–log and semi-log plots because such plots are easier to read. Equation 3.24 can be written accordingly as follows:

$$\eta_s = a + b \log i, \quad (3.25)$$

$$a = \frac{-2.303 RT}{\alpha_a F} \log i_o, \quad b = \frac{2.303 RT}{\alpha_a F}.$$

This equation yields a slope of 59.2 mV per decade for an α_a of 1. The slope of experimental data can be used as a quick reference for determining α. For example, a Tafel slope of about 118 mV per decade would correspond to a single-electron reaction with α_a equal to 0.5. Note that a decade is one x unit on the base$_{10}$ semi-log plot of η_s versus $\log i$.

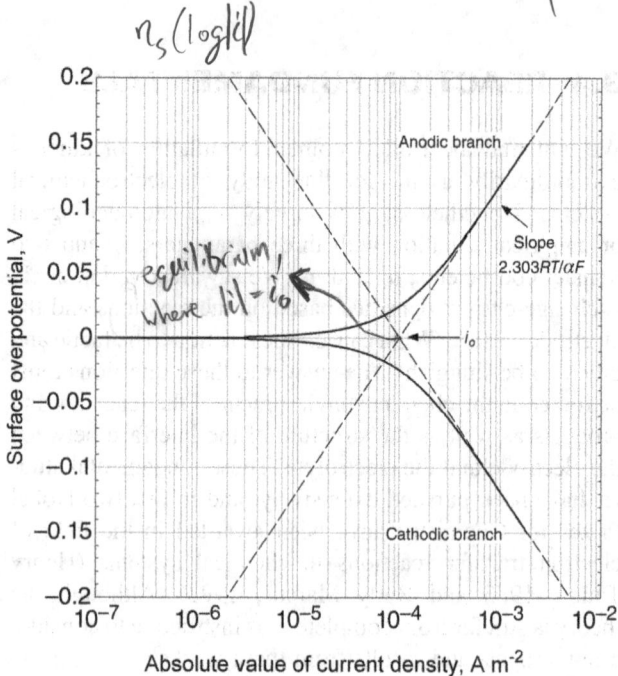

Figure 3.7 Classic Tafel plot. Parameters used are $i_o = 10^{-4}$ A·m^{-2}, 25 °C, $\alpha = 0.5$.

A similar development can be used to obtain a Tafel expression for the cathodic reaction at large negative values of η_s:

$$i = -i_o \exp\left(-\frac{\alpha_c F}{RT} \eta_s\right). \quad (3.26)$$

$$\eta_s = -\frac{RT}{\alpha_c F} \ln(-i) + \frac{RT}{\alpha_c F} \ln i_o = -\frac{RT}{\alpha_c F} \ln|i| + \frac{RT}{\alpha_c F} \ln i_o. \quad (3.27)$$

Note that $i < 0$ for cathodic current so that $-i$ is positive as it must be inside the logarithm. To emphasize this, we have included the absolute value in the last term of Equation 3.27 and on graphical representations of the equation. The slope of the cathodic curve is negative. A classical Tafel plot is shown in Figure 3.7 for η_s as a function of $\log|i|$. Both the anodic and cathodic Tafel lines have been included, which meet at $\eta_s = 0$, where $|i| = i_o$. The absolute value of the current from the full BV equation is also shown. The Tafel approximation and the full BV equation are identical at high overpotentials, but diverge as the overpotential approaches zero. As mentioned above, such plots are useful for fitting parameters to experimental data. These types of plots have also been used extensively to analyze corrosion systems (see Chapter 16), among others.

How large must the overpotential be in order for the Tafel expression to be valid? To address this question, we

general rule for using Tafel eqs:
η_s should be greater than ≈ 100 mV

compare the current from the Tafel expression with that from the corresponding full BV equation. That comparison yields the following:

$$|\eta_s| > \frac{B}{(\alpha_a + \alpha_c)}\left(\frac{T(K)}{298.15}\right), \quad (3.28)$$

where B is a constant that is equal to 0.12 V for a maximum error less than 1%, and 0.062 V for a maximum error <10%. Equation 3.28 provides an easy way to determine the minimum value of the overpotential that meets these error specifications. However, it assumes that α_a and α_c are known. This, of course, does not help if you are using the Tafel expression to fit data in order to estimate these parameters. A general rule of thumb is that the overpotential should be greater than about 100 mV for use of the Tafel simplification. Graphical examination of the data that you are fitting will also allow identification of the linear region (semi-log plot) that can be used for parameter fitting with the Tafel equation.

ILLUSTRATION 3.3

Kinetic data are provided for the evolution of chlorine in 5 M NaCl at 20 °C (*J. Electrochem. Soc.*, **120**, 231 (1973)). a) Write the reaction for the evolution of chlorine. Is this anodic or cathodic? b) What is the Tafel slope for the reaction? c) Determine the exchange-current density for this reaction. d) Find the transfer coefficient α_a.

Current density [A·m^{-2}]	Overpotential [V]
60,300	0.2091
41,300	0.1994
22,300	0.1703
12,400	0.1515
8,080	0.1359
6,180	0.1208
4,110	0.1036
2,070	0.07934
1,230	0.05512
817	0.04727
621	0.04047
427	0.03433
240	0.02547
173	0.02204
116	0.01658
58.4	0.01175
30.1	0.00489

a. The reaction is

$$2Cl^- \rightarrow Cl_2 + 2e^-$$

Because electrons are released, this is an anodic reaction.

b. According to the Tafel equation, the current density is

$$i = i_o \exp\left\{\frac{\alpha_a F}{RT}\eta_s\right\}.$$

Taking the logarithm

$$\ln(10)\{\log(i) - \log(i_o)\} = \left\{\frac{\alpha_a F}{RT}\eta_s\right\}$$

and rearranging results, an equation of the form $y = ax + b$,

$$\eta_s = 2.303\frac{RT}{\alpha_a F}\log(i) - \frac{RT}{\alpha_a F}\ln(i_o).$$

The overpotential data are plotted as a function of the logarithm base 10 of the current density. As expected, the relationship is not linear over the entire range. At larger overpotentials, the Tafel equation should apply and the slope is constant. A line is fitted through the data points at overpotentials greater than 50 mV. The slope of the fitted curve is 92 mV per decade.

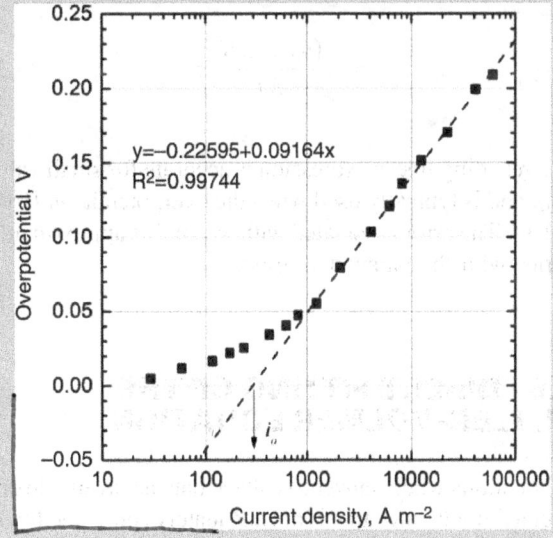

c. The Tafel equation reduces to $i = i_o$ when the overpotential is zero. Graphically, the fit is extrapolated to $\eta_s = 0$, and the value for the exchange-current density is read off the graph. Alternatively, the value is calculated from the regression fit:

$$\eta_s = -0.22595 + 0.09164 \log(i_o) = 0.$$

Solving, $i_o = 292 \text{ A·m}^{-2}$.

d. Using the slope fitted to the data,

$$\text{slope} = 0.09164 = 2.303 \frac{RT}{\alpha_a F}.$$

Thus,

$$\alpha_a = \frac{1}{0.09164} 2.303 \frac{RT}{F} = 0.63.$$

Linear Approximation

The BV equation can also be approximated at low overpotentials by expanding the full BV expression in a Taylor series around $\eta_s = 0$, and keeping the first two terms of the series for both the anodic and cathodic portions:

$$i \approx \left[i_o + i_o \frac{\alpha_a F}{RT}(\eta_s - 0)\right] - \left[i_o - i_o \frac{\alpha_c F}{RT}(\eta_s - 0)\right].$$

(3.29)

$$= i_o \frac{(\alpha_a + \alpha_c)F}{RT} \eta_s.$$

(3.30)

The resulting linear expression is accurate for small values of η_s and is typically used when the overpotential is 10 mV or less. The error associated with use of this approximation varies with the parameters used.

3.6 DIRECT FITTING OF THE BUTLER–VOLMER EQUATION

In situations where current–voltage data are available only over a limited range that is not adequately addressed by one of the limiting cases above, a direct fit of the data to the full BV equation may be appropriate. This fitting can be done in a straightforward manner with use of a nonlinear solver or optimization routine to minimize the error between the data and the desired BV expression. This process is illustrated for Microsoft Excel in the example that follows:

ILLUSTRATION 3.4

The following data were taken for a NiOOH electrode, which involves a single-electron reaction. Please fit the data to the appropriate kinetic expression.

Overpotential [V]	Current density [A·m^{-2}]
−0.1	−4.20
−0.09	−3.36
−0.08	−2.40
−0.07	−2.30
−0.06	−1.80
−0.05	−1.25
−0.04	−1.00
−0.03	−0.80
−0.02	−0.50
−0.01	−0.22
0.01	0.24
0.02	0.45
0.03	0.80
0.04	1.00
0.05	1.45
0.06	1.80
0.07	2.10
0.08	2.80
0.09	3.50
0.1	4.10

Solution:

We first plot the data to get a feel for what we have. Note that the maximum overpotential is 100 mV, so we would not expect to be able to fit the data with just Tafel expressions for the anodic and cathodic currents.

We first plot the data using linear scales for both the current and potential. We next plot the same data with the absolute value of the current on a log scale.

From the semi-log plot, we see that most of the data are not in the Tafel region, although we can get an estimate of what the Tafel slopes might be from the last few data points that correspond to the highest values of either the anodic or the cathodic current. Therefore, we will fit the data to the full BV equation. We will do this by using the Solver in Microsoft Excel to determine i_o, α_a, and α_c that best fit the data. Please note that the Solver may not come loaded by default, although it is included within the software as delivered.

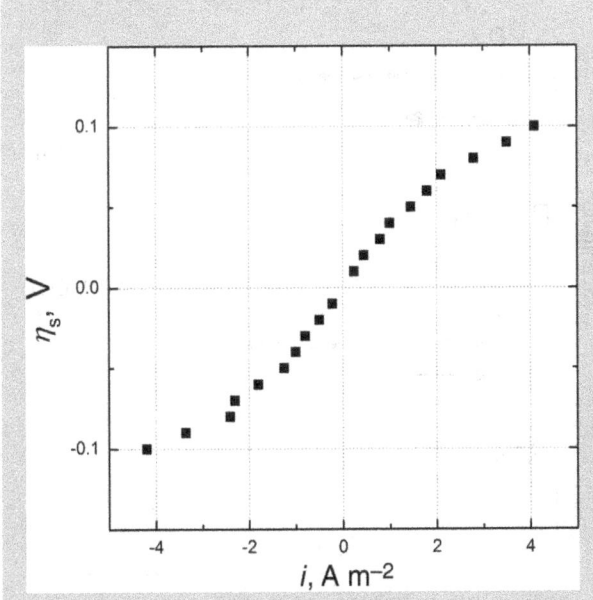

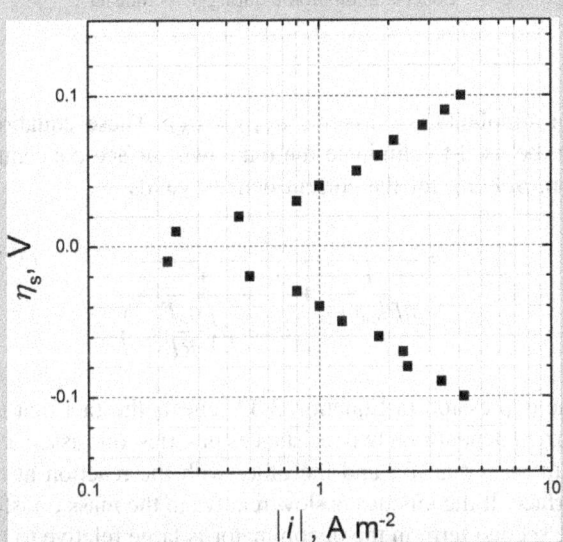

When loaded, it is found under the "Data" ribbon in Excel. Before demonstrating, we will use the last few points of the cathodic curve to get an estimate of α_c and i_o. Note that the α values are bounded and relatively easy to guess with reasonable accuracy, but i_o can vary over orders of magnitude. Consequently, we will fit the last few points to a Tafel expression to get starting values for the desired parameters. A linear fit of $\ln |i|$ versus η_s for the last four points of the cathodic curve yields a slope of -0.0435 and an intercept of -0.0376. The corresponding values of α_c and i_o are 0.59 and 0.42, respectively, where i_o has the same units as the current (A·m^{-2}) (see Equation 3.24). These will be our starting estimates for fitting the full BV equation. Since we have a single-electron reaction, we will assume $\alpha_a = 1 - \alpha_c$ as the starting value. The setup on the spreadsheet is as follows:

Excel Solver Fit of BV Equation

Physical Constants (25 C)

F/RT   38.924

Fitting Parameters

α_c 0.5900
i_o 0.4200
α_a 0.4100

Overpotential	Measured Current	Calculated Current	Error
−0.1	−4.20	−4.0894	0.111
−0.09	−3.36	−3.2181	0.142
−0.08	−2.40	−2.5200	−0.120
−0.07	−2.30	−1.9586	0.341
−0.06	−1.80	−1.5047	0.295
−0.05	−1.25	−1.1350	0.115
−0.04	−1.00	−0.8306	0.169
−0.03	−0.80	−0.5763	0.224
−0.02	−0.50	−0.3596	0.140
−0.01	−0.22	−0.1704	0.050
0.01	0.240	0.1589	0.081
0.02	0.450	0.3126	0.137
0.03	0.800	0.4670	0.333
0.04	1.000	0.6276	0.372
0.05	1.450	0.7996	0.650
0.06	1.800	0.9883	0.812
0.07	2.100	1.1994	0.901
0.08	2.800	1.4387	1.361
0.09	3.500	1.7130	1.787
0.1	4.100	2.0295	2.071
		11.866	Sum of Error Squared

The "Calculated Current" is calculated with the desired BV equation using the parameters above. The "Error" is simply the difference between the calculated value and the measured value. The Sum of Error Squared was calculated with the SUMSQ function in Excel. Solver will minimize this value by changing the fitting parameters. The fitting is done by selecting the cell containing the sum of the squared errors (initial value 11.866), choosing Solver under Data in Excel, and then using the Solver to minimize the value of this computational cell by varying the three fitting parameters. The resulting values are shown in the following table:

α_c	0.5035
i_o	0.5893
α_a	0.5087

The Solver fit reduced the error from 11.866 to a final value of 0.186. Note that the fitting did not constrain $\alpha_a + \alpha_c = 1$. Therefore, the two parameters were fit independently and validated the fact that we have a single-electron reaction. The above process works well as long as the values of the current do not vary widely over orders of magnitude. In those cases, errors associated with the high current values are weighted more heavily and bias the fit. The bias can be offset through the use of a normalized error, but it is better to just use a Tafel fit, since you would undoubtedly have data in the Tafel region and a Tafel fit adequately handles the wide range of data.

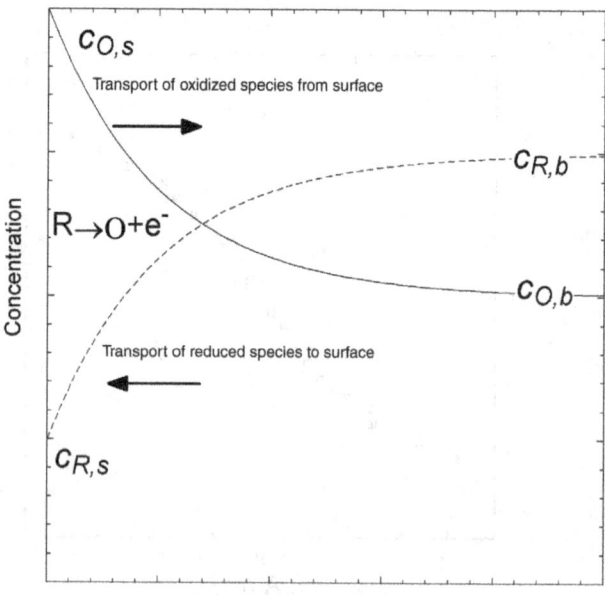

Figure 3.8 Concentration profile during mass transfer.

3.7 THE INFLUENCE OF MASS TRANSFER ON THE REACTION RATE

As noted above, the exchange-current density and, hence, the current depends on concentration at the surface (see Equation 3.20). Species can be consumed and/or generated at the surface, leading to surface concentrations that can vary significantly from the bulk concentrations, as illustrated in Figure 3.8. In this section, we consider the situation where the reaction rate depends linearly on the reactant concentration at the surface. As the applied potential is increased, the reaction rate increases, the surface concentration decreases, and the mass transfer resistance to the surface becomes important. A simplified expression for mass transfer can be used in the presence of supporting electrolyte, as will be discussed in the next chapter, or for a reactant that is not charged. Under such conditions, the rate of mass transfer of the reactant to the surface can be approximated as

$$N_R = k_m (c_{R,b} - c_{R,s}). \quad (3.31)$$

Normally the mass-transfer coefficient is represented by k_c. In this chapter, k_m is used for clarity. At steady state, the rate of mass transfer must equal the reaction rate at the surface, where the Tafel equation is assumed to apply:

$$i = nFk_m(c_{R,b} - c_{R,s}) = i_{o,ref}\left(\frac{c_{R,s}}{c_{R,ref}}\right) \exp\left(\frac{\alpha_a F}{RT}\eta_s\right). \quad (3.32)$$

For simplicity, we assume $c_{R,ref} = c_{R,b}$. These equations can be used to eliminate the unknown surface concentration. Solving for the current density yields

$$i = \frac{1}{\dfrac{1}{nFk_m c_{R,b}} + \dfrac{1}{i_{o,b}\exp\left(\dfrac{\alpha_a F}{RT}\eta_s\right)}}. \quad (3.33)$$

The expression in Equation 3.33 reflects the fact that the current depends on two resistances in series, one associated with mass transfer and the other with the reaction at the surface. If the kinetics is slow relative to the mass transfer, the second term in the denominator is large relative to the first, and i is limited by kinetics. In contrast, if the mass transfer is slow relative to the kinetics, the first term in the denominator dominates (k_m is small, so its reciprocal is large) and the current depends only on mass transfer. In between those extremes, both resistances influence the current density. The overall impact is shown in Figure 3.9, where the current rises exponentially (kinetically dominated) and then levels off due to mass transport limitations. At high overpotentials, the reaction is no longer limiting, and the current depends only on mass transfer, which is not affected by the overpotential. At high values of surface overpotential, the reaction at the surface is so fast relative to the rate of mass transfer that the reactant is consumed at the surface as quickly as it arrives, yielding a surface concentration of the reactant that is essentially zero. The mass transfer limited or *limiting*

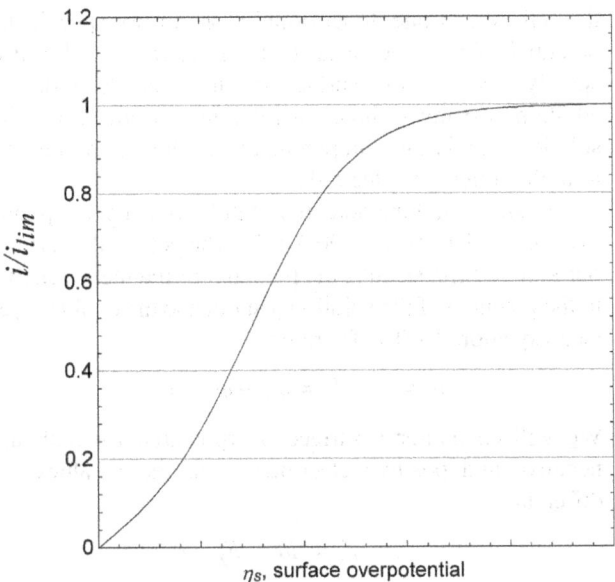

Figure 3.9 Mass transfer limiting current.

current can be calculated from the transport expression by assuming a zero concentration at the surface:

$$i_{lim} = nFk_m c_{R,b}, \quad (3.34)$$

where n is the number of electrons transferred per mole of c_R reacted as determined from the electrochemical reaction. In practice, there are many industrial reactions that are carried out at the limiting current in supporting electrolyte.

The above problem is difficult to solve explicitly when the dependency of i_o on concentration is nonlinear (e.g., a fractional power). We also allow i_o to be known at a concentration other than the bulk concentration. In such cases, the following equation can be solved numerically for the surface concentration over a range of overpotentials:

$$nFk_m(c_{R,b} - c_{R,s}) = i_{o,ref}\left(\frac{c_{R,s}}{c_{R,ref}}\right)^{\gamma}\left[\exp\left(\frac{\alpha_a F}{RT}\eta_s\right) - \exp\left(-\frac{\alpha_c F}{RT}\eta_s\right)\right]. \quad (3.35)$$

Once the surface concentration is known, either the mass transport expression or the rate expression can be used to calculate the current.

Although the expression for mass transfer has intentionally been kept simple, this section provides an initial exposure to the *limiting current*, critical to the design and operation of electrochemical processes. A more detailed treatment of transport is found in the next chapter.

3.8 USE OF KINETIC EXPRESSIONS IN FULL CELLS

The purpose of this section is to help develop some initial experience and intuition with full electrochemical cells. By a full cell, we simply mean an electrochemical cell with the anode and cathode separated by some distance by an electrolyte. What's more, we will consider the potential of the cell, namely, the potential of the positive electrode relative to the negative, rather than the potential of a single electrode relative to a reference electrode. The focus will still be on the rate expressions, although we will account for the potential drop in solution in a simple way in order to illustrate the influence of the potential field across the entire cell. Additionally, we will introduce some nomenclature that will be used extensively in future chapters on electrochemical cells. Concentration effects will not be treated in this section.

Let's consider a system consisting of a Zn electrode and a NiOOH electrode in an alkaline solution. This is a galvanic cell where Zn will oxidize, and NiOOH will be reduced according to the following reactions characteristic of a Ni/Zn battery:

$$Zn + 2OH^- \rightarrow Zn(OH)_2 + 2e^-$$

$$NiOOH + H_2O + e^- \rightarrow Ni(OH)_2 + OH^-$$

Given the geometric and physical property data in Table 3.3, we will find the following:

a. The difference in potential across the cell that corresponds to a current density of 1000 A·m^{-2}.

b. The discharge current that corresponds to a cell voltage of 1.3 V.

Table 3.3 Parameters Used to Generate Figure 3.10

Symbol	Value	Explanation
$R_{a,Ni}$	100	Surface roughness factor for Ni electrode
$R_{a,Zn}$	2	Surface roughness factor for Zn electrode
$i_{o,Ni}$	0.61	Exchange-current density for Ni [A·m^{-2}]
$i_{o,Zn}$	60	Exchange-current density for Zn [A·m^{-2}]
T	298.15	Temperature [K]
U_{Ni}	1.74	Equilibrium potential of Ni relative to Zn reference electrode, V
$\alpha_{a,Ni}$	0.5	Anodic transfer coefficient for Ni electrode
$\alpha_{a,Zn}$	1.5	Anodic transfer coefficient for Zn electrode
$\alpha_{c,Ni}$	0.5	Cathodic transfer coefficient for Ni electrode
$\alpha_{c,Zn}$	0.5	Cathodic transfer coefficient for Zn electrode
κ	60	Conductivity of electrolyte [S·m^{-1}]
L	2	Thickness of electrolyte (mm) (cell gap)

56 Electrochemical Engineering

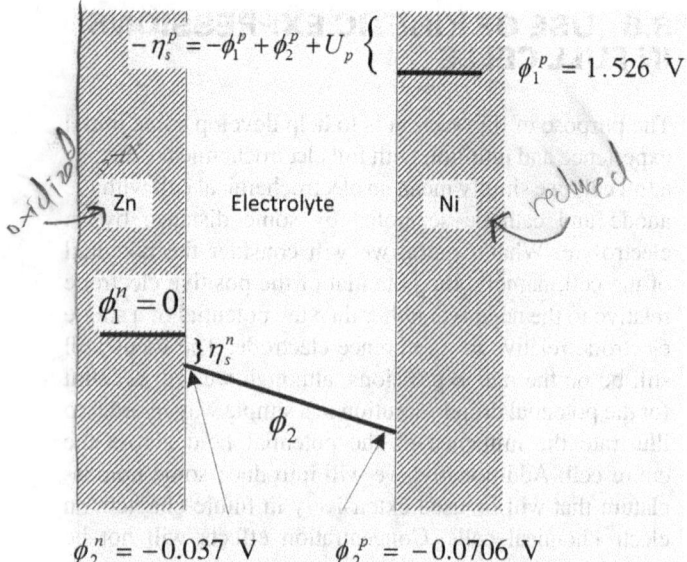

Figure 3.10 Metal and solution potentials for a current density of 1000 A·m^{-2}. The potentials are not to scale.

c. The charge current that corresponds to a cell voltage of 2.0 V.

For our purposes here we will define the discharge current to be positive and the charge current to be negative. We treat the problem as one dimensional, with current flow between two flat electrodes of equal size and electrolyte in the gap L between the two electrodes. The potentials corresponding to 1000 A·m^{-2} are shown in Figure 3.10 (left part). The zinc electrode is the negative electrode and the nickel is the positive one; these are designated with the superscripts n and p. There are three potentials that need to be considered: ϕ_2, the potential of the solution defined with respect to a reference electrode, ϕ_1^p and ϕ_1^n, the metal potentials of the positive and negative electrodes, respectively. Note that ϕ_2 varies in the electrolyte between the two electrodes; therefore, the potential in solution at the zinc electrode is different than that at the nickel electrode.

The potential drop in solution obeys Ohm's law, where

$$i = -\kappa \frac{\Delta \phi_2}{\Delta x}, \qquad (3.36)$$

and κ is the conductivity of the electrolyte. The cell potential is the potential that you would measure if you connected the positive lead (usually the red wire) of your voltmeter to the nickel electrode and the negative lead (usually black) to the zinc electrode. This potential is defined as $V_{cell} = \phi_1^p - \phi_1^n$ and is always found as a difference rather than an absolute value. We may therefore arbitrarily designate a value for the potential at a point in the cell. In this one-dimensional problem, we specify the potential of the zinc metal to be zero, $\phi_1^n = 0$. We also specify $x = 0$ at the surface of the zinc electrode. In addition, we must choose a reference electrode for the solution. In this case, all potentials in solution are referred to a zinc reference electrode.

First, we find the potential of the cell that corresponds to a current density of 1000 A·m^{-2}. The potential V used in our kinetic expression is the potential of the metal relative to the potential of the solution just outside the double layer (see Equation 3.15b). Therefore,

$$\eta_s = V - U = \phi_1 - \phi_2 - U.$$

We will encounter a surface overpotential for both the negative and positive electrodes, and these values are different:

$$\eta_s^n = \phi_1^n - \phi_2^n - U_n$$
$$\text{and} \qquad \eta_s^p = \phi_1^p - \phi_2^p - U_p.$$

Since both the anode and the reference electrode are Zn, the equilibrium potential $U_n = 0$. When the current density i is known, we can move from one side of the cell to the other in steps as we solve for the potentials. We begin at the zinc electrode where the potential of the metal has been specified as $\phi_1^n = 0$. Writing the BV equation for the zinc electrode, the only unknown is the surface overpotential. Note that we have multiplied $i_{o,Zn}$ by $R_{a,Zn}$ in order to account for surface roughness that makes the actual area larger than the superficial area upon which the current density is based. Solving numerically yields $\eta_s^n = 0.0373$ V. As expected, along the anodic branch, the overpotential is positive. This value is also $-\phi_2^n$, the potential in the solution at the surface of the zinc electrode just outside the double layer, because $\phi_1^n = U_n = 0$. Knowing the potential in solution at the zinc electrode interface, we can next calculate the potential drop in solution and hence the potential at the nickel–electrolyte interface.

$$i = -\kappa \frac{\Delta \phi}{\Delta x} = -\kappa \frac{\phi_2^p - \phi_2^n}{L} \qquad (3.37)$$

Therefore, it follows that

$$\phi_2^p = \phi_2^n - \frac{iL}{\kappa} = -0.0706 \text{ V}. \qquad (3.38)$$

Note that positive current flow is from the anode to the cathode and is accompanied by a drop in potential in the solution. The last step is to use the BV equation for the Ni electrode to calculate the potential of the solid phase, ϕ_1^p for the Ni electrode. U_p is the equilibrium potential of the nickel electrode relative to a zinc reference electrode that is provided in Table 3.3. At 1000 A·m^{-2}, $\eta_s^p = -0.1438$ V. On the cathodic branch, the overpotential is negative, and since the exchange-current density is smaller than for the zinc electrode, the magnitude of the overpotential is greater. Finally, we solve for the potential of the positive electrode, $\phi_1^p = U_p + \eta_s^p + \phi_2^p = 1.74 - 0.1438 - 0.0706$ V $= 1.526$. The cell potential is $V_{cell} = \phi_1^p - \phi_1^n = 1.526$ V. This is the potential that would be measured between the electrodes for the cell operating at the specified current density of $1000\,\text{A}\,\text{m}^{-2}$. These potentials are shown in Figure 3.10. Please study this figure carefully in order to help you understand how the potential varies in an electrochemical cell. Note that because the standard potentials of the two electrodes differ by almost 2 V, this figure is not to scale.

Next, in the right part of the figure, we find the discharge current that corresponds to a cell voltage of 1.3 V. Since the cell voltage is $V_{cell} = \phi_1^p - \phi_1^n$ and we can specify the absolute value of the potential at one point, we arbitrarily set $\phi_1^n = 0$. It follows that $\phi_1^p = 1.3$ V. Solution of this problem is a little more complicated when the current density is not known. In this case, we must solve the problem iteratively. The problem can be solved by manual iteration, guessing a current density, calculating a cell potential, and adjusting our guess until the calculated cell voltage matches the specified value. It is much more efficient to simultaneously solve three equations for the *three* unknowns numerically.

Note that the current density in solution is defined as positive when moving from anode to cathode. In contrast, the cathodic current density, which flows in the same direction, is defined as negative by convention. Hence, the current in solution must be set equal to the negative of the cathodic current.

This procedure works equally well for charge as for discharge without the need to modify the equations. However, for charging, the specified potential between the electrodes must be greater than the open-circuit voltage. Thus, solving the same set of equations for an applied potential of 2.0 V yields a charging current of $-821\,\text{A}\cdot\text{m}^{-2}$.

The total voltage loss, or polarization, across the cell is equal to $\eta_{Zn} + \Delta\phi_{solution} - \eta_{Ni}$. The model that we have

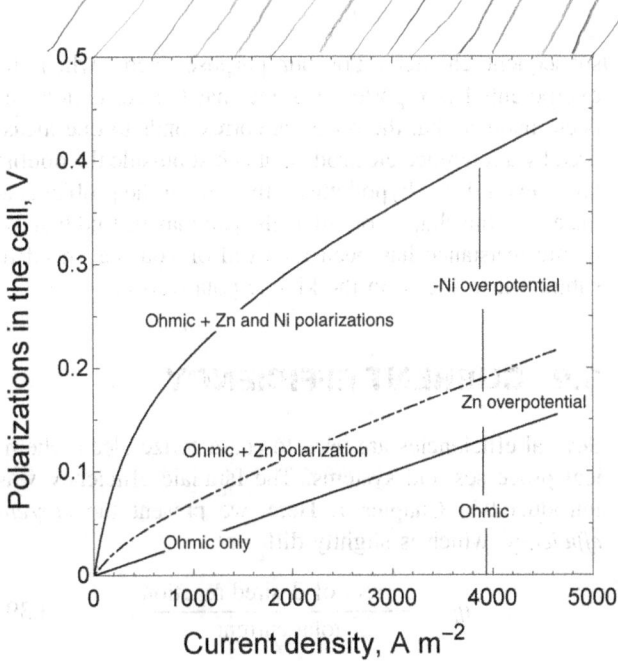

Figure 3.11 Kinetic and ohmic polarizations for a Ni–Zn cell.

developed can be used to evaluate the relative magnitude of each of these losses as a function of current, as shown in Figure 3.11. Examination of the figure shows that the overpotential at the zinc electrode is much less than that at the nickel electrode. Note also that the potential drop in solution increases linearly with current in the absence of concentration gradients, and that its relative importance is greater at the higher currents.

One final note, the reference electrode described here is unlikely to be physically "just outside the double layer." Recall, the typical double layer is only a few nanometers thick. In fact, the reference electrode may be far removed from the electrode surface. As a result, there will be a difference in the potential that is measured by this reference electrode and what is needed for the kinetic expression. Placement of the reference electrode and correcting for this difference in potential will be covered in

Specified variables	Unknown variables	Equations
$\phi_1^n = 0$ V	$i = 4640\,\text{A}\cdot\text{m}^{-2}$	$\phi_2^p = \phi_2^n - \dfrac{iL}{\kappa}$, Ohm's law for electrolyte
$\phi_1^p = 1.3$ V	$\phi_2^n = -0.0627$ V	$i = R_{a,Zn}i_{o,Zn}\left\{\exp\dfrac{\alpha_{a,Zn}F(\phi_1^n - \phi_2^n - U_n)}{RT} - \exp\dfrac{-\alpha_{c,Zn}F(\phi_1^n - \phi_2^n - U_n)}{RT}\right\}$
	$\phi_2^p = -0.218$ V	$-i = R_{a,Ni}i_{o,Ni}\left\{\exp\dfrac{\alpha_{a,Ni}F(\phi_1^p - \phi_2^p - U_p)}{RT} - \exp\dfrac{-\alpha_{c,Ni}F(\phi_1^p - \phi_2^p - U_p)}{RT}\right\}$

subsequent chapters. For our purposes here, when the overpotential is reported, assume that the correction has been made so that the potential corresponds to one measured by a reference electrode that is just outside the double layer, even if it is hypothetical. In some of the problems at the end of the chapter on full cells, you may be told that the ohmic resistance has been removed or you may need to subtract it to focus on the kinetic polarization.

3.9 CURRENT EFFICIENCY

Several efficiencies are used to characterize electrochemical processes and systems. The faradaic efficiency was introduced in Chapter 1. Here, we present the *current efficiency*, which is slightly different:

$$\eta_c \equiv \frac{\text{current of desired reaction}}{\text{total current}}. \quad (3.39)$$

Undesired reactions occur in both electrolytic and galvanic cells. These unwanted side reactions reduce the current efficiency. We can explore this concept with the charging of a lead–acid battery. Two reactions at the positive electrode are possible. The first is the sought-after reaction for the oxidation of the solid lead sulfate:

$$PbSO_4 + 2H_2O \rightarrow PbO_2 + SO_4^{2-} + 4H^+ + 2e^-$$
$$(U^\theta = 1.685 \text{ V})$$

A second, undesired, reaction is the evolution of oxygen:

$$2H_2O \rightarrow O_2 + 4H^+ + 4e^- \quad (U^\theta = 1.229 \text{ V})$$

If a certain current is used to charge the cell, what fraction of the electrons produced result from the desired reaction? The answer to this question is determined by the electrode kinetics. Notice that the equilibrium potential for the positive electrode of the lead–acid cell is more than 0.4 V higher that of the oxygen reaction. Thus, we can expect the overpotential for the oxygen reaction to be much larger.

ILLUSTRATION 3.5

Calculate the current efficiency for the positive electrode during charging of the lead–acid cell. At a particular point in the charging process, the overpotential for the oxidation of lead is 0.05 V, and the overpotential of the oxygen evolution reaction is 0.6 V. Use Butler–Volmer kinetics for the reaction with lead.

Kinetic parameters

Lead acid (1)	Oxygen evolution (2)
Butler–Volmer	Tafel
$\alpha_a = 1.0$, $\alpha_c = 1.0$	$\alpha_a = 0.5$
$i_o = 1 \text{ A} \cdot \text{m}^{-2}$	$i_o = 2 \times 10^{-6} \text{ A} \cdot \text{m}^{-2}$
$\eta_{s,1} = 0.05 \text{ V}$	$\eta_{s,2} = 0.60 \text{ V}$

$$\eta_c = \frac{i_1}{i_1 + i_2}$$

$$= \frac{i_{o,1}\left\{\exp\left[\frac{\alpha_{a,1}F}{RT}\eta_{s,1}\right] - \exp\left[\frac{-\alpha_{c,1}F}{RT}\eta_{s,1}\right]\right\}}{i_{o,1}\left\{\exp\left[\frac{\alpha_{a,1}F}{RT}\eta_{s,1}\right] - \exp\left[\frac{-\alpha_{c,1}F}{RT}\eta_{s,1}\right]\right\} + i_{o,2}\left\{\exp\left[\frac{\alpha_{a,1}F}{RT}\eta_{s,2}\right]\right\}}$$

$$= 0.97.$$

Therefore, 3% of the current is going toward the evolution of oxygen rather than recharging of the battery. Successful operation of this battery relies on sluggish kinetics for oxygen evolution.

Because the current efficiency is not 100%, more current and therefore more energy must be used to charge the battery to the same point. Also, note that current efficiency as defined applies to a single electrode. The current efficiency for the negative electrode of the same battery will be different. Here, the undesired side reaction is the evolution of hydrogen. Because of different transfer coefficients, and different dependencies on temperature, current efficiency can change with polarization.

In subsequent chapters, the current efficiency and some variants will be used for analysis of different electrochemical systems. The focus may be on an electrode, a cell, or even an entire system. As a consequence, it is necessary to modify or even add new definitions. For instance, consider the electrolysis of water to produce hydrogen and oxygen. Even if the current efficiencies of the two electrodes are unity, some oxygen and hydrogen may recombine in the cell and revert to water. This process is clearly undesired and represents a loss that is not captured by the defined current efficiency. The nature of the inefficiency is not electrokinetic, and its discussion is deferred to subsequent chapters.

CLOSURE

The main objective of this chapter was to introduce the relationship between current density and the potential of the electrode. The critical parameter is the surface overpotential, the difference in potential between the metal and the solution just outside of the double layer. In order to remove ambiguity, these potential are measured with a reference electrode. Most of the time, this current–voltage relationship is described with the Butler–Volmer equations. However, it is important to remember that this expression is semiempirical and cannot be expected to apply to every situation. Finally, the potential of full cell that includes kinetic resistance was developed.

FURTHER READING

Bard, A.J. and Faulkner, L.R. (2001) *Electrochemical Methods*, John Wiley & Sons, Inc., Hoboken, NJ.

Delahay, P. (1965) *Double Layer and Electrode Kinetics*, John Wiley & Sons, Inc., New York.

Vetter, K.J. (1967) *Electrochemical Kinetics: Theoretical Aspects*, Academic Press, New York.

PROBLEMS

3.1 The following reaction occurs rapidly at an inert electrode.

$$[Fe(CN)_6]^{3-} + e^- \leftrightarrow [Fe(CN)_6]^{4-}$$

The standard potential for the reaction is 0.26 V; 10 mg of potassium ferricyanide is added to 100 cm³ of water.

(a) Determine the concentration of K^+, $[Fe(CN)_6]^{3-}$, and $[Fe(CN)_6]^{4-}$ at equilibrium if the measured potential relative to a hydrogen reference electrode at 25 °C is 0.26 V.

(b) Calculate the Debye length and sketch the potential near the electrode.

(c) What is the effect of adding KCl to a concentration of 0.1 M on the Debye length and the electric field across the double layer?

3.2 Repeat Illustration 3.2(a) for the situation where the potential (0.10 V) is relative to a Ag/AgCl reference electrode. The equilibrium potential for the saturated Ag/AgCl electrode is 0.197 V. Please comment on any differences that you observe between the two solutions.

3.3 Hydrogen gassing can be a serious problem for lead–acid batteries. Consider two reactions on the negative electrode: the desired reaction for charging

$$PbSO_4 + 2e^- \rightarrow Pb + SO_4^{2-}, \quad U^\theta = -0.356 \text{ V}$$

and an undesired side reaction

$$2H^+ + 2e^- \rightarrow H_2. \quad U^\theta = 0 \text{ V}$$

(a) The exchange-current densities for the two reactions are $i_{o,PbSO_4} = 100$ A·m⁻², and $i_{o,H_2} = 6.6 \times 10^{-10}$ A·m⁻², where the exchange-current density for the hydrogen reaction is on pure lead. Calculate the current density for each reaction if the electrode is held at a potential of -0.44 V relative to a hydrogen reference electrode. The temperature is 25 °C, and the transfer coefficients are 0.5.

(b) With Sb impurity in the lead, the exchange-current density of the hydrogen reaction increases to 3.7×10^{-4} A·m⁻². Repeat the calculation of part (a) in the presence of antimony. Assuming that all impurities cannot be eliminated, what implications do these results have for the operation of the battery?

3.4 Data for the exchange-current density for oxygen evolution on a lead oxide surface are provided as a function of temperature. Develop an expression for i_o as a function of temperature. What is the activation energy? If the transfer coefficient is a constant at 0.5 and the overpotential is 0.7 V, at what temperature will the current density for oxygen reduction be 5 A·m⁻²?

T [°C]	i_o [A·m⁻²]
15	6.9×10^{-7}
25	1.7×10^{-6}
35	7.6×10^{-6}
45	1.35×10^{-5}

3.5 The following reaction is an outer sphere reaction that occurs in KOH.

$$[MnO_4]^- + e^- \leftrightarrow [MnO_4]^{2-}$$

Would you expect the reaction to have a larger or small reorganizational energy compared to an isotope exchange reaction for manganese? What does this imply about the reaction rate?

3.6 On the right find a portion of Julius Tafel's original data for the evolution of hydrogen on mercury (*Zeit. Physik. Chem.* **50**, 641–712 (1905)). Create a Tafel plot and find the Tafel slope for these data. The experiments were conducted in 2 N sulfuric acid at 26.4 °C.

Potential	Current density [A m⁻²]
1.665	4
1.713	10
1.7465	20
1.7665	30
1.777	40
1.824	100
1.858	200
1.878	300
1.891	400
1.912	600
1.94	1000
1.963	1400
1.989	2000

It's interesting that these data were plotted on a linear scale in the manuscript, so Tafel never created the plot named in his honor, but he did report the slope. What additional information would be necessary to determine the exchange-current density?

3.7 The evolution of oxygen is an important process in the lead–acid battery. Assuming that the positive electrode of the flooded lead–acid battery is at its standard potential (entry 2 in Appendix A), calculate the overpotential for the oxygen evolution reaction. It is reported that the Tafel slope for this reaction is 120 mV per decade at 15 °C. What is the transfer coefficient α_a? If the exchange-current density is 6.9×10^{-7} A·m^{-2}, what is the current density for oxygen evolution? You may neglect the small change in equilibrium potential with temperature.

3.8 The tabulated data are for the dissolution of zinc in a concentrated alkaline solution. These data are measured using a Hg/HgO reference electrode. Under the conditions of the experiment, the equilibrium potential for the zinc electrode is −1.345 V relative to the HgO electrode.

Potential [V]	i [A·m^{-2}]
−1.335	58
−1.325	150
−1.315	300
−1.305	600
−1.295	1100
−1.285	1970
−1.275	3560
−1.265	6300
−1.255	11,500
−1.245	20,000
−1.235	36,800
−1.225	66,000

(a) Please determine the exchange-current density and Tafel slope that best represent these data.

(b) For the same electrolyte, the potential of a SCE electrode is 0.2 V more positive than the Hg/HgO reference. If the potential of the zinc is held at −1.43 V relative to the SCE electrode, what is the current density for the oxidation of zinc?

3.9 Derive Equation 3.28. For a reaction at 80 °C, and with $(\alpha_a + \alpha_c) = 2$, at what value for the overpotential will the error with the Tafel equation be less than 1%?

3.10 In order to model the deposition of copper, the following data have been collected for 1 M CuSO$_4$ in 1 M H$_2$SO$_4$. The temperature is 25 °C, and a copper-plated reference electrode was used. From these data determine the parameters for the Butler–Volmer kinetic expression. The reaction is

$$Cu \rightarrow Cu^{2+} + 2e^-, \quad U^\theta = 0.337 \text{ V}$$

I [A m^{-2}]	η_s [V]	I [A·m^{-2}]	η_s [V]
−300	−0.1004	−40	−0.0192
−250	−0.0919	−2.3	−0.00101
−200	−0.0818	67	0.0192
−150	−0.0717	135	0.0293
−125	−0.0616	250	0.0394

(Continued)

I [A m^{-2}]	η_s [V]	I [A·m^{-2}]	η_s [V]
−100	−0.0515	450	0.0495
−75	−0.0414	1100	0.0657

3.11 In Section 3.7, Equation 3.33 was developed using Tafel kinetics, illustrating the effect of mass transfer on the current density. Derive the equivalent expression for linear kinetics. Explain the difference in shape of graphs for i versus η_s between linear and Tafel kinetics.

3.12 Metal is deposited via a two-electron reaction with current voltage data as shown at the right. The equilibrium potential is −0.5 V versus the same reference electrode used to measure the data. The bulk concentration is 100 mol·m^{-3}. You may assume that the exchange-current density is linearly dependent on the concentration of the reactant (i.e., you may use Equation 3.33). Assume 25 °C.

Applied potential (versus reference electrode)	Deposition current [A·m^{-2}]
−0.6	−1.2
−0.7	−2.1
−0.8	−13.7
−0.9	−82.2
−1	−245
−1.1	−330
−1.2	−350
−1.3	−360
−1.4	−362

(a) Is mass transfer important? If so, please determine a value for the mass transfer coefficient.

(b) Assuming Tafel kinetics, find the values of α_c and i_o. Comment on the applicability of this assumption. Can the normal Tafel fitting procedure be used for this part? Why or why not?

(c) If the mass transfer coefficient were reduced by a factor of 2 (cut in half), please predict the current that would correspond to an applied potential of −0.9 V.

Additional Hints: You need to consider carefully which points you use in fitting the kinetic parameters. Also, it is a good idea to normalize the error when fitting the current.

3.13 Use the following property data for a hydrogen–oxygen fuel cell at 25 °C.

Oxygen electrode	Hydrogen electrode	Electrolyte
$i_o = 9 \times 10^{-7}$ A·m^{-2}	$i_o = 14{,}000$ A·m^{-2}	$\kappa = 10$ S·m^{-1}
$\alpha_a = 3$, $\alpha_c = 1$	$\alpha_a =$, $\alpha_c = 1$	$L = 40$ μm
$U = 1.229$ V	$U = 0$ V	

(a) Operating at a current density of 10,000 A·m^{-2}, calculate the potential of the full cell, V_{cell}.

(b) Can the kinetics be simplified for either electrode, or is the complete Butler–Volmer kinetics needed?

3.14 For the Ni/Zn cell analyzed in Section 3.8, determine an I–V curve for the cell by solving for the current as a function of the voltage over the range from 1.0 to 2.2 V. Please comment on the shape of the resulting curve.

3.15 The following data are provided for the oxygen reduction reaction in acid media at 80 °C. The potential of the positive electrode, ϕ_1^p, is measured with respect to a hydrogen reference electrode, which in this case also serves as the counter electrode. Additionally, any ohmic resistance has been removed from the potentials tabulated.

(a) Plot these data on a semi-log plot (potential versus log i). You may assume that the kinetics for the hydrogen reaction is fast, and thus the anode polarization is small. What is the Tafel slope (mV per decade) in the mid-current range?

(b) Even though ohmic polarizations have been removed, at both low and high currents, the slope is not linear on the semi-log plot. Suggest reasons why this may be the case.

I [A·m^2]	ϕ_1^p [V]	I [A·m^{-2}]	ϕ_1^p [V]	I [A·m^{-2}]	ϕ_1^p [V]
1	0.933	112.1	0.9087	5815	0.7722
5.83	0.929	262.3	0.8825	8081	0.7521
10.58	0.93	582.2	0.8576	9442	0.7391
22.26	0.922	1317	0.831	10905	0.7223
46.64	0.92	3201	0.799	13247	0.6664

3.16 One common fuel cell type is the solid oxide fuel cell, which uses a solid oxygen conductor in place of an aqueous solution for the electrolyte. The two reactions are as follows:

I [A·m^{-2}]	V_{cell} [V]
−6981	1.5006
−4871	1.3554
−2946	1.2074
−967	1.0549
930	0.9047
2799	0.7545
4753	0.6020
6836	0.4518
9328	0.3000

$$O_2 + 4e^- \leftrightarrow 2O^{2-}, \quad U = 0.99 \text{ V}$$

and at the positive electrode

$$H_2 + O^{2-} \leftrightarrow H_2O + 2e^-, \quad U = 0 \text{ V}$$

Data for the polarization of a solid-oxide fuel cell/electrolyzer are provided in the table. These potentials are the measured cell potentials, although the anodic overpotential is small and can be neglected. The temperature of operation is 973 K. The ohmic resistance of the cell is 0.067 Ω·cm^2. After removing ohmic polarization, how well can the reaction rate for oxygen be represented by a Butler–Volmer kinetic expression? Comment on the values obtained.

3.17 Using the data from Problem 3.3(b), calculate the current efficiency for the negative electrode of the lead–acid battery in the presence of the Sb impurity. A general rule of thumb is that the reaction rate doubles for each 10° increase in temperature. If the current density is held constant (use the value calculated at −0.44 V), what happens to the potential of the electrode and the current efficiency if the temperature is raised to 35 °C? You may neglect any changes in the equilibrium potential with temperature.

3.18 Referring back to Problem 3.1, when the potential of the electrode is changed away from 0.26 V, current will flow due to faradaic reactions. These reactions will continue until a new equilibrium is established. What is the effect of raising the electrode potential 10 mV above 0.26 V on the following:

(a) The energy of an electron in the electrode?

(b) During the period where the system is evolving to a new equilibrium, will oxidation or reduction take place?

(c) What are the new values for the equilibrium concentration of ferri- and ferrocyanide?

CARL WAGNER

Carl Wagner was born on May 25, 1901 in Leipzig, Germany. His father was a chemist; in fact, an assistant to the renowned Wilhelm Ostwald. Wagner also studied chemistry, receiving his doctorate in 1924 from the University of Leipzig. He then served as a research assistant at the University of Munich.

His interest in solid-state materials and defects in ionic materials stemmed from his time in Berlin at the Max Bodenstein Institute, beginning in 1927. He met Walter Schottky during this period. Schottky, who played an important role in the development of semiconductors, was apparently so impressed with the young Wagner that Schottky invited him to coauthor a book on thermodynamics. The book was published just 2 years later. In 1930, Wagner and Schottky published a seminal paper on point defects in ionic crystals—"Theory of Ordered Alloys," *Z. Phys. Chem.*, **B11**, 163 (1930). Unquestionably, Schottky, almost 20 years Wagner's senior, had an important influence on Wagner's research.

In 1934, Wagner was appointed Professor of Chemistry at the Technical University Darmstadt. His main focus was on solid-state ionics, corrosion, and catalysis, but he was also involved in the development of rocket controls. Anticipating a long period of reconstruction at the end of World War II, he happily accepted an opportunity to work with the von Braun group on the thermodynamics of rocket fuels. He relocated to the United States, where from 1945 to 1949 he was a technical advisor to the Army at Fort Bliss. Subsequently, he became a U.S. citizen; and in 1949 he took a faculty position at the Massachusetts Institute of Technology.

Throughout these many transitions, Wagner continued to publish important works in physical chemistry, corrosion, and metallurgy. Today he is considered to be one of the fathers of solid-state ionics. He clearly showed the effect of impurity atoms in crystal defects, and this work had an important impact on the development of semiconductors. This work also explained oxide conductors that are important for high-temperature fuel cells and sensors. His work inspired many others. In 1958, he returned to Germany to direct the Max Planck Institute for Physical Chemistry in Göttingen.

In 1951, Wagner was the first selectee for the Olin Palladium Medal of the Electrochemical Society, its highest technical award. In 1957, he was selected for the Whitney Award of the National Association of Corrosion Engineers (NACE). The Wagner number, Wa, is a dimensionless quantity that determines the uniformity of current density and is named in his honor. Wagner died on December 10, 1977 in Göttingen, Germany. In 1980, The Electrochemical Society established the Carl Wagner Medal for mid-career researchers.

Image Source: Courtesy of ECS.

Chapter 4

Transport

As our analysis moves away from conditions of equilibrium, kinetic and transport phenomena must be addressed. The focus in the previous chapter was on the kinetics of electron-transfer reactions; here, the effects of transport are considered. The role of mass transfer is illustrated in Figure 4.1. As an example, the reduction at the electrode is a multistep process. Before the oxidant can be reduced, it must be transported from the bulk to a point close to the surface of the electrode. Next, the reactant is frequently adsorbed on the surface. This adsorption is followed by the electron-transfer reaction at the surface. The reduced product then desorbs and is transported from the surface to the bulk. One or more of these steps may control the rate of reaction (or current density) at the surface.

Our interest is to describe quantitatively the rate of transport of reactants and products to and from electrode surfaces. For the same reduction process, what drives the transport to and from the electrode surface? As shown in Figure 4.2, the concentration of the oxidized species is lower at the surface than in the bulk due to consumption of the reactant at the electrode. Similarly, the concentration of the reduced species is higher at the surface where it is generated. These differences in concentration cause material to diffuse in the solution.

For electrochemical systems, charged species are influenced by the presence of an electric field; and the transport of charged species due to the electric field is called *migration*. In addition to migration, typical modes of transport such as diffusion and convection are important. In this chapter, we consider transport by all three modes—migration, diffusion, and convection—as they apply to electrochemical systems.

4.1 FICK'S LAW

The transport of material by diffusion is due to the random thermal movement of molecules and is described by Fick's law:

$$\mathbf{J}_i = -D_i \nabla c_i, \quad (4.1)$$

where $\mathbf{J}_i$ is the molar flux [mol m^{-2} s^{-1}] of species i. The flux represents the rate at which material passes through a plane of unit area. It is a vector quantity with both direction and magnitude. In one dimension, the gradient is simply the derivative, and the flux is

$$J_{i,x} = -D_i \frac{dc_i}{dx}.$$

The driving force for transport by diffusion is the concentration gradient, and material is transported in the x-direction from a region of high concentration to low concentration. D_i is the proportionality constant called the diffusion coefficient or diffusivity [m$^2 \cdot$s^{-1}]. These gradients in concentration, shown in Figure 4.2, drive diffusion of O and R to and from the surface.

The flux defined here, $\mathbf{J}$, is the flux relative to the molar average velocity. It is generally more convenient to work with a flux relative to a fixed frame of reference. This quantity is also a flux, but will be designated with $\mathbf{N}$. For our purposes, we can relate the two quantities by a combination of molecular diffusion and a contribution from bulk flow

$$\mathbf{N}_i = \mathbf{J}_i + c_i \mathbf{v}, \quad (4.2)$$

where $\mathbf{v}$ is the molar average velocity of the fluid.

4.2 NERNST–PLANCK EQUATION

The most widely used expression for the flux in electrochemical systems is the Nernst–Planck equation,

$$\mathbf{N}_i = \underbrace{-z_i u_i F c_i \nabla \phi}_{migration} \underbrace{- D_i \nabla c_i}_{diffusion} + \underbrace{c_i \mathbf{v}}_{convection}. \quad (4.3)$$

The flux of species, i, is the combination of three terms: migration, diffusion, and convection. The Nernst–Planck equation is similar to Equation 4.2 but adds a contribution that arises from the gradient in electrical potential called migration. For charged species, the force due to the electric field is equal to $-z_i F \nabla \phi$. The charge on a species, z_i, can be

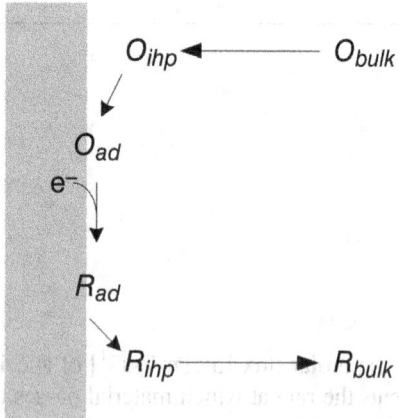

Figure 4.1 Multistep process for reduction.

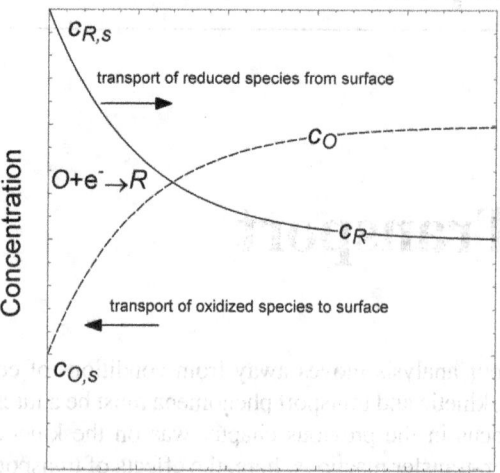

Figure 4.2 Concentration gradients associated with reduction of a solution-phase reactant at an electrode surface to form a soluble product.

positive or negative. The direction of the force changes for anions and cations because their charges are of opposite sign. Multiplying this driving force by the mobility of the ion, u_i, and the concentration, c_i, results in the net flux ascribed to migration. Migration is defined so that positive ions move from high to low potentials. We'll see later that the mobility is related to the diffusivity of the ion.

The second contribution to the flux is the molecular diffusion as described by Fick's law, which was discussed above. The last term is from the bulk movement of the electrolyte or convection. Equation 4.3 is frequently written for dilute solutions and is often referred to as *dilute solution theory*. When the concentration of ionic species is relatively small and the majority of the electrolyte is the solvent, the predominant interactions are between the ions and the solvent; ion–ion interactions are not important. Also, under dilute conditions, the average velocity in Equation 4.3 is well defined and independent of the minor species in solution. Under such conditions, $\mathbf{v}_{mass} = \mathbf{v}_{molar} = \mathbf{v}_{solvent}$, and the balance equation for the solvent is not written since the concentration of the solvent is essentially constant.

The Nernst–Planck equation can also be used for more concentrated solutions. However, as concentration levels increase, it is necessary for transport coefficients to include the effect of ion–ion interactions (in addition to ion–solvent interactions) and for the reference velocity to be defined carefully. The potential in solution is perhaps best defined against a reference electrode under such conditions. As concentration increases, activity effects become more important since it is really the gradient of the electrochemical potential and not the concentration gradient that is the primary driving force for transport. Nevertheless, gradients in the activity coefficients can typically be neglected, even in moderately concentrated solutions.

More commonly, in electrochemical systems we are interested in the current density rather than fluxes of individual species. The electrical current in solution can be related to the species fluxes by Faraday's law:

$$\mathbf{i} = F \sum_i z_i \mathbf{N}_i. \tag{4.4}$$

Substituting the Nernst–Planck equation into Equation 4.4 gives

$$\mathbf{i} = -F^2 \nabla \phi \sum_i z_i^2 u_i c_i - F \sum_i z_i D_i \nabla c_i + F\mathbf{v} \sum_i z_i c_i^{\,0} \tag{4.5}$$

The last term in Equation 4.5 is zero because we assume that the solution is electrically neutral. That is, the charge of anions is exactly balanced by the charge of the cations, which is a good assumption in most cases. It is evident from Equation 4.5 that the current density [A·m^{-2}], like the species fluxes, is a vector. Also, since the current is the result of the transport of ions in solution, it includes contributions from both migration and diffusion.

Let's consider a special case where there are no concentration gradients. Therefore, the second term on the right side of Equation 4.5 is also zero. In the absence of concentration gradients, the current density is proportional to the potential gradient multiplied by a term involving the mobility, concentration, and charge of the ions. Thus, Equation 4.5 reduces to Ohm's law. Note that Ohm's law is synonymous with the statement that the current is proportional to the gradient of the potential.

$$\mathbf{i} = -F^2 \nabla \phi \sum_i z_i^2 u_i c_i = -\kappa \nabla \phi, \tag{4.6}$$

where the electrical conductivity is defined by

$$F^2 \sum_i z_i^2 u_i c_i \equiv \kappa. \tag{4.7}$$

The SI unit for conductivity is [S·m^{-1}], where a siemen [S] is an inverse ohm [Ω^{-1}]. Note that the charge on each ion is squared so that conductivity is always a positive quantity. The electrical conductivity is the sum of the mobility of each ion weighted by its concentration in solution and represents the ability for charge to be transported in the presence of an electrical field.

In the absence of concentration gradients, the conductivity can be directly related to the resistance of the electrolyte. For example, consider one-dimensional transport between two flat plates of area A and separated by a distance L. Application of Equation 4.6 yields

$$i = -\kappa \nabla \phi = \kappa \frac{\phi|_{x=0} - \phi|_{x=L}}{L} = \frac{I}{A}, \quad (4.8a)$$

$$\frac{\phi|_{x=0} - \phi|_{x=L}}{I} = \frac{V}{I} = \frac{L}{\kappa A}, \quad (4.8b)$$

$$\therefore R_\Omega \equiv \frac{L}{\kappa A} \equiv \frac{1}{G}, \quad (4.8c)$$

where R_Ω relates the total current to the difference in the potential. Also note that the conductance, G, is simply the inverse of resistance and has units of siemens.

Finally, remember that Ohm's law only applies in the absence of concentration gradients; it must be modified if concentration gradients are present. Regardless, the electrical conductivity is an important transport property of the electrolyte.

4.3 CONSERVATION OF MATERIAL

In order to solve most transport problems, expressions for the flux, such as the Nernst–Planck equation, are incorporated into material balances or conservation equations. Here we derive a balance for a single species over a control volume of size $\Delta z \Delta x \Delta y$ as shown in Figure 4.3. The balance takes the form

$$\text{accumulation} = \text{in} - \text{out} + \text{generation}. \quad (4.9)$$

The rate of accumulation in the control volume is

$$\frac{\Delta(c_i \Delta x \Delta y \Delta z)}{\Delta t}.$$

The net rate of material entering the control volume in the x-direction is

$$\left(N_{i,x}|_x - N_{i,x}|_{x+\Delta x} \right) \Delta y \Delta z.$$

Similar equations can be written for the y- and z-directions. If R_i is the homogeneous rate of production per volume of species i, the rate of generation is

$$\mathcal{R}_i \Delta x \Delta y \Delta z.$$

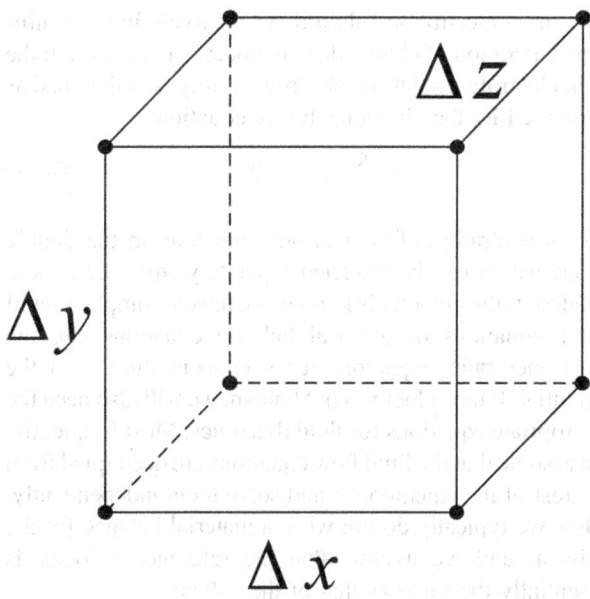

Figure 4.3 Illustration of control volume for material balances.

These equations combine to

$$\frac{\Delta(c_i \Delta x \Delta y \Delta z)}{\Delta t} = \left(N_{i,x}|_x - N_{i,x}|_{x+\Delta x} \right) \Delta y \Delta z$$
$$+ \left(N_{i,y}|_y - N_{i,y}|_{y+\Delta y} \right) \Delta x \Delta z + \left(N_{i,z}|_z - N_{i,z}|_{z+\Delta z} \right) \Delta x \Delta y$$
$$+ \mathcal{R}_i \Delta x \Delta y \Delta z.$$

In the limit of an infinitely small control volume,

$$\frac{\partial c_i}{\partial t} = -\left(\frac{\partial N_{i,x}}{\partial x} + \frac{\partial N_{i,y}}{\partial y} + \frac{\partial N_{i,z}}{\partial z} \right) + \mathcal{R}_i.$$

The species material balance may be written more compactly in vector form as

$$\frac{\partial c_i}{\partial t} = -\nabla \cdot N_i + \mathcal{R}_i. \quad (4.10)$$

Equation 4.10 represents a differential material balance on species i. Equation 4.10 along with the appropriate initial and boundary conditions will be used frequently and should be committed to memory. For a one-dimensional geometry, one initial condition and two boundary conditions must be specified to define completely a problem.

Let's suppose for a moment that you would like to solve an electrochemical transport problem involving multiple ionic species in solution. How would you use Equation 4.10? What other equations are needed? A complete set of equations for n species in solution would include the material balance (Equation 4.10) for each of the species in solution, into which the corresponding flux equation (Equation 4.3) for each species has been substituted. Also, since the electrical forces between charged species in solution are quite large, and the mobilities of ions in an electrolyte solution are relatively high, significant separation of charge does not occur. Therefore, in the bulk electrolyte solution, electroneutrality is maintained as expressed by the electroneutrality equation:

$$\sum_i z_i c_i = 0. \quad (4.11)$$

Electroneutrality, of course, does not hold in the double layer, which can be modeled separately from the bulk if needed. If the velocity is known, we have a complete set of $n+1$ equations (n material balance equations and the electroneutrality equation) for n concentrations and the potential. If the velocity is not known, we will also need the appropriate equations for fluid dynamics. Most frequently, we assume that the fluid flow equations are decoupled from the rest of the equation set and solve them independently. Also, we typically do not write a material balance for the solvent, and we assume that the reference velocity is essentially the same as that of the solvent.

The current density, i, is most often of great interest in the simulation of electrochemical systems. Consequently, a charge balance

$$\nabla \cdot i = 0 \quad (4.12)$$

can be substituted for one of the material balance equations. Equation 4.12 is not independent, but a weighted sum of the other material balance equations as follows:

$$\frac{\partial}{\partial t} F \sum_i z_i c_i = -\nabla \cdot \left(F \sum_i z_i N_i \right)$$

$$= -\nabla \cdot \left(-F^2 \nabla \phi \sum_i z_i^2 u_i c_i - F \sum_i z_i D_i \nabla c_i \right)$$

$$= -\nabla \cdot i = 0.$$

Alternatively, one can solve the equation set above without the charge balance and then calculate the current density "postprocess" with Equation 4.4.

The Nernst–Planck description of the flux, when combined with material balances just described, results in a set of nonlinear differential equations that often do not have an analytical solution. However, some important simplifications are possible. In the absence of concentration variations, the potential follows Laplace's equation, which is examined below. The most common simplifications for systems of varying composition involve elimination of the migration term in the flux equation. These, too, are considered below and include binary electrolytes, where the electric field is eliminated from the conservation equation through mathematical manipulation, and "excess supporting electrolyte," where the effect of the electric field can be neglected. A form of the convective diffusion equation results for each of these cases, but for different reasons and with different implications. These and other classifications of electrochemical problems are summarized in Figure 4.4.

No Concentration Gradients

In situations where the concentration gradients can be neglected, the charge balance in Equation 4.12 can be simplified to yield

$$-\nabla \cdot i = 0 = -\nabla \cdot \left(-F^2 \sum_i z_i^2 u_i c_i \nabla \phi \right) = \kappa \nabla^2 \phi$$

or $\nabla^2 \phi = 0.$

$$(4.13)$$

Equation 4.13, Laplace's equation, is the only equation that needs to be solved since the concentrations and

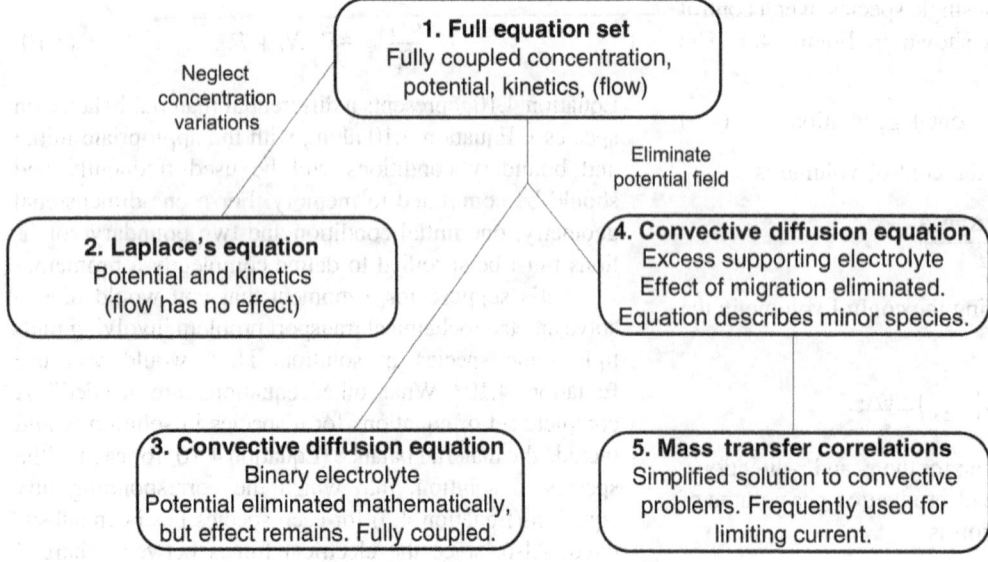

Figure 4.4 Classification of electrochemical problems.

conductivity are known and constant. For a one-dimensional problem, the potential drop in solution for a system at constant concentration can be readily calculated if the solution potential at the two electrodes is known or if the current is known. This calculation is demonstrated in Illustration 4.1.

ILLUSTRATION 4.1

In Section 3.4, we used the equation $i = -\kappa(\Delta\phi/\Delta x)$ to calculate the potential drop in solution. What is the source of this equation, and what assumptions are implicit in its use? In that problem we assumed that the solution concentration was constant. The Laplace equation in one dimension is

$$\frac{d^2\phi}{dx^2} = 0.$$

This equation can be integrated twice to yield

$$\phi = Ax + B,$$

where $x=0$ at the zinc electrode (negative electrode: n) and $x=L$ at the nickel electrode (positive electrode: p). Since we are looking for the potential difference, we arbitrarily set the potential in solution at $x=0$ to be ϕ_2^n. Therefore, $B = \phi_2^n$. In general, we will at a minimum need to specify the potential at one point in the domain when solving Laplace's equation. To find A, we note that

$$A = \frac{d\phi}{dx} = -\frac{i}{\kappa}.$$

The expression for ϕ becomes

$$\phi = -\frac{i}{\kappa}x + \phi_2^n.$$

We can now use this expression to determine the potential difference across the electrode:

$$\Delta\phi = \phi_2^p - \phi_2^n = -\frac{i}{\kappa}L.$$

This result is equivalent to the expression we used in Chapter 3, where $\Delta x = L$. Note that Laplace's equation here relates the potentials in solution, ϕ_2.

In Chapter 3, we showed how to use the kinetic expressions to relate the solution potential, current, and potential of the metal at each electrode. What if the electrode reaction is very fast? For a large exchange-current density, the surface overpotential is zero and the potential difference between the metal and the solution is at the equilibrium value. Using the nomenclature from Chapter 3 relative to a zinc reference electrode,

$$\eta_s^n = \phi_1^n - \phi_2^n - U_n,$$
$$0 = 0 - \phi_2^n - 0,$$
$$\phi_2^n = 0,$$
$$\phi_2^p = \phi_2^n - \frac{iL}{\kappa} = -\frac{iL}{\kappa},$$
$$\eta_s^p = \phi_1^p - \phi_2^p - U_p,$$
$$0 = \phi_1^p - \left(-\frac{iL}{\kappa}\right) - U_p,$$
$$\phi_1^p = U_p - \frac{iL}{\kappa},$$
$$V_{cell} = \phi_1^p - \phi_1^n = U_{cell} - \frac{iL}{\kappa}.$$

In other words, the cell voltage is equal to the equilibrium voltage of the cell minus the ohmic drop (or overpotential). This expression was derived for fast kinetics.

We can write an expression for the potential of the cell, V_{cell}, similar to the one shown in the illustration above, which includes the surface overpotentials. For a galvanic cell, the expression is

$$V_{cell} = U_{cell} - |\eta_{anode}| - |\eta_{cathode}| - |\eta_{ohmic}|. \quad (4.14a)$$

Thus, for a galvanic cell, the maximum voltage is the equilibrium voltage, and the actual cell voltage is reduced from the equilibrium voltage by the amount needed to drive the surface reactions and the flow of current in solution. For an electrolytic cell,

$$V_{cell} = U_{cell} + |\eta_{anode}| + |\eta_{cathode}| + |\eta_{ohmic}|. \quad (4.14b)$$

Here, the cell voltage is higher than the equilibrium voltage as additional power must be added to drive the reactions and the flow of current in solution. Integration of Laplace's equation in one dimension is straightforward and is something that you should be able to do in Cartesian, cylindrical, and spherical coordinates. For two- and three-dimensional problems, solution of Equation 4.13 is more difficult and frequently results in a nonuniform current density at a surface. This issue and the use of kinetic expressions as boundary conditions are treated in more detail in Section 4.6.

Binary Electrolyte

A binary electrolyte is an electrolyte with only one type of anion and one type of cation. It is formed by dissolving a salt in a solvent to create an electrolyte solution. The neutral salt $M_{\nu_+}X_{\nu_-}$ dissolves in the solvent as follows:

$$M_{\nu_+}X_{\nu_-} \rightarrow \nu_+ M^{z+} + \nu_- X^{z-}, \quad (2-51)$$

where
$$\sum_i \nu_i z_i = \nu_+ z_+ + \nu_- z_- = 0. \quad (4.15)$$

As an example, for $CaCl_2$, $\nu_+ = 1$ and $\nu_- = 2$. The resulting binary electrolyte is electrically neutral and, therefore, satisfies the electroneutrality equation (Equation 4.11). Maintaining electroneutrality during transport means that the fluxes of the two ions are not independent. In fact, electroneutrality couples the concentration of the two ions together and allows us to express concentration in terms of a single salt concentration,

$$c = \frac{c_+}{\nu_+} = \frac{c_-}{\nu_-}. \quad (4.16)$$

Using this salt concentration, the Nernst–Planck equation, and the assumption of incompressible fluid ($\nabla \cdot \mathbf{v} = 0$), one can write the material balance equations for the positive and negative species as follows:

$$\frac{\partial c}{\partial t} + \mathbf{v} \cdot \nabla c = z_+ u_+ F \nabla \cdot (c \nabla \phi) + D_+ \nabla^2 c, \quad (4.17a)$$

$$\frac{\partial c}{\partial t} + \mathbf{v} \cdot \nabla c = z_- u_- F \nabla \cdot (c \nabla \phi) + D_- \nabla^2 c. \quad (4.17b)$$

We have assumed the absence of a homogeneous reaction. Equation 4.17b is subtracted from Equation 4.17a to yield

$$(z_+ u_+ - z_- u_-) F \nabla \cdot (c \nabla \phi) + (D_+ - D_-) \nabla^2 c = 0. \quad (4.18)$$

Equation 4.18 provides a relationship between the migration and diffusion terms in the material balance. This equation can be used to eliminate the potential from either one of the species balance equations to yield

$$\frac{\partial c}{\partial t} + \mathbf{v} \cdot \nabla c = D \nabla^2 c, \quad (4.19)$$

where c is the concentration of the salt. Thus, the concentration follows the same equation as that for a neutral species. Here, the equivalent diffusion coefficient of the salt is given by

$$D = \frac{z_+ u_+ D_- - z_- u_- D_+}{z_+ u_+ - z_- u_-} = \frac{D_+ D_- (z_+ - z_-)}{z_+ D_+ - z_- D_-}. \quad (4.20)$$

Equation 4.20 defines the diffusivity of the salt, which may often be more readily measured and reported than the diffusivities of the individual ions. Although the diffusivities may be quite different, the concentration of the two species will always be related to the stoichiometry of the neutral salt because of electroneutrality. The diffusivity of the salt, given by Equation 4.20, can be thought of as the weighted average of the diffusivities of the two ionic species.

Although the electric field has been eliminated mathematically in Equation 4.19 to facilitate solution of the

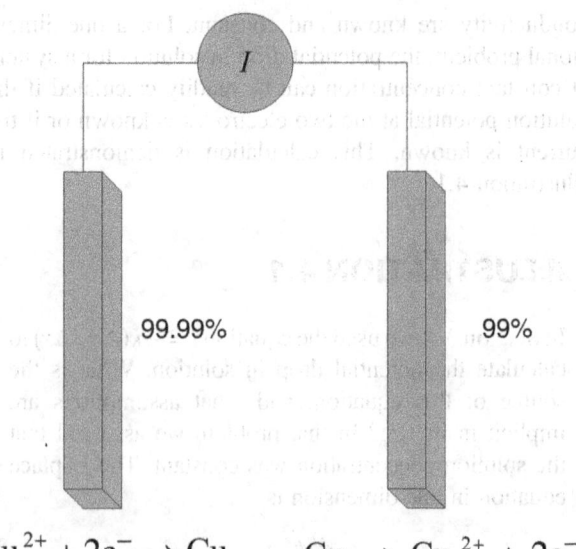

$Cu^{2+} + 2e^- \rightarrow Cu \qquad Cu \rightarrow Cu^{2+} + 2e^-$

Figure 4.5 Electrorefining of copper.

problem, the physics have not changed; therefore, there is still a potential gradient in solution that leads to transport by migration. The effect of the potential can most readily be seen at the boundary. To illustrate this, consider the steady-state electrorefining of copper shown in Figure 4.5. Here Cu is oxidized at the anode and deposited on the cathode. The electrolyte is $CuSO_4$ and there is no convection. The solution contains only $CuSO_4$ and is therefore a binary electrolyte. For a one-dimensional system, Equation 4.19 reduces to

$$\frac{d^2 c}{dx^2} = 0. \quad (4.21)$$

From Equation 4.21, it is clear that the concentration profile is linear between electrodes. Since only the copper reacts, the current is related by Faraday's law to the flux of Cu^{2+} as follows:

$$i = 2F N_{Cu^{2+}} = 2F(-z_+ u_+ \nu_+ F c \nabla \phi - \nu_+ D_+ \nabla c). \quad (4.22)$$

As you can see, the copper(II) flux includes both migration and diffusion. At first glance, this creates a problem since we have not solved explicitly for the potential field. However, because sulfate does not react at the electrodes, its net flux must be zero; therefore, the contributions of migration and molecular diffusion must exactly offset each other. This balance allows us to use Equation 4.3 for the anion to relate the potential and concentration gradients:

$$z_- u_- \nu_- F c \nabla \phi = -\nu_- D_- \nabla c. \quad (4.23)$$

We can now use Equation 4.23 to eliminate the potential gradient in Equation 4.22 in order to provide the boundary relation in terms of just the concentration gradient as needed. This process is illustrated with an example in Section 4.4. From this discussion, we see that solution of a problem involving a binary electrolyte can be facilitated by mathematically eliminating the electric field and using electroneutrality to express the equations in terms of a single salt concentration. The resulting equations can be solved directly for the concentration profile of the salt and appropriate fluxes. Migration does not show up explicitly in the equations; nonetheless, it is completely and properly accounted for in the above treatment.

Excess Supporting Electrolyte

In instances where a large amount of a salt that is not involved in the electrochemical reaction is added to the solution, the effect of the potential gradient on transport is diminished. The added salt that is not involved in the reaction is called a *supporting electrolyte*. There are several reasons for using a supporting electrolyte. Here we are concerned with the effect of the supporting electrolyte on migration. Returning to the electrorefining of copper discussed previously, we modify the system with the addition of sulfuric acid, H_2SO_4, which is a supporting electrolyte. If enough sulfuric acid is added to the solution so that migration can be neglected, then we have an *excess* supporting electrolyte. To repeat, the effect of the excess, nonreacting electrolyte is to reduce the electric field so that migration no longer contributes significantly to the transport of the reacting species. In contrast to the binary electrolyte, use of a supporting electrolyte implies that there must be at least three ionic species in solution.

It is frequently reported or implied that the role of the supporting electrolyte is to reduce the electric field by carrying the current. This conclusion is in error since, at steady state, the flux of any nonreacting species must be zero, and the supporting electrolyte is by definition non-reacting. Therefore, only the fluxes of the reacting species contribute to the current (see Equation 4.4). That said, the presence of the supporting electrolyte does diminish the magnitude of the potential gradient. To explore this further, consider a system at steady state with no bulk flow where any transport of the supporting electrolyte due to migration must be offset by diffusion. Concentration gradients of the nonreacting species must be connected to those of the minor reacting species in order to preserve electroneutrality. Therefore, concentration gradients of the supporting species cannot be large. Furthermore, the conductivity of the supporting electrolyte is high. These two facts together make it impossible to support a large potential gradient in solution. As a result, migration can be neglected for the reacting species when its concentration is significantly less than that of the supporting electrolyte. Convection can be used to minimize concentration gradients and to increase the rate of transport of the reacting species.

Returning to the Nernst–Planck equation, we neglect migration due to the supporting electrolyte and express the flux as

$$\mathbf{N}_i = -z_i u_i F c_i \nabla \phi - D_i \nabla c_i + c_i \mathbf{v} \qquad (4.24)$$

If we further assume that the diffusivity is constant, the resulting material balance is

$$\frac{\partial c_i}{\partial t} + \mathbf{v} \cdot \nabla c_i = D_i \nabla^2 c_i. \qquad (4.25)$$

Finally, in the absence of bulk flow, we are left with Fick's second law:

$$\frac{\partial c_i}{\partial t} = D_i \nabla^2 c_i. \qquad (4.26)$$

Note that, in contrast to the binary electrolyte, Equations 4.25 and 4.26 include both the diffusivity and the concentration of the individual minor species of interest. There are many techniques available for solution of Equation 4.26. What's more, there are many physical situations where Fick's second law is appropriate.

Figure 4.6 shows results from a simulation for a problem similar to the copper electrorefining considered earlier. In both cases, copper is consumed at the anode and deposited at the cathode. This particular simulation only considers transport by diffusion and migration. Because there is no convection, mass transport limitations occur at very low current densities, limiting the range of results. In spite of the limited range, the results illustrate the influence of the supporting electrolyte. The current density at a given cell voltage is higher for the system with the supporting electrolyte when mass transfer is not limiting. This increase is because the potential drop in solution is very small in the presence of the supporting electrolyte, and the entire applied voltage goes to driving the reactions at the electrodes (see Figure 4.6c). In contrast, a considerable ohmic drop is observed without the supporting electrolyte (Figure 4.6b). At just under 0.04 V in this simulation, the curves cross as the mass-transfer limiting current (~10 $A \cdot m^{-2}$) is approached for the system with the excess electrolyte. The mass-transfer limit in the presence of the sulfuric acid is half of the limiting value without the acid because migration does not contribute to the transport when there is excess electrolyte. As the mass-transfer limit is approached, the concentration of copper(II) ions at the cathode decreases, leading to a sharp increase in the cathodic overpotential (Figure 4.6c). As illustrated by this simulation, excess electrolyte increases the current that can be achieved at a particular cell voltage as long

70 Electrochemical Engineering

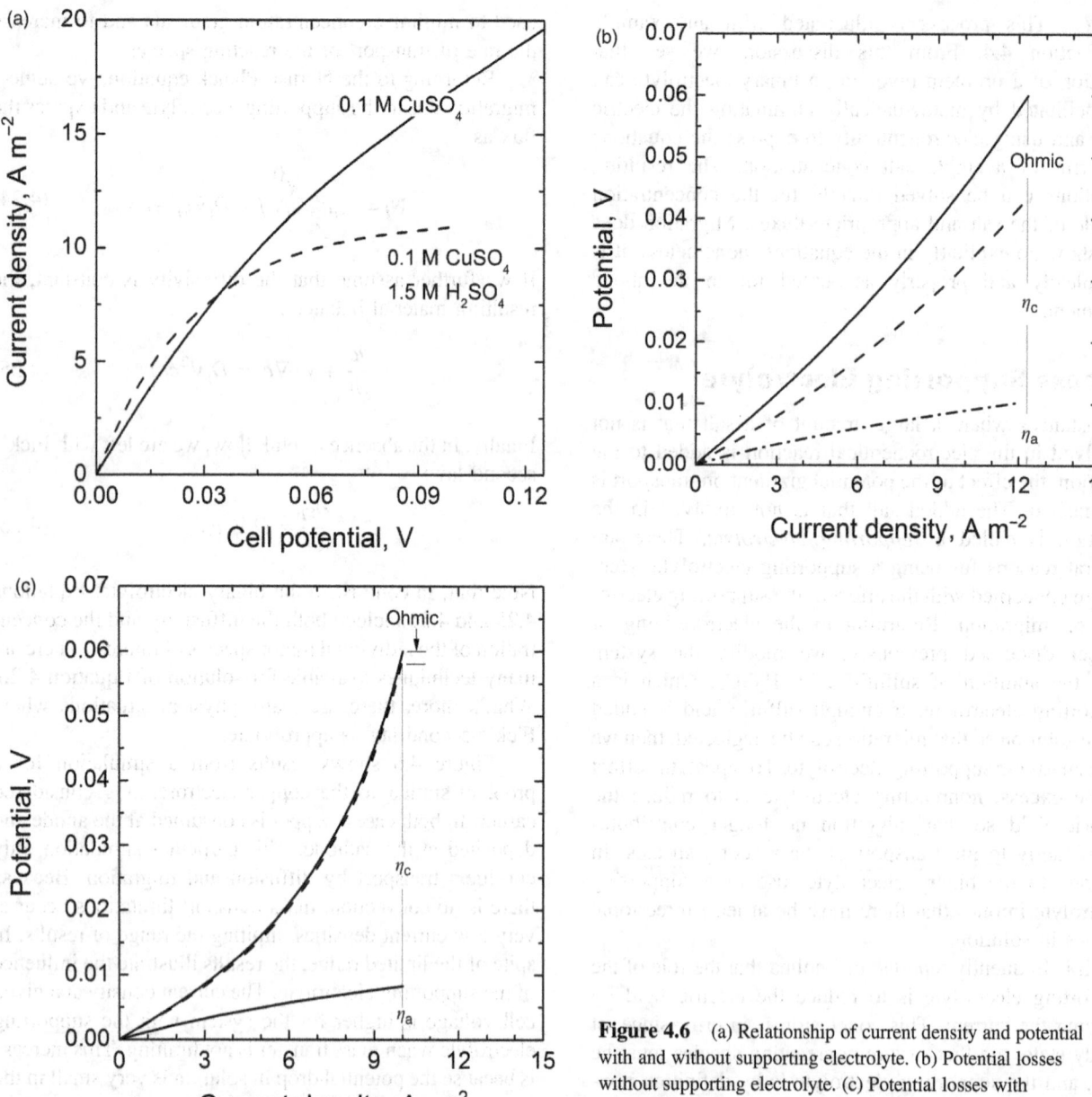

Figure 4.6 (a) Relationship of current density and potential with and without supporting electrolyte. (b) Potential losses without supporting electrolyte. (c) Potential losses with supporting electrolyte.

as the system is operated below the mass-transfer limit. Note that the mass-transfer limit in a practical system will be much higher than the values that pertain to this simulation due to convection in the practical system. Excess supporting electrolytes are used in both industrial processes and analytical experiments. For industrial processes, a reduction in ohmic loss is the principal reason. Supporting electrolyte makes it possible to achieve the same current at a lower cell potential. This result is true in the region of operation where kinetics is important. The supporting electrolyte may actually reduce the magnitude of the limiting current by eliminating the contribution of migration to the transport of the minor species. Industrial use of supporting electrolytes may also be driven by the lower cost of the supporting electrolyte relative to the minor salt of interest. In addition, supporting electrolyte can be used to overcome difficulties associated with a low solubility of the reacting component. For electroanalytical methods (Chapter 6), the goal is the elimination of migration so that transport is controlled by diffusion and analysis is greatly simplified.

ILLUSTRATION 4.2

A common experiment is a potential step study under diffusion control, where the diffusion of a reactant is limiting. As shown in the figure below, the concentration at the surface approaches zero at time zero when the potential step is applied. Assuming no convection and neglecting migration, Fick's law applies (Equation 4.1). Substituting this into the material balance equation, Equation 4.10, gives

Fick's second law $\quad \dfrac{\partial c_i}{\partial t} = D_i \nabla^2 c_i = D_i \dfrac{\partial^2 c_i}{\partial x^2}$

With initial conditions $\quad t = 0,\ c_i = c_i^o$

and boundary conditions $\quad x = 0,\ c_i = 0$
$\quad x \to \infty,\ c_i = c_i^o$

There are several ways to solve this equation, including the use of Laplace transforms or Fourier analysis. Here we pursue a similarity solution. For convenience, we drop the i subscript since we are only dealing with the transport of a single species. We begin by introducing a new variable, $\eta = x/\sqrt{4Dt}$, which transforms the partial differential equation to an ordinary differential equation:

$$\dfrac{d^2 c}{d\eta^2} + 2\eta \dfrac{dc}{d\eta} = 0,$$

with the following BCs, $\eta = 0,\ c = 0;\ \eta \to \infty,\ c = c^o$. This ODE may be readily solved using reduction in order. The solution is

$$\dfrac{c}{c^o} = \mathrm{erf}(\eta^2),$$

where the error function is defined as

$$\mathrm{erf}(x) \equiv \dfrac{2}{\sqrt{\pi}} \int_0^x \exp^{-\eta^2} d\eta.$$

The concentration profiles under these conditions are shown in the figure. For electrochemical systems, we are generally interested in the response of the current or potential. If we consider that the diffusing species undergoes reduction at the surface, the flux of the oxidized species to the surface can be related to the current with Faraday's law:

$$i = nFN_o|_{x=0} = nFD\dfrac{dc}{dx}\bigg|_{x=0}.$$

The error-function solution above can be differentiated to obtain an expression for the concentration gradient at the electrode surface. Substituting that gradient into the expression for the current density, i, yields

$$i = \dfrac{nF\sqrt{D_i}c_i^o}{\sqrt{\pi t}}.$$

Here we see that the current depends inversely on the square root of time. This relationship is known as the Cottrell equation (see Chapter 6), and this dependence on time is indicative of mass-transfer control. This simple analytical relationship applies in situations where migration can be neglected.

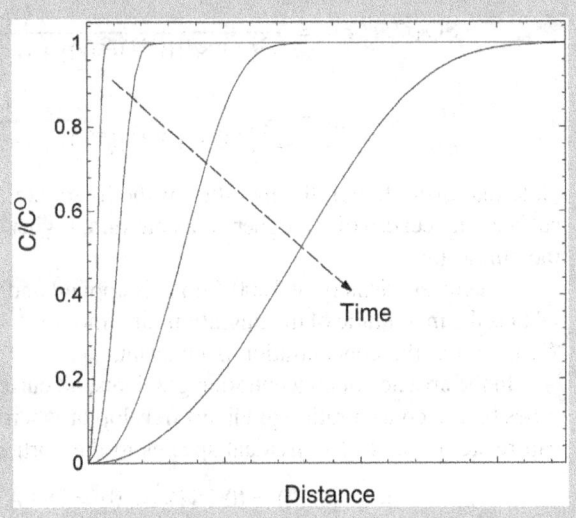

4.4 TRANSFERENCE NUMBERS, MOBILITIES, AND MIGRATION

In Equation 4.3, both the ion mobility and a diffusion coefficient appear. Both the concentration and the potential impact the electrochemical potential (μ_i). While the details are beyond the scope of this text, it is the gradient of the electrochemical potential that is the true driving force for transport. Therefore, we might expect the diffusivity and mobility to be related. That relationship is

$$D_i = RTu_i, \qquad (4.27)$$

which is the Nernst–Einstein relationship for dilute solutions. This equation provides us with a convenient way of estimating one of these parameters from knowledge of the other.

Now let's take a second look at the Nernst–Planck equation in the absence of concentration gradients:

$$\mathbf{i} = -F^2 \nabla\phi \sum_i z_i^2 u_i c_i = -\kappa \nabla \phi.$$

ILLUSTRATION 4.3

Calculate the transference number of the ions present in an electrolyte that is 0.5 M CuSO$_4$ and 0.1 M H$_2$SO$_4$. Assume complete dissociation of the ionic species.

$$u_{Cu^{2+}} = 2.8 \times 10^{-13} \text{ m}^2\text{mol J}^{-1}\text{s}^{-1},$$
$$u_{H^+} = 1.9 \times 10^{-12} \text{ m}^2\text{mol J}^{-1}\text{s}^{-1},$$
$$u_{SO_4^{2-}} = 4.3 \times 10^{-13} \text{ m}^2\text{mol J}^{-1}\text{s}^{-1}.$$

Cu^{2+}: $t_{Cu^{2+}} = \dfrac{2^2(2.8 \times 10^{-13} \text{ m}^2\text{mol/Js})(500 \text{ mol/m}^3)}{2^2(2.8 \times 10^{-13})(500) + (-2)^2(4.3 \times 10^{-13})(600) + 1^2(1.9 \times 10^{-12})(200)} = 0.284,$

SO$_4^{2-}$: $t_{SO_4^{2-}} = \dfrac{(-2)^2(4.3 \times 10^{-13})(600)}{2^2(2.8 \times 10^{-13})(500) + (-2)^2(4.3 \times 10^{-13})(600) + 1^2(1.9 \times 10^{-12})(200)} = 0.523,$

H$^+$: $t_{H^+} = \dfrac{1^2(1.9 \times 10^{-12})(200)}{2^2(2.8 \times 10^{-13})(500) + (-2)^2(4.3 \times 10^{-13})(600) + 1^2(1.9 \times 10^{-12})(200)} = 0.193.$

Note that even though the mobility of the hydrogen ion is much higher, the majority of the current is carried by the sulfate ion because of its higher concentration, higher charge per ion than H$^+$, and a mobility that is greater than that of the cupric ion.

If a current density of 1000 A·m^{-2} is applied and current is positive when it flows from left to right in the system, what is the magnitude of the migration flux (mol·m^{-2}·s^{-1}) for each of the ions, and in which direction do the ions move? Assume that the concentration is constant.

In the absence of concentration gradients, all current flow is due to migration. Under these conditions (e.g., at short times before concentration gradients develop or when convection is sufficient to keep the concentrations constant), we can relate the flux of individual species to the portion of the current that is carried by that species:

$i_{Cu^{2+}} = (t_{Cu^{2+}})(i) = (0.284)(1000) = 284 \text{ A m}^{-2},$

$\dfrac{i_{Cu^{2+}}}{z_{Cu^{2+}}F} = N_{Cu^{2+}} = \dfrac{284 \text{ A m}^{-2}}{(2 \text{ equiv mol}^{-1})(96485 \text{ C equiv}^{-1})} = 0.00147 \text{ mol m}^{-2}\text{s}^{-1}$ (flux : left to right),

$\dfrac{i_{SO_4^{2-}}}{z_{SO_4^{2-}}F} = N_{SO_4^{2-}} = \dfrac{523 \text{ A m}^{-2}}{(-2 \text{ equiv mol}^{-1})(96485 \text{ C equiv}^{-1})} = -0.00271 \text{ mol m}^{-1}\text{s}^{-1}$ (flux : right to left),

$\dfrac{i_{H^+}}{z_{H^+}F} = N_{H^+} = \dfrac{193 \text{ A m}^{-2}}{(1 \text{ equiv mol}^{-1})(96485 \text{ C equiv}^{-1})} = 0.00200 \text{ mol m}^{-2}\text{s}^{-1}.$

We see that the total current is the sum of contributions from movement of the individual species. This observation leads to the definition of a second convenient transport property, the *transference number*,

$$t_i \equiv \frac{z_i^2 u_i c_i}{\sum_j z_j^2 u_j c_j}. \qquad (4.28)$$

The quantity can be thought of as the fraction of current carried by species i in the absence of a concentration gradient. Similar to Ohm's law, it has this intuitive physical meaning only when the concentration is uniform; nonetheless, the transference number represents a key transport property of the electrolyte. From its definition, it can be seen that the sum of the transference numbers for the individual ions is unity,

$$\sum_i t_i = 1. \qquad (4.29)$$

Remember, this calculation is valid only for systems without concentration gradients. It is interesting to see that the flux of the hydrogen ion is higher than that of copper, but it still does not carry as much current. The equivalent ionic conductance is often reported in the literature rather than the mobility (u_i) or diffusivity (D_i) of individual ions. The symbol for the equivalent ionic conductance is λ_i, with units of S·cm²·equiv⁻¹. Thus, λ_i is the conductance per equivalent. Ionic conductance is related to the mobility as follows:

$$\lambda_i = |z_i|F^2 u_i = \frac{|z_i|F^2 D_i}{RT}. \qquad (4.30)$$

The values for λ_i will depend on the solvent, the species, the temperature, and the concentration. However, one typically finds values tabulated for common ions in water at room temperature at infinite dilution. There are theories that attempt to predict the dependence on concentration, for example; but, for the most part, these transport properties must be measured for practical electrochemical systems. Here we will only look at a simple salt that dissociates into one anion and one cation of equal charge. The conductance is given by the sum of the conductance of the two ions.

$$\Lambda = \lambda_+ + \lambda_- = \frac{\kappa}{z_+ \nu_+ c} = \frac{\kappa}{c_{equiv}}. \qquad (4.31)$$

Cation	Equivalent conductance, λ_i [S·cm²·equiv⁻¹]	Anion	Equivalent conductance, λ_i [S·cm²·equiv⁻¹]
H⁺	349.8	OH⁻	197.6
Li⁺	38.69	Cl⁻	76.34
Zn²⁺	53	HSO₄⁻	50
Na⁺	50.11	SO₄²⁻	80
Cu²⁺	54	Br⁻	78.3
K⁺	73.52	NO³⁻	71.44

ILLUSTRATION 4.4

Calculate the (a) conductivity, (b) transference number of Na⁺, and (c) diffusivity of 0.1 M NaCl from values for equivalent conductance.

a. $\Lambda = \lambda_+ + \lambda_- = 50.11 + 76.34 = \dfrac{\kappa}{c_{equiv}}.$

The concentration is 0.1 M, which is equal to 0.1/1000 equiv·cm⁻³. Substituting into Equation 4.31 gives $\kappa = 0.01265\,\text{S}\,\text{cm}^{-1}$. The experimental value is 0.01067.

b. Using Equations 4.28 and 4.30, the transference number of the sodium ion is simply

$$t_+ = \frac{\lambda_+}{\lambda_+ + \lambda_-} = \frac{50.11}{50.11 + 76.34} = 0.40.$$

c. The diffusion coefficient of the salt is given by Equation 4.20:

$$D = \frac{D_+ D_-(z_+ - z_-)}{z_+ D_+ - z_- D_-} = \frac{\lambda_+ \lambda_-(z_+ - z_-)}{z_+ \lambda_+ - z_- \lambda_-} \frac{RT}{F^2}$$

$$= \frac{(50.11)(76.34)2}{(50.11 + 76.34)} \frac{(8.314)(298)}{(96485)^2} = 1.6 \times 10^{-5}\,\text{cm}^2\,\text{s}^{-1}.$$

Binary Electrolyte: Lithium-Ion Battery

Now that we have examined transport properties, let's consider an additional example that involves a binary electrolyte. Specifically, let's look at transport in the separator of a lithium-ion battery. The electrolyte consists of an organic solvent into which a salt of lithium is dissolved (e.g., LiPF₆). The salt is assumed to dissociate completely to form a binary 1:1 electrolyte. For simplicity, we will treat the electrodes as planar even though porous electrodes are used in practice (see Chapter 5). During charging of the battery, lithium ions are reduced at the negative (left electrode) and lithium oxidized at the positive electrode (right side). The anion does not react at either electrode. Thus, in effect, we are shuttling lithium ions from one electrode to the other. Graphite is a typical negative electrode, and metal oxides are commonly used for the positive electrode. The half-cell reactions of interest are

$$\text{LiC}_6 \leftrightarrow \text{Li}^+ + e^- + C_6$$
$$\text{Li}^+ + e^- + \text{CoO}_2 \leftrightarrow \text{LiCoO}_2$$

Convection is neglected. We can write the flux of cations and anions using the Nernst–Planck equation for one-dimensional transport in the x-direction as

$$N_+ = -z_+ u_+ F c_+ \frac{d\phi}{dx} - D_+ \frac{dc_+}{dx}, \qquad (4.32)$$

$$N_- = -z_- u_- F c_- \frac{d\phi}{dx} - D_- \frac{dc_-}{dx}. \qquad (4.33)$$

Since only lithium ions can react, these are the only ions responsible for current, which can be expressed as

$$N_+ = \frac{i}{F}. \qquad (4.34)$$

For the anion, however, the flux must be equal to zero. Thus, we can use Equation 4.33 to express the potential gradient in terms of a concentration gradient.

$$\frac{d\phi}{dx} = -\frac{D_-}{z_- u_- F c_-} \frac{dc_-}{dx}. \qquad (4.35)$$

This expression can be substituted back into Equation 4.32 to eliminate the potential. Also note that although the concentration of the salt will change, because of electroneutrality, $c_- = c_+ = c$ since $\nu_+ = \nu_- = 1$. Thus,

$$N_+ = \frac{i}{F} = \left(\frac{z_+ u_+ D_- - z_- u_- D_+}{z_- u_-}\right)\frac{dc}{dx}. \quad (4.36)$$

If we now introduce the following for the transference number and the diffusivity of the salt:

$$t_+ = \frac{z_+ u_+}{z_+ u_+ - z_- u_-} \quad \text{and} \quad D = \frac{z_+ u_+ D_- - z_- u_- D_+}{z_+ u_+ - z_- u_-},$$

we arrive at

$$N_+ = \frac{i}{F} = \frac{-D}{1-t_+}\frac{dc}{dx}. \quad (4.37)$$

Equation 4.37 can be generalized rather simply by adding ν_+ so that it applies to any binary electrolyte where only the cation reacts, a situation that is quite common. The general expression is

$$N_+ = \nu_+\left(\frac{z_+ u_+ D_- - z_- u_- D_+}{z_- u_-}\right)\frac{dc}{dx} = \nu_+ D_+\left(\frac{z_+ - z_-}{z_-}\right)\frac{dc}{dx}$$

$$= \nu_+\left(\frac{-D}{1-t_+}\right)\frac{dc}{dx}, \quad (4.38)$$

where the simplification in the middle expression was obtained via the Nernst–Einstein relation.

Now that we have the needed expression for the flux as a function of the salt concentration, we can move forward to solve for the concentration profile in the lithium-ion battery containing a binary salt. As we saw earlier, for a binary electrolyte, the equation for convective diffusion applies. Neglecting convection, the material balance for the salt is $\partial c/\partial t = D\nabla^2 c$.

A battery is inherently a transient device whose state-of-charge changes continuously during charge and discharge (see Chapter 7). However, batteries are often operated at conditions where the time required for the transport processes to reach steady state, as characterized by the diffusion time (L^2/D), is short relative to the time scale that characterizes the chemical conversion. Physically this means that the concentration gradient changes relatively quickly compared to time it takes to charge or discharge the battery. Under such conditions, a pseudo steady-state is maintained. Consequently, if we assume steady-state for a one-dimensional case with constant diffusivity, we are left with Laplace's equation for the concentration, which we can integrate to show that the concentration varies linearly with position,

$$c = Ax + B.$$

We can now use Equation 4.37 to find the constant A. Also, since the electrolyte is not involved in the reactions, lithium ions are simply shuttled between electrodes, the amount of salt is constant. This conservation of lithium allows for the evaluation of the second constant B. The net result is

$$c = \frac{i}{F}\frac{(1-t_+)}{D}\left[\frac{L}{2} - x\right] + c_o. \quad (4.39)$$

Figure 4.7 shows the concentration graphically for discharge (i is positive). Since the amount of salt is fixed, the average concentration cannot change. Therefore, the concentration is constant in the middle of the separator. As the current density increases, two undesired results are possible. The concentration of the salt and therefore lithium can go to zero at the electrode surface ($x=0$). This case represents a limiting-current behavior. A second possibility is seen at the opposite electrode where the concentration increases with current density. There is the risk of exceeding solubility limits at that electrode, which would lead to salt precipitation.

The impact of the transference number for the lithium ions can be seen in Equation 4.39. An increase in the transference number for the lithium ion leads to a lower concentration gradient for a particular flux (or current). As the transference number approaches one, the concentration gradient of the salt goes to zero. In general, it is desirable to have a high lithium-ion transference number in these batteries.

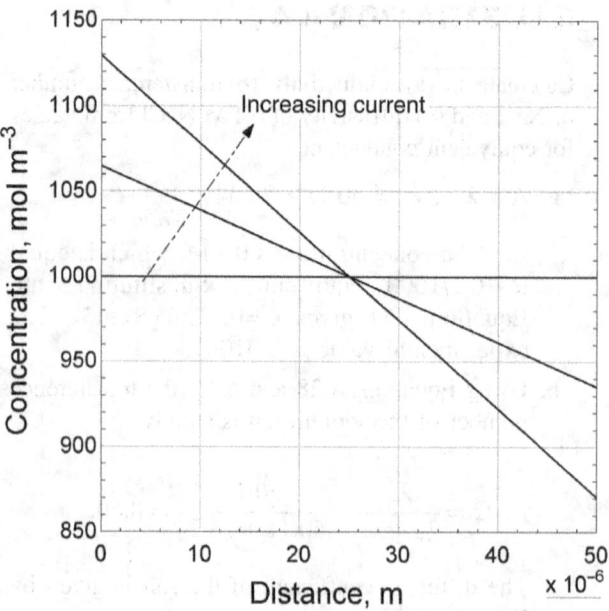

Figure 4.7 Concentration profile in separator.

ILLUSTRATION 4.5

Calculate the following quantities given data for conductivity and transference number of an electrolyte in a separator of thickness 200 μm. The initial concentration of the salt is 800 mol m^{-3}. Also, $\kappa = 1.3$ S m^{-1} and $t_+ = 0.4$.

a. *Diffusion time.* The conductivity and transference number can be expressed in terms of mobilities (Equations 4.7 and 4.28). Solving for the mobilities gives

$u_+ = 6.98 \times 10^{-14}$ m^2 mol J^{-1} s^{-1};

$u_- = 1.05 \times 10^{-13}$ m^2 mol J^{-1} s^{-1};

the salt diffusivity $D = 2.08 \times 10^{-10}$ m^2 s^{-1}.

The time for diffusion is $\tau_{diffusion} = \frac{L^2}{D} = \frac{[2 \times 10^{-4}]^2}{2.08 \times 10^{-10}} = 200$ s, which is short relative to the time typically required to charge or discharge a battery.

b. *Limiting current*, corresponding to zero concentration at the electrode surface at $x = L$, is given by

$$i_{\lim} = \frac{FD}{(1-t_+)} \frac{2c_o}{L}$$

$$= \frac{96485 \text{ C}}{\text{mol}} \left| \frac{2.08 \times 10^{-10} \text{ m}^2}{(1-0.4) \text{ s}} \right| \frac{(2)800 \text{ mol}}{(2 \times 10^{-4} \text{ m}) \text{m}^3}$$

$$= 267 \text{ A m}^{-2}.$$

c. *Potential drop at half of limiting current* measured instantly after the start of the current flow. Before there is time for the concentration gradient to develop, the potential drop will be given by Ohm's law.

$$\Delta\phi = \frac{iL}{\kappa} = \frac{(2 \times 10^{-4} \text{ m})(133.5 \text{ A}) \text{ m}}{\text{m}^2 1.3 \text{ S}} \left(\frac{1000 \text{ mV}}{\text{V}}\right)$$

$$= 21 \text{ mV}.$$

d. Once sufficient time elapses, a concentration gradient will develop and the modified form of Ohm's law is used, Equation 4.5, which is integrated to give

$$\Delta\phi = \frac{iL}{\kappa} - \frac{F}{\kappa} \sum_i z_i D_i \Delta c = 15 \text{ mV}.$$

Thus, as the concentration gradient develops, the potential drop *decreases* by about 5 mV.

4.5 CONVECTIVE MASS TRANSFER

In cases where migration can be neglected (e.g., excess supporting electrolyte), traditional methods and correlations for mass transfer can be used to determine reaction rates and the current density, which of course are related through Faraday's law. In this section, we consider convective mass transfer for geometries that are of interest for electrochemical systems.

It is frequently convenient to describe mass transfer with a convective mass-transfer coefficient rather than molecular diffusion and a detailed description of the fluid motion. If, as mentioned previously, we restrict ourselves to cases where transport by convection is important and where migration can be neglected, we can use the multitude of correlations developed in other fields of engineering to describe mass and heat transfer. Generally, the flux is expressed as

$$N_i = k_c(c_{i,bulk} - c_{i,surf}), \quad (4.40)$$

where k_c is the mass-transfer coefficient with units of m·s^{-1}. Please note that although the flux is still a vector quantity, this quality has been lost with Equation 4.40. Therefore, you need to be careful to ensure that the sign is correct, and that transport is still from high concentration to low concentration. The difference in concentration between the bulk, $c_{i,bulk}$, and the surface, $c_{i,surf}$, is the driving force for transport.

Our interest in convective mass transfer is twofold. First there are a number of problems of interest where one needs to describe transport of reactants/products to or from the surface. Second, one can use electrochemical methods as a means of measuring mass-transfer coefficients and transport properties.

Typically, mass-transfer coefficients are correlated with dimensionless groups. A dimensional analysis of forced convection suggests that the Sherwood number is a function of the Reynolds and Schmidt numbers.

$$\text{Sh} = f(\text{Re}, \text{Sc}). \quad (4.41)$$

The three dimensionless quantities are

$$\text{Sh} = \frac{k_c L}{D}; \quad \text{Re} = \frac{\rho v L}{\mu}; \quad \text{Sc} = \frac{\nu}{D}.$$

Note that the v in Re is the characteristic velocity. The Sh number is a dimensionless mass-transfer coefficient defined in terms of the characteristic length, L, and the diffusivity of the species being transported. For internal flow through a pipe or channel, the characteristic length is the diameter of the pipe or the hydraulic radius of the channel. Flow through an electrochemical cell is often represented as flow through a rectangular channel. For external flow such as flow over a flat plate or a cylinder, the characteristic length is the length of the plate or the diameter of the cylinder.

Unfortunately, dimensional analysis does not provide any help for the functional form of Equation 4.41. There are a number of cases from heat and mass transfer where analytical solutions are possible, but mostly the correlations of interest for electrochemical systems are empirical. Whether the result of analytical solution or experimental measurements, the correlations are of the form

$$\text{Sh} = \text{constant } \text{Re}^a \text{Sc}^b, \quad (4.42)$$

where a and b are constants.

The definition of Sh can be combined with Equation 4.40 as follows:

$$N_i = k_c(c_{i,bulk} - c_{i,surf}) = \text{Sh}\frac{D_i}{L}(c_{i,bulk} - c_{i,surf}). \quad (4.43)$$

It is sometimes advantageous to run an electrochemical cell at the limiting current. The mass-transfer limiting current represents the highest rate at which a cell can be operated and, therefore, the rate that maximizes production. At limiting current, Equation 4.43 becomes

$$i_{\lim} = \frac{nF\text{Sh}D_i c_{i,bulk}}{L}. \quad (4.44)$$

There are also important problems where the fluid motion is driven by density differences in the fluid rather than forced flow. Because of the gravitational field, buoyancy forces result from these differences in density. This effect is similar to flow induced by temperature differences, which is commonly studied in heat transfer and is called natural or free convection. As we saw earlier, electrochemical reactions often result in concentration gradients. These concentration differences can result in variations in density, which culminate in fluid flow. The boundary layer developed from a vertical electrode is sketched in Figure 4.8. For natural convection, the dimensionless mass-transfer coefficient is expressed as

$$\text{Sh} = f(\text{Gr}, \text{Sc}), \quad (4.45)$$

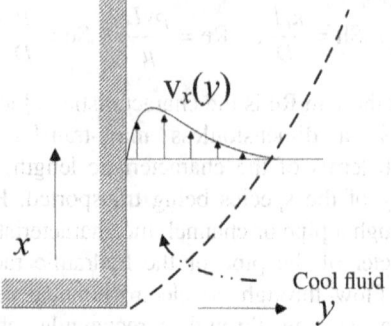

Figure 4.8 Velocity profile in boundary layer.

where the Grashof number, Gr, replaces the Re number. The Grashof number is given by

$$\text{Gr} \equiv \frac{gL^3 \Delta\rho}{\rho_{bulk} \nu^2},$$

where g is the acceleration due to gravity, $\Delta\rho$ is the difference between the density of the bulk liquid and that at the surface of the electrode, ρ_{bulk} is the density of the bulk electrolyte and ν is the kinematic viscosity. The Re number is the ratio of inertial and viscous forces; similarly, the Gr number may be thought of as the ratio of buoyancy and viscous forces. The characteristic length for Gr is typically the height of a vertical plate or electrode, which is of particular interest to us.

For the most part, one can use these correlations without a comprehensive knowledge of their origin, which will be our approach here. The interested reader is directed to any number of texts that treat mass transfer in depth. Since these correlations are by and large empirical, it is important to know what characteristic length is used in the definition of Re, Sh, and Gr. Similarly, the velocity used in Re is essential. In contrast, Sc is a property of the fluid and thus independent of the flow.

The methodology of solving these convection problems will be illustrated by the following two examples. The first involves mass transfer in a parallel plate electrode; the second is an electrorefining example.

ILLUSTRATION 4.6

A divalent metal contaminant in water is removed by electrodeposition as it flows past parallel plate electrodes. The average velocity of the fluid is 0.05 m·s^{-1}, and the plate electrodes are separated by a distance of 1.5 cm and are 20 cm long. The counter electrode is located downstream of the working electrodes. Estimate the limiting current density. The concentration of contaminant is 0.1 M and its diffusivity is 7.1×10^{-10} m^2·s^{-1}. Use a density of 1000 kg·m^{-3} and a viscosity of 1 mPa.

Calculate the Reynolds number, $\text{Re} = \frac{v\rho d_h}{\mu}$. For the characteristic length, we use the hydraulic diameter, $d_h = \frac{4A_c}{P} = \frac{4 \text{ cross-sectional area of flow}}{\text{perimeter}} \sim 2h$

$$\text{Re} = \frac{(0.05 \text{ m s}^{-1})(1000 \text{ kg m}^{-3})(2 \times 0.015 \text{ m})}{0.001 \text{ Pa}} = 1500$$

indicating laminar flow.

$$\text{Sc} = \frac{0.001 \text{ Pa}}{(1000 \text{ kg m}^{-3})(7.1 \times 10^{-10} \text{ m}^2\text{s}^{-1})} = 1408.$$

The following correlation for the Sherwood number is provided, assuming that the flow is laminar and well developed before reaching the electrodes. Since the electrodes are of finite length, there is an additional dimensionless parameter in the correlation that accounts for entrance effects, which are important in this problem.

$$Sh_{avg} = 1.8488 \, Re^{0.333} Sc^{0.333} \left(\frac{d_h}{L}\right)^{0.333},$$

$$Sh_{avg} = 1.8488 \left[(1500)(1408)\frac{0.03}{0.2}\right]^{0.333} = 126,$$

$$k_c = Sh\frac{D}{d_h} = 126\frac{7.1 \times 10^{-10}}{0.03} = 3 \times 10^{-6} \, m\,s^{-1},$$

$$i_{lim} = nFN_i = nFk_c\left(c_{i,bulk} - c_{i,surf}\right)$$
$$= (2)(96485)(3 \times 10^{-6})(100 - 0) = 58 \, A\,m^{-2}.$$

ILLUSTRATION 4.7

Copper is purified by a process known as electrorefining. Here Cu with some impurities is oxidized at one electrode and reduced at the other. The purity can be improved from 99 to 99.99% by this method. This is a huge operation—world production of copper is in the millions of metric tons per year.

The concentration of CuSO$_4$ is about 0.25 M and the supporting electrolyte is sulfuric acid with a concentration of about 1.5 M. Impurities more noble than Cu (see Appendix A), will not oxidize appreciably, and therefore do not enter the electrolyte. Metals, such as Ni that are less noble than Cu will not be reduced. Thus, if the potential is controlled carefully, the cathode can be significantly purer than the anode starting material.

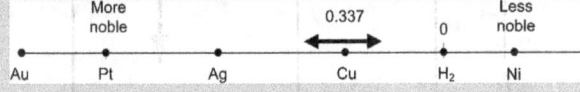

Estimate the limiting current and the electrode area required to produce 1000 metric tons per day of copper. For a vertical electrode, free convection is assumed and the following correlation is suggested.

$$Sh = 0.31(ScGr)^{0.28}.$$

The following data are provided:

$\rho_e = 1094 \, kg\,m^{-3}$, $\quad \Delta\rho = 32 \, kg\,m^{-3}$,
$D = 5.33 \times 10^{-10} \, m^2\,s^{-1}$, $\quad \nu = 1.27 \times 10^{-6} \, m^2\,s^{-1}$,
$L = 0.96 \, m.$

The Grashof and Schmidt numbers are calculated from the data provided:

$$Sc = \frac{\nu}{D} = 2383; \quad Gr = \frac{gL^3\Delta\rho}{\rho\nu^2} = 1.573 \times 10^{11}.$$

Next, the Sherwood number is estimated from the correlation; Sh = 3739.

The limiting current is when $c_i = 0$ at the surface.
Therefore, $Sh = \dfrac{Li_{lim}}{nFD_ic_i^\infty}.$

Solving for $i_{lim} = 100 \, A\,m^{-2}$,

$$Area = \frac{1000\,ton}{day}\bigg|\frac{day}{86,400s}\bigg|\frac{10^6\,g}{ton}\bigg|\frac{mol\,Cu}{63.5g}\bigg|\frac{2(96485)\,C}{mol\,Cu}\bigg|\frac{s\,m^2}{100\,C}\bigg|\frac{1\,km^2}{10^6\,m^2}$$
$$= 0.35\,km^2.$$

There are certain mass-transfer situations that are particularly relevant to electrochemical processes. For some systems gases are evolved on the electrode; for example, chlorine is evolved on the positive electrode in a chlor-alkali cell. Similar to boiling heat transfer, the bubbles disrupt the boundary layer near the electrode and dramatically enhance the rate of mass transfer. Additionally, the bubble can create macroscopic circulation of the electrolyte. Thus, as shown in Figure 4.9, gas evolution at one electrode may enhance mass transfer at the counter electrode.

To describe mass transfer at gas-evolving electrodes, the expansion of gas bubbles at the surface and its similarity to nucleate boiling was used as the basis of a correlation developed by Stephak and Vogt:

$$Sh = 0.93 \, Re^{0.5}Sc^{0.487}, \quad (4.46)$$

where the characteristic length in the Sherwood and Reynolds numbers is the break-off diameter of the bubbles. Measurements of the break-off diameter for hydrogen and oxygen evolution in both acidic and alkaline solutions yielded an average value of about 50 μm. This value of the diameter should be used unless more detailed information is available. Equation 4.46 was tested against a large number of experiments over a wide set of conditions involving hydrogen, oxygen, or chlorine evolution from acidic or alkaline electrolytes at temperatures ranging from 0 to 80 °C, and found to represent the data well in most cases. However, 15% of the data were found to lie outside an error margin spanning −50 to +100%, which is higher than the typical error found for correlations that do not involve formation of a second phase. The correlation found in Equation 4.46 is recommended for the calculation of mass-transfer rates at gas-evolving electrodes in the absence of experimental data for the specific system of interest.

The above correlation is for the mass-transfer rate at the gas-evolving electrode itself. In addition, Mohanta and

78 Electrochemical Engineering

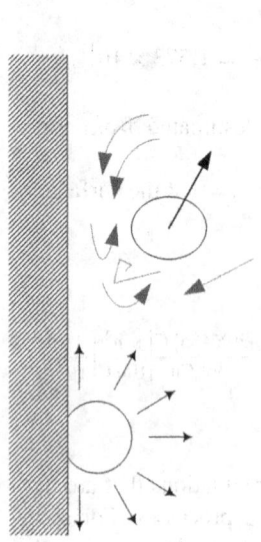
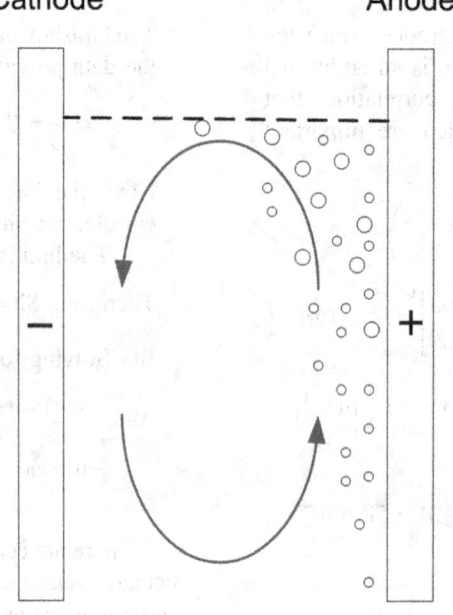

Figure 4.9 Evolution of gases on electrodes showing the flow patterns and the disruption of the stagnant fluid near the surface. On the right, bulk electrolyte motion shown.

Fahidy[1] studied two vertical, parallel plate electrodes shown in Figure 4.10 where the evolution of gas at the anode affected the rate of mass transfer at the cathode. The gas flow will be upward, and the Reynolds number is based on the hydraulic diameter (see Illustration 4.6) and the superficial velocity of the gas:

$$v_s = \frac{\dot{V}}{A_c} = \frac{\dot{V}}{Wh}, \quad (4.47)$$

where $\dot{V}$ is the volumetric flow of gas in m^3 s^{-1}. Therefore,

$$\text{Re} = \frac{v_s \rho d_h}{\mu} = \frac{2\dot{V}\rho}{(W+h)\mu}. \quad (4.48)$$

Fahidy et al. provided the following empirical correlation for mass transfer at the electrode opposite that of the gas-evolving electrode:

$$\text{Sh} = \text{Sh}_o + 3.088\,\text{Re}^{0.77}\text{Sc}^{0.25}\left(\frac{L}{h}\right)^{0.336}. \quad (4.49)$$

Sh$_o$ is the Sherwood number in the absence of bubbles; thus, the second term on the right side is the increase in mass transfer due to the bubbles at the opposite electrode. These and other correlations should be used cautiously as many are based on a small set of data and have only a weak theoretical basis. Nonetheless, Equation 4.49, for example, is consistent with the expected behavior. Specifically, if the distance between electrodes, h, is reduced, then the superficial velocity and Re increase, resulting in a higher mass-transfer coefficient. Additionally, if L is the length from the bottom of the electrode, as L increases, the mass-transfer coefficient increases because the bubbles reach further across the electrode (Figure 4.9).

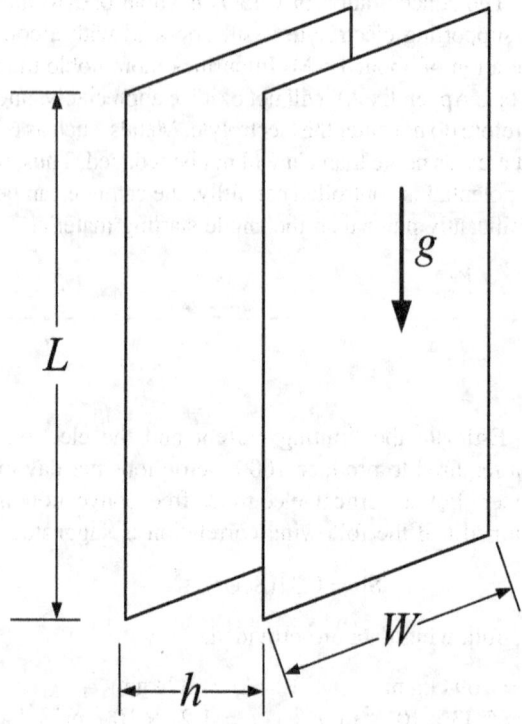

Figure 4.10 Parallel plate vertical electrodes. The acceleration of gravity is downward.

[1] Mohanta, S. and Fahidy, T.Z. (1977) *J. Appl. Electrochem.*, **7**, 235.

> The void fraction of the gas, ε, is also referred to as the "gas holdup," φ_g. The superficial velocity of the gas is defined as the volumetric flow rate of the gas divided by the cross-sectional area of flow:
>
> $$v_s = \frac{\dot{V}}{A_c}.$$
>
> Often, the velocity of the gas relative to the velocity of the liquid can be determined. For example, the terminal velocity of a bubble in liquid for a diameter less than 0.7 mm is given by Stoke's law:
>
> $$v_t = \frac{g d_b^2 \Delta \rho}{18 \mu}.$$
>
> From the terminal velocity, the gas holdup can be estimated
>
> $$\varepsilon = \frac{v_s}{v_t} = \varphi_g.$$

Finally, mass transfer can be improved by aggressively bubbling (sparging) a nonreacting gas through the electrolyte. This phenomenon is similar to natural circulation in that buoyancy forces are critical. Sigrist et al.[2] have proposed the following correlation:

$$\text{Sh} = 0.19(\text{Ar}^* \text{Sc})^{0.333}. \quad (4.50)$$

The Archimedes number is similar to the Gr number, with the principal distinction being that in gas sparging there is a second phase present.

$$\text{Ar} = \frac{L^3 g (\rho_L - \rho_g)}{\rho_L \nu^2}. \quad (4.51)$$

Ar characterizes the buoyancy and viscous forces on the bubble.

Rather than a single bubble, swarms of bubbles would be used to enhance mass transfer. In this case, the fraction of the volume occupied by the bubbles, ε, is important. Assuming that the density of the gas is small compared to the density of the liquid, the Ar number is modified to

$$\text{Ar}^* = \frac{g L^3}{\nu^2} \cdot \frac{\varepsilon}{1 - \varepsilon}. \quad (4.52)$$

In the empirical data, no dependence between the mass-transfer coefficient and either the height of the electrode, L, or the space between electrodes, h, was observed. Therefore, the correlation given by Equation 4.50 uses the bubble diameter, d_b, as the characteristic length in the Sh and Ar*. As above, assume a bubble diameter in the electrolyte of 50 μm, unless additional information is given.

4.6 CONCENTRATION OVERPOTENTIAL

In Section 4.3, we presented the following expression for the cell potential during discharge in the absence of concentration gradients:

$$V_{cell} = U_{cell} - |\eta_{anode}| - |\eta_{cathode}| - |\eta_{ohmic}|. \quad (4.53)$$

We also examined how to solve problems with concentration gradients at various levels of approximation. The cell voltage comes naturally from the solution of the coupled equations for the potential field and concentrations. However, there are situations where we would like to estimate the cell voltage in the presence of concentration gradients without solving the fully coupled equations. In doing so, we assume that the other voltage drops associated with the reactions and the ohmic drop in solution still apply. We also assume that concentration gradients are limited to a mass-transfer boundary near the electrode surfaces. The task, then, is to determine the cell voltage that pertains to a certain current. Most frequently, the Tafel approximation is used to facilitate calculation of the surface overpotential for each reaction at the desired current. The ohmic drop between the electrodes is calculated assuming a constant conductivity at the bulk concentration, which accurately represents most of the domain between the electrodes. Following the treatment presented by Newman and Thomas, we now define a concentration overpotential to calculate the impact of the concentration gradients near each of the electrodes on the cell voltage:

$$\eta_{conc} = i \int_0^\infty \left(\frac{1}{\kappa} - \frac{1}{\kappa_\infty} \right) dy + \frac{RT}{nF} \ln \left(\prod_i \frac{c_{i,\infty}^{s_i}}{c_{i,0}^{s_i}} \right)$$
$$+ F \int_0^\infty \sum_j \frac{z_j D_j}{\kappa} \frac{\partial c_j}{\partial y} dy, \quad (4.54)$$

where y is the distance from the electrode of interest, ∞ represents the bulk solution, i is positive for anodic current and negative for cathodic current, and activity coefficients have been neglected. The first term on the right side accounts for the error associated with use of the bulk concentration to calculate the ohmic drop near the electrode surface. The second term is the equilibrium potential difference between the electrode of interest and a reference electrode of the same type located in the bulk electrolyte. It is thermodynamic in nature and is commonly referred to as a concentration cell. The species of interest are only those

---

[2] Sigrist, L. Dossenbach, O., and Ibl, N. (1979) Int. J. Heat Mass Transfer., 22, 1393.

that participate in the electrochemical reactions. Note that the stoichiometric coefficients in this term correspond to the cathodic reaction when calculating the concentration overpotential associated with either the cathode or the anode, consistent with the convention described in Chapter 3. The last term is the diffusion potential, which accounts for the fact that differences in the transport properties of the ions in solution impact the potential field. This last term is summed over all the charged species in solution. Because we have used the normal from the surface for both electrodes, concentration gradients that are in the same physical direction have the opposite sign at the anode and cathode.

It is possible for the terms in Equation 4.54 to have different signs, and hence contribute in opposite directions to the concentration overpotential. The sign of the cathodic and anodic concentration overpotentials, themselves, have significance. The concentration overpotential contributes in a negative way to the overall cell potential when its value at the cathode is negative and its value at the anode is positive.

If we assume a stagnant diffusion layer of thickness δ with linear variation of the concentration, the concentration gradients in the last term can be approximated as

$$\frac{\partial c_j}{\partial y} \approx \frac{c_{j,\infty} - c_{j,0}}{\delta}. \quad (4.55)$$

Equation 4.54 then becomes

$$\eta_{conc} = i \int_0^\infty \left(\frac{1}{\kappa} - \frac{1}{\kappa_\infty}\right) dy + \frac{RT}{nF} \ln\left(\prod_i \frac{c_{i,\infty}^{s_i}}{c_{i,0}^{s_i}}\right)$$
$$+ F \int_0^\infty \sum_j \frac{z_j D_j}{\kappa}\left(\frac{c_{j,\infty} - c_{j,0}}{\delta}\right) dy. \quad (4.56)$$

When there is excess supporting electrolyte, the conductivity in the boundary layer does not change significantly, and the first term in Equation 4.56 can be neglected. Also, since the conductivity scales linearly with concentration, the last term scales as the concentration of the minor species divided by that of the supporting electrolyte and is likewise small. With these assumptions, the concentration overpotential in the presence of a supporting electrolyte becomes

$$\eta_{conc} \approx \frac{RT}{nF} \ln\left(\prod_i \frac{c_{i,\infty}^{s_i}}{c_{i,0}^{s_i}}\right). \quad (4.57)$$

This is the equation most commonly found in the literature and the one that we will use most frequently in this text. The concentration overpotential expressed in Equation 4.57 will always be negative at the cathode and positive at the anode since the concentration of reactants and products at the surface will always be depleted or enriched, respectively, relative to the bulk. Consequently, Equation 4.57 represents a potential loss for both the anode and cathode. Therefore, the appropriate expressions for the cell potential during galvanic operation (discharge) and electrolysis (charge) become

$$V_{cell} = U_{cell} - |\eta_{s,anode}| - |\eta_{s,cathode}| - |\eta_{conc,anode}|$$
$$- |\eta_{conc,cathode}| - |\eta_{ohmic}|, \quad (4.58a)$$

$$V_{cell} = U_{cell} + |\eta_{s,anode}| + |\eta_{s,cathode}| + |\eta_{conc,anode}|$$
$$+ |\eta_{conc,cathode}| + |\eta_{ohmic}|. \quad (4.58b)$$

One of the challenges with all of the above expressions for the concentration overpotential is that they require knowledge of the concentration at the surface of the electrode, which is a quantity that is not readily accessible. In the presence of excess supporting electrolyte where the bulk concentration is known, the current is related to transport across the mass-transfer boundary layer as follows:

$$i = nFk_c(c_{bulk} - c_{surf}) = nF \, \text{Sh} \frac{D_i}{L}(c_{bulk} - c_{surf}). \quad (4.59)$$

Equation 4.59 applies to all species that participate in the reaction. As before, we treat this equation as a scalar rather than a vector. In doing so, we note that reactants are transported to the surface and have a surface concentration that is less than the bulk, while the opposite is true for products of the reaction. The n in this equation represents the number of equivalents per mole of the species of interest. Once the surface concentrations are known, Equation 4.57 can be used to calculate the concentration overpotential.

Finally, if there is only one species in the electrolyte that participates in the reaction (e.g., a metal cation that is deposited on the electrode) and the limiting current is known, the surface concentration can be written as

$$c_{i,0} = c_{i,\infty}\left(1 - \frac{i}{i_{\lim}}\right) + c_{i,0,\lim} \frac{i}{i_{\lim}}, \quad (4.60)$$

where $c_{i,0,\lim}$ is most often zero, but can be the saturation concentration in situations where the surface concentration is higher than the bulk.

ILLUSTRATION 4.8

Consider an electrorefining cell similar to that of Illustration 4.7. In this case, however, there is forced convection (instead of natural convection) and Sh is dependent on the hydraulic diameter. The following data are known:

Electrolyte composition: 0.25 M $CuSO_4$ and 1.5 M H_2SO_4

Sherwood number (based on hydraulic diameter): 1200

Electrode dimensions: cell gap = 0.05 m, width = 0.5 m, height = 0.96 m

Operating current density: 200 A·m^{-2}

Cupric ion diffusivity: 5.33×10^{-10} m^2·s^{-1}

Kinetic parameters for copper reaction (apply at both electrodes):

$i_o = 0.001$ A·cm^{-2}

$\gamma = 0.42$

$c_{Cu^{2+},ref} = 0.1$ M

$\alpha_a = 1.5$

$\alpha_c = 0.5$

Determine the surface and concentration overpotentials at both the anode and the cathode. Also determine the cell potential at the specified current density.

Solution:

1. We need to find the concentration of cupric ions at both electrode surfaces. To do this, we will use Equation 4.59 to relate the current to the surface concentration, assuming that the bulk concentration is known.

   The characteristic length for Sh is
   $d_h = 4A_c/P = 4(0.05 \text{ m})(0.5 \text{ m})/(2(0.05 \text{ m} + 0.5 \text{ m})) = 0.0909$ m

   For the cathode:
   $$i = 200 \text{ A m}^{-2} = nF \text{ Sh} \frac{D_i}{L}\left(c_{bulk} - c_{surf}\right)$$
   $$= \frac{(2)(96485)(1200)\left(5.33 \times 10^{-10}\right)}{0.0909}\left(250 - c_{surf}\right),$$
   $$c_{surf,c} = 102.7 \text{ mol m}^{-3}.$$

   Assuming that the same mass-transfer coefficient and properties apply at the anode, $c_{surf,a} = 397.3$ mol·m^{-3}.

2. To calculate the concentration overpotential, we need the value of s_i. The reaction written in the cathodic direction is
   $$Cu^{2+} + 2e^- \rightarrow Cu$$
   Therefore, $s_{Cu^{2+}} = -1$. This same value is used at both the anode and cathode since, in both cases, we are calculating the difference between the potential at the surface and that of a reference electrode of the same type as the working electrode located in the bulk solution.

3. The concentration overpotentials can now be determined:

   Cathode: $\eta_{conc} = \frac{RT}{nF}\ln\left(\frac{c_{i,\infty}^{-1}}{c_{i,0}^{-1}}\right) = \frac{RT}{nF}\ln\left(\frac{c_{surf}}{c_{bulk}}\right)$
   $= -0.0114$ V.

   Similarly, for the anode: $\eta_{conc} = 0.00595$ V. Neither of these values is very large.

4. We can also calculate the surface overpotentials:

   Cathode: $i_o = i_{o,ref}\left(\frac{c_1}{c_{1,ref}}\right)^{\gamma_1}$
   $= 0.01 \text{ A m}^{-2}\left(\frac{102.7}{100}\right)^{0.42} = 0.0101$ A m^{-2}.

   From the BV equation with use of a nonlinear solver, $\eta_{s,c} = -0.1533$ V.

   Similarly, for the anode, $i_o = i_{o,ref}\left(\frac{c_1}{c_{1,ref}}\right)\gamma_1$
   $= 0.01 \text{ A m}^{-2}\left(\frac{397.3}{100}\right)^{0.42} = 0.0178$ A m^{-2}.

   From the BV equation with use of a nonlinear solver, $\eta_{s,a} = 0.0420$ V.

5. We can now calculate the potential of the cell.
   From Equation 4.58b, $V_{cell} = U_{cell} + |\eta_{s,anode}| + |\eta_{s,cathode}| + |\eta_{conc,anode}| + |\eta_{conc,cathode}| + |\eta_{ohmic}|.$

   Equation 4.58b is used because this is an electrolytic cell. We will neglect the ohmic drop because of the supporting electrolyte. In practice, there will be a small ohmic drop, depending on the relative concentrations of the species in solution.

   $V_{cell} = 0 + 0.0420 + (0.1533) + 0.00595 + (0.0114) = 0.213$ V.

In summary, in this section we developed an expression for the concentration overpotential that can be used in conjunction with the surface overpotentials and ohmic loss to estimate the potential of a cell at a given current density, as illustrated previously. It applies in the presence of flow since it assumes that the concentration gradients occur near the electrodes across the respective boundary layers.

4.7 CURRENT DISTRIBUTION

Current distribution refers to how the current density varies across the surface of an electrode. A current density that is constant over the electrode surface corresponds to a *uniform current distribution*. Imagine you are electroplating a metal object such as a door handle or a dental instrument. From Faraday's law, the thickness of the deposit will vary if the current density is not uniform, which is clearly not desirable as it could result in incomplete coverage in some areas and/or overplating in others. Whether we are designing an electrode, a corrosion protection system, or an electrodeposition process, it is usually advantageous to have a uniform current distribution. The consequences of a nonuniform distribution can be an uneven deposition, unwanted side reactions resulting in low efficiencies, oxidation of cell components, and poor yield. Specific examples will be considered in more detail in subsequent chapters. Our purpose in this section is twofold: (1) to develop a method for assessing the nonuniformity of the current distribution and (2) to explore quantitatively the current distribution for some simplified cases.

Geometric Symmetry

Geometries that are one-dimensional naturally lead to a uniform current distribution. For example, consider a rectangular cell with electrodes at opposite faces that occupy the entire face area. All of the transport between the electrodes is inherently one dimensional because of the geometry. Concentric cylinders with insulating planes at the two ends would also be one-dimensional due to symmetry. Although less common, spherical symmetry can also lead to a uniform current density. Since the uniformity is a result of the geometry, the current distribution will always be uniform, independent of the complexity of the problem and the parameters involved. Care should be taken with problems that include convective flow since flow may disrupt the inherent geometric symmetry.

Electrochemical systems are frequently not geometrically symmetric and may therefore have a current distribution that is not uniform. Consider the geometry of different types of electrochemical cells and see if you can predict where you expect the current between the two electrodes to be highest. What is the rationale for your prediction? Your intuition may have told you that the current will likely be highest at locations where the electrodes are closest together. It turns out that you are mostly correct. Let's explore this issue further.

Constant Concentration

When concentration gradients can be neglected, we are left with Laplace's equation for the potential as shown in Section 4.3. In that section, we solved for the potential drop in a one-dimensional (symmetric) geometry for fast kinetics. Previously, we examined the impact of the surface reaction on the potential losses at each of the two electrodes, in addition to losses associated with the ohmic drop in solution. Rapid kinetics leads to small surface overpotentials and voltage losses that are dominated by the solution. In contrast, surface overpotentials dominate the voltage losses for slow kinetics. How are these observations related to the current distribution in systems that are not geometrically symmetric?

Figure 4.11 provides a conceptual view of the situation in terms of two types of resistances: ohmic resistance and

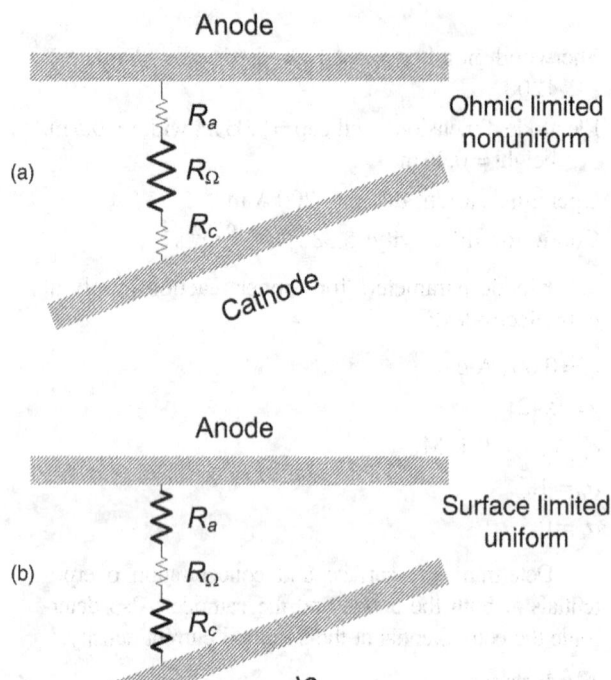

Figure 4.11 Ohmic and charge-transfer (surface) resistances.

charge-transfer resistance. The charge-transfer resistance occurs at the surface and does not depend on the geometry. In contrast, the ohmic resistance is strongly dependent on the geometry and is responsible for nonuniformities in the current distribution. Therefore, situations where the charge-transfer resistance dominates over the ohmic resistance result in a uniform current distribution. Conversely, situations where the ohmic resistance is dominant lead to a nonuniform current distribution (except, of course, in cases where the problem is geometrically symmetric). In Figure 4.11a, the current would be higher on the right where the electrodes are closer together. In contrast, the current would be nearly uniform over both electrodes when the surface resistance dominates as in Figure 4.11b.

Consider current flowing between two large, parallel plates separated by a distance L. The resistance between the plates is (see Equation 4.8c)

$$R_\Omega = \frac{L}{A\kappa}. \quad (4.61)$$

The charge-transfer resistance can be expressed as

$$R_{ct} = \frac{1}{A}\frac{d\eta_s}{di}, \quad (4.62)$$

where η_s is the surface overpotential. The ratio of these two resistances is the dimensionless quantity known as the *Wagner number*, which is a measure of the uniformity of the current density.

$$Wa = \frac{R_{ct}}{R_\Omega} = \frac{\kappa}{L}\frac{d\eta_s}{di}, \quad (4.63)$$

The change in the surface overpotential with respect to the current can be obtained from an expression for the kinetics. Beginning with the Butler–Volmer equation introduced in Chapter 3,

$$i = i_o \left[\exp\frac{\alpha_a F}{RT} \eta_s - \exp\frac{-\alpha_c F}{RT} \eta_s \right],$$

we consider the simplified expressions for linear and Tafel kinetics. The results are as follows (see Problem 4.20):

Linear kinetics : $\text{Wa} = \dfrac{\kappa}{L_c} \dfrac{RT}{F} \dfrac{1}{i_o(\alpha_a + \alpha_c)}.$ (4.64)

Tafel kinetics : $\text{Wa} = \dfrac{RT\kappa}{FL_c} \dfrac{1}{|i_{avg}|\alpha_c}.$ (4.65)

For Tafel kinetics, the average current density appears in the denominator of the Wagner number. Under these conditions, the charge-transfer resistance decreases with increasing current density due to the exponential nature of the current. In contrast, the exchange-current density appears in Equation 4.64 for linear kinetics. The resistance is independent of the current density for the linear region. In general, the current distribution becomes less uniform as the Wagner number decreases. When the Wa approaches zero, the kinetic resistance is no longer significant and the current distribution is controlled completely by the ohmic drop in solution. To summarize:

- In the absence of concentration gradients, the current distribution is influenced by kinetic and ohmic resistances.
- In situations where the kinetic resistance dominates, the current distribution tends to be uniform (Wa >> 1).
- In situations where ohmic losses dominate, the current distribution tends to be nonuniform (Wa → 0).

Equations 4.64 and 4.65 also highlight the important factors in controlling the distribution of current. The current distribution can be made more uniform by (1) increasing the conductivity of the solution, by adding a supporting electrolyte for instance, (2) reducing the characteristic length L, (3) lowering the exchange-current density, perhaps by adding inhibitors to the electrolyte, and (4) keeping the average current density low. All of these techniques are used in practice.

The choice of characteristic length for use in the Wa number is often not clear since there is frequently more than one possibility to consider. A couple of generalizations are helpful in making this choice. If the electrodes are similar in size, then the resistance is best characterized by the distance between them. In contrast, for a situation where the electrode of interest is smaller than the counter electrode and where spherical or cylindrical symmetry leads to decay of the potential field with increasing distance away from the electrode, the most appropriate length is the diameter or length of the small electrode. In such cases, the precise size and location of the counter electrode are not important since nearly all of the potential drop occurs within 5–10 radii of the small electrode. In all cases, if the electrodes are very far apart (>5L), the potential distribution near the electrode is not affected by the other electrode and a dimension characteristic of the electrode (L) should be used as the characteristic length. These generalizations are illustrated in Figure 4.12.

We can now provide some important definitions:

Primary current distribution: The current distribution where Wa = 0 and kinetic resistances are not important. The distribution of the current depends entirely on ohmic losses in the electrolyte. Constant concentration is assumed and Laplace's equation is solved for the potential field with constant potential boundary conditions at the electrodes.

Secondary current distribution: The current distribution that is determined by including kinetics as a boundary condition when solving Laplace's equation for the potential field. Also assumes constant concentration. This distribution approaches the primary current distribution when kinetics are extremely rapid.

Tertiary current distribution. This distribution includes mass transfer, kinetics, and potential field effects. A fully coupled solution of the equations that describe all of these effects is required.

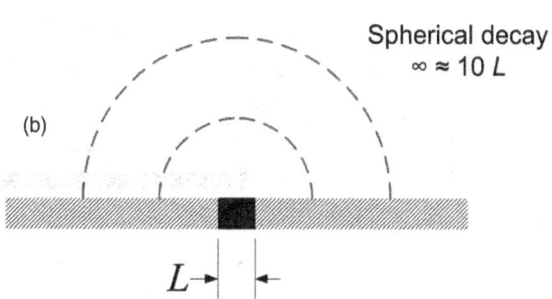

Figure 4.12 Characteristic length.

Primary Current Distribution Example (Wa = 0)

For the primary current distribution, the kinetic resistance is negligible. Under these conditions, the kinetics are sufficiently fast to be considered at equilibrium. Since the metal potential of the electrode is constant, this assumption implies that the solution just beyond the metal surface is also at a constant potential. Ohmic losses control the distribution and, since the conductivity is constant, the geometry dictates the current distribution as determined from Laplace's equation.

For the primary current distribution we have two types of boundary conditions. The first is the constant potential boundary condition at the electrode as just described. The second boundary condition applies at insulating boundaries where there is no current flow. From Ohm's law, the gradient of the potential normal to an insulating boundary is equal to zero.

Figure 4.13 shows the potential distribution for two parallel electrodes embedded in insulating walls, where the potential of the electrodes has been specified and the gradient of the potential at the insulating walls is zero. This problem can be solved analytically with use of the Schwarz–Christoffel transformation to yield the current distribution. A number of different analytical and numerical methods for solving Laplace's equation for the primary current distribution are available, and solutions for a number of different geometries have been published. While analytical solutions are useful and perhaps even elegant, numerical solution with use of modern computational tools is quite straightforward and is probably the quickest way to solve a problem whose primary current distribution is not readily available. The results shown in figures 4.13 and 4.14 were determined numerically.

The current flows normal to the lines of constant potential. Therefore, the current flows out from the edges of the electrodes at right angles to the potential lines. At insulating surfaces, the lines of constant potential are perpendicular to

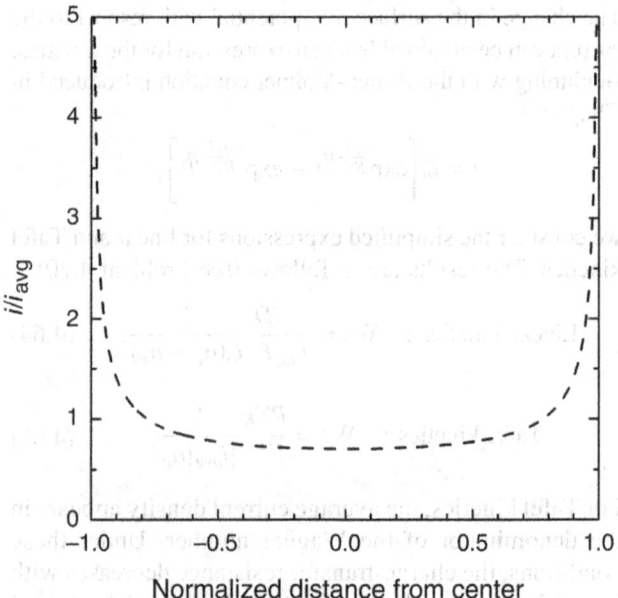

Figure 4.14 Primary current distribution between two parallel plate electrodes. The distance between the electrodes (h) is 1.25 times the electrode width (L).

the surface, corresponding to a zero gradient and hence zero current at those surfaces. Why is the current distribution so nonuniform (see Figure 4.14)? The answer is that there is much more area in which current can flow at the edges of the electrodes, and the current takes the path of least resistance. This path is a bit longer but less crowded. To illustrate, imagine that you and thousands of other people are trying to get into the stadium entrance to a sporting event. What do you do if you are in a hurry? You run along the outside of the crowd where things are not all jammed up with people, and then head for the outside edge of the entrance. That is what the current does and that is why the current at the edges is much higher. Basically, a much larger cross section of the electrolyte feeds the current on the outside of the electrodes.

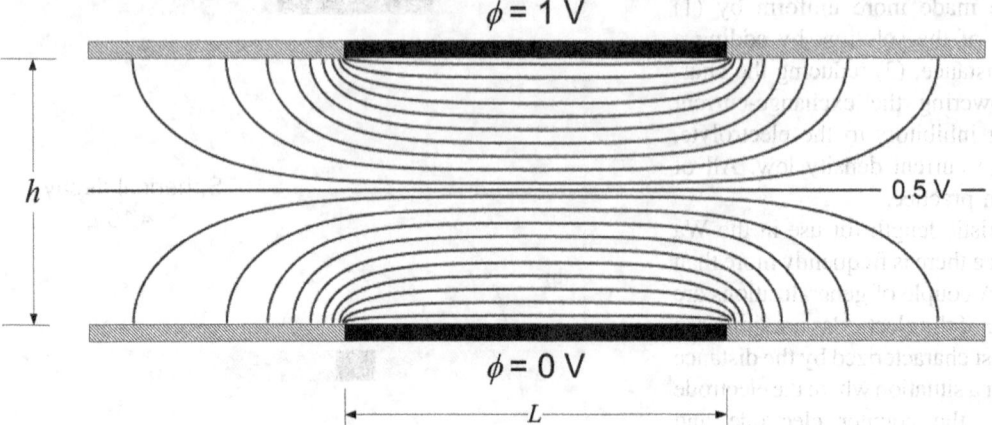

Figure 4.13 Lines of constant potential for the primary current distribution between two parallel plate electrodes. The distance between the electrodes (h) is 0.67 times the electrode width (L).

Figure 4.15 Current density between insulators and electrodes for primary current distributions.

A few general rules are illustrated by Figure 4.15. When the angle between an insulating surface and the electrode is less than 90°, the current density in the corner is zero. For an angle of 90°, the current is finite; for angles equal to or greater than 90°, the current density is infinite. This behavior has important implications for electrodeposition and electropolishing.

Secondary and Tertiary Current Distributions

For the secondary distribution, kinetic resistances of the electrode reactions are included. As we discussed previously, the finite kinetics has the effect of smoothing out the current distribution. Since we still assume a uniform concentration in the electrolyte, Ohm's law and Laplace's equation apply. The difference is that the boundary conditions treat finite kinetics for the electrochemical reactions. There are analytical solutions for a few secondary current distributions that take advantage of simplified kinetics. However, most problems for the secondary current distribution involve nonlinear boundary conditions that are more easily addressed numerically. While the details of the solution are beyond the scope of this section, it is useful to examine the results for the parallel plate problem as a function of Wa number (Figure 4.16). For a uniform current distribution, the local current density and the average current density are equal ($i/i_{avg} = 1$). As seen in the figure, the results for Wa = 30 are close to uniform. The figure also shows very little difference between the results for a Wagner number of 30 and those for Wa = 2.7. Thus, from an engineering perspective, efforts to increase Wa from 2.7 to 30 would not be warranted for this system. The nonuniformity of the current distribution at low Wa values is also apparent from the figure.

The normalized data in Figure 4.16 provide an accurate view of the current distribution. However, the price paid for uniformity is perhaps not obvious. Figure 4.17 shows the average current density that corresponds to each value of Wa. In this case, the average current was reduced by orders of magnitude in order to get a current distribution that was nearly uniform, with rapid changes occurring at very low Wagner numbers. The reduction in the average current means that, for example, the rate of electrodeposition has been dramatically reduced in order to make the current distribution more uniform. In situations where it is possible, it is advisable to start with a primary current distribution that is more nearly uniform in order to avoid the severe reductions in current observed here.

Finally, we have tertiary current distributions, which include concentration gradients and mass-transfer effects, which also impact the current distribution. Ohm's law can no longer be used and, for all intents and purposes, numerical solutions are required.

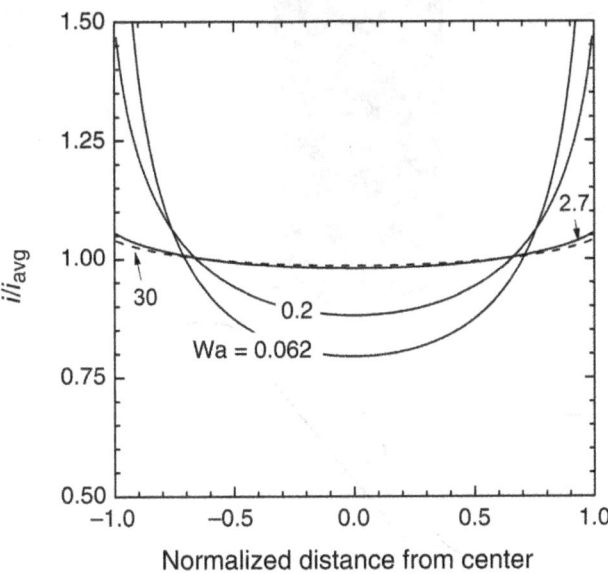

Figure 4.16 Primary current distribution between two parallel plate electrodes. The distance between the electrodes (h) is 1.25 times the electrode width (L). Geometry and conductivity.

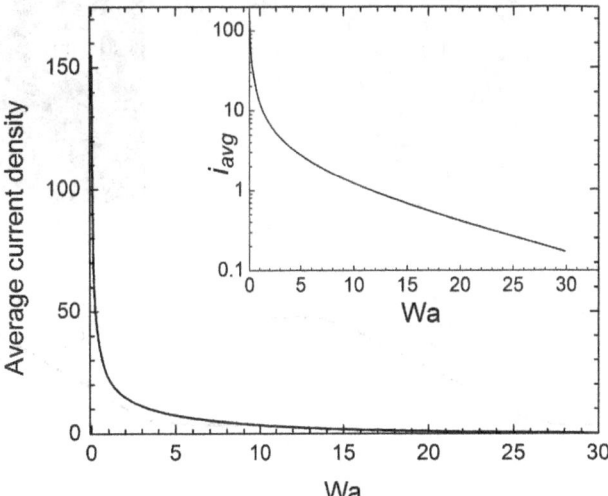

Figure 4.17 Impact of Wa on the average current for the conditions of Figure 4.16.

4.8 MEMBRANE TRANSPORT

Electrolyte and Electrode Separation

The electrodes in electrochemical cells are physically separated, and electrolyte occupies the space between the two electrodes. The manner in which this separation is implemented can vary and has significant impact on transport. Let's inspect the space between electrodes for three electrochemical systems to highlight the differences. The first system we consider is the electrorefining of copper as discussed earlier in this chapter. In this system, there is no physical separator between the electrodes, and the space between them is filled with electrolyte. Since the electrolyte does not conduct electrons, the electrodes are insulated from each other as they must be. Current flows in solution by the transport of ions as we have discussed throughout this chapter (Figure 4.18a).

The second system is a lithium-ion battery. For battery operation, it is critical to minimize losses in order to get as much power as possible from the battery. Lithium ions are shuttled between the two electrodes during charge and discharge. In order to minimize losses related to transport, we want the distance between electrodes to be as small as possible, say 25 μm. At the same time, the two electrodes cannot touch each other. We cannot reliably construct a large area cell with the two electrodes so close together without having them touch if we only have liquid electrolyte between the electrodes. Instead, a porous separator made from a nonconductive material (e.g., polymer) is used (Figure 4.18b). The electrolyte fills the pores and enables ionic transport, while the polymer separator material prevents contact between the two electrodes. Movement through the pores restricts transport; therefore, transport properties, conductivity, and diffusivity must be adjusted to account for volume fraction filled by the inert polymer and for the tortuous path through the separator. These adjustments are discussed in detail in Chapter 5.

Finally, the third type of system we consider is one that uses a *membrane* between the electrodes. An important example of a membrane-based system is the chlor-alkali process used to produce chlorine gas and caustic. Many fuel-cell systems also use membranes. In contrast to a porous separator film, a membrane has no open porosity. Rather, the molecular structure of the membrane permits selective transport of one or more species through the membrane while blocking other species. For example, the chlor-alkali membrane selectively allows cations to move through the membrane, but excludes anions. This feature is critical to an efficient chlor-alkali process (Chapter 14).

Ion Transport Through Membranes

When used as the separator in an electrochemical process, the membrane itself serves as an electrolyte to allow ionic current to flow between the electrodes. It represents an additional phase in our electrochemical system, and there is selective partitioning of species between the bulk

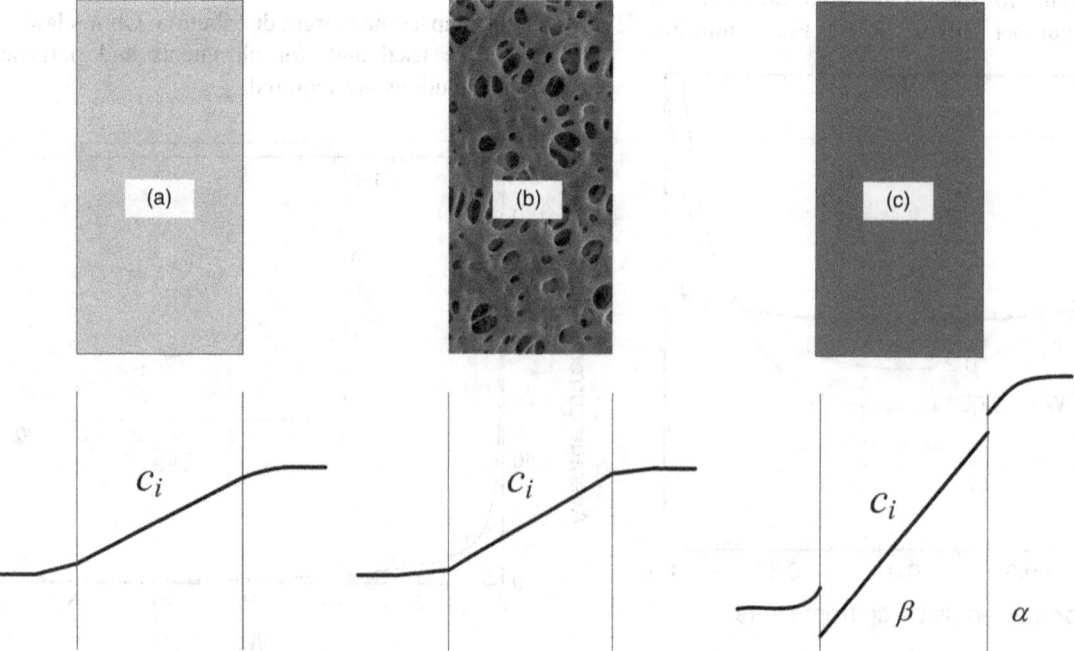

Figure 4.18 (a) Transport across a simple electrode gap filled with electrolyte. (b) Porous inert separator filled with electrolyte. (c) Solid membrane representing a separate phase.

electrolyte and separator. In other words, the concentration of a species in the membrane is different from, but related to, the concentration of that species in the bulk electrolyte at the interface between the membrane and the electrolyte. We will only consider situations where the interfacial compositions are in equilibrium and can be described by a thermodynamic relationship. A partition or distribution coefficient is used to provide the required relationship as described below.

The partition coefficient is simply the ratio of the concentration of species i in phase α divided by its concentration in phase β (see Figure 4.18c):

$$K_i = \frac{c_i^{\alpha}}{c_i^{\beta}}. \quad (4.66)$$

These values are usually determined experimentally and are different for different species. In general, as you might expect, concentrations in the nonporous membrane are lower than those in the bulk.

The concentration difference between the c_i^{β} values on either side of the membrane provides a driving force for transport. However, the potential field and charge interactions are also important for the transport of ions. Thus, ionic transport through the membrane can be quite complex and are not usually adequately described by the Nernst–Planck equation. The relationships needed to accurately describe transport in cation-exchange membranes that are commonly used in fuel cells are presented in Chapter 9.

Transport of Gases Through Membrane's Thin Liquid Films

An approach similar to the partition coefficient mentioned previously can be used to describe the transport of gases in films. Such gas-phase transport is much simpler than ion transport since charge interactions are not important. However, transport of gases through membranes can result in important losses of efficiency. For example, hydrogen gas in a fuel cell may transport across the membrane to the cathode and react directly with oxygen instead of reacting separately on the anode to produce power. For gases, equilibrium at the interface between the gas and the membrane is expressed with Henry's law:

$$y_i \equiv \frac{x_i H_i}{p}, \quad (4.67)$$

where x and y are the mole fractions in the membrane and gas phases, p is the total pressure, and H_i is Henry's law constant. As defined, $1/H_i$ describes the solubility of gas i in the membrane. A common physical situation is *solution and diffusion*, where a gas dissolves into and then diffuses across the membrane. To illustrate, we describe the transport of a gas species i. On the right side of the membrane the partial pressure is p_i, and on the left side it is zero.

Equation 4.10 is applied to the membrane, which at steady state reduces to

$$\frac{d^2 c_i}{dz^2} = 0, \quad (4.68)$$

where z is the position in the membrane in the "thickness direction." Integration gives

$$c_i = Az + B. \quad (4.69)$$

As for boundary conditions, we express the concentration of the gas in the membrane with Henry's law to yield

$$z = 0; \quad c_i = 0$$

and

$$z = \delta; \quad c_i = \frac{p_i c}{H_i},$$

where δ is the membrane thickness, c is the total molar concentration or molar density of the membrane, and we have made use of the fact that $x_i = c_i/c$. We can now determine the constants A and B to yield the following expression for the concentration as a function of position

$$c_i(z) = \frac{p_i c}{H_i} \frac{z}{\delta}. \quad (4.70)$$

The flux is determined with Fick's law:

$$J_i = \left[\frac{D_i}{H_i}\right] \frac{p_i c}{\delta}. \quad (4.71)$$

The rate of transport of gas through the membrane will depend both on its solubility ($1/H_i$) and the diffusivity of the species (D_i). Often, these properties are combined into a quantity called the *permeability*,

$$\text{permeability} = P_i = \text{diffusivity} \times \text{solubility} = \frac{D_i}{H_i}. \quad (4.72)$$

We also see that the rate of transport depends linearly on the difference in partial pressure of gas across the membrane, as expected. When the species must first go into solution and then diffuse across the membrane, these processes are combined and described as permeation.

CLOSURE

Transport in electrochemical systems by molecular diffusion and convection was introduced. Distinctive to electrochemical systems, the electric field plays a role in transport of ions. The Nernst–Planck equation was presented and several important simplifications were examined. Two important circumstances are the binary electrolyte and the case of an excess supporting electrolyte. For all intents and purposes, we treat the electrolyte as electrically neutral, which has the effect of coupling ion transport. For a binary electrolyte, there are three transport properties: electrical conductivity, diffusivity of the salt, and the transference number. Additionally, the significance of the current distribution was presented, which is influenced by geometry, kinetics, and mass transfer. The Wagner number is a dimensionless ratio of the ohmic and kinetic resistances. This dimensionless group provides a good starting point for assessing the uniformity of a current distribution. For transport of gases in liquids and membranes, the solubility and diffusivity can be combined to define the permeability; this quantity plays an important role in fuel cells and gas transport in membranes.

FURTHER READING

Bard, A.J. and Faulkner, L.R. (2001) *Electrochemical Methods*, John Wiley & Sons, Inc., Hoboken, NJ.

Cussler, E.L. (2009) *Diffusion: Mass Transfer in Fluid Systems*, Cambridge University Press.

Deen, W.M. (2011) *Analysis of Transport Phenomena*, Oxford University Press.

Levich, V.G. (1962) *Physicochemical Hydrodynamics*, Prentice-Hall, New Jersey.

Newman, J. and Thomas-Alyea, K.E. (2004) *Electrochemical Systems*, John Wiley & Sons, Inc., Hoboken, NJ.

Welty, J.R., Wicks, C.E., Wilson, R.E. and Rorrer, G. (2008) *Fundamentals of Momentum, Heat, and Mass Transfer*, John Wiley & Sons, Inc., Hoboken, NJ

PROBLEMS

4.1 Oxygen diffuses through a stagnant film as shown in the figure. At the electrode, oxygen is reduced to form water. Estimate the limiting current density, that is, when the concentration of oxygen at the electrode goes to zero, corresponding to the maximum flux of oxygen. The following data are provided: thickness of the film 5 μm, the diffusivity of oxygen is 2.1×10^{-10} m^2·s^{-1}, and the concentration of oxygen at the film surface in contact with gas is 3 mol·m^{-3}.

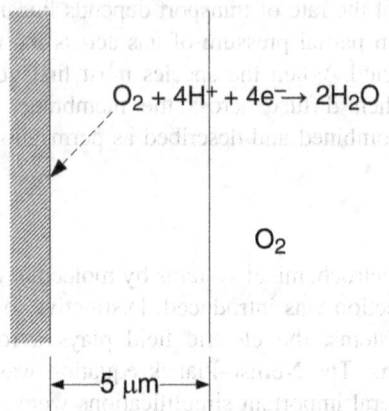

4.2 For a binary electrolyte, show that the electric field can be eliminated and Equation 4.19 results.

4.3 At the positive electrode of a lead–acid battery, the reaction is

$$PbO_2 + SO_4^{2-} + 4H^+ + 2e^- \leftrightarrow PbSO_4 + 2H_2O$$

(a) If the electrolyte is treated as a binary system consisting of H$^+$ and SO$_4^{2-}$, show that at the surface the current is given by

$$\frac{i}{F} = \frac{-2D}{2-t_+} \frac{\partial c}{\partial x}.$$

(b) What issues could arise from the formation of solid lead sulfate on the surface?

4.4 A porous film separates two solutions. The left side contains 2 M sulfuric acid, whereas on the right side there is 2 M Na$_2$SO$_4$. Discuss the rate of transport of sulfate ions across the membrane and the final equilibrium state.

4.5 The purpose of this problem is to compare the limiting current for the electrorefining of copper both with and without supporting electrolyte. Assume that copper is reduced at the left electrode ($x = 0$) and oxidized at the right electrode ($x = L$). You should also assume steady state and no convection. Finally, assume that the current efficiency for both copper oxidation and reduction are 100%. (Note that the assumption of no convection leads to very low values of the limiting current since transport by diffusion is slow over the 5 cm cell gap.)

(a) Derive an expression for the limiting current as a function of the diffusivity, the initial concentration of copper (which is also the average concentration), and the cell gap, L.

(b) Derive an analogous expression for the limiting current in the presence of a supporting electrolyte.

(c) How do the two expressions compare? Why is the limiting current lower in the presence of a supporting electrolyte?

(d) For the binary system, derive an expression for the concentration profile as a function of current for values below the limiting current. What is the sign of the current in this expression?

4.6 One type of Li-ion battery includes a graphite negative electrode and manganese dioxide spinel positive electrode, described by the following reactions (written in the discharge direction):

$$Li_xC_6 \rightarrow xLi^+ + xe^- + C_6 \quad \text{(negative)}$$
$$xLi^+ + xe^- + Mn_2O_4 \rightarrow Li_xMn_2O_4 \quad \text{(positive)}$$

The electrolyte consists of an organic solvent that contains a binary LiPF$_6$ lithium salt, where the anion is PF$_6^-$. Assume a one-dimensional cell with the cathode located at $x = 0$ and the anode located at $x = L$. The following properties are known:

$$D_{Li^+} = 1.8 \times 10^{-10} \text{ m}^2 \text{ s}^{-1},$$
$$D_{PF_6^-} = 2.6 \times 10^{-10} \text{ m}^2 \text{ s}^{-1},$$
$$t_+ = 0.4.$$

For our purposes here, we assume that the electrodes are flat surfaces, and that they are separated by an electrolyte-containing separator that is 25 μm thick. The initial concentration of LiPF$_6$ in the electrolyte is 1.0 M. The parameters given in the question are for transport in the separator. The same expression derived in the text for the concentration also applies to this situation, Equation 4.39.

(a) In which direction does the current in solution flow during discharge? Is this positive or negative relative to the x-direction?

(b) Where is the concentration highest, at $x = 0$ or $x = L$?

(c) What is the concentration difference across the separator if a cell with a 2 cm × 5 cm electrode is operated at a current of 10 mA? Is this difference significant?

(d) Briefly describe how you would calculate the cell potential for a constant current discharge of this cell if the current is known and the concentration variation is known. The potential that would be measured between the current collectors of the cathode and anode during discharge. You do *not* need to include all the equations that you would use. The important thing is that you know and are able to identify the factors that contribute to the measured cell potential, and that you know the process by which you might determine their values.

4.7 A potential step experiment is conducted on a solution of 0.5 M $K_3Fe(CN)_6$, and 0.5 M Na_2CO_3. The potential is large enough so that the reduction of ferricyanide is under diffusion control. Using the data provided, estimate the diffusion coefficient of the ferricyanide.

$$[Fe(CN)_6]^{3-} + e^- \rightarrow [Fe(CN)_6]^{4-} \quad (U^\theta = 0.3704 \text{ V})$$

Time (s)	Current density (A·m^{-2})
1	731
1.7	564
2.8	438
4.6	340
7.7	263
12.9	201
21.5	156
35.9	122
59.9	94
100	72

4.8 Platinum is used as a catalyst for oxygen reduction in low-temperature, acid fuel cells. At high potentials, the platinum is unstable and can dissolve (see Problem 2.17), although the concentration of Pt^{2+} is quite small. The electrolyte conductivity is 10 S·m^{-1}, and the current density, carried by protons, is 100 A·m^{-2}. Assuming that the temperature is 80 °C and the Pt^{2+} is transported over a distance of 20 μm, is it reasonable to neglect migration when analyzing the transport of platinum ions? Why or why not?

4.9 Calculate the diffusivity, transference number, and conductivity of a solution of 0.05 M KOH at room temperature.

4.10 Estimate the conductivity of a 0.1 M solution of $CuSO_4$ at 25 °C.

4.11 An electrochemical process is planned where there is flow between two parallel electrodes. In order to properly design the system, an empirical correlation for the mass-transfer coefficient is sought. An aqueous solution containing 0.05 M $K_4Fe(CN)_6$, 0.1 M, M $K_3Fe(CN)_6$, and 0.5 M Na_2CO_3 is circulated between the electrodes. The reactions at the electrodes are

$$[Fe(CN)_6]^{3-} + e^- \leftrightarrow [Fe(CN)_6]^{4-} \quad (U^\theta = 0.3704 \text{ V})$$

(a) If the potential difference between electrodes is increased slowly from zero, sketch the current–voltage relationship that would result. Include on the graph, the equilibrium potential for the reaction, the open-circuit potential, the limiting current, and the decomposition of water.

(b) For the conditions provided, which electrode would you expect to reach the limiting current first?

(c) Show how to calculate a mass-transfer coefficient, k_c, from these limiting current data.

4.12 For the system described in Problem 4.11,

(a) Use the data provided for I_{lim} versus flow rate to develop a correlation for dimensionless mass-transfer coefficient,

Sh. The Re should be defined with the hydraulic diameter, $d_h = 4A_c/P$. At 298 K, the diffusivity of the ferri/ferrocyanide is 7.2×10^{-10} m^2·s^{-1}. Treat the electrodes as 0.1 m × 0.1 m squares separated by a distance of 1 cm. Use density as 998 kg·m^{-3} and a viscosity of 0.001 Pa·s. Based on the literature, the following correlations are suggested.

Laminar flow $\text{Sh} = a\,\text{Re}^{1/3}\text{Sc}^{1/3}\left(\dfrac{d_h}{L}\right)^{1/3}$,

Turbulent flow $\text{Sh} = b\,\text{Re}^{2/3}\text{Sc}^{1/3}\left(\dfrac{d_h}{L}\right)^{1/4}$.

(b) What is the importance of the sodium carbonate in this experiment?

(c) Justify the need for the additional dimensionless factor (d_h/L).

Flow rate [cm^{3}s^{-1}]	Limiting current [A]
0.96	0.15
2.06	0.19
3.69	0.22
6.41	0.27
9.16	0.30
11.89	0.34
14.62	0.36
17.35	0.39
20.22	0.41
22.83	0.44
25.54	0.46
28.56	0.50

4.13 For the Cu electrorefining problem (see Illustration 4.7),

(a) Calculate the mass-transfer coefficient, k_c [m·s^{-1}], based on natural convection using the height of the electrode as the characteristic length.

(b) It is desired to increase the mass-transfer coefficient by a factor of 10 and thereby raise the limiting current. If forced convection is used, what fluid velocity and Re are required? The following correlations for mass transfer between parallel planes are available. The distance between the electrodes is 3 cm. Assume that the electrodes are square.

Laminar flow $\text{Sh} = 1.85\left(\text{ReSc}\dfrac{d_h}{L}\right)^{0.333}$,

Turbulent flow $\text{Sh} = 0.0789\left[\dfrac{0.079}{\text{Re}^{0.25}}\right]^{0.5}\text{ReSc}^{0.25}$.

Note that in contrast to the correlation for natural convection, the Sh and Re here are based on the equivalent diameter $d_h = 4A_c/P$.

(c) For this arrangement, sparging with air at a rate of 2 L/min per square meter of electrode area results in a mass-transfer coefficient on the order of 2×10^{-5} m·s^{-1}. Calculate the superficial velocity of the air and compare with the velocity obtained in part (b). Why might air sparging be preferred over forced convection?

4.14 Correlations for mass-transfer coefficients for full-sized, commercial, electrowinning cells under the actual operating conditions can be difficult to measure.

(a) What could be some of the challenges with measuring these mass-transfer coefficients?

(b) It has been suggested (*J. Electrochem. Soc.*, **121**, 867 (1974)) that k_c can be estimated by codeposition of a trace element that is more noble. For instance, in the electrowinning of Ni ($U^\theta = 0.26$ V), Ag ($U^\theta = 0.80$ V) could be used. The idea is that because the equilibrium potential for the more noble material is higher, the limiting current will be reached sooner. After a period of deposition, the electrode composition is analyzed to determine the local and average rates of mass transfer. Would the mass-transfer coefficient for Ni be the same as for Ag? If not, how would you propose correcting the measured value?

4.15 A porous flow-through electrode has been suggested for the reduction of bromine in a Zn–Br battery.

$$Br_2 + 2e^- \rightarrow 2Br^-$$

Calculate the limiting current if the bulk concentration of bromine is 5.5 mM using the correlation

$$Sh = 1.29 Re^{0.72}$$

The Re is based on the diameter of the carbon particles, d_p, that make up the porous electrode.

$D_{Br_2} = 6.8 \times 10^{-10}$ m^2 s^{-1}, $v = 0.2$ cm s^{-1}, $d_p = 25\ \mu$m,
$\nu = 9.0 \times 10^{-7}$ m^2 s^{-1}, $c_{Br_2}^\infty = 5.5$ mol m^{-3}.

4.16 Chlorine gas is being evolved on the anode of a cell. $L = 0.5$ m, $W = 0.5$ m, and $h = 0.03$ m. The electrode is operating at a uniform current density of 1000 A·m^{-2}.

(a) Using the geometry from Figure 4.7, relate the superficial velocity of the gas between electrodes as a function of x, distance from bottom of the electrode.

(b) If the bubbles have a diameter of 3 mm, estimate the void fraction of gas, ε, between the electrodes. Use a surface tension of 0.072 J·m^{-2} and a density of 997 kg·m^{-3}. Assume the bubbles travel with a terminal velocity given by

$$v_t = \sqrt{\frac{2\sigma}{d_b \rho_L} + \frac{g d_b}{2}}.$$

(c) Compare estimates for the mass-transfer coefficient from Equations 4.37 and 4.38. Use a diffusivity of 1×10^{-9} m^{-2}·s^{-1}. The viscosity of the solution is 0.00089 Pa·s.

4.17 The Hull cell is used to assess the ability of a plating bath to deposit coatings uniformly; this is known as the "throwing power" of the bath. The cell shown below has two electrodes that are not parallel. The other sides of the cell are insulating. The electrodeposition of a metal is measured on the long side (cathode), and the more uniform the coating, the better the throwing power.

Sketch the potential and current distribution assuming a primary current distribution.

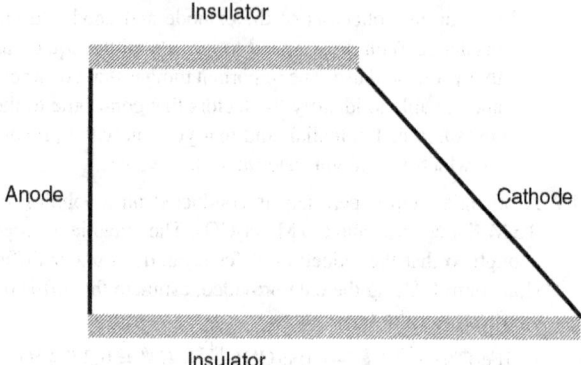

4.18 Please comment on the effect that moving the counter electrode farther away from the working electrode has on the primary and secondary current distributions. Also, name at least one advantage and one disadvantage of doing so.

4.19 Electropolishing is an electrochemical process to improve the surface finish of a metal whereby rough spots are removed anodically. Sometimes this leveling process is characterized as acting on *macroroughness* or *microroughness*. Assume that there is a mass-transfer boundary layer on the surface.

(a) Propose a mechanism for the leveling of macroroughness, specifically where the thickness of the boundary layer is small compared to variations in the roughness.

(b) When silver is electropolished in a cyanide bath, CN$^-$ ions are needed at the surface to form Ag(CN)$_2^-$. Transport of the cyanide ion to the surface may limit the rate of dissolution of silver. Sketch how the current density for anodic dissolution of silver changes with the anode potential.

(c) With microroughness, the variation in thickness of the surface is less than that of the boundary layer. How would the diffusion of the cyanide ion to the surface affect the electropolishing process?

4.20 Derive Equations 4.64 and 4.65.

4.21 Suppose that you have an electrolysis tank that contains 10 pairs of electrodes. The tank is 75 cm long and the gap between electrodes is 4 cm. The tank is 26 cm wide, and the width of the electrodes is 20 cm. The distance between the electrodes and each side of the tank is 3 cm. The temperature is 25 °C.

The tank is used for the electrorefining of copper ($i_o = 0.01$ A·m^{-2}, $\alpha_a = 1.5$, $\alpha_c = 0.5$) at an average current density of 250 A·m^{-2}. The composition of the solution is 0.7 M CuSO$_4$ and 1 M H$_2$SO$_4$. The following properties are known:

$$D_{Cu^{2+}} = 0.72 \times 10^{-9}\ m^2\ s^{-1},$$
$$D_{SO_4^{2-}} = 1.0 \times 10^{-9}\ m^2\ s^{-1},$$
$$D_{H^+} = 9.3 \times 10^{-9}\ m^2\ s^{-1}.$$

Would you expect the secondary current distribution to be uniform or nonuniform? Please support your response quantitatively.

4.22 Use a charge balance on a differential control volume to show that $\nabla \cdot i = 0$ at steady state.

4.23 Permeation of vanadium across an ionomer separator of a redox flow battery should be kept small for high efficiency. In contrast to a simple porous material, for an ionomer membrane separator there can be partitioning between the bulk solution and the ionomer membrane (see Figure 4.18c). Consider the case when the current is zero.

  (a) If this equilibrium is represented by a partition coefficient, $K = c_i^{membrane}/c_i^{bulk}$, what is the expression for the molar flux across the separator analogous to Fick's law?

  (b) The product of solubility and diffusivity is called permeability, here equal to DK. For a permeability of 4×10^{-11} m$^2 \cdot$s^{-1}, estimate the steady-state flux of vanadium across a membrane that is 100 μm thick. On one side, the concentration of V is 1 M, whereas on the other side assume the concentration is zero.

  (c) This flux results in a loss of current efficiency for the cell. If the valence state of vanadium changes by one, what current density does this flux represent?

4.24 It is possible to obtain both the diffusivity and the solubility of a diffusing species from one experiment. The experiment involves establishing the concentration on one side of the membrane and measuring the total amount of solute that is transported across the membrane as a function of time. Assume that at the start of the experiment, there is no solute in the membrane.

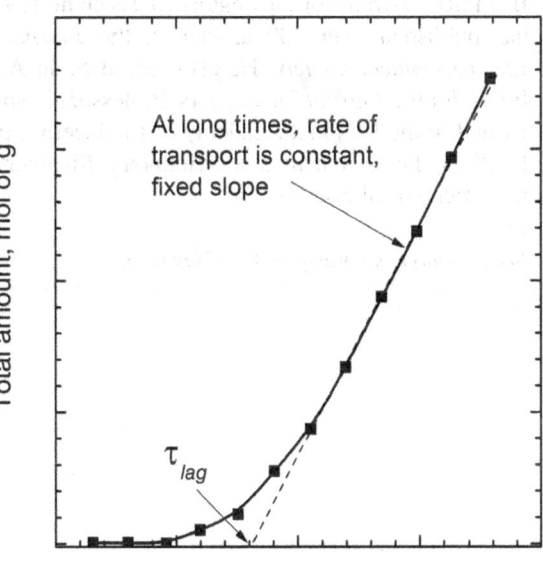

Time [s]	Total flux [mol·m^{-2}]
45.9	0.0
90.0	0.0
136.9	0.0
179.3	0.5
226.0	1.1
270.9	2.8
315.8	4.4
359.0	6.7
403.5	9.4
446.6	11.9
489.6	14.7
536.9	17.7

  (a) Sketch concentration across the membrane as a function of time. Your sketch should include the initial concentration, and the pseudo-steady-state profile, that is, when the flux becomes constant.

  (b) Using the data provided, determine the permeability.

  (c) The time lag can be estimated, and the diffusivity calculated from the formula $\tau_{lag} = L^2/6D$. Using these data, estimate both the diffusion coefficient and the partition coefficient for the solute. The concentration on one side is 1 M, and the thickness of the membrane is 400 μm.

4.25 The relative permeability of oxygen and hydrogen, $\alpha = \frac{(D/H)_{O_2}}{(D/H)_{H_2}}$, in a membrane is measured to be 0.4, where H_i is Henry's law constant, representing the equilibrium between the gas phase and the ionomer defined in Equation 4.67. If air is on one side of the membrane and pure hydrogen on the other, both at atmospheric pressure, and the thickness of the membrane is L,

  (a) Estimate the relative rate of hydrogen permeation and oxygen permeation across the membrane film

  (b) If there is a catalyst in the membrane where oxygen and hydrogen react instantaneously to form water, sketch the concentration of hydrogen and oxygen across the film.

  (c) Derive an expression that describes at what location in the membrane this reaction occurs? How will your answer change if the air is replaced with pure oxygen?

4.26 Calculate the Wagner number for the Zn and Ni electrodes discussed in Section 3.4 ($i = 4640$ A·m^{-2}). Comment on the values that you obtain and the relative magnitude of the kinetic and ohmic resistances that you calculated for this system. Are the values consistent with your expectations? Why or why not?

JOHN NEWMAN

John Newman was born in Richmond, Virginia, in 1938. He earned a B.S. in Chemical Engineering from Northwestern; and during this period he was also a co-op student at Oak Ridge National Laboratory. In 1960, he entered the University of California, Berkeley, to study under Professor Charles Tobias. He obtained his M.S. in 1962 and his Ph.D. in 1963. His master thesis was on the analysis and behavior of porous electrodes, which is the topic of this chapter. His doctoral dissertation investigated laminar flow past a cylinder at high Reynolds number. During his graduate studies, he also helped prepare the English edition of Levich's book, *Physicochemical Hydrodynamics*, published in 1962.

Following his Ph.D., he joined the faculty in Chemical Engineering at U.C. Berkeley. He won the Young Author's prize from the Electrochemical Society for work in current distributions on a rotating disk below the limiting current in 1966 and a *second* time in 1969 with William Parish on modeling channel electrochemical flow cells. Dr. Newman was promoted to full professor in 1970. The bulk of his work is contained in his monograph, *Electrochemical Systems*, published in 1973. The first edition has been translated into Russian and Japanese. In 1975, Professor Newman (along with a long-time collaborator, Bill Tiedemann) published one of his most influential papers, "Porous-Electrode Theory with Battery Applications," *AIChE J.* **21**, 25 (1975). These details were incorporated into the second edition of his book, which was published in 1991; a third edition came out in 2004.

Professor Newman has been the driving force in both educating a generation of electrochemical engineers and the development of mathematical models to describe numerous electrochemical systems. It would be unfair to view his work as being defined only by his additions to the theory of porous electrodes, but without a doubt his role in developing the theory and applying it to electrochemical systems is one of his most important and lasting contributions. These concepts have been instrumental for design and scale-up of fuel cells, batteries, electrochemical double-layer capacitors, and flow-through electrodes.

Professor Newman has advised more than 70 graduate students and about 20 of these have gone on to academic careers. Each of his mentees has had the utmost respect for his penetrating insight into complex problems and the thoroughness of the analysis he demanded of the students and himself.

Professor Newman has received every major award of the Electrochemical Society. Most recently, the Acheson Award, the highest honor of the Society, was presented in 2010. Other significant awards include the Vittorio di Nora Award (2008), Olin Palladium Award (1991), and Henry B. Linford Award for Distinguished Teaching (1990). He has published over 200 articles in the *Journal of the Electrochemical Society*. He also served as an Associate Editor for the *Journal* for 10 years. Professor Newman was elected to the National Academy of Engineering in 1999. In 2013, he was named an Honorary Member of the Electrochemical Society.

Image Source: Courtesy of John Newman.

Chapter 5

Electrode Structures and Configurations

Rather than the planar electrodes that have been the focus of earlier chapters, electrodes for electrochemical systems are often three-dimensional structures. The primary driving force behind three-dimensional electrodes is to increase the reaction area per volume in order to facilitate heterogeneous electrochemical reactions. Other benefits include providing storage volume for solid reactants, supporting dispersed catalysts, establishing a so-called *three-phase boundary*, and efficiently evolving gases, to name but a few.

One example of a porous electrode is the electrochemical double-layer capacitor (Chapter 11). Here, the capacitance is proportional to the surface area, which can be augmented by two or more orders of magnitude with a porous electrode. Without the high surface area made possible by porous electrodes, these double-layer capacitors would be of limited commercial interest.

A second illustration of an effective porous electrode is the positive electrode of the lead–acid battery, which is the battery used for starting automobiles. The discharge reaction for this positive electrode is

$$PbO_2(s) + SO_4^{2-} + 4H^+ + 2e^- \rightarrow PbSO_4(s) + 2H_2O$$

The interfacial area between phases is greatly increased by the use of a porous electrode. A greater surface area lowers the current density at the surface, which reduces the kinetic polarization and increases the power output of these batteries enormously. The porous structure also provides other important advantages for this electrode. During discharge, solid lead oxide reacts with the electrolyte to form another solid, lead sulfate, whose density is greater than that of the original lead oxide. The porous electrode provides the volume needed for accumulation of the reaction product, and accommodates changes in the solid volume of the electrode. The small pore dimensions also reduce the diffusion path length that is important for precipitation (discharge) and subsequent dissolution (charge) of the $PbSO_4$, leading to enhanced performance. Additional information on battery applications is provided in Chapters 7 and 8.

Another example of a porous electrode is the cathode of a phosphoric acid fuel cell. The reaction at the cathode is

$$O_2(g) + 4H^+ + 4e^- \rightarrow 2H_2O(g)$$

where the oxygen is in the gas phase, the protons are found in the electrolyte, and electrons are from the solid phase. In this case, a porous electrode is used to provide the three interpenetrating phases (gas, liquid, and solid) needed for the desired reaction to take place. The actual situation is complicated further by the need for a catalyst to facilitate oxygen reduction reaction in acid, which is notoriously sluggish if not catalyzed. The catalyst is typically supported on high-surface-area carbon to reduce the kinetic polarization. Transport of oxygen to reaction sites can also be important. Finally, creating a stable interface between the gas and the liquid electrolyte is perhaps the principal challenge, and is discussed further in Section 5.5, as well as in Chapter 9 on fuel cells.

The objective of this chapter is to establish the basic principles that govern the behavior of porous electrodes. We first consider a simplified porous electrode, consisting of straight cylindrical pores, and derive the relationships that we will use for analysis of three-dimensional electrodes. This analysis is followed by a discussion of the physical characteristics of porous electrodes. We then examine the current distribution in porous electrodes, which is critical to the design and optimization of these electrodes. Porous electrodes containing a gas phase are then considered. Finally, we consider flow-through electrodes for reaction under mass-transport control.

Electrochemical Engineering, First Edition. Thomas F. Fuller and John N. Harb.
© 2018 Thomas F. Fuller and John N. Harb. Published 2018 by John Wiley & Sons, Inc.
Companion Website: www.wiley.com/go/fuller/electrochemicalengineering

5.1 MATHEMATICAL DESCRIPTION OF POROUS ELECTRODES

As we examine porous electrodes, we first consider a simple geometry consisting of straight cylindrical pores in an electrically conductive matrix as shown in Figure 5.1. The idea behind using a three-dimensional porous structure is to increase the amount of surface area in a given electrode volume. Let's assume that we have an electrode that is 10 cm × 10 cm and 1 mm thick. The electrode is filled with an array of straight pores as shown in the figure. Assume that each pore has a diameter of 10 μm, and that the minimum distance between pores is 2.5 μm. That would give us 64 million pores in the 10 cm × 10 cm area. The internal surface area of each pore is about 0.0004 cm^2. This single pore area results in a total pore surface area greater than 25,000 cm^2, which is more than 250 times greater than the flat area of the original electrode. The area enhancement can be even higher for practical porous electrodes.

Previously, we assumed that the electrochemical reactions took place on the surface of the electrode and we treated the reactions appropriately as boundary conditions. The analysis of a porous electrode, however, is a bit more complex since the reaction takes place throughout the volume of the electrode. We could use our previous approach and perform a three-dimensional calculation on the actual geometry, including the individual pores, and apply the reaction boundary condition on each pore wall. Fortunately, a much more effective method, referred to as *porous electrode theory*, has been developed extensively by Newman. We will illustrate the method in one dimension since in most applications it is only variations in the thickness direction that are important. Situations that require multidimensional treatment are beyond the scope of this chapter.

Let's begin with a discussion of the physical system of interest. As shown in Figure 5.1, we have a porous electrode attached to a *current collector*. We assume that the current collector is at a constant potential, and that the current density at the current collector is uniform. These assumptions are consistent with the one-dimensional treatment mentioned previously and, as a result, we don't need to model the current collector. Our spatial coordinate system is defined as $x = 0$ at the *back* of the porous electrode (the side of the electrode in contact with the current collector) and $x = L$ at the other side of the electrode, where L is the thickness of the electrode. Consistent with the notation introduced previously, i_1 and ϕ_1 are the current density and potential in the solid, electron-conducting phase; and i_2 and ϕ_2 represent the current density and potential in the electrolyte phase. The electrolyte fills the cylindrical pores. For the purposes of discussion, we assume that the electrode is the anode, although the relationships developed apply equally well to a cathode. The current associated with charging of the double layer is not considered in the following treatment. The extension of porous electrodes to electrochemical double-layer capacitors is made in Chapter 11.

At the back of the electrode ($x = 0$), all of the current is in the solid phase. Therefore,

$$i_1 = I/A \text{ at } x = 0, \tag{5.1}$$

where I/A is the specified current density in A·m^{-2}, and i_1 is based on the superficial area, A, of the electrode (100 cm^2 = 0.01 m^2 for this case). At the other end of the electrode (front), the solid phase ends and all the current must be transferred in the electrolyte. Therefore,

$$i_2 = I/A \quad \text{at} \quad x = L. \tag{5.2}$$

In between $x = 0$ and $x = L$, the current is split between the solid and liquid phases. The following relationship must hold at any point in the electrode:

$$i_1 + i_2 = I/A. \tag{5.3}$$

How does the current change from being entirely in the solid at the back of the electrode to being entirely in the electrolyte at the front of the electrode? The current varies due to electrochemical reaction at the interface between the solid and electrolyte phases, just as it did for the flat electrodes that we have considered up to now. In fact, we can use the same type of kinetic expression (e.g., linear, Tafel, or BV) that we used earlier to describe the reaction. The difference is that the reaction area is spread throughout the volume of the electrode in a porous electrode. That area is the surface area of the cylinders in our simplified electrode model (Figure 5.1). As we prepare to write a charge balance for the electrode, it is convenient to define a volumetric charge generation term,

$$\text{Volumetric charge generation rate} = ai_n, \tag{5.4}$$

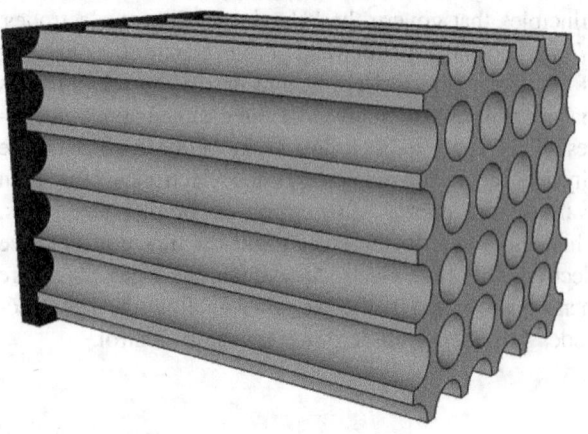

Figure 5.1 Straight pores of an idealized porous electrode.

where i_n is the current density normal to actual surface (A·m^{-2}) and a is the surface area per volume (m^{-1}) or specific interfacial area. The volumetric charge generation rate has units of [C·m^{-3}·s^{-1})] or [A·m^{-3}]. The current density, i_n, is the same current density that we have discussed previously in Chapter 3 and can be represented by, for example, a Tafel or BV expression. The surface area to which i_n corresponds is the interfacial area between the solid and electrolyte, and is equal to the combined surface area of the 64 million cylindrical pores in our example. The specific interfacial surface area, a, in our example is equal to the combined surface area divided by the superficial volume of the electrode:

$$a = \frac{\text{combined suface area}}{\text{superfical volume}} = \frac{2.5 \text{ m}^2}{0.1 \text{ m} \times 0.1 \text{ m} \times 0.001 \text{ m}}$$
$$= 250{,}000 \text{ m}^{-1}$$

We are now prepared to write charge balances for the current in solution (i_2) and the current in the solid phase (i_1). The balance for i_2 is

$$-\nabla \cdot i_2 + a i_n = 0. \quad (5.5)$$

The positive sign on the generation term is consistent with the convention that anodic current is positive and represents the flow of current into the electrolyte. A similar expression can be written for i_1,

$$-\nabla \cdot i_1 - a i_n = 0. \quad (5.6)$$

Note that the generation term is negative since charge leaves the solid phase when i_n is positive. It follows that

$$\nabla \cdot i_1 + \nabla \cdot i_2 = 0. \quad (5.7)$$

It is important to remember that both i_1 and i_2 are based on the same superficial area (10 cm × 10 cm in our example). Also, note that only two of these three equations are independent.

To describe completely the electrode, we also need material balance equations for the species in solution. In Chapter 4, a general material balance was provided.

$$\frac{\partial c_i}{\partial t} = -\nabla \cdot N_i + \mathcal{R}_i. \quad (4.10)$$

This equation needs to be modified since our control volume now includes the electrolyte and solid phases. The term on the left is the rate of accumulation of species i in our control volume. We will continue to let c_i represent the actual concentration in the electrolyte (moles/volume of electrolyte). In the porous medium, only a fraction of the volume is occupied by the electrolyte; therefore, the amount of species i per control volume is $c_i \varepsilon$, where ε is the volume fraction of the electrolyte. Furthermore, since the porosity may change with time and position, we must keep this term in the derivative. The flux, N_i, is the superficial molar flux and does not require a change in the differential material balance. Nevertheless, when the flux is expressed with the Nernst–Planck equation, for instance, effective transport properties must be used. Similar to our treatment of the flux, $\mathcal{R}_i$ is the generation rate based on the superficial volume that includes both solid and electrolyte phases. The form of the material balance for porous electrodes becomes

$$\frac{\partial \varepsilon c_i}{\partial t} = -\nabla \cdot N_i + \mathcal{R}_i. \quad (5.8)$$

Heterogeneous reactions in the porous electrode affect the individual balances of reacting species, which are either consumed or generated by reaction at the surface. In the present macroscopic approach, we account for these terms as volumetric generation terms, similar to what was done above for the charge balance. Therefore, if we neglect generation terms other than those related to the electrochemical reaction:

$$\mathcal{R}_i = a j_n, \quad (5.9)$$

where j_n is the species reaction rate at the surface in moles/(time–area). This rate is related to the current at the surface through Faraday's law:

$$j_n = -\frac{s_i}{nF} i_n. \quad (5.10)$$

Note that j_n is positive when the current is positive (anodic) and the species of interest is a product of the anodic reaction ($s_i < 0$). You should review the definition of s_i and convince yourself that this is true. The species equation now becomes

$$\frac{\partial \varepsilon c_i}{\partial t} = -\nabla \cdot N_i - a \frac{s_i}{nF} i_n = -\nabla \cdot N_i - \frac{s_i}{nF} \nabla \cdot i_2. \quad (5.11)$$

This step completes the balance equations needed to describe a porous electrode. The unknowns are c_i for each of n species, ϕ_1 (potential in solid), and ϕ_2 (potential in electrolyte), for a total of $n + 2$. Thus, $n + 2$ equations are required: two charge balance equations, $n - 1$ species balances, and the electroneutrality equation.

We still need to write the equations that relate the currents and fluxes to their respective driving forces in order to substitute these into the relevant material balances. We will use the Nernst–Planck equation, although a similar procedure is suitable for other descriptions of the species flux. The expression after adjustment for the porous nature of the electrode is

$$N_i = -\varepsilon D_i \nabla c_i - \varepsilon z_i u_i F c_i \nabla \phi + \varepsilon c_i \mathbf{v}, \quad (5.12)$$

where **v** is the actual molar-averaged velocity in the pores. Remember, the flux is based on the superficial area of the porous electrode, and not on the actual pore area. The porosity in the first two terms of the flux expression is usually combined with an additional correction to yield effective transport properties. Finally, ε and **v** are sometimes combined into a superficial velocity, which is based on the superficial area of the electrode and is essentially the volumetric flow rate divided by the superficial area. Effective properties are described below after a brief discussion of the characteristics of porous electrodes in order to broaden our view beyond the simplified electrode geometry considered previously. Note, however, that the material balances derived above are completely general and not restricted to the illustrative geometry consisting of straight cylindrical pores.

We have identified the equations necessary to describe the behavior of the porous electrode. Next, the methods of characterizing porous electrodes are presented, followed by a look at specific examples.

5.2 CHARACTERIZATION OF POROUS ELECTRODES

An idealized porous electrode with straight cylindrical pores was considered above. Next our aim is to describe and characterize porous media in a way that is broadly applicable. The empty space of the electrode available to the electrolyte (or gas phase) is the void volume fraction or porosity, ε. This parameter is the most important feature of any porous media.

$$\varepsilon = \frac{\text{total volume} - \text{volume occupied by solid phase}}{\text{total volume}}. \quad (5.13)$$

Initially, we assume that this void space is completely filled by the electrolyte. In our general description, we treat the porous medium as a continuum; that is, we'll take a macroscopic point of view and do not try to describe the microscopic details. The key to this approach is that the size of the electrode is large compared to the features of the pores. The solid and electrolyte phases are therefore treated as two interpenetrating continua as shown in Figure 5.2. Additional phases can also be considered. For example, a gas phase is important for many fuel cells and for some batteries. Multiple solid phases are also possible. Thus, our two continua model can be readily expanded to include three or more interpenetrating phases.

As mentioned earlier, one key objective of porous electrodes is to increase the interfacial area between the two phases. Thus, a second key characteristic of the porous

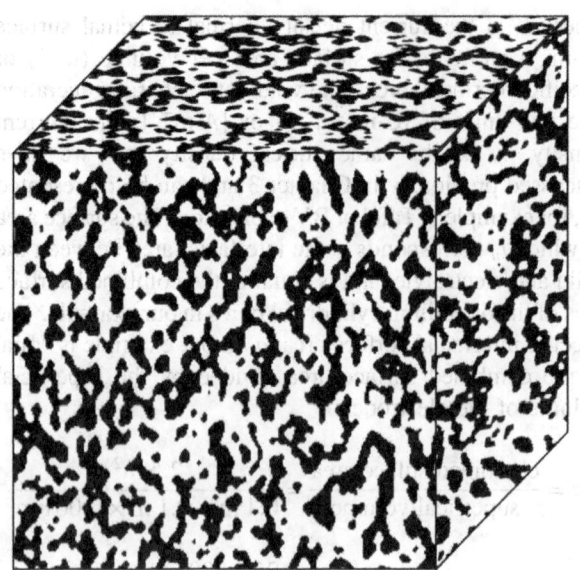

Figure 5.2 Typical porous electrode.

media is the specific interfacial area, a.

$$a \equiv \frac{\text{interfacial area}}{\text{superficial volume}}. \quad (5.14)$$

This quantity, which was introduced previously for our cylindrical pore electrode, has units of [m^{-1}]. An electrode geometry that is more representative of electrodes used in practice is that formed from a packed bed of spherical particles, where the interstitial space between particles forms the void volume, ε. A characteristic length between spheres, analogous to a hydraulic diameter, can be used to describe the pore size. Although still simplified, let's explore this construct in a bit more depth. For a single sphere of radius r, the specific interfacial area is $\frac{4\pi r^2}{4/3 \pi r^3} = \frac{3}{r}$. In general, the specific interfacial area increases as the features of the porous electrode become smaller. Without worrying about the exact details of how the spherical particles pack together, we can express the specific interfacial area with

$$a = \frac{(\text{number of spheres})4\pi r^2}{\text{volume}} = \frac{\frac{V(1-\varepsilon)}{4/3 \pi r^3} 4\pi r^2}{V} = \frac{3(1-\varepsilon)}{r}. \quad (5.15)$$

We see that the specific interfacial area for this packed bed depends not only on the size of the spheres, but also on the number of spheres as reflected in the solid volume fraction $(1-\varepsilon)$.

The characteristic pore size, r_p, can be estimated as follows:

$$r_p = \frac{\text{cross section available for transport}}{\text{wetted perimeter}}$$

$$r_p = \frac{\dfrac{\text{volume of voids}}{\text{bed volume}}}{\dfrac{\text{wetted surface}}{\text{bed volume}}} = \frac{\varepsilon}{a} = \frac{\varepsilon}{1-\varepsilon}\left(\frac{r}{3}\right). \qquad (5.16)$$

$$= \frac{\text{volume available for transport}}{\text{wetted surface}}$$

Although simplified, this analysis helps us to develop some intuition into the behavior of porous media. As the void volume gets small, the pore diameter also decreases. For larger porosity, the pore diameter will be larger.

A typical porous electrode has a distribution of pore sizes rather than the single pore size that characterizes our idealized electrode. Similarly, there is also a distribution of particle sizes in the electrode. The detailed structure of the pores controls the properties; however, in general we won't attempt to treat the geometric minutiae. Instead, we will take the specific interfacial area as a measurable quantity of the porous media or electrode. An example distribution of the pore volume is shown in Figure 5.3, which was measured with Hg-intrusion porosimetry. Here, larger pores are easier to fill than small ones. Since the pores don't fill uniformly, we can relate the volume of Hg and required pressure to pore diameter. Not only don't we have a single pore size, but often there is a relatively broad distribution, perhaps centered on a couple of predominant sizes. Figure 5.3 shows a bimodal distribution, which is also common. The volume labeled primary corresponds to the volume inside the individual particles, and the volume labeled secondary refers to the volume between particles. From data for pore volume, quantities such as the mean pore size and specific interfacial area can be estimated. The size of the pores impacts transport in the electrode and, as we'll see below, is important in establishing a stable interface with electrodes involving gases.

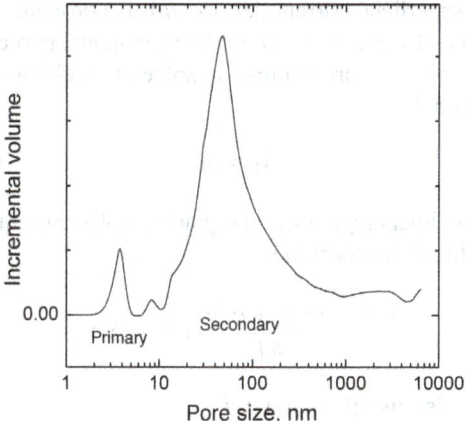

Figure 5.3 Bimodal distribution of pore sizes for a hypothetical electrode containing metal catalyst on carbon.

ILLUSTRATION 5.1

A porous carbon felt is being used for the electrode of a redox-flow battery. The felt is 2 mm thick with an apparent density of 315 kg m^{-3}; the true density of the carbon is 2170 kg m^{-3}. The felt is made up of 20 μm fibers. Calculate the porosity and specific interfacial area.

a. $\varepsilon = 1 - \frac{\rho_A}{\rho_T} = 1 - \frac{315}{2170} = 0.85$.

b. Use a cylindrical filament to model the fibers.
$$a = \frac{4}{d}(1-\varepsilon) = 29{,}032 \text{ m}^{-1}.$$

The superficial current density is 1000 A·m^{-2}. If at 60 °C the exchange-current density is 95 A·m^{-2}, how much is the kinetic polarization reduced because of the porous electrode? Assume $\alpha_a = \alpha_b = 0.5$.

c. The apparent current density is equal to the actual current density times the thickness and specific interfacial area:
$$I/A = i_n aL.$$

The true current density is 17.2 A·m^{-2}, and is assumed to be constant throughout the electrode. The Butler–Volmer kinetic expression is used to evaluate the overpotential, η_s.

$$I/A = i_o\left\{\exp\frac{\alpha_a F \eta_s}{RT} - \exp\frac{-\alpha_c F \eta_s}{RT}\right\}.$$

For the apparent current density, $\eta_s = 136$ mV; for the true current density, $\eta_s = 5$ mV.

5.3 IMPACT OF POROUS ELECTRODE ON TRANSPORT

Clearly, the porous structure has an effect on transport through the media. The way to treat this influence is to define effective transport properties that can be related to the bulk transport properties discussed in Chapter 4. The need for the effective transport properties can be seen with the cartoon illustrated in Figure 5.4. There are two phenomena that we need to consider. First, with a porous medium, the cross-sectional area available for transport is reduced by the presence of the solid phase. The effective area can be approximated as the superficial area multiplied by the porosity. Second, the path length for transport is increased. A molecule diffusing through the electrode must go around the solid obstacles following a tortuous path. This characteristic is known as the *tortuosity*, τ. The tortuosity can have values as high as 6–20, although values of 2–3 are typical for

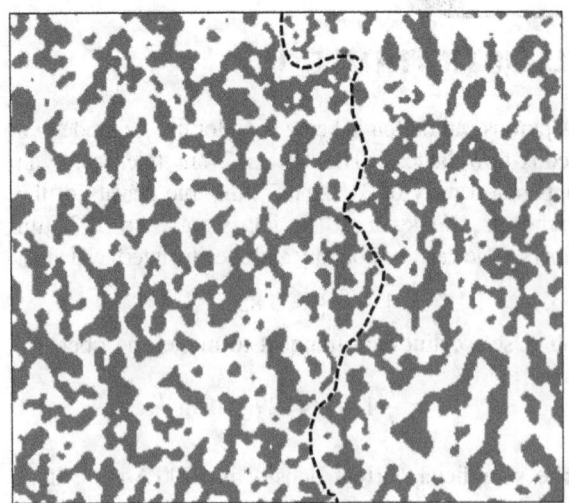

Figure 5.4 Tortuous path through porous media.

many applications. Effective transport properties are related to those in the bulk by the combination of these two effects. This correction is represented below for conductivity, but applies equally to diffusion:

$$\kappa_{\text{eff}} = \kappa_{\text{bulk}} \frac{\varepsilon}{\tau}. \tag{5.17}$$

Thus, a higher tortuosity leads to a lower effective conductivity as would be expected. The effective transport properties can be measured directly, although a common estimate that accounts for porosity and tortuosity is

$$\kappa_{\text{eff}} = \kappa_{\text{bulk}} \varepsilon^{1.5}, \tag{5.18}$$

which is known as the Bruggeman relationship. A similar expression can be used for other transport properties such as the diffusivity and mobility. With these effective transport properties, we can now write the flux equations in terms of the appropriate driving forces. For example, using the definition for conductivity (Equation 4.7), Equation 4.5 can be written for the solution-phase current density:

$$\mathbf{i}_2 = -\kappa \nabla \phi_2 + F \sum_i z_i D_i \nabla c_i. \tag{5.19}$$

All that is needed to apply this relationship to a porous electrode is to replace the transport properties, conductivity, and diffusivities, with effective values that account for the porosity and tortuosity.

$$\mathbf{i}_2 = -\kappa_{\text{eff}} \nabla \phi_2 + F \sum_i z_i D_{i,\text{eff}} \nabla c_i. \tag{5.20}$$

ILLUSTRATION 5.2

The effective conductivity of a porous separator is measured to be $0.035\,\text{S}\cdot\text{m}^{-1}$. The porosity is 0.5. Estimate the tortuosity if the accepted bulk value for conductivity is $0.2\,\text{S}\cdot\text{m}^{-1}$.

$$\kappa_{\text{eff}} = \kappa_{\text{bulk}} \frac{\varepsilon}{\tau}.$$

$$\tau = \kappa_{\text{bulk}} \frac{\varepsilon}{\kappa_{\text{eff}}} = \frac{0.2}{0.035} 0.5 = 2.86.$$

5.4 CURRENT DISTRIBUTIONS IN POROUS ELECTRODES

As noted in Chapter 4, a uniform current distribution is often desired. Generally, the calculation of the current distribution for porous electrodes is not amenable to analytical solutions. There are a few exceptions—most notably the one-dimensional treatment of a porous electrode in the absence of concentration gradients, which is described by the secondary current distribution. In contrast to what was examined in Chapter 4, here we are interested in how the current is distributed through the thickness of the porous electrode.

Consider the porous electrode shown in Figure 5.5. Let κ_{eff} be the conductivity of the electrolyte and σ_{eff} the conductivity of the solid phase. For convenience, we will drop the subscripts in the treatment that follows, but you should remember that these are effective properties. The current density in the electrolyte is i_2 and the current density in the solid phase is i_1. There are also two potentials of interest: ϕ_2, the potential of the electrolyte, and ϕ_1, the potential of the solid. We will simplify the analysis by neglecting concentration gradients, and assuming steady state. We also assume that the superficial current density, I/A, is constant.

In the absence of concentration gradients, two charge balance equations are required to solve the problem. From Equation 5.5,

$$\nabla \cdot \mathbf{i}_2 = ai_n. \tag{5.21}$$

Assuming linear kinetics (see Equation 3.30), this equation in one dimension becomes

$$\frac{di_2}{dx} = \frac{ai_o(\alpha_a + \alpha_c)F}{RT}(\phi_1 - \phi_2). \tag{5.22}$$

We will also use Equation 5.7

$$\nabla \cdot \mathbf{i}_1 + \nabla \cdot \mathbf{i}_2 = 0, \tag{5.23}$$

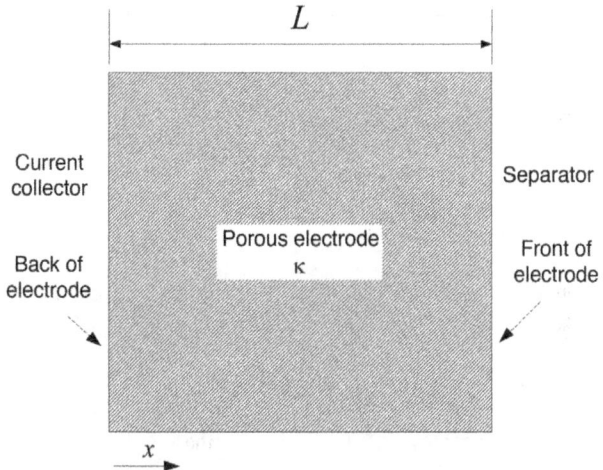

Figure 5.5 One-dimensional porous electrode showing front and back of the electrode.

which can be integrated to yield (see Equation 5.3)

$$i_1 + i_2 = I/A. \quad (5.24)$$

Since there are no concentration gradients, we can use Ohm's law for both the electrolyte and solid phases. As a reminder, σ and κ are effective values for the conductivity.

$$i_1 = -\sigma \frac{d\phi_1}{dx}, \quad (5.25)$$

$$i_2 = -\kappa \frac{d\phi_2}{dx}. \quad (5.26)$$

Initially, we will develop an analytical solution for the current density in solution. To do this, we first differentiate Equation 5.22 to yield

$$\frac{d^2 i_2}{dx^2} = \frac{ai_o(\alpha_a + \alpha_c)F}{RT}\left(\frac{d\phi_1}{dx} - \frac{d\phi_2}{dx}\right). \quad (5.27)$$

Since we are after the current density, we can use Equations 5.24–5.26 to eliminate the potential from this equation and obtain an expression where the only independent variable is i_2:

$$\frac{d^2 i_2}{dx^2} = \frac{ai_o(\alpha_a + \alpha_c)F}{RT}\left(\frac{i_2 - I/A}{\sigma} + \frac{i_2}{\kappa}\right), \quad (5.28)$$

where I/A is constant and equal to the applied superficial current density at the electrode. Equation 5.28 is an ordinary differential equation with i_2 as the only unknown. We now define the following dimensionless variables:

$$i^* \equiv \frac{i_2}{I/A}, \quad (5.29)$$

$$z \equiv \frac{x}{L}, \quad (5.30)$$

$$\nu^2 \equiv \frac{ai_o(\alpha_a + \alpha_c)FL^2}{RT}\left(\frac{1}{\sigma} + \frac{1}{\kappa}\right), \quad (5.31)$$

$$K_r \equiv \frac{\kappa}{\sigma}. \quad (5.32)$$

Substituting these into Equation 5.28 and simplifying yields:

$$\frac{d^2 i^*}{dz^2} = \nu^2\left(i^* - \frac{K_r}{1 + K_r}\right). \quad (5.33)$$

This second order, ordinary differential equation can be solved analytically to yield i^* as a function of z. Two boundary conditions are required. Note that $x = 0$ is the back of the electrode at the current collector. The original boundary conditions for i_2 and the dimensionless equivalents are

$$i_2 = I/A \quad \text{at } x = L; \quad i^* = 1 \quad \text{at } z = 1, \quad (5.34)$$

and

$$i_2 = 0 \quad \text{at } x = 0; \quad i^* = 0 \quad \text{at } z = 0. \quad (5.35)$$

The solution is

$$i^* = \frac{K_r}{1 + K_r} + \frac{\sinh(\nu z) + K_r \sinh(\nu(z-1))}{(1 + K_r)\sinh(\nu)}. \quad (5.36)$$

This current density can also be differentiated to give a dimensionless local reaction rate:

$$\frac{di^*}{dz} = \frac{\nu\cosh(\nu z) + \nu K_r \cosh(\nu(z-1))}{(1 + K_r)\sinh(\nu)} = \frac{aLi_n}{I/A}. \quad (5.37)$$

Use of the dimensionless values allows us to examine the shape of the current distribution and the local reaction rate as a function of two parameters, ν and K_r, without having to insert values for each of the variables in the dimensional equation. For this exercise to be meaningful, we must first examine the physical significance of each of these two parameters. The parameter ν^2 is the ratio of the ohmic and the kinetic resistances, and includes contributions from both the solid and liquid phases to the ohmic resistance. It is essentially the inverse of the Wa for linear kinetics that was presented earlier in Chapter 4 adapted to the porous electrode. The only reason that this ratio is defined as ν^2 rather than ν is to avoid having to write the square root of ν repeatedly in the solution of the differential equation. Small values of ν^2 are controlled by kinetic resistance and result in a uniform or near-uniform distribution of current through the thickness of the porous electrode, similar to what you learned in Chapter 4. In contrast, ohmic resistance controls the current behavior at high values of ν^2, leading to a nonuniform distribution. The second parameter, K_r, is simply the ratio of the conductivity of the electrolyte (κ) to that of the solid matrix (σ). Limiting behavior is reached as $K_r \rightarrow 0$ (electrolyte limited) or $K_r \rightarrow \infty$ (solid limited). It is common to have systems where the solid-phase conductivity is significantly greater than that of the electrolyte. With this background, we will now examine the behavior of i_2 and the local reaction rate in their dimensionless form.

Figure 5.6 shows the dimensionless current density in solution (a) and the local reaction rate (b) as a function of position in the electrode at different values of ν^2 for the case where the conductivity of the solid is an order of magnitude larger than that of the electrolyte, $K_r = 0.1$. For small values of ν^2 (kinetically limited), the reaction rate is flat throughout the electrode (ai_n is constant) and the current in solution varies roughly linearly with position. The linear behavior is expected since, for this case, di^*/dz is equal to a constant. As the value of ν^2 is increased, the reaction rate becomes nonuniform and is preferred at the front of the electrode due to the additional ohmic resistance of the electrolyte associated with penetration of the electrode. The nonuniform reaction rate is reflected in the solution current as shown in the figure. If the solid-phase conductivity is high compared to the solution phase, then the reaction would shift toward the electrode/separator interface ($x = L$). If the conductivities are reversed, ($\sigma < \kappa$), the reaction would shift toward the current collector.

In situations where a local solid-phase reactant is not "consumed," the distribution for high ν^2 values implies that a significant fraction of the electrode is not being used and that similar performance could be obtained with a thinner electrode. In contrast, for a battery electrode with a solid-phase reactant, the results for high ν^2 would indicate, for example, that the outside of the electrode would discharge first, followed by the regions closer to the middle, leading to increased resistance at lower states of charge.

The asymmetry shown in Figure 5.6 is the result of the assumption that $K_r = 0.1$. If the conductivity of the solid and the electrolyte were identical, the reaction distribution would be symmetrical. Figure 5.7 shows the current distribution for a porous electrode with the solid-phase conductivity equal to that of the solution phase. Because of the equal resistance in the solid and

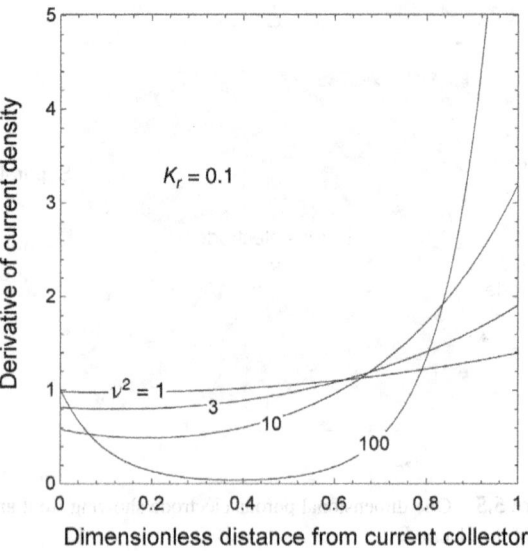
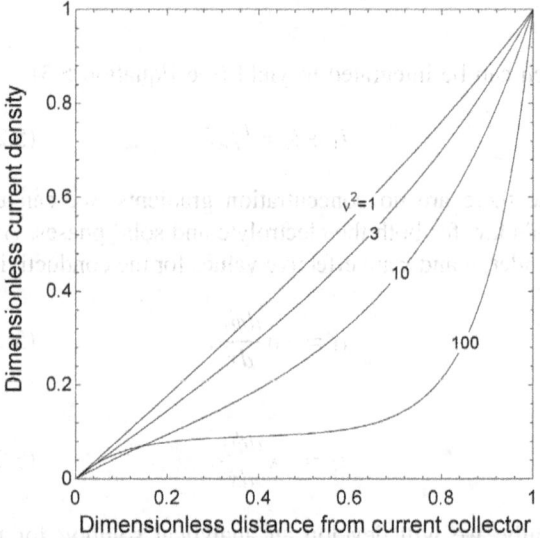

Figure 5.6 Dimensionless current density and its derivative across the electrode. The back of the electrode is a $z = 0$, the front is at $z = 1$, adjacent to the electrolyte. $Kr = 0.1$.

electrolyte phases, the current distribution is symmetrical, neither preferring the front nor the back of the electrode. The reaction is more nonuniform at higher values of ν^2, shifting away from the center to the front and the back of the electrode.

It is useful to estimate a penetration depth for an electrode. Penetration depth is not an issue for small values of ν^2, since such a system is under kinetic control and reaction takes place throughout the entire electrode. Here we consider the situation where $K_r \rightarrow 0$ and behavior is limited by transport of current through the electrolyte. With these assumptions, the current changes monotonically from a high value at the front of the

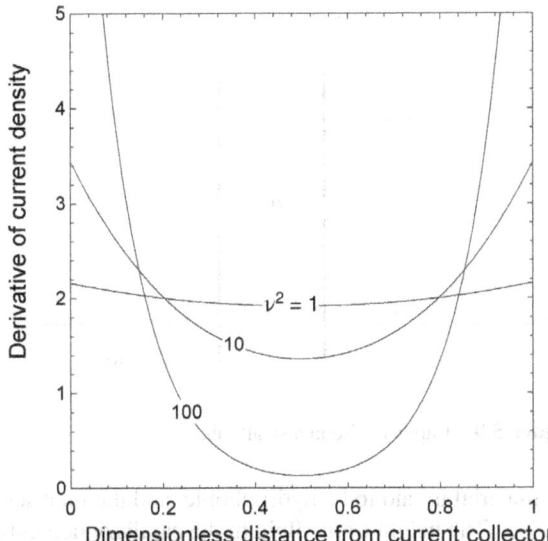

Figure 5.7 Current distribution with equal electronic and ionic conductivity.

electrode to a low value at the back of the electrode. The expression for the local reaction rate, Equation 5.37, can be simplified to

$$\frac{aLi_n}{I/A} \approx \frac{\nu\cosh(\nu z)}{(1+K_r)\sinh(\nu)}. \tag{5.38}$$

We will use a practical approach to the penetration depth by defining it in terms of the reaction rate at the back of the electrode relative to that at the front. The approximate rate at the front of the electrode ($z = 1$) is

$$\left.\frac{aLi_n}{I/A}\right|_{front} \approx \frac{\nu\cosh(\nu)}{(1+K_r)\sinh(\nu)}. \tag{5.39}$$

Similarly, at the back of the electrode ($z = 0$):

$$\left.\frac{aLi_n}{I/A}\right|_{back} \approx \frac{\nu}{(1+K_r)\sinh(\nu)}. \tag{5.40}$$

The ratio of the rate at the back over that at the front is

$$\text{Relative rate} \approx \frac{1}{\cosh(\nu)}. \tag{5.41}$$

A penetration depth can be estimated by solving this equation for the ν value that corresponds to a specified relative rate of reaction. The penetration depth, L_p, can then be back calculated from the definition of ν, remembering that we have assumed $1/\sigma \sim 0$. In other words, given the rest of the parameters, we are seeking the value of L that will yield the ν value calculated from Equation 5.31. This process is shown in Illustration 5.3 that follows.

ILLUSTRATION 5.3

You are designing an electrode for a lithium-ion battery for operation at room temperature. The solid-phase electronic conductivity is much larger than that of the electrolyte, $\kappa = 0.1\,\text{S·m}^{-1}$. If it is desired to keep the reaction rate at the back of the electrode no less than 30% of the front, what is the maximum thickness of the electrode? Additional kinetic data are $\alpha_a = \alpha_c = 0.5$, $i_o = 100\,\text{A·m}^{-2}$, $a = 10^4\,\text{m}^{-1}$. Starting with Equation 5.41,

$$0.3 = \frac{1}{\cosh\nu}, \text{ and therefore, } \nu = 1.874.$$

Equation 5.31 is

$$\nu^2 = \frac{ai_o(\alpha_a + \alpha_c)FL_p^2}{RT}\left(\frac{1}{\sigma} + \frac{1}{\kappa}\right).$$

σ is assumed to be infinity; thus, the only unknown is L_p. Solving for L_p yields

$$L_p = 95\,\mu m.$$

We are also interested in the internal resistance of the electrode. This resistance is nothing more than the change in potential with current density. For linear kinetics the resistance of a porous electrode is

$$R_{int} = \frac{L}{\sigma+\kappa}\left[1 + \frac{2 + \left(\frac{\sigma}{\kappa}+\frac{\kappa}{\sigma}\right)\cosh\nu}{\nu\sinh\nu}\right]. \tag{5.42}$$

The internal resistance of the electrode as a function of thickness is shown in Figure 5.8. It is important to recognize that this resistance is a combination of ohmic drop in the solution and solid phases as well as kinetic resistance. For a large electrode thickness, the value of ν^2 is high, and the reaction is nonuniform. Here the resistance increases with thickness. For thinner electrodes, the reaction is more uniform but there is relatively little surface area available for the reaction. Increasing the thickness makes more surface area available for reaction, and thus, the resistance decreases. There is a value of L, the electrode thickness where the resistance is minimized. These effects are shown in Figure 5.8 for two values of K_r.

Similar analytical solutions are available for Tafel kinetics. Although the use of Tafel kinetics is more appropriate for a large number of problems, the linear and Tafel results are qualitatively similar. We have presented the linear results because they have fewer parameters and are easier to understand. Tafel kinetics is explored in

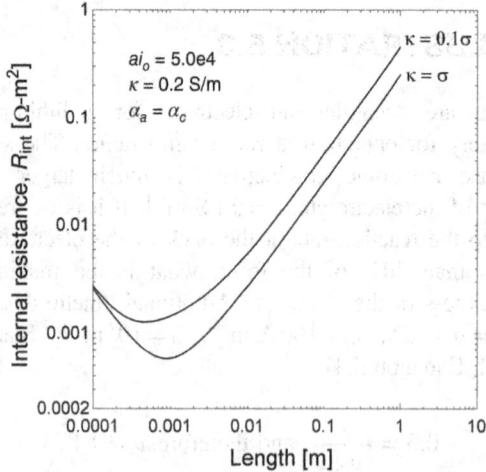

Figure 5.8 Internal resistance of an electrode assuming linear kinetics.

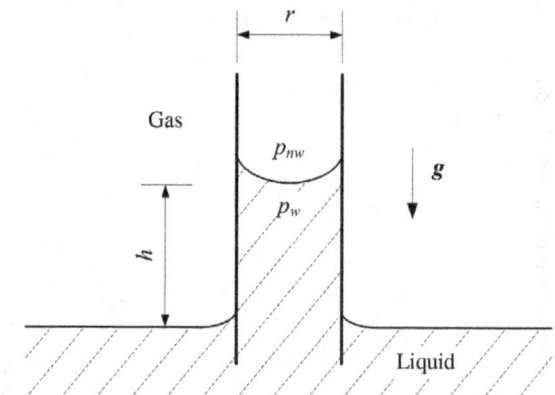

Figure 5.9 Capillary rise in a small tube.

Problem 5.9 at the end of the chapter. There is an important difference between the two solutions. As we can see from Equation 5.36, there are only two parameters for linear kinetics: K_r and ν. Although the exchange-current density is part of ν, the current, I, does not impact the dimensionless current distribution. This result in completely analogous to what we saw for the Wa number in Chapter 4. For linear kinetics, the exchange-current density is a parameter, and the current distribution is independent of the average current density as long as it is small. In contrast, for Tafel kinetics the exchange-current density is not a parameter and has been replaced with the average current density. Since Tafel kinetics applies when the current density is much larger than the exchange-current density, this difference makes sense.

5.5 THE GAS–LIQUID INTERFACE IN POROUS ELECTRODES

There are many instances when both gas and liquid fill the void volume of a porous electrode. Here, we introduce important concepts to describe these two fluids contained in the pores. A key aspect of porous media relevant for our studies is capillarity or capillary action. The capillarity can be understood from the ability of a fluid to flow against a gravitational or other body force. The wetting of a paper towel by wicking of the fluid against gravity is a familiar example. Let's consider a simple situation of a small capillary (thin tube) in a fluid as shown in Figure 5.9. Here surface tension causes the fluid to rise in the capillary.

Figure 5.10 shows the contact angle where there is an intersection of the liquid, gas, and solid. If the contact angle is less than 90°, the material is said to be hydrophilic or wetting. In this case, the meniscus will be concave as shown in Figure 5.9. Conversely, for large contact angles, the material is said to be hydrophobic, and the meniscus is convex. The height of the fluid in the capillary depends on the contact angle, θ, and the surface tension, γ.

$$h = \frac{2\gamma\cos\theta}{\rho g r}. \quad (5.43)$$

Typically, gravitational forces are not significant in porous electrodes; and Equation 5.43 is not applicable. On the other hand, surface tension and the contact angle are important. We define a *capillary pressure*, which represents the difference in pressure between the nonwetting (*nw*) and wetting (*w*) phases. For instance, at the top of the column of water in Figure 5.9, the pressure of the gas (nonwetting) will be higher than the pressure in the adjacent liquid (wetting phase). The liquid pressure can be determined from hydrostatics to be below that of the gas by a value of $\rho g h$. Thus, the capillary pressure, p_c, is

$$p_c \equiv p_{nw} - p_w = \frac{2\gamma\cos\theta}{r}. \quad (5.44)$$

For a wetting fluid with a contact angle less than 90°, the capillary pressure is positive, meaning the gas pressure is

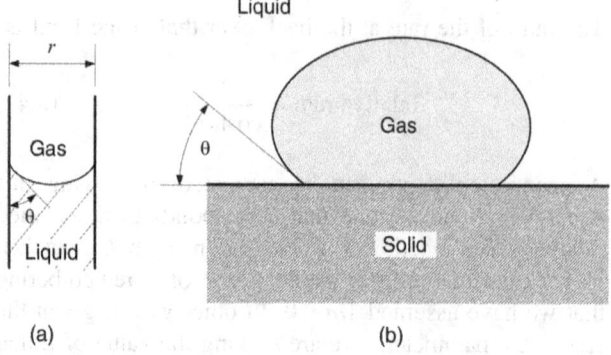

Figure 5.10 (a) Contact angle associated with capillary rise. (b) Contact angle for gas bubble attached to solid surface.

above that of the liquid. Conversely, for contact angles greater than 90°, the liquid pressure will be larger than that of the gas. Hydrophilic materials will have a positive capillary pressure and naturally wick up the fluid. In contrast, hydrophobic materials will have a negative capillary pressure and pressure must be applied to wet the material.

For many electrodes, the entire void volume is filled with the electrolyte. In others, gas occupies some of the volume. Often, this partially filling with fluid is by design, such as with low-temperature fuel cells where the void volume is filled with one or more immiscible fluids. The degree to which these fluids fill the void volume is quantified by the saturation level, given the symbol S_i.

$$S_i = \frac{\text{volume filled by phase } i}{\text{void volume}}. \qquad (5.45)$$

Since the porous medium typically consists of a range of pore sizes, there are a range of capillary pressures as given by Equation 5.44. Again, this is determined principally from the surface tension, contact angle, and pore size distribution. For a partially filled porous body, the distribution of fluid will be described by this capillary pressure. Assuming the contact angle and surface tension are constant, for a wetting fluid the smallest pores will have the largest capillary pressure and will be filled first. As more liquid is made available, larger and larger pores will fill. If the distribution of pore size is known, we can relate the fill level with the capillary pressure as shown in Figure 5.11. The pore size distribution becomes the key factor in determining the shape of this curve.

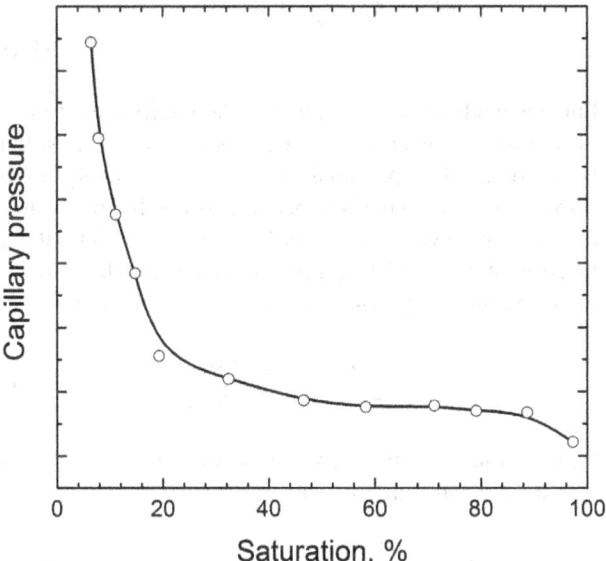

Figure 5.11 Fill as a function of capillary pressure.

ILLUSTRATION 5.4

The separator of a phosphoric acid fuel cell consists of a matrix of micrometer-sized silicon carbide particles filled with phosphoric acid. The function of the separator is to keep the fuel and oxidant separate. Invariably, there will be small pressure differences between these gases. Estimate the capillary pressure in the matrix if the particle size is 1.5 µm and the porosity is 0.5. Assume that the contact angle is zero and the surface tension is $0.02\,\text{N·m}^{-1}$.

$$r_p = \frac{\varepsilon}{1-\varepsilon}\left(\frac{r}{3}\right) = \frac{0.5}{(1-0.5)}\frac{1.5\,\mu\text{m}}{3} = 0.5\,\mu\text{m}.$$

$$p_c = p_{nw} - p_w = \frac{2\gamma\cos\theta}{r_p} = \frac{2(0.2)(1)}{5\times 10^{-5}} = 80\,\text{kPa}.$$

This pressure represents the maximum pressure that the matrix can resist before gas breakthrough occurs.

5.6 THREE-PHASE ELECTRODES

There are situations where a stable three-phase structure is required. For example, the cathode of a low-temperature fuel cell involves three phases (gas, liquid, solid) in intimate contact. Let's examine the reduction of oxygen in acid:

$$4H^+ + 4e^- + O_2 \rightarrow 2H_2O$$

The protons are supplied from the electrolyte, electrons supplied from a solid phase, and oxygen is in the gas phase. The product water may be either in the gas or liquid phase depending on temperature and pressure. The interface of two phases that we have seen before is a surface, but now with three phases the intersection is reduced to the line at which all three phases meet (Figure 5.10b). Thus, the region of intimate contact will be small and represents an obstacle that must be overcome in the development of practical electrodes. This challenge was recognized almost immediately by investigators such as Sir William Grove, the inventor of the fuel cell (1839), who described it as a "notable surface of action." One approach to address the small contact region is with what is known as a flooded-agglomerate electrode.

This approach is illustrated in Figure 5.12, which shows the electrode at three different length scales to clarify its critical features. The porous electrode consists of agglomerate particles that are packed together to form a porous bed. Each agglomerate consists of a porous, solid, electronic conductor that supports catalyst and is completely filled with electrolyte. These agglomerates are

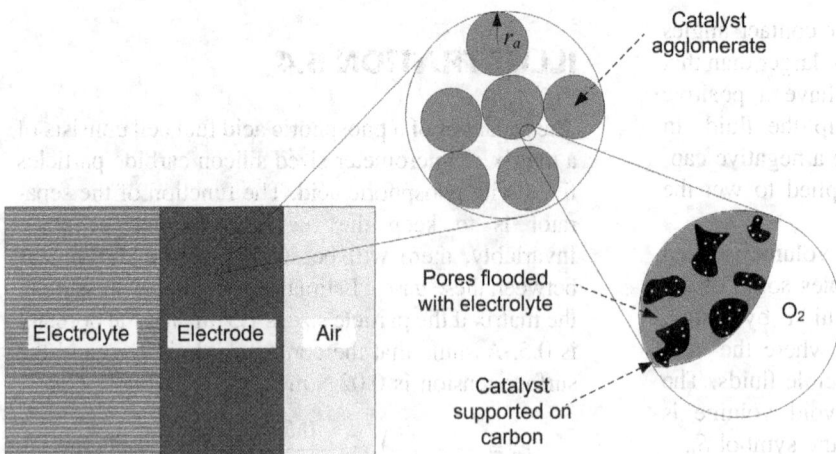

Figure 5.12 Flooded-agglomerate model.

organized to provide both a continuous electronic and ionic path across the electrode. Electrons are transferred through the electrode via contacts between the electrically conducting solid phase (carbon) of the particles. Ionic conduction occurs through liquid-phase connections between the particles. However, in contrast to the electrodes considered earlier, a large fraction of the volume between particles is occupied by a gas phase, which permits rapid gas transport and allows ready access of the small agglomerate particles to the needed oxygen. In analyzing these electrodes using the methodology described previously, we would use the respective volume fractions of gas and liquid in the electrode rather than the electrode porosity since we have added an extra phase (gas).

There is another important difference between this model and the porous electrode treatment that we have considered to the point. In the treatment above, the reaction took place at the interface between the particles and the electrolyte, and there were no physically important processes taking place inside the particles other than electrical conduction of the current. In contrast, the reaction of interest in the flooded-agglomerate model takes place throughout the agglomerate particle as oxygen diffuses into the particle. Therefore, we need to model the processes inside the particle and couple that model with the balance equations previously developed for porous electrodes. Development of the fully coupled model is beyond the scope of this chapter. We will, however, develop the model for the single agglomerate particle in order to illustrate this aspect of the overall approach.

If we now look more closely at a single agglomerate, depicted for a spherical geometry in Figure 5.12, we see that the agglomerate of catalysts dispersed on a support and is completely filled with a liquid electrolyte. Oxygen from the gas phase dissolves into the electrolyte where it simultaneously diffuses and reacts inside the agglomerate.

If the dimension of the agglomerate is small enough, then oxygen can access all of the catalyst.

Starting with Equation 5.11,

$$\frac{\partial \varepsilon c_i}{\partial t}^{0} = -\nabla \cdot \mathbf{N}_i - \frac{s_i}{nF} \nabla \cdot \mathbf{i}_2, \quad (5.46)$$

which is applied to oxygen at steady state. We then will use Fick's law to describe the transport of oxygen and Tafel kinetics for the reduction of oxygen, which is assumed to be first order in oxygen concentration.

$$\mathbf{N}_{O_2} = -D_{eff} \nabla c_{O_2}. \quad (5.47)$$

$$-\frac{s_i \nabla \cdot \mathbf{i}_2}{nF} = \frac{s_i a i_n}{nF} = \frac{a i_o}{4F} \frac{p_{O_2}}{p_{O_2}^*} \exp\left\{\frac{-\alpha_c F}{RT}(\phi_1 - \phi_2 - U^\theta)\right\}. \quad (5.48)$$

For this analysis, we assume that the overpotential inside the particle is constant. This makes sense since the particle is so small that potential losses are not likely to be significant. The analysis is performed for the local partial pressure of oxygen at the surface of the particle at that location in the bed. The particle is assumed to be spherical with a radius of r_p. The boundary conditions are

$$\begin{aligned} r &= 0, \quad \nabla c_{O_2} = 0, \\ r &= r_p, \quad c_{O_2} = H p_{O_2}. \end{aligned} \quad (5.49)$$

Substitution into the differential material balance and solving for concentration gives

$$N_{O_2}\big|_{r=r_p} = \frac{D_{eff} H p_{O_2}}{r_p}(1 - K \coth K), \quad (5.50)$$

where

$$K^2 = \frac{ai_o}{D_{eff}4F} \frac{r_p^2}{p_{O_2}^* H} \exp\left\{\frac{-\alpha_c F}{RT}(\phi_1 - \phi_2 - U^\theta)\right\}. \tag{5.51}$$

Equation 5.50 provides the flux and, through Faraday's law, the current needed for the equations that model the transport and reaction through the porous electrode (as opposed to the particle). In turn, the local potential difference and partial pressure of oxygen are provided by the porous electrode model to the particle model. Hence, the additional physics that take place in the particle can be accounted for while retaining the advantages of the macrohomogeneous porous electrode model.

Exploring the particle behavior further, the concentration is given by

$$\theta = \frac{c_{O_2}}{Hp_{O_2}} = \frac{r_p}{r}\frac{\sinh Kr}{\sinh Kr_p}. \tag{5.52}$$

The dimensionless concentration is shown in Figure 5.13. For low values of K, diffusion is rapid compared to the rate of reaction. Here the concentration is nearly constant. In contrast, for large values of K, the oxygen is consumed near the surface of the agglomerate, and for all intents and purposes no reaction occurs in the middle of the agglomerate. In this case, the effectiveness of the agglomerate is low; that is, most of the volume is not being used. We can define an effectiveness factor, similar to what is done for reaction and diffusion problems in reaction engineering. For a spherical agglomerate,

$$\eta_{eff} = \frac{3(K\coth K - 1)}{K^2}. \tag{5.53}$$

Although analytic solutions are not possible for the current distribution in this electrode, the general rules learned in the previous section are applicable. In this particular instance, the kinetics for oxygen reduction is slow and therefore the reaction tends to be spread out across the thickness of the electrode. As the current density increases, the distribution becomes more nonuniform. Because of the slow kinetics, more concern is applied to the structure of the agglomerate rather than the thickness of the electrode.

5.7 ELECTRODES WITH FLOW

Fluid flow is sometimes used with porous electrodes in order to bring reactants and products in and out of these high-surface-area electrodes. A number of different flow configurations are possible. Two basic categories are the *flow-through* and *flow-by* configurations (Figure 5.14). In the flow-through arrangement, the fluid flow is parallel to the current flow, and the fluid flow actually flows through the current collector. In the flow-by configuration, fluid flow is perpendicular to the current flow. Also shown in Figure 5.14b is a gap between the electrodes. In cases where it is not desirable to mix the two reactant streams, a porous or membrane separator may be used.

One important application for flowing systems is the removal of low concentrations of valuable metals or contaminants by electrochemical deposition or reaction. We will consider the mass-transfer limited deposition of a metal, such as copper in a supporting electrolyte, where the flow and the current are in the same direction (flow-through configuration). The analysis is performed for the cathode (−) where the anode (+) is assumed to be upstream as shown in Figure 5.15.

Because of the supporting electrolyte, it is not necessary to include the migration term in the flux equation (see Section 4.3). The flow velocity is controlled independently and is assumed to be constant. The mass-transfer coefficient is taken to be a function of the flow rate only, and is therefore constant for this analysis. Plug

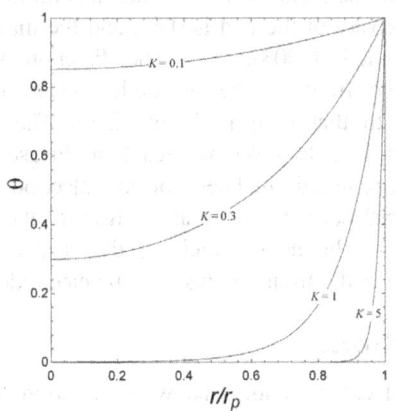

Figure 5.13 Dimensionless concentration in the agglomerate.

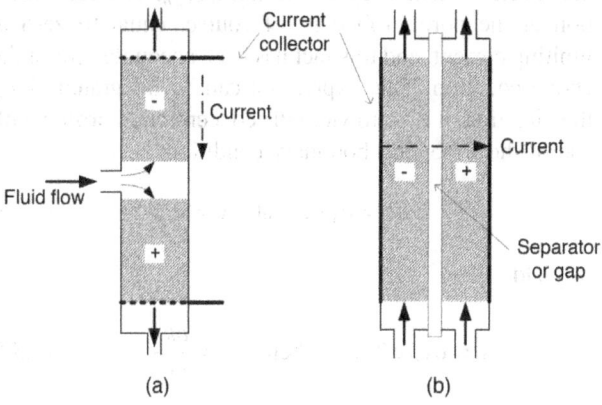

Figure 5.14 Electrode configurations for flowing systems. (a) Flow-through electrode. (b) Flow-by electrode.

106 Electrochemical Engineering

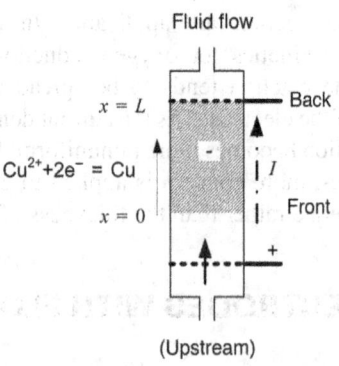

Figure 5.15 Removal of Cu^{2+} from stream.

flow is assumed so that the problem is one dimensional, and changes occur only in the direction of the flow. With these assumptions, the flux equation for the species of interest becomes

$$N_A = -D_{eff}\frac{dc_A}{dx} + \varepsilon c_A v_x, \quad (5.54)$$

where v_x is the velocity of the fluid in the pores in the direction of interest (εv_x would be the superficial velocity). As before, this flux is based on the superficial area of the electrode perpendicular to flow direction. For simplification, we further assume that axial diffusion is small relative to convective transport, which leaves us with just the second term on the right side of Equation 5.54. At limiting current, the potential and concentration fields are decoupled, and we can solve the material balance directly for the concentration profile of species A in the electrode. At steady state, the material balance is (see Equations 5.8 and 5.9)

$$-\frac{dN_A}{dx} - aj_n = -\frac{dN_A}{dx} - ak_c c_A = -\varepsilon v_x \frac{dc_A}{dx} - ak_c c_A = 0, \quad (5.55)$$

where a mass-transfer expression has been used to describe the species reaction rate at the surface, j_n. The concentration at the pore surface is, of course, equal to zero at limiting current, and this fact has been accounted for in the above equation. This expression can be integrated along the direction of flow to yield the concentration profile with use of the following boundary condition:

$$c_A = c_{A,in}, \quad \text{at} \quad x = 0 \quad (5.56)$$

to yield

$$c_A = c_{A,in} e^{-\alpha x}, \quad \text{where} \quad \alpha = \frac{ak_c}{\varepsilon v_x}. \quad (5.57)$$

If we assume that the desired reaction is the only reaction that takes place in the electrode, we can solve the charge balance to obtain the current and potential distributions. The charge balance is

$$-\frac{di_2}{dx} - nFak_c c_A = -\frac{di_2}{dx} - nFak_c c_{A,in} e^{-\alpha x} = 0. \quad (5.58)$$

Integration followed by application of the following auxiliary condition:

$$i_2 = 0, \quad \text{at} \quad x = L, \quad (5.59)$$

yields the expression for the superficial current density,

$$i_2 = nF\varepsilon v_x c_{A,in}\left(e^{-\alpha x} - e^{-\alpha L}\right). \quad (5.60)$$

Examining this equation, we see that the current in solution is directly related to the amount of A that reacts. Specifically, the total current at the front of the electrode ($x=0$) is equal to the difference between the amount of A that enters the electrode and the amount that leaves the electrode, multiplied by nF. As $L \to \infty$, essentially all of the A is reacted in the electrode and the current reaches its maximum value of $nF\varepsilon v_x c_{A,in}$. However, the utilization of the electrode drops as the electrode thickness increases beyond the point at which almost all of A has reacted. Further increases in thickness do not increase the current but result in increased pumping costs and a larger initial capital investment because of the thicker electrode.

ILLUSTRATION 5.5

A packed bed electrode is used to remove metal ions from solution. The inlet concentration of the metal ions in a supporting electrolyte solution is 0.002 M, and the desired outlet concentration is 0.0001 M. The size of the electrode perpendicular to the flow is 25 cm × 25 cm, and the volumetric flow rate of the solution to be treated is 25 L·min^{-1}. The spherical particles that make up the bed are 0.5 mm in diameter. The porosity of the bed is 0.45, and the mass-transfer coefficient is 0.003 cm·s^{-1}. The direction of flow is from the front of the electrode toward the back, similar to that seen in Figure 5.15. The reduction reaction is a two-electron reaction. Please calculate (a) the required thickness of the electrode, (b) the superficial current density at the front of the electrode closest to the anode, and (c) the relative rates of reaction at the front and back of the electrode.

Solution:

Part (a) can be addressed with Equation 5.57. The value of the mass-transfer coefficient was given in the problem statement. However, we need values for the velocity and a, the area per volume. The superficial velocity is

$$\frac{\dot{V}}{A} = \varepsilon v_x = \frac{25 \text{ l/min}}{(25 \text{ cm})^2} \times \frac{1000 \text{ cm}^3}{1} \times \frac{\min}{60 \text{ s}} \times \frac{\text{m}}{100 \text{ cm}}$$

$$= 6.67 \times 10^{-3} \text{ m s}^{-1}.$$

We can use Equation 5.15 to estimate a for a packed bed of spherical particles:

$$a = \frac{3(1 - 0.45)}{0.25 \text{ mm}} \cdot \frac{1000 \text{ mm}}{\text{m}} = 6600 \text{ m}^{-1}$$

From Equation 5.57 for $x = L$,

$$L = -\ln\left(\frac{c_{A,out}}{c_{A,in}}\right) \cdot \frac{\varepsilon v_x}{k_c a} = 0.10 \text{ m}.$$

For part (b), we first determine α:

$$\alpha = \frac{k_c a}{\varepsilon v_x} = 29.7 \text{ m}^{-1}.$$

We then use Equation 5.60 at $x = 0$:

$$i_2 = nF\varepsilon v_x c_{A,in}\left(1 - e^{-\alpha L}\right) = 2442 \text{ A} \cdot \text{m}^{-2}.$$

For part (c), the local reaction rate is proportional to $\exp(-\alpha L)$, as seen in Equation 5.58. Another way of looking at this is to realize that the local reaction rate is directly proportional to the local concentration, which varies exponentially through the electrode. Therefore, for this example,

$$\frac{\text{rate at back}}{\text{rate at front}} = \frac{e^{-\alpha L}}{1} = 0.05.$$

This is also equal to $c_{A,out}/c_{A,in}$. Therefore, the rate at the back of the electrode is only 5% of that at the front. That number would decrease exponentially with increasing L.

In the above treatment, side reactions were not considered. What is the impact of such reactions? To address this question, let's consider perhaps the most straightforward example of a side reaction, hydrogen evolution at cathodic potentials. If the cathodic overpotential becomes too large, hydrogen evolution may dominate over the desired reaction. Thus, there is a practical limit to the potential drop that is permitted across the electrode. The overpotential must be sufficiently large to drive the desired reaction, but not so large that side reactions prevent the desired reaction from occurring. The range of overpotentials at which limiting current conditions prevail is roughly 0.1–0.3 V, depending on the system. Beyond this, the current increases due to side reactions. This means that the change in overpotential across the thickness of the electrode cannot vary by more than this amount without appreciable side reaction.

The most common situation is where $\sigma \gg \kappa$. Under such conditions, the change in overpotential is equal to the change in the potential in solution across the thickness of the electrode. To estimate the change in potential, we can substitute Ohm's law into Equation 5.60, using the effective conductivity of the supporting electrolyte, and integrate to yield the following expression for the potential:

$$\phi_2 = \phi_{2,L} + \frac{\beta}{\alpha}\left(e^{-\alpha x} - e^{-\alpha L}\right) + \beta e^{-\alpha L}(x - L), \quad (5.61)$$

$$\text{where } \beta = \frac{nF\varepsilon v_x c_{A,in}}{\kappa_{eff}}.$$

The total potential difference across the electrode is

$$\Delta\phi_2 = \frac{\beta}{\alpha}\left(1 - e^{-\alpha L}\right) - \beta L e^{-\alpha L}. \quad (5.62)$$

The maximum value occurs for electrodes of large thickness, where Equation 5.62 simplifies to

$$\Delta\phi_2|_{max} = \frac{\beta}{\alpha} = \frac{nF(\varepsilon v_x)^2 c_{A,in}}{\kappa_{eff} k_c a}, \quad (5.63)$$

and εv_x is the superficial velocity. For comparison, if we substitute for the values from the illustration above, $\Delta\phi_2 = 1.38$ V, while $\Delta\phi_2|_{max} = 1.73$ V. Both of these values are too high for practical application. Note that the difference between the calculated potential drop and the maximum value is significant, even when we are removing 95% of the incoming metal-ion concentration. Decreasing our throughput by lowering the volumetric flow rate from 25 L·min^{-1} to 9 L·min^{-1} reduces $\Delta\phi_2$ to 0.18 V, a value that may be feasible. A fraction of this decrease will be offset by a decrease in the mass-transfer coefficient with decreasing velocity, which was not considered in the calculation.

Given that there are voltage losses associated with the anodic reaction and ohmic drop across the channel separating the electrodes, that a sufficient overpotential is needed on the cathode to reach the limiting current, and that impurities may influence behavior, practical systems rarely operate at 100% current efficiency. In fact, in extreme cases current efficiencies for removal systems with low initial concentrations and high removal fractions may dip below 5%, although this is clearly not preferred. It is your job as the electrochemical engineer to understand and quantify the factors that influence electrode design and operation in order to achieve an optimal solution that meets both technical and economic constraints.

CLOSURE

The basic concepts of a porous electrode have been introduced and the principal methods of characterizing these electrodes were discussed. The main terms used in discussing porous electrodes are shown in Figure 5.16. Additional key parameters are the porosity, specific interfacial area, pore size distribution, and surface energy. The most important feature is the increased surface area per volume. Porous electrodes are the mainstays of nearly all electrochemical systems for energy storage and conversion, such as batteries, fuel cells, electrochemical capacitors, and flow cells. Therefore, the information from this chapter is essential for the analysis of these devices in subsequent chapters. The governing equations to describe a porous electrode have been developed. Analytic solutions are possible for either linear or Tafel kinetics when concentration gradients are neglected. The distribution of current through the electrode is determined by the ratio of ohmic and kinetic resistances. Systems with a gas phase were also discussed. In particular, fuel cells require three separate phases in contact. One specific example, the flooded-agglomerate model, was used to illustrate how an effective electrode can be created.

Porous Electrode Terminology

Front: the edge of the electrode next to the electrolyte

Back: the edge of the electrode next to the current collector

Porosity: the empty or void portion of the electrode

Superficial area: the area of a plane cutting through the electrode normal to the direction of superficial current

Specific interfacial area: the actual physical surface area of electrode in contact with the electrolyte divided by the volume of the electrode [m^{-1}]

FURTHER READING

Dullien, F.A.L. (1992) *Porous Media, Fluid Transport and Pore Structure*, Academic Press.

Newman, J. and Thomas-Alyea, K.E. (2004) *Electrochemical Systems*, John Wiley & Sons, Inc., Hoboken, NJ.

PROBLEMS

5.1 The discharge of the lead–acid battery proceeds through a dissolution/precipitation reaction. These two reactions for the negative electrode are

$$Pb \rightarrow Pb^{2+} + 2e^- \quad \text{dissolution}$$

and

$$Pb^{2+} + SO_4^{2-} \rightarrow PbSO_4 \quad \text{precipitation}$$

A key feature is that lead dissolves from one portion of the electrode but precipitates at another nearby spot. The solubility of Pb^{2+} is quite low, around 2 g·m^{-3}. How then can high currents be achieved in the lead–acid battery?

(a) Assume that the dissolution and precipitation locations are separated by a distance of 1 mm with a planar geometry. Using a diffusivity of 10^{-9} m^2·s^{-1} for the lead ions, estimate the maximum current that can be achieved.

(b) Rather than two planar electrodes, imagine a porous electrode that is also 1 mm thick made from particles with a radius 10 μm packed together with a void volume of 0.5. What is the maximum superficial current here based on the pore diameter?

(c) What do these results suggest about the distribution of precipitates in the electrodes?

5.2 A porous electrode is made from solid material with an intrinsic density of ρ_s. When particles of this material are combined to form an electrode, it has an apparent density of ρ_a. What is the relationship between these two densities

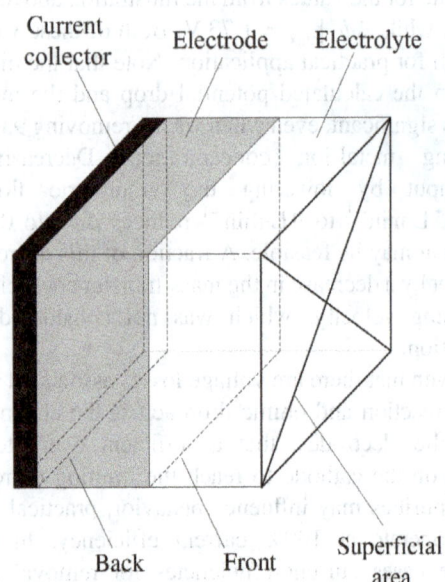

Figure 5.16 Porous electrode terminology.

and the porosity? Assuming the particles are spherical with a diameter of 5.0 μm, what is the specific interfacial area? If the electrode is made from particles with a density of 2100 kg·m^{-3} is 1 mm thick and has an apparent density of 1260 kg·m^{-3}, by what factor has the area increased compared to the superficial area?

5.3 Calculate the pressure required to force water through a hydrophobic gas diffusion layer of PEM fuel cell. The contact angle is 140° and the average pore diameter is 20 μm. Use 0.0627 [N·m^{-1}] for the surface tension of water.

5.4 The separator of a phosphoric acid fuel cell is comprised of micrometer and submicrometer-sized particles of SiC. Capillary forces hold the liquid acid in the interstitial spaces between particles, and this matrix provides the barrier between hydrogen and oxygen. What differential gas pressure across the matrix can be withstood? Assume an average pore size of 1 μm, and use a surface tension of 70 mN·m^{-1}, a contact angle of 10°.

5.5 The separator used in a commercial battery is a porous polymer film with a porosity of 0.39. A series of electrical resistance measurements are made with various numbers of separators filled with electrolyte stacked together. These data are shown in the table. The thickness of each film is 25 μm, the area 2×10^{-4} m^2, and the conductivity of the electrolyte is 0.78 S·m^{-1}. Calculate the tortuosity. Why would it be beneficial to measure the resistances with increasing numbers of layers rather than just a single point?

Number of layers	Measured resistance [Ω]
1	1.91
2	3.41
3	5.17
4	6.65
5	7.79

5.6 To reach the cathode of a proton exchange-membrane fuel cell, oxygen must diffuse through a porous substrate. Normally, the porosity (volume fraction available for the gas) is 0.7 and the limiting current is 3000 A·m^{-2}. However, liquid water is produced at the cathode with the reduction of oxygen. If this water is not removed efficiently, the pores can fill up with water, and the performance decreases dramatically. Use the Bruggeman relationship to estimate the change in limiting current when, because of the build-up of water, only 0.4 and 0.1 volume fractions are available for gas transport.

5.7 Calendering of an electrode is a finishing process used to smooth a surface and to ensure good contact between particles of active material. The electrode is passed under rollers at high pressures. If the initial thickness and porosity were 30 μm and 0.3, what is the new void fraction if the electrode is calendared to a thickness of 25 μm? What effect would this have on transport?

5.8 For a cell where $\sigma \gg \kappa$, the reaction proceeds as a sharp front through the porous electrode. Material near the front of the electrode is consumed before the reaction proceeds toward the back of the electrode. This situation is shown in the figure, L_s is the thickness of the separator, L_e the thickness of the electrode, and x_r is the amount of reacted material.

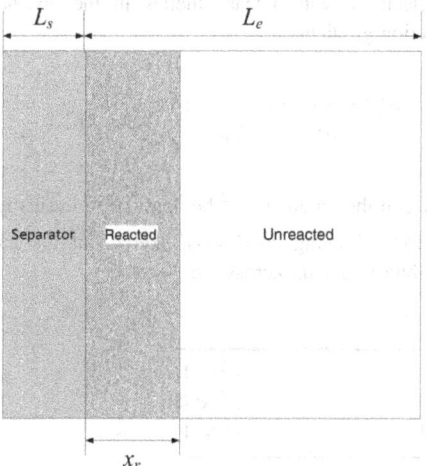

(a) If the cell is discharged at a constant rate, show how the distance x_r depends on time, porosity, and the capacity of the electrode, q, expressed in C·m^{-3}.

(b) What is the internal resistance? Use κ_s and κ for the effective conductivity of the separator and electrode, respectively.

(c) If the cell is ohmically limited, what is the potential during the discharge?

5.9 For Tafel kinetics and a one-dimensional geometry, an analytic solution is also possible analogous to the one developed for linear kinetics. The following is the solution for a cathodic process:

$$\frac{i_1}{I} = 2\theta \tan\left(\theta \frac{x}{L} - \psi\right)$$

and

$$\frac{di_1}{dx} = \frac{I}{L}\left\{\frac{\theta^2}{\delta} \sec^2\left(\theta \frac{x}{L} - \psi\right)\right\},$$

where

$$\tan\theta = \frac{2\delta\theta}{4\theta^2 - \varepsilon(\delta - \varepsilon)}, \quad \tan\psi = \frac{\varepsilon}{2\theta},$$

$$\delta = \frac{\alpha_c FIL}{RT}\left(\frac{1}{\kappa} + \frac{1}{\sigma}\right), \quad \varepsilon = \frac{\alpha_c FIL}{RT}\frac{1}{\kappa}.$$

(a) Make two plots of the dimensionless current distribution (derivative of i_1) for Tafel kinetics: one with $K_r = 0.1$ and the other with $K_r = 1.0$. δ is a parameter, use values of 1, 3, and 10.

Hint: It may be numerically easier to first find the value of θ that corresponds to the desired value for δ.

(b) Compare and contrast the results in part (a) with Figures 5.6 and 5.7 for linear kinetics.

5.10 The two parameters that describe the current distribution in a porous electrode with linear kinetics in the absence of concentration gradients are

$$\nu^2 = \frac{ai_o(\alpha_a+\alpha_b)FL^2}{RT}\left(\frac{1}{\kappa}+\frac{1}{\sigma}\right) \quad \text{and} \quad K_r = \frac{\kappa}{\sigma}.$$

(a) How can the parameter ν^2 be described physically?
(b) For the following conditions, sketch out the current distribution di_2/dx across the electrode:

$K_r = 1$,	$\nu^2 \gg 1$,
$K_r = 1$,	$\nu^2 \ll 1$,
$K_r = 0.01$,	$\nu^2 \gg 1$.

5.11 Problem 5.9 provides the solution for the current distribution in a porous electrode with Tafel kinetics in the absence of concentration gradients. Describe the physical parameters? Compare and contrast these results with the analysis that leads to the Wa for Tafel kinetics found in Chapter 4.

5.12 The figure shows the current distribution in a porous electrode for Tafel kinetics. $\delta = 100$ and K_r is a parameter. As expected, for such a large value of δ, the current is highly nonuniform. Further, for the values of K_r chosen, the reaction is concentrated near the back of the electrode. Here, the scale of the ordinate has been selected to emphasize the behavior at the *front* of the electrode. Note that in all cases, rather than getting ever smaller at the front of the electrode, the derivative of current density always goes through a minimum and is increasing at the current collector. Physically explain this behavior.

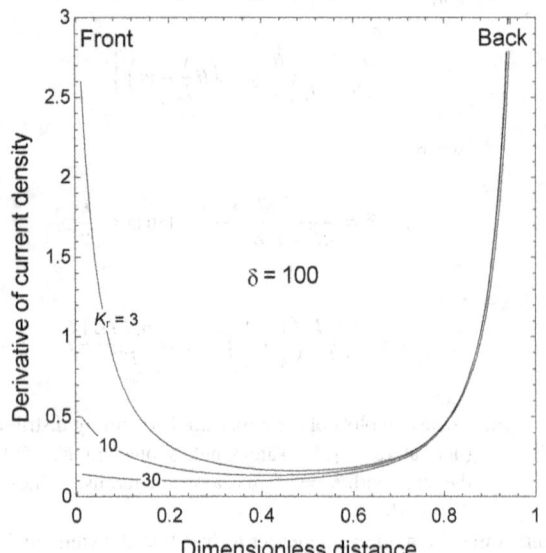

5.13 An electrode is produced with a thickness of 1 mm, $\kappa = 10$ S·m^{-1} and $\kappa = 100$ S·m^{-1}. The reaction follows linear kinetics, $i_o = 2$ A·m^{-2} and the specific interfacial area is 10^4 m^{-1}. It is proposed to use the same electrode for a second reaction where the exchange-current density is much larger, 100 A·m^{-2}. What would be the result of using this same electrode? What changes would you propose?

5.14 The exchange-current density i_o does not appear in the solution for the current distribution for Tafel kinetics (see how ε is defined in Problem 5.9). Why not?

5.15 Rather than the profile shown in Figure 5.7, one might expect that for the case where $\sigma = \kappa$, the distribution would be uniform and not just symmetric. Show that this cannot be correct. Start by assuming that the profile is uniform; then sketch how i_1 and i_2 vary across the electrode. Then sketch the potentials and identify the inconsistency.

5.16 Repeat Illustration 5.3 for a more conductive electrolyte, $\kappa = 100$ S·m^{-1}. If it is desired to keep the reaction rate at the back of the electrode no less than 40% of the front, what is the maximum thickness of the electrode? Additional kinetic data are $\alpha_a = \alpha_c = 0.5$, $i_o = 100$ A·m^{-2}, $a = 10^4$ m^{-1}. Using the thickness calculated, plot the current distribution for solutions with the following conductivities, 100, 10, 1, and 0.1 S·m^{-1}.

5.17 Consider a similar problem to the flooded-agglomerate model developed in Section 5.6, except that now a film of electrolyte covers the agglomerate to a depth of δ. Find the expression for the rate of oxygen transport that would replace Equation 5.50.

5.18 Derive the expression for the effectiveness factor, Equation 5.53. What is the expression for a slab rather than a sphere?

5.19 A porous flow-through electrode was examined in Chapter 4 for the reduction of bromine in a Zn-Br battery.

$$Br_2 + 2e^- \rightarrow 2Br^-$$

The electrode is 0.1 m in length with a porosity of 0.55. What is the maximum superficial velocity that can be used on a 10 mM Br$_2$ solution if the exit concentration is limited to 0.1 mM? Use the following mass-transfer correlation:

$$Sh = 1.29\, Re^{0.72}$$

The Re is based on the diameter of the carbon particles, d_p, and the superficial velocity, which make up the porous electrode.

$$D_{Br_2} = 6.8 \times 10^{-10}\ m^2\ s^{-1}, \quad d_p = 40\ \mu m,$$
$$v = 9.0 \times 10^{-7}\ m^2\ s^{-1}.$$

5.20 Derive Equation 5.42. Start with Equation 5.36.

5.21 Rearrange Equation 5.57 to provide a *design equation* for a flow-through reactor operating at limiting current. Specifically, provide an explicit expression for L, the length of the reactor, in terms of flow rate, mass-transfer coefficient, and the desired separation.

5.22 A direct method of removing heavy metals, such as Ni^{2+}, from a waste stream is the electrochemical deposition of the metal on a particulate bed. The goal is to achieve as low concentration of Ni at the exit for as high a flow rate as possible. A flow-through configuration is proposed. However, here only the negative electrode is porous, and the counter electrode (+) is a simple metal sheet. Would you recommend placing the counter electrode upstream or downstream of the working electrode? Why? For this analysis, assume $\sigma \gg \kappa$, and that the reaction at the electrode is mass-transfer limited. *Hint:* Develop an expression for the change in solution potential similar to Equation 5.63.

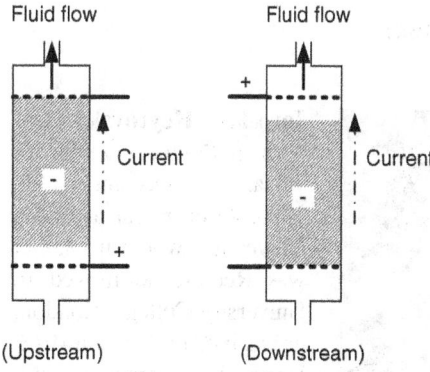

JAROSLAV HEYROVSKÝ

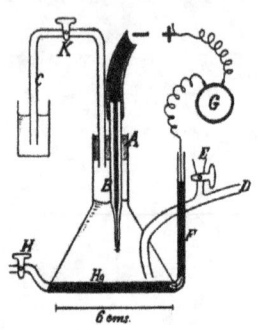

Heyrovsky's original drawing of the hanging mercury drop electrode apparatus.
Source: Heyrovsky 1924. Reproduced with permission of Royal Society of Chemistry.

Jaroslav Heyrovský was born in Prague, Czechoslovakia, on December 20, 1890. After a year at Czech University, where his father was Rector, he moved to University College, London; he received his B.Sc. in 1913. He continued graduate studies on the electrochemistry of aluminum under the direction of Frederick G. Donnan, the eminent Irish physical chemist. One of the challenges for these studies was the oxidation of aluminum during the experiment. Their innovation was to study flowing aluminum so that the surface was continually refreshed. The advantages of a liquid metal surface that is continually renewed foreshadowed his subsequent work on mercury electrodes.

Like so many of the individuals featured in these biographies, his life was shaken by the World Wars. Perhaps because of his single-minded focus, for Heyrovský these were more distractions than life-altering moments. After just a year conducting research in London, World War I broke out in 1914. Jaroslav happened to be in Prague visiting his family at this time and could not return to London. He was called up to service in the Austro-Hungarian Army, but served his time at a hospital in Prague dispensing chemicals and taking X-rays. Throughout the war, he was able to continue his research and was awarded a Ph.D. from Czech University on September 26, 1918, just before the armistice that ended the war.

Heyrovský devoted more than 40 years to the study of electrochemistry at the dropping mercury electrode. Here, mercury flows through a narrow capillary into an electrolyte solution, and every few seconds a drop falls off. Current flows between the drop and the pool below. He was made director of a newly formed Physical Chemistry Department in 1922, and before long Prague was the center of research on dropping Hg electrodes. The same year he measured the first *polarogram*, a graph showing current from a dropping Hg electrode versus applied potential. The term *polarography* was coined in 1925 while working with Masuzo Shikata, who helped systematize the process and create *polarography*, the first fully automatic analytical instrument. In 1939, the Nazi occupation of Czechoslovakia resulted in the closing of the University; yet Heyrovský was able to continue his work, although without students and coworkers.

He was awarded the Nobel Prize in Chemistry in 1959 for development of polarographic method or polarography, which in his own words is

> ". . . the science of studying the processes occurring around the dropping-mercury electrode. It includes not only the study of current-voltage curves, but also of other relationships, such as the current-time curves for single drops, potential-time curves, electrocapillary phenomena and the streaming of electrolytes, and its tools include besides the polarograph, the microscope, the string galvanometer, and even the cathode ray oscillograph."

He is sometimes referred to as the father of electroanalytical chemistry. Today, methods involving Hg electrodes have largely been replaced by other techniques, such as a variety of spectroscopic methods. Yet, these early studies are the foundation of electroanalytic methods; and Heyrovský was one of the pioneers that recognized the importance of controlling potential and measuring limiting currents. Professor Heyrovský died on March 27, 1967.

Image Source: Archive of J. Heyrovsky Institute of Physical Chemistry, CAS.

Chapter 6

Electroanalytical Techniques and Analysis of Electrochemical Systems

The methods described here are used to screen or evaluate the suitability of electrodes and electrolytes for specific applications, to measure transport and kinetic parameters, and to diagnose the performance of electrochemical systems. As an example, it is often desirable to identify the physical processes that control the behavior of an electrochemical cell. Often the procedures explored in this chapter are the first tests performed when one is confronted with a new electrode, electrolyte, or system. An important goal of this chapter is to give you an opportunity to apply the fundamental principles of thermodynamics, kinetics, and transport that you learned in previous chapters to real systems of interest. Though not exhaustive, the approaches here will help you become familiar with basic electroanalytical methods and provide a foundation for the analysis of more complex systems in the future.

6.1 ELECTROCHEMICAL CELLS, INSTRUMENTATION, AND SOME PRACTICAL ISSUES

The basic electrochemical cell used for analysis of electrochemical systems consists of three electrodes: the working electrode (WE), the counter electrode (CE), and a reference electrode (RE). The three-electrode setup is shown in Figure 6.1. The working electrode is where the electrochemical reaction being studied occurs. Typically, the potential or current of the working electrode is controlled. Of course, an electrochemical cell must contain at least two electrodes. The second electrode is called the counter electrode. Sometimes this electrode is referred to as an auxiliary electrode. The counter electrode carries current and completes the circuit. The detailed reactions that take place on the counter electrode may or may not be of interest, depending on the experiment. Finally, whenever possible, we want to include a reference electrode in an electrochemical cell in order to allow us to separate the processes that occur on the working electrode from those at the counter electrode. The reference electrode does not participate in the reactions (no current is passed through the reference electrode), but provides a reference point for measuring and/or controlling the potential. Because there is no current flow to the reference electrode, its potential stays as close as possible to the reversible potential of the electrode.

There are two basic modes of operation for the three-electrode setup. The first mode is called *potentiostatic control*. Here the current flowing through the WE is adjusted until the desired potential is measured between the working and reference electrodes. Generally, we prescribe the potential as a function of time and measure the current that is required to achieve this potential. The second mode of operation is *galvanostatic control*. In this case, the value of the current is set and the potential of the WE is measured relative to the RE. We can specify the current as constant or as a function of time. There is need for both modes of operation and, fortunately, modern instrumentation makes this control straightforward.

The high-precision instrument used for such measurements is called a potentiostat/galvanostat. The typical potentiostat may be limited to no more than a few amperes of current. Quite often, with laboratory fuel cells, for instance, the engineer may require currents on the order of 50 A or more. A battery for an automobile may draw hundreds of amperes when a short burst of power is needed. In these situations, an electronic load is used rather than a potentiostat. It may be possible to operate in a near-potentiostatic mode, but often there is no RE included and the precision is reduced. Also consider the case when we are cycling batteries or electrochemical capacitors. It is common to have dozens if not hundreds of cells operating at the same time. Providing the level of control that a high-quality

---

Electrochemical Engineering, First Edition. Thomas F. Fuller and John N. Harb.
© 2018 Thomas F. Fuller and John N. Harb. Published 2018 by John Wiley & Sons, Inc.
Companion Website: www.wiley.com/go/fuller/electrochemicalengineering

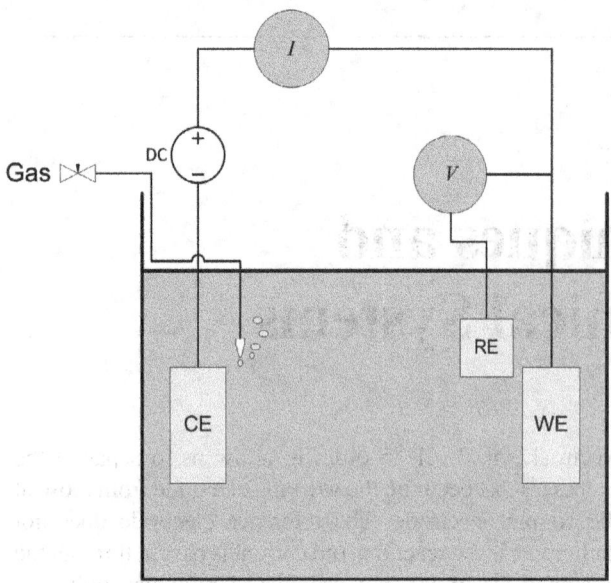

Figure 6.1 Schematic diagram of typical three-electrode setup.

potentiostat offers is not economically feasible. In these instances, automatic cyclers are used.

There are several factors that need to be considered when analyzing an electrochemical system. Suppose that we are examining the oxidation and reduction of a particular species in the electrolyte. In this situation, the WE provides a surface on which the electron-transfer reaction can take place, but does not react itself. Noble metals like Pt or Au that are stable (nonreactive) over a wide range of conditions are often used for this purpose. Conversely, if we investigate the active material of a battery, the WE is the active material, which would participate directly in the electrochemical reaction. Similarly for corrosion studies (Chapter 16), the working electrode participates in the reaction.

As introduced in Chapter 2, a reference electrode should consist of well-defined materials and be capable of carrying out a specific, known electrochemical reaction. Ideally, the reference electrode reaction should involve an ion present in the electrolyte so that liquid junctions are avoided and additional corrections to the potential are not needed. Selection of reference electrodes is explored in Problem 6.1. Another consideration is the placement of the reference electrode. With current flow between the working and counter electrodes, there will be a potential drop through the electrolyte. As already noted, essentially no current flows through the RE; consequently, there is no polarization of the RE (this is just another way of saying that the surface overpotential of the reference electrode is zero). Nevertheless, the measured potential is affected by the potential drop in solution, and therefore depends on the placement of the reference electrode. Generally, we cannot eliminate the impact of the potential drop on the measured potential, but we can correct for it. This correction is examined in Section 6.9. The ideal situation is to have an infinitesimally small reference electrode located just outside of the double layer so that no correction for ohmic drop in solution is necessary.

Similarly, the counter electrode should be selected judiciously. This choice will vary according to what we are trying to accomplish. Frequently, it may be a relatively inert material such as Au or Pt. At times, the reaction at the counter electrode may not be important, and we may electrolyze the solvent or corrode the CE. For example, the evolution of oxygen (anodic) or hydrogen (cathodic) is commonly done at the CE. In all cases, you should know what reaction is occurring at the CE, and you should be aware of issues that may result from these reactions. A minute dissolution of metal from the CE may be transported to and interfere with the reaction at the WE. For example, Pt like Au is generally inert, but when studying oxygen reduction on a nonprecious metal catalyst, Pt would be a poor choice for the CE. Typically, the area of the CE should be much larger than the area of the WE; when this is true, the current density at the CE is significantly lower than that at the WE so that the cell potential and overall behavior of the cell are not limited by processes at the counter electrode.

In addition to the electrodes, there are factors related to the electrolyte that should be considered. In Chapter 4, we introduced the concept of a supporting electrolyte. As we saw before, the mathematical analysis is greatly simplified with a supporting electrolyte because we can neglect migration. In addition, the resistance of the solution is lowered. Supporting electrolyte is used with a number of electroanalytical techniques where it is beneficial and where the introduction of the additional electrolyte does not compromise other aspects of the experiment. It is also common to remove oxygen from the electrolyte. For instance, if oxygen reacts under the conditions (potential and composition) present, it may interfere with the reactions of interest. Oxygen removal is accomplished by bubbling an inert gas such as nitrogen or argon through the electrolyte to strip out any dissolved oxygen in solution (Figure 6.1). In contrast, if we are studying the reduction of oxygen, we would bubble oxygen through the solution to saturate the solution.

What exactly are we doing when we use a potentiostat to apply a potential? The physical situation is shown in Figure 6.2, which shows the CE, RE, and WE connected to a potentiostat. The current through the WE (measured with the I/E converter) and potential difference between the WE and RE (measured with the electrometer) are sensed. To change this potential, the instrument passes current between the working and the counter electrodes. When you select a value of E_{app} on the potentiostat, the control

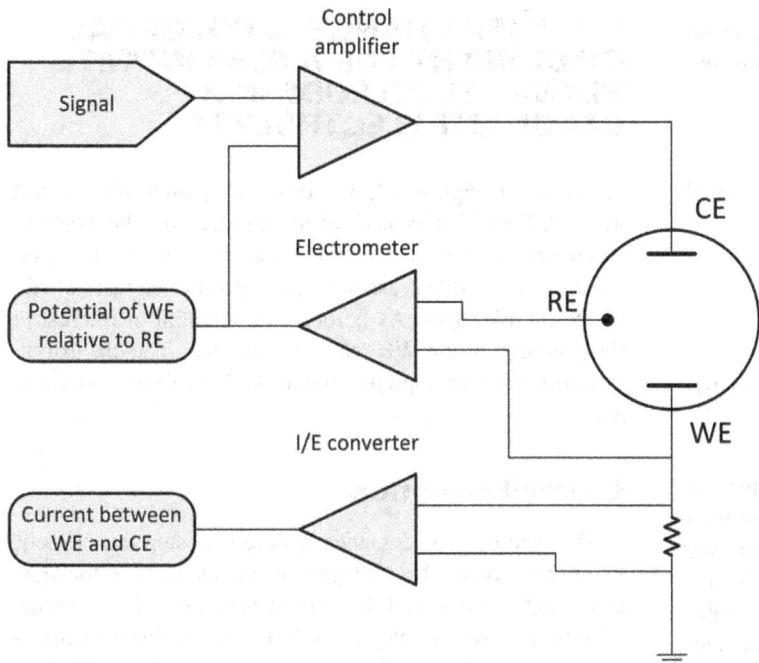

Figure 6.2 Setup and operation of a potentiostat.

amplifier drives whatever current is needed through the WE so that the measured potential matches E_{app}. If that potential is greater than the equilibrium potential, then the resulting current will be anodic at the working electrode. Conversely, application of a potential below the equilibrium potential would drive the cathodic reaction at the working electrode.

What precisely does E_{app} represent? E_{app} can be expressed as follows:

$$E_{app} = U_{working/ref} + \eta_s + \eta_{conc} + \Delta\phi_{iR}. \quad (6.1)$$

$\Delta\phi_{iR}$ is the potential difference between the solution at the location of the reference electrode and a location just outside the double layer of the working electrode. This potential drop is well defined if the current distribution is uniform, or nearly so, and must be accounted for because it is not practical to locate the reference electrode just outside the double layer (see Section 6.9). The influence of the concentration overpotential is often neglected and can be reduced through the use of convection. Finally, a note about the sign of the current. According to the convention that we have used and Equation 6.1, you might expect a positive current when E_{app} is greater than the equilibrium potential. This is not necessarily the case and will depend on the instrument that you are using. A voltage greater than the equilibrium voltage will always yield an anodic current, but that current may be displayed as a negative current by some potentiostats.

6.2 OVERVIEW

There are a multitude of experiments that can be performed with the basic three-electrode setup described above. Through a series of examples, we will explore a few of the more common ones, highlighting the main information that is obtained from each and its advantages and disadvantages. The key features of the any experiment are (1) the geometry of the electrode and the system, (2) the flow of electrolyte, and (3) control of potential or current. The examples that we will treat in this chapter are shown in Table 6.1. Of course, the same principles illustrated in these examples can be applied to different situations. Hence, an understanding of these experimental techniques will help to equip you with the tools needed to analyze any system that may be of interest.

Each of the features will be explored in more detail in the upcoming sections and are only briefly discussed here.

Table 6.1 Experimental Features Considered in This Chapter

Geometry	Fluid flow	Control
Planar	Stagnant	Potential or current step
Spherical	Convection	Potential sweep (CV)
Disk	Infinite rotating disk	Small sinusoidal perturbation (EIS)

Mass transfer has a central role in the analysis of these systems. Two key equations from Chapter 4 are the Nernst–Planck equation and mass conservation:

$$\mathbf{N}_i = \underbrace{-z_i u_i F c_i \nabla \phi}_{\text{migration}} - \underbrace{D_i \nabla c_i}_{\text{diffusion}} + \underbrace{c_i \mathbf{v}}_{\text{convection}}, \quad (4.3)$$

and

$$\frac{\partial c_i}{\partial t} = -\nabla \cdot \mathbf{N}_i + \mathcal{R}_i. \quad (4.10)$$

For the circumstances considered in this chapter, there is no homogeneous reaction ($\mathcal{R}_i = 0$). *We will also assume that there is a supporting electrolyte so that migration can be neglected.* In addition, we assume that the diffusion coefficient is constant and that the fluid is incompressible. With these assumptions, the general material balance, Equation 4.10, reduces to

$$\frac{\partial c_i}{\partial t} + \mathbf{v} \cdot \nabla c_i = D_i \nabla^2 c_i. \quad (4.25)$$

Alternatively, a binary electrolyte yields a similar equation by elimination of the potential gradient. The balance equation in this form can be applied to any geometry and will be starting point for our studies in this chapter. The time-dependent term on the left side is included only for transient experiments. The flow and therefore the velocity field, $\mathbf{v}$, will either be zero or a well-defined known value. An initial condition is required for transient problems. In addition, we must specify boundary conditions. Since electrode reactions occur at the surfaces, electrode kinetics will enter as boundary conditions. Most often the concentration is known far from the electrode and can be used as one of the boundary conditions. For the second boundary condition, the flux is frequently specified at the electrode surface. Our approach is to identify the geometry, characterize the flow, and specify the boundary conditions. As we examine different electrochemical experiments, we will be interested in determining which reactions take place over a specified range of potentials. For these reactions, we want to know whether the reaction is controlled by kinetics or mass transfer or is under mixed control. If under mixed control, how much of the polarization is due to kinetics, ohmic resistance, and concentration overpotentials. Often, the objective is to measure kinetic and transport parameters for the system. Keep these types of things in mind as we describe several different types of experiments with their corresponding analyses and interpretations.

6.3 STEP CHANGE IN POTENTIAL OR CURRENT FOR A SEMI-INFINITE PLANAR ELECTRODE IN A STAGNANT ELECTROLYTE

As shown in Figure 6.3, we consider a planar WE located at $x = 0$. The CE is located sufficiently far from the working electrode so that it has no effect on the results. The reference electrode is assumed to be positioned just outside of the double layer. As before, the potential of interest is the potential of the WE relative to the RE. The electrolyte is assumed to be stagnant so that the bulk fluid velocity is zero.

Cottrell Equation

In this section we consider a potential step experiment where the potential is changed instantaneously at the start of the experiment, and the current is measured as it varies with time. This technique is referred to as *chrono-amperometry*, where "chrono" refers to time and "amperometry" to the measurement of current. We restrict our analysis to a case where the overpotential of an electrode is changed from zero to some large positive or negative value, a magnitude so large that a limiting current is reached.

The case that we examined in Illustration 4.2 is essentially the same as that considered here. For the one-dimensional, planar geometry with no convection, Equation 4.25 reduces to

$$\frac{\partial c_i}{\partial t} = D_i \nabla^2 c_i = D_i \frac{\partial^2 c_i}{\partial x^2}, \quad (6.2)$$

which is also known as Fick's second law. The initial and boundary conditions for this problem are

Initial condition $\quad t = 0, \quad c_i = c_i^\infty,$
Boundary conditions $\quad x = 0, \quad c_i = 0,$
$\quad x \to \infty, \quad c_i = c_i^\infty.$

These conditions assume that the reaction is limited by the transport of a reactant to the surface. One could also consider the case where the reaction is limited by transport away from the surface owing to, for example, a saturated layer on the surface, with similar results. The solution is

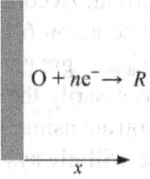

Figure 6.3 One-dimensional planar working electrode.

well known and can be expressed in terms of the similarity variable η, where $\eta = x/\sqrt{4Dt}$. The concentration profile in terms of η is

$$\frac{c_i}{c_i^\infty} = \text{erf}(\eta^2),$$

where the error function is defined as

$$\text{erf}(x) \equiv \frac{2}{\sqrt{\pi}} \int_o^x \exp^{-\eta^2} d\eta.$$

The concentration profiles under these conditions were shown in Figure 4.3. For electrochemical systems, we are generally interested in the current or potential response. If we consider that the diffusing species undergoes an electron-transfer reaction at the surface, the flux of that species to the surface can be related to the current by Faraday's law:

$$i = nFN_i|_{x=0} = nFD_i \frac{dc_i}{dx}\bigg|_{x=0}. \quad (4.17)$$

Differentiating the error-function solution above and substituting yields

$$i = \frac{nF\sqrt{D_i}c_i^\infty}{\sqrt{\pi t}}. \quad (6.3)$$

Here we see that the current depends inversely on the square root of time. This relationship is known as the *Cottrell equation*. Regardless of which electrochemical system is being analyzed, you should always be mindful of how the current changes with time. This dependence on the reciprocal of the square root of time is indicative of mass-transfer control. Also note that with this semi-infinite geometry, a steady-state solution does not exist.

What can we learn by using the potential step method as an analytical technique? A plot of the current versus the reciprocal of $\sqrt{t}$ results in a straight line, with the current density approaching zero at long times. From the slope of the line, either the diffusivity, D_i, the number of electrons transferred, n, or concentration, c_i^∞, can be calculated if the other quantities are known. We'll see later in this chapter that there are other, perhaps better, approaches to determining these quantities. Nonetheless, it is worth taking a minute to examine a couple of our assumptions. First, we assumed that the CE was far enough away to be treated as infinity. Just how far is far enough? The length over which concentration changes occur is proportional to $\sqrt{Dt}$, and can be seen in Figure 4.3. If the distance between electrodes is at least an order of magnitude larger, the supposition is good. Second, we assumed the electrolyte was stagnant. Often it is difficult to achieve a stationary electrolyte with

this configuration. Small vibrations in the laboratory can affect the results. Additionally, the flow of current will cause some heat generation and temperature differences, as well as concentration differences. Both result in density differences and free convection as described in Chapter 4. The net result is that the flow is not zero; and worse yet, the fluid velocity is not well defined. These challenges can be mitigated if the time of measurement is short.

If the overpotential is not sufficiently large so that the reaction is mass-transfer controlled, there is no simple analytical solution. Thus, this experiment depends on achieving a limiting current. Finally, we must consider the possibility of side reactions occurring. The overpotential cannot be so large as to promote other reactions, such as the electrolysis of the solvent.

Sand Equation

Whereas the above analysis describes a *controlled potential* technique, we can easily imagine a *controlled current* method. In this experiment, a step change in current is made for the same semi-infinite, planar system. Generally the equipment for controlled current experiments is cheaper, but unfortunately the analysis is not as straightforward as with a controlled potential method.

Our starting point is again the convective–diffusion equation (Equation 4.25) and, since the geometry and assumptions are the same, it reduces to Equation 6.2. The boundary conditions are different for this galvanostatic experiment. In this case, the current $I(t)$ is prescribed at the surface; therefore, the flux of any species involved in the reaction is known at all times. Here we will only consider a step change to a constant value of current. Figure 6.4 shows the concentration profiles at several

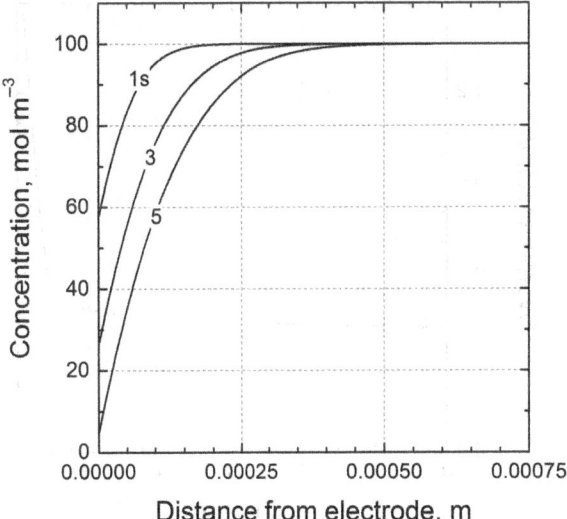

Figure 6.4 Concentration of reactant near electrode surface for step change in current.

different times. Note that the slope at the electrode surface is constant—directly proportional to the current density. At longer times, the concentration of limiting reactant at the surface decreases, and the effects of the step change in current propagate farther away from the electrode surface. The time at which the surface concentration drops to zero is known as the *transition time*, τ, which is given by

$$\frac{i\tau^{0.5}}{c_i^\infty} = \frac{nFD_i^{0.5}\sqrt{\pi}}{2} \,. \qquad (6.4)$$

This relationship is known as the *Sand* equation. For a semi-infinite case, the current cannot be sustained indefinitely because eventually the concentration of the reactant approaches zero at the transition point. As this reactant concentration decreases, the overpotential at the electrode increases to maintain constant current. At some point, the overpotential becomes very large and a second reaction, for example, electrolysis or corrosion, occurs. This behavior is shown in Figure 6.5, which plots the overpotential as a function of time. This overpotential remains relatively constant until the concentration of reactant approaches zero. At that point, the exchange-current density becomes small and the overpotential increases sharply. Similar to the Cottrell equation, this analysis allows one to determine either n, the bulk concentration, or the diffusivity, assuming the other two quantities are known. Like the step change in potential experiment, it can be difficult to prevent unwanted convection, leading to differences between the experimental results and the analysis. Note that, in addition to the utility of this specific analytical technique, it is important to understand the behavior exhibited in Figure 6.5, which is another indication of a system that is mass-transfer limited.

> **The system is likely mass-transfer limited if**
> 1. at constant potential, the current decreases inversely with the square root of time;
> 2. at constant current, the overpotential increases dramatically to sustain the current;
> 3. at steady-state, a change in potential does not result in a change in current.

6.4 ELECTRODE KINETICS AND DOUBLE-LAYER CHARGING

Charging Time

Recall from Chapter 3 that the surface of the electrode is usually charged, and that the structure of the interface is described as a double layer. The charge associated with the double layer depends on the electrode potential, and current must flow to alter this charge as either more or less charge is stored at the interface when the potential is changed. In the above analysis, we neglected this charging current since it was presumably not important for the physical situation and timescales of interest. In contrast, here we explore the current associated with double-layer charging in some detail and identify its effect on our analysis. We start by creating a simple *equivalent circuit* model of the electrode. While there are numerous equivalent circuits developed to describe electrochemical systems, we will limit ourselves to the simplified *Randles* circuit shown in Figure 6.6. This equivalent circuit is an idealized electrical circuit that mimics the electrical response of the actual electrode. In general, there may or may not be a close physical connection between the electrical components of the equivalent circuit and the actual physical processes occurring at the electrode. Here, we consider a simplified system where the physical connection is apparent and important. The first element is R_Ω,

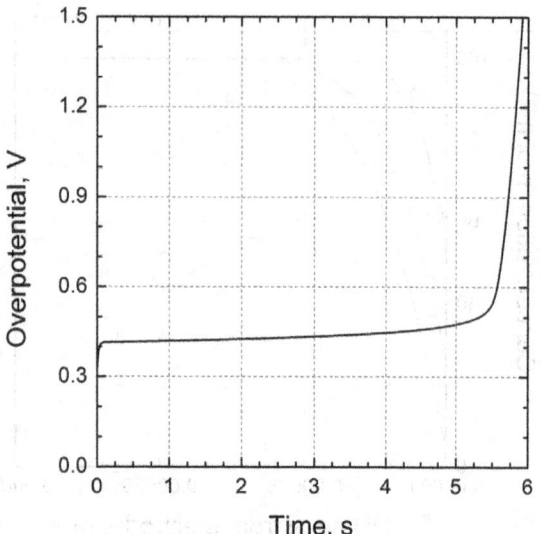

Figure 6.5 Overpotential for step change in current. At long times, the overpotential increases sharply.

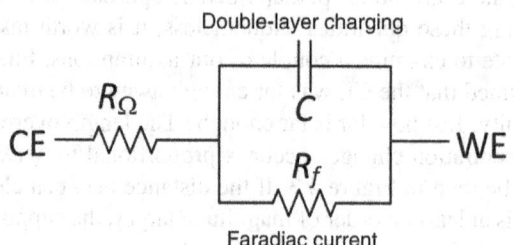

Figure 6.6 Simplified Randles equivalent circuit for an electrode.

which represents the resistance to current flow in the electrolyte. We next see that there are two paths for current to flow to or from the electrode. The first is associated with the charging of the double layer. This current does not involve electron transfer at the surface and is termed *nonfaradaic* or a charging current density, i_c. The second path represents the faradaic current density, i_f, which is the current due to an electron-transfer reaction at the surface. The faradaic current density is generally described with the Butler–Volmer equation discussed in Chapter 3. The sum of these two is the measured current density, i_m:

$$i_m = i_f + i_c. \qquad (6.5)$$

Devices using the charge stored in the double layer are treated in detail in Chapter 11 on electrochemical double-layer capacitors. Here we simply want to describe the double layer as a capacitor in parallel with an element representing the resistance to charge transfer, R_f. This equivalent circuit concept will allow us to understand the response of the systems considered here and, more specifically, how the charging current can affect our analysis of electrochemical systems.

The behavior of an ideal resistor is familiar. We can describe the relationship between current and potential with Ohm's law, namely, that the voltage across the resistor is directly proportional to the current flow ($\Delta V = IR_\Omega$). The ideal capacitor behaves differently, which can be expressed mathematically as

$$I = C \frac{dV}{dt}. \qquad (6.6)$$

At steady state, $dV/dt = 0$. Therefore, the current is zero and the capacitor has a fixed charge that depends on the potential across the capacitor, $Q = CV$. Conversely, for a step change in potential or current, the capacitor first behaves as short circuit, providing no initial resistance to the flow of current. In response to a change in current between the WE and CE in Figure 6.4, at first all of the current would flow through the capacitor. Because of the resistance of the electrolyte (R_Ω) in series with the capacitor, this current is finite even for an ideal capacitor. As the charge builds up on the capacitor, the potential across the capacitor increases, as does the resistance to the flow of current through the capacitor. As a result, current begins to flow through the parallel path representing the faradaic reaction. Once the capacitor is fully charged, no current flows through the capacitor, and all of the current is faradaic.

How long does it take to charge the capacitor that represents the double layer? To address this question, let's consider the response of a capacitor in series with a resistor to a step change in potential. This situation is similar to that just described, although simplified further to only consider

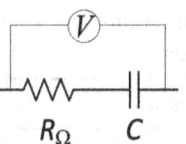

Figure 6.7 Circuit considered for determining double-layer charging time.

the current flow through the capacitor as shown in Figure 6.7. The resistance in series with the capacitor comes from the ohmic resistance of the electrolyte.

As seen in the definition of capacitance, the charge, Q, is proportional to the potential across the capacitor. In practice, we cannot move the charges instantaneously; that is, there is some resistance to current flow. Assume that at time equal to zero, the potential applied to the circuit is changed to V_0. Current will flow in response to the change in voltage, and the charge on the capacitor will vary. After a time, the capacitor will be fully charged, and the current will be zero. We can define Q_0 to be the charge equal to CV_0, which is the charge at the final steady state. The potential across the capacitor is simply Q/C, where Q varies with time, and the potential drop across the resistor in series is IR_Ω. We can use Kirchhoff's voltage law to get

$$V_0 = \frac{Q}{C} + IR_\Omega = \frac{Q}{C} + R_\Omega \frac{dQ}{dt}, \qquad (6.7)$$

where dQ/dt is the current. Equation 6.7 can be integrated to give the fractional charge on the capacitor:

$$\frac{Q(t)}{Q_o} = 1 - e^{-t/R_\Omega C}. \qquad (6.8)$$

The quantity $R_\Omega C$ has units of seconds $\left(\frac{\text{ohm} \cdot \text{coulomb}}{\text{volt}} = \frac{\text{coulomb}}{\text{ampere}} = s\right)$ and is a characteristic time representing the time to charge or discharge the capacitor; it is called the *charging time*. Since the capacitance is the proportionality constant between charge and voltage, Equation 6.8 can also be written as

$$\frac{V(t)}{V_0} = 1 - e^{-t/R_\Omega C}. \qquad (6.9)$$

The current is obtained by differentiation of Equation 6.8:

$$I = \frac{dQ}{dt} = \frac{V_0}{R_\Omega} e^{-t/R_\Omega C}. \qquad (6.10)$$

For a step change in potential, the current decreases exponentially to zero with a time constant $\tau = R_\Omega C$. A large resistance implies that the current is small and, therefore, it takes longer to charge the capacitor. Similarly, as the capacitance increases, it will also take longer to charge or discharge the capacitor.

Table 6.2 Physical Parameters That Affect the Charging and Diffusion Times

Quantity	Description	Value
L	Distance between CE and WE	0.001 m
D_i	Diffusion coefficient	$10^{-9}\ \text{m}^2 \cdot \text{s}^{-1}$
C_{DL}	Electrode capacitance per unit area	$0.2\ \text{F} \cdot \text{m}^{-2}$
κ	Electrical conductivity of solution	$10\ \text{S} \cdot \text{m}^{-1}$

We can gain insight into the timescales that correspond to different physical processes in our electrochemical system by comparing the time constant for capacitor charging with the characteristic time for diffusion:

$$\tau_D = \frac{L^2}{D_i} \quad \text{and} \quad \tau_c = R_\Omega C = \frac{LC_{DL}}{\kappa},$$
diffusion time constant   charging time constant

where C_{DL} is the specific capacitance of the double layer (capacitance per area). If τ_c is much smaller than τ_D, then the current is not affected by the charging process except at very short times following a step change in current or potential. Let's pick some typical values (Table 6.2), and estimate these characteristic times. The resistance in series with the capacitor is taken to be the resistance of the electrolyte between the WE and CE.

The diffusion time is 10^3 s, while the charging time is only 20 μs. This difference tells us that often we don't need to worry about the charging time. There are some cases where this time constant for charging is important as discussed in subsequent sections. Also, charging times can be considerably longer for a porous electrode where the surface area and, therefore, the capacitance is high. The time constant for double-layer charging of porous electrodes is addressed further in Chapter 11, which examines electrochemical double-layer capacitors.

We have assumed that double-layer charging can be represented by a capacitor in series with the solution resistance, and defined a time constant, τ_c, that is equal to the product of the resistance and capacitance. At a time equal to τ_c, the system reaches 63.2% of the steady-state value (see Equation 6.9). It actually takes $3\tau_c$ to reach 95% of steady-state value, which is a reasonable approximation of completion.

It turns out that the resistor and capacitor in series is a simplification that provides a reasonable estimate of the double-layer charging time. In reality, the capacitor is in parallel with the faradaic resistance (see Figure 6.6), which influences the charging time and the extent to which the double layer must be charged since, in the presence of current, not all of the voltage drop needs to be across the capacitor. This issue is explored in more detail in Problem 6.4. A different time constant, $\dfrac{R_\Omega C}{1 + R_\Omega/R_f}$, results that is smaller than that presented above, but approaches the same value as the faradaic resistance becomes large. Consequently, the above treatment overestimates the double-layer charging time and therefore provides a conservative estimate.

Current Interruption

Figure 6.8 shows the voltage response for a step change of current (current step is not shown). At a time of 20 s, the current is increased and held constant. As shown, the potential also changes, but not instantaneously. In fact, multiple time constants are evident. There is an immediate step change in potential associated with the ohmic resistance in the electrolyte. Next we see a relatively rapid increase in potential, shown in more detail in the inset, which is associated with charging of the double layer. Finally, the potential continues to increase. This further increase can be ascribed to diffusion and changes in concentration. Recall from Chapter 4 that there are usually concentration changes associated with the flow of current. Here, the reactant concentration becomes lower, decreasing the exchange-current density; thus, a higher overpotential is required to keep the current constant. Generally, the double-layer charging is fast compared to the time for diffusion. Charging or discharging of the double layer takes place fairly quickly, while concentration changes take significantly longer.

When the cell potential is recorded experimentally for a step change in current, one needs to consider the sampling rate of the measurement. The smooth, continuous curves seen in Figure 6.8 may not always be obtained. This particular challenge is explored in Illustration 6.1.

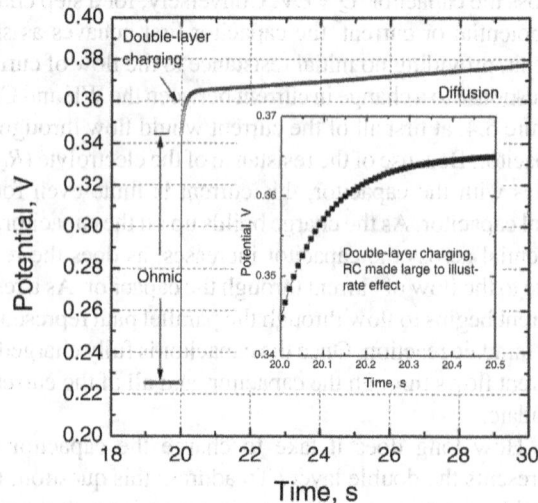

Figure 6.8 Current interrupt showing the charging time and the diffusion time.

A comment regarding the counter electrode is appropriate here. Electroanalytical methods are almost invariably focused on the behavior of one electrode. Therefore, it is important that the other electrode, the counter electrode, does not interfere with the measurement. The counter electrode also has a double-layer capacitance that is in series with that of the working electrode. In order to avoid interference from the counter electrode, we ensure that its capacitance is much larger than that of the working electrode, since it is the smaller capacitor in series that controls the dynamic behavior (Why?). The easiest way to increase the size of the counter electrode capacitance is to increase its active area. Hence, we see a second reason for the counter electrode to be significantly larger than the working electrode in electroanalytical experiments.

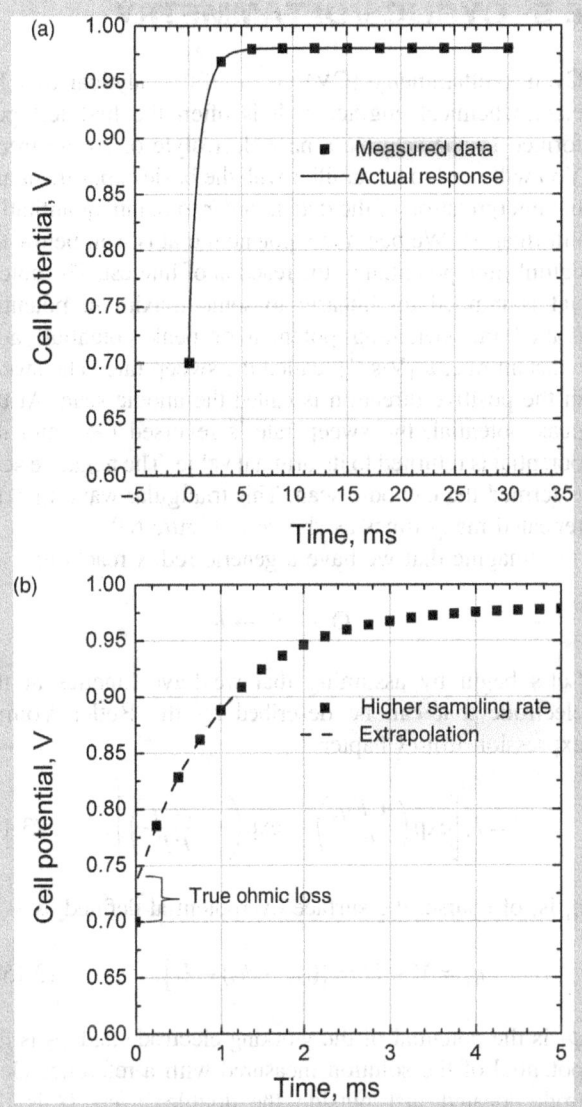

ILLUSTRATION 6.1

During operation of a low temperature H_2-O_2 fuel cell at steady state, it is desired to periodically measure the resistance of the cell. Current interruption is proposed as a means of accomplishing the measurement. It is common to report the resistance in [$\Omega \cdot m^2$] for electrochemical systems. We used the same symbol, R_Ω for both resistance quantities. Here, the current density is used in place of the current:

$$R_\Omega = \frac{\Delta V}{\Delta i}.$$

At 0.4 $A \cdot m^{-2}$, the potential of the cell is 0.7 V. The sampling rate is every 3 ms, resulting in the data shown in part (a) of the following figure for current interruption (step to zero current) at $t = 0$. The estimated resistance is $(0.98 - 0.7)/0.4 = 0.70\ \Omega \cdot m^2$. This value seems large, and your colleague suggests increasing the sampling rate. When the sampling rate is increased to once every 50 μs, the data in part (b) of the following figure are obtained. Clearly, the potential of the cell changes rapidly during the first few milliseconds following interruption of the current. The potential is then extrapolated back to time zero in order to estimate its value at the instant just after the current is interrupted. This value is 0.73 V:

$$R_\Omega = \frac{0.73 - 0.70}{0.40} = 0.075\ \Omega \cdot m^2$$

Why was the first calculation so far off? When the current is interrupted, there are reductions in ohmic losses, kinetic losses, and mass-transfer losses. Although the ohmic loss changes instantaneously when the current is interrupted, other losses have a time constant associated with them. In particular, it takes time for the charge in the double layer to reach a new steady state; this change in the double layer is the main reason for the delayed response of the potential at short times. The sampling time must be short enough to distinguish the ohmic losses from losses associated with charging or discharging of the double layer.

6.5 CYCLIC VOLTAMMETRY

Cyclic voltammetry (CV) is a useful analytical tool for electrochemical engineers. It is often the first test performed to characterize a new electrolyte or an electrode. You will need to be familiar with the basic experiment and the interpretation of the data in order to obtain quantitative information. We begin at a potential that is well below the equilibrium potential of the reaction of interest. The potential is ramped up linearly to some maximum potential (called the switching potential or peak potential) at a constant rate, ν [V·s^{-1}], called the sweep rate. The sweep in the positive direction is called the anodic scan. At the peak potential, the sweep rate is reversed ($-\nu$) and the potential is returned to its original value. The negative scan is termed the cathodic scan. This triangular wave may be repeated many times as shown in Figure 6.9.

Imagine that we have a generic redox reaction:

$$O + ne^- \leftrightarrow R$$

Let's begin by assuming that we have kinetics at the electrode that can be described by the Butler–Volmer expression from Chapter 3:

$$i = i_o \left[\exp\left(\frac{\alpha_a F \eta_s}{RT}\right) - \exp\left(-\frac{\alpha_c F \eta_s}{RT}\right) \right]. \quad (3.17)$$

η_s is, of course, the surface overpotential defined as

$$\eta_s = V - U = [(\phi_1 - \phi_2) - U]. \quad (3.15b)$$

ϕ_1 is the potential of the working electrode and ϕ_2 is the potential of the solution measured with a reference electrode located just outside the double layer. U is the equilibrium potential defined against that same reference electrode.

If we simply use Equation 3.17 to calculate the current when the surface overpotential undergoes a triangle wave as depicted in Figure 6.9, the response is the solid black line shown in Figure 6.10. Note that the current is identical on the anodic and cathodic sweeps and only depends on the overpotential. In other words, there is only one value of the current for each value of the overpotential, and the current does not depend on the direction that the voltage is changing. This response was calculated with what we learned in Chapter 3 for steady-state conditions under kinetic control. However, for the common sweep rates used in CV, too much of the physics of the problem have been neglected. First of all, since the voltage changes constantly during the experiment, the double layer is also changing by either being charged or discharged; consequently, there is current associated with the constant charging or discharging of the electrical double layer. An equivalent circuit representation of this situation was shown in Figure 6.6. The faradaic current is given by the BV equation. The non-faradaic current associated with the double-layer charging is

$$i_c = C_{DL} \frac{dV}{dt} = \nu C_{DL}, \quad (6.11)$$

where C_{DL} is the double-layer capacitance of the electrode [F·m^{-2}]. We also make the simplifying assumption that C_{DL} does not depend on potential. With this assumption, Equation 6.11 tells us that the current associated with the double-layer charging is proportional to the sweep rate. The double-layer current is shown as the solid gray line in Figure 6.10 for the forward and reverse voltage sweeps. It is constant for a given sweep rate and direction, but changes sign when the sweep direction changes from positive to negative. The total current is the sum of the faradaic and double-layer current:

$$i = i_f + i_c. \quad (6.12)$$

The total current response for the CV that includes both the faradaic and charging current is shown as the dashed lines

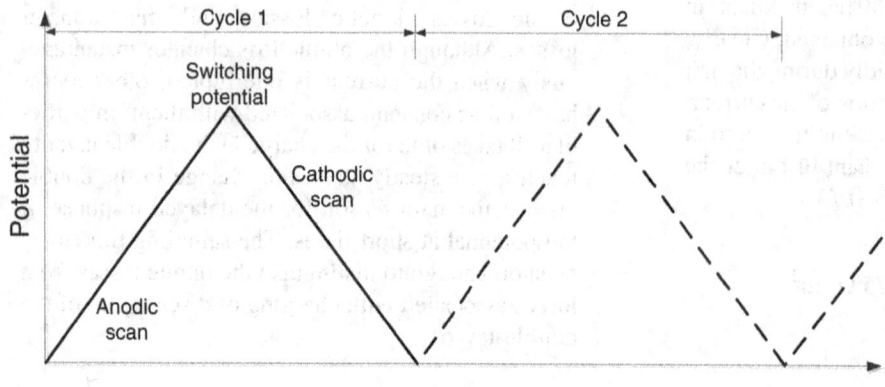

Figure 6.9 Triangle wave used in cyclic voltammetry.

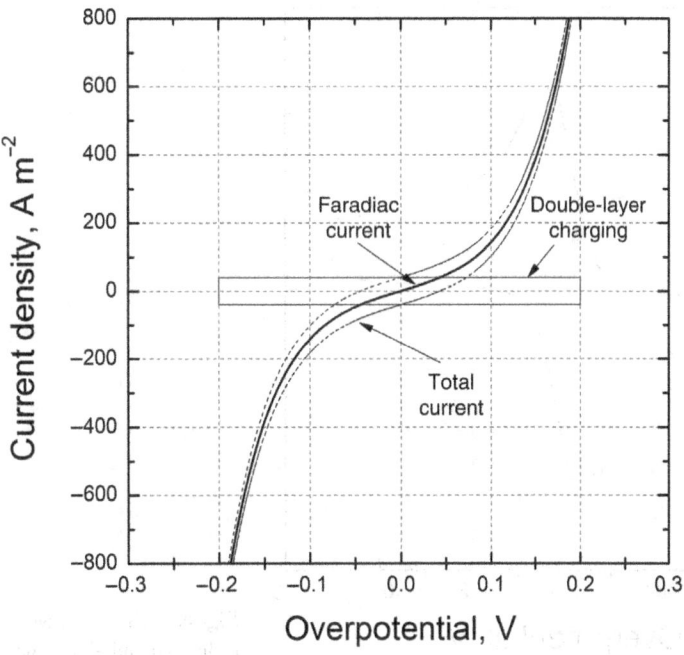

Figure 6.10 Cyclic voltammogram in the absence of mass transfer.

in Figure 6.10. The figure showing the relationship between potential and current is called a cyclic voltammogram, although it doesn't yet have the prototypical shape.

Another important phenomenon that needs to be included is diffusion of products and reactants. So far we have assumed that both species (O and R) are present uniformly at the bulk values in the solution. Recall from Chapter 3 that the exchange-current density depends on the concentration of reactants. Therefore, a changing concentration implies that the exchange-current density, i_o, is not constant. Also, as we have seen from Chapter 4, mass transfer of species to and from an electrode can limit the rate of reaction. The equilibrium potential of the cell can change with composition too—think back to the Nernst equation from Chapter 2.

Before we go through the mathematical analysis of this situation, let's reason through what might happen during the CV when mass transfer is considered. Imagine that our bulk solution has an equal amount of R and O. If the current follows the trajectory as shown in Figure 6.9 for the anodic sweep, we would quickly deplete R from the solution near the electrode. Molecular diffusion would allow transport of R from the bulk to the electrode surface, but there would be a finite maximum rate of replenishment; that is, a limiting current. Above this point, increases in potential would not result in an increase in the current. At the same time, we would be building an excess of O at the electrode surface. When the peak potential in the triangular wave is reached, the potential sweep is reversed. At the time of reversal, there is little R present and a large amount of O at the electrode surface. Instantaneously, the current associated with the double layer reverses sign (since dV/dt changes sign), and the faradaic current also starts to decrease as the overpotential diminishes. When the potential becomes low enough that the cathodic reaction is favored, O is consumed. This cathodic reaction continues until a limiting current is reached again. The current voltage response for this process is shown in Figure 6.11 for a typical voltammogram with fast or reversible electrode kinetics.

Because of changing concentration profiles, the shape of the voltammogram is different for the initial sweep and subsequent sweeps. Multiple sweeps are common experimentally, and sweeping is most often done experimentally from the open-circuit potential. The simulation results shown in this section follow multiple sweeps, where a quasi-stationary condition has been reached.

The steps for calculating this relationship between the current and voltage are analogous to those worked previously, and assume the presence of supporting electrolyte. For one-dimensional transport in the absence of bulk flow, Equation 4.25 once again becomes the well-known diffusion equation:

$$\frac{\partial c_i}{\partial t} = D_i \frac{\partial^2 c_i}{\partial x^2}.$$

As before, we can use Faraday's law to relate the current to the flux of species at the electrode surface by, for example, setting $i = nFN_O$ (see Equation 4.17), where N_O is the flux

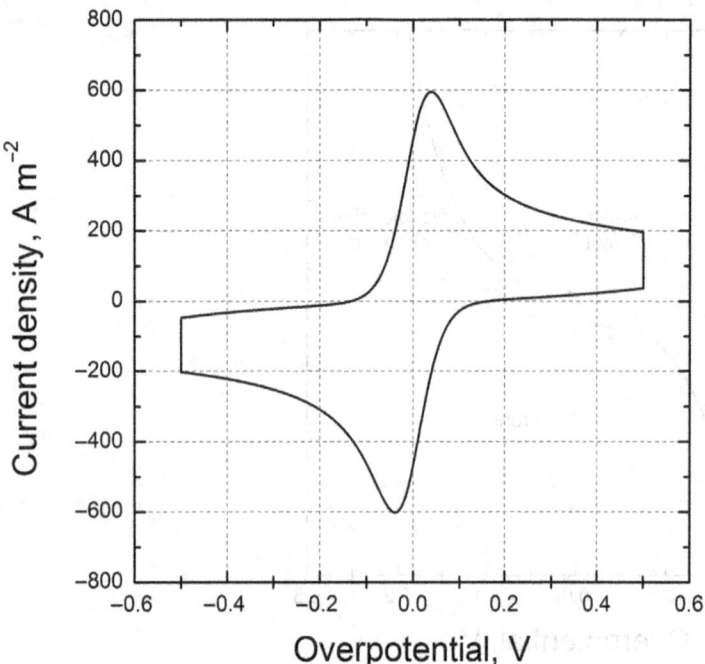

Figure 6.11 Prototypical cyclic voltammogram that includes double-layer charging.

of species O and is positive when O moves away from the surface. Since we are specifying the potential of the WE, this potential combined with the BV equation is one of our boundary conditions. Another boundary condition expresses the fact that the concentrations far from the surface of the WE are equal to the bulk values. Finally, we add an initial condition, which also sets the composition of each species to its bulk value. Both analytical and numerical solutions of the diffusion equation are possible. The results shown here were obtained numerically for multiple voltage sweeps. Note that many of the analytical relationships in the literature, including popular textbooks, apply strictly to a single linear voltage sweep (LSV) rather than to multiple cycles and involve additional simplifying assumptions. Care should be taken to ensure that the assumptions made in deriving a relationship are consistent with your experimental system.

Whether solving the equations analytically or numerically, it is important to express the exchange-current density as a function of concentration. It is that concentration dependence that keeps the current from increasing with potential once the mass-transfer limit is reached. Using Equation 3.20 for the exchange-current density with the reference concentration taken as the bulk concentration and $\gamma_1 = \gamma_2 = 0.5$ yields

$$i_o = i_{o,\text{ref}} \left(\frac{c_R}{c_{R,\text{ref}}}\right)^{0.5} \left(\frac{c_O}{c_{O,\text{ref}}}\right)^{0.5}. \quad (6.13)$$

An additional simplification is possible for reversible systems where kinetic rates are sufficiently rapid that the potential does not move significantly from its equilibrium value. In such cases, the ratio of c_O and c_R at the surface is directly related to the potential through the Nernst equation. This simplification is used in many of the analytical solutions available and is explored as part of a numerical solution in Problem 6.24.

A typical cyclic voltammogram is shown in Figure 6.11 for reversible kinetics. There are two peak currents and two peak potentials—These values are the basis for the primary interpretation of the voltammogram. For reversible kinetics at 25 °C, the distance between the two peaks is

$$\Delta E_p = \frac{0.057}{n} \quad (6.14)$$

as long as the sweep extends ~300 mV above and below the equilibrium potential in either direction. A smaller sweep will lead to a slightly greater distance between the peaks (e.g., $0.059/n$ for a sweep that extends 120 mV). It follows from Equation 6.14 that for reversible kinetics, the number of electrons transferred can be determined from the distance between peaks. Conversely, if n is known, the peak separation can be used to infer whether or not the reaction is reversible.

The concentrations of O and R during CV just before the peak current are displayed in Figure 6.12. Here the concentration of the reduced species is low and the concentration of the oxidized species is high. The penetration depth of this concentration difference depends on the square root of time. Thus, the slower the scan rate, the deeper the penetration, and the lower the limiting current.

Chapter 6 Electroanalytical Techniques and Analysis of Electrochemical Systems 125

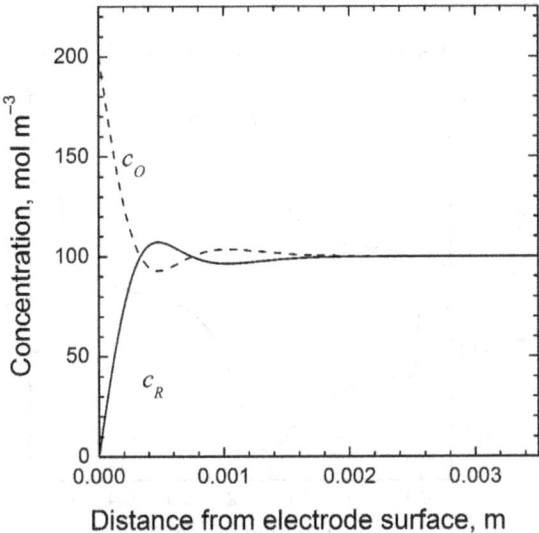

Figure 6.12 Concentration of O and R during potential sweep.

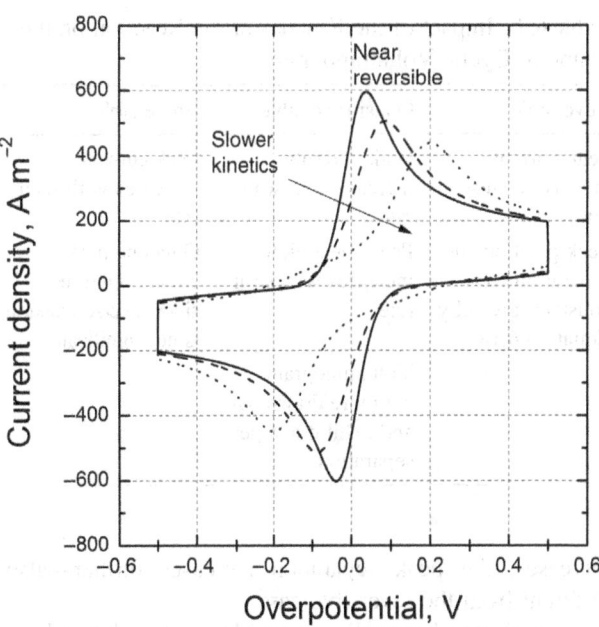

Figure 6.14 Effect of the exchange-current density on the cyclic voltammogram when kinetic limitations are important.

The height of the peaks due to the faradaic current is proportional to $n^{3/2}D^{1/2}\nu^{1/2}c_\infty$, whereas the double-layer charging current is proportional to ν, the scan rate. Figure 6.13 shows the effect of scan rate, ν, on the voltammogram with the double-layer charging removed. The results in Figure 6.13 are for reversible (fast) kinetics. Under such conditions, the peak position is constant and the peak height is a function of the scan rate.

When examining a cyclic voltammogram, the waveforms are frequently described in terms of how facile the electrochemical reactions are: reversible, quasi-reversible, and irreversible kinetics. We've already seen examples of

reversible kinetics (see Figures 6.11 and 6.13). Figure 6.14 provides three CVs for reactions with more and more sluggish kinetics; that is, decreasing exchange-current density, i_o. The peak currents decrease as the kinetics become slower. Also, in contrast to the reversible case, the peak potentials are not constant for finite kinetics. For irreversible or quasi-reversible kinetics, as the kinetics decrease the peak potentials move farther and farther apart. Although not shown in Figure 6.14, as the scan rate

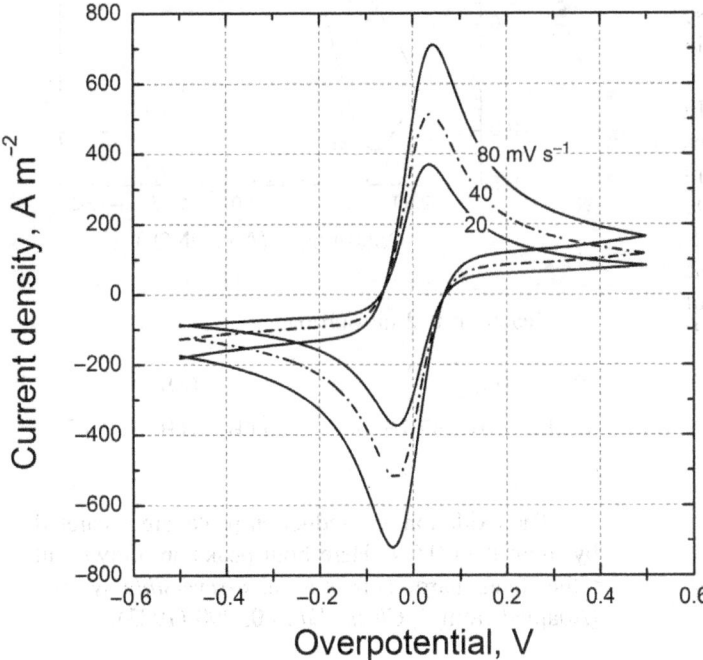

Figure 6.13 Effect of scan rate for a reversible reaction. Double-layer charging has been removed.

Table 6.3 Impact of the Kinetics of the Reaction on the Shape of Cyclic Voltammogram

Reversible	Quasi-reversible	Irreversible
Peak currents increase with scan rate	Peak currents increase with scan rate	Peak current increases with scan rate
Peak position and separation are constant, given by Equation 6.14	Peak separation increases with scan rate	Only one peak current observed since reverse reaction is not significant
	Voltammograms are more drawn out and exhibit a larger separation	

increases, the peak separation increases further—also different from the reversible case.

An electrochemically reversible reaction has a high exchange-current density. As a result, the surface overpotential is small and the potential of the electrode is close to its equilibrium potential as approximated by the Nernst equation, which is the primary condition for reversibility. Quasi-reversible reactions have slower kinetics where the potential differs significantly from the Nernst value. Here we refer to irreversible reactions as reactions for which the reverse reaction (cathodic or anodic) does not take place to any appreciable extent. Unfortunately, these terms are not particularly precise and are not used consistently in the literature. For example, reactions with very slow kinetics are often referred to as electrochemically irreversible reactions, even though they are chemically reversible (the reverse reaction does take place). The behavior is summarized in Table 6.3.

We close this section with one final note. In Equation 6.1 we saw that the applied potential includes the influence of concentration and ohmic drop in solution, due to the fact that the reference electrode is not located just outside the double layer. Consequently, the CV data, as measured, will include these influences and may need to be corrected. Ways for handing these effects during measurement are considered in Section 6.9.

ILLUSTRATION 6.2

Please identify each of the following figures as reversible, quasi-reversible, or irreversible (the answer is below each figure).

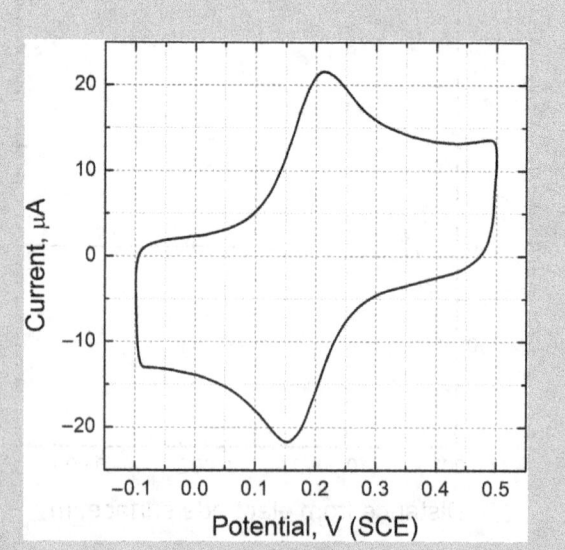

Ferri-Ferro cyanide
$$[Fe(CN)_6]^{3-} + e^- \rightarrow [Fe(CN)_6]^{4-} \quad (U^\theta = 0.55\,V)$$

The peaks are symmetrical and separated by about 60 mV. This reaction is *reversible*. (Adapted from *Phys. Chem. Chem. Phys.*, **15**, 15098 (2013).)

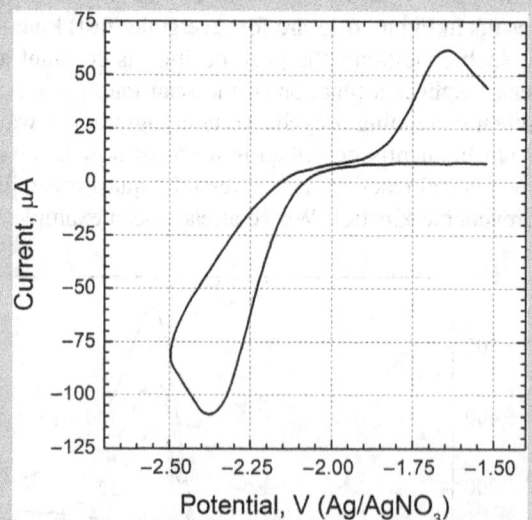

Reaction of 2-nitropropane.

$$CH_3-\underset{|}{\overset{NO_2}{CH}}-CH_3 + e^- \underset{k_b}{\overset{k_f}{\rightleftarrows}} \left[CH_3-\underset{|}{\overset{NO_2}{CH}}-CH_3\right]^-$$

The oxidation and reduction peaks are separated by more than 0.5 V. Here both peaks are drawn out rather than sharp. This is a *quasi-reversible* system. (Adapted from *J. Chem. Ed.*, **60**, 290 (1983).)

6.6 STRIPPING ANALYSES

The underlying principle used in this type of analysis is that a quantitative amount of material is either deposited or removed from a surface or from an amalgam electrode. Because the coulombs passed can be measured very accurately, information about the surface area of an electrode or about the concentration of a metal species in solution can be determined precisely. A couple of illustrative examples will be discussed here, but there are many other applications where a similar analysis is useful.

To illustrate, we consider a method that is used to measure the surface area of a platinum catalyst in a porous electrode. At low potentials and relatively low temperatures, carbon monoxide adsorbs strongly on the surface of the catalyst. If the electrode is held near the potential of the hydrogen electrode, even a small amount of CO (100 ppm, for instance) will completely cover the surface of the electrode. The chemisorption is so strong that, even after removing the exposure to CO, the surface will remain covered until either the temperature is elevated enough to drive the adsorbed CO off or the potential is increased to oxidize the CO:

$$CO_{ads} + H_2O_{ads} \rightarrow CO_{2(g)} + 2H^+ + 2e^- \quad (U^\theta = -0.11 \text{ V}) \quad (6.15)$$

We'll use removal by oxidation to quantitatively determine the amount of adsorbed CO. Specifically, after the surface is covered with CO, a linear sweep of potential is used, up to approximately 1 V relative to hydrogen. Beginning about 0.6 V, the adsorbed CO oxidizes to CO_2 resulting a sharp peak as shown in Figure 6.15. This peak has some resemblance to the peaks seen with cyclic voltammetry, but those were the result of a mass-transfer limitation. Here, the peak occurs because there is a finite amount of CO on the surface. The current versus time can be measured. This current represents both the charge needed for the oxidation of CO and that needed for charging the double layer. In this case, the formation of oxides on the surface of Pt also consumes a small amount of charge, which we will ignore. The double-layer charging (Equation 6.11) must be subtracted and the coulombs associated with CO oxidation are

$$Q_{CO} = \int_{0.4\,V}^{1.0\,V} (I - \nu C) dt.$$

Next we need to relate the charge measured to the surface area of the platinum. Pt has an FCC structure with a lattice parameter of 0.392 nm. From this information, the number of Pt atoms per unit area is calculated. As an example, for the (100) surface, there are 1.3×10^{19} atoms per square meter. We can then relate the area to coulombs of charge passed using the stoichiometry of reaction (6.15) and

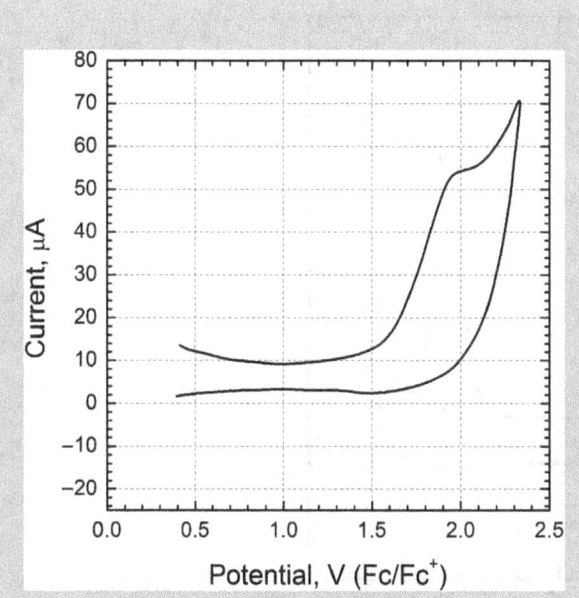

1.5 M LiTFSI in a mixture of ethylene carbonate and diethyl carbonate. There is an oxidation current at high potentials, but no reverse reaction. This system would be described as *irreversible*. In fact, the reaction occurring is the oxidation of the solvent. This test establishes the potential limits for the stability of the solvent. (Adapted from *Phys. Chem. Chem. Phys.*, **15**, 7713 (2013).)

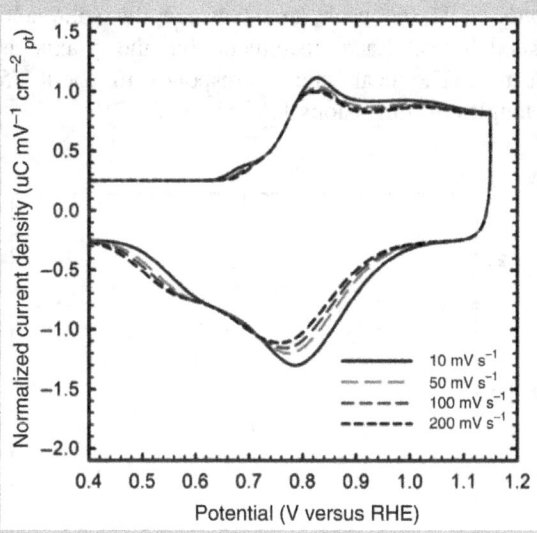

Platinum oxidation

Although the peak potentials are not widely separated, the peaks are extremely drawn out. Furthermore, the peak potentials shift slightly with scan rate This reaction is *quasi-reversible*. (Adapted from *Phys. Chem. Chem. Phys.*, **16**, 5301 (2014).)

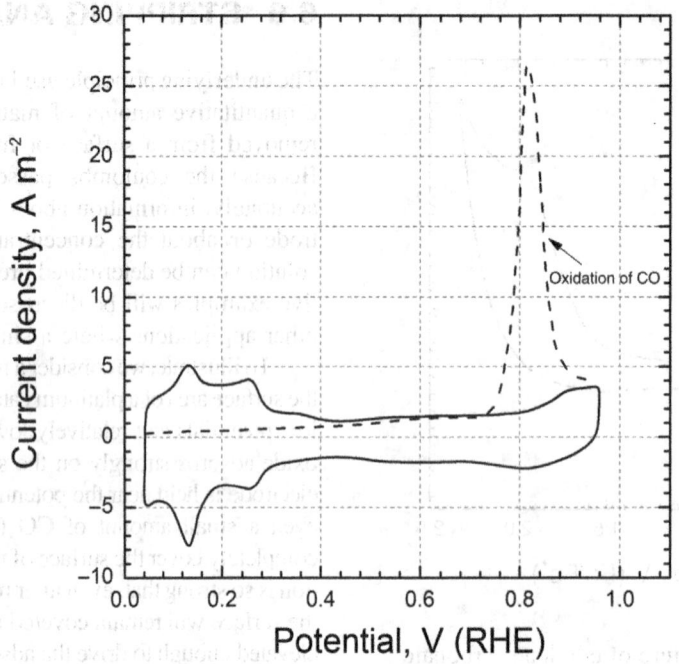

Figure 6.15 Oxidation of adsorbed CO on Pt surface during linear sweep.

assuming one CO molecule per surface atom of Pt:

$$\frac{1.3 \times 10^{19}\,\text{atoms}}{\text{m}^2} \left| \frac{\text{mol Pt}}{6.023 \times 10^{23}\,\text{atoms}} \right| \frac{\text{mol CO}}{\text{mol Pt}} \left| \frac{2\,\text{equiv}}{\text{mol CO}} \right| \frac{96485\,\text{C}}{\text{equiv}} = 4.17\,\text{C m}^{-2}.$$

We have assumed that the CO is linearly bonded to the Pt. If the Pt surface is initially completely covered with CO, and given that there is $4.17\,\text{C·m}^{-2}$ of Pt surface, the surface area of platinum can be determined as simply

$$\text{ECSA} = \frac{Q_{\text{CO}}}{4.17\,\text{C m}^{-2}}. \qquad (6.16)$$

This is described as the electrochemically active surface area (ECSA) because, in contrast to BET, for instance, the metal particles must be in electronic and ionic contact to be detected with this technique. Fortunately, this is the value that is most relevant to electrochemical systems.

ILLUSTRATION 6.3

Using the data from Figure 6.15, calculate the ECSA per gram of Pt (also known as the specific surface area) from the CO adsorption experiment. What is the double-layer capacity? The platinum loading of the electrode is $0.028\,\text{mg·cm}^{-2}$ (based on the superficial area), and the scan rate is $20\,\text{mV·s}^{-1}$.

Solution:

The solid line on the plot corresponds to the current in the absence of adsorbed CO during the potential sweep. The charge associated with the oxidation of CO is the area under the CO peak (centered around 0.8 V) minus the charge associated with double-layer charging and oxidation of the platinum; in other words, it is the peak area between the solid and dashed lines. Each increment on the y-axis is $1\,\text{A·m}^{-2}$. The peak area corresponds to about 75 rectangles of dimensions $1\,\text{A·m}^{-2}$ by 0.02 V.

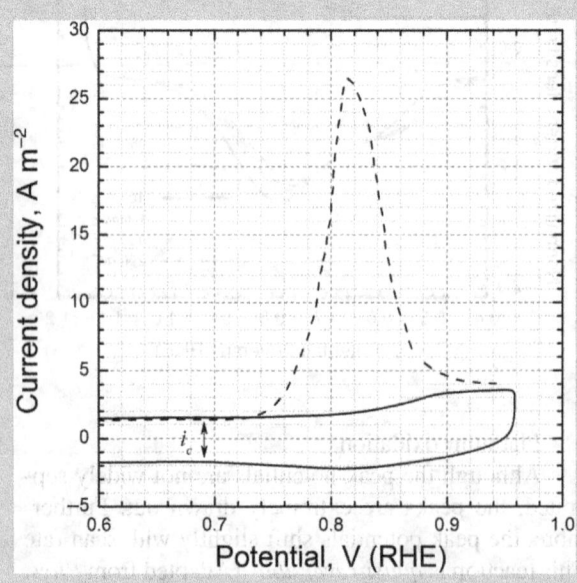

Since $Q = \int I dt$, we need to convert the voltage increment to a time. This can be done readily noting that the voltage is scanned at a constant rate of $20\,\text{mV}\,\text{s}^{-1}$. Therefore, 0.02 V is equivalent to 1 s. The charge associated with CO oxidation is therefore

$$\frac{Q}{A} = 75 \left(\frac{1\,\text{C}}{\text{m}^2\text{s}}\right)(1.0\,\text{s}) = 75\,\text{C}\,\text{m}^{-2}.$$

This is the amount of charge used to oxidize CO per superficial area (same area upon which the current density is based). We can now use the relationship above to calculate the ECSA per superficial area as follows:

$$\frac{\text{ECSA}}{\text{superficial area}} = \frac{75\,\text{C}}{\text{m}^2(\text{superficial})} \frac{\text{m}^2}{4.17\,\text{C}}$$

$$= 18.0 \frac{\text{m}^2}{\text{m}^2(\text{superficial})}.$$

The specific surface area is then

$$\text{specific surface area} = 18.0 \frac{\text{m}^2}{\text{m}^2(\text{superficial})}$$

$$\times \frac{\text{m}^2(\text{superficial})}{0.28\,\text{gPt}} = 64 \frac{\text{m}^2}{\text{gPt}}.$$

The capacitance of the double layer is estimated from the current around 0.5 V, above the hydrogen region but before oxidation of platinum or carbon. The current density is about $14\,\text{A}\cdot\text{m}^2$; therefore, from Equation 6.11,

$$C_{DL} = \frac{i_c}{\nu} = \frac{14}{0.02} = 700\,\text{F}\,\text{m}^{-2}.$$

The area here refers to the superficial electrode area.

6.7 ELECTROCHEMICAL IMPEDANCE

Fundamentals

As mentioned previously, one of the key objectives of electroanalytical methods is to quantify the relationship between the current and voltage and to provide insight into the processes that influence that relationship. The simplest relationship is Ohm's law:

$$V = IR_\Omega, \qquad (6.17)$$

which we have already used frequently. In this relationship, the resistance, R_Ω, is the proportionality constant that relates current and voltage. Of course, the complex physical phenomena occurring in an electrochemical cell cannot be adequately represented with just a resistor. In particular, there are several physical processes that are time dependent: molecular diffusion, adsorption, and charging of the double layer, for example. The time constants associated with these processes vary significantly, making it possible to extract information about the system by examining the relationship between the voltage and current at different timescales. To do this, we need a method for describing and quantifying the relationship between current and voltage under dynamic conditions.

Here we briefly describe a technique that is known as *electrochemical impedance spectroscopy* or EIS. Our objective is to introduce you to the basic elements of the technique. For information beyond this initial treatment, please see references such as those listed at the end of the chapter. The electrochemical cell is the same as that described in Section 6.1, containing a WE, CE, and RE. Starting from steady state, the potential or current is perturbed by superposing an oscillating signal on the steady-state value. The response of the system to the oscillating input for a range of frequencies is then examined. Generally, this means that we look for the ratio of the voltage and current; this quantity is called the impedance, and has dimensions of ohm or ohm·m^2.

Figure 6.16 illustrates the response of the current to an oscillating input potential. For small perturbations, the system is approximately linear and the current varies with the same frequency as the potential. The applied potential can be expressed equivalently as a sine or cosine wave, since both functions describe the same

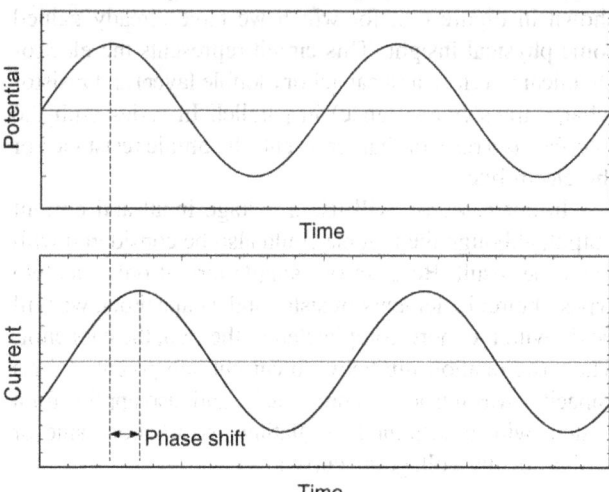

Figure 6.16 Illustration of voltage input and current output showing amplitude difference and phase shift.

sustained oscillation. For our treatment, we use the cosine as follows:

$$V(t) = \Delta V \cos(\omega t), \quad (6.18)$$

where ΔV is the amplitude of the wave, f is its frequency [s^{-1}], and $\omega = 2\pi f$ [rad·s^{-1}]. Note that in our discussion of impedance, the voltage, $V(t)$, and the current, $I(t)$, refer only to the oscillating portion and represent the deviation from the steady-state or baseline value. The current response that corresponds to Equation 6.18 has the same frequency, but may be shifted in phase (see Figure 6.16):

$$I(t) = \Delta I \cos(\omega t - \varphi). \quad (6.19)$$

We seek the relationship between the voltage and the current, which depends on the physical characteristics of the system and changes with the frequency of the oscillating input. This relationship can be used as a "fingerprint" to characterize the physical system of interest.

EIS is performed by perturbing the experimental system with an oscillating input and measuring the corresponding output signal over a broad range of frequencies. The input and output signals are used to determine the impedance, which relates the current and voltage at different frequencies. The impedance data can then be fit to an equivalent circuit to provide quantitative physical insight into processes that control cell behavior at different timescales.

Impedance for Basic Circuit

In order to help you understand what the impedance is and how it represents some of the basic electrochemical processes that we have already identified, we will work this process "backwards," beginning with the equivalent circuit shown in Figure 6.6, for which we have already gained some physical insight. This circuit represents the electrochemical interface as a capacitor (double layer) and resistor (charge-transfer resistance) in parallel. In series with the interface is a resistor that represents the ohmic resistance of the electrolyte.

In our case, we will use a voltage input and current output, although the reverse could also be considered with the same result. Because our simple circuit only has two types of circuit elements (resistor and a capacitor), we will begin with the more complicated of the two, the capacitor. The basic relationship between current and potential for a capacitor was introduced previously, and also applies for a system with a sustained oscillating perturbation (sine or cosine) of the voltage or current:

$$I(t) = C\frac{dV(t)}{dt}. \quad (6.20)$$

There are several ways to solve this equation for an oscillating input voltage. Here we will use complex variables in order to introduce the concept of the complex impedance that is most commonly used. As we saw above,

$$V(t) = \Delta V \cos(\omega t). \quad (6.18)$$

Alternatively, we could have chosen a sine wave

$$V(t) = \Delta V \sin(\omega t), \quad (6.21)$$

which would have yielded exactly the same relationship between the current and voltage (i.e., the same impedance). Because the system is linear, we can use superposition and add two inputs to get a combined output that is equal to the sum of the outputs from each individual input. Proportionality applied to the linear system allows us to multiply an input by a constant to yield a proportional output. Taking advantage of these properties, we define an alternative input that is equal to the sum of a cosine and sine wave,

$$\mathcal{V}(t) = \Delta V \cos(\omega t) + j\Delta V \sin(\omega t) = \Delta V[\cos(\omega t) + j\sin(\omega t)], \quad (6.22)$$

where we multiplied the sine wave by the constant j, which is equal to $\sqrt{-1}$. Our strategy is to solve the problem with $\mathcal{V}(t)$ as the input in order to simplify the math. Solution of the problem with $\mathcal{V}(t)$ as the input will provide a combined output. Once the combined output is determined, we can easily recover the desired portion of the output, which corresponds to only the cosine input, since the real (cosine) input will yield a real output, and the imaginary (sine) input will yield an imaginary output.

The reason why we have introduced this alternative input is to take advantage of Euler's formula, which will simplify the math needed to solve this and more complex equations for circuit elements, and will lead naturally to the complex impedance:

$$e^{j\varphi} = \cos\varphi + j\sin\varphi \quad (6.23)$$

With this formula, we now rewrite the input voltage from Equation 6.22 as

$$\mathcal{V}(t) = \Delta V e^{j\omega t}. \quad (6.24)$$

In this form, we can easily differentiate the complex input potential and use Equation 6.20 for a capacitor to determine the complex current response as follows:

$$\mathcal{I}(t) = C\frac{d\mathcal{V}(t)}{dt} = Cj\omega \Delta V e^{j\omega t}. \quad (6.25)$$

As stated above, we can recover the time-dependent current that results from just the (real) cosine input by taking the real portion of our answer:

$$I(t) = \text{Re}\{\mathcal{I}(t)\} = \text{Re}\{C\Delta V\omega j[\cos(\omega t) + j\sin(\omega t)]\}$$

$$= -C\Delta V\omega \sin(\omega t). \quad (6.26)$$

A key advantage of this approach is that it transforms derivatives with respect to time into algebraic terms, a fact that greatly facilitates solution. Equation 6.26 represents a solution to Equation 6.20 and provides the time-dependent current that results from the cosine input defined in Equation 6.18.

The above problem was readily solved for $I(t)$. However, we would like to be able to solve for the response of the complete circuit rather than for just one element of the circuit. An effective way to approach the circuit problem is to determine the impedance for each element of the circuit, and then to combine the individual impedances to get an equivalent impedance for the entire circuit. The complex impedance is defined as the ratio of the complex voltage and the complex current as follows:

$$Z(\omega) = \frac{\mathcal{V}(t)}{\mathcal{I}(t)} = \frac{\Delta V e^{j\omega t}}{\Delta I e^{j(\omega t - \varphi)}} = \frac{\Delta V}{\Delta I}e^{j\varphi} = r(\cos\varphi + j\sin\varphi), \quad (6.27)$$

where r is the magnitude of Z and φ is the phase angle (see inset). Note that the time dependence cancels out. Both the phase angle and the impedance are a function of the frequency, *but not of time*.

From Equations 6.25 and 6.26, the complex impedance of a capacitor is

$$Z(\omega) = \frac{\mathcal{V}(t)}{\mathcal{I}(t)} = \frac{\Delta V e^{j\omega t}}{C\Delta V j\omega e^{j\omega t}} = \frac{1}{j\omega C} = -\frac{1}{\omega C}j. \quad (6.28)$$

Again, the time dependence cancels out since, for a linear system, it is the same in both the input and the output.

Now that we have the impedance for the capacitor, we need the appropriate values for the resistors to complete the circuit shown in Figure 6.6. For a resistor, the current and voltage are always in phase and there is no imaginary component to the impedance. Therefore, *the impedance for a resistor* is simply equal to the resistance:

$$Z = R_\Omega. \quad (6.29)$$

One of the useful characteristics of impedances is that they add together like resistors. Making use of this, we can derive the equivalent impedance in ohms for the circuit (Figure 6.6) as follows:

$$Z(\omega) = R_\Omega + \frac{1}{\frac{1}{R_f} + j\omega C} = R_\Omega + \frac{R_f}{1 + \omega^2 R_f^2 C^2} + j\left(\frac{-\omega R_f^2 C}{1 + \omega^2 R_f^2 C^2}\right).$$

$$(6.30)$$

The impedances for the faradaic reaction and the double-layer capacitor have been added as parallel resistors in a circuit (Problem 6.4), and the real and imaginary terms have been grouped. The equivalent impedance can be used to relate the voltage and current for the complete circuit, which is much more complicated than the response due to any individual circuit element. You should understand and be able to reproduce Equation 6.30.

Complex Variables

Impedance is typically described with complex quantities. A complex number is expressed as the sum of real and imaginary portions:

$$Z = a + bj,$$

where $j = \sqrt{-1}$. a is the real part of Z, $\text{Re}\{Z\}$; and b is the imaginary part of Z, $\text{Im}\{Z\}$. There are other ways of representing a complex number. Graphically, a complex number can be interpreted with a polar form:

$$Z = r(\cos\varphi + j\sin\varphi),$$
$$a = r\cos\varphi,$$
$$b = r\sin\varphi,$$
$$r = |Z| = \sqrt{a^2 + b^2},$$
$$\tan\varphi = \frac{b}{a}.$$

r represents the magnitude of the number (modulus) and φ is the phase angle in radians (argument). The value Z is a point in the complex plane.

Use of Euler's formula (described in text) permits us to express Z in exponential form as follows:

$$Z = r(\cos\varphi + j\sin\varphi) = re^{j\varphi}.$$

Table 6.4 The Impedance of Common Circuit Elements

Component	Behavior	Impedance
Resistor	$V = IR_\Omega$	R_Ω
Capacitor	$C = (Q/V)$ or $I C dV/dt$	$1/j\omega C$
Inductor	$V = L di/dt$	$j\omega L$

We used the equation for a capacitor to solve for its response to the oscillating input (Equation 6.26) and for its complex impedance (Equation 6.28). However, we need not go through this detailed process every time since the same circuit elements are frequently used. Table 6.4 summarizes the impedance for three common circuit elements.

For a circuit comprised of these elements, we can simply use the impedances of the individual circuit elements from the table and combine them by adding them as we would resistors. The result is the equivalent impedance for the circuit such as that shown in Equation 6.30 for the circuit of Figure 6.6.

How can we graphically represent this complex impedance? One common way is a Nyquist plot, shown in Figure 6.17, where the imaginary portion, $-Z_i$, is on the ordinate and the real part, Z_r, is on the x-axis. Note that since the complex portion is frequently negative, $-Z_i$ rather than Z_i is traditionally plotted on the imaginary axis. The shape of the plot that corresponds to the circuit in Figure 6.6 is a semicircle and is a common feature of EIS impedance spectra measured experimentally. The impedance for this circuit approaches R_Ω as $\omega \to \infty$ and $(R_\Omega + R_f)$ as $\omega \to 0$. Also, at the top of the semicircle, $\omega = 1/R_f C$. Knowing this, we can estimate the value of these parameters from EIS data obtained experimentally. One shortcoming of the Nyquist plot is that frequency is not displayed explicitly on the diagram. The curve represents the whole range of frequencies, with high frequencies toward the left of the diagram and low frequencies toward the right.

It is useful to consider the physical significance of the behavior shown in Figure 6.17. At high frequencies, the capacitor acts as a "short" since the voltage oscillates at a frequency that is fast relative to the time required to charge the double layer. Because the double layer is never charged to any appreciable extent, the capacitor offers essentially no resistance to the oscillating current and the kinetic resistance, in parallel with the capacitor, does not play a role. Therefore, the complete circuit resistance is equal to the ohmic resistance. At lower frequencies where cycle time is comparable to the double-layer charging time, both the capacitor impedance and the faradaic (kinetic) resistance are important.

The result is the semicircular region, with the phase shift caused by the capacitor. Finally, at sufficiently low frequencies (longer times), the cycle time is long relative to the double-layer charging time. Consequently, the double-layer charges and discharges quickly relative to the cycle time and does not contribute to the overall behavior; rather, essentially all of the current flows through the faradaic and ohmic resistors in series (see Figure 6.6). At these low frequencies, the circuit resistance for this simple circuit reaches a constant value of $R_\Omega + R_f$, and the output current is in phase with the input voltage. An understanding of EIS requires the ability to connect the physics, circuit model (equivalent circuit), and the complex impedance.

Connection of Complex Impedance to Time Domain

In the previous section, we derived an expression for the impedance of a capacitor and combined that expression with the faradaic and ohmic resistances shown in Figure 6.6 to determine the equivalent impedance for the circuit. We then demonstrated how to illustrate that complex impedance on a Nyquist plot. The semicircle that we obtained is a characteristic feature of impedance spectra for electrochemical systems and is connected to key physical parameters, as shown in Figure 6.17. In this section, we briefly examine the connection of the impedance to the time domain by using the complex impedance to determine the output current as a function of time.

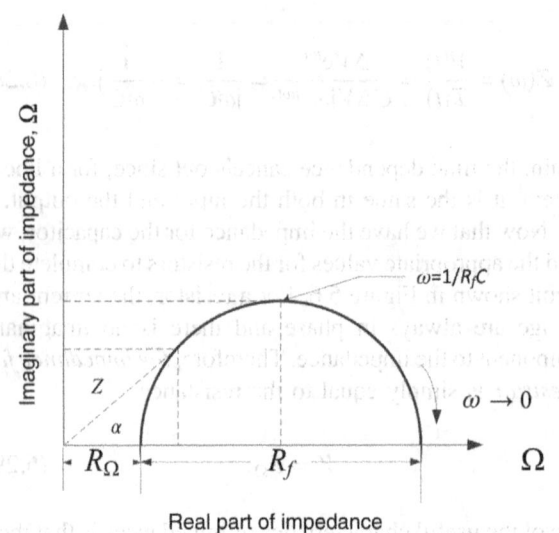

Figure 6.17 Nyquist plot for the equivalent circuit of Figure 6.6.

The equations that we have used thus far are summarized below:

$$\mathcal{V}(t) = \Delta V[\cos(\omega t) + j\sin(\omega t)] = \Delta V e^{j\omega t},$$
$$\mathcal{I}(t) = \Delta I[\cos(\omega t - \varphi) + j\sin(\omega t - \varphi)] = \Delta I e^{j(\omega t - \varphi)},$$
$$V(t) = \text{Re}\{\mathcal{V}(t)\} = \Delta V \cos(\omega t),$$
$$I(t) = \text{Re}\{\mathcal{I}(t)\} = \Delta I \cos(\omega t - \varphi),$$
$$Z(\omega) = \frac{\mathcal{V}(t)}{\mathcal{I}(t)} = \frac{\Delta V e^{j\omega t}}{\Delta I e^{j(\omega t - \varphi)}} = \frac{\Delta V}{\Delta I} e^{j\varphi} = r(\cos\varphi + j\sin\varphi).$$

(6.31)

The process of converting from an arbitrary value of $Z(\omega)$ to the time-dependent expression for the current at that frequency is shown in Illustration 6.4. Its purpose is to show how the complex impedance relates both the magnitude and phase of the input and output signals. The data in the complex domain provide the information needed, as a function of frequency, to quantify the dynamic behavior of the system.

ILLUSTRATION 6.4

The complex impedance of an electrochemical system measured at a frequency of 1000 Hz was found to be $3 - j$. The amplitude of the applied (oscillating voltage) is 5 mV. What is the corresponding current response in the time domain?

To answer this question, we first put $Z = 3 - j$ into polar form $Z = r(\cos\varphi + j\sin\varphi)$.

$$r = \sqrt{3^2 + (-1)^2} = 3.16.$$

$$\varphi = \tan^{-1}\left(\frac{-1}{3}\right) = -0.322 \text{ [rad]} = -18.4 \text{ [degrees]}.$$

Therefore, $Z = 3.16[\cos(-0.322) + j\sin(-0.322)] = 3.16e^{-0.322j} \, [\Omega]$.

$$Z(2\pi \cdot 1000) = \frac{\mathcal{V}(t)}{\mathcal{I}(t)} = 3.16e^{-0.322j} \, [\Omega]$$

$$\mathcal{I}(t) = \frac{\mathcal{V}(t)}{Z(2\pi \cdot 1000)} = \frac{0.005e^{j\,2\pi(1000)t}\,[\text{V}]}{3.16e^{-0.322j}\,[\Omega]}$$

$$= 0.0016 e^{j(2\pi(1000)t + 0.322)} \, [\text{A}]$$

$$I(t) = \text{Re}\{\mathcal{I}(t)\} = 0.0016 \cos[2\pi(1000)t + 0.322] \, [\text{A}]$$

The magnitude of Z (equal to 3.16 in this example) is the ratio of the amplitudes of the voltage and current, $\Delta V/\Delta I$. In other words, it tells you how much the current will change in response to a small change in the voltage, or vice versa. The angle φ (−0.322 rad in this example) provides the phase shift between the input and output. As illustrated by this example, the complex impedance, Z, provides both the ratio of the magnitudes of the voltage and current and the phase shift.

You may be thinking that φ is so small relative to the other term in the brackets that it will not make any difference. However, remember that cosine is a periodic function. The voltage and the resulting current are shown in the following figure, where the phase shift is evident.

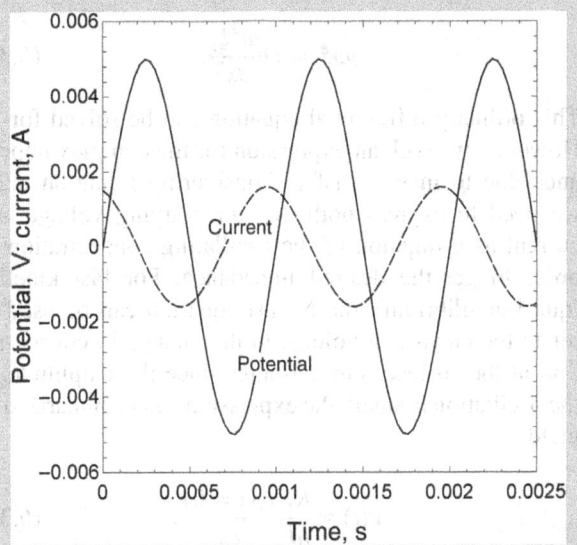

Note that the voltage "lags" the current (reaches its peak at a time later than that at which the current reaches its peak).

Mass Transport

We now consider a common process for electrochemical systems, molecular diffusion, and how it might be included in our EIS analysis. Let's start with one-dimensional diffusion where the process of interest is influenced by the concentration of a single species. The relevant equation is

$$\frac{\partial c_i}{\partial t} = D_i \frac{\partial^2 c_i}{\partial x^2}.$$

(6.32)

Consider a semi-infinite domain with small perturbations around the initial uniform concentration, c_o. The steady-state solution is just a constant; consequently, we are left to solve for the oscillating portion. Similar to what we did previously, we use a combination of cosine and sine terms to define the oscillating input:

$$\xi_i(t) = \Delta c_i[\cos(\omega t) + j\sin(\omega t)] = \Delta c_i e^{j\omega t}, \quad (6.33)$$

where

$$c_i(t) = \text{Re}\{\xi_i(t)\} = \Delta c_i \cos(\omega t). \quad (6.34)$$

Note that the steady-state value has been subtracted off in (6.34), as with other variables in Section 6.7. Equation 6.33 can be differentiated and substituted into (6.32) to yield

$$j\omega \xi_i = D_i \frac{d^2 \xi_i}{dx^2}. \quad (6.35)$$

This ordinary differential equation can be solved for ξ_i. However, we seek an expression for the complex impedance due to mass transfer. Considering Equation 6.27, we need to express both the time-varying voltage and current as a function of the oscillating concentration in order to get the desired impedance. For fast kinetics (quasi-equilibrium), the Nernst equation can be used to relate the change in voltage to the change in concentration at the surface. Furthermore, since the amplitude of the oscillation is small, the expression can be linearized to yield

$$\mathcal{V}(t) \approx \frac{RT}{nF} \frac{\xi_i(x=0)}{c_o}. \quad (6.36)$$

The current is related to the flux at the surface as follows:

$$\mathcal{I}(t) = -nFAD_i \frac{d\xi_i}{dx}\bigg|_{x=0} \quad (6.37)$$

The impedance becomes

$$Z = \frac{\mathcal{V}(t)}{\mathcal{I}(t)} = \frac{RT}{(nF)^2 AD_i c_o} \left(\frac{\xi_i}{-\frac{d\xi_i}{dx}} \right)_{x=0} \quad (6.38)$$

We can now solve Equation 6.35 for the required ratio of the surface concentration to the surface flux, which can then be substituted into Equation 6.38 to yield the desired impedance. The boundary conditions include Equation 6.36 for the concentration and $\xi_i = 0$ as $x \to \infty$. The result is

$$\left(\frac{\xi_i}{-\frac{d\xi_i}{dx}} \right)_{x=0} = \sqrt{\frac{D_i}{j\omega}}. \quad (6.39)$$

$$Z(\omega) = \frac{\mathcal{V}(t)}{\mathcal{I}(t)} = \frac{RT}{(nF)^2 AD_i c_o} \sqrt{\frac{D_i}{j\omega}} = \frac{RT(1-j)}{(nF)^2 A c_o \sqrt{2D_i \omega}}, \quad (6.40)$$

where Z is in ohms. This impedance yields a line with a 45° slope on a Nyquist plot at low frequencies (why?). It is most commonly called the *Warburg impedance*. To get the impedance in ohm·m², which is what results if the current density rather than the current is used, the impedance $Z(\omega)$ should be multiplied by the area.

The Warburg impedance shown in Equation 6.40 assumed that the kinetic resistance was not significant, which permitted the use of the Nernst equation to relate the voltage and concentration. A complete kinetic expression (e.g., Butler–Volmer with concentration dependence) can alternatively be used with the transport equation to obtain the expressions for both the kinetic resistance and Warburg impedance. The interested reader should refer to a more advanced treatment such as the text by Orazem and Tribollet (2008) for additional information.

A procedure similar to that used to derive Equation 6.40 was also used to determine the complex impedance for mass transfer through a stagnant film of finite thickness δ to yield

$$Z(\omega) = \frac{RT\delta}{(nF)^2 AD_i c_o} \frac{\tanh\left[\sqrt{\frac{j\omega\delta^2}{D_i}}\right]}{\sqrt{\frac{j\omega\delta^2}{D_i}}}, \quad (6.41)$$

where Z is again in ohms. The Nyquist plot for the finite thickness impedance with $-Z_i$ is on the ordinate and Z_r is on the x-axis as shown in Figure 6.18, where the impedance has been made dimensionless by multiplying by $\frac{(nF)^2 AD_i c_o}{RT\delta}$. As mentioned previously, each point on the graph represents a different frequency, with high frequencies on the left. The impedance increases with decreasing frequency to yield the 45° slope characteristic of semi-infinite diffusion. The impedance levels off as the frequency decreases further owing to the finite diffusion layer thickness.

Another common representation of the complex impedance is the Bode plot, shown on the right of Figure 6.18. Here frequency is the independent variable and the magnitude and phase angle are plotted on the left and right axes. The key feature here is that the frequency or equivalently the characteristic time constant for different processes is highlighted by a change in magnitude and phase angle. For the diffusion example here, it is not surprising the change occurs at a dimensionless frequency of about 1, where the dimensionless frequency is $\omega\delta^2 / D_i$.

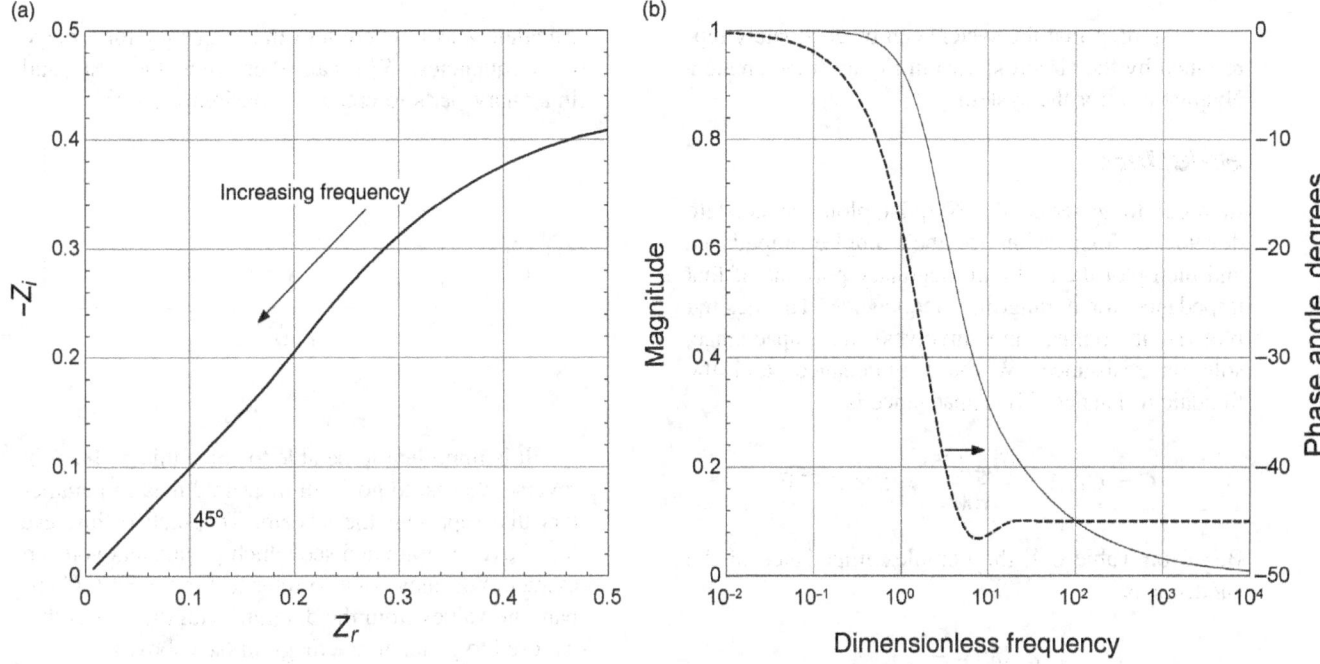

Figure 6.18 Nyquist and Bode plots for impedance due to diffusion through a film of finite thickness.

Interpretation of Data

It is common to measure impedance data and then use software to fit the data to a variety of circuit elements. This approach is intuitive and generally helpful in correlating features of the impedance with physical phenomena in the electrochemical system. It also yields quantitative parameters characteristic of the system of interest. In this section, we explore further the relationship between the impedance data and the key parameters that characterize an electrochemical system. To do so, consider the circuit shown in Figure 6.6, but with a slight modification. Namely, we will add an element that accounts for diffusion, the Warburg impedance, as shown in Figure 6.19.

Finally, we note that the treatment of electrochemical impedance spectroscopy provided in this text is just an introduction. There are many important characteristics of electrochemical systems that influence their dynamic behavior that have not been considered, where additional complexity is required for accurate representation. For example, these characteristics may include the adsorption of species on the surface, the presence of surface films, or even a polymer coating on the surface. Frequently, the circuit representation of complex systems is not unique, and is best coupled with a physical description to enhance understanding. You should refer to more comprehensive resources for EIS analysis of such systems (see Further Reading section).

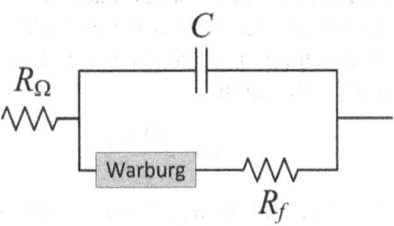

Figure 6.19 Equivalent circuit with Warburg impedance.

ILLUSTRATION 6.5

Suppose that you have an electrochemical system characterized by the following parameters

$C_{DL} = 0.2\,\text{F}\cdot\text{m}^{-2}$

Electrode dimensions: $5\,\text{cm} \times 5\,\text{cm}$ (only one side active)

Conductivity of electrolyte $= 10\,\text{S}\cdot\text{m}^{-1}$

Distance from electrode to reference electrode $= 1\,\text{cm}$

$i_o = 10\,\text{A}\cdot\text{m}^{-2}$, single-electron reaction ($n = 1$)

$D_i = 1 \times 10^{-9}\,\text{m}^2\cdot\text{s}^{-1}$

$c_o = 10\,\text{mol}\cdot\text{m}^{-3}$

$T = 25\,°\text{C}$

Assuming that the system can be adequately represented by the circuit shown in Figure 6.19, create a Nyquist plot for the system.

Solution:

In order to generate the Nyquist plot, we need to develop an expression for the complex impedance and then plot the real and imaginary portions of that impedance for a range of frequencies. The desired plot is in ohms, and involves the capacitance, solution resistance, Warburg impedance, and the faradaic resistance. The capacitance is

$$C = C_{DL}A = \frac{(0.2)(25)}{100^2} = 5 \times 10^{-4} \text{ F}.$$

Based on Table 6.4, the complex impedance of the capacitor is

$$Z_C(\omega) = \frac{1}{\omega C j}.$$

To determine the resistance of the electrolyte, we use Equation 4.8c:

$$R_\Omega = \frac{L}{\kappa A} = \frac{0.01}{(10)(0.0025)} = 0.4 \, \Omega.$$

For the kinetic resistance, we assume open circuit as the steady-state condition, with small oscillations around that point. Because the magnitude of the potential change is small, linear kinetics can be used to determine the resistance according to Equation 4.62:

$$R_f = \frac{1}{A}\frac{d\eta}{di} = \frac{RT}{Fi_oA} = \frac{(8.314)(298)}{(96,485)(10)(0.0025)} = 1.03 \, \Omega.$$

The Warburg impedance is

$$Z_W(\omega) = \frac{RT}{F^2 A D_i c_o}\sqrt{\frac{D_i}{j\omega}}.$$

We now have all that we need to assemble the composite complex impedance:

$$Z(\omega) = R_\Omega + \frac{1}{\dfrac{1}{R_f + Z_W(\omega)} + \dfrac{1}{Z_C(\omega)}},$$

where the only unknown is the frequency. Perhaps the easiest way to generate the desired plot is to enter the above information into a package, such as Python or Matlab, that handles complex arithmetic, and calculate Z as a function of the frequency for a range of frequencies. We can then plot the real and imaginary parts to create the desired figure:

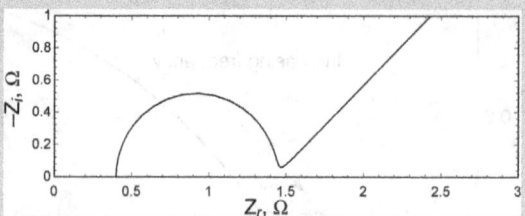

It is important to be able to solve this problem in reverse, that is, to go from measured data to parameters that represent the system. To practice this, use the above diagram and see which parameters you can extract. You may want to refer to Figure 6.17. Compare the values from the diagram with the values that we used to generate the diagram (see above).

6.8 ROTATING DISK ELECTRODES

The rotating disk electrode (RDE) is shown in Figure 6.20. Here an electrode is imbedded in the end of a cylinder, which rotates submerged in the electrolyte. The electrode is typically a smooth surface surrounded by an insulating material such as Teflon. The RDE merits special attention for several reasons. First, the hydrodynamics for the RDE are well known. Second, the velocity toward the surface of the disk depends only on the distance from the surface and not on the radial position or the angle. Thus, the surface is uniformly accessible to mass transfer, and the rate of mass transfer is constant as long as the bulk concentration does not change with time. Finally, compared to other devices with similar characteristics, experiments with the RDE are relatively simple to prepare and conduct. These features allow the RDE and the rotating ring disk electrode (RRDE) to find frequent use in measuring kinetic and mass-transport properties.

We will not go into the detailed development of the hydrodynamics here as that development is available in numerous references. Flow streamlines are illustrated in Figure 6.20 for laminar flow (Re < 200,000). Analytical solution of the flow field yields the following for the velocity in the z-direction:

$$v_z = -az^2 \frac{\Omega^{3/2}}{\nu^{1/2}}, \qquad (6.42)$$

where $a = 0.51023$ and the negative sign indicates that the velocity is toward the surface. Here, ν is the kinematic

Chapter 6 Electroanalytical Techniques and Analysis of Electrochemical Systems   **137**

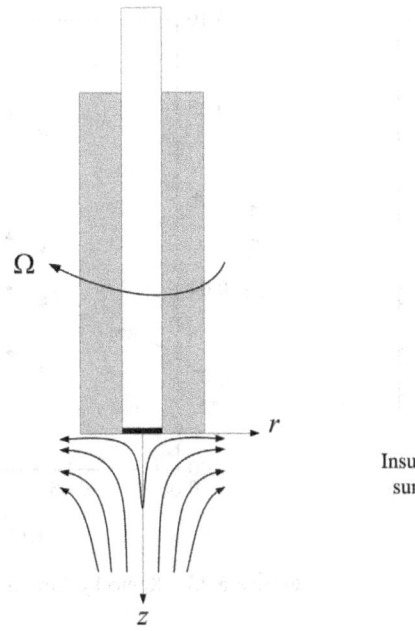

Figure 6.20 Rotating disk electrode (RDE).

viscosity. The key result from this is that the z-velocity depends only on distance from the surface and does not depend on r or θ. Since the z-component of the velocity brings reactant to the surface, we might expect that the concentration also depends only on z. From Equation 4.25, the convective diffusion equation for the case of excess supporting electrolyte is

$$v_z \frac{dc_i}{dz} = D_i \frac{d^2 c_i}{dz^2}. \quad (6.43)$$

This equation has a relatively straightforward solution. The dimensionless flux at the surface can be expressed in terms of the Sherwood number, a result that follows from the solution of Equation 6.43 under laminar flow conditions. The resulting flux at the surface is uniform under mass-transfer control:

$$\text{Sh} = 0.62 \left(\frac{r^2 \Omega}{\nu} \right)^{0.5} \text{Sc}^{1/3}. \quad (6.44)$$

As we saw in Chapter 4, Equation 6.44 can be used to derive an expression for the current density, which is uniform across the surface when mass transfer is controlling:

$$i = 0.62 n F D_i^{\frac{2}{3}} \Omega^{\frac{1}{2}} \nu^{\frac{-1}{6}} \left(c_i^\infty - c_i \right). \quad (6.45)$$

Application of Equation 6.45 at the limiting current shows that i_{lim} varies with $\Omega^{\frac{1}{2}}$. This behavior is illustrated in Figure 6.21, and a plot of i_{lim} versus $\Omega^{\frac{1}{2}}$ yields a line, as

shown in Figure 6.22, from which the diffusivity, for example, can be determined. This type of plot is known as a Levich plot.

At conditions below the limiting current, the RDE can be used to measure kinetic parameters. Typically, the current is high enough so that mass transfer is important, but not so high that the limiting current is reached. An example of such an analysis is that of Koutecký–Levich in which the reaction at the surface is assumed to be first order with respect to the concentration of the reactant. We shall investigate the current as a function of rotation speed when

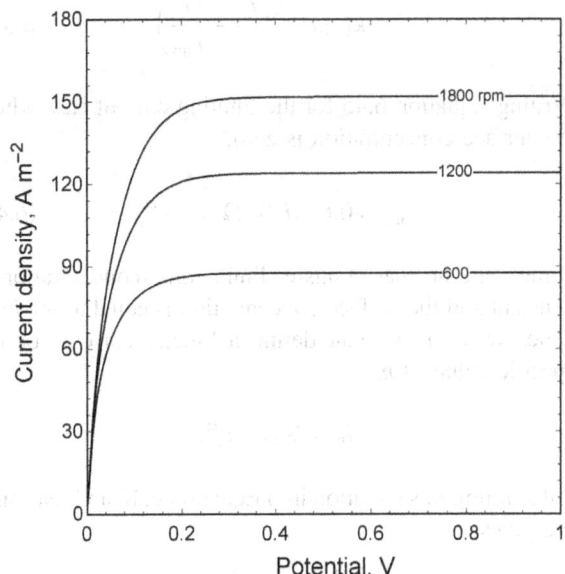

Figure 6.21 Linear potential sweep with RDE.

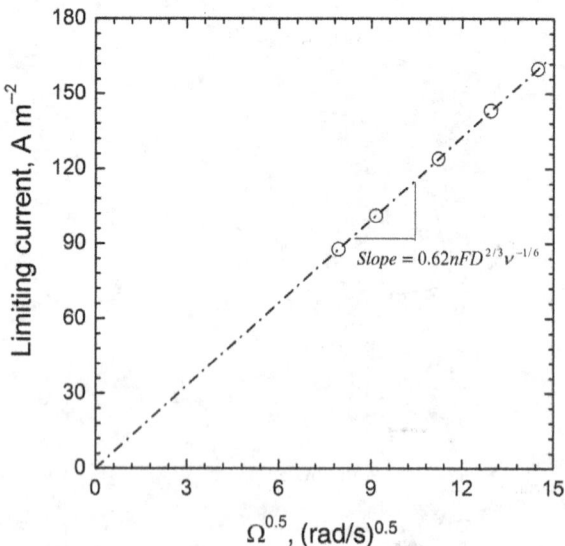

Figure 6.22 Levich plot.

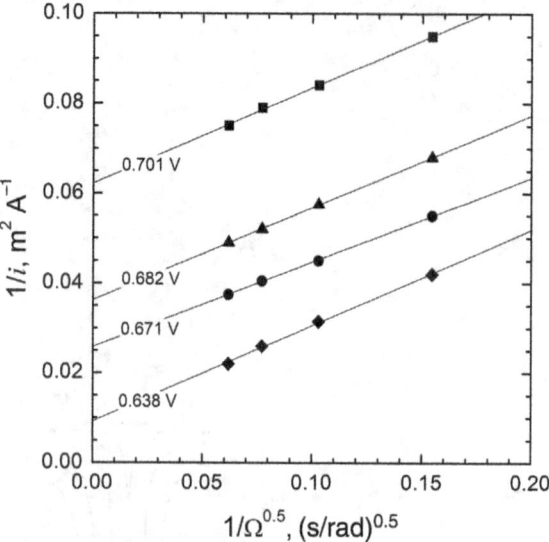

Figure 6.23 Koutecký–Levich plot.

the electrode is held at a constant overpotential. Under these conditions,

$$i = k(\eta)nFc_i,$$

where $k(\eta)$ is the rate constant at the specific value of η. The concentration at the interface can be expressed in terms of the limiting current:

$$c_i = c_i^\infty \left(1 - \frac{i}{i_{\lim}}\right).$$

Thus,

$$i = k(\eta)nFc^\infty \left(1 - \frac{i}{i_{\lim}}\right). \quad (6.46)$$

Writing Equation 6.45 for the limiting current case where the surface concentration is zero,

$$i_{\lim} = 0.62nFD_i^{\frac{2}{3}} \Omega^{\frac{1}{2}} \nu^{\frac{-1}{6}} c_i^\infty. \quad (6.47)$$

If there are no mass-transfer limitations (concentration is constant and the surface concentration is equal to the bulk concentration), we can define a kinetic current for the specific value of η.

$$i_k = k(\eta)nFc_i^\infty.$$

Substituting this equation into Equation 6.46 and rearranging gives

$$\frac{1}{i} = \frac{1}{i_k} - \frac{1}{i_{\lim}}. \quad (6.48)$$

Since the limiting current is proportional to the square root of rotation speed, a plot of $\frac{1}{i}$ versus the reciprocal of the square root of rotation speed gives a straight line with slope of $\left(0.62nFD_i^{\frac{2}{3}}\nu^{\frac{-1}{6}}\right)^{-1}$ and an intercept of $\left(k(\eta)nFc_i^\infty\right)^{-1}$. This relationship is shown in Figure 6.23 where the different lines represent different potentials. The intercept of each of the lines, corresponding to $\Omega \to \infty$, represents the current density in the absence of any mass-transfer limitations. These values can be fit to a kinetic expression such as the Butler–Volmer equation. From this same experiment the slope is determined, which can be used to identify either the diffusivity or n for the reaction.

One of the attractive aspects of the Koutecký–Levich analysis is that the kinetic data obtained from the intercept of each of the lines are all at the bulk concentration, which simplifies the fitting. However, there is nothing about the RDE that limits the coupled kinetic mass-transfer analysis to the first order system considered above. The advantage of the RDE is that it provides a relationship between the bulk and surface concentrations, and provides a mass-transfer configuration that is uniform. Data taken with a RDE over a range of potentials and concentrations can be used to fit a kinetic expression that includes nonlinear concentration-dependent terms. A note of caution is appropriate at this point. While the current is uniform under mass-transfer limited conditions, the primary current distribution is *not* uniform and the secondary distribution may not be uniform. The above kinetic analysis assumes a uniform current distribution, and a nonuniform distribution would introduce error. Consequently, the current distribution should be carefully considered at currents below the limiting current when taking kinetic data.

ILLUSTRATION 6.6

Use the data from Figure 6.23, provided in the table below, to calculate the diffusivity of oxygen. These data represent oxygen reduction in acid media. The solubility of oxygen is $1.21\,\text{mol}\cdot\text{m}^{-3}$, and the kinematic viscosity is $1.008 \times 10^{-6}\,\text{m}^2\cdot\text{s}^{-1}$.

Rotation rate [rpm]	Rotation rate [s^{-1}]	i [A·m^{-2}] 0.701 V	i [A·m^{-2}] 0.682 V	i [A·m^{-2}] 0.671 V	i [A·m^{-2}] 0.638 V
2500	262	13.33	20.41	26.67	45.45
1600	167	12.66	19.23	24.69	38.46
900	94.2	11.90	17.39	22.22	31.75
400	41.9	10.53	14.71	18.18	23.81

Solution:

We first use the data to calculate the slope of each of the lines. Since each set of data is linear, we simply use the first and last points to estimate the slope of each line as follows:

$$\text{Average slope (each curve)} = \frac{\Delta\left(\frac{1}{i}\right)}{\Delta\left(\frac{1}{\sqrt{\Omega}}\right)}.$$

We then average these to get the average slope of the four curves $= 0.216\,\frac{\text{m}^2}{\text{A}\sqrt{\text{s}}}$.

For the Koutecký–Levich plot, the slope is equal to $\left(0.62 n F D_i^{\frac{2}{3}} \nu^{-\frac{1}{6}} c_i^{\infty}\right)^{-1}$, where $n = 4$ for oxygen reduction, and c_i^{∞} is equal to the solubility of oxygen. Substituting in the known values yields

$$D_i = 2.03 \times 10^{-9}\,\text{m}^2\,\text{s}^{-1}.$$

6.9 iR COMPENSATION

Ideally, we would like to measure or control the potential of the WE relative to the RE, where the reference electrode is just outside the diffuse double layer of the working electrode. Unfortunately, it is not possible to do so. There are two principal concerns: first, the degree to which the physical presence of the RE disturbs the current and potential distributions near the electrode, and second, the ohmic losses between the RE and WE. Given the dimensions of the double layer (~nm) and the finite size of any RE, choosing where to place the electrode and how to correct for any iR differences takes some care.

Imagine that we have large planar WE and CEs separated by some distance. The current distribution is uniform, and the potential varies linearly. Figure 6.24 shows isopotential lines around a reference electrode placed far from the WE (a) and close to the WE (b); the WE is on the left. If the reference electrode is placed far from the WE, it does not interfere much with the potential distribution (and current) around the surface of the WE. However, there is clearly a difference between the measured potential and the actual potential just outside the

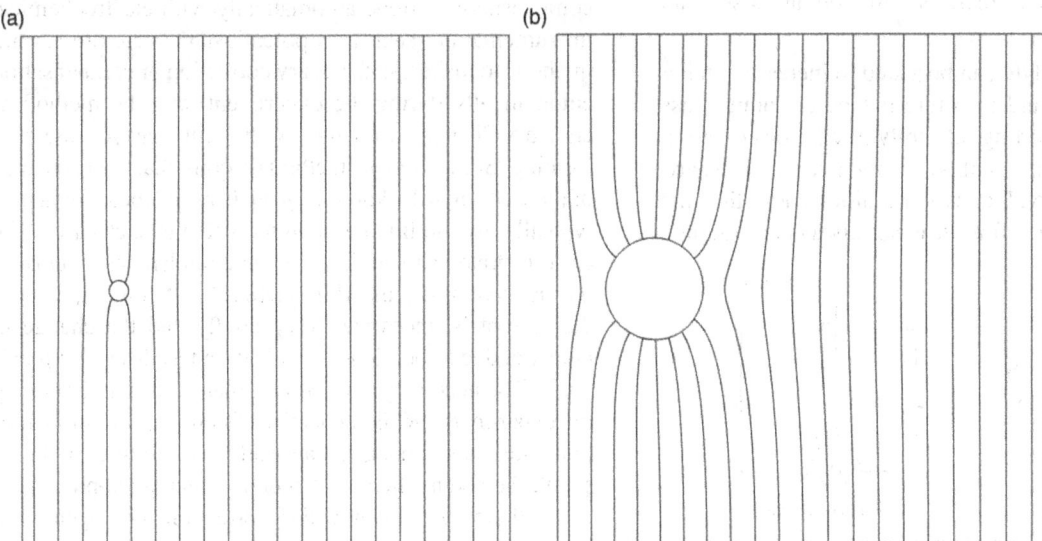

Figure 6.24 Isopotential lines around a reference electrode, the WE is on the left. (a) RE far from WE. (b) RE placed near WE surface.

double layer. The potential difference can be reduced by moving the RE closer to the WE. But what happens if we position the electrode adjacent to the WE? This placement is shown in Figure 6.24b. In contrast to the earlier case, the potential and current distributions are affected by the presence of the RE. Additionally, for systems where fluid flow is important, the reference electrode can change the velocity field. For instance, a RE placed near the disk of the RDE discussed in the previous section would completely alter the fluid flow and negate the key RDE feature of uniform accessibility.

To minimize the disturbance of the current and potential distributions by the RE, the reference electrode should be far away from the surface; however, we noted this may introduce a large error in the measured potential. Fortunately, with some careful attention, this dilemma can be largely addressed.

Figure 6.25 shows an equivalent circuit representation of this situation. The RC element on the right symbolizes the resistance to electron transfer (faradaic) and the double-layer capacitance (non-faradaic). The RE electrode is a finite distance away from the CE and, of course, we control the potential between the RE and WE. Compared to Figure 6.6, there exists an additional resistance between the RE and WE. This resistance is called the uncompensated resistance, R_U. The true overpotential will be either higher or lower depending on the direction of current. Adjusting for this potential drop is called *iR* correction.

The approach to dealing with this resistance is first to minimize it as much as possible, and then to compensate for any *iR* that cannot be eliminated. There are several ways to minimize the uncompensated resistance:

- Use small electrodes. Microelectrodes are discussed in next section—essentially with small electrodes, the currents are small and therefore the ohmic losses are reduced.
- Supporting electrolyte can be added to increase conductivity of solution and therefore reduce the ohmic loss. The use of supporting electrolyte was discussed in Chapter 4, and this method is often used in electroanalytical chemistry. It does add additional ions that may be detrimental. Therefore, it is not always an option.

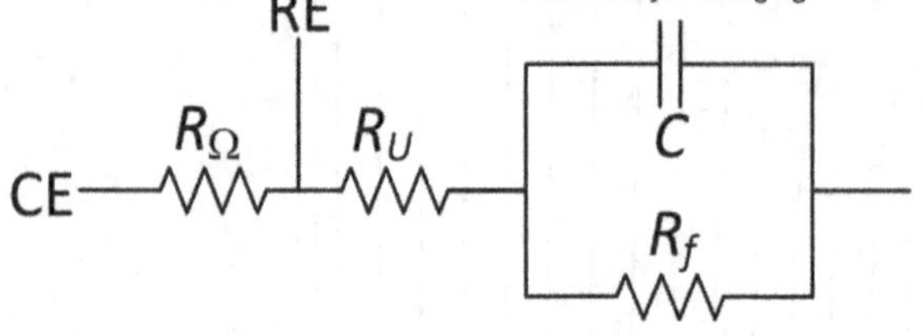

Figure 6.25 The placement of the RE away from the electrode results in an uncompensated resistance.

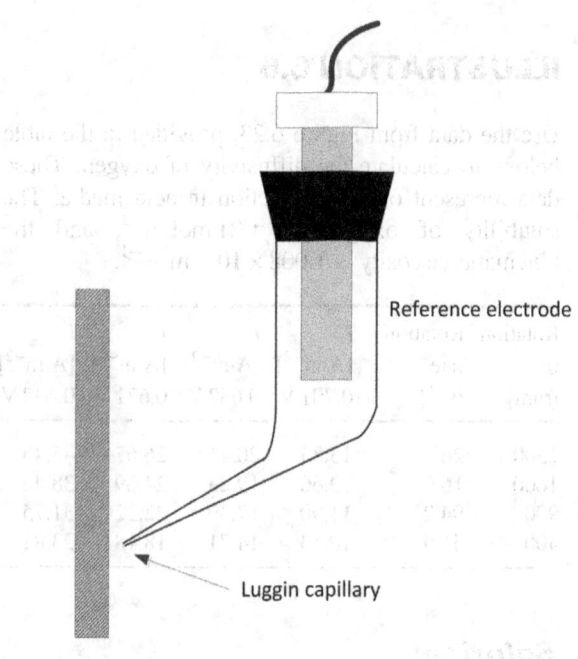

Figure 6.26 Luggin capillary.

- Move the RE closer to WE by using a Luggin capillary, for example. A Luggin capillary (see Figure 6.26) is typically made of glass and consists of a long thin neck that is open at the end. The capillary is filled with the same electrolyte as in the system.

After we have reduced the uncompensated *iR* as much as possible, we can try to either correct for any remaining resistance or to compensate for the resistance. Both attempt to accomplish the same goal, but the key difference is that compensation is done automatically with electrochemical instrumentation, such as a potentiostat. There are several methods to make the necessary correction or compensation automatically during the experiment, and the method of choice will depend on the experiment that you are performing. Some of these methods require that the resistance of the electrolyte be known *a priori*; an AC measurement is typically used to find the required value in such cases. The easiest strategy to understand and implement is *current interruption* as discussed in Section 6.4. During operation, the current is interrupted very briefly and the change in potential due to ohmic loss is identified as discussed previously. The necessary correction is then made while running the experiment. While generally effective in *iR* compensation, automatic strategies can lead to instabilities and care should be taken when implementing such an approach.

Is there a way to know if *iR* correction or compensation is being done correctly? It's not always straightforward, but let's examine a typical cyclic voltammogram collected

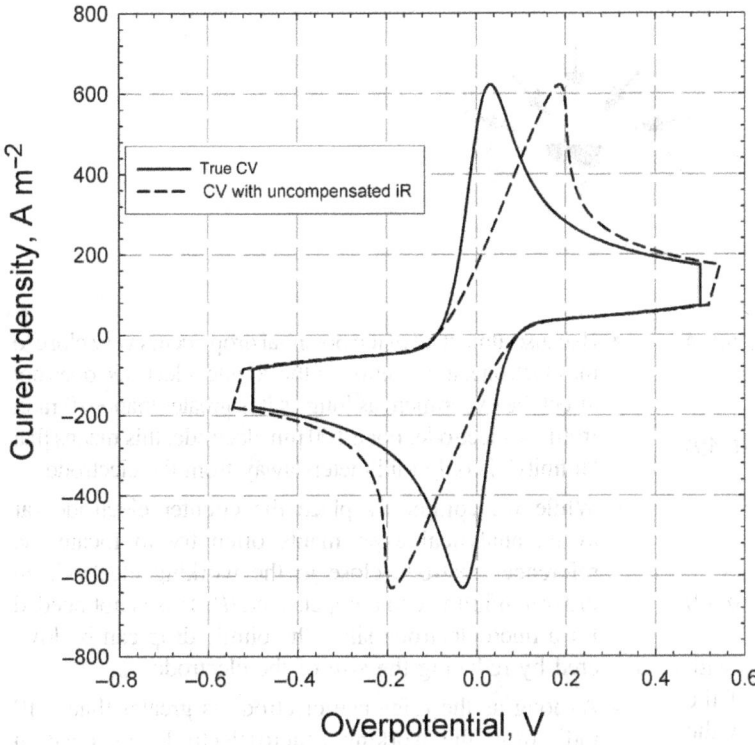

Figure 6.27 Cyclic voltammogram for a reversible system with and without *iR* compensation.

where there is significant *iR* drop between the working and reference electrodes. Figure 6.27 reproduces the CV shown in Figure 6.11, the true electrochemical response is shown with the solid line. Also plotted in this figure (dashed line) is the same CV, but where there is significant uncompensated *iR*. Note that the CV with uncompensated *iR* is skewed, and the peak potentials are no longer separated as predicted by Equation 6.20. One might be tempted to conclude from the uncompensated data that the reaction is quasi-reversible because of the large peak separation. Carefully compare the data for the CV with uncompensated *iR* with one for a quasi-reversible reaction (Figure 6.14) to convince yourself that the behaviors are different.

---

iR, Correction, and Compensation

iR drop refers to the potential difference measured between two points in an electrolyte caused by the flow of current. *iR* is also called ohmic drop.

iR correction is a means of adjusting the potential of the WE relative to the RE to account for *iR* drop in the electrolyte. The correction may be positive or negative.

iR compensation is a technique that is used by typical potentiostats to automatically correct for *iR* drop in the electrolyte. Often, full compensation is not accomplished because of stability problems with the instruments.

6.10 MICROELECTRODES

Microelectrodes are electrodes whose characteristic dimension is only a fraction of a millimeter. They represent an important tool for electroanalytical measurements. In this section, we examine some of the advantages and disadvantages of these electrodes.

Two examples of microelectrode geometries are shown in Figure 6.28, but the general concepts developed here can be applied to other geometries. The simplest type of electrode to analyze is the hemispherical microelectrode. However, as you might imagine, fabricating such an electrode is challenging. The second geometry we will consider is the disk electrode shown in Figure 6.28b. It is more complicated to analyze, but has many features in common with the hemispherical electrode. Owing to the simplicity of fabrication, the disk microelectrode is the one that is most commonly used.

Potential Drop

What is it that makes microelectrodes so useful? In order to answer this question, let's examine the potential drop associated with a microelectrode in solution. As is typically the case, we assume that the counter electrode is large and located far away from the microelectrode. For simplicity, we will consider a hemispherical electrode and neglect concentration gradients in solution. First, let's determine the potential drop in solution for a specified current density at the electrode surface. The problem is spherically

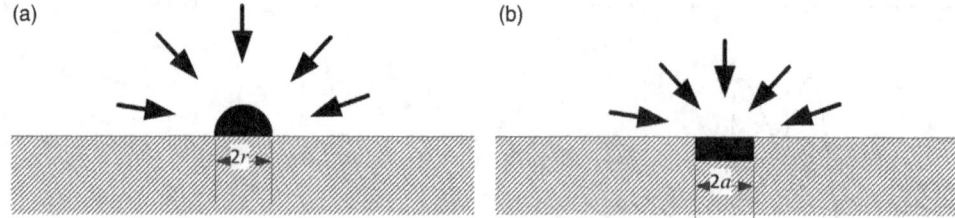

Figure 6.28 Hemispherical and disk microelectrodes.

symmetric and the relevant equation is Laplace's equation with the following boundary conditions:

$$-\kappa \frac{d\phi}{dr}\bigg|_{r=a} = i_{\text{surf}}, \quad (6.49)$$
$$\phi = 0 \text{ as } r \to \infty.$$

Solving for the potential yields

$$\phi(r) = \frac{i_{\text{surf}} a^2}{\kappa r}. \quad (6.50)$$

Because the area available for current flow increases with r^2 as you move further from the electrode, most of the voltage drop occurs close to the electrode. In fact, by the time you are 10 radii ($10a$) away from the electrode, 90% of the total voltage drop has occurred. The total potential drop associated with the microelectrode is

$$\Delta V_\Omega = \phi(a) - \phi(\infty) = \frac{i_{\text{surf}} a^2}{\kappa a} - 0 = \frac{i_{\text{surf}} a}{\kappa}. \quad (6.51)$$

Now that we have determined the voltage drop for a given current density at a microelectrode, it is useful to look at the absolute value of the current:

$$I = i_{\text{surf}} A = 2\pi a^2 i_{\text{surf}}. \quad (6.52)$$

Thus, the total current corresponding to a specific current density at the surface scales with a^2. With the total current and voltage, we can calculate the total ohmic resistance for the hemispherical electrode:

$$R_{\Omega,\text{hemisphere}} = \frac{V}{I} = \frac{1}{2\pi \kappa a}, \quad (6.53)$$

where the resistance is in ohms. With the above relations, we can now explore some of the important characteristics of microelectrodes. Because of spherical decay, the potential drop is centered at the electrode and there is no substantial potential drop far away from the electrode. The total ohmic drop scales with the electrode radius, a, and, therefore, becomes smaller as the size of the electrode is reduced. What are the practical implications? For an experiment at a given current density, i_{surf},

- We can reduce the ohmic drop in solution by reducing the size of the electrode.

- Because almost all of the potential drop occurs very close to the electrode, the position of the counter electrode does not affect the experiment as long as it is greater than ~10 radii from the electrode. For a 100 μm electrode, this means that "infinity" is only millimeters away from the electrode.

- While we commonly place the counter electrode far away, analytical experiments often try to locate the reference electrode close to the working electrode in order to minimize uncompensated iR. This is not needed for a microelectrode since the ohmic drop can be lowered by reducing the size of the electrode.

- As long as the reference electrode is greater than ~10 radii from the working (micro)electrode, its position does not matter and the ohmic drop between the working and reference electrodes can be easily calculated from Equation 6.51.

- Because the total current required to obtain a desired current density is much less for the microelectrode, much higher current densities can be achieved with a given instrument when using a microelectrode.

Because of these characteristics, and others that we will show shortly, microelectrodes have become an important tool for electroanalytical chemistry.

The above characteristics derived for a hemispherical electrode also apply to a disk microelectrode. Analogous to the resistance presented above, the resistance for a disk electrode is

$$R_{\Omega,\text{disk}} = \frac{V}{I} = \frac{1}{4\kappa a}, \quad (6.54)$$

which is also in ohms. Besides the ease of use, there is an important difference between a hemispherical electrode and a disk electrode: The current density at the disk electrode is not uniform. This, of course, is a problem when you want to measure, for example, kinetic data, which describes the relationship between the overpotential and the current density. How does the size of the electrode affect the current distribution at the electrode? The answer is found in the Wagner number, introduced previously in Chapter 4 and shown here for Tafel kinetics.

$$\text{Wa} = \frac{R_{ct}}{R_\Omega} = \frac{\frac{1}{A}\frac{d\eta}{di}}{\frac{1}{4\kappa a}} = \frac{4\kappa RT}{\pi a F}\frac{1}{i_{\text{avg}} \alpha_c}. \quad (6.55)$$

As you can see, the Wagner number is inversely proportional to the disk radius, a. Therefore, as the disk gets smaller, Wa increases and the current density is more uniform. In other words, another advantage of microelectrodes is that they yield a more uniform current density.

Mass-Transfer Control

There are also some very important advantages of microelectrodes with respect to mass transfer. To illustrate these, let's return to the hemispherical electrode and examine mass transfer by diffusion for a stagnant fluid. For simplicity, we neglect the influence of bulk flow caused by the current, which is only important at high concentrations and at high rates. With no variations in the θ or ϕ directions, Equation 4.25 becomes

$$\frac{\partial c_i}{\partial t} = D_i \frac{\partial^2 c_i}{\partial r^2} + \frac{2 D_i}{r} \frac{\partial c_i}{\partial r}. \quad (6.56)$$

The following initial and boundary conditions apply:

$$\begin{aligned} t &= 0, \ c_i = c_i^\infty, \\ r &= \infty, \ c_i = c_i^\infty, \\ r &= a, \ c_i = 0. \end{aligned} \quad (6.57)$$

The solution for this mass-transfer limited case for the current density is

$$i(t) = \frac{nFD_i c_i^\infty}{a} + \frac{nFD_i^{0.5} c_i^\infty}{\sqrt{\pi t}}. \quad (6.58)$$

Here because the hemisphere is uniformly accessible, the current density at the surface does not vary spatially and depends only on time. In contrast to large planar electrodes, we see that there are two parts to the solution. The first term on the right side represents the steady-state solution for diffusion to a sphere. The second part is transient. At short times, the transient term is much larger of the two and the solution is identical to the Cottrell equation for the planar geometry. At these short times, the concentration is depleted only in a very small region near the surface. If this region is thin enough, then the curvature of the sphere is not important. At long times, the transient term goes to zero and we are left with the steady-state solution.

As already noted, Equation 6.58 describes the transient behavior of a system that is mass-transfer controlled. A plot of $i(t)$ versus $\frac{1}{\sqrt{t}}$ yields a straight line whose slope and intercept are a function of both the diffusivity and the bulk concentration, as illustrated in Figure 6.29. Consequently, values of the diffusion coefficient and the concentration can be obtained from one experiment with use of the slope and intercept.

If we want to get an idea of when each of the two terms Equation 6.58 dominates, we simply equate the two terms

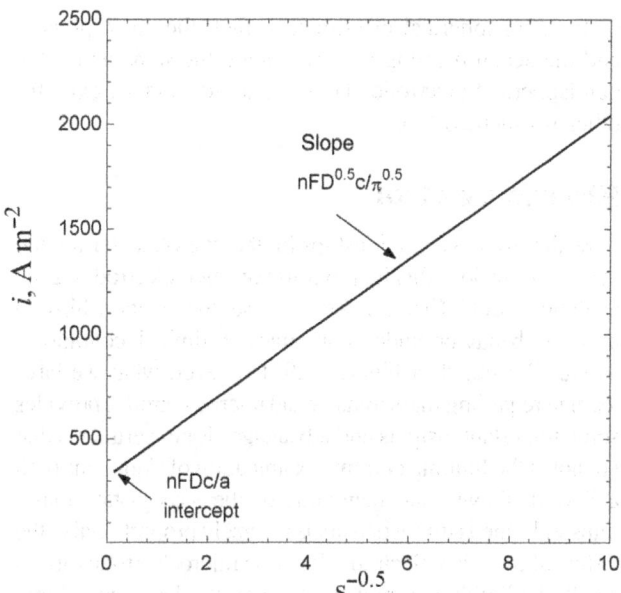

Figure 6.29 Transient behavior at microelectrode under mass-transfer control.

and solve for time. The resulting characteristic time for spherical diffusion is denoted by τ_D:

$$\tau_D = \frac{a^2}{\pi D_i}. \quad (6.59)$$

How does this compare with the time required to charge the double layer? As noted in Section 6.4, the time constant for charging of the double layer, τ_c, is $R_\Omega C$. The time constant for double-layer charging of the spherical electrode is

$$\tau_c = \frac{a}{\kappa} C_{DL}, \quad (6.60)$$

where C_{DL} is the double-layer capacitance per unit electrode area. As the dimension of the microelectrode, a, gets smaller, the charging time decreases. Using a value of $10\,\text{S·m}^{-1}$ for the conductivity, $0.2\,\text{F·m}^{-2}$ for the double-layer capacitance, and $20\,\mu\text{m}$ for the dimension of the electrode gives a charging time of $0.4\,\mu\text{s}$. Referring back to Figure 6.29, one doesn't need to worry about the effect of this non-faradaic current except for very short times, and measuring a good slope is not a problem.

As mentioned previously, the spherical or hemispherical electrode is difficult to implement experimentally. The embedded disk is much preferred. The solution for the disk geometry is not as straightforward as that for the spherical geometry, but we arrive at

$$i(t) = \frac{4nFD_i c_i^\infty}{\pi a} + \frac{\frac{8}{\pi^2} nFD_i^{0.5} c_i^\infty}{\sqrt{\pi t}}. \quad (6.61)$$

Comparison of this equation with Equation 6.58 shows that the transient behavior of the disk is very similar to that of the hemispherical electrode. In fact, the basic physics and the accompanying advantages are the same as for the hemispherical electrode. These same advantages exist for other geometries too.

Kinetic Control

Now that we have relationships for the mass-transfer limited current, how does this help our use of microelectrodes as an analytical tool? Clearly, we can use the relationships to describe behavior under mass-transfer limited conditions, as was illustrated in Figure 6.29. However, what we have learned regarding mass transfer and microelectrodes provides some important insights and advantages for experiments that are not at the limiting current. Examination of Equations 6.58 and 6.61 shows that magnitude of the steady-state mass-transfer limited current density is inversely proportional to the radius of the microelectrode. Because microelectrodes are so small, the limiting current for these electrodes is quite high, much greater than that for large electrodes. This means that microelectrodes make it possible to measure, for example, kinetic data at much higher rates without reaching the mass-transfer limit. In experiments where both rate and transport limitations are important, mass transfer reaches a steady state and does so rather quickly, facilitating both experimentation and analysis of the resulting data.

To illustrate, we'll stick with the spherical geometry because of its simplicity. The governing differential equation remains unchanged (Equation 6.56), but the initial and boundary conditions are modified. Specifically, at the surface we replace the zero concentration boundary condition by equating the flux at the surface with the current density. The current density is then given by the Butler–Volmer equation, and the problem can be solved for a given value of the potential or, equivalently, the surface overpotential at $r = a$.

Assuming steady state, the boundary conditions are

$$r = \infty, \quad c_i = c_i^\infty,$$
$$r = a, \quad -D_i \frac{\partial c_i}{\partial r} = \frac{i}{nF}, \qquad (6.62)$$

where i is calculated for the specified potential. The solution to Equation 6.56 for these boundary conditions is shown as the solid line in Figure 6.30. The concentration at the surface of the electrode is given by

$$c_i = c_i^\infty - \frac{ia}{nFD_i} = c_i^\infty\left(1 - \frac{i}{i_{lim}}\right), \qquad (6.63)$$

where a is the radius and

$$i_{lim} = \frac{c_i^\infty n F D_i}{a}, \qquad (6.64)$$

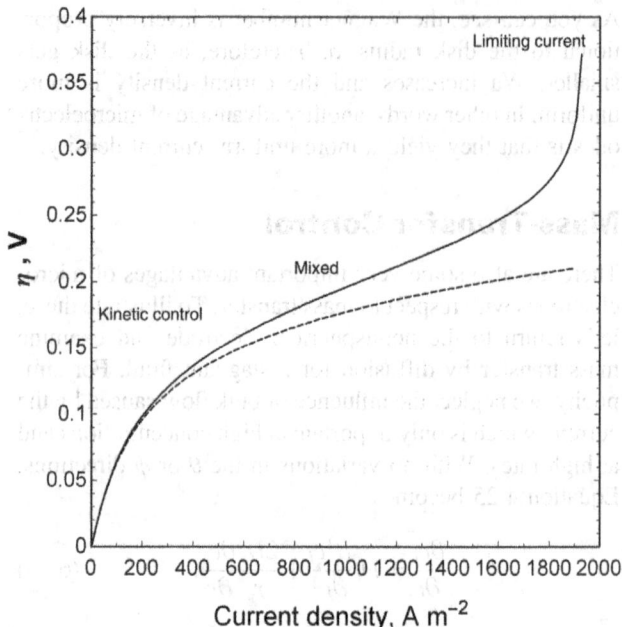

Figure 6.30 Current–voltage relationship for a spherical electrode showing regions of kinetic, mixed, and mass-transfer control. The dashed line is for pure kinetic control with no mass-transfer effect.

which is the steady-state solution from Equation 6.58 under mass-transfer control (surface concentration equal to zero). At low current densities (e.g., less than 10% of the limiting current), the electrode is under kinetic control; under these conditions, the concentration at the surface is close to that of the bulk. As the current density increases, the concentration at the surface decreases and the overpotential increases due to both concentration and kinetic effects; this is the mixed control region. At large current densities, a limiting current is reached; here the electrode is under mass-transfer control.

How is this behavior affected by electrode size? In particular, how is the behavior different with use of a microelectrode? These questions are examined in Illustration 6.7.

ILLUSTRATION 6.7

Compare the limiting current and surface concentration at a current density of $10\,\text{A}\cdot\text{m}^{-2}$ for two hemispherical electrodes, one *with* a radius of 1 mm and the other with a radius of 10 μm. The bulk concentration is 0.1 M, the diffusivity is $1\times10^{-9}\,\text{m}^2\cdot\text{s}^{-1}$, and the reaction is a two-electron reaction.

a (mm)	i_{lim} (A·m^{-2})	% i_{lim}	c_{surf}
1	19.30	52	0.0482
0.01	1930	0.52	0.0995

> The difference between the two electrodes is clear from these numbers. At the same current density, the microelectrode is operating under kinetic control at a very small fraction of the limiting current (less than 1%), and its surface concentration is essentially equal to the bulk concentration. In contrast, the larger electrode is under mixed control at 52% of the limiting current, with a surface concentration that is considerably lower than the bulk.

In summary, microelectrodes represent a powerful electroanalytical tool that allows us to reduce and correct for uncompensated iR, provide a more uniform current density, easily analyze behavior at the limiting current, and make measurements at much higher rates without reaching the mass-transfer limit. These advantages are explored further in homework problems. Care must be taken, however, to accurately measure the current, which can be quite small for very small electrodes, even when the current density is high. Fortunately, modern instrumentation can measure currents in the nanoampere range quite readily.

CLOSURE

In this chapter, we have examined several different analytical methods applicable to electrochemical systems. We began with the description of an electrochemical cell and a description of a three-electrode system for electrochemical measurements. Then in the remaining part of the chapter we examined a variety of different analytical techniques at an introductory level, exploring the influence of geometry, flow condition (transport), and control of potential or current on the output from the techniques. Time-dependent behavior and the time constant associated with different physical processes represent important themes in the chapter. Our goal for this chapter was to provide a foundation upon which a more detailed understanding of electroanalytical techniques may be based.

FURTHER READING

Bard, A.J. and Faulkner, L.R. (2001) *Electrochemical Methods*, John Wiley & Sons, Inc., New Jersey.

MacDonald, J.R. and Barsoukov, E. (2005) *Impedance Spectroscopy: Theory, Experiment, and Applications*, Wiley Interscience.

Orazem, M.E. and Tribollet, B. (2008) *Electrochemical Impedance Spectroscopy*, John Wiley & Sons, Inc., New Jersey.

Wang, J. (2006) *Analytical Electrochemistry*, Wiley-VCH Verlag GmbH, Hoboken, NJ.

PROBLEMS

6.1 You have been asked to measure the kinetics of *nickel* deposition from a *Watts* nickel plating bath. The conductivity of the plating solution is $3.5 \text{ S} \cdot \text{m}^{-1}$. The reference electrode is located 2 cm from the working electrode. The electrode area is 5 cm^2 (with only one side of the electrode active). You may neglect the impact of the concentration overpotential. You may also assume that the current density is nearly uniform.

(a) Recommend a reference electrode for use in this system. (*Hint:* What is in the Watts bath?)

(b) You apply a potential of 1.25 V and measure an average current density of $5 \text{ mA} \cdot \text{cm}^{-2}$. What is the surface overpotential? Is the IR drop in solution important? (Assume a 1D uniform current density for this part.)

(c) How good is the assumption of uniform current density? What type of cell geometry would satisfy this assumption? How would your results be impacted if the current density were not uniform?

6.2 One of your lab colleagues is attempting to measure the kinetics of the following reaction:

$$VO_2^+ + 2H^+ + e^- \rightarrow VO^{2+} + H_2O$$

which is used in the cathode of vanadium-based redox flow batteries. To simplify things, he is making the measurement at constant current. He finds that the potential decreases slightly with time, followed by an abrupt decrease and substantial bubbling.

(a) Qualitatively explain the observed behavior.

(b) Given the following parameters, how long will the experiment proceed until the abrupt change in potential is observed? Assume that there is excess H^+ in solution, and that the electrode area is 2 cm^2.

$$D_{VO_2^+} = 4 \times 10^{-10} \text{ m}^2 \text{ s}^{-1},$$

$$c_{VO_2^+} = 25 \text{ mM},$$

$$I = 1 \text{ mA}.$$

6.3 An estimate of the diffusivity can be obtained by stepping the potential so that the reaction is mass-transfer limited as described in the chapter. From the following data for V^{2+} in acidic solution, please estimate the diffusivity. The reaction is as follows:

t (s)	I (mA)
0.5	6.2
1.0	4.1
5.0	1.7
10.0	1.28
25.0	0.86
60.0	0.58
600	0.17
6,000	0.052
10,000	0.043

$$V^{2+} \rightarrow V^{3+} + e^-$$

The bulk concentration of V^{2+} is 0.01 M, and the area of the electrode is 1 cm^2.

6.4 In Section 6.4 we examined the time constant associated with charging of the double layer. In doing so, we assumed that the physical situation could be represented by a resistor (ohmic resistance of the solution) and a capacitor (the double layer) in series. However, the actual situation is a bit more complex since there is a faradaic resistance in parallel with the double-layer capacitance as shown in Figures 6.6 and 6.25. This problem explores the impact of the faradaic resistance on double-layer charging. Our objectives are twofold: (1) Determine the time constant for double-layer charging in the presence of the faradaic resistance, and (2) determine an expression for the charge across the capacitor as a function of time.

(a) Initially, there is no applied voltage, no current, and the capacitor is not charged.

(b) At time zero, a voltage V is applied.

(c) Your task is to derive an expression for the charge across the double layer as a function of time, and report the appropriate time constant. Use the symbols shown in Figure 6.17 for the circuit components.

Approach: The general approach is identical to that used in the chapter with the simpler model. In this case, you will need to write a voltage balance for each of the two legs, noting that the voltage drop must be the same. Remember that $I = \frac{dQ}{dt} = \frac{dQ_1}{dt} + \frac{dQ_2}{dt}$ where "1" is the capacitor leg and "2" is the faradaic leg. Once you have written the required balances, you can combine them into a single ODE and solve that equation for the desired relationship and time constant.

Finally, please explain physically how the characteristic time that you derived can be smaller than that determined for the simpler situation explored in the chapter.

6.5 GITT (galvanostatic intermittent titration technique) uses short current pulses to determine the diffusivity of solid-phase species in, for example, battery electrodes where the rate of reaction is limited by diffusion in the solid phase. This situation occurs for several electrodes of commercial importance. The concept behind the method is to insert a known amount of material into the surface of the electrode (hence the short time), and then monitor the potential as it relaxes with time due to diffusion of the inserted species into the electrode. In order for the method to be accurate, the amount of material inserted into the solid must be known. For this reason, the method uses a galvanostatic pulse for a specified time, which permits determination of the amount of material with use of Faraday's law assuming that all of the current is faradaic (due to the reaction).

(a) While it is sometimes desirable to use very short current pulses, what factor limits accuracy for short pulses?

(b) Assuming that you have a battery cathode, how does the voltage change during a current pulse?

(c) For a current of 1 mA, what is the shortest pulse width (s) that you would recommend? Assume that you have a small battery cathode at open circuit, and that the drop in voltage associated with the pulse is 0.15 V. The voltage during the pulse can be assumed to be constant. The error associated with the pulse width should be no greater than 1%.

6.6 Assume that you have 50 mM of A^{2+} in solution, which can be reduced to form the soluble species A^+. Assume that the reaction is reversible with a standard potential of 0.2 V. There is essentially no A^+ in the starting solution. Please qualitatively sketch the following:

(a) The *IV* curve that results from scanning the potential from a high value (0.5 V above the standard potential of the reaction) to a low value (0.5 V below the standard potential of the reaction).

(b) The *IV* curve that results from scanning the potential from a low value (0.5 V below the standard potential of the reaction) to a high value (0.5 V above the standard potential of the reaction).

(c) Why are the curves in (a) and (b) different?

(d) Assuming that you started from the open-circuit potential, in which direction would you recommend scanning first? Why?

6.7 The following CV data were taken relative to a Ag/AgCl reference electrode located 1 cm from the working electrode. You suspect that the results may be impacted by IR losses in solution. The conductivity of the solution is $10\,\text{S}\cdot\text{m}^{-1}$.

(a) Determine whether or not IR losses are important and, if needed, correct the data to account for IR losses.

(b) Is it possible to determine n for the reaction from the data? If so, please report the value. If not, please explain why not.

$100\,\text{mV}\cdot\text{s}^{-1}$ Potential [V]	Current density [mA·cm^{-2}]	$10\,\text{mV}\cdot\text{s}^{-1}$ Potential [V]	Current density [mA·cm^{-2}]
0.815	11.49	0.755	5.48
0.928	18.78	0.813	7.28
0.952	17.27	0.841	6.07
0.964	14.44	0.868	4.83
1.008	10.87	0.900	4.04
1.069	9.04	0.972	3.17
1.137	7.91	1.047	2.70
1.173	7.48	1.124	2.39
1.246	6.81	1.201	2.17
1.248	6.54	1.201	2.08
1.164	6.08	1.120	1.94
1.081	5.71	1.039	1.82
0.998	5.41	0.958	1.72
0.915	5.09	0.876	1.60
0.821	3.68	0.790	0.91
0.742	−0.23	0.730	−1.02
0.611	−9.44	0.652	−4.82
0.504	−16.08	0.597	−6.32
0.477	−14.84	0.569	−5.11
0.467	−11.87	0.542	−3.87
0.450	−9.61	0.510	−3.08
0.396	−6.98	0.438	−2.22
0.331	−5.54	0.363	−1.75
0.261	−4.61	0.286	−1.46
0.188	−3.95	0.208	−1.25
0.176	−3.68	0.208	−1.16
0.260	−3.25	0.289	−1.02
0.343	−2.90	0.370	−0.92
0.425	−2.62	0.451	−0.82
0.507	−2.34	0.532	−0.72

(Continued)

100 mV·s⁻¹ Potential [V]	Current density [mA·cm⁻²]	10 mV·s⁻¹ Potential [V]	Current density [mA·cm⁻²]
0.599	−1.16	0.619	−0.05
0.672	2.20	0.677	1.85
0.815	11.49	0.755	5.48

6.8 For hydrogen adsorption on polycrystalline platinum, the accepted loading is 2.1 C·m⁻². Using the (100) face, calculate the amount of H adsorbed on this FCC surface assuming one H per Pt atom. Then, convert this number to the corresponding amount of charge per area. Assume a pure platinum surface with an FCC lattice parameter of 0.392 nm, and compare your results with the polycrystalline number. Provide a possible explanation for any differences between the calculated and accepted values.

6.9 The behavior of an inductor is described by the following differential equation:

$$V = L\frac{dI}{dt},$$

where L is the inductance. Use this equation and the procedure illustrated in Section 6.7 to derive an expression for the complex impedance, Z. Compare your answer to that found in Table 6.4.

6.10 Create a Nyquist plot for the following system considering only electrolyte resistance, kinetic resistance for a single-electron reaction, and double-layer capacitance:

$C_{DL} = 10\,\mu\text{F·cm}^{-2}$.

$R_\Omega = \frac{1}{4\kappa r_0}$ for disk electrode with counter electrode located far away.

Counter electrode is large and located more than 10 radii from the disk electrode.

$i_0 = 0.001\,\text{A·cm}^{-2}$

$\alpha_a = \alpha_c = 0.5$.

$r_0 = 1$ mm.

$\kappa = 10\,\text{S·m}^{-1}$

6.11 Please examine your response to the Problem 6.10 and address the following:

(a) How does the magnitude of the kinetic and ohmic resistances compare to those calculated in Illustration 6.5? Please rationalize the differences and/or similarities.

(b) How is it possible to use just the formula for the disk electrode to estimate the ohmic resistance? Do you expect this to be accurate? Why or why not?

(c) In what ways does a large counter electrode influence the impedance results?

6.12 Please include the influence of the semi-infinite Warburg impedance on the system above (Problem 6.10) for which you generated the Nyquist plot. Provide a Nyquist plot and a Bode plot of the results.

$D_i = 1 \times 10^{-9}\,\text{m}^2\,\text{s}^{-1}$, $c_i^\infty = 10\,\text{mol m}^{-3}$.

6.13 EIS data were taken for a system at the open-circuit potential. Given the Nyquist diagram below:

(a) Estimate the ohmic resistance.

(b) Estimate the kinetic resistance.

(c) Is it likely that the experimental system included convection? Why or why not?

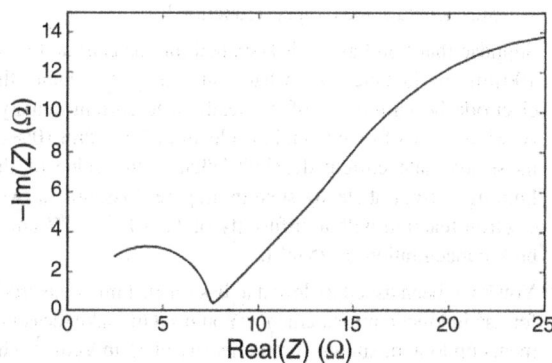

6.14 When measured about the open-circuit potential, the kinetic resistance is frequently larger than the ohmic resistance. However, for systems where mass transfer is not limiting, the ohmic drop inevitably controls at high current densities.

(a) Given that the relative magnitude of the ohmic and kinetic resistance at high current densities has changed, is this because the ohmic resistance has increased or because the kinetic resistance has decreased? Please justify your response.

(b) For the resistance that changed (kinetic or ohmic), please derive a relationship that describes how that resistance depends on the value of the current density.

6.15 The following data were taken with a RDE operating at the limiting current for a range of rotation speeds. The radius of the disk is 1 mm, and the reaction is a two-electron reaction. Assume a kinematic viscosity of $1.0 \times 10^{-6}\,\text{m}^2\,\text{s}^{-1}$. The concentration of the limiting reactant is 25 mol m⁻³. Please use a Levich plot to determine the diffusivity from the data given. Make sure that all quantities are in consistent units.

Rotation speed (rpm)	I (µA)
100	104
500	230
1000	325
1500	404
2000	470
2500	520
3000	565
3500	607
4000	660

6.16 Illustration 6.6 is a Koutecký–Levich for oxygen reduction in water, where the bulk concentration is the solubility of oxygen in water as given in the problem. These data represent oxygen reduction in acid media, and the potential values given are relative to SHE. The equilibrium potential of oxygen is 1.23 V versus SHE under the conditions of interest.

(a) Using the data from the illustration, calculate the rate of reaction for oxygen at the bulk concentration at each value of the overpotential given in the illustration.

(b) Determine the exchange-current density and Tafel slope assuming Tafel kinetics.

(c) What assumption was made regarding the concentration dependence of i_0 in the analysis above? Is the assumption accurate for oxygen reduction?

6.17 Suppose that you have a disk-shaped microelectrode that is 100 μm in diameter. At what value of time would the electrode be within 1% of its steady-state current density? At what value of time would the electrode be within 10% of its steady-state current density? What is the value of the limiting current at steady state in amperes? Assume a two-electron reaction with a diffusivity of $1 \times 10^{-9}\,\text{m}^2\,\text{s}^{-1}$ and a bulk concentration of 25 mM.

6.18 You have been asked to design a disk-shaped microelectrode for use in kinetic measurements. You need to make measurements up to a maximum current density of 15 mA cm^{-2}. The concentration of the limiting reactant in the bulk is 50 mol m^{-3}, and its diffusivity is $1.2 \times 10^{-9}\,\text{m}^2\,\text{s}^{-1}$. The conductivity of the solution is 10 S·m^{-1}. Assume a single-electron reaction.

(a) What size of microelectrode would you recommend? Please consider the impact of the limiting current and the uniformity of the current distribution.

(b) What would the measured current be at the maximum current density for the recommended electrode?

Hint: Can you do kinetic measurements at the mass-transfer limit? How does this affect your response to this problem?

6.19 Derive an expression for the ratio of the iR drop associated with a microelectrode to that associated with a large electrode. Each of these two working electrodes (the microelectrode and the large electrode) is tested in a cell with the same current density at the electrode surface, and with the same reference electrode and counter electrode. Assume that any concentration effects can be neglected and that the current distribution is one-dimensional for the large electrode. Also assume that the distance L from the working electrode to the reference electrode is the same in both cases, and that L is large enough to be considered at infinity relative to the microelectrode.

6.20 Qualitatively sketch the current response of a microelectrode to a slow voltage scan in the positive direction from the open-circuit potential. Assume that the solution contains an equal concentration of the reduced and oxidized species in solution. How does this response differ from that of a typically sized electrode? Please explain. *Hint:* What is the steady-state behavior of a microelectrode and how might this impact the shape of the CV curve?

6.21 You need to measure the reduction kinetics of a reaction where the reactant is a soluble species. The reaction is a single-electron reaction. The diffusivity is not known. As you answer the following, please include the equations that you would use and consider the implications of both mass-transfer and the current distribution.

(a) Can a rotating disk electrode be effectively used to make the desired measurements? If so, how would you proceed? If not, why not?

(b) Is it possible to use a microelectrode to measure the quantities needed to determine the reduction kinetics? If so, how would you proceed? If not, why not?

(c) What are the advantages and disadvantages of the two methods? Which would you recommend? Please justify your response.

(d) What role, if any, does a supporting electrolyte play in the above experiments?

6.22 A CV experiment is performed using a microelectrode with a diameter of 100 μm at room temperature. The potential is swept anodically at $\nu = 10\,\text{mV·s}^{-1}$. The double-layer capacitance is 0.2 F·m^{-2}. Recall that the charging current is $i_c = \nu C_{DL}$. The diffusivity of the electroactive species is $3 \times 10^{-9}\,\text{m}^2\cdot\text{s}^{-1}$. Assume that the fluid is stagnant. The concentration of the redox species is 100 mol·m^{-3} and the solution conductivity is 10 S·m^{-1}. From the data for a sweep in the positive direction, determine the exchange-current density and the anodic-transfer coefficient. The potentials are measured relative to a SCE reference electrode located far away from the microelectrode. The equilibrium potential of the reaction relative to SHE is 0.75 V.

V (SCE)	I (nA)
0.600	0.50
0.650	1.35
0.700	3.30
0.750	9.10
0.800	25.0
0.850	62.5
0.900	168
0.950	425
1.001	1200
1.052	3150
1.104	8000

6.23 Given an elementary single-electron reaction described by the following kinetic expression:

$$i\left[\frac{\text{A}}{\text{m}^2}\right] = 10.0 \left(\frac{c_{ox,\text{surf}}}{c_{ox,\text{bulk}}}\right)^{0.5} \left(\frac{c_{red,\text{surf}}}{c_{red,\text{bulk}}}\right)^{0.5} \left[\exp\left(\frac{0.5F\eta_s}{RT}\right) - \exp\left(-\frac{0.5F\eta_s}{RT}\right)\right],$$

where the bulk concentration of each of the two reactants is 50 mM. You are to use a rotating disk electrode to measure the current density as a function of V for two different disk sizes, one with a 10 mm diameter and a second with a diameter of 1 mm. V is measured against a SCE reference electrode located more than 5 cm from the disk, and the standard potential of the reaction is 0.1 V SCE. Plot the i versus V curve for each of the two electrodes for a range of current densities from −150 to 150 A·m^{-2} at a rotation speed of 500 rpm. Comment on any similarities and differences between the two curves. How does the size of the disk impact the mass transfer and the ohmic losses? You should account for the difference between the surface and bulk concentrations, including its impact on the equilibrium potential. *Hint*: It is easier to start with the current than it is with the voltage.

6.24 Our goal in this problem is to numerically generate a portion of the CV curve for a reversible system, where there are no kinetic limitations. Under these conditions, the potential at the surface remains essentially at equilibrium due to the very small overpotential needed to drive the reversible reaction. The surface potential does not remain constant, however, and *the ratio of concentrations at the surface changes so that the equilibrium potential at the surface matches the applied potential*. At the concentrations considered, the rate of reaction and hence the current depends completely on mass transfer. We will consider a generic single-electron reaction where both the oxidized and reduced species are in solution and there is excess supporting electrolyte (migration is not important). Given this, the following relations apply:

$$O + ne^- \leftrightarrow R$$

$$\frac{c_{ox}}{c_{red}} = \varepsilon = \exp\left(\frac{nF}{RT}(E_{app} - U^\theta)\right).$$

$$E_{app} = E_{start} + vt.$$

$$i = -FD_{ox}\frac{dc_{ox}}{dx} = FD_{red}\frac{dc_{red}}{dx}.$$

These expressions for the current can be combined with the concentration ratio above to yield the required boundary conditions as we will see below. Your task is to solve numerically the diffusion equation (Fick's second law) for the concentration and the current as a function of time (and, therefore, potential). The required time can be determined from the starting and ending potentials and the scan rate. We will only do a single scan, although you should code the scan so that it can be run in either the positive or negative direction.

The discrete form of the diffusion equation needed for numerical solution is

$$\frac{c_{i,new}[n] - c_{i,old}[n]}{\Delta t} = D_i \frac{c_{i,old}[n+1] - 2c_{i,old}[n] + c_{i,old}[n-1]}{\Delta x^2},$$

where explicit time integration has been assumed. This equation can be written for both "O" and "R" to solve explicitly for the concentrations at the internal node points at each time step in terms of the other values at the previous time step. In the discrete equation, $c_{i,old}[n]$ represents the concentration of i at node $[n]$ at the previous time step, where $n = 0$ at the surface. We recommend using about 200 spatial grid points uniformly spaced by Δx, the distance between grid points. Assume that $x = 0$ at the surface, and that x increases as you move away from the surface. Explicit time integration simplifies the problem by eliminating iteration, but is stable only when

$$\frac{D_i \Delta t}{\Delta x^2} < 0.5.$$

This equation can be used to determine the time-step size that is needed for stability. Far away from the electrode surface (e.g., 1 cm), the concentrations remain at their initial values. Use a simple (approximate) boundary condition at the surface where the surface gradient is approximated by

$$\frac{dc_i}{dx} = D_i \frac{c_i[1] - c_i[0]}{\Delta x}.$$

Using this approximation for the gradients of both "O" and "R," it can be shown that

$$c_{red}[0] = \frac{c_{red}[1] + \frac{D_{ox}}{D_{red}}c_{ox}[1]}{1 + \varepsilon\frac{D_{ox}}{D_{red}}},$$

$$c_{ox}[0] = \varepsilon c_{red}[0],$$

where ε is the ratio of concentrations calculated from the applied potential by the equation listed previously. Your program will need to specify your x grid points, determine the time-step size needed for stability, define your parameters and the number of time steps required, and define and initialize the required arrays. In your time loop, the concentration values at the internal nodes can be updated at each time step from the values at the previous time step (no iteration required). Subsequently, the equations above can be used to calculate the surface concentrations of c_{ox} and c_{red} at the new time step, where the ratio is determined from the value of the applied potential at that time step. The current at the new time step can then be calculated. Please use the following parameters:

$$U^\theta = 0.7\ V,\quad D_{ox} = 1.0\times 10^{-9}\ \frac{m^2}{s},$$

$$D_{red} = 2.0\times 10^{-9}\ \frac{m^2}{s}.$$

(a) Please provide the current–voltage (CV) plot for a voltage sweep from 0.4 to 1.1 V at 5 mV·s^{-1}. Assume initial concentrations of $c_{red} = 100$ mM and $c_{ox} = 0$ mM. Also provide the concentration profiles at the ending voltage (1.1 V).

(b) Repeat (a) for scan rates of 10 and 20 mV·s^{-1}. How does the peak current change? Why?

(c) Why is the voltage profile flat at the beginning of the scan in (a)? *Hint*: Look at concentration.

(d) Repeat (a) for initial concentrations of $c_{red} = 10$ mM and $c_{ox} = 10$ mM. Before running the simulation, predict and record what you think will happen. How accurate were your predictions?

(e) Provide the CV plot for a voltage sweep from 0.7 to 1.1 V with an initial concentration of $c_{red} = 10$ mM and $c_{ox} = 10$ mM. Why is the initial current equal to zero?

(f) Provide the CV plot for a voltage sweep in the negative direction from 0.7 to 0.4 V with the same initial conditions as (e). How does the curve generated compare with that from (e)?

(g) Change the value of D_{red} equal to that of D_{ox}. How does this change the curve generated in (f)?

John B. Goodenough

John B. Goodenough was born on July 25, 1922. He grew up in Connecticut where his father taught at Yale, specializing in the history of religion. During his childhood Goodenough struggled with reading—at the time dyslexia was largely undiagnosed, poorly understood, and rarely treated. To his own surprise, Goodenough received a scholarship to a private boarding school and subsequently entered Yale. Initially, he had no plans to study science. In his second year at Yale, World War II began for the United States with the bombing of Pearl Harbor, and John enlisted in the Army Air Force. He served as a meteorologist dispatching tactical aircraft across the Atlantic. Goodenough was awarded a degree in Mathematics from Yale in 1944. Two years later, still in the Army and stationed in the Azores, he was ordered to return to the states to study either mathematics or physics in Chicago. John chose physics and immediately began his graduate studies. He left the Army in 1948 and in 1952 received a Ph.D. from the University of Chicago. Rather than taking an academic position, Dr. Goodenough joined MIT's Lincoln Laboratory, where he worked on the development of ferrite core memory for computers. Subsequently, he chose to focus on solid-state chemistry. For the next decade or so his work was theoretical, and he published two books on oxide materials.

Two events put Dr. Goodenough's career on a new trajectory. First, in 1969 the controversial Mansfield Amendment was passed, prohibiting federal funding for labs like the Lincoln Lab unless the research was directed at a specific military application. This change, of course, discouraged fundamental studies. Second, in 1973, the oil embargo thrust energy into the spotlight. Dr. Goodenough turned his attention to clean energy and was involved in photoelectrolysis, the development of yttria-stabilized zirconia for solid oxide fuel cells, and fast solid-state sodium-ion conductors. He was elected Professor and Head of the Inorganic Chemistry Laboratory at Oxford University in 1976. It was here where he developed the positive electrode materials for lithium-ion batteries. These cobalt oxides were paired with a carbon negative electrode and commercialized by Sony in 1991. Facing a mandatory retirement at the age of 67, Dr. Goodenough left Oxford and joined the faculty of the University of Texas in 1986. In the 1990s, he developed an alternative positive electrode for lithium-ion batteries based on materials with olivine crystal structures, such as iron phosphate Li_xFePO_4. Iron phosphate is both less costly and safer than the cobalt oxide, but operates at a lower potential.

Professor Goodenough has published more than 500 articles and five books. He is a member of the National Academy of Engineering and the French Academy of Sciences and a Foreign Member of the Royal Society. He received the Electrochemical Society's Olin Palladium award in 1999. This award is the Society's highest honor for scientific achievement. In 2001, he received the Japan Prize for his discoveries of the materials critical to the development of lightweight rechargeable batteries. In 2008, the Royal Society of Chemistry established an award in his name that recognizes "exceptional and sustained contributions to the area of materials chemistry." In 2009, Dr. Goodenough was corecipient of the Enrico Fermi Award, and more recently (2012) he was awarded the National Medal of Science for "groundbreaking cathode research that led to the first commercial lithium-ion battery, which has since revolutionized consumer electronics with technical application for portable and stationary power." In 2013, Professor Goodenough was awarded an honorary membership in the Electrochemical Society.

Much of the early information was obtained from a 2001 interview found at http://authors.library.caltech.edu/5456/1/hrst.mit.edu/hrs/materials/public/Goodenough/Goodenough_interview.htm

Image Source: Courtesy of Cockrell School of Engineering.

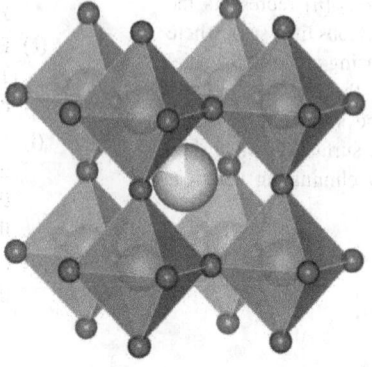

Chapter 7

Battery Fundamentals

Energy storage using batteries is an essential feature for numerous technologies, and applications of batteries are widespread in society. The necessity of energy storage includes cases where portability is desired, emergency power is required, or simply when electrical demand and supply are not matched. The battery is such an energy storage device—converting between chemical and electrical energy. For batteries to be practical, the conversion must be nearly reversible; that is, the reactions must occur with a low overpotential. This near reversibility allows for most of the chemical energy in the battery to be converted into electrical energy that performs useful work. In this chapter, the fundamental characteristics of batteries are covered. Applications and engineering design are discussed in the next chapter. As with other electrochemical systems, our convention will be to put the more negative electrode on the left. Generally, the reactions will be written as discharge reactions, but remember that for rechargeable batteries the reactions are reversed on charge.

7.1 COMPONENTS OF A CELL

Figure 7.1 shows a simple electrochemical cell known as the Daniell cell (1839). Here zinc is oxidized at one electrode and copper ions reduced at the other:

$$Zn \rightarrow Zn^{2+} + 2e^-$$
$$Cu^{2+} + 2e^- \rightarrow Cu$$

The main components of a cell are the negative electrode (zinc bar), the positive electrode (copper bar), and the electrolyte. The electrolyte is an ion conductor and an electronic insulator. As mentioned in Chapter 4, batteries frequently have a physical separator to permit placement of the electrodes in close proximity without shorting. That separator can also serve as the electrolyte (e.g., a solid Li-ion conductor) or can house the electrolyte in a porous structure where current flows by ion movement through the pores. In the cell shown (Figure 7.1), the two electrodes are isolated electronically by a separator (here shown as two beakers connected by a salt bridge of KNO_3). This, of course, is not a configuration practical for commercial use, but it does illustrate the important components of a battery cell. During discharge, electrons travel from the anode to the cathode through the external circuit.

We can estimate the thermodynamic or equilibrium potential for this cell using the methods of Chapter 2.

$$U = U_{cell}^{\theta} - \frac{RT}{nF} \ln \prod a_i^{s_i} = U_{cell}^{\theta} - \frac{RT}{2F} \ln \frac{a_{Zn^{2+}}}{a_{Cu^{2+}}}.$$

From Appendix A, the standard potential is

$$U_{cell}^{\theta} = U_{Cu}^{\theta} - U_{Zn}^{\theta} = 0.337 - (-0.763) = 1.10 \text{ V}. \quad (7.1)$$

The oxidation of zinc and reduction of copper takes place spontaneously at standard conditions; and electrons travel through the external circuit. The chemical nature of the materials used in the cell determines its thermodynamic potential. The chemical energy is stored in the two electrodes and sometimes in the electrolyte. The size of these components dictates the amount of energy stored in the cell.

Often the electrode that undergoes oxidation during discharge is referred to as the anode, and the one where reduction occurs during discharge is called the cathode. This practice can be confusing because for a rechargeable battery the process is reversed on charge, and the electrode that was the anode during discharge is now the cathode during charge. In contrast, it is unambiguous to refer to the electrode that is the anode upon discharge (e.g., zinc) as the *negative electrode* and the electrode that is the cathode on discharge (e.g., copper) as the *positive electrode*. Thus, the terms negative and positive electrode will be used, and the terms anode and cathode will be avoided for batteries.

Electrochemical Engineering, First Edition. Thomas F. Fuller and John N. Harb.
© 2018 Thomas F. Fuller and John N. Harb. Published 2018 by John Wiley & Sons, Inc.
Companion Website: www.wiley.com/go/fuller/electrochemicalengineering

152 Electrochemical Engineering

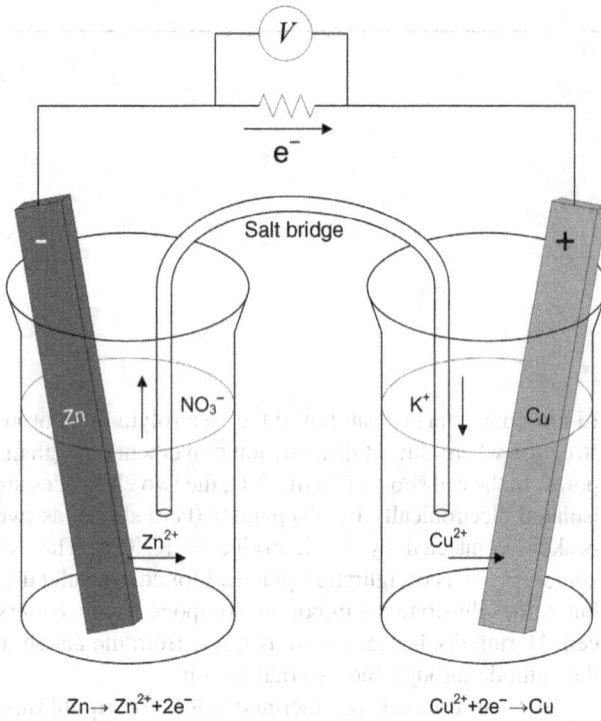

Figure 7.1 The Daniell cell is a simple battery, shown undergoing discharge.

The construction of a commercial cell is represented better by Figure 7.2, which shows a silver–zinc coin or button cell. The separator is a porous film or matrix that electronically isolates the positive and negative electrodes. The electrolyte fills the pores of the separator. This thin separator reduces the ohmic resistance of the cell. The electrolyte is ionically conductive and contains ions that are involved in the two electrochemical reactions. The electrode contains the active material. Quite often it also contains conductive materials to improve the electronic conductivity over that of the active material itself, and a binder to provide structural integrity, as well as other support materials. Typically, the electrodes are porous structures that increase the active area in contact with

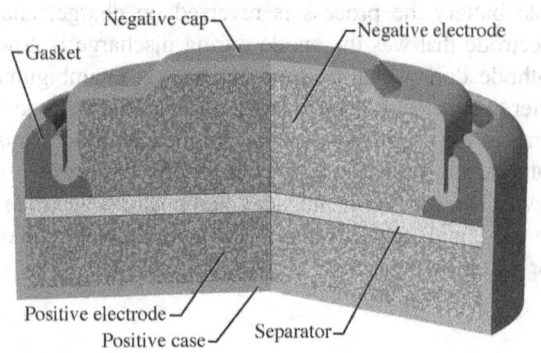

Figure 7.2 Silver–zinc coin cell.

the electrolyte, as was introduced in Chapter 5. Current passes to and from the electrodes through their respective current collectors. For the silver–zinc cell, also known as a silver oxide cell, the overall reaction is

$$Zn + Ag_2O \rightarrow ZnO + 2Ag \quad (7.2)$$

The methods of Chapter 2 can be used to determine the standard potential, U^θ, from the Gibbs energies of formation, found in Appendix C.

$$U^\theta = \frac{-\Delta G}{nF} = \frac{-11210 - (-318300)}{(2)96485} = 1.591 \text{ V}. \quad (7.3)$$

These electrodes are separated from each other with a porous cellophane film. Silver oxide is reduced at the positive electrode on discharge:

$$Ag_2O + H_2O + 2e^- \rightarrow 2Ag + 2OH^- \quad (0.340 \text{ V}) \quad (7.4)$$

The other electrode is a powder of Zn, which reacts as follows during discharge:

$$Zn + 2OH^- \rightarrow H_2O + 2e^- + ZnO \quad (-1.251 \text{ V}) \quad (7.5)$$

The standard potential for each of these reactions is shown to the right of the corresponding half-cell reaction. Because the silver electrode has a higher potential, it is referred to as the positive electrode; the zinc electrode is the negative electrode. The pores of the electrodes and the separator are filled with concentrated KOH to provide a conduction path and to transport OH^-. Also, for this cell, hydroxide ions are involved in the reactions at both electrodes. At the same time, water is consumed at the positive electrode and produced at the negative electrode during discharge. As the reaction proceeds, the net amount of water and hydroxide ions remains constant. This characteristic of the silver–zinc cell will influence its operation as we will see in subsequent sections.

7.2 CLASSIFICATION OF BATTERIES AND CELL CHEMISTRIES

Batteries are classified as either *primary* or *secondary*. The easiest way of thinking about this is that secondary batteries are rechargeable. By reversing the current through the cell, a secondary battery can be recharged and used again. That is, electrical energy is used to restore a secondary cell back to its original state, whereas a primary battery is used only once—the reactions cannot be easily reversed. Both types are common, as you will recognize from your experience with batteries. Primary batteries usually have higher specific energy [W·h·kg^{-1}] and power [W·kg^{-1}] than secondary batteries, although that is not always the case. Also, keep in mind that the distinction between primary and secondary batteries is often based on design rather than on chemistry alone. For instance, the silver–

zinc and alkaline battery chemistries are each used in both primary and secondary (rechargeable) batteries.

> ### *Cell versus Battery*
> A cell is the basic electrochemical device that stores energy. Two or more cells connected together form a battery. It is common to refer to a single cell as a battery even though there is only one cell.

A special category of primary batteries is the *reserve* battery. In these batteries, a key component of the cell, such as the electrolyte, is kept separate from the rest of the components or is otherwise inactive. By doing this, most *self-discharge* and *side reactions* (Section 7.8) are eliminated, leading to an extremely long shelf life under even severe environmental conditions. When it is needed, the reserve battery is activated. Only then are all of the components able to function together. The most common reserve batteries are thermally triggered. When the cell is heated, the electrolyte, which is a solid at room temperature, melts and forms an ionically conductive media that enables operation of the battery.

The chemistries for a number of common batteries are listed in Table 7.1. There is a rich diversity of reactions used in batteries, and a full discussion is beyond the scope of this chapter. Nonetheless, we will find it instructive to consider a few battery chemistries in order to develop a physical understanding of what is happening during the operation of a battery cell. This understanding will help us describe the performance of these cells and will provide a foundation for electrode design (see Chapter 8). Two important classes of electrode reactions for batteries are described by Huggins: *reconstruction* and *insertion*.

Reconstruction (Also Referred to as Conversion Reactions)

As the reaction proceeds, new phases are formed and grow, while other phases shrink and disappear. One or more electrons are transferred, and the reactants and products are distinct stoichiometric species. Reconstruction reactions can further be designated as *formation* or *displacement* reactions.

$$A + B \rightarrow AB \quad \text{formation}$$

An example of a reconstruction/formation reaction is that found in the lithium sulfur dioxide primary cell. The overall reaction is lithium and sulfur dioxide reacting to form solid lithium dithionite.

$$2Li + 2SO_2 \rightarrow Li_2S_2O_4$$

Of course, in a battery the overall reaction proceeds as two separate electron-transfer reactions. At the negative electrode, lithium metal is oxidized; and at the positive electrode, lithium ions react with sulfur dioxide dissolved in an organic electrolyte.

$$Li \rightarrow Li^+ + e^-$$

$$2Li^+ + 2SO_2 + 2e^- \rightarrow Li_2S_2O_4$$

Thus, a new solid phase, lithium dithionite, forms and grows during the discharge process. Note that a key reactant, SO_2, is found in the electrolyte, a feature that is present in a number of batteries.

Displacement is a second kind of reconstruction reaction.

$$A + BX \rightarrow AX + B \quad \text{displacement}$$

An example of a displacement reaction is the silver oxide cell discussed earlier.

$$Zn + Ag_2O \rightarrow ZnO + 2Ag$$

Referring back to the half-cell reactions discussed previously, the equilibrium potential for the Zn (−1.251 V) reaction is more negative than that of Ag (0.340 V). Therefore, Zn will displace the Ag in the oxide spontaneously. For the electrochemical cell, of course, these reactions occur as separate electron-transfer reactions as we have already seen.

Although useful in categorizing chemistry of the cell, these designations of formation or displacement do not provide the mechanistic details of the reaction. Often it is necessary to have a more in-depth picture of the electrode reactions in order to explain the behavior of cells. The physical phenomena associated with reconstruction reactions can be complicated and include dissolution and precipitation, solid-state ion transport, and film formation. Because of its importance in a number of batteries, it is worth taking time to explore the *dissolution–precipitation mechanism* in more detail. In this mechanism, a reaction product is formed that may subsequently react with a species in solution to form a precipitate. For example, a metal (M) used as the negative electrode of a cell (see Table 7.2) may dissolve as a result of electron transfer:

$$M \leftrightarrow M^{z+} + ze^- \quad \text{dissolution}$$

This electrochemical reaction may be accompanied by the subsequent precipitation of a salt from the solution

$$M^{z+} + zX^- \leftrightarrow MX_z(s) \quad \text{precipitation}$$

The amount precipitated depends on the solubility of MX_z in the solvent. Other reactions besides metal dissolution can result in the dissolved product and subsequent precipitation. We will examine the behavior under three solubility conditions: (1) an insoluble salt, (2) a highly soluble salt, and (3) an intermediate or sparingly soluble salt. Each case will be explored through specific chemistries.

Table 7.1 Example Chemistries for a Number of Common Batteries

Cell	Electrochemical reactions	Nominal Cell potential	Comments
Lead–acid	$Pb(s) + HSO_4^- + H_2O \rightarrow PbSO_4(s) + H_3O^+ + 2e^-$ $PbO_2(s) + 3H_3O^+ + HSO_4^- + 2e^- \rightarrow PbSO_4(s) + 5H_2O(l)$ $Pb(s) + PbO_2(s) + 2H_3O^+ + 2HSO_4^- \rightarrow 2PbSO_4(s) + 4H_2O(l)$	1.8 V	Secondary During discharge, lead sulfate is formed at both electrodes. The electrolyte is part of the active material
Alkaline	$Zn(s) + 2OH^- = ZnO(s) + H_2O(l) + 2e^-$ $2MnO_2(s) + H_2O(l) + 2e^- \rightarrow Mn_2O_3 + 2OH^-$ $Zn(s) + 2MnO_2(s) = ZnO(s) + Mn_2O_3$	1.5 V	Primary Common battery
Lithium-ion	$Li_xC_6 \rightarrow xLi^+ + xe^- + 6C$ $xLi^+ + xe^- + Mn_2O_4 \rightarrow Li_xMn_2O_4$ $Li_xC_6 + Mn_2O_4 \rightarrow Li_xMn_2O_4 + 6C$	3.5–4.2 V	Secondary The spinel manganese dioxide is just one of the chemistries of rechargeable lithium-ion batteries
Silver–zinc	$Zn + 2OH^- \rightarrow H_2O + 2e^- + ZnO$ $Ag_2O + H_2O + 2e^- \rightarrow 2Ag + 2OH^-$ $Zn + Ag_2O \rightarrow ZnO + 2Ag$	1.5 V	Manufactured as both a primary and a secondary battery
Ni/Cd	$Cd(s) + 2OH^- \rightarrow Cd(OH)_2 + 2e^-$ $NiOOH + e^- + H_2O \rightarrow Ni(OH)_2 + OH^-$ $Cd(s) + 2NiOOH + 2H_2O \rightarrow Cd(OH)_2 + Ni(OH)_2$	1.2 V	Secondary, out of favor because of environmental issue with cadmium
Ni/Fe	$Fe + 2OH^- \rightarrow Fe(OH)_2 + 2e^-$ $NiOOH + e^- + H_2O \rightarrow Ni(OH)_2 + OH^-$ $Fe + 2NiOOH + 2H_2O \rightarrow Fe(OH)_2 + 2Ni(OH)_2$	1.2 V	Secondary, also called the Edison cell
NiMH	$MH + OH^- \rightarrow M + H_2O + e^-$ $NiOOH + e^- + H_2O \rightarrow Ni(OH)_2 + OH^-$ $MH + NiOOH \rightarrow M + Ni(OH)_2$	1.2 V	Secondary
Na-S	$Na \rightarrow Na^+ + e^-$ $xS + 2e^- \rightarrow S_x^{2-}$ $2Na + xS \rightarrow Na_2S_x$	1.9 V	Secondary high temperature
Li-SO$_2$	$Li \rightarrow Li^+ + e^-$ $2Li^+ + 2SO_2 + 2e^- \rightarrow Li_2S_2O_4$ $2Li + 2SO_2 \rightarrow Li_2S_2O_4$	3 V	Primary Lithium–sulfur dioxide
Li-SOCl	$Li \rightarrow Li^+ + e^-$ $2SOCl_2 + 4e^- \rightarrow 4Cl^- + S + SO_2$ $4Li + 2SOCl_2 \rightarrow 4LiCl + S + SO_2$	3.4 V	Primary Lithium thionyl chloride
Li/FeS$_2$	$Li \rightarrow Li^+ + e^-$ $4Li^+ + 4e^- + 3FeS_2 \rightarrow Li_4Fe_2S_5 + FeS$ $4Li + 3FeS_2 \rightarrow Li_4Fe_2S_5 + FeS$	1.5 V	Reserve Thermally activated LiCl/KCl eutectic for electrolyte
Mg/AgCl	$Mg \rightarrow Mg^{2+} + 2e^-$ $2AgCl + 2e^- \rightarrow 2Ag + 2Cl^-$ $Mg + 2AgCl \rightarrow 2Ag + MgCl_2$	1.4 V	Reserve water activated

Table 7.2 Common Negative Electrode Materials for Batteries. Values Are Theoretical

Element	MW [g·mol^{-1}]	Standard potential [V]	Density [g·cm^{-3}]	Valence	Specific capacity [A·h·g^{-1}]	Volumetric capacity [A·h·cm^{-3}]
Li	6.941	−3.01	0.534	1	3.86	2.06
Na	22.99	−2.714	0.971	1	1.16	1.12
Mg	24.305	−2.4	1.738	2	2.20	3.8
Al	26.98	−1.7	2.699	3	2.98	8.1
Fe	55.845	−0.44	7.86	2	0.96	7.5
Zn	65.39	−0.763	7.13	2	0.82	5.8
Cd	112.41	−0.403	8.65	2	0.48	4.1
Pb	207.21	−0.126	11.34	2	0.26	2.9

For an example of an insoluble salt, we examine the positive electrode of a lithium thionyl chloride battery. From Table 7.1, the reaction is

$$2SOCl_2 + 4e^- \rightarrow 4Cl^- + S + SO_2 \quad \text{dissolution}$$

$$Li^+ + Cl^- \rightarrow LiCl(s) \quad \text{precipitation}$$

Chloride is produced as a result of $SOCl_2$ reduction. The electrolyte is a mixture of $LiAlCl_4$ (lithium tetrachloroaluminate) and $SOCl_2$ (thionyl chloride), which serves as both a reactant and part of the electrolyte for the cell. Each time a chloride ion is produced at the positive electrode, it will encounter a lithium ion in the electrolyte. For all intents and purposes, LiCl is insoluble in the electrolyte and precipitates out. Consequently, LiCl(s) is likely to be found close to where the thionyl chloride is reduced. Because of the insolubility of the salt, there is essentially no Cl^- in solution; therefore, the reverse of the dissolution reaction cannot occur. Thus, because of the poor solubility of the discharge products, the reactions are not easily reversed and the $LiSOCl_2$ is a primary battery.

Next, let's consider the zinc electrode, which provides an example of dissolution to form a highly soluble salt. The reaction takes place in an alkaline electrolyte and follows a dissolution–precipitation mechanism:

$$Zn + 4OH^- \leftrightarrow Zn(OH)_4^{2-} + 2e^- \quad \text{dissolution}$$

$$Zn(OH)_4^{2-} \leftrightarrow ZnO + 2OH^- + H_2O \quad \text{precipitation}$$

The ultimate product ZnO(s) has a relatively high solubility in aqueous KOH. For now, let's imagine that the solubility is so high that the precipitation reaction doesn't occur at all. At first, this doesn't appear to be a problem. When we want to recharge the cell, zincate ions are readily available in solution to be reduced back to zinc. However, in contrast to the LiCl in the previous example, which is expected to be located at the site where the reduction occurs, the soluble zincate can diffuse or be transported by convection to another part of the cell. Now imagine, as is inevitably the case, that the current distribution on the electrode is nonuniform. The sites where deposition is higher will preferentially grow in size leading to changes in the shape of the electrode and possible dendrites (branching tree-like crystals). The shape changes and dendrites can cause shorting in the cell and lead to premature failure. Consequently, cells using a zinc negative electrode behave in a mediocre fashion as rechargeable batteries.

The final example is the negative electrode of the lead–acid battery, where lead reacts to form lead sulfate, a nonconducting ionic solid that is sparingly soluble.

$$Pb \leftrightarrow Pb^{2+} + 2e^- \quad \text{dissolution}$$

$$Pb^{2+} + SO_4^{2-} \leftrightarrow PbSO_4(s) \quad \text{precipitation}$$

As in the previous examples, the first step on discharge is an electron-transfer reaction and dissolution. In this instance, the lead sulfate is sparingly soluble in the electrolyte. The solubility is small enough so that during cycling the lead sulfate product is repeatedly formed and dissolved at roughly the same location. Yet, the solubility is large enough so that the reverse reaction can occur. On charging, not only are the kinetics for the electron-transfer reaction important, but the rate can be limited by dissolution of the lead sulfate and mass transfer of Pb^{2+} to the lead surface. This situation is explored in Problem 7.26.

Insertion

A second important class of electrode reactions found in batteries is the *insertion* reaction. Here there is a relatively stable host material, and the guest material is found in unoccupied sites of the host material. When the host material has a layered structure, the insertion is called *intercalation*. In the case of intercalation, the host material retains its essential crystalline structure and the process can be highly reversible. This process is illustrated by the intercalation of lithium into titanium disulfide (Figure 7.3).

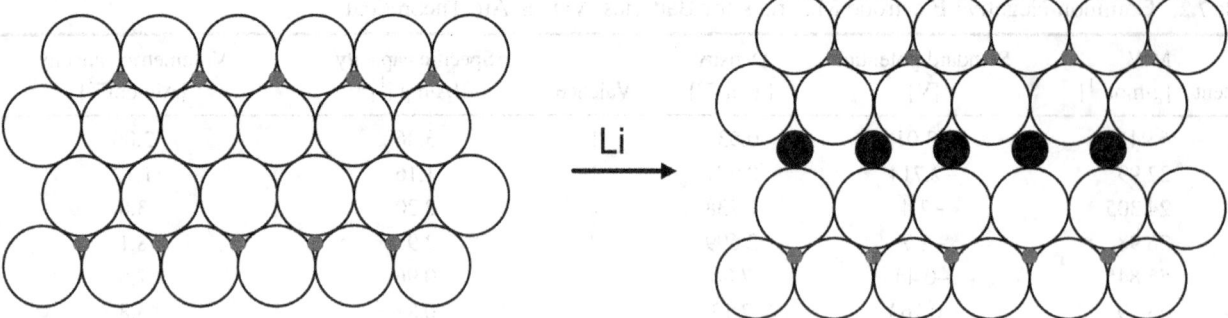

Figure 7.3 Insertion of lithium (solid black circles) into the gallery space of titanium disulfide.

TiS$_2$ has a layered structure, where individual sheets of TiS$_2$ are held together by van der Waals forces. In contrast to the reconstruction reactions, the lithium occupies the space between the layers of titanium disulfide, which is called the *gallery space*, rather than creating a new phase. The intercalation of lithium into the host titanium disulfide, a positive electrode material for a lithium-ion cell, is written as

$$x\text{Li}^+ + xe^- + \text{TiS}_2 \rightarrow \text{Li}_x\text{TiS}_2$$

where x varies from zero to one. When x is zero, the TiS$_2$ is delithiated and the electrode is considered charged; when x is one, the host is lithiated, representing a discharged state. The product of this insertion process results in a kind of solid solution of the guest species in the host, where lithium occupies sites in the gallery. In these instances, a characteristic sloping discharge curve results (Figure 7.5).

There are many types of intercalation and insertion reactions. The interested reader should consult the Further Reading section at the end of the chapter for additional information. Insertion sometimes occurs in *stages*, where there are different sites within a layer or even preferential filling of a stack of layers. Here, the guest material is not randomly distributed between the layers; instead, occupation of specific low-energy interstitial sites is favored. In these cases, the equilibrium potential can have multiple plateaus and steps. We write the reaction for the lithiation of the material as if lithium were reduced to a valence of zero; however, it may be the oxidation state of the metal (e.g., Mn) in the host material that changes.

The development of lithium rechargeable batteries has brought the importance of insertion reactions to the forefront. For lithium-ion cells, positive electrode materials are generally transition metal oxides or phosphates. The predominant negative electrode material is carbon. Finally, it should also be noted that metals other than lithium can undergo insertion reactions.

In this section, we have examined reconstruction and insertion reactions, both of which are important for practical battery systems. You have also been introduced to several different types of battery chemistries. This short introduction is by no means complete, but it is hoped that it has helped you to gain new insights into batteries and battery chemistries. We encourage you to consult other sources, such as those listed at the end of the chapter, for additional information.

Some standard metrics provide a basis for comparison of different electrode and battery systems. For example, Table 7.2 contains information for elements commonly used as negative electrode materials. Key metrics are found in the last two columns: the specific capacity [A·h·g^{-1}] and the volumetric capacity [A·h·cm^{-3}]. The potential of the electrode is also critical, and low values for the negative electrode potential contribute to higher cell voltages. Given its small mass and low potential, it is clear why lithium is the customary choice for high-energy batteries. When the volumetric capacity is considered, other materials compare more favorably to lithium, for example, magnesium and aluminum. Note that both of these metals have a valence greater than one. The standard potentials in this table refer to the oxidation of the elements, and do not necessarily reflect the actual reactions in a practical battery.

7.3 THEORETICAL CAPACITY AND STATE OF CHARGE

The capacity is a rating of the charge or energy stored in the cell. This value is expressed in either ampere-hours [A·h] or watt-hours [W·h]. The first represents the capacity in terms of coulombs of charge available, the second in terms of energy available. The two are related simply by the average voltage of the cell, W·h = A·h × V_{avg}. Clearly, if one wishes to increase the capacity of the cell, more active material is added. Thus, the capacity is directly related to the amount of active material and is a good indication of the size of the cell.

Since the current in amperes has units of C·s^{-1}, the capacity or size of a cell expressed in terms of A·h is really a measure of the coulombs available from the cell

(1 A·h = 3600 C). The theoretical capacity of an electrode for different chemistries can be determined from the electrode reactions and the mass of active material according to the following equation:

$$\text{theoretical capacity of electrode} = \frac{m_i n F}{M_i} \text{ [C]}, \quad (7.6)$$

where m_i is the mass of active material and M_i is its molecular weight. This calculated charge in A·h is often nomalized by the mass of active material to create a theoretical specific capacity that can be expressed in units such as [A·h·kg^{-1}], [A·h·g^{-1}], or [mA·h·g^{-1}]. The theoretical value represents the maximum possible capacity from the active material itself, and does not include the mass of other electrode components such as the current collector, conductive additives, or packaging. Other practical issues related to extraction of the energy are also not taken into account in the theoretical capacity.

Similarly, the theoretical capacity can be calculated for the cell, rather than for just one electrode. In this calculation, the charge capacity of each of the two electrodes is set to the same value. The coulombs of charge are determined from the stoichiometery and normalized by the combined mass of the two electrodes. This calculation is shown in Illustration 7.1. Again, the masses of the other cell components are neglected. The useable capacity of a commercial cell may only be one-fourth of its theoretical capacity. In practice, the mass and volume of the other cell components must be considered, as well as the extent to which the active material can be converted according to its stoichiometry. A number of factors influence the extent of conversion, such as side reactions and limitations on the maximum and minimum potential of the cell. These real-world effects will be discussed over much of the remainder of this chapter and in Chapter 8.

ILLUSTRATION 7.1

a. Calculate the theoretical specific capacity of the following electrodes:
$NiOOH + e^- + H_2O \rightarrow Ni(OH)_2 + OH^-$

$$1 \text{ g NiOOH} \left| \frac{\text{mole NiOOH}}{91.71 \text{ g NiOOH}} \right| \frac{1 \text{ equiv}}{\text{mole NiOOH}}$$

$$\frac{96485 \text{ C}}{\text{equiv}} \left| \frac{1000 \text{ mAh}}{3600 \text{ C}} \right. = 292 \text{ mA·h·g}^{-1}$$

$Li^+ + e^- + 6C \rightarrow LiC_6 \quad 372 \text{ mA·h·g}^{-1}$

$H_2O + 2e^- + ZnO \rightarrow Zn + 2OH^-,$
$819 \text{ mA·h·(per g Zn)}$

---

b. The last example from part (a) was for the negative electrode of the silver–zinc cell. What is the theoretical specific capacity of this full cell?

Using the stoichiometry for the silver–zinc cell from Table 7.1,

$$1 \text{ g Zn} \left| \frac{\text{mole Zn}}{65.38 \text{g Zn}} \right| \frac{\text{mole Ag}_2\text{O}}{\text{mole Zn}} \right|$$

$$\frac{231.7 \text{ g}}{\text{mole Ag}_2\text{O}} = 3.544 \text{ g Ag}_2\text{O}.$$

The total active material mass (positive electrode and negative electrode) = 3.544 g + 1.00 g = 4.544 g

$$\frac{819 \text{ mAh}}{\text{g Zn}} \left| \frac{1 \text{ g Zn}}{4.544 \text{ g active material}} \right.$$

$$= 180 \text{ mA·h·g}^{-1}\text{-active material}.$$

There is a need to quantify the available capacity remaining in a cell while in use. It is common to refer to either the *state of charge* (SOC) or the *state of discharge* (SOD, or depth of discharge, DOD) of a battery to identify the amount of unconverted active material in the cell. This concept is illustrated in Figure 7.4, again for the silver–zinc battery. When the cell is fully charged or at 100% SOC, pure Zn and Ag$_2$O are present. As the reaction proceeds, charge flows through the cell converting the reactants to ZnO and Ag. If the reactants are present in stoichiometric amounts, then the fraction (or percentage) of either Zn or Ag$_2$O that is in the initial state rather than the product state corresponds to the state of charge.

Later we'll see that a measure of the SOC is critical for managing a battery system, as will be discussed in Chapter 8. However, as you might imagine, once a battery is assembled and operated, determining the state of charge is not so easy. One well-known exception is the flooded lead–acid cell. Since the electrolyte is involved in the discharge reaction, its density changes with SOC. Where there is access to the electrolyte, its density can be measured and used to indicate the SOC. Unfortunately, most often we do not have a direct measure of active material in the battery. A further difficulty with most cells is that the two electrodes may have different capacities. Consequently, even if the amount of unconverted material in each electrode were known, we would still need to decide which electrode to use for the SOC. Finally, the full capacity of a battery based on the amount of active material is most often not accessible. If we take SOC as

$$\text{SOC} = \frac{\text{available capacity remaining}}{\text{total capacity}} \times 100\%, \quad (7.7)$$

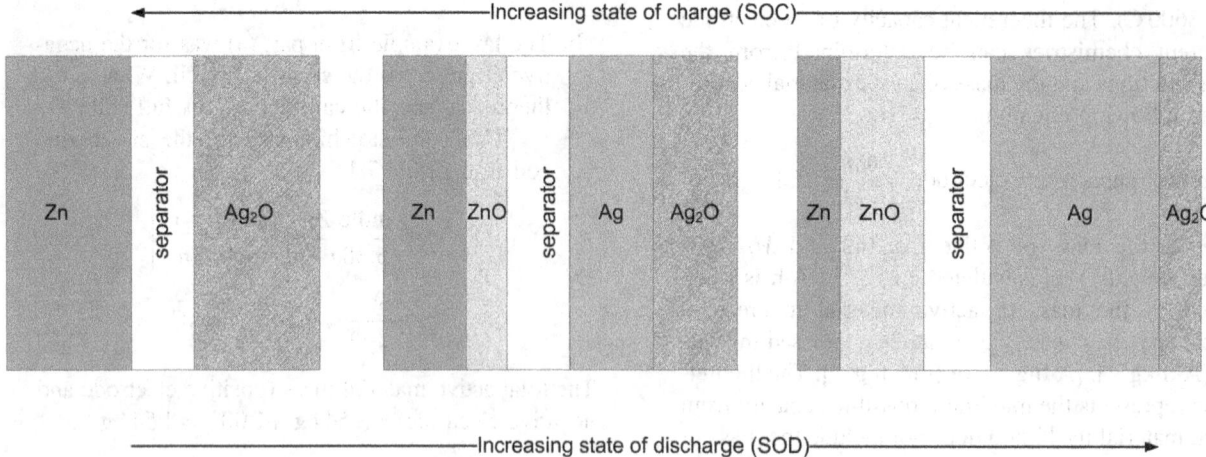

Figure 7.4 State of charge as conversion of active material.

we have two challenges. First, how is the total capacity defined? Total capacity can be expressed in A·h; therefore, a detailed knowledge of the mass of all the individual components in the battery is not required in order to specify its capacity. Still, several choices remain. For example, should the total capacity be the theoretical capacity, the nominal measured capacity for a new cell, or the capacity of the latest charge/discharge cycle? Second, the available capacity depends on temperature and rate of discharge as discussed below. How do these factors impact our definition of the SOC? As the previously mentioned challenges can be addressed in different ways, we must accept that the term SOC is not precise. For now, we'll defer detailed discussion, and we take the SOC to be an indication of the remaining capacity of the cell. As mentioned previously, SOC is used commonly with the SOD where

$$SOD = 100 - SOC. \quad (7.8)$$

7.4 CELL CHARACTERISTICS AND ELECTROCHEMICAL PERFORMANCE

In order to understand how a battery will perform in service, it is important to understand how the potential of the cell is impacted by factors such as the rate at which the cell is charged or discharged, the cell temperature, and the SOC of the cell. Thermodynamics, electrode kinetics, and transport phenomena all have a role in determining the operating potential of the cell. Since the battery is inherently a transient device, the history of the cell can also influence its performance.

The most important factor influencing the cell potential is the battery chemistry. The starting point in determining the potential of a cell is the equilibrium or thermodynamic value calculated from methods discussed in Chapter 2. Figure 7.5 shows the potential of some common batteries as a function of capacity (SOD). These are equilibrium potentials, meaning that there is no current flow. Cell potentials for different types of cells range from 1 to 4 V. The potential of a lithium-ion cell, for instance, is up to two times that of a lead–acid cell and three times that of an alkaline cell. Also note that the potential of the cell varies with the state of charge.

In general, the cell potential decreases with increasing SOD (decreasing SOC). Depending on the cell chemistry, the slopes of the curves shown in Figure 7.5 can vary. For instance, the potential of the nickel–cadmium (NiCd) battery is nearly flat between 20 and 80% SOD, whereas the potential for the lead–acid cell decreases steadily over the entire range. The change of potential with state of

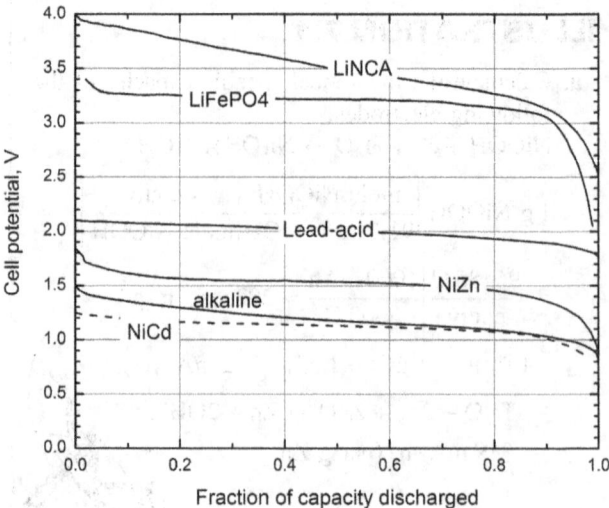

Figure 7.5 Potentials of several battery chemistries.

charge can, in large part, be understood from thermodynamics. To illustrate, let's explore the lead–acid battery.

The overall reaction for discharge of the lead–acid cell is

$$Pb(s) + PbO_2(s) + 2H_2SO_4 \xrightarrow{discharge} 2PbSO_4(s) + 2H_2O \quad (7.9)$$

As the cell is discharged (SOC decreases), solid lead sulfate is formed as a result of the reconstruction reaction. In addition, one of the reactants and one of the products are part of the electrolyte; specifically, water is produced and sulfuric acid is consumed as the cell is discharged. Therefore, the concentration of sulfuric acid decreases and the electrolyte becomes more dilute with increasing DOD. Using the methods of Chapter 2, the following equation for the equilibrium potential of the lead-acid cell results.

$$U = U^\theta + \frac{RT}{F} \ln \frac{a_{H_2SO_4}}{a_{H_2O}}, \quad (7.10)$$

where the activity of the solid components is assumed to be one. As the lead–acid cell is discharged, lead sulfate precipitates and fills the voids of the porous electrode. Because the activity of the solids is taken as unity, precipitation of lead sulfate does not affect the equilibrium potential. However, changes in the activity of the water and that of the acid both impact the equilibrium potential as is evident from Equation 7.10. The result is a sloping potential shown in Figure 7.5, which in this instance is related to the dilution of the electrolyte during discharge.

Insertion reactions are also common in rechargeable cells as typified by the familiar lithium-ion battery. At the negative electrode, lithium intercalates between the graphene-like layers of carbon as shown in Figure 7.6a. One lithium atom can be inserted for every six carbon atoms, and during discharge

$$Li_{1-x}C_6 \xrightarrow{discharge} (1-x)Li^+ + (1-x)e^- + 6C \quad (7.11)$$

In this instance, the change in volume with insertion of lithium is only a few percent. Thus, in a porous graphite electrode, the pore volume filled with electrolyte is more or less constant. As discussed earlier, lithium intercalates and de-intercalates from the carbon in stages, and the equilibrium potential varies with the amount of lithium inserted as shown in the lower line in Figure 7.6.

One type of active material for the positive electrode is $LiMn_2O_4$. During discharge, lithium inserts into the manganese oxide of the positive electrode as follows:

$$xLi^+ + xe^- + Mn_2O_4 \xrightarrow{discharge} Li_xMn_2O_4 \quad (7.12)$$

$LiMn_2O_4$ has a spinel structure, and the insertion is not the same as the layered insertion discussed earlier. The variable x is a measure of the state of discharge of the cell. In the positive electrode, x is zero when the battery is fully charged, and increases as the cell is discharged. This nomenclature is commonly used for scientific studies of lithium-ion cells. Each one of the equilibrium potentials in the figure is relative to a lithium reference electrode. The difference between the potentials of the positive and negative electrodes is the overall equilibrium potential of the cell, which is represented by the upper solid line in the figure.

The Rate of Charging or Discharging as Expressed in Terms of the "C-Rate"

C- rate, h^{-1}	Discharge time, hours
C/20	20 hours
C/5	5 hours
C	1 hour
2C	30 minutes
10C	6 minutes

The above two examples illustrated how the equilibrium voltage may change during discharge for two different types of cells. A second key factor in determining the potential of the cell is the rate at which current is drawn from the cell. The rate at which charge is removed from or added to the cell is frequently normalized in terms of the capacity of the cell in A·h. The reason for this is that the

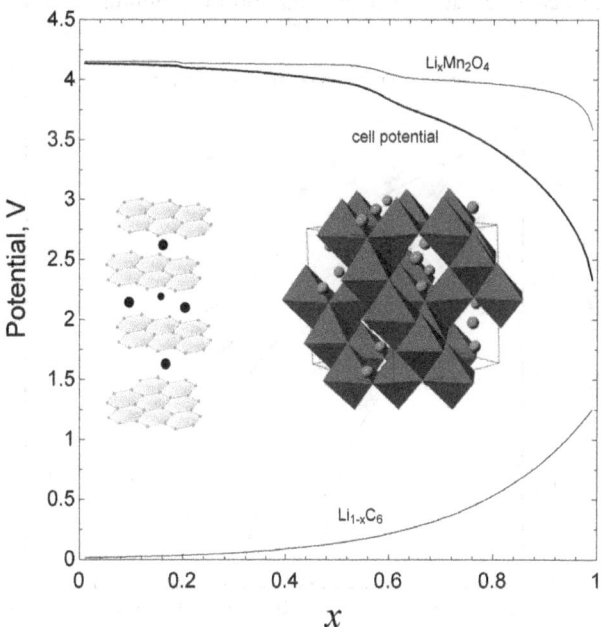

Figure 7.6 Potential versus the fraction discharged for a lithium-ion cell. Also shown are the structural aspects of the negative electrode (a) and positive electrode (b).

same discharge current may represent a very high rate of discharge for a small battery and a low rate of discharge for a large-capacity battery. The normalized rate of charge or discharge is expressed as a *C-rate*, which is a multiple of the rated capacity. A rate of "1C" draws a current [A] that is equal in magnitude to the capacity of the battery expressed in A·h. As such, the 1C rate is the current at which it would require one hour to utilize the capacity of the battery. Thus, a 5 A·h battery delivers 5 A for 1 hour, and 5 A would correspond to the 1C-rate. The same 5 A·h battery delivers roughly 0.5 A for 10 hours, and the matching C-rate would be C/10.

Figure 7.7 shows how the rate at which current is drawn from a lead–acid cell affects the cell potential. During discharge, the potential of the cell is always less than the equilibrium or thermodynamic potential. This difference is called the polarization of the cell and for discharge is given by

$$U - V_{cell} = \eta = |\eta_{ohmic}| + |\eta_{neg}| + |\eta_{pos}| + |\eta_{conc}|. \quad (7.13)$$

This polarization or deviation from the equilibrium potential is a result of contributions from ohmic, kinetic, and concentration effects. The breakdown of polarization in Equation 7.13 is similar to that shown in Figure 3.11, for ohmic and kinetic polarization. Complete separation of these effects is not strictly possible; nonetheless, it is common to ascribe polarization losses roughly to ohmic, kinetic, and concentration effects. During discharge, the potential of the cell decreases with increasing current due to increased polarization. When the cell is charged, the current is reversed. Under these conditions, the potential of the cell is greater than the equilibrium potential. Before delving into the details of the curves shown in Figure 7.7, it is useful to provide a few definitions.

Equilibrium potential (U): Potential of cell described by thermodynamics. The potential depends on the materials, temperature, pressure, and composition.

Open-circuit voltage (V_{OCV}): Potential of the cell when no current is flowing, not necessarily equal to the equilibrium potential.

Nominal voltage: Typical potential of the cell during operation, this will be less than the thermodynamic potential when discharging.

Average voltage: Potential averaged over either the discharge or charge.

End or cutoff voltage (V_{co}): Potential of the cell when the discharge is terminated.

In a well-designed cell, ohmic loss is almost invariably the largest contributor to the overall polarization of the cell. The ohmic polarization is a result of the resistance of the electrolyte, the electronic resistance of the porous electrodes (see Chapter 5), and any contact resistance. Recall that Ohm's law applies for electrolytes in the absence of concentration gradients. Making this assumption, we see that the potential drop across the separator, for instance, increases with increasing current density.

$$i = -\kappa \frac{d\phi}{dx}. \quad (7.14)$$

Thus, as the cell is discharged more rapidly, the potential at the same SOC will be lower. This ohmic polarization is the principal reason for the change in potential with rate in Figure 7.7 (i.e., the reason why, for example, the 2C curve is lower than the C/20 curve).

ILLUSTRATION 7.2

In the absence of concentration gradients, calculate the potential across the separator of a lithium-ion battery. The thickness of the separator is 30 μm, with a porosity of 0.38, and tortuosity of 4.1. The conductivity of the electrolyte is 9 mS·cm^{-1} and the current density is 50 A m^{-2}.

$$\eta_\Omega = \frac{iL}{\kappa_{eff}} = 0.00003 \text{ m} \left| 50 \frac{A}{m^2} \right| \frac{4.1/0.38}{0.9} \frac{m}{S} = 18 \text{ mV}.$$

If the cell area is 0.14 m^2, the current density of 5 mA·cm^{-2} corresponds to the 1C rate for a nominal 7 A·h battery. The change in voltage from 1C to 18 C is estimated as

$$\Delta V = (18 - 1)0.018 = 0.31 \text{ V}.$$

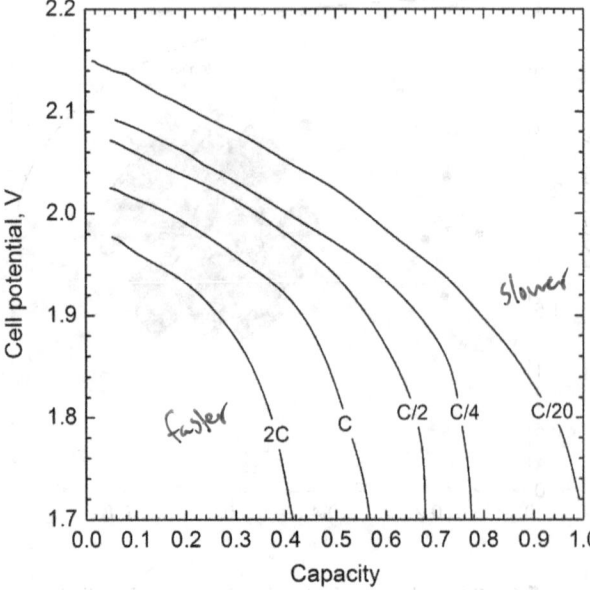

Figure 7.7 Effect of rate on potential as a function of the fractional capacity for a lead–acid cell.

Even in the absence of concentration gradients, the resistance can change during discharge. For example, Figure 7.8 shows resistance data for both a lead–acid battery and a nickel–cadmium battery. As discussed previously, the concentration of sulfuric acid decreases during discharge; therefore, we would expect the electrolyte resistance to increase. Figure 7.8 shows a strong dependence of the internal resistance on SOC for the lead–acid cell due largely to the change in the electrolyte conductivity. Thus, ohmic losses will increase during a constant current discharge. In contrast, hydroxyl ions are not consumed in a NiCd battery, even though water is produced. Consequently, we would not expect the resistance of the electrolyte to vary as strongly with SOC, consistent with the observed behavior. Note that the internal resistance of a cell can be measured with impedance and current interruption techniques as outlined in Chapter 6.

Another key factor is the activation or kinetic polarization at each electrode. The kinetics can frequently be described by the Butler–Volmer expression introduced in Chapter 3. In contrast to ohmic polarization, kinetic losses typically do not vary linearly with the current density. These losses are most significant at low currents, relative to other types of polarization.

Finally, concentration polarization results from concentration gradients that develop due to the passage of current in the cell. Let's consider a lithium-ion cell where lithium ions are shuttled between electrodes. For a concentrated binary electrolyte, applicable to many battery systems, the flux of anions is

$$N_- = -\nu_- D \nabla c + \frac{it_-^o}{z_- F}. \quad (7.15)$$

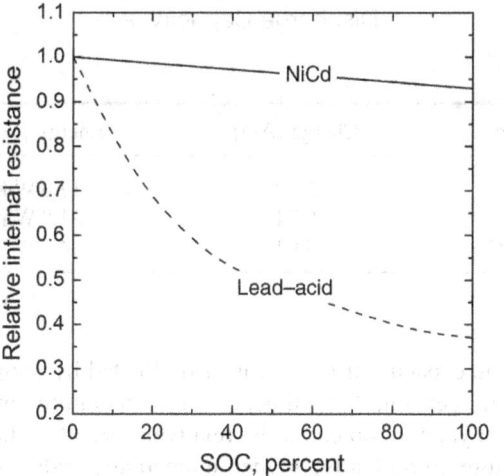

Figure 7.8 Change in cell resistance with state of charge. A fully discharged cell has a value of unity.

Because anions are not involved in the electrode reactions, the steady-state flux of anions is zero. Since there is an electric field that exerts a migration force on the anions, a concentration gradient will be established during the passage of current to counter this force and keep the flux zero. The magnitude of that gradient is

$$\nabla c = \frac{it_-^o}{z_- F \nu_- D}. \quad (7.16)$$

Although the average concentration does not change since there is no net addition or removal of ions from the electrolyte in a lithium-ion cell, the presence of the concentration gradient can have a significant influence on the operation of the cell. For example, the change in concentration will affect the local conductivity of the electrolyte. Let's also consider the effect on kinetics. As the current density is increased, the concentration gradient grows. This development means that the lithium-ion concentration increases at one electrode and decreases at the other. The reaction rate at each electrode depends on concentration, as described by a concentration-dependent kinetic expression. Therefore, the kinetic polarization at each electrode will change as a result of the change in the local concentration. As we continue to increase the current, the concentration at one electrode will eventually drop to zero and the cell will have reached its limiting current. Similar concentration effects are seen in other types of batteries.

In all cases, cell polarization increases with increasing rates of discharge. Since a battery is typically discharged to a specified cutoff voltage, that voltage will be reached sooner (after fewer coulombs have passed) at higher discharge rates due to the increased polarization. This effect is shown in Figure 7.7, which illustrates discharge curves for a lead–acid battery cell to a cutoff voltage of 1.7 V. At higher rates of discharge, the cutoff voltage is reached sooner due to the increased polarization that lowers the cell voltage. The result is that the capacity that can be accessed above the cutoff voltage decreases with increasing discharge rate. For example, a discharge rate of C/20 corresponds to a normalized capacity of one in Figure 7.7. In contrast, the capacity accessible above the cutoff voltage at 2C is only about 40% of the C/20 value. Thus, the same cell with a nominal rating of 20 A·h will have a lower useable capacity at higher rates. A correction factor called the *capacity offset* is frequently applied to the rating of the battery when the cell is discharged at a current other than the one at which it was rated.

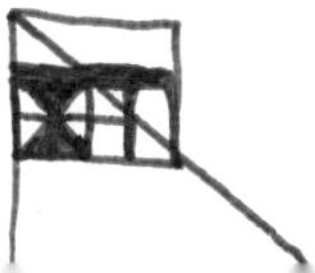

The relationship between cell potential and capacity can be complex and generally requires extensive experimental data or detailed physical models. One well-known empirical relationship for relating capacity and current (rate) is the Peukert equation for capacity offset,

$$C_p = \left[\frac{I_{sp}}{I}\right]^{k-1}, \quad (7.17)$$

where C_p is the capacity offset, I is the current, I_{sp} is the current where the capacity is specified, and k is an empirical coefficient with a value between 1.1 and 1.5. The Peukert equation, developed for lead–acid batteries, is empirical and does not necessarily apply to other battery chemistries. Furthermore, while information about the capacity offset is useful, it is typically not sufficient. Rather, system design most frequently requires knowledge of the actual potential of the cell as a function of the rate and extent of discharge.

An alternative approach that is more useful for design is to fit the voltage of the cell to an expression that includes the influence of both rate and DOD. A common model used to do this is the semiempirical *Shepherd equation*,

$$V_{cell} = U - IR_{int} - K\left(\frac{Q}{Q-It}\right)I + A\exp\{BQ^{-1}It\},$$

$$(7.18)$$

where R_{int} is the internal resistance, Q is the nominal capacity in A·h, and K is a polarization constant in ohms. The first term on the right side is the equilibrium or thermodynamic potential. From this, the polarization due to the internal resistance of the cell is subtracted, as shown in the second term on the right side of Equation 7.18. The third term approximates the kinetic polarization where the current density in the porous electrode is assumed to be inversely proportional to the amount of unused material, and the voltage losses are proportional to the current. The last term accounts for the more rapid "exponential" initial decrease in the voltage that is observed for some cells.

ILLUSTRATION 7.3

The rate capability of a lithium-ion cell is described with the Shepherd equation. Calculate the theoretical energy ($U \cdot Q$) and compare it with the energy available at discharge rates of 0.1 C, 1.0 C, and 3 C. Assume that the battery has a capacity of 6 A·h and is discharged to a cutoff potential of 2 V.

$$V_{cell} = 4.131 - 0.00388I - 0.015138\left(\frac{Q}{Q-It}\right)I$$
$$- 0.02\exp\{(0.1Q^{-1})It\}.$$

a. The theoretical energy is approximately equal to the charge capacity multiplied by the equilibrium voltage 6 A·h × 4.131 V = 24.8 W·h.

b. At other rates, the corresponding energy can be obtained by integrating the power with respect to time as follows

$$\text{Energy} = \int_0^{t_d} IV(t)\,dt = I\int_0^{t_d} V(t)\,dt = \int_0^{V_{co}} V(Q)\,dQ.$$

Note that the current is constant and has been moved outside of the integral. The time t_d is the time at which the voltage reaches its cutoff value (2 V). The cell potential calculated from the Shepherd equation is shown in the figure for discharge at the 3C rate. Energy below the cutoff potential is not included in the calculation.

c. Data for three rates are as follows:

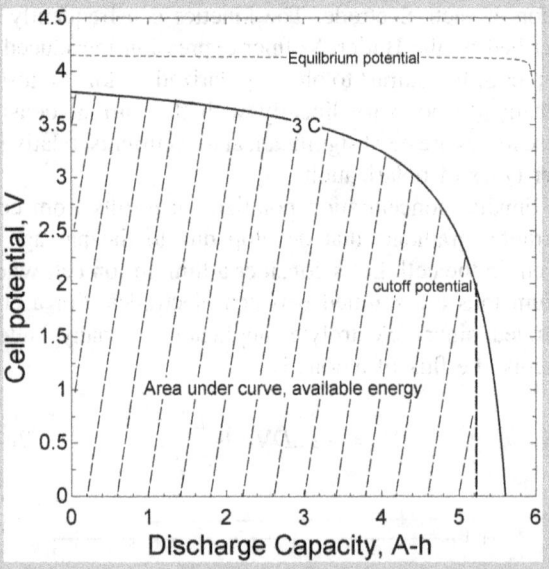

Rate	Charge [A·h]	Energy [W·h]
3 C	5.20	17.8 W·h
C	5.74	21.9 W·h
0.1 C	5.94	24.2 W·h

The capacity of the cell is also affected by temperature. In general, the influence of temperature on the thermodynamic potential is relatively small (Chapter 2). More importantly, higher temperature reduces the ohmic, concentration, and kinetic polarizations. Thus, as the temperature is increased, the potential of the

Chapter 7 Battery Fundamentals 163

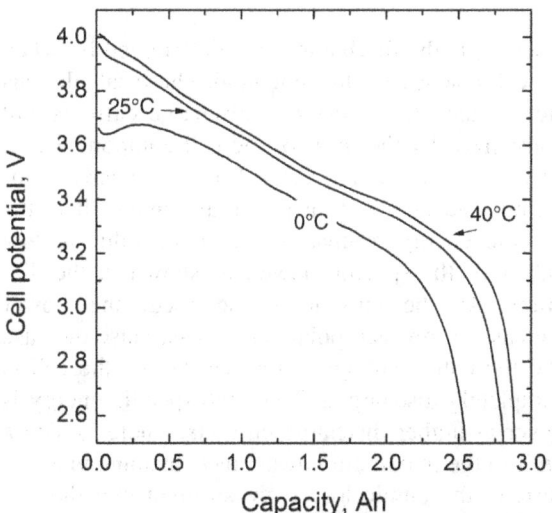

Figure 7.9 Effect of temperature on the cell voltage during discharge for a lithium-ion cell.

cell increases and the capacity increases. The effect of temperature on a commercial lithium-ion cell is shown in Figure 7.9.

In some cases, the *self-discharge* (discussed later in the chapter) may increase so much that the capacity actually decreases with temperature, but this is not common. That being noted, temperature is generally the enemy of long battery life as will be discussed shortly.

In this section, we have examined factors that affect battery voltage as a function of SOC. Battery chemistry is, naturally, the primary factor. However, the voltage profile is also impacted by the rate of discharge or charge, as characterized by the C-rate, due to the impact of the SOC and rate on polarization losses. Methods to describe the capacity offset and the voltage behavior were also presented.

7.5 RAGONE PLOTS

Power and energy are often key design aims of an electrochemical system for energy storage and conversion. The instantaneous power produced by a cell is simply the current multiplied by the potential of the cell. Even if the current is constant, the potential is not generally flat and changes during discharge. Therefore, integration is required to determine the average power.

$$P_{avg} = \frac{1}{t_d} \int_0^{t_d} IV(t)dt. \quad (7.19)$$

The power is measured in W or kW. Similarly, the electrical energy available during discharge can be obtained by integration. The most common ways of reporting the power and energy available are to normalize these quantities to the mass or volume of the cell. The four commonly used terms are *specific power* [kW·kg^{-1}], *specific energy* [kW·h·kg^{-1}], *power density* [kW·L^{-1}], and *energy density* [kW·h·L^{-1}].

The trade-off between power and energy can be represented with a Ragone plot. Here the specific energy is plotted against the specific power of a battery. Figure 7.10 shows the range of values for batteries, as well as for fuel cells, double-layer capacitors, and the internal combustion engine. The large rectangle for batteries reflects the important impact of battery chemistry, and the large number of possible types of batteries. Also, since cell design can be tailored for either capacity or power, each battery chemistry would itself be represented by a range of values on the plot, as shown in Illustration 7.4. A similar plot can be constructed using energy and power density.

Also of note on the log–log Ragone plot are the 45° lines, where each line corresponds to a constant ratio of specific energy to specific power. As shown, the lines are identified as the battery *run time* and represent the time that would be required to fully discharge the battery. Battery run time is the inverse of the C-rate, and is one of the most important characteristics used in selecting or designing a battery.

As discussed previously, battery chemistry is perhaps the foremost design factor. But for a particular chemistry, there can be a large difference in performance due to differences in cell and battery design to address the needs of specific applications. For example, the lead–acid battery

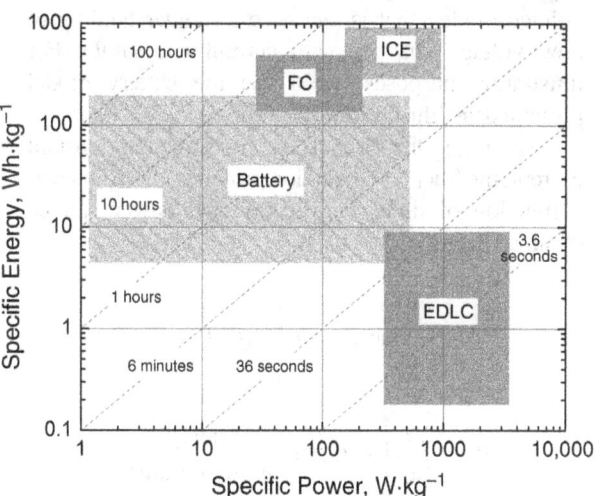

Figure 7.10 Ragone plot illustrating the strengths of different energy storage and conversion devices.

used to start your car engine needs to produce high currents for a short period of time, but does not need to have a high energy density. Therefore, these batteries are designed to optimize power. In fact, because of the importance of power, the required cold-cranking amps (CCA) is typically specified rather than the capacity in A·h. CCA is the maximum current that the battery can sustain for 30 seconds at $-18\,°C$. A typical automotive starting battery may have a capacity of 35 A·h and a CCA of 540 A. The corresponding discharge rate is 540/35, or 15 C, which corresponds to a 4-minute discharge time. In contrast, lead–acid batteries can also be used as backup power for a telecommunication system where they may need to provide power for hours. Lead–acid batteries are also commonly used as the energy storage component for off-grid solar power systems. In both of these stationary applications, the capacity is more important than the rate capability. The cell design for these last two applications, specifically the design of the separator and electrodes, is very different from that used for the starting battery discussed previously. The relationship between battery design and performance is discussed in more detail in Chapter 8.

ILLUSTRATION 7.4

The relationship between specific power and energy is shown on a Ragone plot. In Figure 7.10, the different technologies are represented loosely by rectangles. That single-frame labeled "battery" encompasses everything from an implantable primary lithium battery to the lead–acid battery for starting your car. Here we explore the behavior of a specific battery as depicted on a Ragone plot. To do this, we need a voltage model; that is, we need an understanding of how voltage changes with current and DOD. For illustration purposes, we'll use the battery model presented in Illustration 7.3.

Assuming the discharge is performed at constant current, the Shepherd equation provides cell voltage as a function of time. The energy and power can be obtained from

$$\text{Energy} = I \int_0^{t_d} V(t)dt.$$

$$P_{avg} = \frac{I}{t_d} \int_0^{t_d} V(t)dt = \frac{\text{Energy}}{\text{discharge time}},$$

where t_d is the discharge time, determined by when the cell reaches a cutoff potential. These calculations are repeated for a series of discharge currents and normalized by the mass of the cell components. At very low current, the cells discharge nearly completely and polarization losses are small; thus, the specific energy reaches a maximum value at low values of the specific power, as shown in the diagram. As the current is increased, the power increases. However, polarization losses also increase and the cutoff voltage is reached before the cell is completely discharged. Thus, the specific energy is lower at higher discharge currents. There is also a limit to the maximum power since, at some value of current, the ohmic losses are so great that the cell reaches the cutoff potential instantaneously without any appreciable capacity.

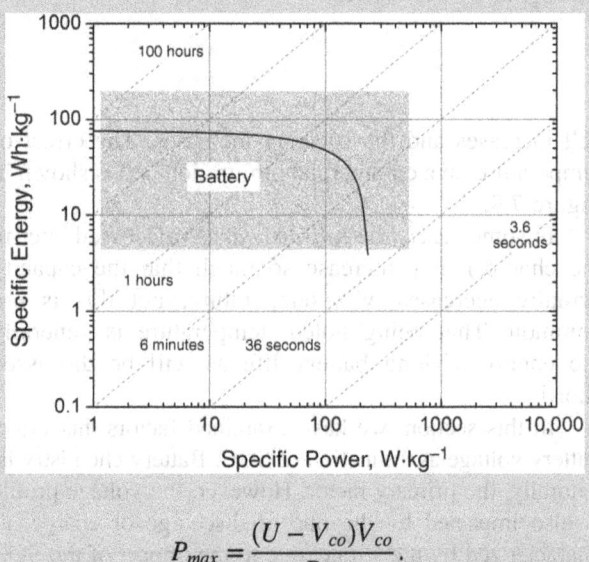

$$P_{max} = \frac{(U - V_{co})V_{co}}{R_{int}}.$$

Clearly, high power requires a low internal resistance.

7.6 HEAT GENERATION

For a commercial secondary cell, the polarization losses discussed in Section 7.4 are usually quite low when the cell is used as designed. Hence, most of the available chemical energy is converted into electrical work rather than heat. Nonetheless, heat generation plays an important role in cell performance, system design, battery safety, and useable cell life. This importance is partly due to the fact that, although the heat generation rates may be low, the thermal resistances of many cells and batteries are high. Thus, it is relatively difficult to remove heat from the interior of the cell. This challenge will be discussed in more detail as it relates to battery design in the next chapter. Here we address the fundamental processes.

Heat can be generated in electrochemical cells due to irreversible losses associated with kinetic and concentration overpotentials, by resistive or Joule heating associated with the flow of current in solution, and by reversible changes due to differences in the entropy of reactants and products. The expression used for heat generation in an electrochemical cell is

$$\dot{q} = I(U - V_{cell}) - I\left(T\frac{\partial U}{\partial T}\right) \, [\text{W}]. \quad (7.20)$$

The first term on the right side is the current density multiplied by difference between the equilibrium potential and the cell voltage. All of the losses in the cell (ohmic, kinetic, and concentration) affect the cell voltage and therefore the rate of heat generation. As we have learned, these polarizations depend on the cell current and on the state of charge. The current is taken to be positive during discharge, and the cell potential is smaller than U. Therefore, the heat generation from this term is positive. During charge, the sign of the current changes (now negative) and the cell potential is larger than U, resulting again in a positive rate of heat generation. In other words, polarization losses in the cell always result in heat generation. The second term on the right side of Equation 7.20 is the entropic contribution, which can be negative or positive. This quantity can be determined from the change in the equilibrium potential of a cell with temperature.

Heat generation data, such as those shown in Figure 7.11, are generally the result of detailed experiments or models. Such data are critical for accurate simulation of battery performance. Another important point is that the heat generation rate increases nonlinearly. Heat generation is small at low rates of charge or discharge, but increases dramatically with current.

ILLUSTRATION 7.5

A small 1.6 A·h cell has a volume of 1.654×10^{-5} m^3 and an internal resistance of 50 mΩ. Assuming that the cell is ohmically limited, calculate the rate of heat generation at 0.25, 1, and 5 C neglecting the entropic term.

If the cell is ohmically limited, then

$$U - V_{cell} = IR_\Omega.$$

The current is simply the capacity in A·h, Q, times the C-rate

$$I = Q(\text{C-rate}).$$

Thus, the heat generation normalized by the volume of the cell is

$$\dot{q}''' = \frac{I(U - V_{cell})}{\text{Volume}} = \frac{I^2 R_\Omega}{\mathbb{V}}.$$

Values for 0.25, 1, and 5 C are 0.48, 7.74, and 194 W L^{-1}. The rate of heat generation is quadratic function of the current or C-rate.

ILLUSTRATION 7.6

Calculate the cell potential that corresponds to zero heat generation for the LiSOCl$_2$ battery.

Data for the equilibrium potential for a primary lithium thionyl chloride battery are shown in the figure. A small decrease in U with increasing temperature is typical of most batteries. Based on a linear fit of the data, the slope of the curve is -0.228 mV·K^{-1}. Using Equation 7.20,

$$\dot{q} = I(U - V) - I\left(T\frac{\partial U}{\partial T}\right) = 0.$$

Therefore,

$$V_{therm} = U - \left(T\frac{\partial U}{\partial T}\right).$$

The potential is 3.723 V. This potential is called the *thermoneutral potential* and represents as condition where there is some current, but no heat flows into or out of the cell. At 25 °C, U is 3.657 V. This equilibrium potential is less than the thermoneutral potential, therefore a small charging current would be required to achieve the condition of zero heat generation. Since this is a primary cell, this condition would not occur.

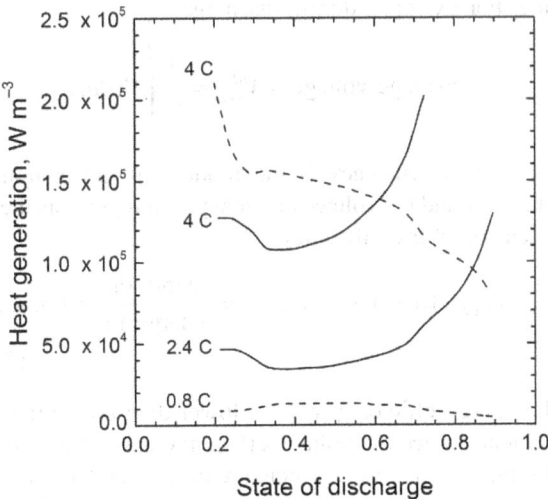

Figure 7.11 Rates of heat generation for a lithium-ion cell. Solid lines are for discharges and dashed lines are for charges. (Adapted from *J. Power Sources*, **247**, 618 (2014).)

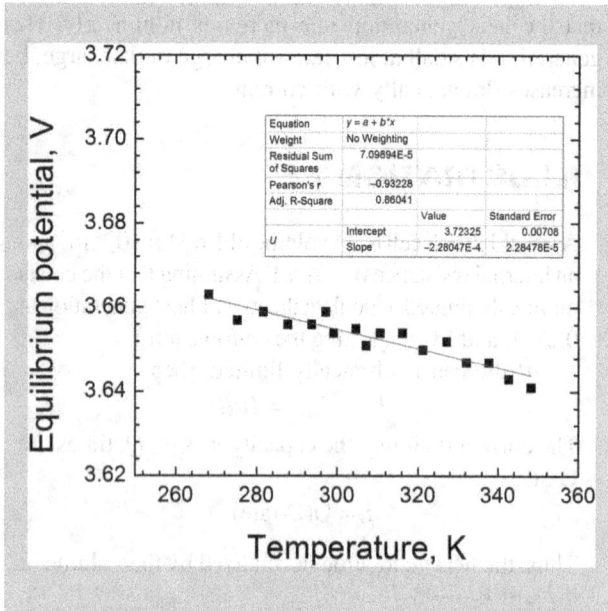

7.7 EFFICIENCY OF SECONDARY CELLS

The most basic efficiency of a rechargeable cell is the coulombic efficiency. Note that in contrast to the faradaic and current efficiencies that were defined in Chapters 1 and 3, this coulombic efficiency refers to a battery that undergoes a complete charge–discharge cycle.

$$\eta_{coul} = \text{coulombic efficiency}$$
$$= \frac{\text{number of coulombs on discharge}}{\text{number of coulombs on charge}} \times 100\%. \quad (7.21)$$

Why would the number of coulombs be different for charge and discharge? The principal reason is that secondary reactions occur; that is, some of the current goes to undesired reactions. For example, when charging a NiCd battery, gases may be evolved as a side reaction. Although it may be possible to recombine the gases (generated from electrolysis of water) in order to avoid loss of electrolyte, the coulombs used to generate the gases are not available for power generation on discharge. There can also be irreversible losses such as the formation of the solid electrode interphase (SEI) in lithium cells. Here, the electrolyte reacts with lithium in the negative electrode, making the reacted lithium unavailable for cycling.

Typical values for coulombic efficiency are 95% or greater for many rechargeable cells; and acceptable values depend on the particular chemistry involved. The coulombic efficiency must be very close to one to achieve reasonable battery life for rechargeable cells when the side reactions are irreversible and result in a loss of active material (see Problem 7.18 for the lithium-ion cell).

In some cases, side reactions can provide overcharge protection for a battery. For example, for an acid-starved lead–acid battery, the dominant side reactions during overcharge are

$$\text{positive: } 2H_2O \rightarrow O_2 + 4H^+ + 4e^-$$

and

$$\text{negative: } O_2 + 4H^+ + 4e^- \rightarrow 2H_2O$$

The cell is designed so that the evolved oxygen can move rapidly through the separator in the gas phase to the other electrode where it reacts to form water. This recombination prevents loss of electrolyte and damage to the cell due to overcharging.

A second efficiency of interest is the voltage efficiency. The difference between the potential during charging and discharging determines the voltage efficiency of a rechargeable cell. Any polarization of the cell due to ohmic, kinetic, or concentration effects will cause the difference to increase and the efficiency to decrease. Therefore, the voltage efficiency decreases with increasing current density due to the increased irreversible losses at the higher current. Similar to the coulombic efficiency, the voltage efficiency for a rechargeable cell applies to a cell undergoing a complete charge–discharge cycle. Voltage efficiency is defined as the ratio of the average cell voltage during discharge and charge.

$$\text{Voltage efficiency} = \eta_V = \frac{V_{cell}^d}{V_{cell}^c} \times 100\%. \quad (7.22)$$

The average voltage is determined by integrating over time. For example, during discharge,

$$\text{average voltage} = V_{cell}^d = \frac{I}{t_d} \int_0^{t_d} V(t)dt.$$

The energy efficiency is the product of the coulombic efficiency and the voltage efficiency, and represents the so-called round-trip efficiency,

$$\text{Energy efficiency} = \eta_{coul}\eta_V = \frac{\text{energy out}}{\text{energy in}} \times 100\%.$$
$$(7.23)$$

The energy efficiency will be lower than the coulombic efficiency since the voltage efficiency is less than 100%. The typical round trip efficiency of a secondary cell can be near 90% for rechargeable batteries. This efficiency will, of course, depend on the rates of charge and discharge.

7.8 CHARGE RETENTION AND SELF-DISCHARGE

Charge retention refers to the amount of charge, usually expressed as a percentage of capacity, remaining after a cell is stored for a period of time and not connected to an external circuit. Self-discharge describes the mechanism by which the capacity of the cell is reduced. Rates of self-discharge can vary dramatically—a 10% loss in capacity may take 5 years for a lithium primary cell, but only 24 hours for some nickel-based cells. Comparing two secondary cells, we see that self-discharge rates also vary widely with chemistry. For example, a Ni-MH cell may discharge 30% per month, whereas lithium-ion cells have rates closer to 5% per month. For a secondary rechargeable battery, self-discharge can be further characterized as either reversible or irreversible. Irreversible self-discharge contributes to *capacity fade*, which is discussed further in Section 7.9.

The rate of self-discharge is an important design requirement for any battery. A low rate of self-discharge is a critical feature of a primary cell. The rate determines the amount of time that the battery remains useable. Many primary batteries are stored for long periods of time before being put into service. Others are in use, but only needed for backup power, for example, memory in electronic circuits. A service life of 10 or 20 years is not uncommon. The need of some applications for extremely long shelf life is the principal reason that reserve batteries have been developed. Some reserve batteries may be required to be *in place*, but not *in use* for a decade or more. As noted previously, the rates of self-discharge are strongly dependent on the cell chemistry, but cell design and manufacturing processes can be specified to reduce the rate of self-discharge—that's the role of the engineer. There are many mechanisms that cause a cell to self-discharge. A few physical causes are discussed below as examples.

The easiest one to understand is an *electronic short* across the cell. A small short, though clearly undesirable, can be introduced in the manufacturing process of both primary and secondary cells. More often shorts result from use. For instance, the plates of a lead–acid cell can swell and shrink during operation, putting pressure on the separator. Mechanical shock could also cause the two plates to touch. In extreme cases, dendrites of lead (tree formation) can penetrate through the separator. In each instance, an electronic path between the two electrodes is established inside the cell. The rate of self-discharge is simply

$$I_{self-discharge} = \frac{V_{cell}}{R_{short}}. \quad (7.24)$$

A key difference between an external short and an internal short is that all of the resistive heating associated with an internal short is inside the cell. This feature has important implications for thermal runaway, a safety concern for lithium-ion and some lead–acid cells.

A second source of self-discharge is a *shuttle* mechanism. For instance, in either NiCd or Ni metal-hydride cells, ammonium hydroxide can be formed from the breakdown of a polyamide separator. This impurity contributes to self-discharge through a chemical shuttle mechanism as described below. At the positive electrode of the Ni-MH cell,

$$NH_4OH + 6NiOOH + OH^- \rightarrow 6Ni(OH)_2 + NO_2^-. \quad (7.25)$$

Transport of nitrite to the negative electrode yields

$$NO_2^- + 6MH \rightarrow NH_4OH + 6M + OH^- \quad (7.26)$$

These two reactions combine to yield

$$MH + NiOOH \rightarrow M + Ni(OH)_2 \quad (7.27)$$

which is the overall reaction for the discharge of a Ni-MH cell (see Table 7.1). Ammonium ions produced at the negative electrode diffuse back through the separator to the positive electrode, where they are converted to nitrite as the cycle continues. These reactions and transport between electrodes result in a chemical short, which discharges the cell.

An analogous shuttle involving the redox reaction of an impurity metal, M,

$$M^{2+} + e^- \leftrightarrow M^+$$

can also contribute to self-discharge if the potential for this reaction lies between that of the positive and negative electrodes. Under such conditions, the metal will be oxidized at one electrode and reduced at the other, forming a chemical short (see Problem 7.25). This possibility emphasizes again the need to minimize impurities in cells during manufacturing.

A third and common mechanism for self-discharge is *corrosion*. Often one or both of the electrodes are thermodynamically unstable in the charged state. For instance, two of the many possible self-discharge reactions for the lead–acid cell are grid corrosion on the positive electrode and evolution of hydrogen on the negative electrode. Note that the lead oxide in the positive plate is supported on a grid of Pb.

$$Pb(s) + PbO_2(s) + 2H_2SO_4 \xrightarrow{grid\ corrosion} 2PbSO_4(s) + 2H_2O(l)$$

$$Pb(s) + H_2SO_4 \xrightarrow{hydrogen\ evolution} PbSO_4(s) + H_2$$

Whether controlled by transport or kinetics, rates of self-discharge are expected to increase strongly with temperature, and this phenomenon is generally observed.

ILLUSTRATION 7.7

A Ni-MH cell is constructed with a separator thickness of 200 μm. The effective diffusion coefficient of ammonium hydroxide in the electrolyte is $3 \times 10^{-10}\ m^2 \cdot s^{-1}$, and the concentration of ammonium hydroxide is 1mM.

a. Sketch the concentration profiles of NH_4^+ and NO_2^- across the separator assuming the reactions at the electrodes are fast.

If reactions are fast, the concentration of reactants will be zero at electrode. The profile is linear with an average of 1 mM.

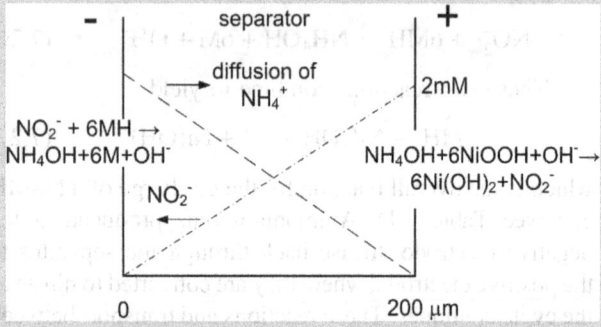

b. Calculate the rate of self-discharge using Fick's law for diffusion.

The integrated form of Fick's law is

$$N_{NH_4^+} = \frac{D2c}{L} = 3 \times 10^{-6}\ mol\ m^{-2} s^{-1}.$$

The flux is converted to a current density with Faraday's law:

$$\frac{3 \times 10^{-6}\ mol}{m^2 s} \left|\frac{6\ equiv}{mole\ NH_4^+}\right| \frac{96485\ C}{equiv} = 1.74\ A\ m^{-2}.$$

c. For a 0.6 A·h cell and a separator area of 0.05 m^2, estimate the time required to fully discharge the battery.

$$0.6\ A \cdot h \left|\frac{3600\ C}{1\ Ah}\right| \frac{s\ m^2}{1.74\ C} \left|\frac{1}{0.05\ m^2}\right| \frac{h}{3600\ s} = 6.9\ hours.$$

7.9 CAPACITY FADE IN SECONDARY CELLS

As is well known by any user of rechargeable batteries, performance degrades over time. Although a new battery might last for 3 or 4 hours, with extended use, the working time before recharging may now be just 1 or 2 hours. This reduction in operating time, also known as aging, represents a loss in the capacity of the cell and is typically quantified in ampere-hours. Loss of capacity is a combination of cycle (use) life and calendar life, since the battery can lose capacity even while being stored due to irreversible self-discharge as discussed in the previous section. Figure 7.12 shows the *capacity fade* of a typical lithium-ion cell. The capacity in A·h that is available before the cutoff voltage is reached decreases with increasing cycle number. The loss of capacity for the lithium-ion cell is in large part due to a decrease in the amount of cyclable lithium in the cell. Specifically, some of the lithium in the graphite negative electrode reacts with the electrolyte to form a solid electrolyte interphase (SEI). The lithium that reacts becomes unavailable for subsequent cycling.

SEI growth is just one of many causes of capacity fade in secondary lithium cells. The predominant mechanisms are different for other cell chemistries, but all secondary cells degrade with operation. Understanding the capacity fade and power fade is essential for designing battery systems. There are a number of mechanisms that can contribute to capacity fade. The first is loss of active material. SEI formation is such an example, where an irreversible side reaction consumes lithium from the cell. Active material can also be disconnected electrically from the electrode, typically due to volume changes associated with restructuring or even from excessive gassing in aqueous systems. In addition to capacity loss, larger polarization of the cell either from increased charge transfer or internal resistance will cause the cutoff potential to

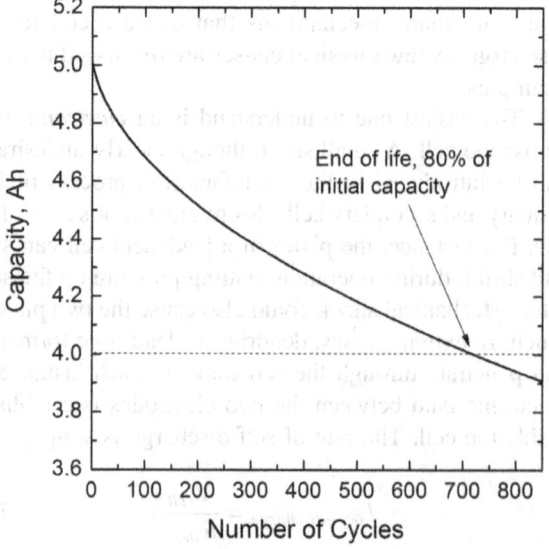

Figure 7.12 Typical behavior for capacity fade of a lithium-ion cell.

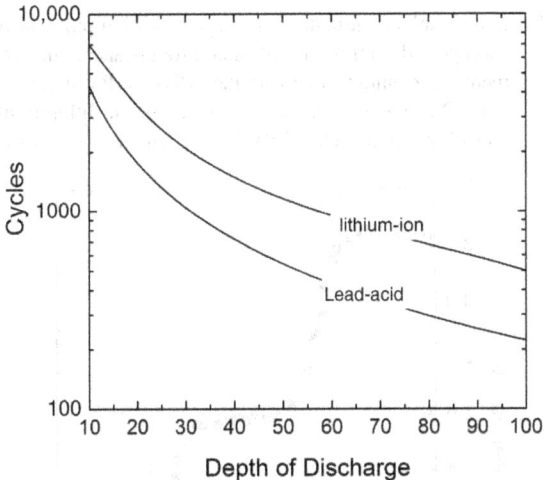

Figure 7.13 Effect of depth of discharge on cycle life.

be reached sooner. This behavior effectively results in a lower useable capacity of the battery. Such increases in internal resistance with time are common for many battery systems. There can also be chemical shorts in cells. For instance, the transport of Ag from the positive to negative electrode in rechargeable silver cells, or the transport of lithium polysulfide in lithium–sulfur cells both represent such shorts (see Problem 7.28). Self-discharge, discussed in the previous section, can also contribute to reduction of battery life.

The end of life for a cell depends on the application, but batteries are often considered unusable when the capacity is reduced to less than 70 or 80% of rated capacity. A key factor that influences capacity fade is the depth of discharge or the extent to which a battery is discharged during cycling. Figure 7.13 shows the impact of DOD on the number of cycles that can be achieved, for example, lead–acid and lithium-ion cells. It is rare for most rechargeable batteries to be repeatedly cycled from a fully charged to fully discharged state. More often than not, the state of charge is maintained in a narrower window of say 40–70%. A principal reason for doing this is to reduce capacity fade.

Temperature also has a strong effect on battery life. Sometimes higher temperatures initiate new mechanisms for failure, but for the most part the main impact is on accelerating existing failure modes.

CLOSURE

In this chapter, a basic description of a battery has been presented and battery operation has been discussed. Cells are characterized as either *primary* or *secondary*, with the two main types of electrode reactions being *reconstruction* and *insertion*. Batteries are rated in terms of nominal voltage and ampere-hour capacity. Methods for calculating the theoretical capacity have been developed. The current–voltage response of the battery is described using the principles of thermodynamics, kinetics, and mass transfer. Other important phenomena that occur in batteries have been introduced: heat generation, self-discharge, and capacity fade. The information presented in this chapter will prepare the reader for the analysis and design of batteries found in Chapter 8.

FURTHER READING

Besenhard, J. O. and Daniel, C. (eds) (2011) *Handbook of Battery Materials*, Wiley-VCH Verlag GmbH.
Huggins, R.A. (2009) *Advanced Batteries: Materials Science Aspects*, New York, Springer.
Reddy, T. and Linden, D. (2010) *Linden's Handbook of Batteries*, McGraw Hill.
Vincent, C.A. and Scrosati. B. (1997) *Modern Batteries: An Introduction to Electrochemical Power Sources*, Butterworth, Oxford.

PROBLEMS

7.1 Use data from Appendix A or Appendix C to determine values of U^θ for the:

(a) A lead–acid battery (both lead and lead oxide react to form lead sulfate).

(b) A zinc–air battery in alkaline media.

7.2 Sodium is far more abundant in the earth's crust than lithium. Consequently, there is interest in replacing lithium as the negative electrode material with sodium in batteries. Consider the overall reaction of lithium with cobalt as an example for a new secondary battery:

$$CoO + 2Li \leftrightarrow Co + Li_2O$$

(a) Write the equivalent reaction where sodium replaces lithium. Categorize this reaction based on the discussion from Section 7.2.

(b) Using the thermodynamic data provided in Appendix C, calculate the equilibrium potential, capacity in $A \cdot h \cdot g^{-1}$, and specific energy for lithium and sodium versions of this battery.

7.3 A common primary battery for pacemakers is the lithium–iodine cell. The negative electrode is lithium metal, the positive electrode is a paste made with I_2 and a small amount of polyvinylpyridine (PVP), and the separator is the ionic salt LiI. The overall reaction is

$$2Li + I_2 \leftrightarrow 2LiI(s)$$

Write out the half-cell reactions and using the data from Appendix A, calculate the equilibrium potential and the theoretical capacity in $A \cdot h \cdot g^{-1}$. You may treat the positive electrode paste as pure iodine. Categorize this reaction based on the discussion from Section 7.2.

170 Electrochemical Engineering

7.4 The lithium–iodine cell described in Problem 7.3 is used for an implantable pacemaker. Note that LiI is produced during discharge, and this salt adds to the thickness of the separator. The nominal current is 28 μA and is assumed constant over the life of the cell. How much active material is needed for a 5-year life? At 37 °C, the LiI electrolyte has an ionic conductivity of 4×10^{-5} S·m^{-1}. If the separator is formed in place from the overall reaction, and LiI has a density of 3494 kg·m^{-3}, what is the voltage drop across the separator due to ohmic losses in the separator after 2.5 years? The cell area is 13 cm^2. Please comment on the magnitude of the voltage drop. Is it important? Why or why not?

7.5 Most high-energy cells use lithium metal for the negative electrode; furthermore, rechargeable lithium systems rely on intercalation for reversible reactions at the cathode. Discuss the idea of replacing Li with Mg for future rechargeable cells. Specifically contrast and compare Li and Mg commenting on the following:

(a) Specific capacity [A·h g-metal^{-1}]
(b) Volumetric capacity [A·h cm^3-metal^{-1}]
(c) Earth abundance
(d) Specific energy and energy density
(e) Ionic radii
(f) Charge/radius ratio (*Hint:* How is this likely to affect intercalation?)

7.6 The carbon monofluoride primary cell consists of a lithium metal negative electrode, and a carbon monofluoride CF$_x$ as the positive electrode. The carbon monofluoride is produced by the direct fluorination of coke or another carbon. The fluorine expands the carbon structure, creating a nonstoichiometric intercalation material; the value of x is about 1. The overall reaction is expressed as

$$xLi + CF_x \leftrightarrow xLiF + C$$

(a) If the equilibrium potential of this cell is about 3.0 V and $x = 0.95$, determine the theoretical specific energy of this battery. How does this value compare to the capacity of a commercial cell, which is about 450 W·h kg^{-1}? Why are they different?

(b) Calculate the theoretical specific energy of the lithium sulfur dioxide battery (Table 7.1) and compare it to that of the CF$_x$ cell.

7.7 Calculate the standard potential, U^θ, for the Ni/Fe (Edison cell) from the following information for the half-cell reactions.

Fe + 2OH$^-$ ↔ Fe(OH)$_2$ + 2e$^-$ (−0.89 V versus SHE)

NiOOH + e$^-$ + H$_2$O ↔ Ni(OH)$_2$ + OH$^-$

(0.290 V versus Ag/AgCl)

7.8 Calculate the theoretical specific energy of the aluminum–air battery. The two electrode reactions are

Al + 3OH$^-$ ↔ Al(OH)$_3$ + 3e$^-$ (−2.31 V)
O$_2$ + 2H$_2$O + 4e$^-$ ↔ 4OH$^-$ (0.401 V)

7.9 Some implantable batteries must provide high-pulse power for short periods of time, a defibrillator for instance. This requirement that cannot be met with the Li/I$_2$ cell (Problems 7.3 and 7.4). One battery for such a device is the lithium silver–vanadium–oxide cell (Li/SVO) cell. The overall reaction is

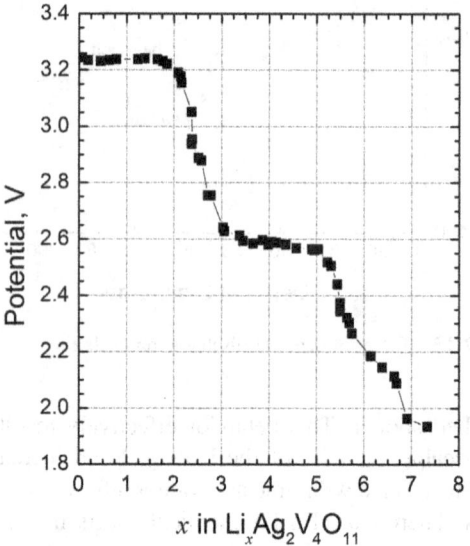

$$xLi + Ag_2V_4O_{11} \rightarrow Li_xAg_2V_4O_{11}$$

Ag$_2$V$_4$O$_{11}$ is a highly ordered crystalline material consisting of vanadium oxide sheets alternating with silver ions. These layers persist with the lithiation of the material. The equilibrium potential is shown on the right. There are two plateaus followed by a sloping decrease in potential at $x > 5$. What does this behavior suggest about the phases of the products? Assuming that the potential must be greater than or equal to that of the second plateau, calculate the theoretical energy density and specific energy of this battery. Problem 7.12 explores power.

7.10 Li ions are shuttled between electrodes of a lithium-ion battery during operation. During the charging process, lithium ions are transported from the positive electrode to the negative electrode. For a binary electrolyte (lithium salt, LiX, in an organic solvent), sketch the concentration of the salt in the separator of the cell. Explain the profile and comment on how it would change with changes in the magnitude and/or direction of the current density, i.

7.11 For an ohmically limited battery, the potential of the cell is given by $V_{cell} = U - IR_{int}$, where R_{int} is the internal resistance of the cell. Derive an expression for the maximum power. At what current and cell potential is the maximum power achieved? How are the results changed if there is a cutoff potential, V_{co}, below which operation of the cell is not recommended, that is reached first (i.e., before the maximum power)?

7.12 For the SVO cell (described in Problem 7.9), in addition to the low average power (~50 μW), periodically a 1 W pulse is needed for 10 seconds. The resistance of the cell initially decreases but then increases as the reaction proceeds (see data in table). Assuming that the cell is *ohmically limited* and that the separator area is 0.0015 m^2, at what value of x in Li$_x$Ag$_2$V$_4$O$_{11}$ is the cell no longer able to provide the required pulse power?

x	U [V]	R_{int} [$\Omega \cdot$cm^2]
0.0	3.2	20
0.6	3.2	17
1.1	3.2	17
1.8	3.2	16
2.4	2.85	14
3	2.65	16
3.5	2.6	20
4.1	2.6	25
4.3	2.5	27

SVO data at 37 °C.

7.13 Fit the data for a small lead–acid cell with the Peukert equation and determine the value of k. Use the capacity at 1.09 C as the reference. How well does the model fit the data?

Rate [C]	Capacity [Ah]
5.45	0.52
1.68	0.68
1.09	0.72
0.31	1.01
0.17	1.20
0.09	1.30

7.14 Fit the data (taken from Figure 7.6) to the Peukert equation and determine the value of k. Use the capacity at 1 C as the reference. How well does the model fit the data? What would be the effect of an increase in the temperature during discharge of the cell due to joule heating?

Rate [C]	Capacity [Ah]
1	7.80
4.5	7.60
9	7.32
14	6.95
18	6.45

7.15 What is the theoretical specific capacity and energy for the Leclanché cell? The overall reaction is

$$Zn + 2MnO_2 \rightarrow ZnO + Mn_2O_3$$

Use 1.6 V as the average potential of the cell. A practical battery of this chemistry has a specific energy of about 85 W·h·kg^{-1}. Please explain the difference between the theoretical and practical values.

7.16 Calculate theoretical specific energy for the lithium–air cell. Base your answer on the mass of lithium only. There are three possible reactions:

$$4Li + O_2 + 2H_2O \rightarrow 4LiOH_{(aq)}$$

$$4Li + O_2 \rightarrow 2Li_2O$$

$$2Li + O_2 \rightarrow 2Li_2O_2$$

7.17 The effective conductivity of a 25 μm separator filled with electrolyte is measured to be 2 mS·m^{-1}. The material has a porosity of 0.55. If the bulk conductivity of the electrolyte is 18 mS·m^{-1}, what is the tortuosity of the separator material? When operating at current density of 4 A·m^{-2}, calculate the ohmic loss in the separator.

7.18 A principal mechanism for power fade in lithium-ion cell is the irreversible reaction of lithium with the electrolyte to form the SEI. This rate of side reaction must be very small in order to have good cycle life. If the end of life is taken as when the capacity is 80% of initial capacity, what is the minimum coulombic efficiency to achieve 100 cycles? 1000 cycles?

7.19 During discharge of the lead–acid battery, the following reaction takes place at the positive electrode:

$$PbO_2 + 2H^+ + 2e^- + H_2SO_4 \rightarrow PbSO_4 + 2H_2O$$

Discharge of a vertical PbO$_2$ electrode results in flow due to natural convection. Briefly explain the natural convection process on the electrode. Sketch the velocity profile near the positive electrode.

7.20 A lithium-ion battery is being discharged with a current density of i mA·cm^{-2}. The positive electrode has a porous structure, and the electronic conductivity is much greater than the ionic conductivity, $\sigma \gg \kappa$. Assume an open-circuit potential where U^+ is essentially flat, but increases for high SOC and drops for low SOC.

(a) Sketch the ionic current density, i_2, across the separator and porous electrode at the start of the discharge.

(b) Sketch the divergence of the current density; physically explain the shape of this curve.

(c) Repeat (a) and (b) when the cell has nearly reached the end of its capacity. Again explain the shape.

(d) How would the internal resistance change with depth of discharge for this cell?

7.21 Develop a simple model for growth of SEI formation in lithium-ion cells. Assume the rate-limiting step is the diffusion of solvent through the film. Show that the thickness of the film is proportional to the square root of time. Discuss how capacity and power fade would evolve under these conditions.

7.22 Starting with Equation 7.20, which gives the rate of heat generation in the absence of any side reactions or short circuits, develop an expression for the rate of heat generation as a function of current. Treat the cell as being ohmically limited with a resistance R_Ω. You may also consider that the entropic contribution, $\partial U/\partial T$, is constant. Finally, assume that there is an additional, constant rate of heat generation due to self-discharge, $\dot{q}_{sd}$. Sketch the rate of heat generation as a function of current for the cell.

7.23 The rate of self-discharge is critical design parameter for primary batteries.

  (a) Assume that a primary battery is designed to last for 5 years at a constant average discharge rate. At what C-rate does this battery operate?

  (b) For this same cell, what is the equivalent C-rate for the self-discharge process if the current efficiency is to be kept above 90%? Assume that the self-discharge reaction operates as a chemical short in parallel with the main electrochemical reaction.

  (c) Because of the extremely long lives of some batteries, microcalorimetry is used to measure the rate of self-discharge. Data for a Li/I$_2$ cell (described in Problem 7.3) are shown on the right. Estimate the current efficiency of the discharge. The equilibrium potential is 2.80 V, the cell resistance is 650 Ω, and the entropic contribution is 0.0092 J·C^{-1} at the cell temperature. Assume a nominal operating current of 70 µA.

I [µA]	Q [µW]
0.047	5.95
14.5	5.79
31.3	6.63
56.5	7.97
125	15.02

7.24 During charging, oxygen can be evolved at the positive electrode of a lead–acid cell. In order to avoid adding water, this oxygen must be reduced back to water. In the so-called *starved* cell design, the electrolyte is limited so that there is some open porosity in the glass-mat separator. Therefore, oxygen can diffuse to the negative electrode. One set of proposed reactions at the negative electrode is

$$Pb + 0.5O_2 \rightarrow PbO$$
$$PbO + H^+ + HSO_4^- \rightarrow PbSO_4 + H_2O$$
$$PbSO_4 + H^+ + 2e^- + H_2SO_4 \rightarrow Pb + HSO_4^-$$

What is the net reaction? Describe how the evolution of oxygen at the positive electrode and its reaction at the negative electrode is in effect a shuttle mechanism with oxygen for the lead–acid cell. How does the oxygen reaction impact battery performance during charging? How does it impact performance during overcharge? In other words, what is the impact of overcharging these starved lead–acid cells? Finally, these cells are designed to be sealed from the atmosphere. What is the impact of having the cell open to the atmosphere on the rate of self-discharge of the starved cell?

7.25 It has been proposed that a small contamination of iron in the electrolyte can result in a shuttle mechanism of self-discharge of nickel–cadmium cells. What is the standard potential for this reaction?

$$Fe^{3+} + e^- \rightarrow Fe^{2+}$$

The two electrode reactions for the NiCd cell can be represented by

$$NiOOH + e^- + H_2O \rightarrow Ni(OH)_2 + OH^- \quad U^\theta = 0.49 \text{ V}$$
$$Cd(OH)_2 + 2e^- \rightarrow Cd + 2OH^- \quad U^\theta = -0.81 \text{ V}$$

Comment on the plausibility of such a self-discharge mechanism.

7.26 The discharge reaction for the lead–acid battery proceeds through a dissolution/precipitation reaction. The two reactions for the negative electrode are

$$Pb \rightarrow Pb^{2+} + 2e^-$$
$$Pb^{2+} + SO_4^{2-} \rightarrow PbSO_4$$

This mechanism is depicted in the figure. A key feature is that lead dissolves from one portion of the electrode but precipitates at another. The solubility of Pb^{2+} is relatively low, around 2 g·m^{-3}. How then can high currents be achieved in the lead–acid battery?

  (a) Assume that the dissolution and precipitation locations are separated by a distance of 1 mm. Using a diffusivity of 10^{-9} m^2·s^{-1} for the lead ions, estimate the maximum current that can be achieved.

  (b) Rather than two planar electrodes, imagine a porous electrode that is also 1 mm thick and made from particles with a radius 10 µm packed together with a void volume of 0.5. What is the maximum superficial current here based on the pore diameter?

  (c) What do these results suggest about the distribution of precipitates in the electrodes?

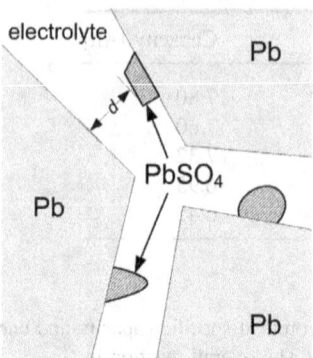

7.27 An 8 A·h Ni–MH cell is charged and discharged adiabatically. The data for the temperature rise are shown in the figure (adapted from *J. Therm. Anal. Calorim.*, **112**, 997 (2013)). Explain the effect of rate on the temperature rise. Comparing the differences between the charging and discharging temperature rise, what can be inferred about the entropic contribution to heat generation? Notice that only during charging a bit above 30 °C, the temperature rise increases sharply. Your colleague suggests that the side reactions of oxygen evolution increase rapidly at high temperatures. Can this evolution and recombination explain the results?

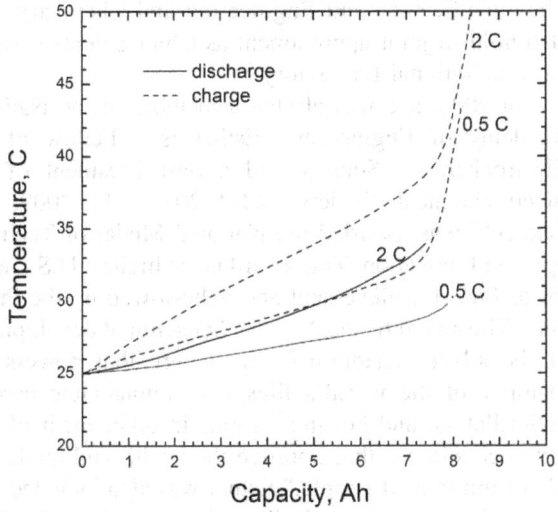

7.28 The lithium–sulfur battery uses lithium metal for the negative electrode and sulfur with carbon for the positive electrode. The overall reaction is

$$16Li + S_8 \rightarrow 8Li_2S$$

The electrochemical process at the positive electrode goes through a series of sequential formation of lithium sulfides (Li_2S_x), specifically

$$Li_2S_8 \rightarrow Li_2S_6 \rightarrow Li_2S_4 \rightarrow Li_2S_2 \rightarrow Li_2S$$

What are the half-cell reactions associated with this mechanism? The higher order polysulfides ($x = 8, 6, 4$) are soluble in the electrolyte. In contrast, Li_2S_2 and Li_2S are much less soluble. How could this situation lead to self-discharge in these cells? Identify some options to mitigate this self-discharge.

7.29 The theoretical specific capacity of an electrode was introduced in Section 7.4. Of course, to make a full cell, a positive and negative electrode must be combined. If SC^+ and SC^- represent the specific capacity of the positive and negative electrodes, show that the specific capacity of the full cell is given by

$$SC = \frac{SC^+ \times SC^-}{SC^+ + SC^-}.$$

Here, it is assumed that the capacities in A·h of the two electrodes are the same; that is, the electrodes are matched. If the specific capacity of the positive electrode is 140 mA·h·g^{-1} and that of the negative electrode is 300 mA·h·g^{-1}, what is the specific capacity of the full cell? If the specific capacity of the negative electrode were doubled to 1000 mA·h·g^{-1}, how much improvement in the specific capacity of the full cell is achieved?

Esther Sans Takeuchi

Esther Sans Takeuchi was born on September 8, 1953 in Kansas City. She is the daughter of Latvian immigrants, who, after the Soviet Union occupied Latvia following World War II, fled to a refugee camp in Germany. Her parents stayed in Europe for six more years, moving to the United States in 1951. When she started school in Ohio, Esther spoke the Latvian she was familiar with from home and barely any English. Her father was an electrical engineer. Esther graduated from the University of Pennsylvania with bachelor's degrees in both chemistry and history. Subsequently, she earned her Ph.D. in organic chemistry from Ohio State University. Dr. Takeuchi learned electrochemistry during two postdoctoral positions, the first at the University of North Carolina with Royce Murray and then with Professor Janet Osteryoung at SUNY Buffalo. For most of her career Esther worked at Greatbatch, Inc., a medical device company founded by the inventor of the implantable pacemaker. Providing power to these implantable devices is a difficult challenge. Esther's first project at Greatbatch was working on the implantable cardiac defibrillator, which is used to treat irregular heartbeats. The battery for the defibrillator is required to provide short pulses of high power in addition to lasting up to 10 years. Today hundreds of thousands of these devices are implanted annually. Over the years, Esther has proved to be an unusually prolific innovator—she holds more than 150 patents. After 22 years with Greatbatch, she took a faculty position at the University of Buffalo. In 2012, she moved to Stony Brook University as a Professor in the Departments of Materials Science and Engineering and Chemistry. She also holds a joint appointment as Chief Scientist, Brookhaven National Laboratory.

In 2004, she was elected a member of the National Academy of Engineering. Esther is a Fellow of the Electrochemical Society and a past President of the Electrochemical Society (2011–2012). In 2009, Dr. Takeuchi was awarded the National Medal of Technology and Innovation. This award is the highest U.S. award for technical achievement and is bestowed by the President. The award reads, ". . . for her seminal development of the silver vanadium oxide battery that powers the majority of the world's lifesaving implantable cardiac defibrillators, and her innovations in other medical battery technologies that improve the health and quality of life of millions of people." Esther was also inducted into the National Inventors Hall of Fame in 2011. In 2013, Professor Takeuchi received the E.V. Murphree Award in Industrial & Engineering Chemistry from the American Chemical Society.

Image Source: Courtesy of Esther Sans Takeuchi.

Chapter 8

Battery Applications: Cell and Battery Pack Design

8.1 INTRODUCTION TO BATTERY DESIGN

Given the critical role that the battery plays in many applications, the specifications or requirements that a battery must meet can be numerous and detailed. For example, an engineering specification for an automotive battery may be 30 pages or more in length. In contrast, we will only consider design of the main features of a battery in this introductory treatment. For simplification, we assume that the cell chemistry has already been selected. Our task is to take a cell of known chemistry and determine the size, number, and arrangement of cells required to assemble a battery that meets a desired set of basic specifications. Some of the more important requirements are tabulated in Table 8.1. Note that three items, **discharge time**, **nominal voltage**, and **energy**, are highlighted in bold. These represent the three most critical design specifications for a battery once the desired chemistry has been determined. Other important characteristics of the battery can, for the most part, be derived from these. For example, battery size (mass and volume) is largely determined by the energy and the battery chemistry. The average power, which is important for many applications, can be determined from the energy and the discharge time. Peak power, on the other hand, requires an additional specification.

To illustrate, let's consider a lithium-ion battery for an electric vehicle application that requires a discharge time of 2 hours, 24 kW·h of energy, and a targeted battery voltage of 360 V. The nominal single-cell voltage depends on the cell chemistry and has a value of 3.75 V for the lithium-ion cell considered here. The number of cells in series is simply,

$$\text{number of cells in series} = m = \frac{V_{batt}}{V_{cell}} = \frac{360}{3.75} = 96. \quad (8.1)$$

Thus, a minimum of 96 cells connected in series is needed. Later we will explore the advantage of using a high battery voltage. Cells connected in series are referred to as a *string*. Multiple strings of cells can be placed in parallel to increase the capacity of the battery or to reduce the single-cell capacity needed to meet the performance requirements. In this example, we will use two parallel strings, so that $n = 2$. The total number of cells is

$$N_c = mn = 192, \quad (8.2)$$

which is the product of the number of cells in series (m) and the number of cell strings connected in parallel (n). Next, we can determine the capacity of each individual cell based on the energy storage requirement for the battery. The energy of an individual cell is just its capacity times its voltage ($E_{cell} = Q_{cell} V_{cell}$). The total energy of the battery (E_{batt}) is the sum of the energy from each individual cell ($E_{batt} = N_c E_{cell}$). It follows that

$$Q_{cell} = \frac{E_{cell}}{V_{cell}} = \frac{E_{batt}}{N_c} \frac{1}{V_{cell}} = \frac{24000}{(192)(3.75)} = 33 \text{ A·h}. \quad (8.3)$$

Thus, a configuration of a total of 192 cells, each with a capacity of 33 A·h, is required. The cells in the battery are arranged in two parallel strings to achieve a nominal voltage of 360 V. Note that the capacity per cell would have been twice as large if only one series string had been used.

Electrochemical Engineering, First Edition. Thomas F. Fuller and John N. Harb.
© 2018 Thomas F. Fuller and John N. Harb. Published 2018 by John Wiley & Sons, Inc.
Companion Website: www.wiley.com/go/fuller/electrochemicalengineering

Table 8.1 Important Battery Requirements

Requirement	Units	Comments
Discharge time	hours	Nominal operation time for application, inversely proportional to rate capability
Nominal Voltage maximum, and minimum voltages	V	Output voltage of the battery string, not an individual cell
Energy	W·h	Capacity of the battery. Linked to average power and discharge time
Weight or mass	kg	Closely related to energy stored in battery
Volume	m³	Closely related to energy stored in battery
Peak power	W	Power for a short pulse of fixed time, 30 seconds, for instance
Cycle life	–	Number of charge/discharge cycles before capacity or power capability is reduced by 20%, for example
Temperature of operation, minimum, maximum	°C	Expected nominal, minimum, and maximum environmental temperatures
Calendar life	years	Beyond the scope of this text

ILLUSTRATION 8.1

The purpose of this illustration is to compare the energy, capacity, and voltage of a string of three cells in series with that of three cells in parallel.

a. For three cells in series, determine V_{batt}, E_{batt}, and Q_{batt}.
b. Repeat Part (a) for three cells in parallel.
c. Based on the results from Parts (a) and (b), compare the voltage and capacity for the two configurations.

Solution:

a. The voltage in series is: $V_{batt} = 3V_{cell} = N_c V_{cell}$.
   The total energy is $E_{batt} = 3E_{cell} = N_c E_{cell}$.
   For the capacity, we note that $V_{batt}Q_{batt} = N_c E_{cell} = N_c V_{cell} Q_{cell} = V_{batt} Q_{cell}$. Therefore, $Q_{batt} = Q_{cell}$.

b. The voltage of the battery with cells in parallel is just the cell voltage. Therefore, $V_{batt} = V_{cell}$.
   The total energy is $E_{batt} = 3E_{cell} = N_c E_{cell}$.
   For the capacity, $E_{batt} = V_{batt}Q_{batt} = N_c E_{cell} = N_c V_{cell} Q_{cell} = V_{cell}(N_c Q_{cell})$. Therefore, since $V_{batt} = V_{cell}$, $Q_{batt} = N_c Q_{cell}$.

c. The energy is the same for the two cases. For the series case we have higher voltage and lower capacity, and for the parallel configuration we have lower voltage and higher capacity.

8.2 BATTERY LAYOUT USING A SPECIFIC CELL DESIGN

A battery consists of a collection of cells that are electrically connected with series and parallel combinations. The general nomenclature is (mS-nP), which means that m cells are connected in series and n of these series strings are connected in parallel. The total number of cells, N_c, is $m \times n$. All of the cells together make up the battery (also referred to as a battery pack or simply pack). Sometimes, a subset of the cells is housed together in a *module*. The primary question we want to answer in this section is, for a given cell design, what is the best way to combine cells together to achieve the design objectives?

For a fixed number of cells, N_c, there are many layouts or different ways of electrically connecting the cells. A generic layout is shown in Figure 8.1. In the previous section, we saw how the voltage requirement determines the number of cells that must be placed in series. Here, we consider the resistance of the battery, which has a crucial impact on its peak power. In principle, any combination of n cells in parallel with m cells in series is acceptable so long as $N_c = mn$. Assuming that all cells are identical, each with an open-circuit potential of V_{ocv} and an internal resistance of R_{int}, the total resistance of the battery can be calculated:

$$R_{tot} = \frac{m}{n} R_{int}. \qquad (8.4)$$

Series connections (m) increase the resistance of the battery, and parallel connections (n) reduce the resistance of the battery. Similarly, series connections cause the battery voltage to increase, and parallel connections increase the capacity and current. Therefore, $V_{batt} = mV_{cell}$, and $I_{batt} = nI_{cell}$.

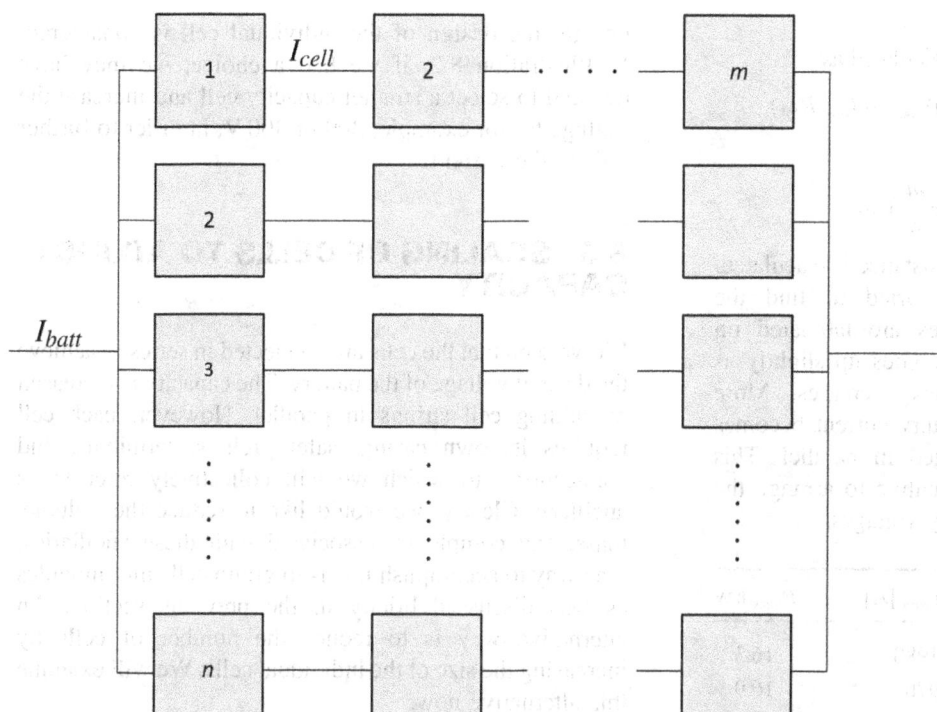

Figure 8.1 Possible series–parallel layout for a total of mn cells.

Assuming that the battery is ohmically limited,

$$V_{batt} = m[V_{ocv} - I_{cell}R_{int}]. \quad (8.5)$$

Power from the battery is

$$P_{batt} = I_{batt}V_{batt} = mnI_{cell}[V_{ocv} - I_{cell}R_{int}]. \quad (8.6)$$

Thus, we see the power from the battery is just the power from an individual cell times the number of cells, and is independent of how the cells are connected together. However, we must have a sufficient number of cells in series to meet the voltage requirements of the application as we have seen previously. Also, there is typically a maximum current specified for the system, since wires, connections, and electrical devices must be sized to handle the system current without, for example, excessive heating. A key advantage of higher voltages is that they result in lower currents and therefore smaller wires and smaller motors for a given power.

The above analysis only included the internal resistance of the cells in the battery pack. How does the situation change if, in addition to the internal resistance, the resistance of the connections between cells is incorporated? There is some resistance in the wires between cells and a contact resistance associated with each electrical connection. For this situation, we need to add the resistance external to the cells to the combined internal resistance calculated from Equation 8.4:

$$R_{tot} = R_{ext} + \frac{m}{n}R_{int} \approx \frac{R_w(1+m)}{n} + \frac{m}{n}R_{int}, \quad (8.7)$$

Resistance of battery accounting for resistances of electrical connections

where R_w is the combined wire and connection resistance between each cell and on each end of the string. Clearly, the lower the external resistance, the higher the maximum power and the greater the energy obtained from the battery. As indicated in Equation 8.7, the external resistance can be calculated from the resistance of the connecting wires and the individual contact resistances. These connections, and therefore the external resistance, vary according to the battery layout. The effect of the battery layout is explored in Illustration 8.2.

ILLUSTRATION 8.2

A 48 kW·h battery is needed for 4 hours of energy storage. The cells available have an open-circuit potential of 2.0 V, a nominal capacity of 1 kW·h (C/4 rate), and an internal resistance of 2 mΩ. R_w is equal to 0.75 mΩ. Compare the nominal voltage, nominal current, and maximum power for four configurations: (4S-12P), (8S-6P), (12S-4P), and (48S-1P).

Solution:

Note that each of the possible configurations has 48 cells in order to meet the total energy requirement with 1 kW·h cells.

Based on the discharge time of 4 hours, 12 kW of power is needed on average. Therefore, the nominal battery current and voltage are calculated as

$$P_{batt} = I_{nom}V_{nom} = (I_{nom})(mV_{ocv} - I_{nom}R_{tot}),$$

where

$$R_{tot} = \frac{R_w(1+m)}{n} + \frac{m}{n}R_{int}.$$

For each layout, the total resistance is calculated. The battery current can be varied to find the maximum power. These values are tabulated on the right. The maximum power goes up slightly as the number of cells in series increases. More importantly, notice that the battery current becomes very large as cells are arranged in parallel. This dramatic increase is strong incentive to arrange the cells in series with higher battery voltages.

m	n	V_{nom} [V]	I_{nom} [A]	P_{max} kW
4	12	6.1	1980	16.3
8	6	12.3	976	16.9
12	4	18.4	647	17.1
48	1	74.7	161	17.3

There is one other term that comes up frequently in describing batteries: the *module*. In Section 8.1, we considered a 24 kWh device where 192 cells were combined to form the battery, with two strings in parallel. However, in practice, a battery design such as this would not likely be implemented as simply a 96S-2P arrangement. Rather, cells would be grouped into modules. For example, in this instance, four cells could be combined to form a module, where each module is 2S-2P and housed together in a single case with one pair of terminals. Forty-eight of these modules are then strung together in series to form the battery (Figure 8.2).

In this section, we have assumed that the cell design was fixed as a constraint, and we examined how cells of that type might be connected to meet the desired specifications. In the next sections, rather than use a fixed design, the design of the individual cell is considered. In Illustration 8.2, if we had a choice, we may have decided to select a smaller capacity cell and increase the voltage to, for example, 200 or 300 V, in order to further reduce the current.

8.3 SCALING OF CELLS TO ADJUST CAPACITY

We've seen that the cells are connected in series to achieve the desired voltage of the battery. The capacity is increased by adding cell strings in parallel. However, each cell requires its own casing, safety reliefs, terminals, and connections, to which we will collectively refer to as ancillary. Clearly, we would like to reduce the volume, mass, and complexity associated with these ancillaries. One way to accomplish this is to group cells into modules as was discussed briefly in the previous section. An alternative way is to reduce the number of cells by increasing the size of the individual cells. We will examine this alternative now.

The reason for increasing cell size is to increase its capacity, recognizing that the capacity of a battery is a function of the amount of reactive material present. Consequently, we begin by developing a quantitative relationship between size (volume) and capacity. To do this, we use a primary lithium thionyl chloride battery as an example. The overall reaction for this battery is

$$4Li_{(s)} + 2SOCl_2 \rightarrow 4LiCl_{(s)} + S_{(s)} + SO_{2(g)}$$

Two of the reaction products are solids that deposit at the positive electrode; the other product is sulfur dioxide gas. Also, note that the $SOCl_2$ electrolyte participates in the reaction. Therefore, the cell volume must include space for the gas that is formed, and should ideally account for the decrease in electrolyte volume as the reaction proceeds. For our purposes here, we adopt the battery construction shown in Figure 8.3. The negative electrode is a foil of lithium metal, and the positive electrode is porous carbon. The volume of the cell is the sum of the volume of individual components: negative electrode, positive

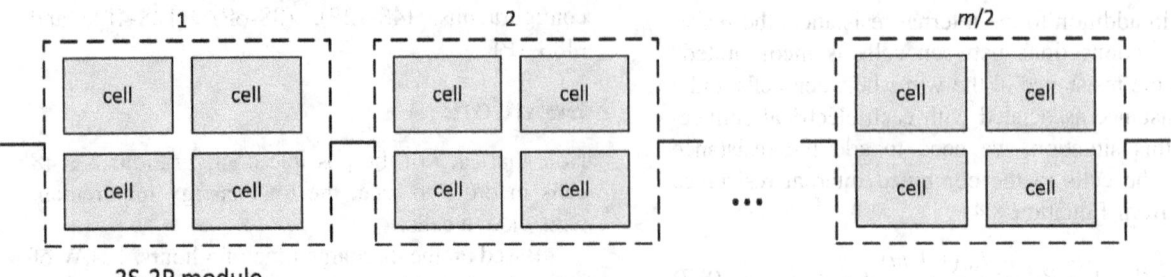

Figure 8.2 Creation of a battery from modules. Each module consists of four cells combined, two in series and two in parallel.

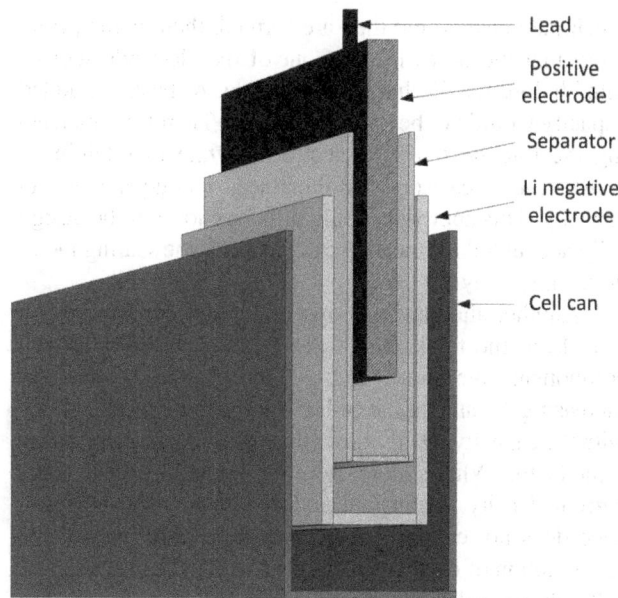

Figure 8.3 Construction of LiSOCl₂ cell.

electrode, separator, positive current collector, negative current collector, and the excess volume. The total cell volume is therefore

$$\mathbb{V}_{cell} = \mathbb{V}_- + \mathbb{V}_+ + \mathbb{V}_s + \mathbb{V}_{-cc} + \mathbb{V}_{+cc} + \mathbb{V}_{ex}. \quad (8.8)$$

For a given capacity Q [A·h], the negative electrode (lithium) volume is given by Faraday's law

$$\mathbb{V}_- = \frac{(3600)QM_{Li}f_a}{F\rho_{Li}}. \quad (8.9)$$

The quantity 3600 converts A·h to coulombs. f_a is a design factor that is included because excess lithium is needed. Since current is also collected by the lithium foil, we don't want to consume all of the lithium, even at full discharge.

The positive electrode is a porous carbon material. It must have an initial pore volume that is in excess of the volume of solid reaction products resulting from the discharge. The volume of solids produced is the sum of the sulfur and lithium chloride, and both are proportional to the capacity.

$$\text{volume of solid products} = (3600)Q\left[\frac{M_{LiCl}}{F\rho_{LiCl}} + \frac{M_S}{4F\rho_S}\right]. \quad (8.10)$$

$$\mathbb{V}_+ = \frac{\text{(volume of solid products)}f_c}{\varepsilon}$$
$$= \frac{(3600)Qf_c}{\varepsilon}\left[\frac{M_{LiCl}}{F\rho_{LiCl}} + \frac{M_S}{4F\rho_S}\right]. \quad (8.11)$$

ε is the initial void volume fraction of the positive electrode. f_c is a second design factor that represents additional porous volume in the positive electrode. Similar to the excess lithium required, some pore volume must remain, even at full discharge. The next three volumes are, respectively, those associated with the separator, negative current collector, and positive current collector.

$$\mathbb{V}_s + \mathbb{V}_{-cc} + \mathbb{V}_{+cc} = A_s\left(\delta_s + \delta_{-cc} + \frac{\delta_{+cc}}{2}\right), \quad (8.12)$$

where δ represents thickness. Each of these volumes is proportional to the separator area, which wraps around both sides of the positive electrode for the design shown in Figure 8.3; therefore, the separator area is divided by two to get the area for the positive current collector. Volume is also needed to accommodate the gas that is formed. We won't go into details for calculating the excess volume needed, but it stands to reason that it also should be proportional to the capacity, which we will express as $\mathbb{V}_{ex} = Qf_{ex}$. Substituting this relationship and Equations 8.9, 8.11, and 8.12 into Equation 8.8 gives

$$\mathbb{V}_{cell} = Q(3600)\left\{\frac{M_{Li}f_a}{F\rho_{Li}} + \frac{f_c}{\varepsilon}\left[\frac{M_{LiCl}}{F\rho_{LiCl}} + \frac{M_S}{4F\rho_S}\right] + f_{ex}\right\}$$
$$+ A_s\left(\delta_s + \delta_{-cc} + \frac{\delta_{+cc}}{2}\right). \quad (8.13)$$

Equation 8.13 provides the desired relationship between volume and capacity. It would seem that we can now easily determine the change in volume necessary to increase the capacity by a desired amount. However, this relationship does not tell the entire story. To explore the issue further, let's imagine that we have an existing cell design with capacity Q_0 and volume $\mathbb{V}_0$. We want to increase the capacity, but keep the discharge time unchanged. If the separator area is held constant, the nominal current density is directly proportional to the capacity; therefore, an increase in capacity by a factor of 2 would require that the current density be doubled. From Equation 8.13, the smallest increase in volume needed to double the capacity results from keeping the separator area, separator thickness, and the thickness of the current collectors constant. This is shown as Option 1 in Figure 8.4 where the separator area, A_s, is held constant and the thickness of each of the two electrodes is increased. The new volume can be calculated from Equation 8.13 and is shown as O1 in Figure 8.5. Unfortunately, if you followed this procedure in practice, you would likely be disappointed. Why? To answer this question, let's examine what happens to the nominal current density as the cell size is changed.

$$i = \frac{Q}{t_d A_s}. \quad (8.14)$$

With the separator area held constant, for two times the capacity, the nominal current density is twice as large. This

180 Electrochemical Engineering

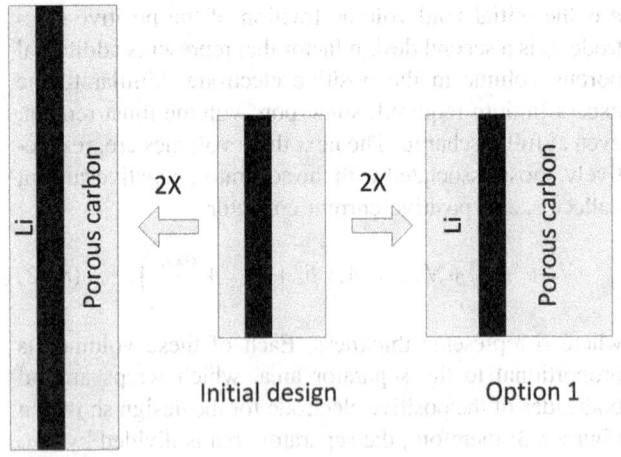

Figure 8.4 Two methods to increase capacity of a cell by a factor of 2.

immediately leads to a couple of concerns. First, with a larger current density, the ohmic, activation, and concentration polarizations of the cell increase, resulting in a lower cell potential for a given current. At this higher current density, the cutoff voltage for the cell will be reached sooner, leading to a lower effective capacity. Second, as we saw in Chapter 5, as the current density increases, the distribution of current through the thickness of a porous electrode becomes increasingly nonuniform (Figure 5.6). Assuming that the reaction products

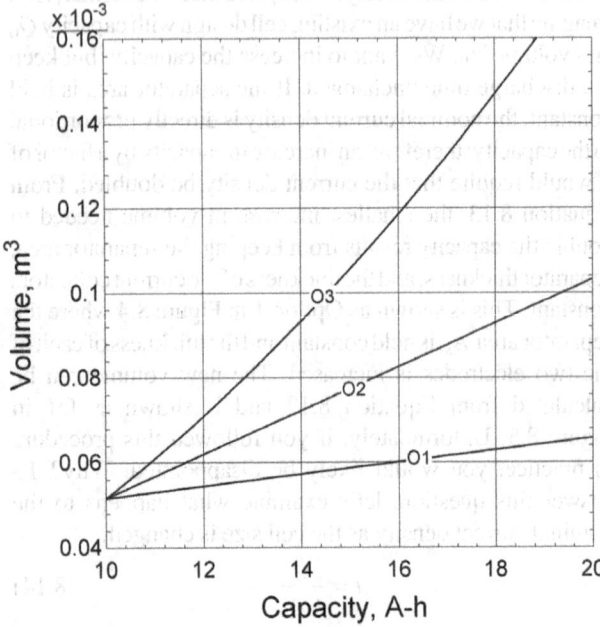

Figure 8.5 Three approaches to the scaling of 10 A·h thionyl chloride primary cell.

precipitate near where they are formed, there is the possibility that the pores in the front of the electrode will be filled before the discharge is complete. At best, the added capacity would not be used effectively. Therefore, increasing the thickness of the electrodes while maintaining a constant cell area is not recommended for large increases in capacity. This approach would only be chosen if the design fully accounted for the above effects or if the scaling factor were small, say 10–15%.

Another alternative, shown as Option 2 in Figure 8.4, is to keep the thickness of the electrodes and other cell components the same, and to increase the capacity by increasing the area of the cell. If the separator area is scaled with the capacity, Q, it follows that the volume scales by the same factor. What's more, from Equation 8.14 the average current density is constant and electrochemical performance does not change appreciably. The resulting volume as a function of capacity is shown as O2 in Figure 8.5. The cell volume scales linearly with capacity, but is larger than the volume calculated for the previous case (Option 1). This is because the volume occupied by the separator and current collectors remained constant for Option 1, but scales with the capacity for the second option.

Can we continue to scale the capacity of the battery indefinitely by increasing its area? No, because other factors will eventually become important. One of these factors is the resistance associated with the two current collectors and their connection tabs as shown in Figure 8.6. This resistance, if important, can be reduced by increasing the thickness of the current collectors. Problem 8.7 explores the effect of scaling the thickness of the current collector by the cell area. This third option, shown as O3 in Figure 8.5, results in an even larger volume. However, it does lead to more confidence regarding cell performance.

Another alternative to increase capacity is to stack several electrodes or plates together in a single assembly. This configuration is shown in Figure 8.7 and represents a common approach used in both lead–acid and lithium-ion

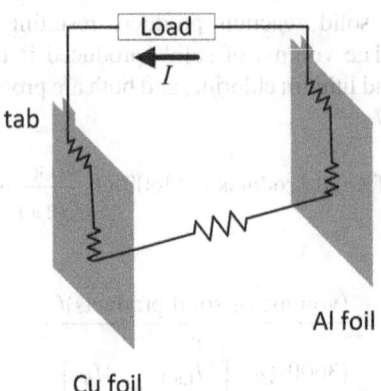

Figure 8.6 Resistance of the current collectors for lithium-ion cell.

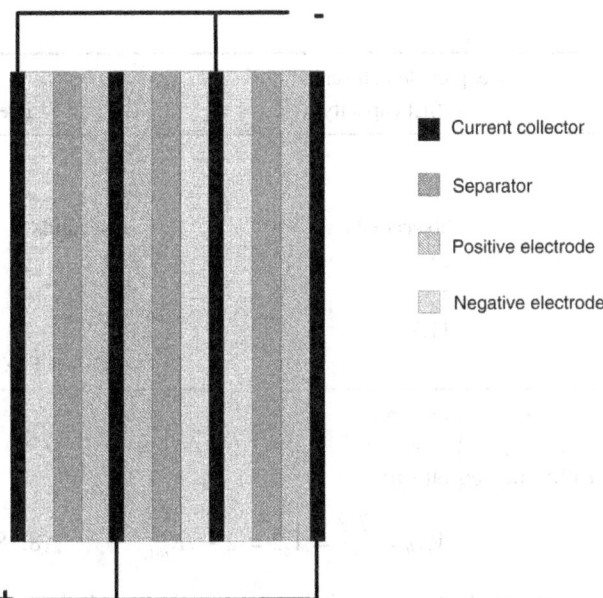

Figure 8.7 Stacked assembly of plates or cells, connected in parallel.

cells. The same electrode active material is placed on both sides of a current collector. This assembly is not a bipolar configuration—the individual electrodes or plates are electrically connected in parallel within a single cell. Additional plates with the same cross-sectional area are added to increase capacity. With the configuration shown in Figure 8.7, it is possible to keep the separator area and electrode thicknesses constant while varying the planform size independently by the addition or removal of plates to yield the desired capacity. The key differences between the configuration shown in Figure 8.7 and simply increasing the size of the electrode are manifest in the current collection, which avoids problems associated with large electrodes, and in planform size. Note that a single current collector serves two electrodes and will need to be sized accordingly. The use of plates to form a battery cell most closely represents case O2.

Finally, we have not considered the volume of the battery casing, tabs, and vents. Volume or mass package factors can range from 1.5 to 1.8. However, the volume of these ancillaries will not change the sizing principles discussed in this section.

8.4 ELECTRODE AND CELL DESIGN TO ACHIEVE RATE CAPABILITY

In the previous section, we examined ways to increase cell capacity in order to decrease the number of cells required in a battery pack. In this section, we examine the impact of cell design on the rate capability or power of a cell. For a given cell capacity, the current and current density are inversely proportional to the time of discharge, which is a key parameter for our discussion in this section:

$$I_{cell}[\text{A}] = \frac{Q_{cell}}{t_d} \quad \text{and} \quad i_{cell}[\text{A} \cdot \text{m}^{-2}] = \frac{Q_{cell}}{A_s t_d}. \quad (8.15)$$

Thus, long discharge times correspond to low-rate capability, and short discharge times denote high-rate capability. We'll explore the impact of discharge time on electrode design through a lead–acid cell example. Before doing that, it may be useful to describe qualitatively the process that will be followed in the example. We begin with a cell that is designed for high capacity rather than high rate. We then scale the capacity of that cell down to the capacity that we need for a high-rate application (starting your car). That scaling is done according to the principles of the previous section where the idea is to keep the electrochemical characteristics constant while changing capacity. Therefore, the scaled cell is still designed for capacity rather than rate. Finally, we examine how to modify the cell, keeping the capacity constant, in order to improve the rate performance.

The lead–acid chemistry is almost universally used to start the automotive engine. Colloquially, it is called a SLI (starting–lighting–ignition) battery. High power is needed, but only for a short period of time; the total capacity is less important. Under normal operation, the state-of-charge of an SLI battery does not change substantially. SLI batteries typically have a capacity of about 60 A·h. Other designs of lead–acid batteries are used for back-up power where they deliver power over much longer periods (on the order of hours or tens of hours). In contrast to SLI applications, batteries for backup power applications undergo a large change in SOC during routine operation, operate at a lower specific power, and have a substantially higher capacity. Often these types of cells are referred to as high-capacity or deep-cycle cells. How are the designs of the SLI and deep-cycle cells different? To answer this question, our approach will be to start with a deep-cycle lead–acid cell and then to discuss changes needed to create a SLI cell. The procedure was described above and results are found in Table 8.2, where the bolded numbers represent SLI specifications.

First, we scale the existing deep-cycle cell from its capacity (1700 A·h) to the desired capacity (60 A·h) of the SLI battery using the cell area. In other words, we decrease the cell area in proportion to the capacity. The internal resistance scales approximately as,

$$R_{int}^{SLI} = R_{int}^{dc} \frac{\text{capacity of the deep cycle}}{\text{capacity of the SLI}}$$
$$= 0.4 \frac{1700}{60} = 11.3 \text{ m}\Omega. \quad (8.16)$$

Table 8.2 Design Parameters for SLI and Deep-Cycle Cells

Feature	Existing deep-cycle cell	Deep-cycle cell scaled for SLI capacity	Desired SLI cell
Nominal voltage [V]	2	2	2
Capacity [A·h]	1700	60	**60**
Discharge time	10 hours	30 seconds	**30 seconds**
Cold cranking amps	NA	71	**560**
Mass [kg]	125	4	16
Internal resistance [mΩ]	0.4	11.3	1.4
Cycle life	500 (50% SOC)		2000 (3% SOC)

Quantities in bold are fixed for the SLI cell.

The resistance increases with decreasing cell area since a higher voltage drop is required to drive the same current in the smaller cell. We further assume that the mass of the cell will scale directly with the capacity or area.

$$m_{SLI} = m_{dc} \frac{\text{capacity of the SLI}}{\text{capacity of the deep cycle}} \quad (8.17)$$
$$= 125 \frac{60}{1700} = 4.4 \text{ kg}.$$

SLI batteries are rated in cold-cranking amps (CCA). SAE J537 (Society of Automotive Engineers publication) specifies that this value is the maximum current that can be sustained for 30 seconds at −18 °C without having the voltage drop below 7.2 V (this assumes a nominal 12 V battery that consists of six cells in series); the corresponding voltage for a single cell is 1.2 V. We would like to estimate the CCA for the scaled cell and compare it to the SLI specification. To do this, we use the following equation:

$$V_{cell} = \frac{7.2}{6} = 1.2 = 2 - IR_{int}^{SLI}, \quad (8.18)$$

which neglects the effect of temperature. Using the calculated value for the internal resistance of our scaled cell, we estimate the current corresponding to the CCA as 71 A. This current is about eight times lower than the required value of 560 A. Clearly, simple scaling of the deep-cycle cell to the desired SLI capacity will not meet the performance requirements for the SLI battery. Consequently, the electrode design must be changed in order to lower the resistance and increase the power. In doing so, we do not need to nor want to increase the capacity of the battery.

A strategy for increasing the power at constant capacity is to make the individual electrodes or plates thinner as illustrated in Figure 8.8. The amount of active

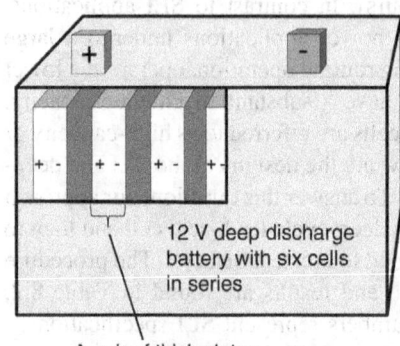

A pair of thick plates

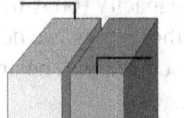

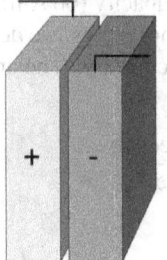

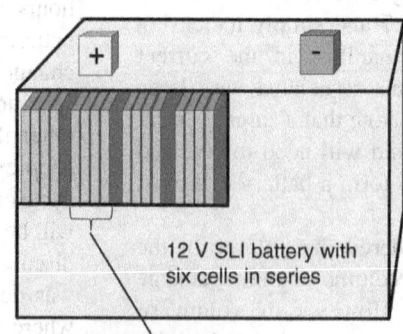

Four pairs of thin plates in parallel

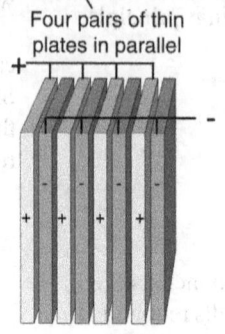

Figure 8.8 Comparison showing the internals of a deep discharge and an SLI lead–acid battery. Both are 12 V and have the same amount of active material. The electrodes for the SLI battery are thinner, with multiple plates connected in parallel to reduce the internal resistance of the cell.

material is the same, but we have used many more (thinner) plates in the SLI design. This strategy does two important things that both increase the rate capability of the cell. First, the electrode thickness is reduced, making it easier to access the available active material. Second, the cell area is increased, leading to a lower current density for a given battery current. Ignoring the benefits of the thinner electrodes for a moment, we can estimate the cell area that is needed to adequately lower the internal resistance.

$$V_{cell} = \frac{7.2}{6} = 2 - (560)R_{int}^{SLI}. \qquad (8.19)$$

An internal resistance of 1.4 mΩ is calculated; therefore, the cell area must be increased by a factor of about 8 (11.3/1.4). Thus, to achieve the desired power, the separator area for the SLI battery would need to be about eight times that of the deep-cycle battery, which would result in electrodes that are approximately eight times thinner. In practice, the electrodes are designed to be thinner and also more porous to further reduce the resistance. Thus, the rate performance benefit is significantly greater than that due to just the increased area.

What about the specific capacity? If we discharged the SLI battery at a rate comparable to that of the deep cycle (C/10), would it have a similar specific capacity? The rudimentary analysis just presented suggests that the specific capacities would be the same. However, for each plate, a current collector is needed, and additional separator area is required. When we account for this additional mass, the specific capacity of the SLI cell is lower. More important than the specific capacity is the cycle life. SLI batteries perform very well at the limited DOD for which they are designed. However, while the details are beyond the scope of this chapter, commercial SLI batteries show rapid degradation and a short lifetime when cycled over a large SOC window.

8.5 CELL CONSTRUCTION

To this point in the chapter, we have examined several aspects of cell and battery design. This section takes a brief look at ways in which cells and modules are constructed. The mechanical designs can be complex, and often manufacturers differentiate their products on these attributes.

Almost invariably, it is desirable to have cells and modules in a sealed container. In the case of lithium batteries, for instance, contact of the active materials with water and oxygen must be avoided; thus, a hermetic seal is required. The internal components of some cells can be exposed to the atmosphere, but it is still generally preferable to have a sealed design in order to prevent escape of electrolyte and any gases generated during normal

operation. The container or casing is either metal or plastic, depending on the application. Gases can be generated due to normal operation or due to an internal cell failure. Uncontrolled venting of gases as a result of cell faults varies from mild to violent, and depends on the chemistry and size of the cell. Thus, with a sealed casing, a means of safety pressure relief (Figure 8.9a) is needed to avoid injury and equipment damage. The complexity of the pressure relief system depends on the needs of the particular application. Each case must also include a positive and a negative terminal. Any current that enters or leaves the cell or module must pass through the terminals. Terminals come in a wide variety of sizes and shapes, and have features that manifest little uniformity among manufacturers.

Cells may take many forms. Here we will discuss only the more prevalent types: prismatic, plate designs, cylindrical, and coin. *Prismatic* cells take the shape of a rectangular prism, or simply a three-dimensional rectangular object with six faces that are rectangles. Because they can pack together efficiently, prismatic cells are preferred

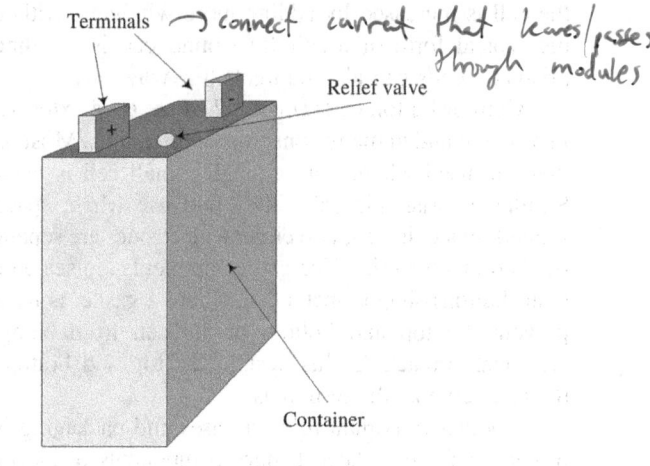

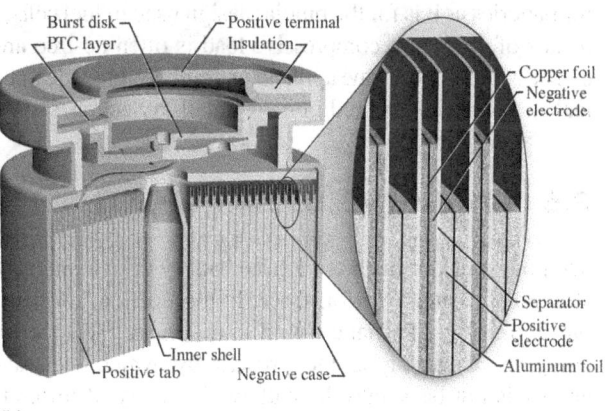

Figure 8.9 Cell configurations (a) A prismatic cell showing terminals and relief valve. (b) Spirally wound cylindrical cell.

where space is limited. One method of forming a prismatic cell is to stack electrodes (plates) together as shown in Figures 8.7. The size and shape of the electrodes can be controlled independently. Multiple cells can also be combined and connected in series to create a battery as shown in Figure 8.8. The details of a prismatic design will depend on the cell chemistry. For instance, both lead–acid and lithium-ion cells can be made in a prismatic form, but the design details are quite different.

The common alkaline cell is *cylindrical* and probably already familiar to you. Construction starts with a steel can, which also serves at the current collector for the positive electrode. Active material is then molded to the inside wall, a separated introduced, and the negative material and electrolyte added. A metal pin is inserted to collect the current for the negative electrode before sealing the can. An alternative method of creating cylindrical cells is to wind two electrodes around a core. The winding consists of current collectors coated with active material on both sides and two separator layers, as shown in Figure 8.9b. The winding is then inserted into a casing. The capacity of the cell is increased by adding more windings. Although the typical form of a spirally wound cell is a cylinder, prismatic cells can also be made this way.

Coin or button cells (Figure 7.2) are used extensively in research and in many consumer applications. Most often they are used where only a single, small cell is needed. Similar to prismatic cells, they find use where space is limited. In a coin cell, two circular electrodes are separated by a porous material. The cell components are sealed in a coin (button)-shaped metal can, where a gasket is used to prevent the top and bottom of the can from being in electrical contact. In this design, the top and bottom of the can serve as the terminals.

Another important role for cases and packaging is to provide a force to keep battery components in electrical contact and to reduce the redistribution and loss of material during cycling. Although this force is generally not as critical for batteries as it is for the bipolar design used in fuel cells, a means of providing compressive load is often desired and can be essential for some applications and designs. Mechanical issues are considered in more detail in Section 8.10.

8.6 CHARGING OF BATTERIES

We now shift our discussion from battery design and construction to battery performance. In this section, we begin our look at performance with a description of battery charging. Charging is the process by which electrical energy is put back into the battery. A variety of different procedures or protocols are used for charging. These protocols depend on the battery chemistry, the application, and, to some extent, the battery manufacturer. We can broadly differentiate the methods used for charging based on cell chemistry, and specifically, whether or not the cell can tolerate overcharging. Lead–acid and NiCd cells are both tolerant of overcharge and, in some instances, even benefit from some overcharge. In contrast, lithium-ion cells cannot be overcharged without causing permanent damage. When a cell is overcharged, this means that more coulombs are passed with charging than are required to fully charge the cell. The ratio is known as the *charge coefficient*:

$$\text{charge coefficient} \equiv \frac{\text{coulombs passed to fully recharge}}{\text{coulombs drawn during discharge}}$$
$$= \frac{1}{\eta_{coul}}. \quad (8.20)$$

Note that this charge coefficient is related to the coulombic efficiency introduced in Chapter 7. Batteries with a coulombic efficiency less than 100% will have a charge coefficient greater than 1. The extra coulombs indicate that a side reaction is occurring, such as hydrogen and oxygen evolution in aqueous systems or SEI formation and electrolyte decomposition for lithium-ion cells.

The two basic methods of adding charge are charging at constant current (CC) and charging at constant voltage. In the first of these, the cell is charged at a constant rate (e.g., C/2–C/8 depending on the battery). During the charge, the potential of the cell rises, and charging is allowed to continue until a specified voltage is reached. The second basic method is to hold the potential of the cell constant and allow the current to vary. Constant voltage (CV) is seldom used as the sole means of charging a cell because very large currents are possible at the beginning of the charge if the voltage is held constant at its final value.

Often, these two basic methods are combined: constant current charge followed by a constant voltage charge (CCCV) as shown in Figure 8.10. In this combined method, charging is done at constant current until a specified maximum voltage is reached. Once that voltage is reached, charging continues at a fixed potential and the charging current decreases with time. This decrease in current with time is known as a current taper. The current is allowed to decrease until it reaches a specified value at which the device is considered to be fully charged. Using this protocol, the majority of charge is added in the constant current mode, but the time spent charging in each mode is roughly equal. For a lithium-ion cell, which cannot tolerate overcharge, charging is stopped once the specified current is reached (C/20 for example), and the charge coefficient is very close to one.

Two other charging protocols deserve mention. The first is *pulse-charging* (Figure 8.11). Here the charging current is not constant, but rather is interrupted by brief periods of rest. In some cases, the cell may even be discharged briefly. The pulse-charging protocol has been

Chapter 8 Battery Applications: Cell and Battery Pack Design 185

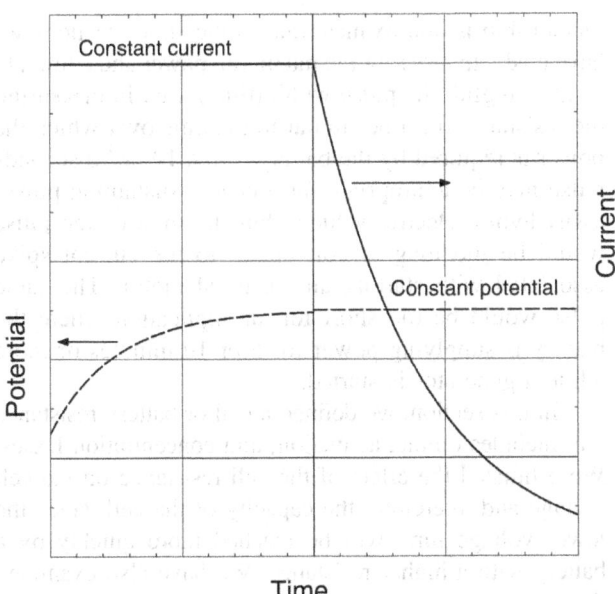

Figure 8.10 Constant current–constant voltage charge (CCCV).

widely applied to different cell chemistries. Although more complex and not without controversy, there are some reported advantages, namely, more rapid charging and a reduced extent of sulfation for lead–acid cells.

The final charging protocol we will consider relates to the use of charging to counteract the effects of self-discharge. Although its magnitude varies greatly, all cells exhibit self-discharge. For some secondary battery applications, for example, standby power applications, the user expects that the battery will always be available and near a full state-of-charge. The most obvious strategy for addressing self-discharge is to occasionally recharge the batteries after a substantial standby period has passed. This strategy is used for lithium-ion cells and other cells that do not tolerate overcharge. An alternative that works for cells that can be overcharged is to put the cell on a *float charge* following the constant voltage portion of a normal charge cycle. The float charge, commonly used with lead–acid batteries, continues to pass a small current through the cell indefinitely in order to maintain a full state-of-charge. Note that the float charge portion of the cycle is not used in the charge-coefficient calculation.

This section has examined a few of the issues related to battery charging as a subset of battery performance. We next examine battery resistance as a way of evaluating the health of the battery, which is a comparison of its current performance with that achieved initially.

8.7 USE OF RESISTANCE TO CHARACTERIZE BATTERY PEFORMANCE

The laboratory tests to determine the capacity and rate capability of a cell that were described in Chapter 7 can also be applied to a battery and are not repeated here. Rather, the focus in this section is on the use of resistance to provide information on the power capability and health of the battery. Health in this context is a measure of the condition of the battery relative to its initial "full performance" state.

There are a number of resistances defined and used by specialists, and some care is needed in identifying the specific quantity of interest. We begin with the ohmic resistance, which was used extensively in Chapters 6 and 7, and throughout the book. Ohmic resistance is determined from the high-frequency intercept of an electrochemical impedance spectrum or, equivalently, from current interruption. Its measurement and meaning are unambiguous. Although this ohmic resistance is used extensively in research, it is not the cell or battery resistance commonly used in practice. The resistance used commercially is a more loosely defined term that depends on the specific application. While there are many variations, the general idea is the same. The concept and its relationship to the ohmic resistance are illustrated in Figure 8.12. After a period of rest, two

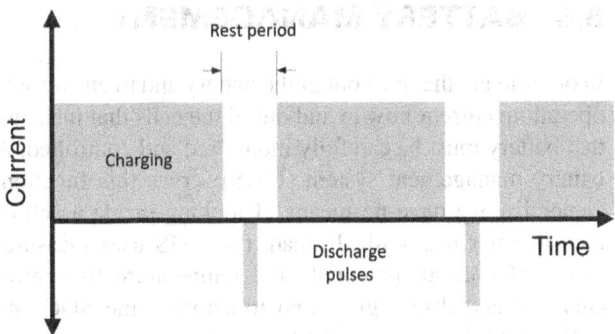

Figure 8.11 Pulse charging.

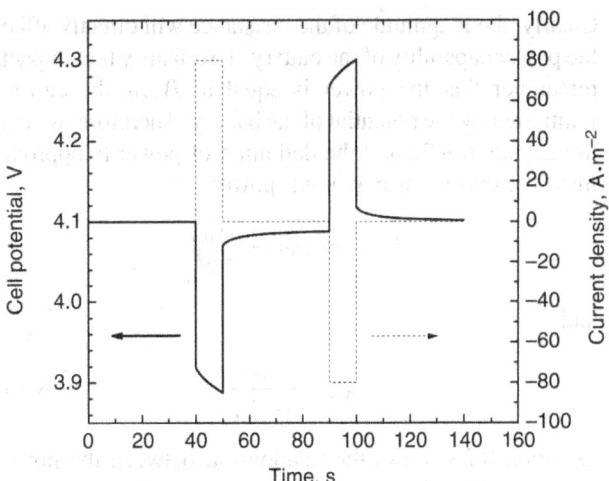

Figure 8.12 Pulse power test.

short current pulses are applied, one for charging and one for discharging. During discharge, the potential of the cell drops instantaneously and then decreases with time. The immediate drop represents the ohmic resistance of the cell as described in Chapter 6. The further decrease that occurs in the time interval from t_0^+ (just after the instantaneous drop) to t_1 is due to activation and concentration polarizations in the cell. The discharge resistance for the cell or battery includes all of these polarizations (ohmic, activation, and concentration) and is defined as the change in potential divided by the change in current:

$$R_{cell}^d = \frac{\Delta V}{\Delta I} = \frac{V(t_0) - V(t_1)}{\Delta I}, \qquad (8.21)$$

where the superscript d refers to discharge. Whereas the ohmic resistance depends on temperature and slightly on the SOC, the battery resistance defined by Equation 8.21 can have a strong dependence on SOC. It is also evident from Figure 8.12 that the resistance may vary with changes in the width or magnitude of the discharge pulse. Analogous to Equation 8.21, a resistance for charging can be defined, R_{cell}^c. These two values are different because the activation and concentration polarizations will change depending on whether a cell is being charged or discharged. To account for differences in battery size, the resistance is often normalized by the separator area of the cell. This quantity is the area-specific resistance (ASR) with units of $\Omega \cdot m^2$. Again, as measured, this quantity includes ohmic, activation, and concentration polarizations.

For preliminary analysis of batteries, we assume that the measured cell resistance is constant so that voltage losses vary linearly with the current. In fact, this practice is used extensively in this chapter. With use of the newly defined cell resistance, we can estimate the cell voltage as

$$V_{cell} = V_{ocv} - IR_{cell}^d. \qquad (8.22)$$

Clearly, the magnitude of the resistance will directly affect the power capability of the battery. To quantify that impact, remember that the power is equal to IV_{cell}, the current multiplied by the potential of the battery. Therefore, we can use Equation 8.22 and the definition of power to approximate the current at maximum power.

$$I(\text{max power}) = \frac{V_{ocv}}{2R_{cell}^d}$$

and

$$P_{max} = \frac{(V_{ocv})^2}{4R_{cell}^d}. \qquad (8.23)$$

Equation 8.23 shows the relationship between the power and the cell resistance. It must be remembered that this relationship is approximate due to the simplifications we have made. In order for the maximum power and current to be meaningful, the pulse width (time) used in measuring the resistance must be similar to the time over which the power is required by the battery. So, a 10 or 30 seconds pulse may be appropriate for a charge-sustaining power assist hybrid-electric vehicle, but the time of the pulse would be too long to correspond to the current spike associated with starting an electrical motor. The same pulse would be too short for an application where the battery is supplying power for 5 or 10 minutes or more while a generator is started.

In this section, we defined a cell or battery resistance that includes ohmic, activation, and concentration losses. We estimated the effect of the cell resistance on the cell voltage and, therefore, the capacity of the cell, since the lower voltage limit will be reached more quickly by a battery with a higher resistance. We have also examined the impact of the resistance on the power or rate capability of the cell. Given its impact, resistance is a convenient way of characterizing battery performance, and the change in resistance with time as the battery is cycled provides a measure of the state of health (SOH) of the battery, defined generically as

$$\text{SOH} = \frac{\text{present capability}}{\text{design capability}}. \qquad (8.24)$$

SOH relates to the ability of the cell to meet its specified performance ratings. From the manufacturer, we assume that the battery meets its design goals, and therefore the "as received" SOH is 100%. There are numerous physical processes that cause the performance of the cell to degrade over time. Many of these processes also cause an increase in the cell resistance. Therefore, the cell resistance is one of the best ways to assess the SOH of the cell. While it is not the only method, it represents a common and effective way to make the desired assessment. Finally, we note that cell resistance changes with temperature, an effect that must be accounted for in order to make an accurate assessment.

8.8 BATTERY MANAGEMENT

In order to get the most out of the battery and to ensure safe operation, current flow in and out of the cells that make up that battery must be carefully monitored and controlled. A battery management system (BMS) serves this function. Generally, we have no means of looking inside a cell or module after it is built. Instead, the BMS uses measurements of current, potential, and temperature to control charging and discharging, and to estimate the SOC and SOH. Additionally, the BMS must communicate with other systems that interact with the battery pack. It must also provide the electrical hardware and software to accomplish cell balancing. Our purpose in this chapter is not to examine the architecture of the BMS and describe how it works; these topics are beyond the scope of this text. Here, we are concerned with the electrochemical processes that drive the need for a battery management system.

The importance of and the sophistication required in the BMS depend on the cell chemistry, the size of the battery, and the application. With regard to cell chemistry, it is critical to understand the consequences of overcharging and overdischarging a cell. As you know, lithium-ion cells cannot be overcharged without incurring damage, whereas overcharging can help to prolong the life of lead–acid cells. An important function of the BMS is to manage overcharging and overdischarging of individual cells.

One of the most important issues in the management of a multicell battery is keeping the individual cells in balance. Cell balance is another way of stating that, to the extent possible, the SOC of all of the cells should be the same. Despite our best efforts, the cells in a string will not behave identically. There can be variations from the manufacturing process that introduce small differences in the cell resistance or in its capacity. Even if these nonuniformities are eliminated, cells assembled into a battery can experience different temperatures, depending on their location in the pack. Cells can age differently too. Although the same current passes through each cell in series, this does not ensure that the state-of-charge of a given cell remains synchronized with that of the other cells in the string. Cells will not have exactly the same coulombic efficiency, which will affect the usable capacity of the battery. For example, every time the pack is charged, cells with lower coulombic efficiencies will not charge to the same extent as other cells in the pack. In spite of their poorer charge performance, these weak cells are likely to be discharged to the same extent as the other cells. This means that their SOC relative to the other cells drops each cycle. Eventually, the SOC of the weak cells will drop to zero. Prior to that, the cell is likely to be damaged, perhaps irreversibly. Damage of other cells in the string is also possible. Illustration 8.3 demonstrates the impact of lower charge efficiency on the SOC.

ILLUSTRATION 8.3

A string of 1 A·h NiCd batteries are cycled between 30 and 80% SOC. If the nominal current efficiency for charging is 80%, but one weak cell has a coulombic efficiency of only 70%, what happens to the SOC of the weak cell with cycling?

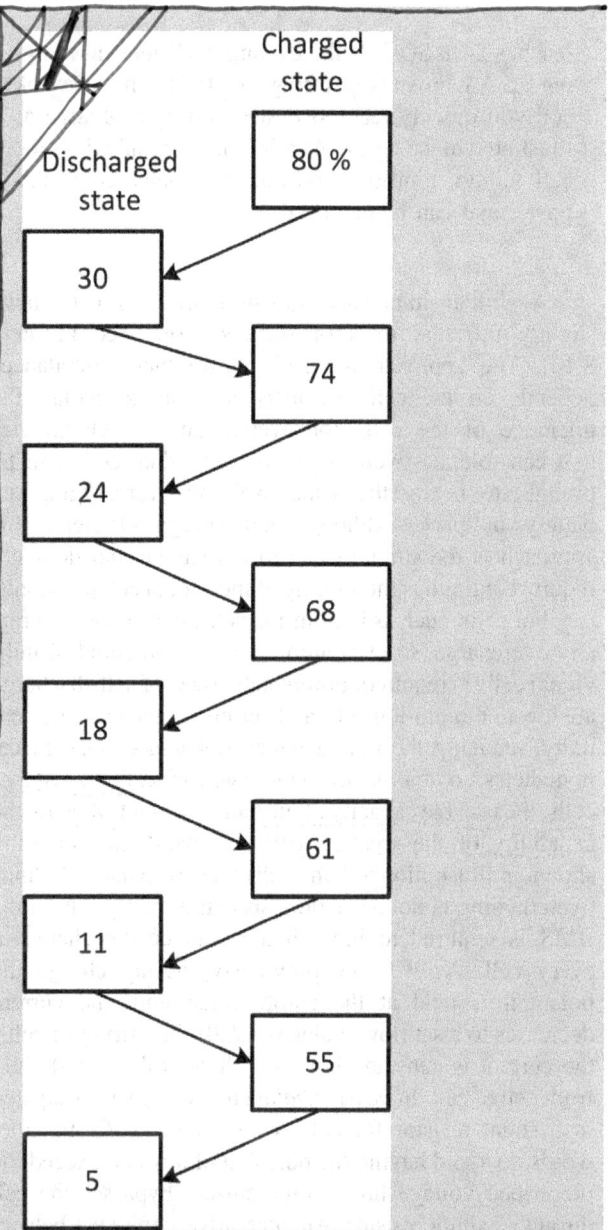

The coulombs needed to restore the SOC of the nominal cells is

$$Q = \frac{\text{capacity} \times \Delta \text{SOC}}{\eta_{coul}^{nominal}} = \frac{1\text{A} \cdot \text{h} \times 0.5}{0.8} = 0.625 \text{ A} \cdot \text{h}.$$

The weak cell receives the same number of coulombs, but a smaller fraction goes to restoring its SOC. Instead of 0.5, the change in SOC of the weak cell is

$$\Delta \text{SOC} = \frac{Q \eta_{coul}^{weak}}{\text{capacity}} = \frac{0.625 \text{A} \cdot \text{h} \times 0.70}{1.0 \text{A} \cdot \text{h}} = 0.4375.$$

Thus, when discharged 0.5 A·h and subsequently charged, the new SOC of the weak cell is 0.3 + 0.44 = 0.74. After just one cycle, this cell only reaches 74% SOC. If the cycling continues discharging 0.5 A·h but only restoring with 0.44 A·h, this weak cell will quickly reach 0% SOC during discharge as illustrated in the figure. At this point, not only does the cell fail to contribute power, it actually consumes power and can be damaged.

A similar imbalance situation arises if cells have slightly different rates of self-discharge (see Problem 8.13). The approach to dealing with these imbalances depends on the cell chemistry and, in particular, the tolerance of the cells for overcharge. For chemistries that can tolerate overcharge, the individual cells can be brought to nearly the same SOC by overcharging the battery or pack. Although not energy efficient, this approach is the simplest one to implement and does not require continuous monitoring of individual cell potentials.

For cells, such as lithium-ion cells, that cannot tolerate any overcharge, strict monitoring of the potential of individual cells is required. Fortunately, rates of self-discharge are low in lithium-ion cells and current efficiencies are near unity, reducing the rate at which imbalances take place; nonetheless, over time imbalances will arise. For a string of cells, the useable capacity of the battery is controlled by the capability of the weakest cell. To avoid the condition shown in Illustration 8.3, the cells must be balanced. Since overcharging is not an option, special circuitry within the BMS is required to individually balance the charge in every cell. As we saw previously, during charge the potential is held at the cutoff value until the current decreases to a set (low) value (~C/20). In a string of cells, the current is constant, but the voltage will vary slightly from cell to cell. In *passive balancing*, energy is dissipated in a shunt resistor for cells with excess SOC. In other words, to avoid having the potential of any cell exceed the prescribed voltage limit, some current bypasses the cell through a shunt resistor. An alternative to passive balancing is a*ctive balancing*. Here, energy from a cell with a high SOC is moved to a cell with a lower SOC. Active balancing requires more complicated electrical circuitry, but has greater energy efficiency.

Knowledge of the SOC is an important aspect of cell balancing. There are two approaches for determining the SOC. The first means is to simply count the coulombs passed relative to a known condition. In doing so, current is taken as positive during charge and negative during discharge.

$$\text{SOC} = \text{SOC}(t_o) + \frac{\int_t^{t_o} I dt}{\text{capacity}}, \quad (8.25)$$

where $\text{SOC}(t_0)$ represents the known condition. Let's consider some challenges with this approach. First, this method is only accurate if the current efficiency is close to one, although it can be corrected with use of an efficiency, if known. The SOC calculated by Equation 8.25 will also not be accurate if there are appreciable rates of self-discharge. Finally, since the capacity of the battery changes with time due to aging, the capacity used in Equation 8.25 should be the capacity at time t_0. Temperature also influences capacity and must be considered in determining the SOC.

A second alternative to SOC determination is to measure the open-circuit potential of the cell or battery. With some chemistries, the potential change with SOC is minimal (Figure 7.4); with others it is substantial. When there is an appreciable change in the open-circuit potential with SOC, measuring the voltage of the cell is a quick way of estimating its SOC. Often, algorithms are used to combine these two approaches to estimate the SOC.

Finally, we close with a couple of comments about temperature. At either excessively high or low temperatures, the current in or out of the battery may need to be restricted by the BMS in order to avoid damage. Concerns include the possibility for thermal runaway at high temperatures, and the possibility of cell damage at temperatures that are too low for acceptable operation of the battery. Often individual lithium-ion cells have a positive temperature coefficient (PTC) current-limiting device that acts as a fuse to prevent thermal runaway. Thermal management is an important aspect of battery operation and will be considered in more depth in the next section.

8.9 THERMAL MANAGEMENT SYSTEMS

The spirally wound cylindrical cell is particularly convenient to manufacture. Why then are these cells typically small? To answer this question, heat removal must be considered. A number of simplifying assumptions will be made, but the basic physics will provide clear guidance for the engineer. As was discussed in the previous chapter, during operation heat is generated in the cell. Here we will assume a resistive cell with a uniform rate of heat generation, $\dot{q}$ (see Illustration 7.5). To solve for the temperature in the battery, we start with a differential energy balance:

$$\nabla^2 T + \frac{\dot{q}'''}{k_{eff}} = 0. \quad (8.26)$$

This equation is applied to a cylindrical battery. k_{eff} is an effective thermal conductivity in the radial direction. There are no variations in the theta direction and we assume that the cell is long in the z-direction so that the problem becomes one dimensional, and in cylindrical coordinates,

$$\frac{1}{r}\frac{\partial}{\partial r}\left(r\frac{\partial T}{\partial r}\right) + \frac{\dot{q}'''}{k_{eff}} = 0. \quad (8.27)$$

For the spirally wound cell, a sheet of current collectors, electrodes, and separators are wound around a shaft of radius r_i, and the final radius of the winding is r_0 (Figure 8.9b). Two boundary conditions are needed for Equation 8.27. We assume that no heat is conducted to the shaft and that the temperature is specified at r_0:

$$r = r_i, \quad \frac{\partial T}{\partial r} = 0,$$

$$r = r_0, \quad T = T_0.$$

Equation 8.27 is separable and upon integration, the solution is

$$T = T_0 + \frac{\dot{q}'''}{4k_{eff}}\left[(r_0^2 - r^2) + 2r_i^2 \ln \frac{r}{r_0}\right]. \quad (8.28)$$

Figure 8.13 shows the variation in temperature with radial position. As seen in Illustration 7.5, the rate of heat generation is proportional to the current, and therefore proportional to the C-rate.

Next, we consider what happens when more windings are added to the cell. The capacity of the cell is directly proportional to the length, L, of the winding around the shaft. The diameter or radius of the winding can be determined from basic geometric considerations:

$$r = \sqrt{r_i^2 + \frac{\delta L}{\pi}}. \quad (8.29)$$

δ is the thickness of the sheet that is being wound.

The results are shown in Figure 8.13. As the capacity of the cell is increased (by adding windings), the temperature near the center, r_i, increases. This relationship is shown in Figure 8.14. The combination of increased capacity and higher rates is limited by the maximum allowable temperature of the cell. Even though it is efficient to manufacture larger cells by winding more material, heat removal constrains this approach to adding capacity. This illustration also assumed that the temperature on the outside of the cell is fixed. A more realistic boundary condition would relate the rate of heat removal by conduction to the heat removal rate from the surface by convection. This refinement would result in even higher temperatures inside the battery.

Storing large amounts of energy has inherent safety concerns. Although the challenge is best known with lithium-ion cells, many different types of cells can undergo thermal runaway. As illustrated previously, the operating temperature of a cell is determined by the heat generation inside the cell and the heat removal from the cell. This analysis assumed that the rate of heat generation was equal to the rate of heat removal; that is, a kind of steady state. As the rate of heat generation increases, the cell temperature rises until the rate of heat removal balances the rate of generation. Thermal runaway, in contrast, refers to a transient condition where the internal cell temperature rises

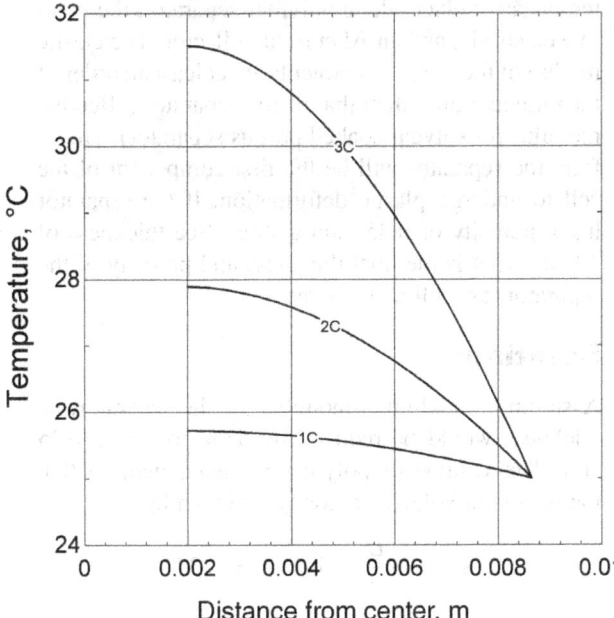

Figure 8.13 Temperature profile in cell with C-rate as a parameter. Cell capacity is 1.6 A·h.

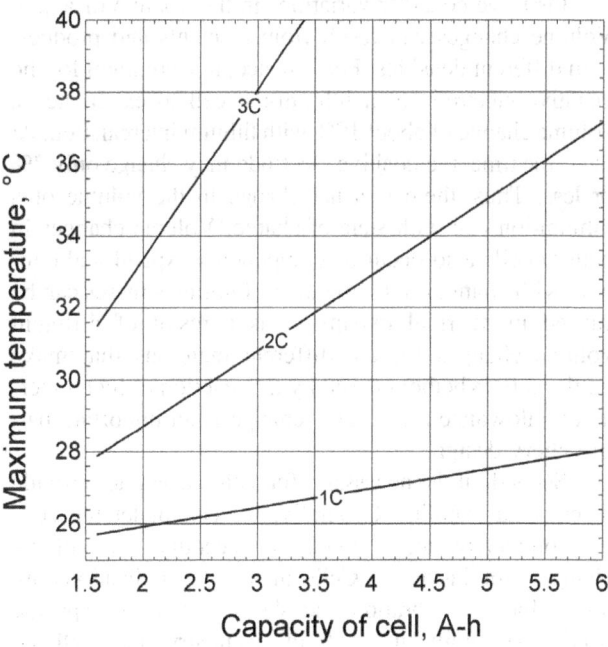

Figure 8.14 Effect of rate and capacity on maximum temperature.

uncontrollably. For aqueous batteries, thermal runaway is usually associated with a constant voltage or float charge. At constant voltage, the current decreases during the taper charge. If the temperature were to rise in one area, the resistance of the cell decreases (lower ohmic and activation overpotentials) and thus the local current density increases. There is, however, negative feedback in this system. As more charge is passed, the local equilibrium potential increases; and therefore, the overpotential decreases, causing the current density to go lower. What happens, during float charge or when most of the current is going into gas evolution? The current flowing does not change the local SOC of the electrode, and therefore the negative feedback is removed. Now more current results in more heat generation and a temperature rise. Higher temperature lowers the polarization, leading to more current and more heat generation. If not controlled, thermal runaway can occur.

8.10 MECHANICAL CONSIDERATIONS

Mechanical effects are important in battery design and operation, and are often underappreciated. They impact battery performance and longevity at a range of length scales from the stress that develops in submicrometer-sized particles of an active material to the macroscale structural requirements of a large battery pack such as the 2-ton battery pack used in a submarine. Frequently, the effects are specific to a particular chemistry and cell design. A few of these effects will be mentioned in this brief introduction, but the analysis is by no means complete or in-depth.

First, we consider variations in the volume of a cell. Volume changes can result from reactants and products with different densities. For instance, it is common for the negative electrode of a lithium-ion cell to experience a volume change of about 10% with lithium intercalation. At the same time, the positive electrode may change only 2% or less. Thus, there is a net change in the volume of a lithium-ion cell with state-of-charge. Volume changes in battery cells also occur as components expand and contract with changes in temperature. Internal stresses can be caused by thermal expansion as a result of different volume changes for the different materials that make up the cell. Whether caused by the reaction or by temperature, allowance for volume change is an important part of battery design.

Second, it is necessary for cell design to provide mechanical stability. Generally, we cannot let the cells be completely unrestrained (to account for volume changes, for instance). Cells in mobile applications are also subject to vibration and shock. Again, the specific needs vary significantly with the chemistry of the cell and with the particular application. For example, a small amount of compression is desired in order to prevent delamination (connected to volume change) in cells where the electrodes are coated onto a current collector. What if the compression is too high? Excessive compression can lead to creep of the polymer separator, causing a reduction in its porosity and resulting in increased resistance or increased mass-transfer overpotentials. This possibility is explored further in Illustration 8.4. Other cells will have different needs. Our purpose in this short section was to briefly introduce you to the mechanical issues that are an essential aspect of cell design and operation.

ILLUSTRATION 8.4

In this illustration we consider the impact of compression on the separator of a lithium-ion cell. The strain for two electrodes of a cell sandwich is given by

$$\epsilon = \frac{h - h_0}{h_0},$$

where h_0 is the thickness in the assembled state (fully discharged). The maximum thickness occurs when the electrode is fully charged, which for the lithium-ion cell under consideration is 105 μm. If the assembled thickness of the electrodes is 100 μm, the strain is calculated as

$$\epsilon = \frac{105 - 100}{100} = 0.05.$$

The cell sandwich consists of a Cu current collector, the negative electrode, a polymer separator, the positive electrode, and an Al current collector. The elastic moduli of the other components are at least an order of magnitude higher than that of the separator (effective modulus for solvent-soaked porous separator). Therefore, the separator will be the first component of the cell to undergo plastic deformation. If the separator has a porosity of 0.45, and a stress-free thickness of 30 μm, what is the final thickness and porosity of the separator, as well as the stress?

Solution:

Assuming a negligible modulus for the separator, its thickness would be reduced by 5 μm to $30 - 5 = 25$ μm. The volume of polymer is unchanged, so that the new void volume (porosity) is given by

$$\epsilon = 1 - \frac{L_{s,0}}{L_s}(1 - \epsilon_0) = 0.34.$$

Under these assumptions, there would be no stress in the assembly.

Let's check to see if the assumption of plastic deformation is reasonable. The strain experienced by the polymer is

$$\epsilon = \frac{25-30}{30} = \frac{\sigma}{E}.$$

Using an effective modulus, E, of 1 GPa, the stress is 170 MPa, which is well above the yield stress of 20 MPa. This would result in plastic deformation of the separator.

CLOSURE

In this chapter, the process of selection and design of batteries was introduced. The key factors are discharge time, energy, and voltage. The discharge time profoundly influences the electrode and cell design. The energy, or capacity, determines the size of the battery. The desired voltage is achieved by connecting multiple cells together in a string. Capacity is changed by adding more active material, and several means of accomplishing this addition were considered, particularly with regard to rate capability. Proper monitoring and control of the current to the individual cells in the battery is critical during operation, but particularly during charging. Finally, the importance of thermal and mechanical aspects of battery design was introduced.

FURTHER READING

Crompton., T.P.J. (2000) *Battery Reference Book*, Reed Educational and Professional Publishing Ltd.
Keihne, H.A. ed., (2003) *Battery Technology Handbook*, CRC Press.
Reddy, T. and Linden, D. (2010) *Linden's Handbook of Batteries*, McGraw Hill.

PROBLEMS

8.1 You are asked to configure a battery with the following requirements: 1512 W·h, 200–400 VDC, and a discharge time of 3 hours. A lithium-ion battery is available with the following specifications at a rate of C/3: nominal voltage of 3.5 V and is manufactured in two sizes: 3.3 and 4.5 A·h. The cutoff potential of the cells is 2.75 V, and the maximum charging potential is 4.1 V. What configuration would you recommend?

8.2 For the same battery specified in Problem 8.1, what would be the ideal capacity in A·h of the individual cells to achieve as close as possible to 300 VDC for the nominal voltage of the battery? Compare this answer to the configuration found in Problem 8.11. Which would be preferred and why?

8.3 Derive Equation 8.23 for the maximum power for an ohmically limited cell. What is the expression for maximum power if there is a cutoff potential greater than one half of the open-circuit potential?

8.4 You have a battery with a configuration of 50S-3P. The specifications for individual cells are open-circuit potential 3.1 V, an internal resistance of 2 mΩ, and a capacity at 1C of 7 A·h. Assuming the cells are ohmically limited, what is the maximum power that can be achieved? If individual cells have a cutoff voltage of 2.75 V, what is the maximum power? What must the value of the external resistance be to ensure that less than 5% of the available power is lost because of the connections?

8.5 The Tesla Model S electric car uses 6831 cells for the battery. The cells are the so-called 18650 (cylindrical, 18 × 65 mm) lithium-ion cells, each with a capacity of 3.1 A·h and a nominal voltage of 3.6 V. What is the capacity in kW·h for the battery? How might you configure these cells to make a pack? Discuss the challenges with packaging the large number of cylindrical cells. A battery voltage of 300–400 V is generally desired.

8.6 The nominal design of a 20 A·h cell is shown below. The tabs for current collection are on the same side and the dimensions of the cell ($L \times W \times H$) are $140 \times 100 \times 15$ mm³. Alternative designs are being considered.

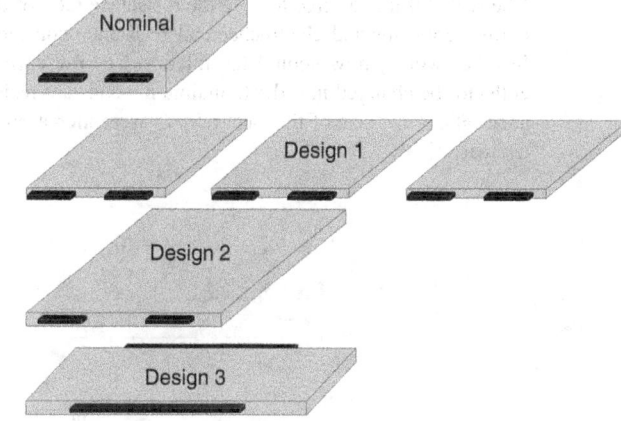

Option	Capacity [A·h]	L [mm]	W [mm]	Tabs
Nominal	20	140	100	Narrow, same side
1	6.67	140	100	Narrow, same side
2	20	200	140	Narrow, same side
3	20	250	120	Opposite sides, wide

(a) Assuming that the electrodes and current collectors are unchanged and that the thickness of the current collector is small relative to the cell thickness, what are the cell thicknesses of the alternate designs?

(b) Discuss the advantages and disadvantages of these alternatives. For option 1, three cells are required to keep the capacity the same. Consider the following in your answer:

   i) heat removal
   ii) uniformity of current density across the planform
   iii) rate capability, resistance of current collectors and tabs

8.7 L is the characteristic dimension of the electrode, δ is the thickness of the current collector, and σ is its electrical conductivity. The width of the electrode, perpendicular to the section illustrated below, can also be assumed to have a length equal to L so that the area of the electrode exposed to the electrolyte is L^2. Show that if the current density over the electrode is constant (i_y), the resistance to current flow in the current collector is $\frac{1}{2\sigma\delta}$. This "average" resistance is defined as the total current that enters the current collector divided by the voltage drop across the current collector. Assume that electrical connection to the current collector is made at $x = L$, and treat current flow in the current collector as one-dimensional. Under these conditions, $i_x = i_y \frac{x}{\delta}$. How should δ be scaled if it is desired to keep the resistance ratio of the current collector and electrochemical resistance constant? In other words, how should the thickness of the current collector be changed in order to maintain a constant resistance ratio if the size of the electrode, L, were increased or decreased?

8.8 For current collectors in lithium-ion cells, copper foil is used for the negative electrode and aluminum foil for the positive. Often the aluminum foil is about 1.5 times as thick as the copper. Why is this done?

8.9 Problems 7.3 and 7.4 examined the LiI battery that is used for implantable pacemakers. There is also a desire to have a defibrillator capability, where a short high pulse of power is occasionally required. Recall that the separator is formed in place from the overall reaction. Assume that all of the resistance is in the separator. At 37 °C, the LiI electrolyte has an ionic conductivity of 4×10^{-5} S·m^{-1}. The open-circuit potential of the LiI cell is 2.8 V. What is the maximum thickness of LiI allowed so that the power requirements for the defibrillator (3 W) can be met? LiI has a density of 3494 kg·m^{-3}. The cell area is 13 cm^2. Is it feasible to redesign this LiI cell to include the defibrillator feature and still meet the energy and life requirements? Explain why or why not.

8.10 Using data from Figure 8.12, determine charging and discharging resistance of the cell. The answer should be in $\Omega \cdot m^2$. Compare these values with the ohmic resistance of the same cell. Discuss why the values are different.

8.11 For a 125 A·h lead–acid cell, after a discharge from 100 to 20% SOC, 119 A·h was required to restore the cell to a full state-of-charge. Calculate the charge coefficient. How might this charge coefficient change with the rate of discharge? Temperature?

8.12 During charging of a lithium-ion battery, lithium ions are transported to the negative electrode, where they are reduced and then intercalate into the graphite active material. One limitation on the rate of charging is the concentration of lithium at the interface. If the rate is too high, then lithium metal can plate, which is a dangerous situation. This level of lithium is sometimes referred to as the saturation level.

(a) Qualitatively sketch the concentration profile of lithium in the electrolyte and in the graphite. How do these profiles change with the rate of charging?

(b) Discuss differences that correspond to the following charging protocols: (1) constant current density of 20 A·m^{-2} until the saturation level of lithium is reached, and (2) repeated pulses of charging at 25 A·m^{-2} for 3 seconds followed by a lower rate of 5 A·m^{-2} for 1 second.

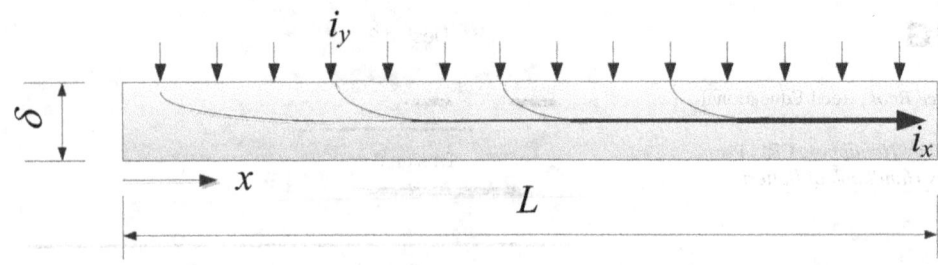

8.13 Lithium-ion batteries have self-discharge rates of 1–2% per month. If two adjacent cells in a long string connected in series have rates of self-discharge of 1 and 2% per month, respectively, and the battery is fully charged each month, how long before the SOCs of these two cells vary by 5%? The rate of self-discharge, however, can be as high as 5% in the first 24 hours. If the initial rates of self-discharge for the two cells are 3 and 5%, respectively, how does the answer change? What role would the battery management system play in this scenario?

8.14 A cylindrical cell is fabricated by winding the electrodes around a shaft with a diameter of 2 mm. The length of the winding is 1.8 m and its thickness is 0.5 mm. If the rate of heat generation is 50 kW·m^{-3} and the effective conductivity in the radial direction is 0.15 W·m^{-1}·K^{-1}, what is the maximum temperature of the cell at steady state?

8.15 Rather than specifying the temperature at the outside of the cell as was done in Section 8.9, in practice heat is removed by forced convection. What is the appropriate boundary condition? Use h for a heat-transfer coefficient and T_∞ for the temperature of the fluid. Solve the differential equation to come up with an equation equivalent to (8.28). In general, would liquid or air cooling be more effective? Why?

8.16 The analysis in Section 8.9 is for a cylindrical cell. Develop a similar analysis for a prismatic cell. Assume that all the heat is removed from the top and bottom of the cell (i.e., assume there is no heat loss from the sides). Furthermore, heat is removed from the bottom of the cell using convection through a cold plate (h_c, T_c) and from the top to ambient, also by convection (h_a, T_a).

8.17 Electrodes for lithium-ion cells are made by coating active material onto both sides of metal foils. The coated electrodes are then wound with a polymer separator or stacked to form a cell sandwich. Given the dimensions and physical properties in the following table, calculate the effective conductivity in the in-plane ($\parallel$) and through-plane ($\perp$) directions for a prismatic cell sandwich. What are the implications for the design of a cell and heat removal? Use the following formulas:

$$k_\perp = \frac{\sum t_i}{\sum t_i/k_i} \quad \text{and} \quad k_\parallel = \frac{\sum k_i t_i}{\sum t_i}.$$

Component	Thickness, t_i [µm]	Conductivity, k_i [W·m^{-1}·K^{-1}]
Al current collector	45	238
Positive electrode material	66	1.5
Cu current collector	32	398
Negative electrode material	96	1.0
Separator	50	0.33

8.18 Negative feedback was mentioned in Section 8.9 as an important feature in preventing thermal runaway of cells. For the following instances, explain the role of feedback in either reducing or exacerbating temperature variations.

(a) Because of ohmic resistance of the current collectors, the current density over the electrode is not uniform. The current density is higher near the tabs.

(b) During normal discharge of a single cell, heat removal is not uniform and a local temperature rise occurs at one spot in the planform of a cell.

(c) For valve-regulated lead–acid battery thermal runaway is a possibility during float charge at a fixed voltage. In these cells, rather than being vented, oxygen generated at the positive electrode is directed to the negative electrode where it reacts exothermically. Up to 90% of the current during float charge goes to this oxygen recombination.

8.19 Consumers desire to charge their electric vehicles as quickly as a conventional car can be refueled. If time for refueling with gasoline is about 2 minutes, at what C-rate would the battery need to be recharged in the same period? For a 50 kWh battery, what power corresponds to this rate? Frequently, researchers report extremely high rates of charge and discharge for tiny experimental cells, typically with very thin electrodes. What challenges exist in translating these results to a full-sized electrical vehicle battery?

8.20 A coin cell is prepared by sealing layers of material in a can. A spring washer is included to apply a controlled force on the assembly. Using the data for the layers below, if the final thickness of the sandwich is 2.4 mm, what is the normal stress? The spring has an initial thickness of 2 mm and a spring constant of 120 kN·m^{-1}. The diameter of the layers is 1.4 mm.

Cell component	Initial thickness [mm]	Elastic modulus [GPa]
Anode spacer	0.50	210
Anode	0.070	15
Separator	0.025	1
Cathode	0.070	70
Cathode spacer	0.50	210

8.21 Derive Equation 8.29. (*Hint:* How might you express the volume of the wound and unwound cell?)

Supramaniam Srinivasan

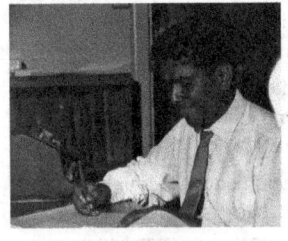

Supramaniam Srinivasan was born on August 12, 1932, in Sri Lanka. After receiving his Bachelor of Science degree in Chemistry with Honors from the University of Ceylon in 1955, he went to the University of Pennsylvania to study electrochemistry with John O'Mara Bockris, an innovator in physical electrochemistry (also the mentor for Brian Conway, Chapter 11). His thesis was entitled "Mechanism of Electrolyte Hydrogen Evolution: An Isotope Effect Study." "Srini", as he was called by his friends and colleagues, continued to work in Bockris' laboratory as a postdoc for a few years after completing his Ph.D. in 1963. He and Bockris coauthored *Fuel Cells: Their Electrochemistry*, which was published in 1969.

Srini next went to the Downstate Medical Center, State University of New York, Brooklyn, where he applied electrochemical techniques to medical problems. He subsequently took a position at Brookhaven National Laboratory. There, he led a hydrogen energy technology group and resumed his studies of fuel cells. His research covered a wide range of fuel-cell types: alkaline, phosphoric acid, and solid oxide, as well as electrolysis and electrochemical energy storage. Srini was instrumental in the formation of the Energy Technology Division of the Electrochemical Society, which emerged in response to the Energy Crisis of the 1970s. Srini continued to nurture and support the Energy Technology Division throughout his career. He moved to Los Alamos National Laboratory (LANL) in 1981 and led a group focused on proton-exchange membrane fuel cells (PEMFCs). This group published a series of influential articles in the late 1980s that helped to spur the renaissance of PEMFC technology. Although first used in the Gemini space program in the 1960s, PEMFCs were displaced by alkaline fuel cells for manned space missions of the United States. Two key developments, the introduction of more stable polymers and the development of electrodes with low loadings of precious metals capable of high power, changed the landscape. In 1988 Srini left LANL to serve as Deputy Director of the newly formed Center for Electrochemical Systems and Hydrogen Research (CESHR) at Texas A&M University. After retiring from Texas A&M, Srini continued to pursue his passions, but now at Princeton University where he was a Visiting Research Collaborator at the Center for Energy and Environmental Studies.

Those who worked with Srini were inspired with his tireless dedication to electrochemistry and the energy he devoted to the development of young engineers and scientists. Dr. Srinivasan was a Divisional Editor of the *Journal of the Electrochemical Society* for a little more than a decade (1980–1991). Srini was presented with the Research Award of the Energy Technology Division in 1996 and named a Fellow of the Electrochemical Society in 2001. Srini died in 2004, and *Fuel Cells: From Fundamentals to Applications* was published posthumously in 2006. In 2011, the Energy Technology Division that he held dear established a Young Investigator Award in his honor. This annual award recognizes and rewards an outstanding young researcher in the field of energy technology.

Image Source: Reproduced with kind permission of Dr Mangai Srinivasan.

Chapter 9

Fuel-Cell Fundamentals

Fuel cells are energy conversion devices. In contrast to batteries, fuel cells do not store energy, but are used to convert the chemical energy of a fuel directly into electricity. This chapter will help you to gain a general understanding of the operation of fuel cells and an appreciation for the role that key components, such as the electrolyte, play with respect to fuel-cell design and operation. An important objective of the chapter is to guide you to a detailed understanding of the current–voltage relationship, or polarization curve, for a fuel cell. There are many varieties of fuel cells. After introducing some common types and providing the principles of fuel-cell operation, we will take a more in-depth look at two types of fuel cells: the proton-exchange membrane fuel cell (PEMFC) and the solid oxide fuel cell (SOFC). The concepts learned here will be applicable to other fuel-cell systems that an electrochemical engineer may encounter. The principal reason for considering fuel cells for energy conversion is efficiency. Therefore, a system-level treatment of fuel-cell systems is critical. After examining the more fundamental electrochemical aspects in this chapter, broader systems implications will be considered in Chapter 10.

9.1 INTRODUCTION

Operation of a methanol fuel cell is illustrated in Figure 9.1. This particular fuel cell uses a proton-exchange membrane separator in which the ionic current is carried by protons. In contrast to batteries, where we were careful to use the terms negative and positive for electrodes rather than anode and cathode, the terms anode and cathode are appropriate and unambiguous when applied to fuel cells. In fact, while either set of terms is acceptable, anode and cathode are more commonly used. At the anode (negative electrode), methanol is oxidized to produce carbon dioxide and protons. The protons move toward the cathode (positive electrode) through the electrolyte. At the cathode, oxygen is reduced to form water. Electrons move through an external circuit from the anode to the cathode:

$CH_3OH + H_2O \rightarrow CO_2 + 6H^+ + 6e^-$	anode
$O_2 + 4H^+ + 4e^- \rightarrow 2H_2O$	cathode
$CH_3OH + 1.5O_2 \rightarrow CO_2 + 2H_2O$	overall

The overall reaction is the reaction of methanol and oxygen to form water and carbon dioxide, which is the same reaction that would have resulted from burning the methanol. The increased efficiency of the fuel cell comes from using the fuel directly to produce electrical work rather than combusting the fuel in a thermal process, such as is done in an internal combustion engine.

Methanol is just one of a number of possible fuels, the most common of which is hydrogen gas. The fuel cell in Figure 9.1 is known as a *direct* methanol fuel cell because the fuel is oxidized at the anode without first being converted to hydrogen by a chemical process called *reformation*. Figure 9.1 also identifies a number of basic components used in fuel cells. These components and their functions will be discussed in this and the next chapter. In contrast to a combustion process that only releases heat, the conversion of methanol to carbon dioxide and water occurs in the fuel cell through two separate electrochemical reactions. As a consequence, fuel cells are not thermal devices and are not limited by the Carnot efficiency of heat engines.

Thermodynamics (Chapter 2) informs us about the equilibrium potential of the half-cell reactions and overall cell, as well as the enthalpy of the overall reaction. The equilibrium potential, U, is related to the change in Gibbs energy according to Equation 2.3. For the methanol fuel cell at standard conditions,

$$U^\theta = \frac{-\Delta G^o}{nF} = \frac{702,3306}{6F} = 1.21 \text{ V}.$$

A second important thermodynamic quantity is the change in enthalpy associated with the reaction. For the methanol reaction, this change in enthalpy at standard conditions is

---

Electrochemical Engineering, First Edition. Thomas F. Fuller and John N. Harb.
© 2018 Thomas F. Fuller and John N. Harb. Published 2018 by John Wiley & Sons, Inc.
Companion Website: www.wiley.com/go/fuller/electrochemicalengineering

196 Electrochemical Engineering

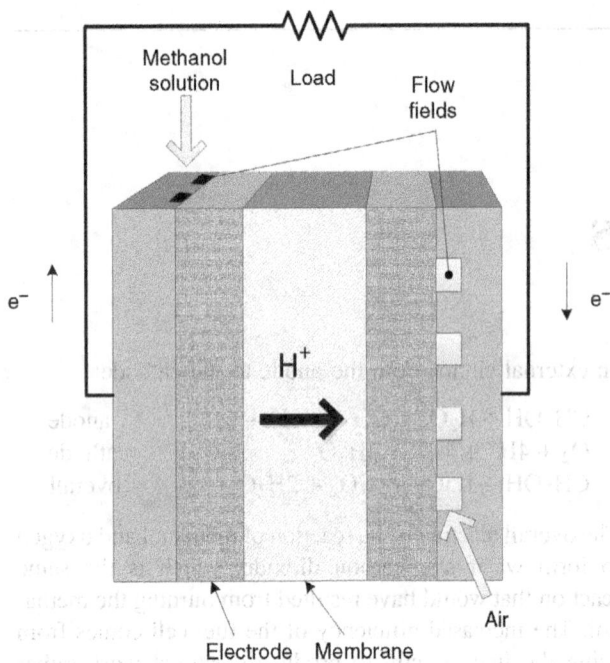

Figure 9.1 Direct methanol fuel cell that uses an acid electrolyte.

calculated with the data from Appendix C:

$$\Delta H^\circ_{f,CO_2} + 2\Delta H^\circ_{f,H_2O} - \Delta H^\circ_{f,CH_3OH} = -726.5 \text{ kJ mol}^{-1}.$$

In calculating both ΔG° and ΔH°, liquid water and liquid methanol were assumed. Since the overall reaction is the same as the combustion of methanol, we expect the reaction to be highly exothermic with a change in enthalpy that is a large negative value. Let's use the first law of thermodynamics to examine the fuel cell, as depicted in Figure 9.2. For an open system, the change in enthalpy is equal to the heat transfer to the system, Q, minus the electrical work done by the system:

$$\Delta H = Q - W. \qquad (9.1)$$

Note that the signs for Q and W depend on an arbitrary convention; here, positive indicates heat transfer to the system and work done by the system. ΔG, and therefore U, represent the maximum electrical work that can be obtained from the cell corresponding to a reversible process occurring at an infinitesimally slow rate. The electrochemical engineer is interested in designing a practical device where rate, size, and cost are important considerations. Much of this chapter will focus on how the rate (think of current density) affects the useful electrical work that can be extracted. Chapter 10 will expand this discussion and apply the principles to more complete systems. Mass and energy balances will illustrate many key ideas that will be addressed in more detail in the subsequent sections.

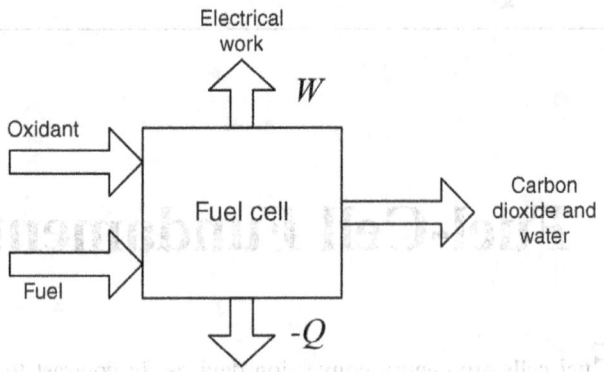

Figure 9.2 Fuel cells are open systems with reactant streams entering and exiting the fuel cell.

The efficiency of the fuel cell can be expressed as the electrical work produced divided by the energy available in the reactants. In contrast to the typical battery, fuel cells are not cycled but operate indefinitely with the continuous flow of fuel and oxidant. Thus, our analysis will involve rates of energy (power). Assuming that we provide $\dot{m}_i$ [kg·s^{-1}] of reactants, the voltage efficiency of a fuel cell is

$$\eta_V^{fc} = \frac{\text{rate of electrical energy from cell}}{\text{maximum rate of electrical energy}}$$
$$= \frac{\text{electrical power from cell}}{\text{theoretical power from cell}} \qquad (9.2)$$
$$= \frac{IV}{\left(\dfrac{\dot{m}_i}{M_i}\right)(-\Delta G_{rx})} = \frac{IV}{\left(\dfrac{I}{nF}\right)(nFU)} = \frac{V}{U}.$$

We have used both Equation 2.3 and Faraday's law to eliminate the current. The net electrical work done by the cell is simply the cell current multiplied by the operating potential of the cell. The efficiency we have defined, η_V^{fc}, is a type of voltage efficiency. Here, we have presumed that the reactants are supplied in exactly stoichiometric quantities. Clearly, we'll want the cell voltage to be as high as possible in order to maximize efficiency. Thus, minimization of ohmic, kinetic, and mass-transfer losses in the cell is desired, just as was with batteries. Compare this definition with the voltage efficiency used for batteries. Why are they different?

ILLUSTRATION 9.1

A hydrogen oxygen fuel cell is operating at 0.7 V. What is the fuel-cell voltage efficiency? You may assume standard conditions.

Solution:

Assuming that the hydrogen and oxygen are supplied in exactly stoichiometric amounts and at standard conditions at 25 °C, the change in Gibbs energy is given by Equation 2.3. For the H$_2$/O$_2$ fuel cell this value is 1.229 V. The efficiency of this specific fuel cell is therefore

$$\eta_V^{fc} = \frac{0.7}{1.229} = 0.57$$

or 57%. In general, efficiency is a function of the specific operating conditions, which differ from the standard state.

We'll see in Chapter 10 that the analysis of efficiency done here, while appropriate from a fundamental perspective, is not adequate for the characterization of fuel-cell systems. For example, it does not account for the fraction of the current or power from the fuel cell that goes to operating ancillary equipment, such as a blower to supply air; this power is subtracted from the gross power, IV, to get the net electrical power produced. Also, by convention, the change in enthalpy (ΔH) is often used as the benchmark rather than the change in Gibbs energy. This convention is rooted in history more than science. The analysis of efficiency for fuel cells is explored further in Chapter 10. In this chapter we next describe several types of fuel cells. We will then look in detail at the polarization curve from a fuel cell.

9.2 TYPES OF FUEL CELLS

There are many different kinds of fuel cells, which are distinguished primarily by the electrolyte used and also by the type of fuel that is consumed. The most common fuel is hydrogen gas; hydrocarbons are frequently reformed to produce hydrogen for consumption in the fuel cell. Fuels other than hydrogen can also be used directly, such as in the methanol fuel cell discussed earlier. The electrolyte is important because, more than anything else, the nature of the electrolyte determines the structure of the electrode, the design of the individual cell, how individual cells are combined to form the cell-stack assembly (CSA), and even the architecture and control of the fuel-cell system. In all instances, the electrolyte should have a high ionic conductivity and a low electronic conductivity. It should also provide a reasonably good barrier to the transport of reactants between electrodes. Referring back to Figure 9.1, without a barrier the methanol and oxygen would mix and combust on the electrocatalyst, releasing heat rather than undergoing separate electrochemical reactions to produce electrical work.

Some of the more common fuel-cell types are identified in Table 9.1. Note that the temperature of operation ranges from room temperature up to 1000 °C. When evaluating a new fuel cell, understanding the physical and chemical properties of the electrolyte is a natural place to begin with. The choice of electrolyte and temperature of operation are closely linked. The properties of the electrolyte influence the temperature of operation, the electrocatalysts selected, the approaches to water and thermal management, and the system design. Ultimately, these electrolyte properties also drive the selection of materials.

Each type of fuel cell has its own advantages and limitations. For example, a low-temperature fuel cell may offer the advantages of rapid start-up and shutdown, as well as the possibility of using a broader set of materials for fuel-cell construction. However, such fuel cells may be

Table 9.1 Types of Fuel Cells

Fuel-cell type	Main application	Operating temperature [°C]	Comments
Direct methanol	Portable power	25–90	Uses same membrane as PEM FC
Proton-exchange membrane (PEMFC)	Automotive, buses portable	60–90	Tolerant to carbon dioxide in air Requires precious metal catalysts Rapid start-up and shutdown
Alkaline (AFC)	Space	80–100	Requires pure hydrogen and oxygen Nonprecious metal catalysts possible
Phosphoric acid (PAFC)	Stationary, combined heat and power	180–220	Operates on reformed fuels Long life Some cogeneration possible
Molten carbonate (MCFC)	Stationary, combined heat and power	600–650	High efficiency Good cogeneration
Solid oxide (SOFC)	Stationary, combined heat and power	650–1000	High efficiency High temperature limits materials available and makes thermal cycles challenging

Table 9.2 Anode and Cathode Reactions for Different Types of Fuel Cells Operating on Hydrogen and Oxygen

Fuel-cell type	Anode	Cathode
Alkaline	$H_2 + 2OH^- \rightarrow 2H_2O + 2e^-$ (−0.828 V)	$O_2 + 2H_2O + 4e^- \rightarrow 4OH^-$ (0.401 V)
Acid	$H_2 \rightarrow 2H^+ + 2e^-$ (0 V)	$4H^+ + 4e^- + O_2 \rightarrow 2H_2O$ (1.229 V)
Oxygen conducting ceramic, solid oxide	$O^{2-} + H_2 \rightarrow H_2O + 2e^-$	$O_2 + 4e^- \rightarrow 2O^{2-}$

challenged to find the electrocatalysts needed to carry out the oxidation and reduction reactions effectively at the lower operating temperature. The appropriate type of fuel cell depends on the application, and the references available at the end of the chapter provide information on fuel cells suitable for a wide variety of applications. For our purposes here, we narrow the focus of the rest of this chapter to fuel cells that use hydrogen as the fuel. Natural gas, higher alcohols, and a variety of hydrocarbons can also serve as fuels and will be treated in the next chapter. While it is possible to use other oxidants, oxygen is used almost invariably, and air is likely the source of the oxygen.

Let's now consider the operation of hydrogen fuel cells in more detail. Operating with hydrogen as the fuel, Table 9.2 shows the electrode reactions for three electrolytes: alkaline, acid, and an oxygen-conducting ceramic. Note that the half-cell reactions are different in different electrolytes. However, the overall reaction obtained from each pair of half-cell reactions is the same, namely, the reaction of hydrogen and oxygen to form water:

$$2H_2 + O_2 \rightarrow 2H_2O$$

Additionally, the half-cell potentials sum to the same value, 1.229 V, under standard conditions. Nonetheless, the operation and design of the fuel cell depend dramatically on the nature of the electrolyte. For example, while hydrogen always reacts at the anode and oxygen at the cathode, water can be produced at either the anode or the cathode, depending on the electrolyte. The current carrier is different for each of the three electrolytes considered. In the acid case, the ionic current is conducted by protons that move from the anode to the cathode. In the case of the alkaline cell, hydroxyl ions carry the current in solution and move in the opposite direction (cathode to anode), although the direction of the ionic current does not change. Finally, for the oxygen-conducting ceramic, oxygen ions move from the cathode to the anode. In all circumstances, electrons flow through the external circuit from anode to cathode.

In spite of important differences, such as those just mentioned, the electrochemical principles that govern the operation of fuel cells are much the same. The application of those principles is explored in the next section.

9.3 CURRENT–VOLTAGE CHARACTERISTICS AND POLARIZATIONS

A key objective in this chapter is to gain a detailed understanding of what is known as the *polarization curve*. This curve represents the steady-state relationship between the potential of the cell, V_{cell}, and its current density, i. It is typically measured experimentally. During this measurement, the temperature and pressure are held constant, and the flow of reactants is either fixed or proportional to the current density. Figure 9.3 shows example curves for three types of fuel cells: AFC, PEMFC, and SOFC. As we expect for a galvanic cell, the potential decreases as the current density increases. The current–voltage characteristics of a fuel cell can be understood with use of the principles that we have already learned, namely, thermodynamics, kinetics, ohmic losses, and mass transfer. It turns out that, to a good approximation, each of these determines the behavior of a part or section of the polarization curve as follows: (i) open-circuit potential (thermodynamics), (ii) low-current behavior (kinetics), (iii) behavior at moderate currents (ohmic losses), and (iv) the behavior at high currents (mass transfer).

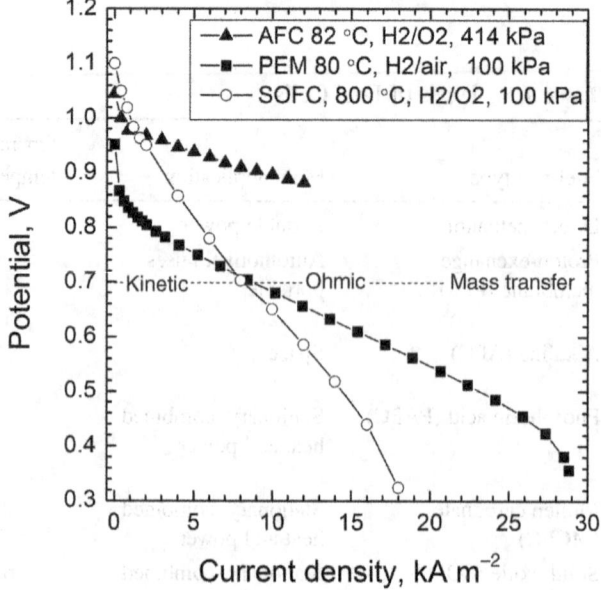

Figure 9.3 Example polarization curves for PEM, AFC, and SOFC.

Thermodynamics

Looking at Figure 9.3, we see that the potentials at open circuit (zero current density) vary. Also, note that none of the curves reaches 1.229 V (the equilibrium value found in Appendix A) at zero current. There are several factors that can explain these data. First, let's consider the impact of temperature, which is different than 25 °C for all three cases. The influence of the temperature on the equilibrium potential can be estimated by integrating Equation 2.16b. Assuming that the change in enthalpy does not vary with temperature, the resulting expression is

$$U^o(T) \approx U^\theta(T_\theta)\frac{T}{T_\theta} + \frac{\Delta H^\theta}{nF}\left\{\frac{T}{T_\theta} - 1\right\}, \quad (9.3)$$

where the temperature ratios must be in absolute units. Figure 9.4 shows the equilibrium potential, U^o, as a function of temperature as given by Equation 9.3. Also included are the results of a more accurate calculation that does not assume a constant enthalpy change and is needed for increased fidelity at high temperatures. This graph was created assuming that the product water is formed as a vapor rather than a liquid, so the standard potential at 25 °C is 1.184 instead of 1.229 V. The equilibrium potential is 1.171 V at 80 °C, and drops to 0.9794 V at 800 °C.

Let's compare the results of these calculations with the open-circuit potentials observed in Figure 9.3 for each type of fuel cell. Starting with the SOFC data, we see that the potential at zero current is actually higher than the value calculated at 800 °C. How can these data be explained? In addition to the influence of temperature on the equilibrium potential, the effect of gas composition must be considered.

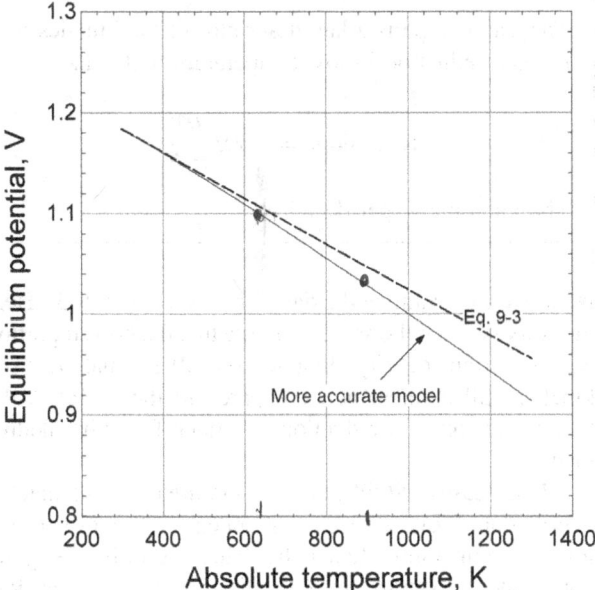

Figure 9.4 Effect of temperature on the equilibrium potential of a hydrogen/oxygen fuel cell. Water is assumed to be a vapor.

Using the methodology of Chapter 2, we arrive at the following equation for a hydrogen–oxygen fuel cell, where the activity has been retained:

$$U = U^0 + \frac{RT}{2F}\ln\frac{a_{H_2}a_{O_2}^{0.5}}{a_{H_2O}}. \quad (9.4)$$

Note that the activity correction term is also a function of temperature. The equilibrium potential, U^0, is at the operating temperature. The nature of the electrolyte does not affect this equation, which applies to all three types of fuel cells. At 100 kPa, the activity of each gas species can be taken simply as its mole fraction, and the equilibrium potential can be calculated. For the SOFC data, the anode gas stream is 94% hydrogen and 6% water. The cathode is pure oxygen. Thus, at 800 °C

$$U = 0.9794 + \frac{RT}{2F}\ln\frac{(0.94)(1)}{(0.06)} = 1.11\text{ V},$$

which is close to the value found in Figure 9.3. It is typical that for many high-temperature fuel cells, the open-circuit potential is close to the thermodynamic value.

For the alkaline system operating with pure oxygen, the mole fractions of water, hydrogen, and oxygen are all about 0.5. However, the pressure is above standard conditions ($p = 100$ kPa), and the activity for an ideal gas is $a_i = p_i/p^0$. Thus, at 80 °C and 414 kPa,

$$U = 1.171 + \frac{RT}{2F}\ln\frac{(2.07)\sqrt{2.07}}{(2.07)} = 1.172\text{ V}.$$

In this case, the OCV is a bit less than the thermodynamic value. The reduced value can be explained by permeation of oxygen and hydrogen across the separator (see Problem 9.16). Finally, we examine the PEM fuel cell. Here, the cell is operating on air. Both the air and fuel are humidified, containing 47 mol% water vapor. Thus, the mole fractions are roughly $y_{H_2O} = 0.47$, $y_{H_2} = 0.53$, and $y_{O_2} = 0.21 \times 0.53 = 0.11$.

$$U = 1.171 + \frac{RT}{2F}\ln\frac{(0.53)\sqrt{0.11}}{(0.47)} = 1.156\text{ V}.$$

The data in Figure 9.3 for the PEMFC are well below this value. There are two important reasons for the observed difference in potential. As in the AFC example, there is some permeation of reactants across the electrolyte, which results in a mixed-potential and depression of the OCV. A second, more important, reason is the sluggishness of the oxygen reduction reaction in acid at low temperatures. The reaction is so slow that even minute impurities and contaminants can compete with the oxygen reduction reaction. Therefore, in low-temperature acid systems, the thermodynamic potential is difficult to achieve experimentally. To summarize, the open-circuit potential (zero current) is predicted theoretically by thermodynamics. However, quite often these thermodynamic values are not observed because of finite permeation of reactants through the separator, impurities, and side reactions. In particular, the reaction for oxygen in acid media at low temperatures is highly irreversible, and the thermodynamic potential is hard to achieve in practice. Finally, we note that the equilibrium potential is about 1 V for a fuel cell, independent of the chemistry. This uniformity is in contrast to batteries, where large variations in the equilibrium potential are observed for different chemistries.

Kinetics

As we know from Chapter 3, the kinetics of electrochemical reactions depends strongly on overpotential, catalysts, and temperature. We will focus initially on the low-temperature fuel cells: PEM and AFC. Referring to Figure 9.3, drawing a small current from these cells causes the cell potential to decrease rapidly, followed by a more gradual decline. An alternative way to examine these same data is with a *Tafel plot* as shown in Figure 9.5. There are two main differences between the polarization curve and the Tafel plot. First, the cell potential is plotted as a function of the logarithm of the current density. Second, the ohmic resistance is removed as discussed in Chapter 6. Figure 9.5 shows the original data for the PEMFC along with the iR-corrected data. At high current densities, mass-transfer effects are present; but at lower current densities, the iR-corrected curve clearly shows the kinetics and the corresponding surface overpotential. Generally, the kinetics of low-temperature fuel cells is described well by the Butler–Volmer equation. Recall that the oxygen reduction reaction (ORR) is highly irreversible, whereas the hydrogen oxidation is fast. Consistent with this, the kinetic polarization at the hydrogen anode is negligible, and we only need to consider the polarization at the oxygen electrode. With the slow ORR kinetics, a large overpotential is needed to achieve any appreciable current; therefore, the reaction operates in the Tafel regime and polarization losses vary linearly with the logarithm of current density. As a result, we can readily obtain basic kinetic information for the oxygen reduction reaction from the data as plotted in Figure 9.5. The slope of the curve can be measured and is described as the *Tafel* slope in V per decade (see Section 3.5 and Illustration 9.2). Often it is assumed that for ORR $\alpha_c \approx 1$, and the value of the Tafel slope is estimated as follows:

$$\text{Tafel slope} = \ln(10)\frac{RT}{\alpha_c F} = 2.303\frac{(8.314)353}{(1)96485} = 0.07\ \frac{\text{V}}{\text{dec}}.$$
(9.5)

By comparison, the data in Figure 9.5 indicate a value of 0.06 V/decade. Referring back to Figure 9.3, when plotted on a linear scale, the polarization curve is not straight. At low current densities, ohmic and mass-transfer polarizations are small and the cell potential is dominated by the reaction kinetics. Ohmic losses, which scale linearly with the current density, increase in relative importance as the current increases. Thus, strong curvature at low current densities is an indication of sluggish reaction kinetics.

The Tafel slope is a key descriptor of the kinetics of oxygen reduction in low-temperature fuel cells.

$$\text{Tafel slope} = 2.303\frac{RT}{\alpha_c F}.$$

The units are V per decade.

Next, we can examine the data for AFC (Figure 9.3). Here the story is much the same, but note that the bowed portion at low current density (kinetic control) is smaller. Tafel kinetics still applies; but, compared to the PEMFC, the kinetics for oxygen reduction are much faster in alkaline media.

In the case of SOFC, the temperature is so high that the overpotential for the reaction is much less important than for low-temperature fuel cells. There is only a slightly noticeable curvature in the polarization curve at low current densities. Use of the Butler–Volmer equation for solid oxide fuel cells is explored in Problem 9.20.

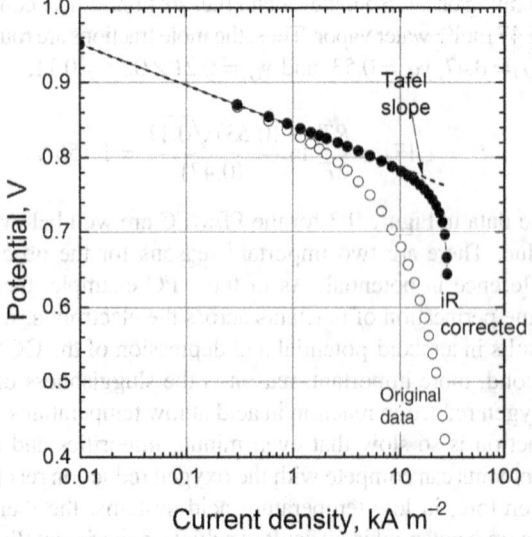

Figure 9.5 Tafel plot used for low-temperature PEM fuel cell from Figure 9.3.

ILLUSTRATION 9.2

If oxygen reduction kinetics are described by a Tafel equation that is first-order in oxygen pressure:

$$i = -i_0 \frac{p_{O_2}}{p_{ref}} \exp\left\{\frac{-\alpha_c F(V-U)}{RT}\right\},$$

estimate the change in potential for a one-decade change in current density assuming that the fuel cell is operating in the kinetic region at 25°C. Write the Tafel equation for i_1 and i_2, which are rearranged to give

$$\ln\left(\frac{i_2}{i_1}\right) = \frac{-\alpha_c F(V_2 - V_1)}{RT}.$$

If $\alpha_c = 1$, then for a change of 10 in the current density, the potential change is

$$\Delta V = V_2 - V_1$$
$$= -\ln(10)\frac{(8.314)(298\text{ K})}{96485} \times 1000 = -59 \frac{\text{mV}}{\text{decade}}.$$

This is equivalent to the Tafel slope calculation from Chapter 3. The slope is negative since it is a cathodic reaction and a lower voltage gives a higher current. In practice, only the magnitude of the Tafel slope is typically reported since the sign is understood from the reaction.

Ohmic Region

At moderate current densities, the importance of ohmic polarization increases compared to the activation polarization. Ohmic losses increase linearly with current, whereas kinetic losses are proportional to the logarithm of current density. The absolute magnitude of the kinetic polarization is still large for the PEMFC, but it increases much more slowly than the ohmic polarization at moderate current densities, resulting in a linear decrease in cell voltage with increasing current density. This linear region of the polarization curve is the *ohmic region*. Referring to Figure 9.3 again, we see that above a few thousand $A \cdot m^{-2}$, the polarization curves for all three cells are roughly linear. However, the slopes of the curves are not the same. This slope, which we will refer to as the ohmic resistance, is largely dependent on the conductivity and thickness of the electrolyte.

$$\text{slope} = \frac{\Delta V}{\Delta i} \approx \frac{L}{\kappa} = R_\Omega \; [\Omega \cdot m^2]. \qquad (9.6)$$

Thus, either decreasing the thickness of the separator or increasing its conductivity will reduce the slope of the polarization curve and lower ohmic losses. Using the example polarization curves from Figure 9.3, the resistance of this SOFC (0.04 $\Omega \cdot m^2$) is a little more than three times higher than the resistance of the PEMFC.

Mass Transfer

Finally, as the current density increases further, mass-transfer effects become important. Reactants and products are transported to and from the catalyst sites due to molecular diffusion and bulk fluid motion. For instance, consider oxygen at the cathode. Assuming that the source of oxygen is atmospheric air, the initial concentration of oxygen is only 21%, with the balance mostly N_2. Depending on the type of fuel cell, water vapor or other gases can be present. For instance, gases are saturated with water vapor in a typical PEM fuel cell, which at 80°C corresponds to almost 50% water vapor. As oxygen is consumed at the electrode surface, a concentration gradient develops between the bulk stream and the surface. In the extreme, the concentration of oxygen approaches zero at the surface and the limiting current is reached. Of the three examples shown in Figure 9.3, the PEMFC shows the most obvious effect of mass transfer near 25 $kA \cdot m^{-2}$, where the potential of the cell drops rapidly with current density. The SOFC appears to have small additional polarization due to mass transfer near 18 $kA \cdot m^{-2}$, but this is largely masked by the large ohmic polarization. Unfortunately, the data for the AFC do not extend much beyond 10 $kA \cdot m^{-2}$. However, the use of pure oxygen in the AFC will reduce the mass-transfer resistance relative to that of the PEMFC, which uses air.

Recall that the efficiency of a fuel cell is proportional to the potential of the cell as shown in Equation 9.1. Thus, the fuel cell is more efficient at low power or low current densities. This behavior is a key feature of fuel cells. In a well-designed system, operation in or near the mass-transfer region is avoided.

The polarization of SOFC and PEMFC are summarized pictorially in Figure 9.6. For the data in Figure 9.3, the operating potentials of the two cells are nearly the same, ~ 0.7 V, at a current density of 8 $kA \cdot m^{-2}$. The thermodynamic potentials are close too. However, the individual contributions to the overall polarization are quite different. It is important that the electrochemical engineer know the dominant source of polarization in order to determine the types of changes in cell design that are likely to improve performance. For instance, looking at Figure 9.6 and holding the operating potential at 0.7 V, we would conclude that the current density of the SOFC could be improved dramatically with a thinner electrolyte. In contrast, most of the polarization for the

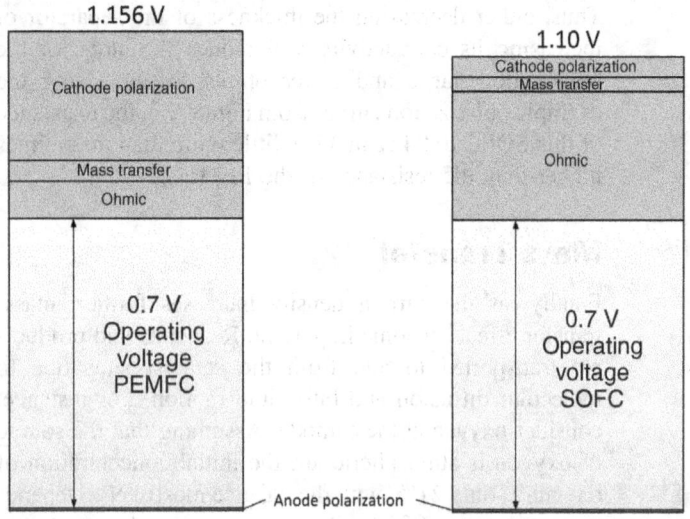

Figure 9.6 Polarizations of the PEM and solid oxide fuel cells shown in Figure 9.3 at a cell potential of 0.7 V.

PEMFC is at the cathode, and a thinner electrolyte would cause minimal improvement in current density at 0.7 V.

9.4 EFFECT OF OPERATING CONDITIONS AND MAXIMUM POWER

It is important to understand how changes in operating conditions influence the efficiency of and power available from a fuel cell. Two important variables that influence fuel-cell operation are temperature and reactant composition (partial pressure of gases). In this section, we examine the effect of these variables on the ohmic, kinetic, and concentration (mass-transfer) overpotentials that directly impact the operating voltage and efficiency of the fuel cell.

$$V_{cell} = U - iR_\Omega - |\eta_{s,anode}| - |\eta_{s,cathode}| \\ - |\eta_{conc,anode}| - |\eta_{conc,cathode}|. \quad (4.58a)$$

A thorough understanding of these coupled effects can be rather involved; our approach here is to simply provide some insight into important cause-and-effect relationships. Broadly speaking, increasing temperature lowers the equilibrium potential slightly and reduces the key overpotentials. Thus, if other factors are kept constant, higher temperature results in a higher cell potential and a more efficient fuel cell. When analyzing the effect of temperature and composition, it is important to have an understanding of the relative magnitudes of the major polarizations or losses in the system, Figure 9.6 for instance, in order to guide one's focus.

Let's start with ohmic losses. The major source of ohmic losses is the resistance of the electrolyte, $R_\Omega \approx L/\kappa$. Temperature has a strong effect on conductivity. The behavior of conductivity of the electrolyte with temperature can be complex. For our purposes, think of it as an activated process, and the electrical conductivity increases with temperature according to an Arrhenius relationship: $\kappa \propto \exp\{\frac{-E_{a,\kappa}}{RT}\}$, where $E_{a,\kappa}$ is the activation energy for the conduction process. Values of the activation energy can vary widely, but are on the order of a kJ mol^{-1}. Thus, as temperature increases, the conductivity of the electrolyte increases and the ohmic loss decreases. Pressure, on the other hand, has a negligible effect on the conduction process.

Temperature influences the kinetic overpotential in several ways, namely, through the exchange-current density, the equilibrium potential, and via the explicit temperature dependence in the exponential term of the kinetic expression (e.g., Butler–Volmer equation). Temperature may also cause the reaction mechanism to change, which can appear as a change in a parameter that is normally not a function of temperature. Perhaps the most important of these is the influence of temperature on the exchange-current density, i_0, as introduced in Chapter 3. This parameter represents the rate per area of the forward and reverse reactions at equilibrium. Since this parameter is associated with the energy barrier between reactants and products, it is not surprising that it also follows an Arrhenius relationship. Specifically,

$$i_o(T) = A \exp\left\{\frac{-E_{a,i}}{RT}\right\}. \quad (3.18)$$

The activation energy, $E_{a,i}$ does not have the same value as the one that describes the conductivity; the proportionality constant is also different. As shown in Equation 3.18, the exchange-current density increases with increasing temperature, the kinetics for the reaction improves, and the kinetic polarization decreases.

We also saw in Chapter 3 that the exchange-current density depends on the composition of reactants and products. When considering the effect of pressure on the kinetic overpotential, this relationship must be known. Based on the simple analysis shown in Chapter 3, we anticipate that the compositional dependence of the exchange-current density can be approximated as

$$i_o = i_{o,ref}\left(\frac{c_1}{c_{1,ref}}\right)^{\gamma_1}\left(\frac{c_2}{c_{2,ref}}\right)^{\gamma_2} \quad (3.20b)$$

Since many of the reactions of interest in fuel cells are complex, experimental data are usually needed to determine the exponents in the above equation. Alternatively, data can be used to determine in more detail the actual mechanism as a basis for an improved description of the composition dependence. Finally, we note that instances in which the kinetic polarization is significant or dominant are invariably connected with a large kinetic overpotential and can frequently be described with Tafel kinetics. This is true for the oxygen reaction at the air electrode in a PEM fuel cell where, even at very modest current densities, the kinetics are highly nonlinear and far from equilibrium. In contrast, the hydrogen electrode (anode) in a PEM fuel cell has rapid kinetics and, as a result, remains close to its equilibrium potential. Thus, for the hydrogen electrode, we can use the thermodynamic relationships from Chapter 2 to estimate changes in its potential with pressure.

The last overpotential to consider is that due to concentration polarization. If we consider the limiting current associated with gas transport by diffusion,

$$i_{lim} = nF\frac{D_i c_i}{\delta} = nF\frac{D_i p_i}{\delta RT}, \quad (9.7)$$

where δ is the thickness of the diffusion layer. For an ideal gas, the concentration is proportional to the partial pressure. This change in concentration is the dominant effect of pressure. Simply, as the partial pressure increases, the limiting current increases and the concentration polarization decreases. For a change in temperature, we can see that the gas-phase concentration decreases with increasing temperature as $1/T$. In contrast, the dependence of the diffusivity on temperature is greater than that of the concentration, and leads to an increase in the diffusivity with increasing temperature. The net result is a small increase in the limiting current with increasing temperature.

The above discussion can be summarized to provide the general guidelines shown in Table 9.3. Again, these are only general guidelines based on simplification of the actual coupled effects. In addition, there are factors that limit the temperature of operation and, therefore, counter the benefits of higher operating temperatures. The most important of these are materials considerations as most materials of construction are not thermodynamically stable in the fuel-cell environment. A common example is carbon, which is used extensively in low-temperature fuel cells. The standard potential for the corrosion of carbon is 0.207 V, well below typical operating potentials found in fuel cells. Therefore, carbon would be expected to oxidize under fuel-cell conditions. Fortunately, the poor kinetics allows for operation with carbon, but this becomes more

Table 9.3 Effect of Temperature and Pressure on Fuel-Cell Performance

Factor	Temperature	Pressure
U	$\Delta H < 0$ for fuel-cell reactions. Therefore U decreases with increasing temperature; effect is not large, Equation 2.18	$\approx U^\circ + \frac{RT}{2F}\ln\frac{\left(\frac{p_{H_2}}{p^\circ}\right)\left(\frac{p_{O_2}}{p^\circ}\right)^{0.5}}{a_{H_2O}}$
R_Ω	Resistance is strongly lowered as the temperature is increased. The conductivity is an activated process that increases exponentially with temperature: $\kappa = \kappa_o \exp\left\{\frac{-E_{a,\kappa}}{RT}\right\}$	Negligible effect
η_s	Exchange-current density is an activated process and increases exponentially with temperature: $i_0 = A\exp\left\{\frac{-E_{a,i}}{RT}\right\}$	Generally experimental data are needed to determine pressure dependence because reactions are not elementary
η_{conc}	Diffusion coefficient in the gas phase depends roughly on the temperature to the 3/2 power, but concentration decreases with temperature. Net result is a small reduction in mass-transfer polarization with temperature	Affects gaseous reactants, for an ideal gas $c_i = p_i/RT$ and $i_{lim} \sim c_i$. Thus, mass-transfer polarization decreases at higher pressure

and more difficult as the temperature increases. Similar to the effect of temperature, pressure is almost always going to help the performance of the cell because of better kinetics and reduced mass-transfer losses. However, it is not immediately clear if there is a net benefit to increasing pressure when the entire fuel-cell system is considered. This topic is discussed in Chapter 10.

Now that we have examined the effect of temperature and concentration on cell polarization, we turn to the relationship between the cell current, operating voltage, and the power available from the fuel cell. As you are aware, power (IV) increases with increasing current and cell voltage. However, the cell voltage and current are not independent; specifically, the cell voltage decreases with increasing current due to increased polarization losses. As a result, the power available from the cell is zero at both $I=0$ and at high currents where the entire equilibrium voltage is needed to drive the current, yielding a cell voltage of zero. Between these two extremes there exists a current and voltage at which the power is a maximum.

In order to estimate the maximum power, we note that the potential of a well-designed fuel cell changes linearly with current density over a relatively large range of current densities. The linear variation is due to ohmic losses, and the cell resistance can be determined from the slope of the current–voltage curve. As mentioned above, there may also be additional losses that do not vary significantly with current over the range of interest. These additional losses have been grouped together in a single overpotential, which is assumed to be constant, η_{other}. The value of this overpotential can be determined from the cell voltage at a given current, assuming that the equilibrium potential and the cell resistance are also known. This information can be used to estimate the maximum power as follows:

$$V \approx U - \eta_{other} - iR_\Omega. \quad (9.8)$$

Therefore, the power per unit area is

$$P = Vi = (U - \eta_{other})i - i^2 R_\Omega. \quad (9.9)$$

Taking the derivative of this power with respect to current density and setting it equal to zero allows us to determine the current density and voltage at maximum power:

$$V(\text{max power}) = \frac{(U - \eta_{other})}{2}, \quad (9.10)$$

$$i(\text{max power}) = \frac{(U - \eta_{other})}{2R_\Omega}, \quad (9.11)$$

$$P_{max} = \frac{(U - \eta_{other})^2}{4R_\Omega}. \quad (9.12)$$

Because $(U - \eta_{other})$ appears as a group, it is sometimes more convenient to determine its value directly from the data without the need to explicitly calculate U. Therefore, for a cell with an open-circuit potential of 0.9 V and a resistance of 0.10 m$\Omega\cdot$m^2, the maximum power is roughly 20 kW·m^{-2}. Later, we'll refine this calculation (Illustration 9.5). Nonetheless, Equation 9.12 shows the importance of low cell resistance in achieving high power from a fuel cell.

ILLUSTRATION 9.3

If the kinetics for oxygen reduction is described by the Tafel equation with a first-order dependence on the oxygen partial pressure:

$$i = -i_o \frac{p_{O_2}}{p_{ref}} \exp\left\{\frac{-\alpha_c F \eta_s}{RT}\right\}.$$

a. For a fixed current density, estimate the change in potential when operating on pure oxygen compared to air ($y_{O_2} = 0.21$). While your initial thoughts might be to consider equilibrium, the oxygen reaction is far from equilibrium during operation. Therefore, we write the Tafel expression twice, once for oxygen and once for air as the oxidant. We then set them equal to each other since the comparison is made at the same current density. The result is

$$\Delta\eta_s = \eta_{s,air} - \eta_{s,ox} = \frac{RT}{\alpha_c F} \ln\left(\frac{p_{air}}{p_{ox}}\right)$$

$$= \frac{(8.314)(353\,\text{K})}{(1)96485} \ln\left(\frac{0.21}{1}\right) \times 1000$$

$$= -47\,\text{mV},$$

where an operating temperature of 80 °C has been assumed. Therefore, the use of air changes the overpotential by 47 mV in the negative direction, which is an increase in the overpotential since the sign of the overpotential is negative for a cathodic reaction. In other words, the overpotential increases by becoming more negative.

b. Estimate the change in the hydrogen electrode polarization for reformate with 40% H$_2$ versus pure hydrogen. Given the fast kinetics of the hydrogen reaction, we can assume the electrode is at equilibrium and calculate the change in potential from the thermodynamic relations. The temperature is assumed to be 80 °C.

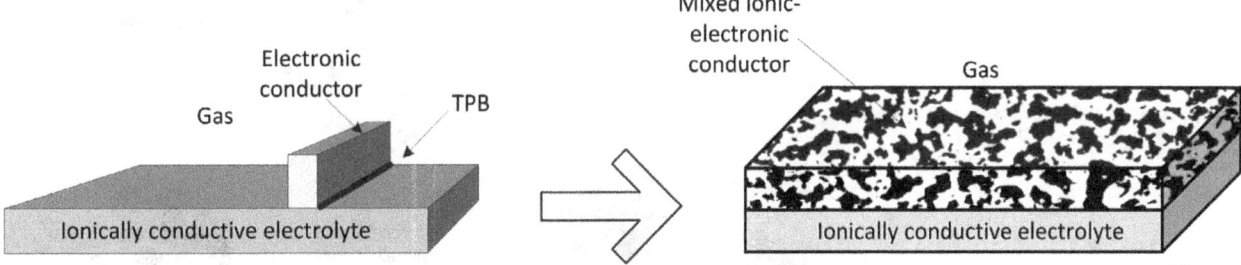

Figure 9.7 Porous, mixed electronic ionic conductor used to increase the triple-phase boundary.

From Chapter 2, assuming that the hydrogen concentration is the only thing that changes,

$$\Delta V = U_r - U_h = \frac{RT}{nF}\ln\left(\frac{p_h}{p_r}\right)$$
$$= \frac{(8.314)(353K)}{(2)96485}\ln\left(\frac{1}{0.4}\right) \times 1000$$
$$= 14 \text{ mV}.$$

The electrode potential increases (moves toward the O_2 potential). Therefore, the cell potential decreases.

9.5 ELECTRODE STRUCTURE

Electrode structures were introduced in Chapter 5. One of the distinguishing features of a fuel-cell electrode is that three phases are present. First, there is a solid phase that is electronically conductive to supply or remove electrons. Second, an electrolyte phase that conducts ions is needed. The electrolyte may be solid or liquid. So far, this arrangement is similar to that of a battery. The difference is that the reactants and products flow to and from the electrode to allow continuous operation. What's more, the reactants and products are distinct from the electrolyte, and invariably a gas phase is present in practical devices. Thus, contact between three phases is needed to carry out the electrochemical reactions. This region of contact is known as a *triple-phase boundary* (TPB). Of course, the intersection of three phases in three dimensions defines a line. As we have learned from Chapter 5 on porous electrodes, one key to good performance is to achieve a high surface area for reaction. In fact, from their inception, this precise challenge for fuel cells was identified by Grove (he is credited with the discovery of the fuel cell in 1839), which he referred to as the "notable surface action." Let's look at two examples of fuel-cell electrode structures that are widely used to get around the difficulties of bringing three phases together while providing adequate surface area for the reactions.

Let's first examine the SOFC. Shown on the left in Figure 9.7 is the line that defines the intersection of the three phases. The reaction is limited to a very small region (TPB) around the line, and only very small currents can be drawn from this cell.

$$\underset{\text{electrolyte}}{O^{2-}} + \underset{\text{gas}}{H_2} \rightarrow \underset{\text{gas}}{H_2O} + \underset{\text{electronic conductor}}{2e^-}$$

A common approach for solid oxide fuel cells is to make a porous electrode that is both ionically and electronically conductive. This region is called a mixed ionic–electronic conductor (MIEC). More specifically, it is a composite of two materials, but still porous so that the reactant gas can enter the region. Typically, the effective conductivities for the electronic and ionic phases are within a couple orders of magnitude of each other. With this composite material, the reaction zone goes from a line to a three-dimensional region where the reactions can occur throughout the region, which is electrochemically active.

The second example is used in cells with liquid electrolytes (Figure 9.8):

$$\underset{\text{gas}}{H_2} + \underset{\text{electrolyte}}{2OH^-} \rightarrow \underset{\text{gas}}{2H_2O} + \underset{\text{electronic conductor}}{2e^-}$$

On the left, imagine we have a planar electrode partially immersed in aqueous KOH. The region indicated at the surface of the liquid is the only place where all three phases are present. Because there is some small solubility of oxygen in the electrolyte, the reaction zone is not strictly just the line; even so, the performance of such a system would be poor. The *flooded-agglomerate model*, introduced in Chapter 5, describes the approach to overcome this limitation by taking advantage of the solubility of the gases. On the right, small agglomerates of carbon and catalyst are seen that form an electrode structure. The particles are in intimate contact ensuring electronic conductivity, and at the same time a porous structure is formed to allow gas access. The microporosity of the agglomerate is filled with electrolyte to form a continuous ionic path

206 Electrochemical Engineering

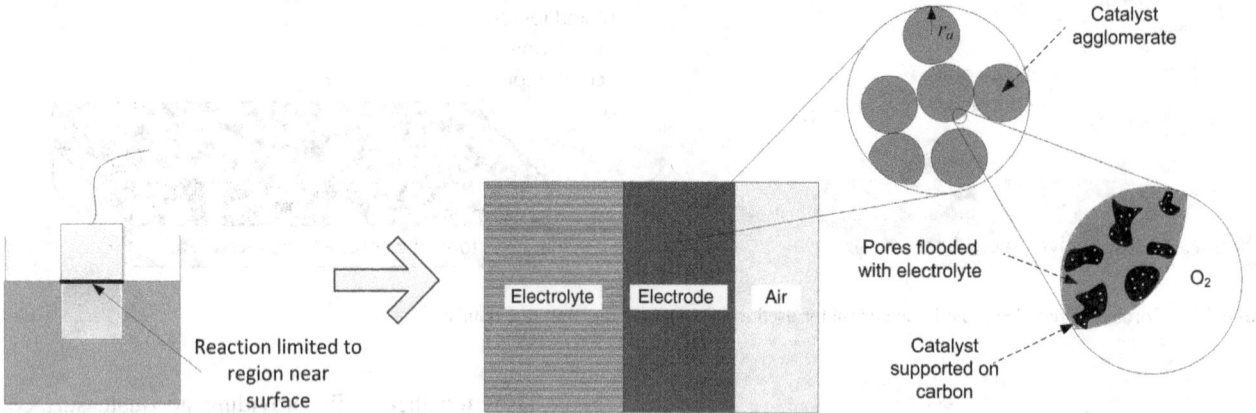

Figure 9.8 Flooded-agglomerate electrode, which is designed to increase the TPB.

through the electrode. The agglomerates are made small enough so that oxygen can dissolve and diffuse through the agglomerate, extending the reaction zone through the diameter of the agglomerate.

About 100 years after Grove's discovery, Francis Bacon took up the hydrogen/oxygen fuel cell. Bacon, the father of the modern fuel cell, made seminal contributions to improving the electrode structure. Previously, the electrochemical reactions had been limited to a small interfacial area, resulting in low limiting currents because of the poor access of gases to the catalysts. Bacon developed alkaline fuel-cell electrodes with controlled porosity and almost immediately obtained current densities in excess of $10 \, kA \cdot m^{-2}$ at 0.6 V at temperatures around 230 °C. This concept was subsequently extended to phosphoric acid fuel cells (PAFCs). The theme of controlled porosity, highlighted in Chapter 5, has proved to be critical for the development of all fuel cells.

9.6 PROTON-EXCHANGE MEMBRANE (PEM) FUEL CELLS

The electrolyte in PEM fuel cells is a solid polymer material with covalently bonded sulfonic acid groups. These materials are similar to ion-exchange resins. A cation is associated with each negatively charged sulfonic acid group. For fuel-cell applications, the cations are protons. In other applications, see Chapter 14 and the chlor-alkali process, sodium ions are present rather than protons. PEM is almost synonymous with perfluorinated ionomers known under the trade name Nafion®. Figure 9.9 shows the structure of the ionomer material. Nafion® is a copolymer of tetrafluoroethylene (Teflon) and sulfonyl fluoride vinyl ether. The incorporation of ionic groups into the polymer has a dramatic effect on its physicochemical properties. A bicontinuous nanostructure is formed. There are two domains: one that is Teflon like that forms from the backbone of the material, and a second that contains the sulfonic acid groups. These materials segregate into hydrophobic and hydrophilic regions. The sulfonic acid groups arrange themselves to form the hydrophilic regions that are strongly acidic; and the two domains are randomly connected. Equivalent weight, the mass in grams of the polymer per mole of sulfonic acid group, is a key measure of the ion-exchange capacity of the membrane.

The sulfonic acid groups represent fixed anions in the polymer membrane since they are covalently bound to the polymer backbone. When current flows in a PEM fuel cell, protons are transported via two mechanisms as illustrated in Figure 9.10. In the vehicular mechanism, protons are dissociated from the sulfonic groups and hydrated. These move by molecular diffusion. A second mechanism is proton hopping. Here, the proton is closely associated with the sulfonic acid group. During transport, protons hop from acid group to acid group. The only charge carrier is the protons; thus, the transference number of protons is unity.

In addition to the ability of cations to move through these ionomers, a second important aspect of perfluorinated ionomers is their ability to take up large amounts of water and other solvents. It was recognized early on that the water content in ionomer membranes is important for the operation of PEMFCs. The amount of water can be

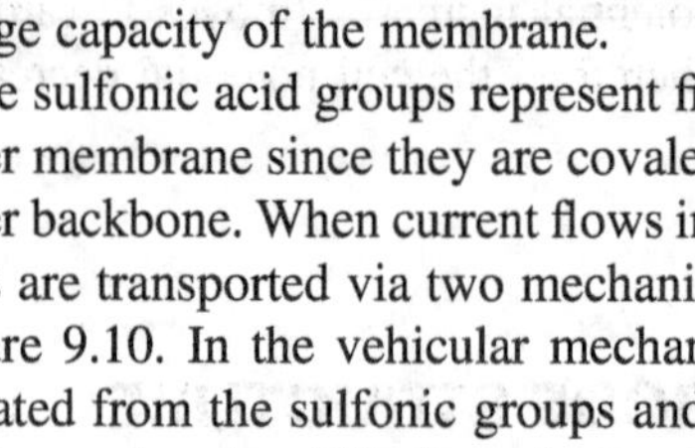

Figure 9.9 Basic structure of Nafion® co-polymer.

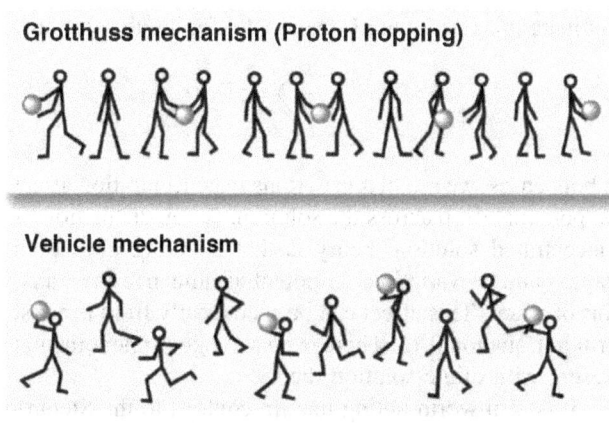

Figure 9.10 Scheme of proton-transfer mechanisms (vehicle and Grotthuss mechanisms). *Source:* Ueki 2008. Reproduced with permission of American Chemical Society.

expressed in several ways, but the most common approach is the ratio:

$$\lambda = \frac{\text{number of water molecules}}{\text{number of sulfonic acid groups}}. \quad (9.13)$$

When exposed to water vapor, a relationship exists between the activity of water in the vapor phase and λ. This equilibrium for Nafion is shown in Figure 9.11. When plotted against activity, $\sim p_w/p_o$, the curve is nearly independent of temperature. The first few molecules of water are associated with high enthalpic changes and are tightly bound to the protons.

An important characteristic of these materials is that the ionomer must be hydrated to allow efficient conduction of protons. The conductivity depends roughly linearly with water content or λ.

$$\kappa = A + B\lambda. \quad (9.15)$$

Conductivity is the first key transport property of the electrolyte. Compared to other electrolytes used for fuel cells, PEM shows a particularly strong dependence for its conductivity on water content. This behavior has critical implications for PEM cell and system design. For all intents and purposes, the membrane must be close to fully hydrated for good cell performance. Lastly, we note that because the sulfonic acid groups are relatively close to each other in the hydrophilic regions, electrostatic forces repel anions. Effectively, these cation-exchange membranes exclude anions.

Although protons are the only charge carriers in ionomer membranes such as Nafion, the water in the membrane is also mobile. The transport of water influences the design and operation of the PEMFC. A second key transport property is the *electroosmotic drag coefficient*

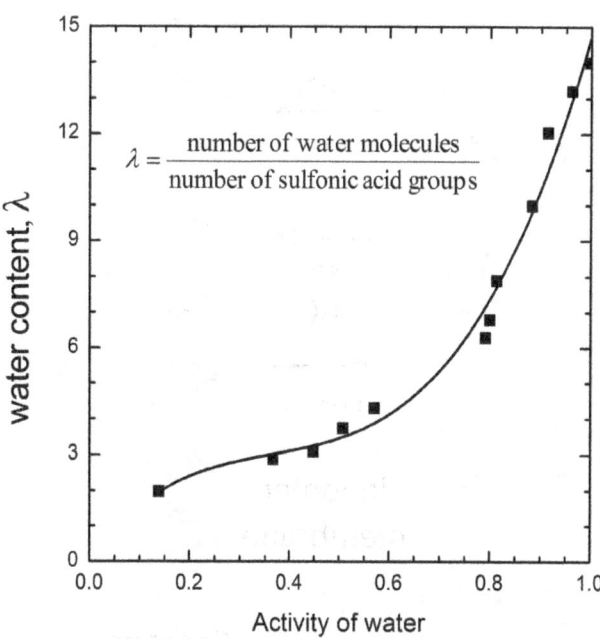

Figure 9.11 Water content in membrane as a function of partial pressure of water vapor. *Source:* Adapted from Zawodzinski 1993.

of water, ξ. This drag may be thought of as the number of water molecules that move with each proton in the absence of concentration gradients. This property is most easily measured with a concentration cell (Chapter 2), which can provide the electroosmotic drag as function of water content, λ. The third transport property is the diffusion coefficient of water in the proton-exchange membrane, D_o.

Accordingly, there are three transport processes that occur in the membrane as highlighted in Figure 9.12. During fuel-cell operation, protons move from the anode to the cathode. Because these protons are hydrated, they "drag" water with them. Finally, concentration gradients can develop, leading to molecular diffusion of water. Thus, we have three species (water, protons, and ionomer) and three transport properties, κ, ξ, and D_o.

Because the ionic current carried by the protons also significantly influences the transport of water (uncharged species), the two processes are coupled. This coupling between ionic current and water transport is not described well with dilute solution theory. The transport described in Chapter 4 focused on the interaction of each species with the solvent. Here, there is no clear choice for the solvent, and further, binary interactions among all three species can be important. Hence, concentrated solution theory is used to derive the equations below. Because of electroneutrality, the concentrations of sulfonic acid groups and protons are equal, and only one composition needs to be specified. Taking the amount of water as the compositional variable,

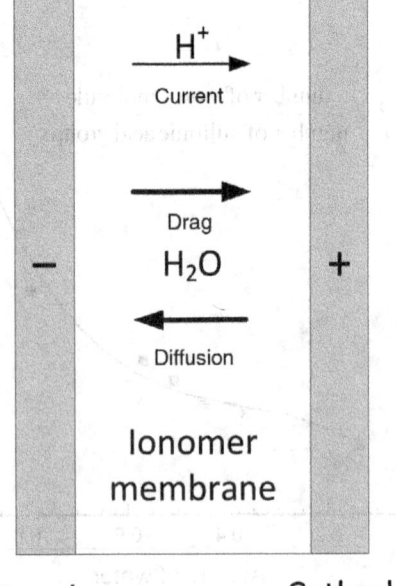

Figure 9.12 Transport processes in PEM fuel cell.

the relevant transport equations are

$$\mathbf{N}_0 = -\frac{\kappa \xi}{F} \nabla \phi_2 - \left(\alpha + \frac{\kappa \xi^2}{F^2} \right) \nabla \mu_0, \quad (9.15)$$

$$\mathbf{i} = -\kappa \nabla \phi_2 - \frac{\kappa \xi}{F} \nabla \mu_0. \quad (9.16)$$

Equation 9.15 expresses the flux of water as a function of two independent gradients: the chemical potential of water, μ_o, and the electrical potential, ϕ_2. α is a diffusion coefficient of water based on the thermodynamic driving force (i.e., the gradient of the chemical potential of the water rather than its concentration gradient). In addition to molecular diffusion due to a gradient in the chemical potential of water, we see from Equation 9.15 that water movement is also associated with a gradient in potential.

Since protons are the only mobile charged species, the current density is proportional to the molar flux of protons. Therefore, Equation 9.16 is written in terms of current density rather than molar flux. The first term on the right side of Equation 9.16 would be analogous to Ohm's law. Here we note that there is a second term arising from the gradient in chemical potential of water. ξ is the electro-osmotic coefficient, which represents the number of water molecules that move with each proton in the absence of concentration gradients. Equation 9.16 may be rearranged to

$$\nabla \phi_2 = -\frac{\mathbf{i}}{\kappa} - \frac{\xi}{F} \nabla \mu_0. \quad (9.17)$$

Compare this result with Equation 4.6 for dilute solutions:

$$\nabla \phi = -\frac{\mathbf{i}}{\kappa} - \frac{-F}{\kappa} \sum_i z_i D_i \nabla c. \quad (4.6)$$

In both cases, we see that variations in concentration affect the potential drop across the solution. A key distinction of concentrated solution theory is that coupling goes both ways: namely, variations in potential influence the transport of water. This effect can be seen clearly from the first term in Equation 9.15. There is no analogous phenomenon present with dilute solution theory.

It is also worth noting that, in contrast to the Stefan–Maxwell formulation for concentrated systems to which you may have been previously exposed, nonidealities in the solution are likely to be important, which significantly complicates obtaining the necessary physical parameters. On the positive side, for this three-component system, the three independent binary interaction parameters can be related to three straightforward transport properties: that is, the electrical conductivity (κ), the diffusion coefficient of water (D_0), and the electro osmotic drag coefficient, (ξ).

Equations 9.15 and 9.16 can be simplified to illustrate better the coupling of current flow and water transport in PEM fuel cells.

$$\mathbf{i} \approx -\kappa \nabla \phi_2, \quad (9.18)$$

which is just Ohm's law, and

$$N_o \approx \underbrace{\frac{i}{F} \xi}_{\text{electroosmotic drag}} - \underbrace{D_0 \nabla c_o}_{\text{molecular diffusion}}. \quad (9.19)$$

Thus, we can see more clearly that water moves due to a concentration gradient, but also from the electroosmotic drag. Let's examine this behavior in a bit more detail. Imagine that we have a proton-exchange membrane in contact with water vapor on both sides at the same activity. Initially, there is no current flowing and the water content, λ, is determined with the data from Figure 9.11 and is constant across the membrane. Next, we apply a fixed current density of i. What happens to water? Assuming that a positive current is in the direction of left to right, electroosmotic drag will carry water to the right. Unless water can be supplied at the left side and removed at the right side instantaneously, a concentration gradient will develop. Of course, the rate of water supply and removal would be limited by mass transfer of water between the membrane and the vapor. Once a concentration gradient is established, water diffuses down the gradient. Molecular diffusion occurs to counteract the effect of water drag. Depending on the current density, physical properties, and mass transfer on each side, it is possible to have a steady net water

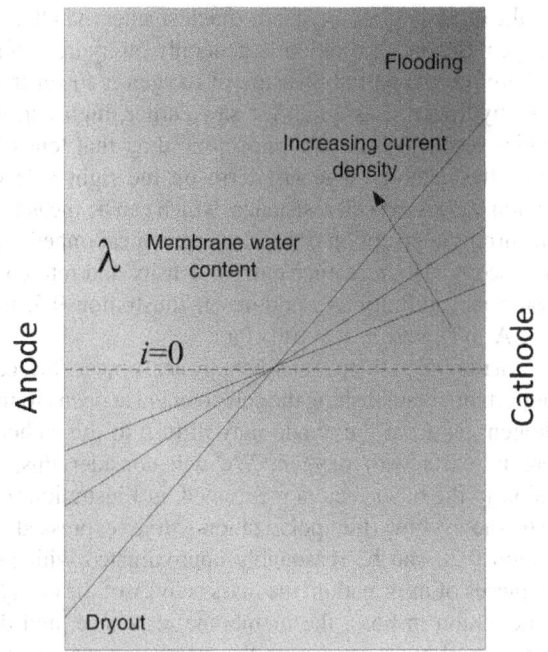

Figure 9.13 Water content in ionomer membrane during current flow.

movement across the membrane. These phenomena are depicted in Figure 9.13.

What are the implications for the PEMFC? Keep in mind that there is also production of water at the cathode of a PEMFC. This production is an additional source of water that would need to be considered to arrive at a detailed picture of the water content in the membrane. For our purposes, we can refer back to Figure 9.11 to illustrate two possible issues for the operation of PEMFCs. First, note that as the current density increases, the water content on the left side (anode) of the membrane decreases. As we have seen before, the conductivity of the membrane depends on water content. Thus, with increasing current density the anode can "dry out," resulting in poor performance. At the same time, the water content at the right side grows with increasing current density. What's more, water is being produced here at the cathode. What if the combined rate transport of water and the rate of production are greater than the rate at which water is removed? In this instance, conductivity is not an issue, but the cathode can "flood", resulting in severe mass-transfer limitations and poor performance. The excess water can form drops that cover the catalyst sites and prevent access by oxygen.

The PEM material is formed into a thin film, which in addition to serving as the electrolyte, also forms a barrier between the fuel and oxidant. Water balance at the cell sandwich level becomes a key challenge for PEM fuel cells. Too little water, and the membrane conductivity drops; too much water, and the electrode floods. The ideal situation is for the gas in contact with the membrane to be saturated with water vapor. It is impractical to completely avoid liquid water and simultaneously keep the membrane hydrated; therefore, some means of dealing with liquid water is needed for PEMFCs. Further, because of the unique importance of water content in PEMFCs, λ has taken on a special significance to the design of cells and complete systems. The system level analysis is discussed in Chapter 10.

ILLUSTRATION 9.4

Estimate the limiting current in a PEM fuel cell assuming that the conductivity goes to zero as the concentration of water approaches zero. Referring to Figure 9.13, this implies that the minimum concentration of water on the left-hand side of the diagram is zero, since any further increase in the current is not possible because the membrane conductivity without water goes to zero. The maximum current is the current that corresponds to a net water flux of zero. Under limiting current conditions at steady state, Equation 9.19 becomes

$$\frac{i}{F}\xi = D\nabla c_0.$$

$$i = \frac{F}{\xi}D\frac{\Delta c_0}{L}.$$

The concentration of water can be estimated from the density, equivalent weight, and λ, the ratio of water to sulfonic acid. Taking this concentration to be 15 kmol m^{-3} and using a value for ξ of 1.0 with a membrane thickness of 100 μm, we arrive at

$$i = \frac{96485}{1.0}10^{-9}\frac{15000}{0.0001} = 14500 \text{ A} \cdot \text{m}^{-2}.$$

It is clear from the above equation that the limiting current can be increased by reducing the thickness of the separator, which also reduces ohmic losses in the cell.

Whereas Nafion has an electroosmotic drag coefficient of near unity over a wide range of levels of hydration, other membranes show about half the value. If the electroosmotic drag were reduced to 0.4 for an alternative membrane, the limiting current would increase to 36,200 A m^{-2}.

As noted earlier, a three-dimensional electrode structure is needed in PEMFCs. The flooded-agglomerate model is a reasonable description of electrodes found in PEMFCs. The electrodes for PEMFCs require electrocatalysts to carry out the reactions. At these low temperatures and in acid conditions, platinum and alloys of platinum are the only presently acceptable materials. Given that platinum is a precious metal and therefore expensive, the amount of platinum used in the fuel cell is of critical importance. First, the platinum must be part of the triple phase boundary, and therefore in contact with the gas and the electrolyte, as well as connected electronically with the current collector. Platinum that fails to achieve this is not active toward the desired reactions in the cell. The degree to which the electrode achieves the desired connections defines the gross utilization of platinum. In a well-designed electrode this value is near one. Second, consider a particle of platinum. Only the atoms at the surface participate in the electrochemical reactions. Therefore, it is desirable to have a high surface area to volume ratio; this is achieved by creating small particles of platinum and is characterized by the dispersion of the catalyst, which is defined as the fraction of catalyst atoms that reside at the surface (as opposed to inside the particle). The small catalyst particles are placed on a support material with a high surface area. The most common choice of support material in a fuel-cell electrode is carbon. There are a large variety of carbon materials with high surface area; what's more, the carbon is conductive and reasonably stable at the conditions found in PEMFCs.

A detailed accounting of the effects of water transport in the membrane, the mechanisms for electrocatalytic reactions, and multicomponent gas diffusion on the polarization curve for PEM fuel cells is beyond the scope of this text. For our purposes we will assume that the polarization curve for a PEM fuel cell can be represented by

$$V_{cell} = \text{const} - 2.303 \frac{RT}{F} \left\{ \underbrace{\log\left(\frac{i}{i_{ref}}\right)}_{\substack{\text{Tafel slope} \\ \text{for ORR}}} - \underbrace{\log\left(\frac{p_{O_2}}{p_{ref}}\right)}_{\substack{\text{effect of} \\ O_2 \text{ partial pressure}}} \right.$$

$$\left. + \underbrace{\log\left(1 - \frac{i_{load}}{j_D}\right)}_{\substack{\text{mass} \\ \text{transfer}}} \right\} - \underbrace{R_\Omega i_{load}}_{\substack{\text{ohmic} \\ \text{resistance}}} . \quad (9.20)$$

As noted earlier, the kinetic polarization of the cathode is much larger than that of the anode for PEMFCs; thus, the kinetic polarization for the hydrogen electrode is neglected. Equation 9.20 includes a term associated with the Tafel slope for the oxygen reduction reaction (ORR) and a second term for the effect of oxygen partial pressure on the kinetics of ORR. All mass-transfer limitations (anode, cathode, membrane) are lumped together with a single term. Again, most often mass-transfer limitations are associated with the cathode. The reasons are as follows.

First, the diffusivity of oxygen is much smaller than that of hydrogen. Second, operation is generally on hydrogen/air and therefore the partial pressure of oxygen is lower than that of hydrogen. Lastly, as we saw earlier, high current densities result in high electroosmotic drag that tends to "flood" the cathode. The last term on the right side of Equation 9.20 is the cell resistance, which can be measured from current interruption or by electrochemical impedance spectroscopy. The reference current density and reference pressure are arbitrary. As is done in Illustration 9.5, use $i_{ref} = 1 \text{ A m}^{-2}$, and $p_{ref} = 100 \text{ kPa}$.

As a reminder, there is another physical process that can be important in establishing the cell potential at open circuit; hydrogen gas from the anode may diffuse to the cathode where it reacts with oxygen. We will consider this, in addition to the other factors mentioned, in Illustration 9.5, which shows how the polarization curve expressed in Equation 9.20 can be reasonably approximated with just three pieces of information: the mass activity of the catalyst (see definition in box), the membrane resistance, and the permeation of hydrogen across the membrane separator.

Mass Activity of Pt Catalysts

The most common way to describe the effectiveness of the oxygen reduction catalysts for PEM fuel cell is with the mass activity. The current per unit mass of platinum in the cathode is reported. This current is recorded at a cell potential of 0.9 V (iR free) at 80 °C on humidified hydrogen and oxygen at 150 kPa.

ILLUSTRATION 9.5

The mass activity of the catalyst for a PEM fuel cell is 60 A·g^{-1} (O_2 at 0.9 V). The platinum loading is 5 g·m^{-2}. The rate of hydrogen crossover is equivalent to 10 A·m^{-2}, and the membrane separator is 50 μm with a conductivity of 5 S·m^{-1}. The fuel cell will operate at 100 kPa using air at the cathode. Calculate the current–voltage curve, the open-circuit potential, and the maximum power assuming no mass-transfer limitations.

Solution:

We first use the mass activity, measured at 150 kPa, to determine the needed constant in Equation 9.20. The vapor pressure of water at 80 °C is 47 kPa; therefore, the partial pressure of oxygen is 103 kPa and the ratio in Equation 9.20 is close to one. Using the conditions and definition for mass activity, we can determine the constant in Equation 9.20. The potential of 0.9 V (iR free) assumes that the iR contribution has already been accounted for as described on the previous page. Thus,

$$\text{const} = 0.9\,\text{V} + 2.303\frac{RT}{F}\left\{\log\left(\frac{i}{i_{ref}}\right)\right\}$$

$$= 0.9\,\text{V} + 2.303\frac{RT}{F}\left\{\log\left(\frac{(60)(5)}{1}\right)\right\}$$

$$= 1.074\,\text{V}.$$

The resistance is simply

$$R_\Omega = \frac{L}{\kappa} = \frac{5\times 10^{-5}}{5} = 10^{-5}\,\Omega\cdot\text{m}^2.$$

The current density that is used to determine the polarization of the cathode is the sum of the load current density and the crossover of hydrogen.

$$i = i_{load} + i_{H_2\text{ crossover}}.$$

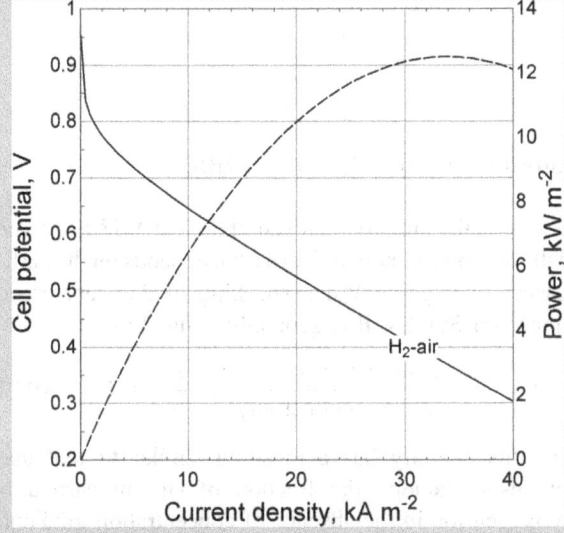

Equation 9.20 can now be used to yield a reasonable approximation to the polarization curve. In doing so, we include the effect of polarization, the influence of the operating pressure (21 kPa O$_2$), and the ohmic resistance. We neglect the mass-transfer limitations as specified in the problem statement. The iR resistance is included as shown in Equation 9.20. For each value of the load current we can now calculate the cell potential, V_{cell}. The results are shown in the plot. The corresponding power curve is also shown. The current–voltage curve is linear at the higher current densities considered, consistent with ohmically limited operation.

The open-circuit potential for this cell corresponds to a load current of zero and is determined by the reaction of oxygen at the cathode with hydrogen that crosses over from the anode; it is not equal to the equilibrium potential. Assuming that the Tafel approximation holds due to sluggish oxygen kinetics, and that the hydrogen crossover rate remains constant, the open-circuit potential is

$$V_{OC} = 1.074\,\text{V} - 2.303\frac{RT}{F}\left\{\log\left(\frac{10}{1}\right) - \log\left(\frac{21}{100}\right)\right\}$$

$$- R_\Omega(0) = 0.96\,\text{V}.$$

9.7 SOLID OXIDE FUEL CELLS

A second major type of fuel cell is the SOFC. Here the electrolyte is a ceramic material that conducts oxygen ions. The electrode reactions for a SOFC were provided in Table 9.2. Note that in contrast to the PEMFC, although the overall reaction is the same, water is produced at the anode rather than the cathode. Additionally, the charge carrier is the oxygen ion and not a proton.

Just as in PEMFCs, we desire the electrolyte to be a fast ion conductor, an electronic insulator, and a barrier to prevent mixing of the fuel and oxidant gases. There are three general classes of electrolytes used in SOFCs as shown in Table 9.4. We will discuss only the most common one, zirconia, in any detail.

Zirconia is a ceramic with low intrinsic conductivity. The crystal structure of zirconia changes with temperature: monoclinic at room temperature, tetragonal, and finally cubic at high temperatures. It is the cubic structure that is of interest because of its ionic conductivity. The cubic fluorite structure is face centered cubic (fcc) with O^{2-} filling the tetrahedral interstitial sites. Zirconia is typically doped with yttria. Since Y has three valence electrons compared to four for Zr, O^{2-} vacancies are created to preserve charge neutrality as shown in Figure 9.14. This doping behavior is analogous to doping of semiconductors (Chapter 15). The replacement of two Zr^{4+} with two Y^{3+} creates an oxygen vacancy. The vacancies created by these substitutional cations with a different charge provide a means for more facile charge transport in the ceramic. Yttria also serves a second function. The crystal structure of zirconia changes with temperature, as noted previously, through displacive transformations; that is, no bonds are broken when changing from cubic to monoclinic structures. However, large volume changes accompany these phase changes and cause zirconia to crack during cooling. The

212 Electrochemical Engineering

Table 9.4 Types of Electrolytes Used in SOFCs

ZrO_2, zirconia based	YSZ (yttria-stabilized zirconia), fluorite structure, most common electrolyte	700–1000 °C
CeO_2, ceria based	Lower temperature, fluorite structure	600 °C
$LaGaO_3$	Perovskite structure	800 °C

addition of a small amount of yttria stabilizes the tetragonal and cubic structures so that a metastable cubic structure is preserved even at low temperatures. Hence, these materials are called *yttria-stabilized zirconia* (YSZ).

The existence of these vacancies provides a conduction mechanism for the ceramic. The ability of the oxygen ions to move depends on whether an adjacent site is empty. Oxygen hops through these vacancy sites by a process called *vacancy diffusion* as shown in Figure 9.15. The addition of Y creates more vacancies, and oxygen now has room to move more easily, much faster than by interchange or interstitial mechanisms. This phenomenon is typically expressed in terms of ionic conductivity, κ. Recall our definition for conductivity (Equation 4.7):

$$\kappa = F^2 \sum_i z_i^2 u_i c_i. \quad (4.7)$$

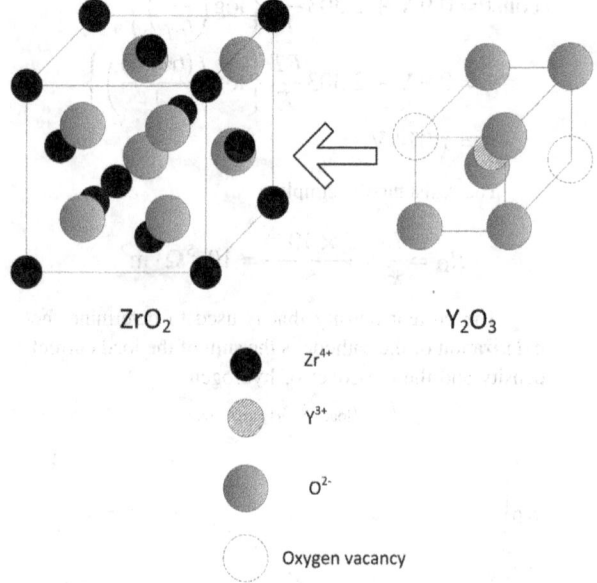

Figure 9.14 Creation of vacancies in ZrO_2.

u_i is the mobility of the ion, z_i its charge, and c_i is the concentration of the species. The ability of oxygen to move depends on the number of vacancies (concentration) and the energy barrier or activation energy for exchange. The vacancy concentration increases with increased levels of doping (Y addition), and this simple model predicts that the conductivity increases linearly with doping. The actual behavior is more complex, and the percent doping used in YSZ is about 8%, near the point of maximum conductivity. The ionic conductivity is strongly dependent on temperature, which is shown in Figure 9.16. This dependence is again explained in terms of the mobility of the oxygen. There is an energy barrier to the exchange of atoms with the vacancies, and with increasing temperature the mobility follows an Arrhenius relationship. Temperatures in the range of 700–1000 °C are needed for a practical YSZ-based fuel cell.

O^{2-} is the only ion that is mobile, but YSZ has a very small electronic conductivity (σ) that depends on the partial pressure of oxygen. When speaking of the transference number for SOFCs, it is generally defined as

$$t_{O^{2-}} = \frac{\text{ionic conductivy}}{\text{electrical conductivity}} = \frac{\kappa}{\kappa + \sigma} \approx 1. \quad (9.21)$$

Here this quantity has a meaning similar to that used previously, namely, the fraction of current carried by the oxygen ion in the absence of concentration gradients. In contrast to liquid electrolytes, the concern is that the minor electronic conductivity is effectively a small short in the cell. In contrast to YSZ, ceria has an appreciable electronic conductivity.

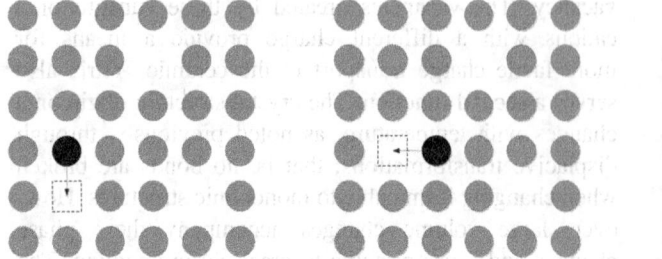
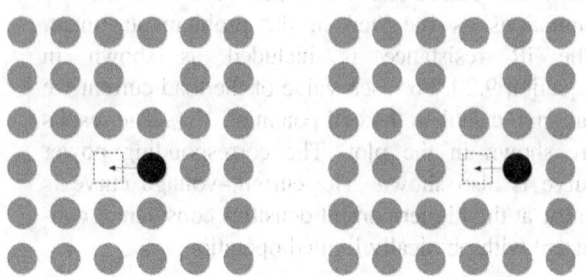

Figure 9.15 Oxygen vacancies created by doping allow more facile diffusion of vacancies.

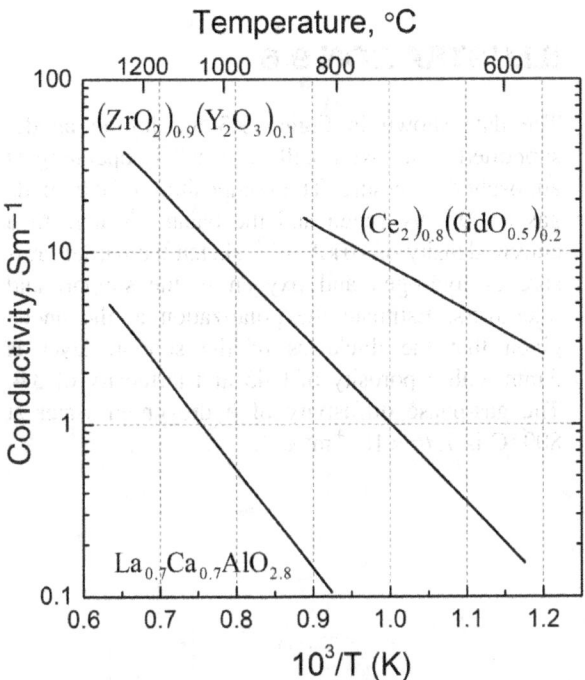

Figure 9.16 Conductivity of common oxide materials used in SOFCs.

As with any electrochemical system, the voltage of a solid oxide cell can be expressed as

$$V_{cell} = U - iR_\Omega - |\eta_{s,anode}| - |\eta_{s,cathode}| \\ - |\eta_{conc,anode}| - |\eta_{conc,cathode}|. \quad (4.58a)$$

The first polarization is the ohmic loss, and it is proportional to current density. This includes the resistance of the electrolyte, electrodes, and any interfacial or contact resistances. Activation and concentration polarizations exist at each electrode. We introduced the idea of a charge-transfer resistance in Chapter 4.

$$R_{ct} = \frac{1}{A}\frac{d\eta}{di}. \quad (4.62)$$

For Butler–Volmer kinetics (Chapter 3), this charge-transfer resistance decreases as the current density increases; however, for many SOFCs, the kinetic resistance is nearly constant with current. Therefore, we can adequately express the cell potential for a SOFC with

$$V_{cell} = U - i(R_\Omega + R_{anode} + R_{cathode}) \\ - |\eta_{conc,anode}| - |\eta_{conc,cathode}|. \quad (9.22)$$

The concentration polarizations arise due to changes in reactant composition at the electrodes. In this SOFC, we only need concern ourselves with variations in concentration of the gaseous reactants.

A good understanding of the materials of construction, including their microstructure, and the design of a solid oxide cell provide a basis for understanding SOFC polarization curves. Before exploring this, three typical configurations of cells are introduced: anode supported, electrolyte supported, and cathode supported, as shown in Figure 9.17. In short, one of the components is made thicker so that it can provide structural integrity to the cell.

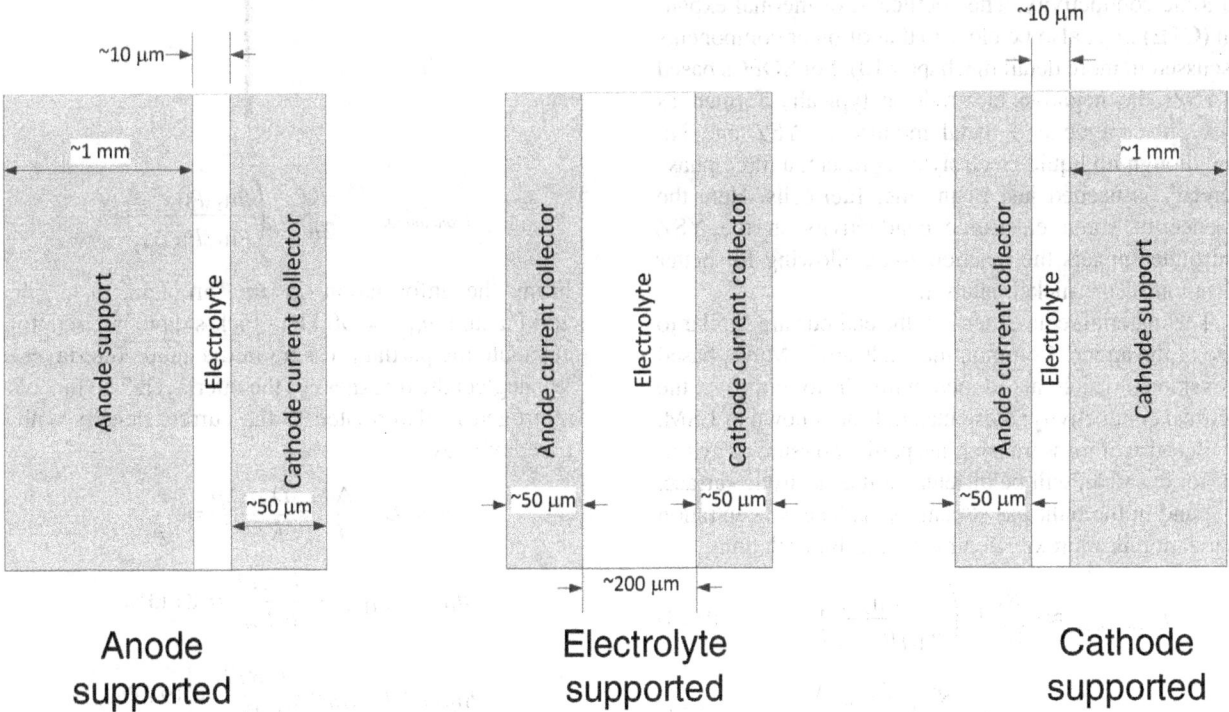

Figure 9.17 Anode-, electrolyte-, and cathode-supported designs used in SOFCs. Drawing is not to scale.

214 Electrochemical Engineering

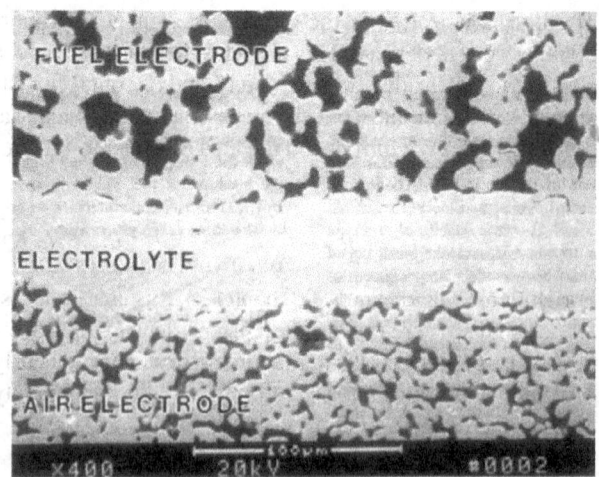

Figure 9.18 Cross section of a SOFC showing the porous nature of the two electrodes and the dense separator. *Source:* Ormerod 2003. Reproduced with permission of Royal Society of Chemistry.

The electrolyte is dense, but the other components are porous as shown in Figure 9.18.

The high operating temperature, required for sufficient ionic conductivity, has some advantages. The reaction kinetics at high temperature are typically facile; thus, precious metal catalysts are not essential. What we have referred to as the electrodes are commonly called cathode and anode *interlayers* within the SOFC community. The anode materials must be catalytically active, chemically compatible, porous for gas access, and have both electronic and ionic conductivity. The coefficient of thermal expansion (CTE) must also be close to that of other components (discussed in more detail in Chapter 10). For SOFCs based on YSZ, the negative electrode is typically formed as *cermet*, a ceramic and metal mixture of YSZ and Ni. Even though no liquid electrolyte is present, a three-phase boundary is needed just as in other fuel cells. Here the presence of some electronic conductivity in the YSZ electrolyte spreads the reaction zone, allowing for better utilization of the nickel catalyst.

The material requirements of the cathode are similar to those of the anode. Common materials are LaMnO$_3$-based perovskites, which are doped with Sr to improve the electrical conductivity; these materials are known as LSM.

Because of mass transfer, the partial pressure of hydrogen, for example, will be different at the electrode surface, $p_{H_2,s}$, and in the bulk gas stream, $p_{H_2,b}$. The concentration polarization is approximated by equilibrium relations,

$$\eta_{conc,anode} \approx \frac{-RT}{2F} \ln\left(\frac{p_{H_2,s} p_{H_2O,b}}{p_{H_2,b} p_{H_2O,s}}\right). \tag{9.23}$$

$$\eta_{conc,cathode} \approx \frac{-RT}{4F} \ln\left(\frac{p_{O_2,s}}{p_{O_2,b}}\right). \tag{9.24}$$

ILLUSTRATION 9.6

The data shown in Figure 9.3 are for an anode-supported solid oxide cell. The cell is operating at atmospheric pressure. The composition of the anode gas is 90% hydrogen and the balance water. At a current density of 10 kA m^{-2}, sketch the partial pressure of hydrogen and oxygen in the support and interlayers. Estimate the polarization at the anode given that the thickness of the support layer is 2 mm with a porosity of 0.45 and tortuosity of 3.4. The gas-phase diffusivity of hydrogen in water at 800 °C is 7.76×10^{-4} m^2 s^{-1}.

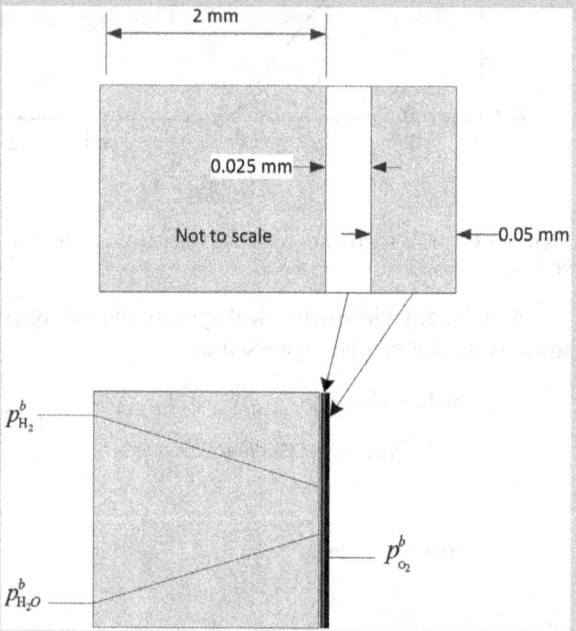

$$\eta_{conc,anode} \approx \frac{-RT}{2F} \ln\left(\frac{p_{H_2,s} p_{H_2O,b}}{p_{H_2,b} p_{H_2O,s}}\right).$$

From the information in the problem, $p_{H_2O,b} \approx$ 10 kPa and $p_{H_2,b} \approx$ 90 kPa. Fick's law is used to calculate the partial pressure in the anode interlayer. We neglect the thickness of the interlayer. The flux of hydrogen is also related to the current density with Faraday's law:

$$N = D_{eff} \frac{\Delta c}{L} = \frac{D_{eff}}{RT} \frac{\Delta p}{L} = \frac{i}{nF}.$$

$$p_{H_2,s} = p_{H_2,b} - \frac{i}{2F} \frac{RTL}{D_{eff}} = 81 \text{ kPa}.$$

$$p_{H_2O,s} = p_{H_2O,b} + \frac{i}{2F} \frac{RTL}{D_{eff}} = 19 \text{ kPa}.$$

The effective diffusion coefficient is $D_{eff} = \dfrac{D\varepsilon}{\tau} = 1.03 \times 10^{-4}$ m^{2}s^{-1}:

$$\eta_{conc,anode} \approx \dfrac{-RT}{2F} \ln\left(\dfrac{81 \times 10}{90 \times 19}\right) = 34 \text{ mV}.$$

Interconnects

The primary function of interconnects is to electrically join the anode of one cell with the cathode of another. This term is a bit broader than what is called a bipolar or separator plate for PEMFCs, but the function is identical. The interconnect term is likely more common in SOFC because of the importance of the tubular design, which is not bipolar (see Problem 9.9).

Refractory metals, such as W, Nb, and Ta, are prohibitively expensive; thus only ceramics are practical at typical operating temperatures of 800–1000 °C. Most often perovskite-type oxide ceramics are used for the interconnects; and these are based on rare-earth chromites. If good oxygen conductors can be found that operate at lower temperatures, say 600 °C, then more common metals and alloys, such as steel, can be used.

The high quality of the waste heat from a SOFC integrates well with a fuel processor, where the reformation reaction is endothermic. There is a good temperature match between the reformer and the waste heat. These high temperatures make the SOFC tolerant of CO in fuel. In fact, only partial reformation is required. Finally, the high-quality waste heat can be used to drive a Rankine bottoming cycle for instance.

The high temperature of SOFCs brings challenges too. The system takes a relatively long time to start up, and SOFCs are generally not well suited to applications that require frequent starting and stopping. There are a relatively small number of materials that can be used at these high temperatures. Sealing is also a major issue and is critical in order to keep the fuel and oxidant gases from mixing. Different materials used in the SOFC expand at different rates with temperature. This expansion is reflected in the coefficient of thermal expansion (CTE) of the material. The CTEs must be matched to prevent large stresses from developing and causing damage to the components.

The effect of temperature on the polarization curve of a SOFC is shown in Figure 9.18. Note that the open-circuit potential doesn't change much when the temperature is raised from 600 to 800 °C. The largest effect is on the resistance of the electrolyte, which decreases sharply as the temperature is increased.

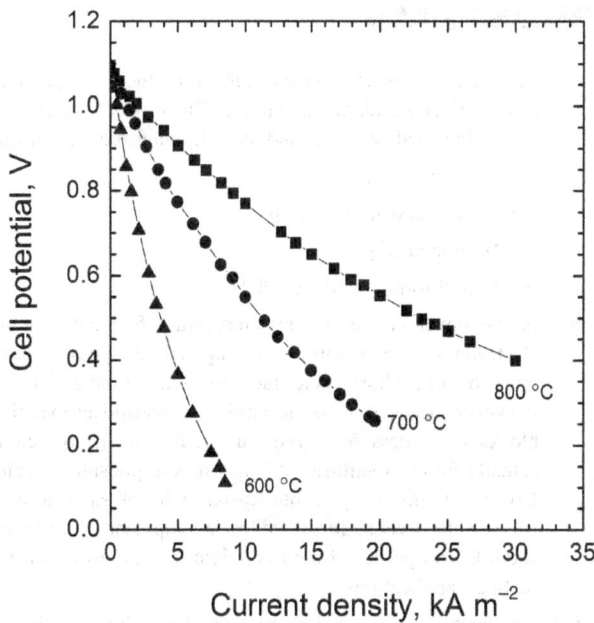

Figure 9.19 Effect of temperature on SOFC performance. *Source:* Adapted from Zhao 2005.

CLOSURE

In this chapter, the basic operation of a fuel cell has been examined. We learned that fuel cells are classified by their electrolytes, and this is a good starting point to understanding the behavior of any fuel-cell system. Two electrolytes were discussed in detail: proton-exchange membranes and solid oxide conducting ceramics. The most important feature of a fuel cell is its current–voltage relationship, or polarization curve. These curves can be understood in terms of fundamental electrochemical concepts: equilibrium potential, kinetic polarization, ohmic losses, and mass-transfer limitations.

FURTHER READING

Huang, K. and Goodenough, J.B. (2009) *Solid Oxide Fuel Cell Technology: Principles, Performance and Operations*, Woodhead Publishing.

Larminie, J. and Dicks, A. (2003) *Fuel Cell Systems Explained*, SAE International.

Mench, M. (2008) *Fuel Cell Engines*, John Wiley & Sons, Inc., New York.

Minh, N.Q. and Takahashi, T. (1995) *Science and Technology of Ceramic Fuel Cells*, Elsevier Science.

O'Hayre, R.P. Cha, S.-W., Colella, W.G., and Prinz, F.B. (2009) *Fuel Cell Fundamentals*, John Wiley & Sons, Inc., New York.

Singhal, S.C. and Kendall, K. (2003) *High-Temperature Solid Oxide Fuel Cells: Fundamentals, Design and Applications*, Elsevier Science.

Srinivasan, S. (2006) *Fuel Cells: From Fundamentals to Applications*, Springer.

PROBLEMS

9.1 Calculate the voltage efficiency, η_V^{fc}, for a fuel cell operating at 0.65 V at standard conditions. The product water is a liquid, the oxidant is air, and the following fuels are used:
   (a) Methane, CH_4
   (b) Liquid methanol, CH_3OH
   (c) Hydrogen, H_2
   (d) Liquid formic acid, $HCOOH$

9.2 In the development of low-temperature fuel cells, many electrolytes were explored. For a liquid acid type of electrolyte, the phosphoric acid fuel cell was commercialized. However, because of the adsorption of phosphate ions that blocks the access of oxygen, the reduction of oxygen is actually faster in sulfuric acid than it is in phosphoric acid. Given this, discuss possible reasons why phosphoric acid was selected over sulfuric acid for development. *Hint:* Think about the properties of the electrolyte that are important for fuel-cell applications.

9.3 The electrolyte for a molten carbonate fuel cell is a liquid salt mixture of lithium and potassium carbonate (Li_2CO_3 and K_2CO_3). Suggest the electrode reactions for molten carbonate chemistry. The reactants are hydrogen and oxygen, as is common for fuel cells. In addition, carbon dioxide is consumed at the cathode and produced at the anode. How might these high-temperature cells be designed so that the anode and cathode do not short out and so that an effective triple phase boundary is achieved? Discuss the importance of managing gaseous CO_2 in these cells.

9.4 The performance curves in the figure can be fit with a theoretical curve (*J. Electrochem. Soc.*, **152**, A1290 (1985)). The fitting parameters are shown below. Focus on just two parameters, j_D and R_{int}. What do these parameters represent, and how do their values impact the shape of the performance curves? Discuss the physical changes that were likely made to the cell to achieve the observed changes in performance.

$$V = U - \eta_c - R_{int} i.$$

$$\frac{\eta_c}{b} = \left(1 + \frac{i/j}{1 + i/j}\right) \ln \frac{i}{j} - \ln k - \ln\left(1 - \frac{i}{j_D}\right).$$

Curve	U [V]	b [V]	j [A·m^{-2}]	j_D [A·m^{-2}]	$-\ln(k)$	R_{int} [$\mu\Omega\cdot m^2$]
1	1.05	0.026	2772	9273	10.18	11.0
2	1.05	0.026	1014	14066	10.72	10.2
3	1.05	0.026	1014	17023	11.01	8.0
4	1.05	0.026	1014	28381	7.0	6.8

9.5 Shown are polarization data for a PEM fuel cell operating on hydrogen air at 70 °C and atmospheric pressure. Also shown is the Tafel plot, where ohmic and anodic polarizations have been removed. By means of a sketch, show how these plots would change under the following conditions:

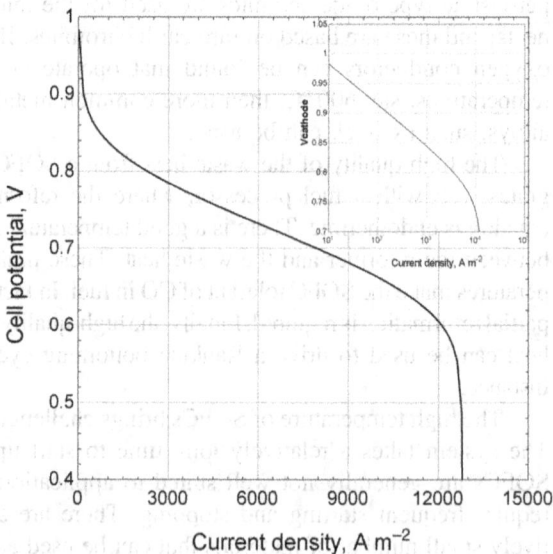

(a) The pressure is raised to 300 kPa.
(b) The oxidant is changed to pure oxygen in place of air.
(c) The platinum loading of the cathode catalyst (mg Pt cm^{-2}) is doubled.

9.6 A polarization curve for a molten carbonate fuel cell is shown in the figure. The temperature is 650 °C, and the electrolyte is a eutectic mixture of lithium and potassium carbonate. Discuss the polarization curve in terms of the four principal factors that influence the shape and magnitude of the curve.

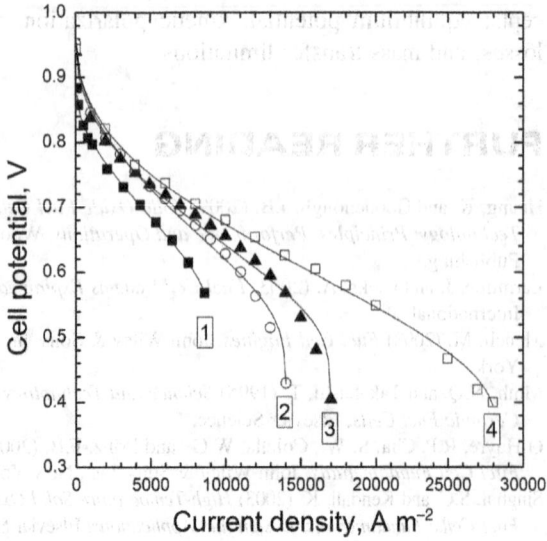

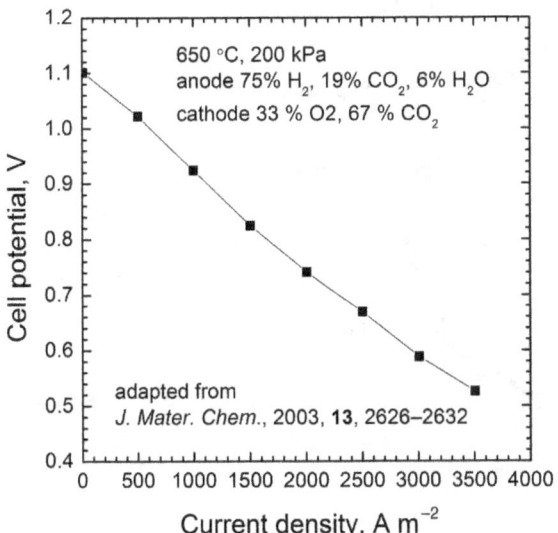

Thickness [μm]	Resistance [Ω·cm²]
4	0.100
8	0.105
14	0.120
20	0.140

9.7 A series of polarization curves at different temperatures for a direct methanol fuel cell are shown in the figure. This cell uses a Nafion separator. It is suggested that Nafion is permeable to methanol. Could this explain why the open-circuit potential is so low? What information about the cell can be inferred from these data specifically?

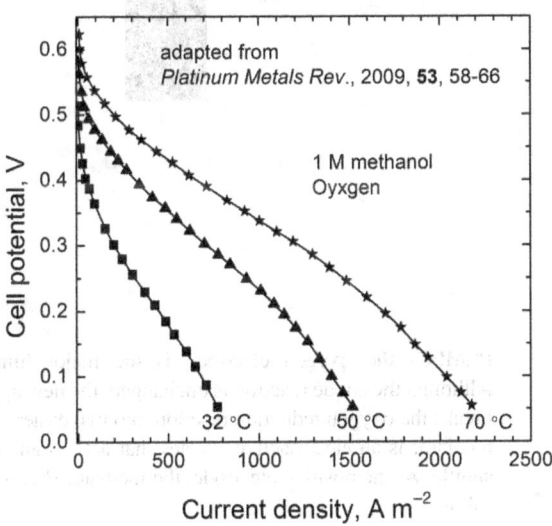

9.8 A series of anode-supported SOFCs were tested at 800 °C. The only parameter that was changed was the thickness of the electrolyte. Data for the cell resistance measured with current interruption are shown in the table. Determine the conductivity of the YSZ electrolyte and the fraction of the resistance that can be ascribed to the interlayers, current collectors, and contact resistances combined.

9.9 The tubular configuration is the most developed design for the solid oxide fuel cell. This design is shown in the figure. Air flows through the center and fuel flows over the outside. The separator is YSZ (yttria-stabilized zirconia), an oxygen ion (O^{2-}) conductor. What is the direction of current flow in the cell? How is the current carried in the cell? Sketch the potential and current distributions in the cell. Use the approximate schematic shown in the figure, where one half of the tube has been flattened out. Why is the performance (current–potential relationship) of the tubular design much lower than that of planar designs?

9.10 A proton-exchange membrane (PEM) fuel cell is fabricated with a separator that is 100 μm thick and has a conductivity of 5 S·m⁻¹. The cell is operating on hydrogen and air.

(a) If the open-circuit potential is 0.96 V, which corresponds to 20 A·m⁻² of crossover current, calculate the maximum power per unit area if only ohmic losses in the separator are considered.

(b) For the ohmically limited cell in part (a), sketch the current–voltage relationship. On the same graph compare the performance of a cell that includes kinetic- and mass-transfer polarization. Explain the curve.

9.11 After a prolonged shutdown of a PEM FC, both the anode and the cathode will contain air. During start-up, a front of hydrogen displaces the air in the fuel channels. This condition is illustrated in the figure and was first reported by Reiser et al., *Electrochem. Solid State Lett.*, **8**, A273 (2005). This situation is clearly transient and 2D in nature. Nonetheless, we can gain insight by examining a one-dimensional, steady-state analog, shown in the figure.

The two cells are electrically connected in parallel to an external load. Cell (1) has air and fuel provided normally, and the second cell (2) has air on both electrodes. At the positive electrode of the cell (2) oxygen evolution or carbon corrosion can occur. At the negative electrode of cell (2) we can expect oxygen reduction. For any reasonable potential, the current through the second cell will be small and we may assume that the solution potential, ϕ_2, is nearly constant between the anode and cathode of that cell. The anodic and cathodic currents for cell (2) must be equal to each other. Assuming Tafel kinetics for oxygen reduction and carbon corrosion, and assuming the kinetics for hydrogen oxidation to be fast, estimate the overpotential for carbon corrosion in cell (2) as a function of the measured potential of the cell V_c.

9.12 For the situation described in Problem 11, sketch current and potential in the separator of the fuel cell during start–stop

218 Electrochemical Engineering

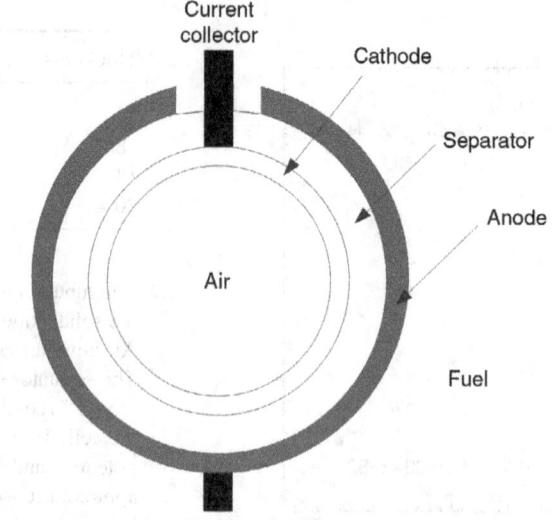

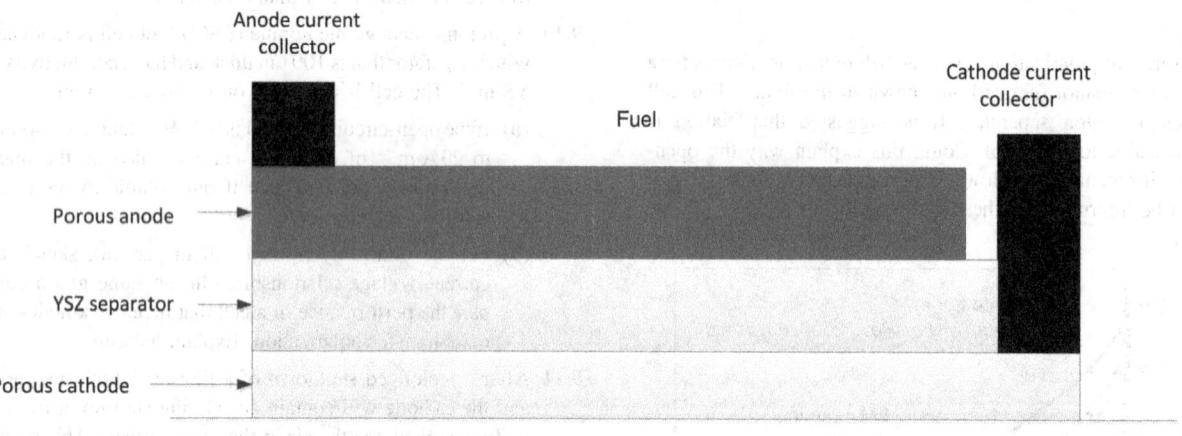

Problem 9.9

phenomena, that is, during the situation illustrated in the figure with Problem 9.11.

9.13 How does the maximum power density improve if the separator is decreased in thickness from 50 to 25 μm. Assume that there are no mass-transfer limitations.

9.14 Redo the maximum power calculation for Illustration 9.5 assuming that there is a mass-transfer limitation of 15,000 A·m^{-2} for the cathode.

9.15 You are evaluating a new technology for a hydrogen-air fuel cell. The incumbent is the traditional proton-exchange membrane fuel cell (PEMFC). For both fuel cells, the overall reaction is

$$H_2 + 0.5\, O_2 \rightarrow H_2O$$

At the anode of the PEMFC, hydrogen is oxidized; and at the cathode, oxygen is reduced. It is well known that for PEMFCs the oxygen electrode is the major limitation. Although the anode reaction is unchanged, the new approach breaks the oxygen reduction reaction into two easier parts. A mediator is an electroactive species that acts as an electron shuttle. At the positive electrode, the mediator (M) reacts as follows:

$$M_{ox}. + e^- \rightarrow M_{red}$$

In a separate nonelectrochemical reaction, the mediator is regenerated outside the cell.

$$O_2 + 4H^+ + 4M_{red} \rightarrow 2H_2O + 4M_{ox}$$

On the right are two polarization curves, one for the PEMFC and one for the new concept. Both curves are taken at 80 °C.

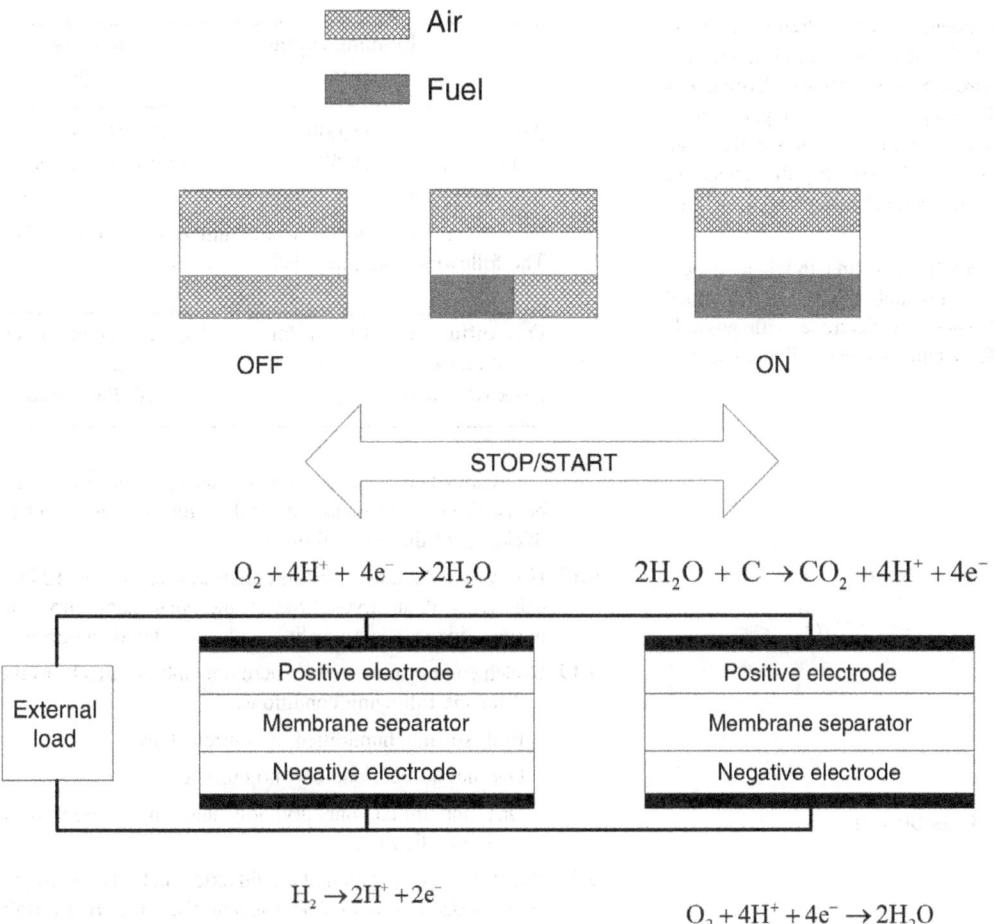

Problem 9.11

Compare and contrast the polarization curves of the two types of fuel cells. Specifically address the open-circuit potential as well as kinetic, ohmic, and mass-transfer losses.

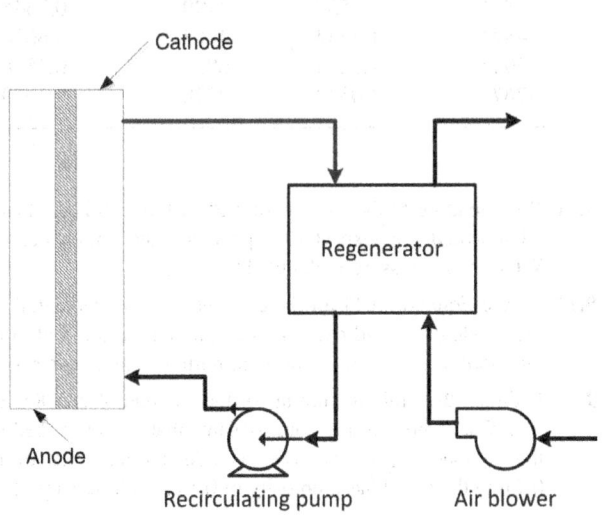

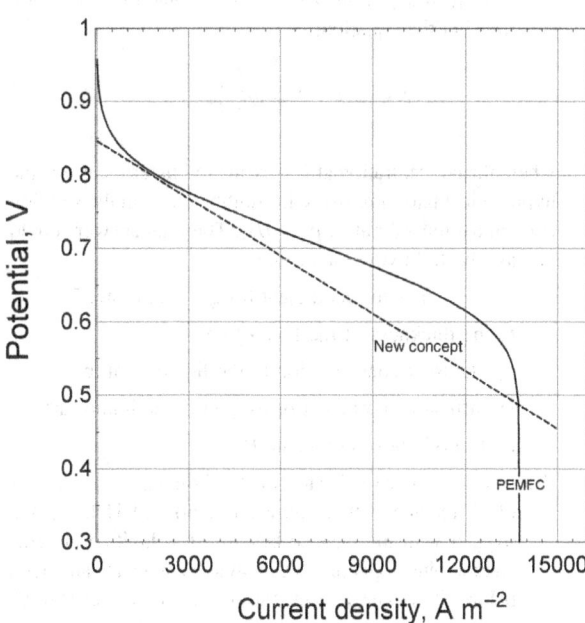

Problem 9.12

9.16 In Illustration 9.5, it was assumed that the hydrogen crossover current was $10 \text{ A} \cdot \text{m}^{-2}$. Estimate the permeability (Equation 4.72) for hydrogen through the membrane. Assume c is $30{,}000 \text{ mol} \cdot \text{m}^{-3}$. There is also permeation of oxygen across the membrane. The oxygen permeation is smaller than hydrogen, but not insignificant. Justify why this crossover is ignored in calculating the open-circuit potential in Illustration 9.5.

9.17 You are investigating mass-transfer limitations in a new cathode structure that your research group has developed for a PEM fuel cell. Consider an electrode with possible limitations to mass transfer in either the gas or liquid phase as illustrated in the cartoon.

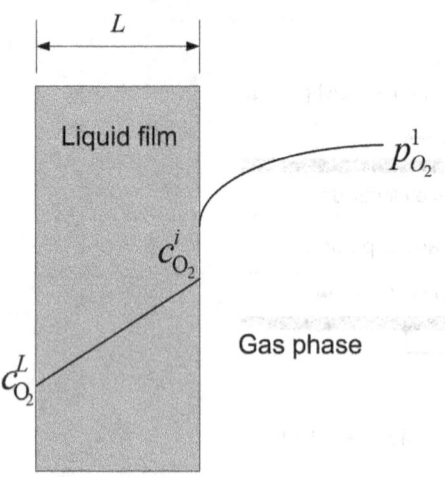

(a) Using the figure as a guide, develop a relationship for an overall mass-transfer coefficient, in terms of a pressure driving force; in other words, determine the expression for K in the equation:

$$N_{O_2} = K\left(p_{O_2}^1 - p_{O_2}^*\right).$$

where the concentration at L is expressed in terms of p_i*, the hypothetical partial pressure in equilibrium with the solution at composition x_i; that is, $p_i^* = H x_i$. The final answer should include the following parameters:

k_c: mass-transfer coefficient for gas phase, $\text{m} \cdot \text{s}^{-1}$

L: the thickness of the liquid film, m

c_T: the total concentration in the liquid, mol m^{-3}

D: diffusion coefficient of oxygen in the liquid, $\text{m}^2 \cdot \text{s}^{-1}$

H: Henry's law coefficient, Pa

(b) Data have been collected for the limiting current on air (21% oxygen with balance nitrogen) and Helox (21% oxygen with the balance helium). The limiting currents and the binary gas-phase diffusivity coefficients are given below. These were collected at ambient pressure and 80 °C.

	Limiting current $[\text{A} \cdot \text{m}^{-2}]$	Diffusivity of oxygen
Air	15,000	$2.47 \times 10^{-5} \text{ m}^2 \cdot \text{s}^{-1}$
Helox	19,500	$8.97 \times 10^{-5} \text{ m}^2 \cdot \text{s}^{-1}$

Assume that the Sherwood number is a constant, 3.66. The following additional data are provided:

D_{O_2}, diffusion coefficient for liquid phase	H_{O_2}/c_T, Henry's law constant
$3.0 \times 10^{-9} \text{ m}^2 \cdot \text{s}^{-1}$	$8.0 \times 10^4 \text{ Pa} \cdot \text{m}^3 \cdot \text{mol}^{-1}$

Estimate the fraction of mass-transfer resistance that can be ascribed to gas phase. What does this suggest about the thickness of the liquid film?

9.18 How would flooding affect the polarization curve of a PEM fuel cell. What about dryout? Sketch the polarization curves for normal, dryout, slight flooding, and severe flooding operation.

9.19 Sketch composition of water across membrane in a PEM fuel cell for the following conditions:

Both streams humidified, no current flow

One humidified, one dry, no current

One humidified, one dry, low and high current (from humidified to dry)

9.20 Data for the polarization of a solid oxide fuel cell/electrolyzer are provided below. (*J. Electrochem. Soc.*, **158**, B514–B525 (2011)). These potentials are reported with respect to a hydrogen reference electrode. The temperature of operation is 973 K. The ohmic resistance of the cell is $0.067 \, \Omega \cdot \text{cm}^2$. After removing ohmic polarization, how well can the data be represented by Butler–Volmer kinetic expression? Discuss whether the BV expression is appropriate for these data.

$i \, [\text{A} \cdot \text{m}^{-2}]$	V_{cell} [V]	$i \, [\text{A} \cdot \text{m}^{-2}]$	V_{cell} [V]
−6981	1.5006	930	0.9047
−4871	1.3554	2799	0.7545
−4871	1.3554	4753	0.6020
−2946	1.2074	6836	0.4518
−967	1.0549	9328	0.3000

9.21 If the loading of the cathode of a PEM fuel cell is doubled, what would you expect to happen to polarization curve? What if the pressure is doubled?

9.22 Derive Equation 9.24 for the concentration overpotential of the anode of a solid oxide fuel cell assuming that only the thermodynamic contribution is important (see Chapter 4).

9.23 Estimate the limiting current for the cathode of a SOFC at 900 °C and ambient pressure. Assume that air is supplied to the cathode: $y_{O_2} = 0.21$. The cathode current collector is 0.7 mm thick and has a porosity of 0.5 and a tortuosity of 6.

9.24 From the data provided, calculate the transference number of oxygen for doped ceria and YSZ. How would open-circuit potentials of the two cells compare?

YSZ $\quad \kappa = 7\ \text{S·m}^{-1} \quad \sigma = 0.1\ \text{S·m}^{-1}$
Doped ceria $\quad \kappa = 15\ \text{S·m}^{-1} \quad \sigma = 10\ \text{S·m}^{-1}$

9.25 Calculate the maximum power for three designs of SOFC: anode supported, cathode supported, and electrolyte supported. Assume operation at 800 °C and 1 bar. The cathode operates on air and the composition of the anode gas is 90% hydrogen, with the balance water. The diffusivity of oxygen in air at this temperature is $1.9 \times 10^{-4}\ \text{m}^2\text{·s}^{-1}$, and that of hydrogen in water is $7.8 \times 10^{-4}\ \text{m}^2\text{·s}^{-1}$. The anode is 45% porous and has a tortuosity of 2.5. The porosity of the cathode is 40% and the tortuosity is 3. Transport in the anode approximates equimolar counter-diffusion. In contrast, transport in the cathode is best described by diffusion of oxygen through stagnant nitrogen. In determining the cell voltage, you should include ohmic, surface, and concentration overpotentials. The conductivity of the solid electrolyte at 800 °C is $10\ \text{S·m}^{-1}$. Anode kinetics are approximately linear with an exchange current density of $5300\ \text{A·m}^{-2}$. The surface overpotential at the cathode, which is negative, can be approximated as

$$\eta = -0.2\ \sinh^{-1}\left(\frac{i}{4000}\right)$$

	Electrolyte supported (μm)	Anode supported (μm)	Cathode supported (μm)
Anode thickness	50	750	50
Cathode thickness	50	50	750
Electrolyte thickness	500	40	40

9.26 A molten carbonate fuel cell uses hydrogen as the fuel and air as oxidant and operates at 650 °C, ambient pressure. The electrolyte layer is 0.5 mm thick and has a conductivity of $100\ \text{S·m}^{-1}$.

(a) If the partial pressure of carbon dioxide on the anode and cathode side were 20 and 10 kPa, respectively, predict the theoretical voltage of the cell.

(b) Estimate the highest possible operating cell voltage for this cell at $2000\ \text{A·m}^{-2}$.

(c) Estimate the reduction in electrolyte layer thickness needed to triple the power density of the cell at the same cell potential. (This problem was suggested by S.R. Narayan).

FRANCIS THOMAS BACON

Francis Thomas Bacon was born on December 21, 1904, and died on May 24, 1992. He was a direct descendent of Sir Francis Bacon (1561–1626), philosopher and sometimes called the father of empiricism and the scientific method. Tom, as he was called by his friends, obtained a bachelor's degree from Trinity College in the Mechanical Sciences Tripos, a predecessor to Engineering. He then worked for C.A. Parsons and Co., a firm named for its founder and a pioneer in the development of the modern steam turbine.

In 1932, a pair of articles in *Engineering* set Tom's professional life on a new course: "The Erren Hydrogen Engine," and "Gas-Operated Vehicle." The idea was to store hydrogen generated from electricity during off-peak hours, and then use it in an internal combustion engine. Rather than directly injecting and burning the hydrogen as originally conceived, Bacon envisioned reversing the electrolysis process and *somehow* recombining the hydrogen and oxygen to generate electrical power. He had yet to hear of the concept of a fuel cell, but immediately started working on the idea. This electrochemical process should be more efficient than a thermal device, which is limited by the Carnot efficiency. Since a fuel cell didn't fit within the core business of C.A. Parsons and Co., he left the company in 1940 after several years of working surreptitiously at the office and at home on fuel cells. Bacon worked for many years with only intermittent support on an alkaline-based hydrogen/oxygen fuel cell. The challenges he had to overcome are familiar to those working in the area today: controlling the porosity of the electrodes and finding materials that are stable in the environment of the oxygen electrode. In 1959, he achieved a 6-kW, 40-cell stack with high efficiency that included automatic controls and a suitable water management system. Despite these developments, there was little commercial interest in the technology. However, this setback was not the end of the story.

In 1957, the launch of the Soviet Sputnik satellite ignited a space race. For manned flight, batteries were too heavy for long endurance missions. In the United States, a competing fuel-cell technology was under development, polymer electrolyte fuel cells, which would be used in NASA's Project Gemini. The goal of that program was to put a pair of astronauts in low-earth orbit for an extended period of time in preparation for the Apollo mission to the moon. These early polymer electrolyte fuel cells were not dependable—alkaline technology offered a higher efficiency and better reliability. Pratt and Whitney (now part of United Technologies Corp.) licensed Bacon's patents and successfully bid a proposal to develop Bacon's alkaline fuel-cell technology. In contrast to Bacon's original small team, at its peak more than a thousand people were advancing Bacon's fuel cell for manned space missions. Of course, the Apollo program was successful, and Bacon was repeatedly and justly recognized for his contribution. When meeting with then President Nixon, he put his arm around Bacon and remarked, "Without you, Tom, we wouldn't have made it to the moon." The same technology was used for the Space Shuttle fleet on a total of 135 missions, the last of which took place in 2011.

In 1978, Bacon received the Vittorio de Nora Award of the Electrochemical Society. Established in 1971, the award recognizes distinguished contributions to the field of electrochemical engineering and technology. In 1991, Bacon was the first winner of the Grove Medal.

Image Source: Courtesy of ECS.

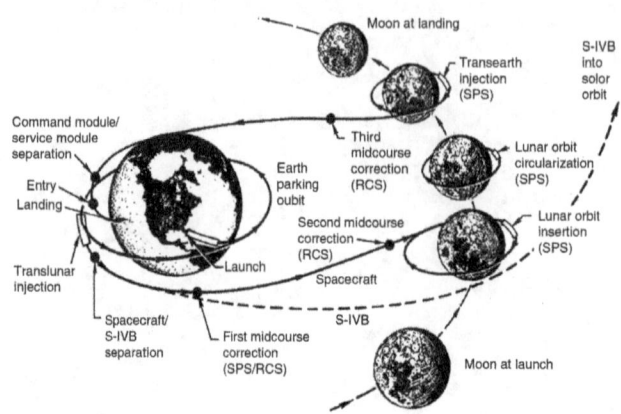

Chapter 10

Fuel-Cell Stack and System Design

Having introduced the basic concepts of fuel cells in the previous chapter, we now turn our attention to engineering challenges with the design and operation of fuel-cell systems. Of course, the current–voltage relationship, or polarization curve, discussed in the previous chapter is of vital importance. A wide range of additional issues must be considered. What is the best way to arrange the individual cells? How will reactants be supplied to the cell stack? How will products be removed? Is a thermal management system needed? As we examine these issues, we will explore different ways in which they might be addressed. We will also examine the impact that the methods selected to resolve these issues have on the polarization curve.

As we noted in the earlier chapters, runtime is the key design factor for batteries. In contrast, fuel cells are typically expected to run more or less indefinitely. Similar to heat engines, fuel cells convert the energy in a chemical fuel to electricity, a more useable form of energy. Efficiency, namely, the fraction of the chemical energy of the fuel that is converted to electricity, often replaces runtime as the key metric in stationary applications. A careful analysis of the entire system that contains the fuel cell is required to assess and maximize the benefits of fuel cells for electric power generation. For mobile devices where the fuel is carried, the system behaves a lot like an energy storage device. We'll see that as the desired operating time of mobile power system increases, fuel cells are favored over batteries. A well-known example is U.S. manned space missions, which have relied on fuel cells rather than batteries because of higher system-specific energy [W·h·kg^{-1}], among other advantages.

10.1 INTRODUCTION AND OVERVIEW OF SYSTEMS ANALYSIS

In contrast to the battery, a fuel cell is typically a steady-state device. As such, the fuel and oxidant are supplied continuously. In the previous chapter, we focused on the electrochemistry of the fuel cell. Now we will examine the entire fuel-cell system. Many cells are combined to form a cell stack assembly (CSA), which is the "heart" of the fuel-cell system. In addition to the cell stack, there can be considerable equipment associated with the proper operation of the system. Figure 10.1 shows a highly simplified arrangement, which will serve to illustrate the main points. Here, the system is divided into three parts: a fuel processor, a cell stack, and a power conditioning section.

As mentioned previously, the oxidant, typically air, must be supplied continuously to sustain the electrochemical reaction at the cathode. Therefore, a compressor or air blower is used to provide the required flow of oxidant. Higher air flow rates lead to a reduction in the polarization at the cathode, but high air flow rates are not always preferred. The optimal flow rate will depend on the type of fuel cell and the application for which it is used. In all cases, some of the power generated in the cell stack must be used to operate the compressor or blower, impacting the efficiency of the system.

Similarly, fuel is supplied constantly to the anode. For many fuel-cell systems, the fuel must be *processed* prior to entering the electrochemical device. Operation on pure hydrogen is relatively simple, but use of a hydrocarbon fuel may require sulfur removal, *reformation* of the fuel (discussed below), and in some cases even the removal of small amounts of carbon monoxide from the reformed fuel. Here, we consider natural gas as the fuel for a low-temperature PEM fuel cell. After sulfur removal, three steps are needed to process the fuel.

$$CH_4 + H_2O \rightarrow CO + 3H_2 \quad \text{reformation}$$

$$CO + H_2O \leftrightarrow CO_2 + H_2 \quad \text{water-gas shift reaction}$$

$$CO + 0.5\,O_2 \rightarrow CO_2 \quad \text{selective oxidation}$$

Reformation is a general term that refers to the conversion of a hydrocarbon fuel to hydrogen and carbon monoxide.

Electrochemical Engineering, First Edition. Thomas F. Fuller and John N. Harb.
© 2018 Thomas F. Fuller and John N. Harb. Published 2018 by John Wiley & Sons, Inc.
Companion Website: www.wiley.com/go/fuller/electrochemicalengineering

Reformation is an endothermic reaction; therefore, heat must be added to drive the reaction. Ideally, reforming is done with close thermal coupling of the chemical (reformation) reactor with the fuel-cell stack in order to use waste heat from the stack to drive the chemical reaction. If the amount or quality of heat available is not sufficient, then some of the fuel may be burned to assist with reformation. Any combustion of the fuel to provide heat will, of course, reduce the efficiency of the system. Reformation is performed at high temperatures. Although some of the CO reacts with water to form hydrogen at the high temperature, a second lower temperature reactor is often needed to shift the equilibrium further toward hydrogen production. Finally, for acid fuel cells below about 150 °C, the remaining carbon monoxide must be removed before delivering the hydrogen to the cell stack. Removal is accomplished by selective oxidation of the carbon monoxide.

As we saw in Chapter 9, a general characteristic of a fuel cell is that the efficiency is highest at low current densities. Efficiency will be considered in greater detail in the next section, but one cannot design a fuel cell with efficiency as the only target. System size, cost, and durability are also important design considerations. Referring back to Figure 10.1, the efficiency of a fuel cell can be expressed as the net electrical work produced divided by the energy available from the reactants:

$$\eta_{sys} = \frac{\text{net electrical output}}{\text{energy available from the fuel}}$$
$$\approx \frac{IV_s - \text{ancillary power} - \text{electrical losses}}{\sum_i \Delta G_{f,i} \frac{\dot{m}_{i,in}}{M_i} - \sum_i \Delta G_{f,i} \frac{\dot{m}_{i,out}}{M_i}}, \quad (10.1)$$

where the free energy change at standard conditions has been used to approximate the free energy change of the system. This is a reasonable assumption since it is likely that streams entering and leaving the overall system will be close to standard conditions. The gross electrical work is simply the current multiplied by the operating potential of the cell

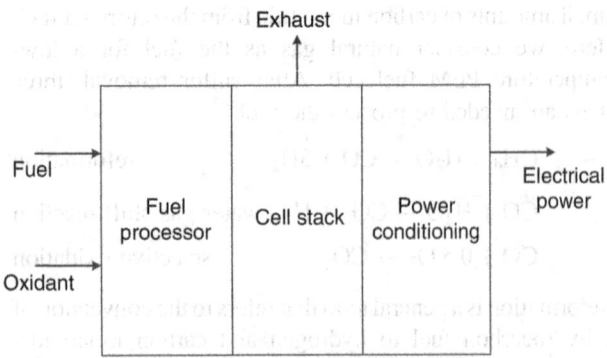

Figure 10.1 Generic fuel-cell system operating on fuel and oxidant. There are three main subsystems: fuel processor, the cell stack, and the power conditioning subsystem.

stack, IV_s. From this value, we subtract any power needed to run ancillary equipment, also called parasitic power, as well as any electrical losses external to the stack (e.g., losses resulting from power conditioning). Clearly, we want the cell potential to be high in order to achieve high efficiency. Thus, minimization of ohmic, kinetic, and mass-transfer polarizations is desired, just as it was for batteries.

The fuel can be thought of as a compound containing carbon, hydrogen, and oxygen, generically represented with the formula $C_xH_yO_z$. We have already seen some examples of fuels, CH_3OH, CH_4, and H_2. When calculating the energy content or availability of the fuel, we assume that all of the fuel supplied to the fuel cell reacts completely to form water and carbon dioxide,

$$C_xH_yO_z + \left(x + \frac{y}{4} - \frac{z}{2}\right)O_2 \rightarrow xCO_2 + \frac{y}{2}H_2O \quad (10.2)$$

This reaction describes the combustion of the fuel; but remember, a fuel cell is not a thermal device. The maximum work per time available from the reactants is given by the change in Gibbs energy for the reaction (10.2) multiplied by the flow rate of the fuel. Ideally, the flow of oxidant would be the exact stoichiometric amount needed for complete combustion of the fuel. It turns out that complete stoichiometric utilization of the fuel is not practical. Rather, reduction of the required fuel flow rate and optimization of the oxidant flow rate represent important design considerations.

For fuel cells, a system efficiency based on the change in enthalpy is more commonly used than the definition above in terms of the Gibbs energy. The enthalpy-based efficiency is called the *system thermal efficiency*, and its use in electrochemical systems is limited mostly to fuel cells. In defining this system efficiency, we assume that water and carbon dioxide are produced according to the stoichiometry of Equation 10.2. Since the enthalpy of formation of molecular oxygen is zero,

$$\eta_{th} = \frac{\text{net electrical output}}{\text{enthalpy of combustion}}$$
$$\approx \frac{IV_s - \text{ancillary power} - \text{electrical losses}}{\frac{\dot{m}_f}{M_f}\left(\Delta H_{f,fuel} - x\Delta H_{f,CO_2} - \frac{y}{2}\Delta H_{f,H_2O(g)}\right)}$$
$$\approx \frac{IN_cV_{cell} - \text{ancillary power} - \text{electrical losses}}{\frac{\dot{m}_f}{M_f}\left(\Delta H_{f,fuel} - x\Delta H_{f,CO_2} - \frac{y}{2}\Delta H_{f,H_2O(g)}\right)}.$$
$$(10.3)$$

The denominator here is the change in enthalpy associated with combustion of the fuel, also termed the heating value of the fuel. In order to calculate the heating value, one must decide whether to treat product water as a vapor or as a liquid. The enthalpy of reaction depends strongly on the phase of the product water. The higher heating value (HHV) assumes that the product water is in the liquid phase; the lower heating value (LHV) assumes it is in the vapor phase. Usually, the lower heating value is used, regardless of the actual state of the water. Can you infer why someone promoting fuel cells might adopt this practice? Ideally, the enthalpy change would account for actual temperatures of the inlet and outlet streams. However, for our purposes, we will assume that the heat of reaction at 25 °C can be used to adequately approximate the required heating value. The system thermal efficiency is illustrated in the following example.

ILLUSTRATION 10.1

A fuel-cell system uses natural gas as the fuel. The gross electrical output is 220 kW, the ancillary power is 20 kW. Neglect any electrical and power conditioning losses. If methane is fed to the fuel-cell system at a rate of 0.011 kg s^{-1}, what is the system thermal efficiency?

The net electrical output is simply $220 - 20 \text{ kW} = 200 \text{ kW}$. The mass flow rate of 0.011 kg s^{-1} is converted to a molar flow rate, using the molecular weight of 16 g·mol^{-1} for methane.

$$\text{molar flow rate} = \frac{11 \text{ g s}^{-1}}{16 \text{ g mol}^{-1}} = 0.6875 \text{ mol s}^{-1}.$$

The enthalpy of formation for methane, water (vapor), and carbon dioxide are found in Appendix C and are used to find the efficiency. For methane, $x=1$, $y=4$.

$$\eta_{th} = \frac{IV_s - \text{ancillary power}}{\frac{\dot{m}_f}{M_f}\left(\Delta H_{f,\text{fuel}} - x\Delta H_{f,CO_2} - \frac{y}{2}\Delta H_{f,H_2O(g)}\right)}$$

$$= \frac{220,000 - 20,000}{\frac{11}{16}[-74,852 + 393,509 + 2(241,572)]} = 0.36.$$

Note that we used the thermodynamic data at 25 °C. How can this be justified? Although the cell may be operating at a much higher temperature, it is likely that the air, fuel, and even exhaust of the complete system won't be far away from these temperatures.

We can write the system thermal efficiency as the product of the fuel efficiency, thermal voltage efficiency, power-conditioning efficiency, and mechanical efficiency:

$$\eta_{th} = \eta_{fuel} \times \eta_{V,t} \times \eta_{pc} \times \eta_{mech}. \quad (10.4)$$

The first term on the right side of Equation 10.4 is the fuel efficiency. It is defined as the fraction of the fuel fed into the system that contributes to the electrical current in the fuel cell. Ideally, this would be one, but that is difficult to achieve.

$$\eta_{fuel} = \frac{\text{fuel that contributes to current in fuel cell}}{\text{total amount of fuel that is supplied}}$$

$$= \frac{I/(4x+y-2z)F}{\dot{m}_{f,cell}/M_f} = \frac{IN_c/(4x+y-2z)F}{\dot{m}_f/M_f}, \quad (10.5)$$

where $\dot{m}_{f,cell}$ is the flow rate through a single cell and $\dot{m}_f$ is the combined flow rate through all of the cells in the stack. Take a moment and think about the term $(4x + y - 2z)$. You should be able to relate this to the number of electrons being transferred per mole of fuel (see Equation 10.2). It may be helpful to think about this in terms of the moles of O_2 required per mole of fuel and the number of equivalents associated with the reduction of each mole of O_2.

Next is the *thermal voltage efficiency*, which describes how effectively the energy from the fuel is converted to electrical power in the fuel cell. This efficiency accounts for voltage losses in the cell; any voltage losses or cell polarizations, explored in Chapter 9, reduces this efficiency. Here, $\eta_{V,t}$ is defined as the operating potential of the cell divided by the change in enthalpy of the reaction expressed in [J·C^{-1}]:

$$\eta_{V,t} \equiv \frac{V_{cell}}{\left(\dfrac{\Delta H_{f,\text{fuel}} - x\Delta H_{f,CO_2} - \dfrac{y}{2}\Delta H_{f,H_2O(g)}}{(4x+y-2z)F}\right)}. \quad (10.6)$$

We can also express thermal voltage efficiency in terms of the voltage efficiency for fuel cells introduced in Chapter 9:

$$\eta_{V,t} \equiv \eta_V^{fc}\left(\frac{\Delta G}{\Delta H}\right) = \frac{V_{cell}}{U}\left(\frac{\Delta G}{\Delta H}\right). \quad (10.7)$$

Note that the voltage efficiency defined here for a fuel cell, η_V^{fc}, is slightly different from the voltage efficiency used for batteries, Equation 7.22. The thermal voltage efficiency is obtained by multiplying η_V^{fc} by the ratio of changes in Gibbs energy and enthalpy. Because we have defined this thermal voltage efficiency in terms of enthalpy instead of Gibbs energy, the efficiency is not unity, even for reversible conditions.

> For fuel-cell systems, efficiencies are defined in terms of the change in enthalpy rather than the change in Gibbs energy. This convention is largely historical. Other methods of producing electricity from fuels use thermal engines that are limited by the Carnot efficiency, where the heat released by the combustion of fuels is important.

The gross electrical power produced by the stack generally requires conditioning, converting DC power to regulated

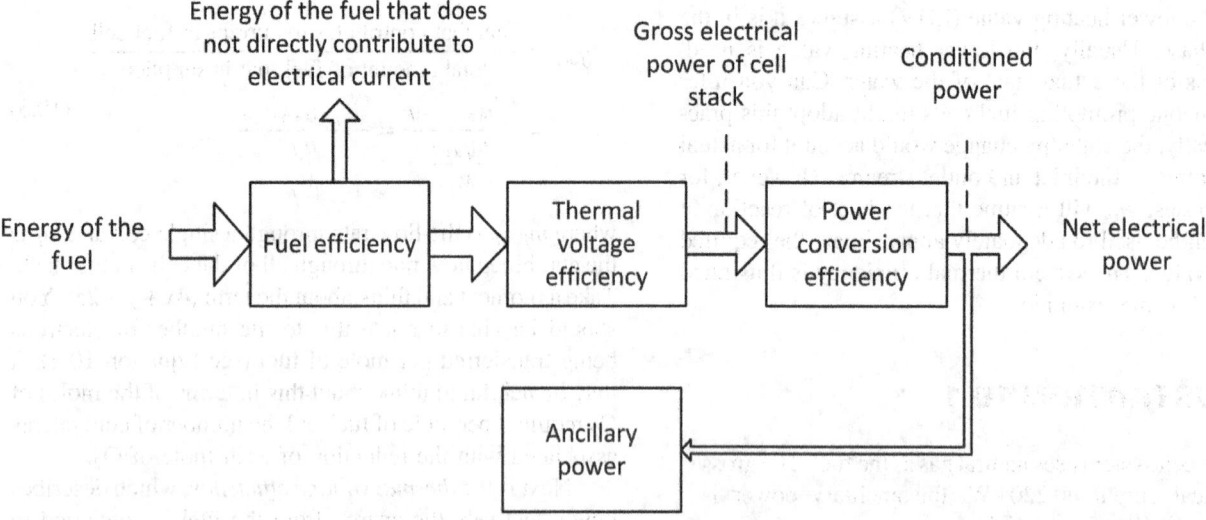

Figure 10.2 Power flow and efficiencies for fuel-cell systems.

AC power for instance. The efficiency for power conditioning is expressed as

$$\eta_{pc} \equiv \frac{\text{conditioned electrical power}}{\text{gross electrical power from stack}}$$
$$= 1 - \frac{\text{electrical losses}}{\text{gross electrical power from stack}}. \quad (10.8)$$

These losses may also come from electrical resistances in the cables and contact resistances, as well as losses from the power conversion. Finally, a mechanical efficiency is defined as

$$\eta_{mech} \equiv \frac{\text{net electrical power}}{\text{conditioned electrical power}}$$
$$= 1 - \frac{\text{ancillary power}}{\text{conditioned electrical power}}. \quad (10.9)$$

Here, the net electrical power is the ancillary power subtracted from the conditioned power. Thus, we see that as the power required to drive any pumps or blowers to operate controllers or other devices in the system is reduced, the mechanical efficiency approaches unity.

The flow of power is depicted in Figure 10.2. Each of these efficiencies will be used over the course of the chapter. In particular, we are interested in how they are interrelated and influenced by the system design.

10.2 BASIC STACK DESIGN CONCEPTS

In this section, we consider the initial or basic design of the fuel-cell stack. The fuel-cell stack is part of the fuel-cell system, which itself is part of a larger system. Therefore, some of the important design specifications and constraints for the fuel-cell stack are derived from the high-level requirements of the larger system. For instance, we might be tasked to generate 400 kW of regulated AC power from natural gas using a SOFC, or to provide electrical power to a transit bus with a PEMFC running on compressed hydrogen, or to supply 100 W of electrical power for 36 hours continuously using a direct methanol fuel-cell system. A list of common high-level requirements is given in Table 10.1. For our purposes here, we will assume that

Table 10.1 Typical High-Level Requirements

Requirement	Units	Comments
Average net power	W	Conditioned electrical power minus ancillary power requirements
Nominal, maximum, and minimum voltages	V	Output voltage of the cell stack, not an individual cell.
System efficiency	–	As is done in practice, we'll use the LHV, Equation 10.3.
Fuel source	–	Examples are compressed hydrogen, industrial natural gas, methanol
Oxidant source	–	Air in all but a few instances
Weight or mass	kg	
Volume	m^3	
Heat sink	–	Available means to reject heat, typically the atmosphere is the sink
Lifetime	years	Important, but beyond the scope of this book

we have already committed to a particular type of fuel cell and have identified the fuel and oxidant.

We now turn our attention to the design of the fuel-cell stack. As described in Chapter 9, the current versus potential or polarization curve establishes the basic performance of a single cell. Assuming that this performance is known and that the individual cell design is fixed, what is left for the fuel-cell designer to specify? It turns out that there are choices that can significantly impact the overall operation and efficiency. In this section, and those that follow, we will introduce a few basic concepts and tradeoffs that are important to fuel-cell design. For example, consider the following:

1. At what point on the polarization curve should the fuel cell be designed to operate? A lower current density results in higher cell potential and greater efficiency. But the lower current density requires a larger cell area, increasing the mass, volume, and cost of the cell stack.
2. If the current density is established, is it preferred to have a few very large cells, or would a larger number of smaller cells be better? What shape should these cells take?
3. How will the required flow rates of air and fuel be distributed to these cells and within each cell?

We begin by developing an initial design for the fuel-cell stack. Table 10.2 lists variables that are critical to this design. Several important relationships exist between these variables that enable us to reduce the number that must be specified. The most important relationship is the polarization curve, written here as a function of the current density:

$$V_{cell} = f(i). \quad (10.10)$$

Table 10.2 Initial Design of a Fuel-Cell Stack Consisting of Eight Key Variables

Variable	Units	Comments
P	W	Power, gross average electrical power from the cell stack
$\eta_{V,t}$	—	Thermal voltage efficiency
V_s	V	Nominal voltage of the cell stack
N_c	—	Number of cells in the stack that are connected in series
V_{cell}	V	Potential of an individual cell
A_c	m^2	Area of individual cell
A	m^2	Total cell area, sometimes referred to as the separator area
i	A·m^{-2}	Current density that corresponds to the nominal voltage

The second and third relationships are also connected to the polarization curve. The thermal voltage efficiency is directly proportional to the cell voltage (see Equation 10.6):

$$\eta_{V,t} = \frac{V_{cell}}{\frac{-\Delta H}{nF}}, \quad (10.11)$$

and the power generated per unit area is simply IV. Thus,

$$P = (iV_{cell})A = IV_s. \quad (10.12)$$

The total area and stack voltage are simply proportional to the number of cells in the stack, N_c:

$$A = N_c A_c, \quad (10.13)$$
$$V_s = N_c V_{cell}. \quad (10.14)$$

where Equation 10.14 assumes that the cells are connected in series. With these five relationships (Equations 10.10–10.14), three of the eight variables must be specified to completely define a solution. Most commonly, the three specifications come from the system-level requirements and are the thermal voltage efficiency, the stack voltage, and the power.

The thermal voltage efficiency ($\eta_{V,t}$) needed for stack design can be estimated from the system thermal efficiency (η_{th}) with use of reasonable estimates for the fuel, mechanical, and power-conditioning efficiencies:

$$\eta_{V,t} = \frac{\eta_{th}}{\eta_{fuel} \times \eta_{mech} \times \eta_{pc}}. \quad (10.15)$$

A similar estimation of the gross power of the stack can be made from the final power requirements for the system:

$$\text{gross power of stack} = P = \frac{P_{req}}{\eta_{mech} \times \eta_{pc}}. \quad (10.16)$$

Use of these relationships is shown in the following illustration, where the number of cells and area are determined.

ILLUSTRATION 10.2

The fuel cell described in Illustration 10.1 requires a voltage for the stack of 190 V; determine the number of cells that must be connected in series and the area of each cell. Assume a power conditioning efficiency of 0.94 and a fuel efficiency of 0.76. The current–voltage performance of the cell is provided in the figure.

Solution:

In the previous illustration, we assumed that the electrical losses were zero. When included, the gross electrical power generated by the stack must increase (Equation 10.16):

$$\text{gross electrical power from stack} = \frac{\text{conditioned electrical power}}{\eta_{pc}}$$

$$= \frac{220}{0.94} = 234 \text{ kW}.$$

The thermal voltage efficiency ($\eta_{V,t}$) is needed for the stack design. We can estimate its value using (η_{th}) from Illustration 10.1, and the additional efficiencies given above. We still need the mechanical efficiency, which can be estimated from the information given in the previous illustration:

$$\eta_{mech} = \frac{\text{conditioned electrical power} - \text{ancillary power}}{\text{conditioned electrical power}}$$

$$= \frac{220 - 20}{220} = 0.91,$$

$$\eta_{V,t} = \frac{\eta_{th}}{\eta_{fuel} \times \eta_{mech} \times \eta_{pc}} = \frac{0.36}{0.76 \times 0.91 \times 0.94} = 0.554.$$

Next, we need to convert this thermal voltage efficiency to a cell potential. Applying Equation 10.11,

$$V_{cell} = \eta_{V,t}\left(\frac{-\Delta H}{nF}\right).$$

For the methane reaction at standard conditions, $\frac{-\Delta H}{nF} = 1.039\,V$. Therefore, the cell voltage is 0.575 V. Using the graph, a cell potential of 0.575 V corresponds to a current density of 9500 A·m^{-2}. The number of cells in the stack is the number of cells in series needed to achieve the specified stack voltage.

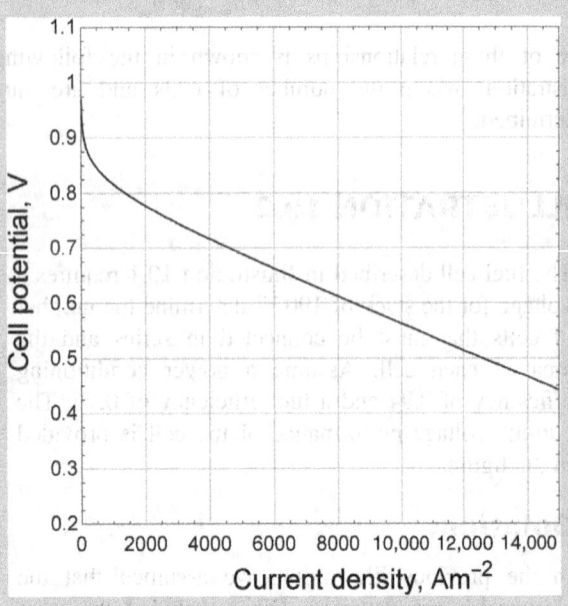

$$N_c = \frac{\text{nominal voltage of stack}}{\text{voltage of individual cell}} = \frac{190}{0.575} = 330 \text{ cells}.$$

Finally, we need the cell area. The gross power from the stack and the stack voltage can be used to determine the current required from the stack:

$$I = \frac{\text{gross power of stack}}{\text{nominal voltage of stack}} = \frac{234,000}{190} = 1232 \text{ A},$$

$$\text{area of individual cells} = A_c = \frac{I}{i} = \frac{1232}{9500} = 0.13 \text{ m}^2.$$

In the next three sections, we discuss how these cells are assembled and explain in more detail how the reactants are supplied. With this knowledge, we can see how volume and mass enter into the analysis.

10.3 CELL STACK CONFIGURATIONS

Potentials of individual cells are about 1 V. Most applications require at least a few volts, but often hundreds of volts are desired. Single cells can be fabricated and then connected in series to increase the voltage, just as is done in batteries. The first approach, which is used commonly in batteries, is referred to as a *monopolar* design as shown Figure 10.3a. The monopolar design consists of distinct positive and negative electrodes. Multiple electrodes are often connected in parallel to form a cell with increased area as shown in the figure. The cells are then connected in series to increase the voltage as required for the application. Series connections are made external to the individual cells as the current is collected from one cell and then directed into the next cell via an external electrical connection.

In contrast to batteries, the most common approach to building voltage in fuel cells is to use what is known as a *bipolar* design or *bipolar stack* as shown in Figure 10.3b. In the bipolar configuration, one positive and one negative electrode are mated together. The reactants are still physically separated, but the negative electrode of one cell is electrically connected to the positive electrode of the next cell in the stack. Thus, there is a single plate that serves as the current collector for two cells—hence the name bipolar plate and bipolar configuration. Note that in the bipolar design the current does not have to be directed in and out of the cells, rather current can pass in a straight line along the axis of the stack. Because external wires are not needed to connect cells in series in order to build voltage, the bipolar configuration is more compact than the monopolar design.

Current collection for the monopolar design has been discussed previously in Chapter 8 for batteries. Similarly, at each end of a bipolar fuel-cell stack, current is collected and

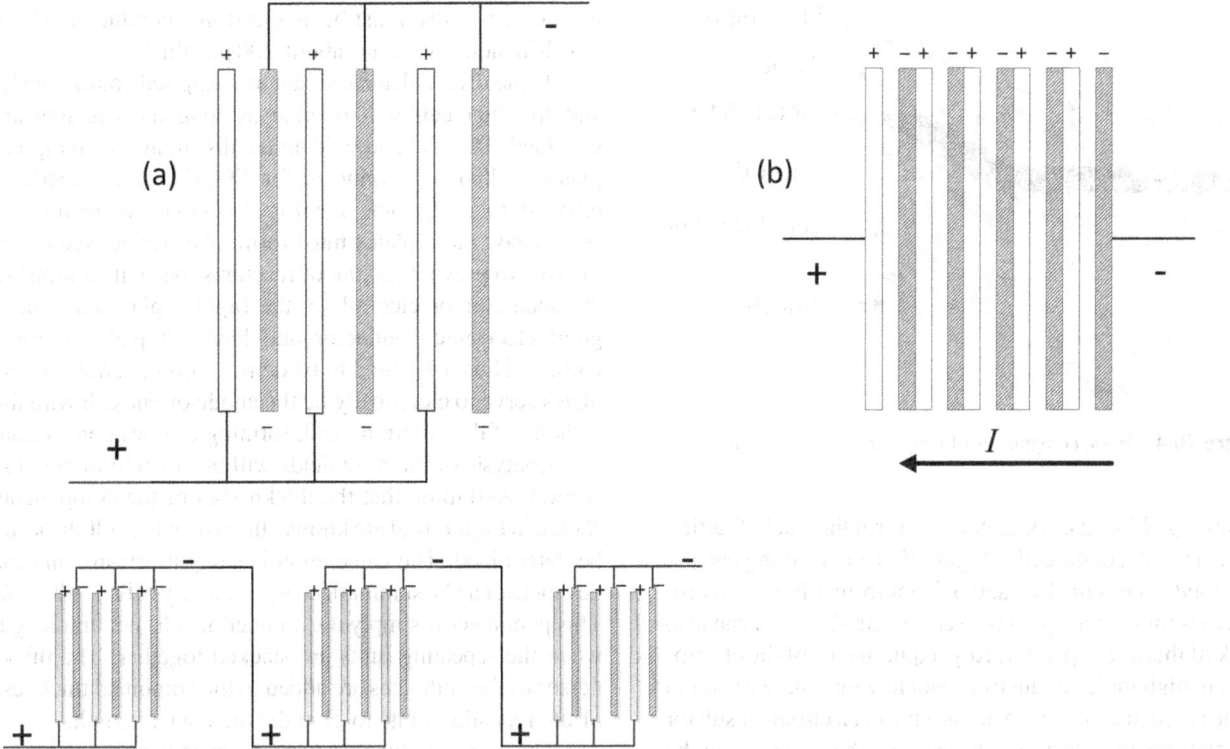

Figure 10.3 Monopolar and bipolar designs. The diagram on the left (a) shows the monopolar design; below this the connections required to build voltage are illustrated. The bipolar design is shown on the right (b).

directed into wires or bus bars. Here, the design challenges are similar to those found in the monopolar design.

The bipolar plates between cells in the fuel-cell stack are critical to our ability to utilize effectively the advantages of the bipolar configuration. Of primary importance is that this bipolar plate serves as a barrier so that reactants from the anode do not mix with those of the cathode from the next cell. The prevention of mixing is essential from both safety and efficiency perspectives. A second important function of the bipolar plate is to collect current from the cathode of one cell and pass it to the adjoining anode of the next cell. A highly conductive bipolar plate will tend to even out any local nonuniformities in the current from an individual cell.

Assuming that the bipolar design is preferred, are we always going to put all of the cells in series in a single stack? The answer is no. As the number of cells increase, it become progressively more difficult to assemble the cell stack. Further, there are times when one long stack just doesn't suit the available space allocated for the fuel cell. In these instances, two, three, or more cell stacks may be fabricated and then the stacks connected together in series with external wires. In some instances, we may face the reverse problem, although this is rare. Namely, the area of the individual cell calculated as in Illustration 10.2 cannot be used because of manufacturing limitations or space constraints. In such cases, it is possible to split the fuel cell in two cells with smaller areas, and then connect them in parallel with external wires.

We have determined the area of individual cells in the stack, but not their shape. For bipolar configurations, *planform* is used to describe the shape of the cell when viewed from above. A rectangular planform is the norm, but how might we decide on the aspect ratio? Often, space limitations dictate the aspect ratio. Assuming that we aren't constrained by the available space, then ease of manufacturing and pressure drops associated with reactant flows are critical factors. As shown in Figure 10.4, another function of bipolar plates is to direct the flow of reactants over the planform area of the cell. Details of these *flow fields* will be addressed in Section 10.6. Of course, the bipolar stack is not the only approach used to build voltage in fuel cells. A notable example is the so-called *tubular design* used in SOFCs. Some of the challenges and advantages of this design were explored in Problem 9.9 of the previous chapter.

10.4 BASIC CONSTRUCTION AND COMPONENTS

Let's now take a more detailed look at the components of a bipolar fuel-cell stack as illustrated in Figure 10.4. Multiple layers are assembled together to form a cell, and these layers

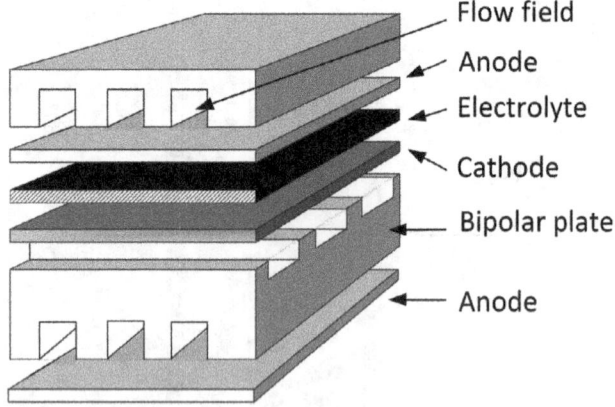

Figure 10.4 Basic components of a bipolar fuel-cell design.

are arranged like a deck of cards to form the stack. Starting from the center of the cell in Figure 10.4 and working out, the electrolyte keeps the fuel and oxidant from mixing and also provides for an ionic path between electrodes. You can also think of this as a separator. Key requirements of the electrolyte are high ionic conductivity and low rates of permeation of the reactants. Of course, it must be an electronic insulator to isolate the two electrodes from each other and prevent the cell from being shorted. In addition to high conductivity, a high transference number of the ion that is involved in the electrochemical reaction is desired.

Next are porous electrodes as described in Chapter 5. Most often a catalyst is used to reduce the overpotential at the electrodes, and generally the electrodes for the anode and the cathode are different. Intimate contact of the reactants, electrolyte, and electronic conducting material, as well as the catalyst, is required to achieve the triple-phase boundary discussed in the previous chapter.

The bipolar plates generally will not be mated directly to the electrode surface. There is an additional porous layer between them as illustrated in Figure 10.5. This layer, called a gas-diffusion layer (GDL) in PEMFCs, allows access by the reactants to the entire electrode surface. Without this layer, the lands (also called ribs) would prevent the reactants from easily reaching the area under the ribs. This gas-diffusion layer should be highly porous to permit easy access to the electrode surface by the reactants, but also must be an electronic conductor. These gas diffusion layers are about 200 μm thick.

Typically, fuel and oxidant are supplied continuously, and the fuel cell will operate as long as reactants are provided. The fuel and oxidant are distributed with bipolar plates (called interconnects for SOFCs, and sometimes referred to as bipolar *separator* plates). As mentioned previously, these plates must form a barrier between cells in order to prevent mixing of reactants. Also, in contrast to the separator or electrolyte, the bipolar plate must be a good electronic conductor and have no path for ionic current. Thus, they tend to be dense solid materials. These plates serve to electrically tie the anode of one cell with the cathode of the adjoining cell, forming a series connection.

Analysis of the flow fields will be covered in the next section. Assuming that the thicknesses of the components shown in Figure 10.4 are known, the size of the cell stack can be determined. The calculation is straightforward, and the key metric can be summarized by a quantity called *cell pitch*. This parameter is simply the number of cells per unit length when the repeating units are stacked together. The thicknesses of the endplates are added to the combined thickness of the repeating units to give the total stack length.

The last quantity we want to introduce is a ratio of active area to planform area. If we examine the planform, we note that a portion of the area is allocated to sealing, and distributing reactants, namely, a manifold. To account for this, we simply use an active area ratio:

$$A_{act} \equiv \frac{\text{active or electrode area}}{\text{planform area}}. \quad (10.17)$$

The connection between the manifold and the individual cells is illustrated in Figure 10.6; the active area ratio is evident from the top (planform) view. Illustration 10.3 demonstrates the calculation of the fuel-cell volume.

ILLUSTRATION 10.3

If the cell in Illustration 10.2 had a cell pitch of 3 cells cm^{-1} and the active area ratio is 0.7, calculate the volume of the cell stack. The endplates may be neglected.

The length of the cell stack is

$$\text{stack length} = 330 \text{ cells} \times \frac{0.01 \text{ m}}{3 \text{ cells}} = 1.1 \text{ m}.$$

The planform area is

$$\text{planform area} = 0.13 \text{ m}^2 \times \frac{1}{0.7} = 0.186 \text{ m}^2.$$

The volume is, therefore,

$$\text{stack volume} = 0.186 \text{ m}^2 \times 1.1 \text{ m} = 0.20 \text{ m}^3.$$

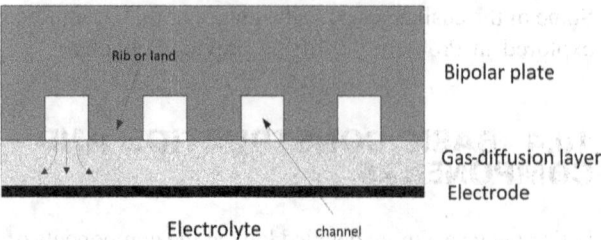

Figure 10.5 Designs used in PEM and SOFC to distribute reactants across the entire electrode.

> The actual volume will be larger still because a loading system is needed to apply a force to compress the stack axially, and hardware may be needed to attach the manifolds and to mount the cell stack assembly in the system.

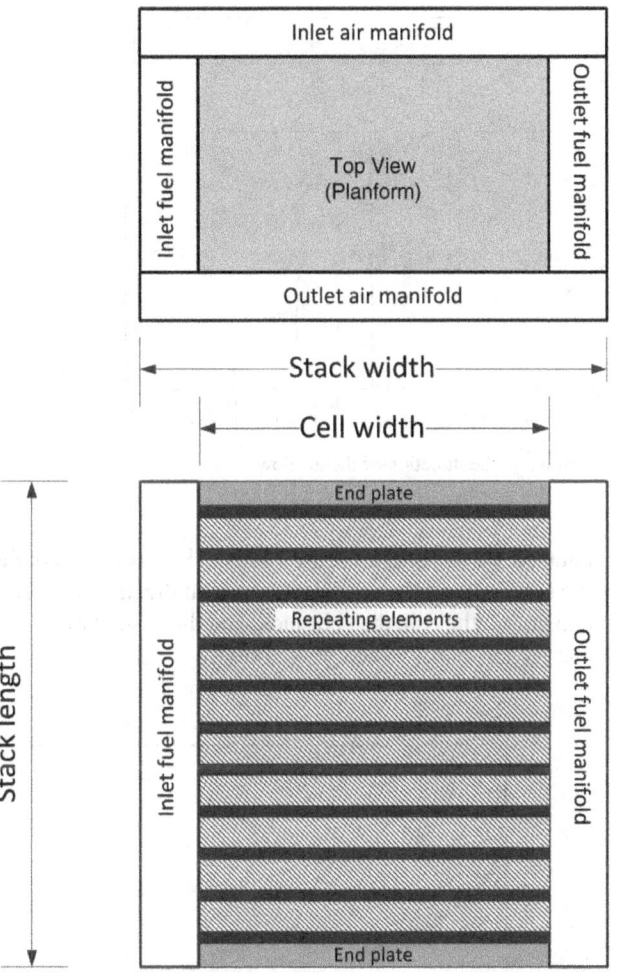

Figure 10.6 Cell stack assembly showing repeating elements, manifolds, and end plates.

10.5 UTILIZATION OF OXIDANT AND FUEL

For both the fuel and oxidant streams, *utilization* is an important design factor. The utilization of fuel plays a key role in the fuel efficiency, and the utilization of the oxidant, typically air, affects the mechanical efficiency of the system and polarization losses in the cell. We'll first consider the oxidant by examining the cathode of a PEMFC to illustrate the concepts. If the oxidant is air, and oxygen is being reduced in an acid medium:

$$O_2 + 4H^+ + 4e^- \rightarrow 2H_2O$$

The utilization of oxygen is defined as

$$u_{O_2} \equiv \frac{\text{rate of oxygen consumption}}{\text{rate of oxygen supplied}} = \frac{(i/nF)A}{\dot{m}_{air} y_{O_2}/M_{air}}, \quad (10.18)$$

where A is the total combined active area of the cells and the current density is the same for all of the cells in series. $\dot{m}_{air}$ is the rate of mass flow of air to the cathodes. n is four for the reduction of oxygen.

As the oxidant is typically air and air is "free," one might conclude that a high flow rate of air, and hence low oxygen utilization, is preferred. However, power is required to supply air to the stack, directly affecting the efficiency of the fuel-cell system. In contrast, a high utilization of oxygen results in a lower flow rate of air and a reduced ancillary power requirement. What then are the disadvantages of high utilization? We can answer this question in part by considering what happens along the direction of air flow (normal to current flow in the stack) as illustrated in Figure 10.7. Imagine that our planform is divided into three segments that are no longer electrically connected to each other. Simultaneously, we apply the same current density to each segment and measure the corresponding voltage. By repeating this procedure, the polarization curve for each segment can be measured while oxygen flows sequentially through the segments. As oxygen is consumed in the first segment, the concentration of oxygen reaching the second segment will be lower, and still lower in the third segment. As we studied in the previous chapter, the lower oxygen partial pressure is expected to increase both kinetic and mass-transfer polarization. Three hypothetical polarization curves are shown in Figure 10.7, one near the air entrance, one in the middle, and one near the air exit. The greater the utilization of oxygen, the lower the partial pressure of oxygen at the exit, and the lower the limiting current.

Of course, rather than isolated segments, the cell is connected to a highly conductive bipolar plate. This connection allows the current density to redistribute to mitigate the effect of oxygen depletion along the flow direction. Thus, the current density would be higher than the average value near the air entrance and lower at the exit. Controlling the total current to the cell and measuring the potential results in single polarization curve. This curve can be thought of as an amalgamation of the polarization curves for the segmented cells. Under special conditions,

232 Electrochemical Engineering

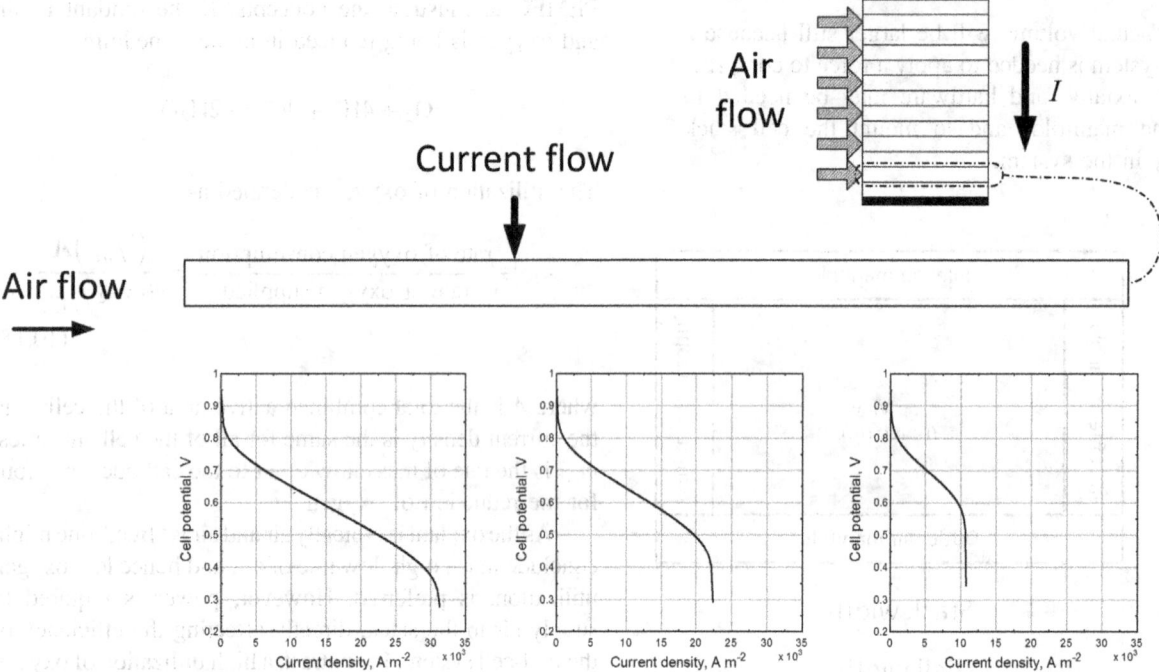

Figure 10.7 Effect of oxygen consumption on the polarization curve as we move in the direction of the air flow.

we can calculate the effect of utilization easily as illustrated in Problem 10.10. However, most of the time, the effect of oxygen utilization is best measured experimentally. The two typical representations of the experimental data are as a set of polarization curves measured at different utilization values (Figure 10.8), or as a utilization sweep, where the current density is held constant and the flow rate varied (Figure 10.9).

Increased polarization at higher oxygen utilization levels is evident from both figures. At low to moderate current densities, there is a kinetic effect due to the lower average partial pressure of oxygen. At higher current densities, mass-transfer effects dominate as evidenced by the reduced limiting current. A well-designed system will avoid operation near the limiting current. In other words, operation at a point with high sensitivity to oxygen utilization should be avoided.

The oxidant flow rate impacts fuel-cell operation in other important ways; these effects will change depending on the type of fuel cell. For a proton exchange membrane fuel cell, water is removed with the vitiated air (exiting air stream). Starting from a steady state, if the flow rate of air increases, the rate of water removal initially increases. However, any new steady state must also be in water balance. This limitation has an important impact on system performance as discussed in Section 10.7. In contrast, for solid oxide fuel cells, water is produced at the anode, and the flow rate of oxidant doesn't influence the water balance. However, air flow plays a key role in cooling the solid oxide cell stack. This situation illustrates a second constraint: the cell, the cell stack, and the system must be in energy balance for steady operation.

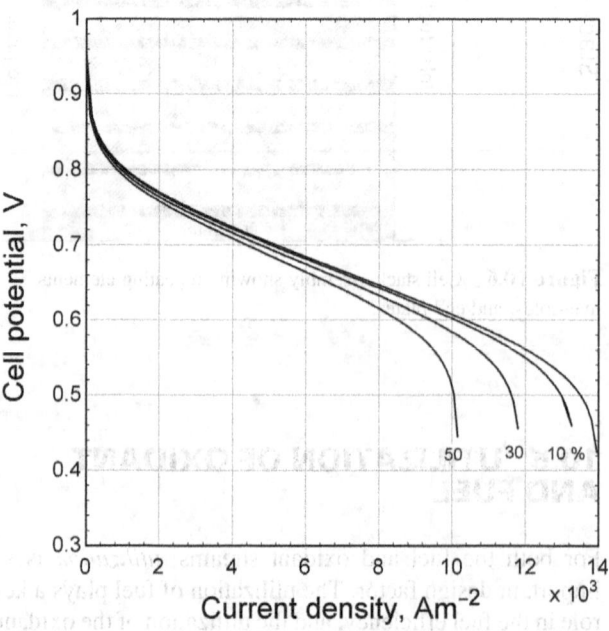

Figure 10.8 Oxygen utilization sweep. The current is held constant and flow of air varied.

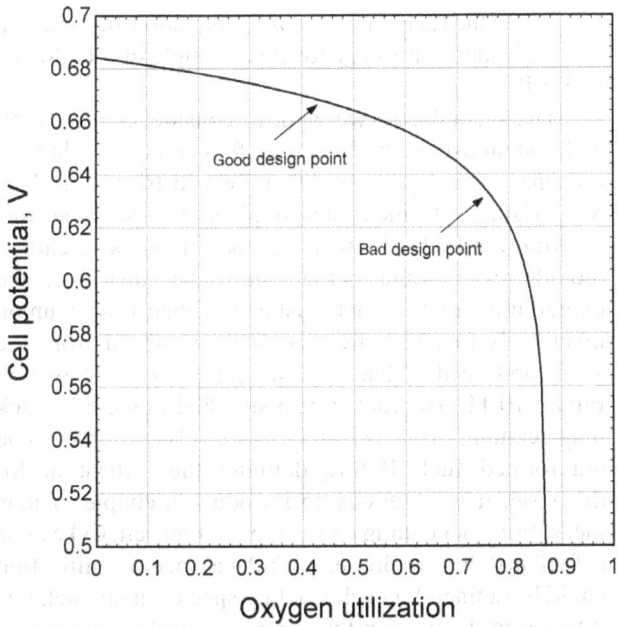

Figure 10.9 Effect of oxygen utilization on the polarization curve.

Let's now consider the fuel. The utilization of fuel in the stack, u_f, is defined similarly to that for oxygen. If the anodic reaction is the oxidation of hydrogen, then

$$H_2 \rightarrow 2H^+ + e^-$$

$$u_f \equiv \frac{\text{rate of hydrogen consumption}}{\text{rate of hydrogen being supplied}} = \frac{(i/nF)A}{\dot{m}_{fuel}y_{H_2}/M_{fuel}}.$$
(10.19)

> Utilization and stoichiometry are two methods of expressing the flow rate of reactants relative to the current in the cells. Both terms are used commonly, and are simply the inverse of each other.
>
> $$\text{utilization} = \frac{1}{\text{stoichiometry}}.$$
>
> High flow rates of reactants mean low utilization or high stoichiometry.

A utilization of 100% means that all of the fuel is used in the cell or, conversely, that just the stoichiometric amount of fuel needed for the required current is provided to the stack. Clearly, the utilization of hydrogen in the stack impacts the fuel efficiency of the system. The consequence of supplying too much fuel is that it is wasted. On the other hand, providing too little fuel can cause severe damage as will be examined more closely in the next section.

ILLUSTRATION 10.4

Calculate the flow rate of hydrogen and air into 5 kW stack operating at a current of 100 A and 47 V. The stack consists of 70 cells (N_c) connected in series with an individual cell area, A_c, of 600 cm². The utilizations of oxygen and fuel are 0.7 and 0.95, respectively.

$$u_f \frac{\dot{m}_{fuel}}{M_{fuel}} y_{H_2} = \frac{IN_c}{n_a F} = \frac{(100)(70)}{2(96485)}.$$

For the oxidation of hydrogen, $n_a = 2$ and we assume pure hydrogen: $y_{H_2} = 1$.

Solving for the mass flow rate, $\dot{m}_{fuel} = 0.077$ g s^{-1}.
For the air, $n_c = 4$ and $y_{O_2} = 0.21$.

$$u_{O_2} \frac{\dot{m}_{air}}{M_{air}} y_{O_2} = \frac{IN_c}{n_c F} = \frac{(100)(70)}{4(96485)}.$$

Hence, $\dot{m}_{air} = 3.57$ g s^{-1}.

In both cases, we have neglected the impact of water vapor in the feed on the inlet mole fraction, which is significant for some types of fuel cells. This is equivalent to doing the calculation on a dry basis.

The utilization we have defined refers to the fraction of reactant consumed in the cell stack. Other definitions of utilization are possible and useful. One can express a more local utilization, for instance, such as the utilization of a single cell in the cell stack. As noted previously, the reactants travel through the stack in parallel. Although the stack is designed to have uniform flow through each cell, there are inevitably variations in flow to the cells. Likewise, within each cell there are often many flow channels in parallel, and the utilization in each channel can vary. Each of these differences in utilization changes the current distribution through the cells.

Looking beyond the cell stack, we can have instances where we are interested in the utilization of fuel in the overall system. As was suggested in Figure 10.2, not all of the fuel provided to the system contributes to electrical current. Some fuel may be combusted, some fuel may be exhausted from the system, and some may permeate across the separator. The overall utilization will depend on the particular design, and there are a great variety of designs and configurations used. Here we consider just two that illustrate the concepts.

To begin, let's imagine we have a bipolar PEMFC stack operating on pure hydrogen. At first thought, you might anticipate that we could easily use all of the hydrogen in the cell stack so that $u_f = 1$. Unfortunately, this proves to be difficult primarily because water vapor and nitrogen dilute the hydrogen. Recall that the membrane requires water for good conductivity, which necessitates that the gas stream be humidified. There is also a small amount of nitrogen from the air side that permeates across the membrane; therefore, there must be a means to remove this nitrogen from the fuel side. Clearly, we don't want to throw away valuable fuel. To compensate for only partial utilization of the fuel entering the stack, we can recycle a portion of the fuel stream that exits the stack as shown in Figure 10.10. With this configuration, we can conceive of two utilizations:

$$u_f = \frac{\text{rate of hydrogen consumed in stack}}{\text{rate of hydrogen supplied to stack}} = \frac{N_c I / 2F}{(\dot{n}_r + \dot{n}_f) y_{H_2}}$$

and

(10.20)

$$u_{f,s} = \frac{\text{rate of hydrogen consumed in stack}}{\text{rate of hydrogen supplied to system}} = \frac{N_c I / 2F}{\dot{n}_f y_{H_2}}.$$

(10.21)

Whereas the consumption terms are identical in Equations 10.20 and 10.21, the supply terms are different (denominators). In this example, we could have pure hydrogen as the feed, but the flow into the cell stack would be diluted with water vapor and a small amount of nitrogen. Additionally, the flow rate through the stack is higher than the feed rate. The recycle ratio, R_r, is the molar flow rate in the recycle loop divided by that of the feed:

$$R_r \equiv \frac{\text{molar flow rate in recycle loop}}{\text{molar flow rate of feed}} = \frac{\dot{n}_r}{\dot{n}_f}. \quad (10.22)$$

Because of recycle, each molecule of hydrogen, on average, may make multiple passes through the fuel-cell stack before either reacting or exiting the system. Use of recycle lowers the concentration of hydrogen in the stack, but enables the utilization of fuel in the system to be increased. Thus, recycle allows us to increase the overall utilization of fuel in the system, $u_{f,s}$, while maintaining a lower single pass utilization of fuel, u_f, in the stack. Note that earlier we

introduced the fuel efficiency, η_{fuel} (Equation 10.5), which we see is identical to $u_{f,s}$ for this example of a hydrogen fuel cell.

The second case we want to consider is for a fuel-cell system using a hydrocarbon fuel and that includes a reformer or fuel processor that we will treat as a black box. Using reformed fuel implies that the reactants entering the fuel-cell stack will be diluted with carbon dioxide, water, and perhaps nitrogen; therefore, the overall utilization of fuel must be less than 1. A common practice is to take unreacted fuel exhausted from the stack and feed it back to the reformer as shown in Figure 10.11. The fuel stream supplied to the cell stack may contain H_2, N_2, CO, CO_2, H_2O, and some unreformed fuel. Before defining the utilization for this case, it is necessary to introduce a couple of new ideas. First, depending on the type of fuel cell, CO can be a fuel just like hydrogen. While a specific utilization could be defined for each reacting species in the fuel, it is difficult to distinguish the contribution of an individual reaction of a multiple reaction set based on the composition of the fuel-cell effluent. The difficulty is complicated further at higher temperatures because reforming reactions may continue to take place inside the fuel cell. We can nonetheless define the utilization for the fuel-cell stack with just a small modification.

$$u_f = \frac{\text{rate of equivalents consumed in stack}}{\text{rate of equivalents supplied to stack}}. \quad (10.23)$$

Think of the equivalents as the amount of electrons that would be supplied by the oxidation of the fuel. For hydrogen, two equivalents per mole of hydrogen are supplied. Earlier, Equation 10.2 was used to describe hydrocarbon oxidation with use of the generic formula $C_x H_y O_z$, where the number of equivalents per mole of fuel is $4x + y - 2z$. This concept can be extended to fuel mixtures. The rate at which fuel is consumed is directly related to the current in the fuel-cell stack. The utilization in Equation 10.23 can be related back to the fuel efficiency used in Equation 10.5 by defining

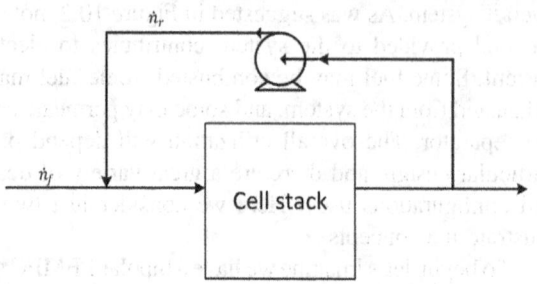

Figure 10.10 Recycle system for hydrogen fuel cell. Utilization of fuel in the stack is different from utilization of fuel in the system.

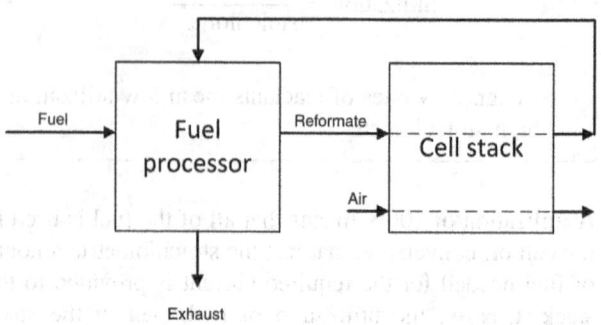

Figure 10.11 One scheme for the integration of cell stack with a fuel processor.

a *fuel processing efficiency*,

$$\eta_{fp} = \frac{\text{rate in equivalents of oxidizable fuel supplied to stack}}{\text{rate in equivalents of fuel supplied to fuel processor}}.$$
(10.24)

Finally, if we multiply these two together,

$$\eta_{fp} \times u_f = \frac{\text{rate in equivalents consumed in stack}}{\text{rate in equivalents of fuel supplied to fuel processor}}$$
$$= \eta_{fuel},$$
(10.25)

we see that this product is the same as the fuel efficiency used earlier. There is an alternative definition for fuel-processing efficiency that is sometimes used:

$$\eta_{fp} = \frac{\text{heating value of the fuel supplied to stack}}{\text{heating value of fuel supplied to fuel processor}},$$

as explored further in Problem 10.12.

ILLUSTRATION 10.5

The fuel-cell system described in Illustration 10.1 uses a steam reformer and a high-temperature SOFC. In a solid oxide fuel cell, both CO and H_2 can be oxidized at the anode. From the $11\,\text{g s}^{-1}$ of methane fed to the system, the output of the fuel processor is a total of $3.678\,\text{mol s}^{-1}$ and has the following composition.

Species	Mole fraction	$(4x+y-2z)$ $C_xH_yO_z$	Equivalent $[\text{mol·s}^{-1}]$
CO	0.1421	2	1.045
CO_2	0.2131	0	0
H_2O	0.06729	0	0
CH_4	0.009346	8	0.275
H_2	0.5682	2	4.180

Assume the methane does not react in the cell stack; that is, the methane does not undergo further reformation and is not oxidized electrochemically. As defined in Equation 10.24, what is the fuel-processor efficiency? If the utilization, u_f, in the stack is 0.80, what is the fuel efficiency?

Solution:

The fuel processor efficiency is (Equation 10.24)

$$\eta_{fp} = \frac{\text{rate in equivalents of oxidizable fuel supplied to stack}}{\text{rate in equivalents of fuel supplied to fuel processor}}.$$

First, we determine the rate in equivalents of oxidizable fuel supplied to the stack (numerator). The methane cannot be oxidized, but hydrogen and carbon monoxide react at the anode according to

$$O^{2-} + H_2 \rightarrow H_2O + 2e^- \qquad O^{2-} + CO \rightarrow CO_2 + 2e^-$$

The numerator can be found using the equivalents (mol·s^{-1}) from the table for H_2 and CO. For the denominator, we need the molar flow rate of methane to the fuel processor, which is $11\,\text{g s}^{-1}/16\,\text{g mol}^{-1} = 0.6875\,\text{mol s}^{-1}$. Methane has 8 equivalents per mole of fuel as shown in the table, where $x=1$, $y=4$, and $z=0$. Therefore, the fuel processor efficiency is

$$\eta_{fp} = \frac{1.045 + 4.180}{0.6875(8)} = 0.95.$$

Note that in the numerator we did not include methane based on the problem statement. Alternatively, we can do the same calculation by subtracting off the methane equivalents exiting the reformer (see table) from the methane equivalents in the original fuel

$$\eta_{fp} = \frac{(8)(0.6875) - 0.275}{(8)(0.6875)} = 0.95.$$

Finally, using Equation 10.25,

$$\eta_{fuel} = \eta_{fp} \times u_f = 0.95 \times 0.8 = 0.76.$$

10.6 FLOW-FIELD DESIGN

As was noted in Section 10.4, the fuel and oxidant must be directed over the surface of anode and cathode in the fuel-cell stack. Most often this is done with rectangular channels. The set of channels make up the *flow field*. Drawing on what we have learned in the previous sections, we now have a good idea of the reactant flow required. The flow to individual cells is in parallel and manifolds are used to distribute flow evenly to individual cells. In order to get the same flow to each cell, we would like the resistance to flow through individual cells to be high relative to the resistance in the manifold. On the other hand, high resistance requires additional pumping power to move the reactants. Remember, efficiency is the key driver in fuel-cell technology. As you might then expect, knowing the resistance to flow or, equivalently, the pressure drop through the flow field is vital.

Let's begin by considering the pressure drop associated with flow through a closed conduit, typically a rectangular channel. The most important parameter is the Reynolds number:

$$\text{Re} = \frac{v_z \rho D_h}{\mu}, \qquad (10.26)$$

where v_z is the average velocity in the channel and the equivalent diameter is

$$D_h = \frac{4(\text{area of flow})}{\text{perimeter}} = \frac{2lw}{(l+w)}, \qquad (10.27)$$

where l and w are the height and width of the channel, respectively. The critical Reynolds number is 2300. As illustrated in the example below, the Re is typically small (about 100) and the flow is laminar. We can calculate the pressure drop through a channel of length L along the direction of flow as

$$\Delta p = L \frac{2f\rho v_z^2}{D_h}. \qquad (10.28)$$

The fanning friction factor, f, is 16/Re for fully developed laminar flow. This equation does not account for entrance effects in the channel as the flow field develops (see Figure 10.12). For laminar flow, the entrance length, L_e, is estimated as

$$\frac{L_e}{D_h} \approx \frac{\text{Re}}{20}. \qquad (10.29)$$

As shown in Figure 10.12, the pressure drop will be a bit higher than that predicted by Equation 10.28 due to entrance effects. We will typically ignore entrance effects for initial calculations, but you should be aware of the implications of this assumption.

Finally, work is required to overcome friction losses in the channel and move the reactants through the fuel cell. The greater the pressure drop, the greater the power needed:

$$\text{Power} = (lw v_z)\Delta p = \dot{V} \Delta p. \qquad (10.30)$$

Changes in the density with pressure have been ignored, a relatively good assumption for the magnitude of pressure drops typically observed. The power required to move the reactants is part of the ancillary power that enters into the mechanical efficiency and directly impacts the overall efficiency of the system.

In addition to the pumping power and its impact on system efficiency, there are some other considerations. As was noted previously, providing equal flow to each cell is important. One of the most important challenges in flow-field design is the distribution of reactants to multiple channels and multiple cells. Reactants are supplied to cells in parallel, and within a single cell there are often many flow channels in parallel. It is important that each receives nearly the same flow of reactants. What are the consequences of supplying too much or too little reactant? The main problem is not providing enough fuel to a cell or to an isolated region of a cell. This phenomenon is called *fuel starvation*. Since all the cells of a bipolar stack are connected electrically in series, the current through each cell is identical. With many cells in series, the voltage differences across the stack can be tens or hundreds of volts. So, even if one cell has too little fuel, the large potential will drive current through the cell anyway. If there is no fuel available to be oxidized, then part of the cell components are oxidized to provide the electrons to allow current to flow. Oxidation of the catalyst carbon support is typical in low-temperature fuel cells,

$$C + 2H_2O \rightarrow CO_2 + 4H^+ + 4e^-$$

Clearly, fuel starvation is a serious problem that must be avoided.

ILLUSTRATION 10.6

The cathode of a fuel cell has straight parallel channels. If the current density is $10\,\text{kA}\,\text{m}^{-2}$, and air is used with a utilization of 0.6, what is the pressure drop through the channels? The dimensions of the channel are 2 mm deep, 2 mm wide (l and w), and the ribs (also called lands) are 1.5 mm wide. The planform size is 30 cm by 30 ($L \times W$) cm.

Solution:

Since the planform is square and the channels are straight, the length of the channels $= L = W$. We will assume that the channels are in along the L dimension. The velocity is determined from the volumetric flow rate of air, $\dot{V}$. The volumetric flow rate of the air can be determined from the cell current density and the utilization value given. The density of air at the operating conditions of the fuel cell is $0.9865\,\text{kg}\,\text{m}^{-3}$. The viscosity of air at the same conditions is $2.096 \times 10^{-5}\,\text{kg} \cdot \text{m}^{-1} \cdot \text{s}^{-1}$.

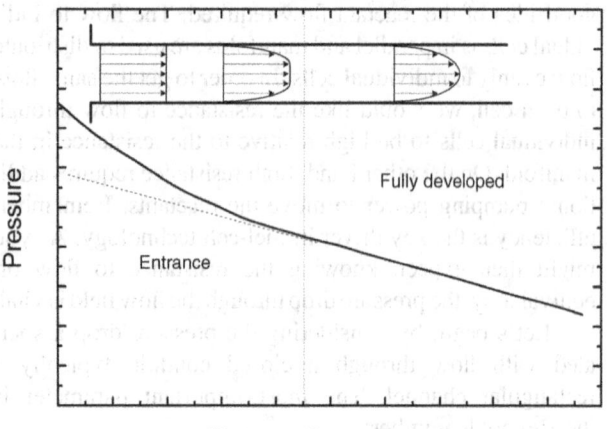

Figure 10.12 Pressure drop for laminar flow including the entrance effect.

Volumetric flow rate $= \dot{V} = i(WL)\dfrac{M_{air}}{nFy_{O_2}\rho_{air}u_{O_2}}$

$= 10000(0.3)(0.3)\dfrac{0.029}{4(96485)0.21(0.9865)0.6}$

$= 5.44 \times 10^{-4} \text{ m}^3 \cdot \text{s}^{-1}.$

The average velocity in the channel is

$v_z = \dfrac{\dot{V}}{Wl}\dfrac{w+rib}{w} = \dfrac{5.44 \times 10^{-4}}{0.3(0.002)}\dfrac{2+1.5}{2} = 1.59 \text{ m} \cdot \text{s}^{-1},$

where a correction has been made for the fact that the air only flows through the channels (rib area is not available for flow). The hydraulic diameter is

$D_h = \dfrac{2lw}{(l+w)} = \dfrac{2(2)(2)}{2+2} = 2.0 \text{ [mm]}.$

And, therefore,

$\text{Re} = \dfrac{v_z \rho D_h}{\mu} = \dfrac{1.59(0.9865)0.002}{2.096 \times 10^{-5}} = 149.$

The flow is laminar, and the fanning friction factor is

$f = \dfrac{16}{\text{Re}} = 0.107,$

$\Delta p = L\dfrac{2f\rho v_z^2}{D_h} = 80 \text{ Pa}.$

The allowable variability (tolerances) in channel dimensions must be specified and carefully controlled when manufacturing these flow fields. Imagine that there are multiple cells connected electrically in series and flow of reactants through these cells is in parallel. What is the impact of having the dimensions of one cell low compared to the others? The overall pressure drop will only be slightly altered by having one of many flow fields out of tolerance, and we will neglect this effect. For a given overall pressure drop, we can use Equation 10.28 to calculate the average velocity:

$$v_z = \dfrac{\Delta p}{L}\dfrac{D_h^2}{32\mu} = \dfrac{\Delta p}{L}\dfrac{1}{8\mu}\dfrac{l^2 w^2}{(l+w)^2}. \qquad (10.31)$$

Of course, the flow rate is proportional to the cross-sectional area, which also changes with the channel dimensions. Accordingly, the volumetric flow through each channel is

$$\dot{V}_{ch} = v_z lw = \dfrac{\Delta p}{L}\dfrac{1}{8\mu}\dfrac{l^3 w^3}{(l+w)^2}. \qquad (10.32)$$

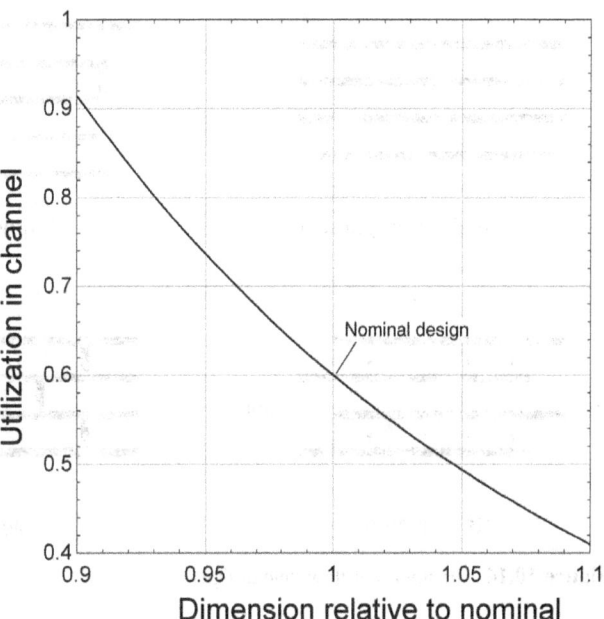

Figure 10.13 Effect of poor tolerance control of channel dimensions in the flow field.

As shown in Equation 10.32, the volumetric flow rate depends on the channel dimensions, and can be significantly different in channels that are out of specification. As a result, the utilization may vary over the planform. Figure 10.13 shows the utilization in a channel if its dimensions are slightly different from the nominal values. Here, the designed utilization of the reactant is 0.6. A 5% change in the dimensions of a channel can cause the utilization to increase from 0.6 to more than 0.7. Severe damage can occur if at any point there is insufficient fuel in the channel. In contrast, channels whose dimensions cause the utilization to decrease relative to the design value are much less of an issue. The flow rate is higher than nominal, but otherwise the performance is good.

There are a large number of flow-field designs. The style chosen depends on the type of fuel cell, the application, and specific design preferences of the manufacturer. These flow-field designs vary greatly in complexity. Several of the approaches are shown in Figure 10.14. You can easily verify that the flow fields with multiple channels will have lower pressure drops; however, the chance of maldistribution is increased. The interdigitated design forces flow through the gas diffusion and electrode layers. The main objective is always to distribute the reactants in a controlled manner without excessive pressure drop.

The simple analysis above can be used to estimate the pressure drop for the straight-through, spiral, and serpentine designs. More complex flow fields require numerical simulation, typically using computational fluid dynamics (CFD), or experiments to characterize their performance.

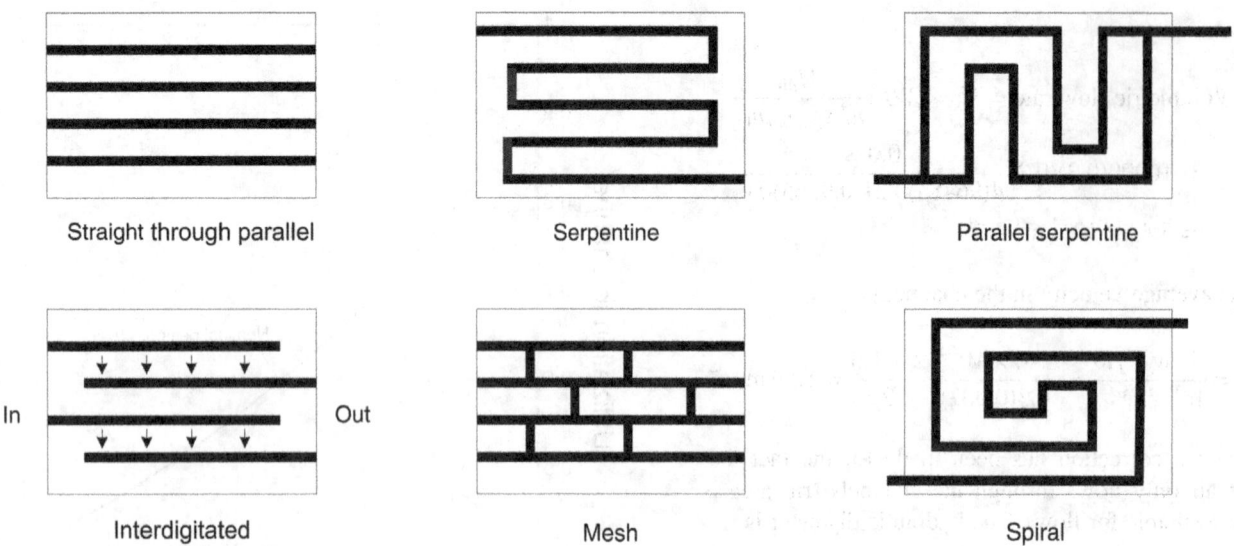

Figure 10.14 Examples of flow-field designs.

10.7 WATER AND THERMAL MANAGEMENT

For any fuel cell system at steady operation, water must be balanced; that is, the amount of water entering the fuel-cell system plus the water produced by the reactions must equal the amount of water in the exhaust. Similarly, water must be in balance around each component of the system. Just as a water balance must be achieved in the system, thermal balance is a condition of steady operation. Although high, the efficiency of a fuel cell is well below 100%; this means that heat must be removed from the cell. The approach for heat rejection depends, as we have come to expect, on the electrolyte and temperature of the system. We'll start by examining a SOFC system integrated with a reformer, and then examine a PEMFC operating on hydrogen. As we'll see, thermal challenges and water balance issues are coupled.

Integration of SOFC Fuel Cell and Steam Reformer

Let's again examine the fuel-cell system with a reformer. For simplicity, we assume that methane is reformed according to

$$CH_4 + H_2O \rightarrow CO + 3H_2 \qquad (10.33)$$

Furthermore, we will assume the reaction goes to completion, in contrast to the assumption made earlier. Reforming reactions are endothermic. For instance, the change in enthalpy for reformation of methane (10.33) is 206 kJ·mol^{-1}-methane. Thus, energy must be added to drive the reaction. This feature will have important implications for systems that use hydrocarbon fuels.

There are several reactions that can lead to the formation of elemental carbon in the reactor during the reforming process. Formation of carbon in a reactor is called *coking* and must be avoided. An example of a coke-forming reaction is shown below:

$$CO + H_2 \leftrightarrow C(s) + H_2O, \qquad (10.34)$$

where solid carbon is formed. Note that the carbon formation reaction is not favored in the presence of excess water. As a result, water in excess of that required for the reformation reaction (10.33) is added. The amount of water is described by the steam to carbon ratio, SC. The steam to carbon ratio may be as high as five in an industrial reformer. Where does this water come from? The clear source is the fuel cell. For the SOFC examined here, water is produced at the anode where hydrogen and carbon monoxide are oxidized,

$$O^{2-} + H_2 \rightarrow H_2O + 2e^- \qquad (10.35)$$

and

$$O^{2-} + CO \rightarrow CO_2 + 2e^- \qquad (10.36)$$

For each mole of methane reformed, there is one mole of water consumed in the reformer and three moles of water produced in the fuel cell. So, in principle, we are in good shape since we produce more water than we consume. As shown in Figure 10.15, water can be recovered from the exhaust of the stack and recycled back to the reformer. However, it turns out that this is not a good approach. To understand why, we need to consider the thermal balance. A solid oxide fuel cell operates at 700–1000 °C. Therefore, the exhaust contains a

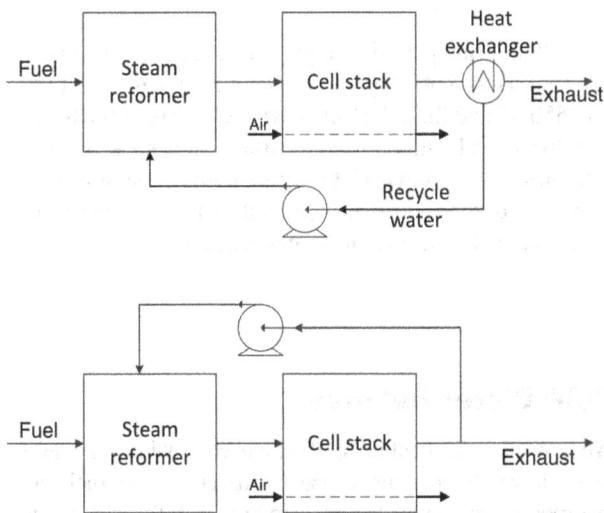

Figure 10.15 Possible configuration to integrate the cell stack and reformer.

large amount of energy that must be removed prior to separation of the water by condensation. Subsequently, heat needs to be added back to the water after separation to vaporize it (raise steam) before feeding it to the reformer. The reforming process is highly endothermic and additional heat is required to drive the reaction. Recall that the fuel-cell stack generates excess heat. An improved method of system integration would be to use excess heat from the fuel-cell stack in the reformer as shown with the lower image of Figure 10.15. Rather than trying to separate out the water, we simply recycle the hot exhaust gas from the outlet of the fuel-cell back to the reformer. Enough of the exhaust is recycled so that the desired SC ratio is achieved; what's more, we have now transferred thermal energy from the fuel cell to the reformer where it is needed. Are there any disadvantages with this approach? First, a high temperature compressor is needed to recycle the hot gas. Second, as a result of the recycle, the concentrations of hydrogen and carbon monoxide are lower in the fuel-cell stack. This situation is explored in Illustration 10.7.

Proper integration with the fuel processing system reduces the cooling load substantially. In addition, heat can be removed with the air stream. Because water is not produced at the cathode in a SOFC and water does not play a key role in electrolyte conductivity, the water balance and the conductivity of the electrolyte are not linked to oxygen utilization, as they are for PEM fuel cells as described below. Therefore, the air flow rate and utilization can be used to control the temperature. Also, the high temperature of the SOFC permits much more heat to be removed with the air compared to that possible with low temperature fuel cells. Controlling the temperature of a SOFC with air utilization is explored in Problem 10.18.

ILLUSTRATION 10.7

Determine the amount of recycle needed to achieve a steam to carbon ratio (SC) of 2 in the reformer. Compare the composition of the anode gas for each of the two configurations shown in Figure 10.15.

In the first case, water is condensed and recycled to provide the correct SC; therefore, the amount of water that is recycled is known. The feed gas is assumed to be pure methane, and the reaction in the reformer is assumed to be complete (see Equation 10.33). Water is the only component recycled, and other components in the anode gas are exhausted from the system. For $\dot{n}$ [mol·s^{-1}] of methane in the feed, the molar flow rates of species exiting the stack can be determined from the reaction stoichiometry and the definition for utilization. To simplify the problem, we have assumed that the utilization of the CO and the H_2 is the same. The recycle stream is 100% water at a flow rate of $\dot{n}SC$.

CH_4	CO	CO_2	H_2	H_2O
0	$\dot{n}(1-u)$	$\dot{n}u$	$3\dot{n}(1-u)$	$\dot{n}(3u+SC-1)$

The composition of hydrogen at the stack exit is plotted as a function of system utilization. For a utilization of 0.85, the mole fraction is 0.09 for H_2 and 0.71 for H_2O.

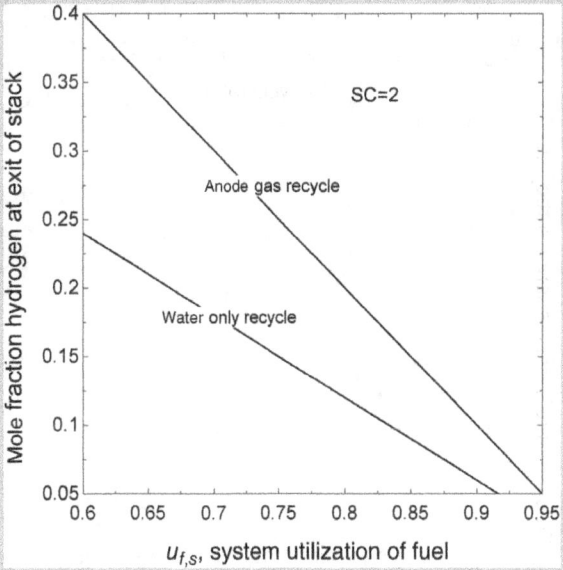

The analysis of second case, where the anode exhaust gas is recycled, still just requires material balances; however, the balances are significantly more complex and the details are left to Problem 10.25. As before, SC, the overall utilization, and the inlet flow rate of methane are specified. The recycle ratio, R_r, defined as the moles being recycled divided by the moles in the feed, $\dot{n}_r/\dot{n}_f$, changes to keep the SC ratio constant. Since SC is based on the feed rate of methane to the reformer, the mole fraction of water decreases as the recycle ratio is increased, as shown in the figure. The relationship between overall utilization and the fuel utilization in the stack is shown in the inset. For an overall utilization of 0.85, the stack utilization is 0.71, and the gas mole fraction exiting the stack is 0.15 for H_2 and 0.52 for H_2O. From the Nernst equation, the change in potential can be determined.

$$U = U^\theta - \frac{RT}{2F}\ln\frac{y_{H_2O}}{y_{H_2}\sqrt{y_{O_2}}}.$$

Thus, as the mole fraction of hydrogen increases and that of water decreases, the potential of the cell

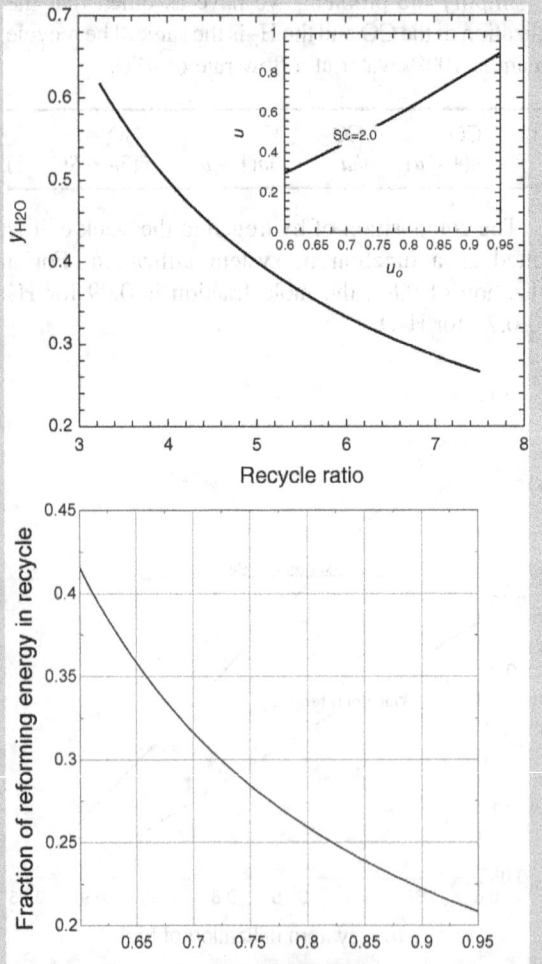

increases. Using the composition at the exit for the two cases, we determine the change in potential. The equilibrium potential is reduced 26 mV for the water-only recycle and 16 mV for the anode gas recycle.

The principal advantage with anode gas recycle though is in heat integration. Assuming the reformer is at 550 °C and the cell stack at 850 °C, the sensible heat in the recycle stream can be transferred back to the reformer. As shown in the third figure, a significant fraction of the energy needed to drive the reformer can be obtained from the anode gas recycle.

PEM Water Balance

We now turn our attention to PEM fuel cells, which operate at much lower temperatures, use hydrogen as the fuel, and incorporate an ion-exchange membrane as the electrolyte. Utilization of the fuel in these fuel cells is typically quite high. Balancing of the water represents a key challenge for PEM fuel cells. Recall from Chapter 9 that, in contrast to most electrolytes, the conductivity of the proton exchange membrane depends strongly on its water content. Too little water and the membrane conductivity drops; too much water, however, and the electrode floods. The ideal situation is for the gas in contact with the membrane to be saturated with water vapor.

To examine the situation further, let's perform a water balance for the fuel-cell stack. Referring to Figure 10.16, we assume that the inlet air has a water mole fraction of $y_{w,in}$, and that the mole fraction of oxygen in the air feed is known. We further assume that the mole fraction of water in the air exit stream, $y_{w,out}$, is at the saturation value. We neglect the water removed by the fuel stream since the utilization of hydrogen is near unity. We will use as a basis the moles of dry air supplied, even though the actual inlet stream also contains water. With $\dot{n}$ (mol·s^{-1} of dry air supplied) and total cell area (A) we have,

$$\text{Water in} = \dot{n}\frac{y_{w,in}}{1-y_{w,in}}, \quad (10.37)$$

$$\text{Water produced} = \frac{iA}{2F}, \quad (10.38)$$

$$\text{Water out} = \left(\dot{n} - \frac{iA}{4F}\right)\frac{y_{w,out}}{1-y_{w,out}}, \quad (10.39)$$

The term in the parentheses is the molar flow rate of dry, vitiated (spent) air leaving, which is now depleted in oxygen due to reaction. If we assume that the spent air stream must

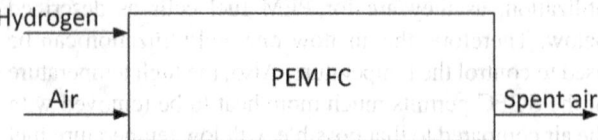

Figure 10.16 Simplified picture of air flow and water balance for a PEMFC.

be saturated with water to keep the membrane conductive, the mole fraction of water out is given simply as

$$y_{w,out} = \frac{p_{vap}(T_{cell})}{p}, \qquad (10.40)$$

where $p_{vap}(T_{cell})$ is the vapor pressure of water at the cell temperature. Next, the utilization of oxygen is used:

$$u_{O_2} = \frac{iA/4F}{\dot{n}y_{O_2}}. \qquad (10.18)$$

Substituting, a material balance on water yields

$$\underbrace{\dot{n}\frac{y_{w,in}}{1-y_{w,in}}}_{\text{water in}} + \underbrace{2u_{O_2}y_{O_2}\dot{n}}_{\substack{\text{water} \\ \text{produced}}} = \underbrace{\dot{n}(1 - u_{O_2}y_{O_2})\frac{y_{w,out}}{1-y_{w,out}}}_{\text{water out}}$$

$$(10.41)$$

$\mathcal{W}$ cancels out, and now we see that the utilization and the outlet mole fraction of water are related. Since we assume that the outlet mole fraction is saturated, we can relate utilization and cell temperature. This connection is illustrated in Figure 10.17. Here it is assumed that the incoming air is near room temperature and as such cannot hold a lot of water vapor. For a given cell temperature there is only one value for utilization of air that perfectly balances water. A higher utilization of oxygen will result in cell flooding and lower u_{O_2} causes dry out of the ionomer.

The perfect balance just described is difficult to achieve. Consequently, it is impractical to avoid liquid water completely while simultaneously ensuring that the membrane is kept fully hydrated. Therefore, some means of dealing with liquid water is needed for PEMFCs. In practice, capillary forces are used to wick out the liquid or, alternatively, high velocities are used in the flow channel to physically expel the water droplets. Exploring these topics, however, is beyond the scope of this book.

Heat removal is important for PEMFCs, just like it was for SOFCs, although the waste heat is at a much cooler temperature. One important way to remove heat is to provide coolers between cells in the cell stack assembly. These coolers may be located between each cell or between groups of cells, such as every six or seven cells. For a bipolar stack, these coolers must allow for electronic current flow through them. Such coolers typically use a liquid as the cooling fluid. The heat is then rejected to the atmosphere in a separate heat exchanger.

10.8 STRUCTURAL–MECHANICAL CONSIDERATIONS

It might seem odd to include mechanical concerns in a text on electrochemical engineering. Experience has taught us that the mechanical design is critical in electrochemical systems. Many times performance difficulties can be traced to poor mechanical design or structural failure. Of course, mechanical engineering of materials is a large field in itself. Here our discussion is restricted to just a few mechanical issues that are relevant to fuel cells. A critical question that you as the engineer must answer is, "What axial load (force applied along the axis of the fuel-cell stack, perpendicular to the planar electrodes) should be applied to the cell stack?" As happens with almost any design decision, this choice involves trade-offs among a number of competing factors. Appendix D reviews some basic concepts of mechanics of materials needed for this section.

Contact Resistance

It is essential that the resistance of electrochemical cells be low for practical use. In the introductory chapters, we discussed many important design factors that affect the cell resistance: the conductivity of the electrolyte, electrode structure, and kinetic parameters to name a few. Now we want to treat the contact resistance between components. Although not electrochemical in nature, it can play a critical role in the performance of the cell. When two surfaces are brought together and the interface between them is examined microscopically, we see that the contact is not perfect (Figure 10.18). In fact, there are generally relatively few points of actual contact. The conductance and resistance of this interface are

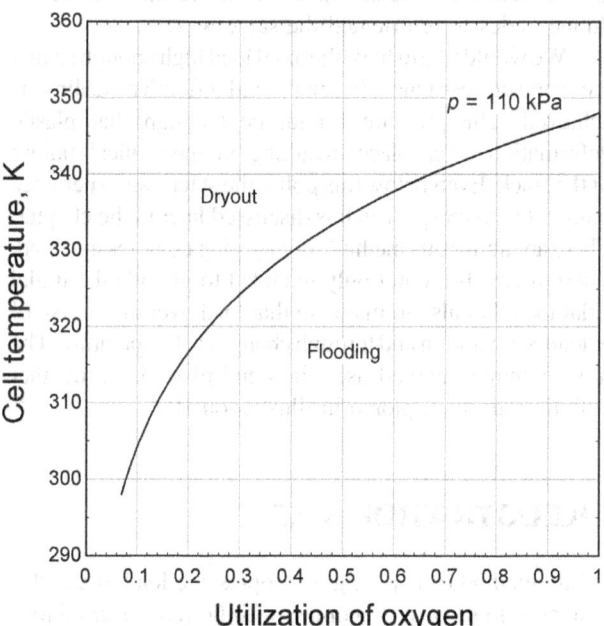

Figure 10.17 Maximum utilization of air consistent with water balance in proton-exchange membrane fuel cell

242 Electrochemical Engineering

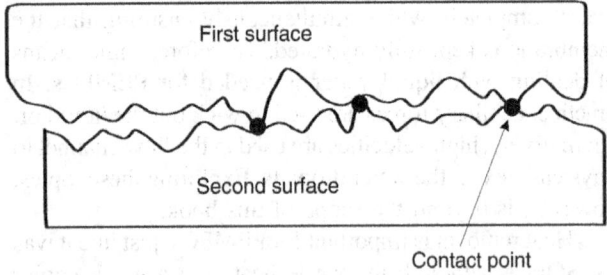

Figure 10.18 Microscopic picture of two surfaces mated together.

described by

$$\text{conductance} = \frac{1}{R_{contact}} = kp^\beta \; [\Omega^{-1} \cdot m^{-2}], \quad (10.42)$$

where p is the contact pressure or what we have called the axial load, and k and β are empirical constants. β has a value between 0.5 and 1. The values of β and k depend on the roughness of the surfaces and the materials of the two components in contact. Figure 10.19 shows an example behavior for smooth and rough surfaces. Note that rough surfaces require a higher axial load to achieve a given resistance. A low axial load results in a high contact resistance. At first, the resistance decreases rapidly with increasing load. However, as the axial load increases further, the reduction in resistance is much smaller. There are disadvantages to high loads, which will be discussed shortly.

Although the basic behavior is well described with Equation 10.42, the parameters are best obtained experimentally. Comparable behavior is seen for thermal conductance because of the similarity in the two conduction mechanisms in metals. One additional factor that is particularly relevant for us is corrosion (Chapter 16). Oxidation of the surfaces can have a significant impact on the contact resistance. A nonconducting oxide on a steel surface, for example, can dramatically increase the contact resistance.

Sealing

Another factor to be considered when establishing the axial load of the cell stack is sealing. A detailed examination of seals is beyond the scope of this chapter. Nonetheless, a little reflection should convince you that escape of fuel overboard, leakage between the fuel and oxidant streams, or dripping of water on the ground must be prevented. Three common sealing approaches used in fuel cells are discussed briefly. First, *wet seals* use controlled porosity to establish a liquid barrier to the transport of gases and are suitable for low-temperature fuel cells—see discussion of capillarity from Chapter 5. Second, materials can be *bonded* together. The specific material used for bonding depends on temperature and compatibility; examples include epoxies, solders and brazes, as well as glass seals. Third, *compression* seals use compliant elastomers to account for unevenness in the surfaces. These seals can also mitigate the effect of tolerance stack-up, which is the cumulative effect of stacking imperfect parts whose dimensions are within specified tolerances. For example, if cells that are slightly thinner on one side, but within tolerances, are stacked, a bowed cell stack would result if all of the thinner sides were oriented the same way.

We would like to have the axial load high enough so that the contact resistance is small and effective sealing is achieved. The pressure cannot be so high that plastic deformation takes place, resulting in mechanical failure of the stack. Even below this point, the stack can experience failure due to *creep*, which is discussed later in the chapter. There are numerous methods of applying compressive load. These methods are not only intended to provide the initial axial load, but also to maintain that load over the course of extended operation and through changes in temperature. The most common method uses rigid end plates held together with tie-rods, as explored in Illustration 10.8.

ILLUSTRATION 10.8

One method of applying a compressive load to a cell stack is to use a stiff endplate with tie-rods external to the stack to apply a compressive force. A tie-rod is nothing more than a metal shaft that joins the two end

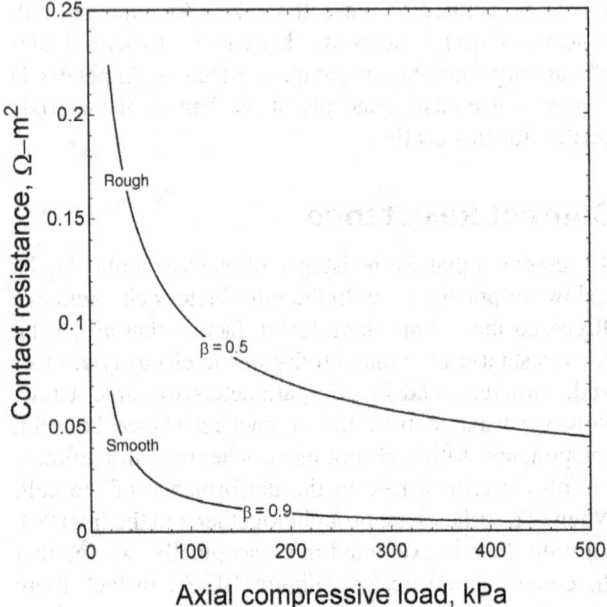

Figure 10.19 Contact resistance of two mated surfaces.

plates. It has been determined that the required force per unit of cell area is 340 kPa. The cell area is 0.1 m². There are 6 tie-rods with a diameter of 3 mm made from carbon steel. Calculate the stress and strain on the tie-rods.

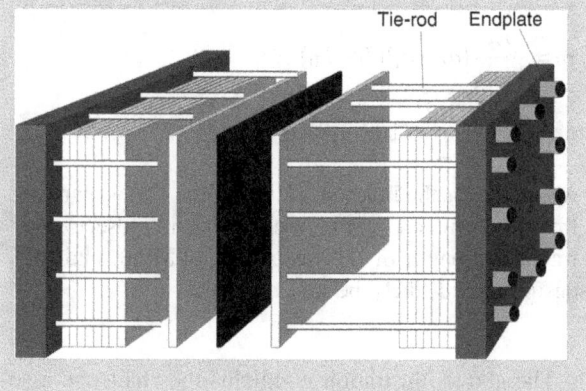

From the axial load, the force per tie-rod can be calculated and converted to a stress:

$$\sigma = \frac{\text{cell area}}{\text{rod area}} \times \frac{\text{axial load}}{\text{number of rods}}$$
$$= \frac{0.01}{7.07 \times 10^{-6}} \frac{340{,}000}{6} = 80.2 \text{ MPa},$$

which is well below the yield strength of steel, 250 GPa. Using Young's modulus for steel, 200 GPa, the strain is

$$\epsilon = \frac{\sigma}{E} = 4.01 \times 10^{-4}.$$

Note that the elongation of the rods is simply $L\epsilon = 0.2$ mm if the height of the stack is $L = 0.5$ m.

Illustration 10.9 examines the load (axial stress) needed to compress elements of the fuel-cell stack.

ILLUSTRATION 10.9

An MEA (membrane electrode assembly) for a PEM fuel cell consists of a membrane coated with catalyst on both sides and two gas-diffusion layers (GDLs). See Figure 10.4. Initially, the membrane is 50 μm thick and each of the two gas diffusion layers is 200 μm thick. What stress is needed to compress these layers from 450 to 445 μm? Assume the materials behave elastically, $E_{\text{GDL}} = 17.1$ MPa, $E_{mem} = 1.0$ GPa.

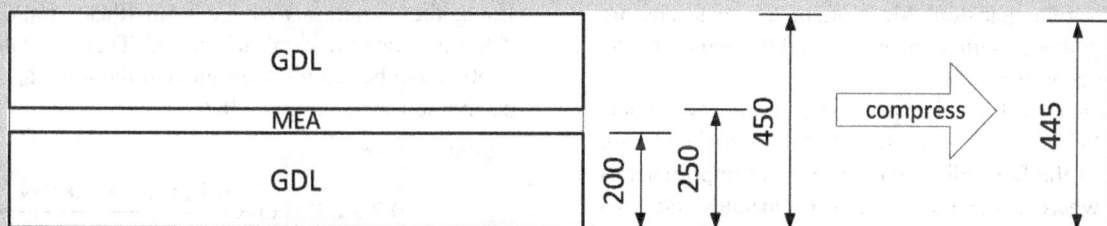

Under load, the stress, σ, in each component is the same. The final compressed thickness for each component is

$$L_{j,f} = L_{j,ini}(1 - \epsilon_j) = L_{j,ini}\left(1 - \frac{\sigma}{E_j}\right),$$

and the overall thickness is

$$L_f = \sum_j L_{j,ini}\left(1 - \frac{\sigma}{E_j}\right).$$

Solving for the stress gives $\sigma = 213$ kPa.

Stress Induced by Temperature Changes

Materials expand when the temperature is raised. The amount of change depends on the material and is quantified by the coefficient of thermal expansion (CTE) (see Appendix D). Thus, a stress is created in a material if it is constrained and the temperature changed from the stress-free state. In our rudimentary analysis of thermally induced stress in a fuel-cell stack, we will assume that one of the components is much stiffer than the other. Think of stiffness as the ability of the component to resist deformation. This assumption means that the deformation of one component is insignificant relative to that of the other. Stiffness is not simply a material property, but is affected by the thickness and shape of the component. For an anode-supported electrode of an SOFC, for instance, the thickness of the anode material may be much greater than that of the separator. In this case, we assume that the strain of the support is determined from the temperature change, and it remains free of stress. The stress generated in the other component is related to the difference in CTE (α) of the two materials.

$$\sigma_i = \frac{E_i}{1-\nu_i}(\alpha_s - \alpha_i)(T - T_0), \qquad (10.43)$$

where the subscript s refers to the support material, the subscript i refers to the material of interest, ν_i is Poisson's ratio, and T_0 is the stress-free temperature. This highly simplified analysis will give the engineer a starting point for understanding the importance of matching thermal coefficients of expansion. Most often, the materials are bonded together at high temperature, and this would be the stress-free condition.

What makes thermal expansion of particular importance for SOFCs is the high temperature associated with operation of the fuel cell. A failure mode of importance is *fracture*, where a material or object separates into two pieces under a mechanical stress. This process is nearly instantaneous and usually catastrophic. This mechanism is of most interest for high temperature cells that use ceramics. The principle of maximum stress applies to brittle materials such as ceramics. If the principal stress reaches its maximum value, the material fractures. This challenge is shown in Illustration 10.10.

ILLUSTRATION 10.10

An anode-supported SOFC is bonded together at 1000 °C. What is the stress in the YSZ separator when cooled to room temperature? The modulus for YSZ is 215 GPa and the coefficients for thermal expansion are 10.5×10^{-6} and $9 \times 10^{-6} \text{ K}^{-1}$ for the support and YSZ separator, respectively. Assume that, because of the thickness of the anode, it is rigid as discussed above. Using Equation 10.43,

Separator
Anode

$$\sigma_i = \frac{E_i}{1-\nu_i}(\alpha_s - \alpha_i)(T - T_0)$$

$$= \frac{2.15 \times 10^{11}}{1 - 0.22}(10.5 - 9)10^{-6}(1000 - 25) = 403 \text{ MPa},$$

where 0.22 is the Poisson's ratio for the YSZ separator from Appendix D. This stress value is close to the fracture strength of the material (416 MPa), and the design would likely be unacceptable.

The above illustration highlights the challenge associated with operation of high-temperature fuel cells. Brittle fracture is not common in low-temperature systems, but a loss of load can occur in such systems with only relatively modest changes in temperature. This situation arises because the CTE for the cell stack materials is often different from the CTE of the materials used in the loading system. Thus, there is a tendency for the axial load to change significantly with temperature.

ILLUSTRATION 10.11

Using the compressive force from Illustration 10.8, determine the torque on each tie-rod. There is a simple relationship between the torque and the force applied by the tie-rod. Taking $c = 0.2$,

Torque $= cDF_x$

$$= 0.2 \times 0.003 \text{ m} \times \frac{0.1 \text{ m}^2 \times 340{,}000 \text{ N} \cdot \text{m}^{-2}}{6}$$

$$= 0.34 \text{ N} \cdot \text{m}.$$

Six tie-rods are used, each of diameter D. If this torque is applied at room temperature, what would you expect to happen to the compressive force on the stack when the system is heated to 150 °C?

To address this question, we need to consider the expansion of the stack and the expansion of the loading system. If the CTEs are the same and the moduli don't change with temperature, then there would be no effect. Our load system is made from steel, and the coefficient of thermal expansion for the steel is $10.8 \times 10^{-6} \text{ K}^{-1}$. Therefore, the length would increase from 0.5 m by

$$\delta_L = 0.5[\alpha_L(150 - 25)] = 0.7 \text{ mm}.$$

The CTE for the stack is $4 \times 10^{-6} \text{ K}^{-1}$; thus, it expands by 0.25 mm. Since the increase in the stack length is less than that of the load system, which is very common, the compressive force will be reduced. To get a quantitative value for the stress at temperature, the specifics of the load system must be considered.

Creep

Below a certain level of stress, the deformation is reversible or elastic; above this stress level, the deformation is permanent. The stress beyond which the deformation becomes plastic is the *yield point*. Design conditions should not exceed the yield point of the materials. Another mechanism of failure is *creep*: the permanent deformation of a material subjected to mechanical stress below its yield point. Yielding can occur at stresses below the yield point if stresses are applied for a sufficiently long time or at elevated temperatures. Unlike *fracture*, creep is a slow process. Furthermore, some creep is generally acceptable as long as it is anticipated and compensated for in the design.

ILLUSTRATION 10.12

For the stack from Illustration 10.8, what impact will there be on the compressive load if, due to long-term creep, the cell components permanently deform by 1 mm? Show how this reduction in load is mitigated by using Belleville washers (see Appendix D) for load follow-up.

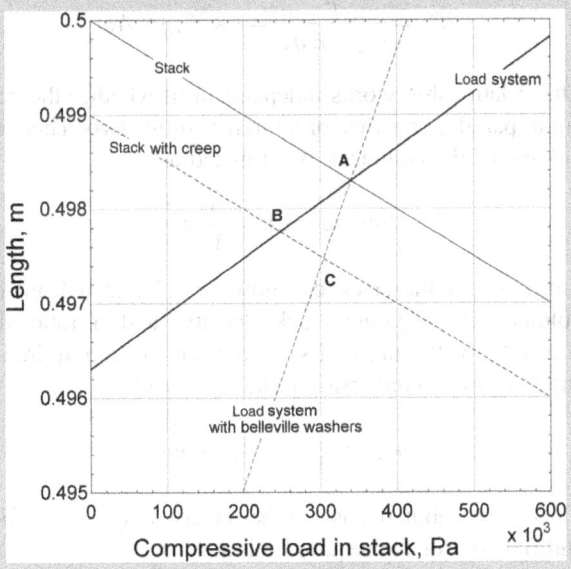

The stress-free height of the stack is 0.5 m. Using a modulus of 100 MPa for the stack material, we can calculate the compressed length as a function of compressive load assuming elastic compression.

$$L = 0.5\left(1 - \sigma_s/E_s\right),$$

which is depicted with the solid line labeled stack. Point A is the nominal design point. To understand the effect of creep, the effect of the load system must be known. The tie-rods are in tension, and their length increases with stack stress. The slope of the line is

$$L_o\left(\sigma_{rod}/E_{rod}\right) = L_o\left(\sigma_s/E_{rod}\right)\left(\frac{A_s}{A_{rod}}\right),$$

where the length of the expanded tie-rods must equal the compressed thickness of the stack at the design stress. The modulus of the tie-rod material is 50 GPa. The line labeled load system represents this response and intersects with the stack curve at the design point A. Permanent deformation of the stack results in a new curve for the stack with creep, where the zero-stress thickness is now 0.499 m. The intersection of this new curve with the load system, B, becomes the new operating point. Here, we see that the compressive load has been reduced from 340 to about 250 kPa, low enough that cell performance may be adversely affected.

Belleville washers (Appendix D) are basically springs that, when added to the end of the tie-rods, reduce the effective modulus of the tie-rods and make the load system less sensitive to changes in the stack length. This is consistent with your intuition that tells you that the use of springs will help to maintain compression on the stack by helping to compensate for the effect of creep. The curve on the plot was made using an effective modulus of 100 MPa for the load system containing the Belleville washers. For the same creep, the compressive load is now 310 kPa.

10.9 CASE STUDY

A fuel cell is being considered for a manned space flight. Because hydrogen and oxygen are needed for propulsion power, these are the preferred fuel and oxidant. The basic cell performance is shown in Figure 10.20. Our job is to recommend the current density and voltage at which the fuel cell operates. A high current density will result in higher power density and therefore a smaller cell stack. However, operation at high power means that the cell voltage is lower, and the efficiency lower, resulting in higher fuel consumption. We will seek a system where the combined mass of the fuel-cell stack, hydrogen, and oxygen is minimized.

We'll need to simplify this optimization, but we can nonetheless illustrate how to approach these challenges. First, let's understand why minimizing the mass is so important. Any object that goes into orbit must be

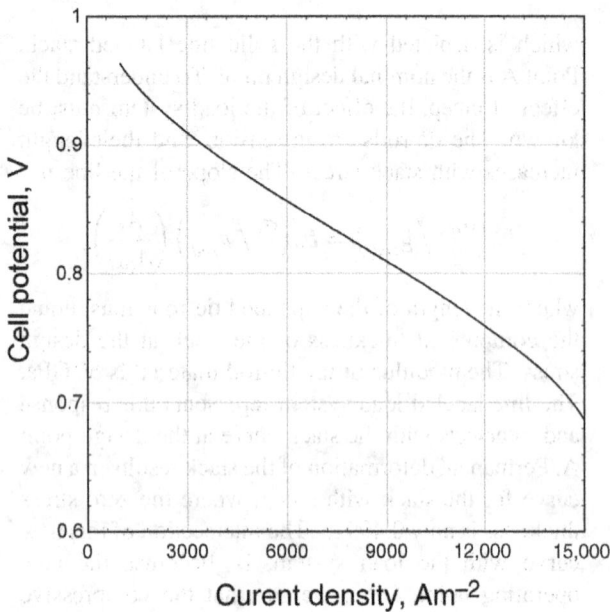

Figure 10.20 Polarization curve for space fuel cell operating on hydrogen and oxygen. Utilization of hydrogen and oxygen are 0.95.

Table 10.3 Baseline Parameters for Case Study

Parameter	Symbol	Value
Power	P	2000 W
Mission length	t	10^6 s
Mechanical efficiency	η_{mech}	0.90
Fuel efficiency, or hydrogen utilization	η_{fuel} u_{H2}	0.98
Power conversion efficiency	η_{pc}	0.95
Oxygen utilization	u_{O2}	0.95
Cell pitch	γ	250 m^{-1}
Active area ratio	A_r	0.7
Apparent density of the fuel-cell stack	ρ_s	2000 kgm^{-3}
System/stack mass ratio	mr	2

accelerated to a high velocity (about 7 km·s^{-1} for low earth orbit). This acceleration requires a large amount of energy and fuel, perhaps 2 kg of fuel for each kg of payload. For our purposes, the fuel-cell system as well as the oxygen and fuel needed for the mission duration can be considered payload. There is a tremendous incentive in keeping this mass low.

Next, let's think about the key factors that go into fuel consumption besides the voltage efficiency. Two essential items are the average net conditioned electrical power and the length of the mission; these items are critical for a space mission because they determine the amount of fuel needed, and it is necessary to carry the fuel for the entire flight with you into space. Other items include those we have already discussed such as the mechanical efficiency, fuel efficiency, and power conversion efficiency. Baseline values for these parameters are given in Table 10.3. From these parameters and the cell potential, we can calculate the mass of hydrogen required.

$$m_{H_2} = \frac{P \times t}{V_{cell}} \times \frac{1}{\eta_{mech}\eta_{pc}} \times \frac{M_{H_2}}{n_a F},$$

where n_a is 2 for the hydrogen oxidation, and the energy (power × time) has been used rather than the power in order to calculate the total mass rather than the mass flow rate. Note that the amount of hydrogen used is inversely proportional to the voltage of the cell.

In contrast to terrestrial applications, the oxidant is pure oxygen not air, and its mass must be calculated and added to the total just like we did for hydrogen. Therefore, we need the utilization of oxygen to determine the amount of oxygen required, which is proportional to the amount of hydrogen.

$$m_{O_2} = m_{H_2} \frac{u_{H_2}}{u_{O_2}} \times \frac{M_{O_2}}{M_{H_2}} \times \frac{n_c}{n_a}.$$

Next, we need the mass of the cell stack. For a given potential of the cell, the current density is determined by the polarization curve. Because a fuel cell is a power conversion device rather than an energy storage device, the total cell area, A, is calculated from the average gross power of the cell stack:

$$P_g = \frac{P}{\eta_{mech} \times \eta_{pc}} = i \times V_{cell} \times A.$$

This relationship works independent of whether the cells are in parallel or series, or a mixture of the two. Then, the volume of the cell stack is estimated as

$$\text{Volume stack} = \frac{A}{A_r} \times \frac{1}{\gamma}.$$

The mass of the stack can now be calculated from the volume, the apparent stack density, and a ratio that accounts for the mass of system components required to support power conversion in the cell stack,

$$m_{fcs} = \frac{A}{A_r} \times \frac{1}{\gamma} \times \rho_s \times mr.$$

Thus, the combined mass for the cell stack, hydrogen (fuel) and oxygen for the required mission, is

$$m_t = m_{fcs} + m_{H_2} + m_{O_2}.$$

Taking the parameters given in Table 10.3, the total mass can be determined as a function of cell potential or, equivalently, current density using the polarization curve. The results of these calculations are shown in Figure 10.21. A minimum mass is obtained for operation at a potential of about 0.88 V. As the cell potential increases, the fuel cell is

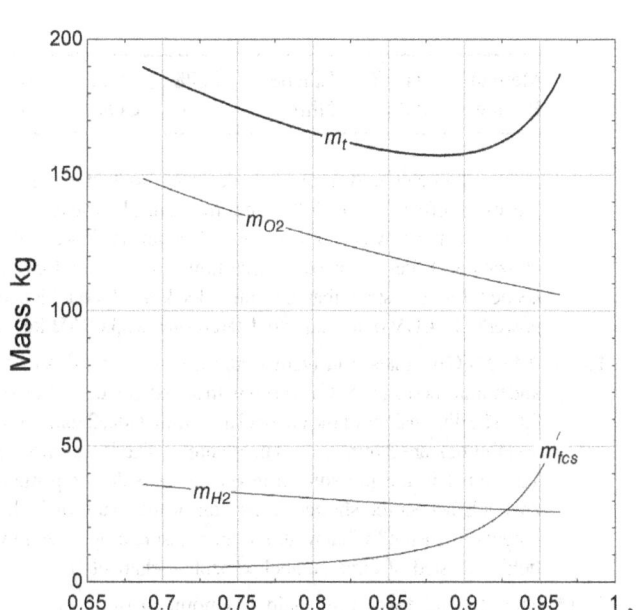

Figure 10.21 Optimized potential for planned space mission, where mass is the most important factor.

more efficient, and therefore the mass of oxygen and hydrogen decrease. At the same time, the mass of the fuel-cell system increases sharply.

CLOSURE

In this chapter, we have extended the analysis of fuel cells to include the overall system. The current–voltage relationship is still fundamental, but the amount of fuel and air provided relative to the stoichiometric value must be established. Material and energy balances are the main tools for understanding and designing full systems. Water and thermal balances have been shown to be critical to these designs. In addition, we have introduced some of the mechanical aspects that are critical to good cell stack and system designs.

FURTHER READING

Larminie, J. and Dicks, A. (2003) *Fuel Cell Systems Explained*, John Wiley & Sons, Inc., New York.
Mench, M. (2008) *Fuel Cell Engines*, John Wiley & Sons, Inc. New York.
Spiegel, C.S. (2007) *Designing & Building Fuel Cells*, McGraw Hill.

PROBLEMS

10.1 Using the definition of efficiency given by Equation 10.3, what is the maximum thermal efficiency of a hydrogen/oxygen fuel cell at 25 °C, standard conditions?

10.2 A fuel cell operating on methane produces 100 kW of gross electrical power. Calculate the voltage and thermal efficiency given the following:
  (1) The individual cells are operating at a potential of 0.65 V and gaseous water is produced at standard conditions.
  (2) 75% of the fuel is converted to electricity (the balance is combusted external to the fuel cell to drive the reformation process).
  (3) 5% of the electrical output is consumed by ancillary equipment and losses in power conversion.

10.3 How would the results of Problem 10.2 change if we assume that the water is produced as a liquid?

10.4 The quantity $\Delta G/\Delta H$ was introduced in Equation 10.7. This quantity is needed because historically with fuel cells the efficiency is based on the heating value of the fuel rather than on the change in Gibbs energy. For a hydrogen/oxygen fuel cell, create a plot of $\Delta G/\Delta H$ as a function of temperature from 25 to 200 °C. For temperatures below 100 °C, use the higher heating value of the fuel (product is liquid water); and above 100 °C, use the lower heating value (product is gaseous water). You'll need to find data for heat capacity as a function of temperature to complete this problem.

10.5 Calculate mass of hydrogen needed to operate PEM FC at an average of 18 kW for a period of 3 hours. Assume the utilization of hydrogen is 0.98 and that each cell is operating at 0.7 V. Roughly 18 kW power is needed to sustainably power a passenger vehicle on the highway. Compare the mass of hydrogen calculated with the mass of gasoline that would be needed. If helpful, you may assume that the average speed for the 3-hour period is 90 kmph.

10.6 A portable hydrogen fuel cell used by the military operates at an overall fuel-cell system thermal efficiency, η_{th}, of 55%. The stack has a specific power of 100 W·kg^{-1} and operates at 50 W, and the system delivers a net power of 35 W continuously. The ancillary systems (plumbing, controls, and electronics) weigh 0.210 kg. The theoretical energy content of a hydrogen storage system is 1200 W·h·kg^{-1}.
  (a) Estimate the mass of the fuel required for providing 35 W of conditioned power for a 72-hour mission.
  (b) How does the mass of the fuel-cell system package (with fuel) compare to a lithium-ion battery with a specific energy of 150 W·h·kg^{-1}?
  (c) Discuss two properties of the fuel-cell system you would improve to achieve 700 W·h·kg^{-1} for the fuel-cell system package for the same mission duration. (This problem was suggested by S.R. Narayan).

10.7 Create a plot of system thermal efficiency versus power level for PEM FC operating on pure hydrogen. Use the LHV of hydrogen, and assume that the utilization of hydrogen is 0.97. Power for ancillary equipment is 500 W + 5% of gross electrical output; that is,

$$P_{ancilary} = 500\,[W] + 0.05 P_{gross}.$$

The cell stack has 100 cells, each of 0.04 m^2 area. The performance curve is represented by the model given in Problem 9.4 with the following parameters:

$b = 0.054$ V, $j = 25,500$ A·m^{-2}, $j_d = 30,000$ A·m^{-2}, $\ln(k) = -9.14$, $R = 5$ μΩ·m^2.

What do these results suggest about how the fuel cell might be best used in a fuel-cell hybrid electric vehicle?

10.8 A direct methanol fuel cell operates at 0.4 V and 2000 A·m^{-2}. If the stack must produce 50 W at 12–24 V, what configuration would you propose? Assume a pitch of 4 mm/cell, an endplate thickness of 7.5 mm, and an active area ratio $A_r = 1.2 + 0.05 \left(\frac{0.002}{A_c}\right)^2$, where A_c is the active area of a single cell [m^2]. What is the volume of the stack?

10.9 Estimate the mass-transfer limiting current density as a function of utilization of oxygen down the channel of a PEM FC. Treat the mass-transfer resistance as simple diffusion through a porous gas diffusion layer. The cell is operating at 70 °C, ambient pressure, and the air is saturated with water vapor. The thickness of the GDL is 170 μm, and the effective diffusivity of oxygen is 6×10^{-6} m^2 s^{-1}. What is the effect of increasing temperature on limiting current density for a PEM FC?

10.10 When analyzing the performance of a low-temperature fuel cell, it is often desirable to include the effect of oxygen utilization with a one-dimensional analysis. If the mole fraction of oxygen changes across the electrode, what value should be used? Assuming that the oxygen reduction reaction is *first order* in oxygen concentration, show that it is appropriate to use a log-mean mole fraction of oxygen as an approximation of the average mole fraction.

$$y_{lm} \equiv \frac{y_{in} - y_{out}}{\ln \frac{y_{in}}{y_{out}}}.$$

10.11 Express the log-mean term in Problem 10.10 in terms of oxygen utilization and the inlet mole fraction of oxygen, y_{in}. Sketch the average current density as a function of utilization, keeping the overpotential for oxygen reduction fixed. How would this change if mass transfer is also included.

10.12 Discuss the advantages and disadvantages of using the alternative definition for fuel processing efficiency based on the heating value of the fuel as shown in Chapter 10 immediately following Equation 10.25.

10.13 The composition of industrial natural gas is as follows:

| Methane | 94.9% | Ethane | 2.5% | Propane | 0.2% |
| Butane | 0.03 | Nitrogen | 1.67 | CO$_2$ | 0.7 |

A fuel-cell power plant is designed so that the fuel-processer efficiency is 78%, the mechanical efficiency is 92%, and the power conditioning efficiency is 96%. If the utilization of fuel (u_f) is 0.85, what flow rate of natural gas is needed for a system that provides 400 kW of conditioned power? The LHV of the industrial natural gas is 53.08 MJ·kg^{-1}.

10.14 A PEMFC operates at an average current density of 16 kA m^{-2}, and humidified air (75 °C) is used with an oxygen utilization of 0.6. The dimensions of the channel are 2 mm wide, 2 mm deep, and the ribs are 3 mm wide. The planform size is 12 cm × 24 cm. What is the pressure drop on the air side for parallel channels across the shorter dimension of the planform? The longer dimension? What would be the pressure drop if the flow field consisted of three parallel serpentine channels?

10.15 The channel for a flow field has nominal dimensions of 2.0 mm × 1.5 mm. If the nominal utilization is 55%, the maximum permissible utilization is 65%, what are the minimum dimensions allowed for the channels?

10.16 Recreate Figure 10.17 for the case where the cell is operating at 200 and then 300 kPa absolute pressure. Explain the effect of operating pressure on the water balance for PEM fuel cells.

10.17 Dry air is fed to a PEMFC that is operating at 10 kA·m^{-2}. The temperature is 55 °C and the pressure is 120 kPa. If the utilization of air is 40%, what fraction of the water produced by the cell is removed as a liquid? Assume that this exit stream is saturated at the cell temperature.

10.18 Whereas liquids are used to cool low-temperature fuel cells, cooling with gas is more appropriate at the high operating temperature of a solid oxide fuel cell. One strategy for cooling is to increase the air flow through the cathode (see the figure that follows). The additional power required to provide air for cooling in excess of what is required for the electrochemical reactions is a drain on the efficiency, but in contrast to PEM fuel cells, air stoichiometry has no effect on the water balance. Since temperature strongly affects the conductivity of the YSZ separator and because thermal gradients can cause failure of the seals, the temperature difference between the inlet air and the cell stack cannot be too large.

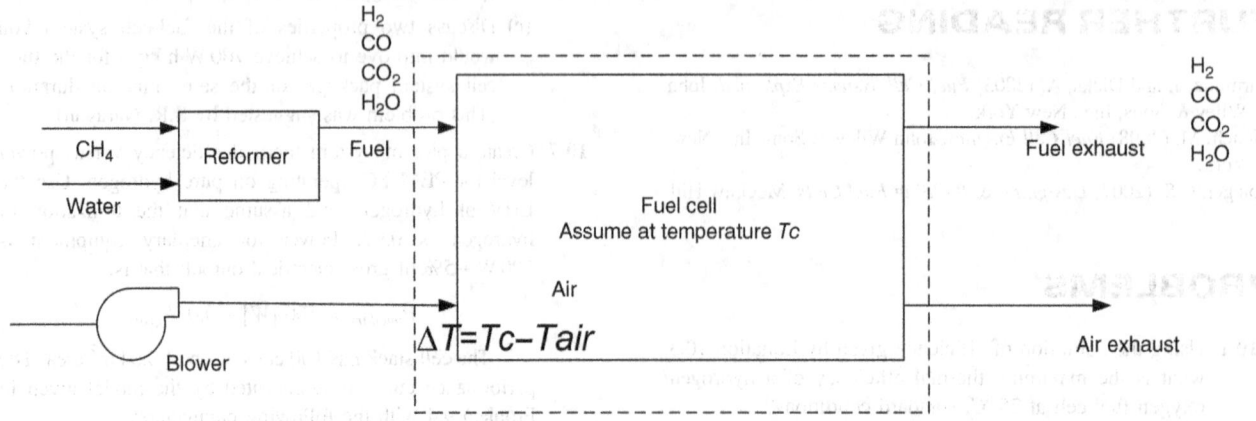

The fuel is reformed methane. Excess water is supplied to the reformer to prevent carbon deposition. Assume that the steam-to-carbon ratio is 2.5. There are two steps to the fuel processing. The first is reformation, which can be assumed to go to completion.

$$CH_4 + H_2O \rightarrow 3H_2 + CO$$

The second reaction is the water-gas shift reaction. Here, assume that 30% of the CO is converted to CO_2.

$$CO + H_2O \leftrightarrow H_2 + CO_2$$

(a) $\Delta T = 200$ K, what is the resulting air utilization? Perform an energy balance on the stack (enclosed by the dashed line). Assume that the fuel, fuel exhaust, and depleted air are at the temperature of the cell stack ($T_c = 900\,°C$). How does this utilization change with a change in the inlet temperature?

(b) Plot the air utilization as a function of the inlet temperature difference and explain the relationship between the air utilization and ΔT.

Both CO and H_2 are oxidized in the fuel cell, assume that the utilization of each of these species is 0.9 and that the potential of each cell is 0.7 V.

$$H_2 + O^{2-} \rightarrow H_2O + 2e^-$$
$$CO + O^{2-} \rightarrow CO_2 + 2e^-$$

Species	Heat capacity [J·mol^{-1}·K^{-1}]	ΔH_f [kJ·mol^{-1}]
H_2	30.132	0
H_2O	42.220	−241.8
Air	33.430	0
CO	29.511	−110.5
CO_2	55.144	−393.5

10.19 One of the simplest models for a low-temperature hydrogen/oxygen fuel cell is

$$V = U + \frac{RT}{F} \ln\left[\frac{p_{O_2}\sqrt{p_{H_2}}}{i}\right] - R_\Omega i.$$

Using this model, which neglects mass transfer, how does the cell voltage change with oxygen utilization if the average current density is fixed? You may assume that the equilibrium potential, U, is constant. You will need to use an average oxygen partial pressure that accounts for the change in oxygen concentration along the length of the electrode.

10.20 In Section 10.5, 100% utilization of hydrogen is noted not to be practical even if the fuel is pure hydrogen. As shown in Figure 10.10, some of the fuel is recycled, and a small amount of fuel is vented through a pulse-width-modulated valve that can be scheduled to discharge gas in proportion to power. Holding the system utilization of fuel, $\eta_{f,s}$, constant at 0.9, plot the composition of the fuel entering the cell stack and the stack utilization, η_f, in the cell stack for recycle rates between 0 and 2.0. Assume the inlet composition to the system is 90% hydrogen.

10.21 By how much is the limiting current reduced when the gas-diffusion layer is compressed from 170 to 100 μm? Assume a Bruggeman relation (Equation 5.18) for tortuosity and an initial porosity of 0.7.

10.22 In the study of fuel consumption and utilization, several factors were neglected. Specifically, there can be leakage or permeation across the separator of a fuel cell. Additionally, there may be small shorts, for instance, due to the small electronic conductivity of a SOFC electrolyte or a small ionic conductivity in the interconnects (bipolar plates). Briefly discuss how these factors would affect the design of a fuel-cell system and propose a definition for the fuel utilization that accounts for these additional factors.

10.23 List advantages and disadvantages of the six flow fields shown in Figure 10.14.

10.24 A PEMFC stack in a fuel-cell vehicle is operating at 80 °C on hydrogen and air at a balanced pressure of 300 kPa and delivers 75 kW (gross power) at 42 V. The individual cells are operating at 0.7 V and 6 kA·m^{-2}.

(a) Determine the number of cells in the stack and the area of each electrode.

(b) Find the minimum flow rates of reactants in g·s^{-1} to sustain this average power. State the reasons why this minimum rate is insufficient in practice?

(c) Calculate the rate of accumulation of water at cathode from the reaction and electro-osmotic drag, assuming that back diffusion from the cathode to the anode is negligible. What are the consequences of not having back diffusion of water from the cathode to the anode and how do you propose to address these consequences to ensure stable operation? (This problem was suggested by S.R. Narayan).

10.25 Work out details for Illustration 10.7.

10.26 How would the results of the case study change if the mission were reduced to just 2 days?

10.27 Methanol can be oxidized in an aqueous fuel cell to carbon dioxide and water as per the following cell reaction:

$$CH_3OH(\ell) + 3/2 O_2 \rightarrow CO_2 + 2H_2O(\ell)$$

(a) Write the individual electrode reactions for the fuel cell.

(b) Calculate the theoretical specific energy for the fuel and report it in W·h·kg^{-1}.

(c) Calculate the maximum thermal efficiency for this reaction at 298 K.

(d) If such a methanol cell were operating at 298 K at 0.5 V and 0.1 A, calculate the voltage efficiency and the rate of heat output from the cell. (This problem was suggested by S.R. Narayan).

BRIAN EVANS CONWAY

Brian Evans Conway was born on January 26, 1927, in Farnborough, England. He studied in London during World War II, receiving his B.S. in 1946 and his Ph.D. from Imperial College in 1949. His graduate work was under the direction of the renowned John O'Mara Bockris. Conway was just 20 years old when he published his first paper with Bockris. His research group at Imperial College also included Roger Parsons and Martin Fleischmann, who along with Conway became preeminent leaders in electrochemistry.

After a short stay at the University of Pennsylvania, he joined the University of Ottawa in 1956 in the newly formed Department of Chemistry. He continued to research at the University of Ottawa for his entire career, some five decades more. During this time, Professor Conway educated more than 100 graduate students and postdoctoral fellows, many of whom became recognized leaders in electrochemistry in their own right.

Conway and Bockris coedited two important series in electrochemistry. These two series exemplified an extraordinary collaboration between a former student and mentor. What's more, these series had an immeasurable impact by revitalizing electrochemistry in the second half of the twentieth century. The first series, *Modern Aspects of Electrochemistry*, started in 1954. Brian Conway was editor or coeditor of 23 volumes. The second series was *Comprehensive Treatise of Electrochemistry*.

In 1999, Conway published a monograph *Electrochemical Supercapacitors: Scientific Fundamentals and Technological Applications*, which is main topic of this chapter. He is reported to have coined the term "supercapacitor." In the 1970s, Conway conducted wide-ranging research on electrochemical capacitors using ruthenium oxide, so-called pseudo-capacitors. His extensive and pioneering work was instrumental in expanding basic knowledge and in clarifying the operation of electrochemical capacitors. In one of his most influential works, the energy-storage characteristics, specifically the transition from battery to capacitor behavior, were clearly described (*J. Electrochem. Soc.*, **138**, 1539 (1991)). All told, Conway authored more than 400 scientific papers.

Professor Conway's interests were broad, and his versatility is demonstrated by his important contributions to fundamental and applied electrochemistry in fields such as oxide film formation, electrode kinetics, fuel cells, electrochemical capacitors, and electrolysis. He was unquestionably the top electrochemist in Canada and one of the top physical chemists of his day. Professor Conway is fondly remembered by his students for his passion for science and for the support and "words of wisdom" he provided them.

Brian Conway received numerous awards over his career. He was made a Fellow of the Royal Society of Canada (1968) and received the Galvani Medal from the Italian Chemical Society in 1991. In 2000, Dr. Conway was awarded the Thomas W. Eadie Medal for contributions in engineering and applied science. He received two major awards from the Electrochemical Society: the Henry B. Linford Award for Distinguished Teaching (1984), and the Society's highest technical honor, the Olin Palladium Award in 1989. He was named a Fellow of the Electrochemical Society in 1995, and Conway remained active in the Electrochemical Society until his death on July 9, 2005.

Image Source: Courtesy of ECS.

Chapter 11

Electrochemical Double-Layer Capacitors

Electrochemical double-layer capacitors (EDLCs) are another important means of storing energy electrochemically. In many ways, EDLCs complement batteries by providing greater power density at the expense of energy density, which is generally lower than that of batteries. Although not nearly as common as batteries, EDLCs are important for several applications. For example, they are often combined with batteries, fuel cells, and other devices in hybrid power systems. In such applications, the EDLC most frequently provides a load-leveling function.

It is important for you to understand the operation of these devices and the conditions where they are a better alternative than batteries, or where hybrid systems may be beneficial. EDLCs have two key advantages or strengths: (i) high power density and (ii) long cycle life. In this chapter, we explore the behavior of these devices and provide the elementary information needed by engineers. In doing so, we will apply the fundamental principles of electrochemistry that we have already learned to help us understand the design and operation of EDLCs. The material in this chapter, however, represents only a brief introduction to the topic, and we encourage you to explore the references at the end of this chapter for additional information.

Before getting into the details of electrochemical double-layer capacitors, we briefly examine more conventional capacitors, which are best exemplified by a parallel plate capacitor. This introduction will provide a foundation for our discussion of EDLCs, and will help to highlight the differences in physics that determine the basic operation of these capacitors.

11.1 CAPACITOR INTRODUCTION

A conventional electrostatic capacitor consists of two conductors separated by a dielectric (electronic insulator). Energy storage is accomplished by charge separation, with positive charge accumulated on one conductor and negative on the other (see Figure 11.1a). The charge, Q, is the amount of charge on either conductor (not the sum of the two). Capacitance is defined as the charge divided by the potential difference and relates the amount of charge stored to the electrical potential required to store that charge.

$$C \equiv \frac{Q}{V}. \qquad (11.1)$$

The unit for capacitance is the farad, [F], which is equivalent to coulomb per volt [C·V^{-1}]. Alternatively, we can express a differential capacitance as

$$C_d = \frac{dQ}{dV}. \qquad (11.2)$$

This differential representation of the capacitance provides additional information and is required in situations where the capacitance, C, varies with the amount of charge. In contrast, Equation 11.1 defines the integral capacitance.

Figure 11.1a shows a conventional parallel plate capacitor consisting of two conductive plates separated by an insulator. If the gap between the plates is a vacuum, the capacitance is constant and equal to

$$C = \frac{\varepsilon_0 A}{d}, \qquad (11.3)$$

where d is the distance between the plates, ε_0 is the permittivity of free space, 8.8542×10^{-12} [F·m^{-1}], and A is the area of the plates. The electrical symbol for this capacitor is also shown in the figure. The positively charged plate is called the anode and the negatively charged plate the cathode. If the vacuum is replaced by a dielectric material, the capacitance is proportional to the

---

Electrochemical Engineering, First Edition. Thomas F. Fuller and John N. Harb.
© 2018 Thomas F. Fuller and John N. Harb. Published 2018 by John Wiley & Sons, Inc.
Companion Website: www.wiley.com/go/fuller/electrochemicalengineering

252 Electrochemical Engineering

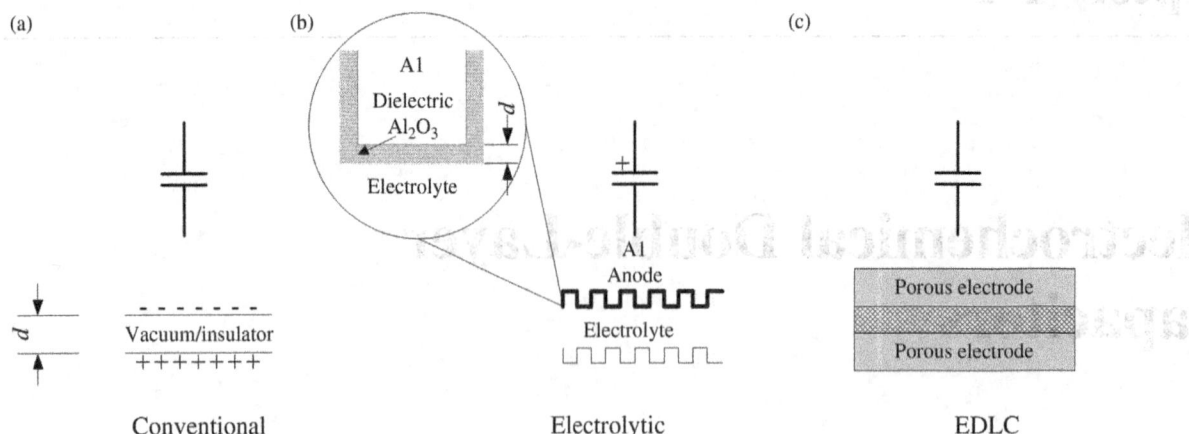

Figure 11.1 Types of capacitors and electrical symbols. (a) Conventional or electrostatic. (b) Electrolytic. (c) EDLC. Please note that, as discussed in the text, the electrical symbols do not accurately reflect the physics in Electrolytic capacitors and EDLCs—both of these have two capacitors in series, one for each electrode.

dielectric constant, which is the ratio of the permittivity of the material to the permittivity of vacuum:

$$\varepsilon_r = \frac{\varepsilon}{\varepsilon_0}. \qquad (11.4)$$

For a parallel plate capacitor using a dielectric material in place of a vacuum, Equation 11.3 is replaced with

$$C = \frac{\varepsilon A}{d} = \frac{\varepsilon_r \varepsilon_0 A}{d}. \qquad (11.5)$$

To this point, we have seen that the capacitance depends on the properties of the material between the conducting plates, as well as on the distance between the plates and the area available for charge storage. We have also defined both a differential and an integral capacitance, which relate the amount of charge stored to the voltage difference across the plates.

The capacitance of a device can be raised by increasing the area available for charge storage and by decreasing the thickness of the dielectric. These strategies are used in the *electrolytic capacitor*, the second type of capacitor, shown in Figure 11.1b. This kind of capacitor is significantly different from and not to be confused with an *electrochemical double-layer capacitor*. In addition to their capacitive properties, electrolytic capacitors are of particular interest because they are manufactured with electrochemical processes. As shown in Figure 11.1b, two metal foils are separated by an electrolyte. In contrast to the parallel plate capacitor, the areas of the two metal foils are enhanced by etching the surfaces. The increase in area is illustrated by the corrugations shown in Figure 11.1b. The etching takes place as the foil is passed through a chloride solution and current is applied to create small, tunnel-like pores, mostly perpendicular to the surface. The surface roughness (Chapter 3) or *foil gain* can be up to 100 times for low-voltage and 25 times for high-voltage electrolytic capacitors. The capacitance of the anode, however, does not result from the interface between the electrolyte and the foil; rather a forming process creates a thin metal oxide layer on the etched surface of one of the metal foils. This oxide is shown as alumina in the enlargement of the corrugation in Figure 11.1b. It is this oxide that serves the same role as that of the dielectric in the parallel plate capacitor. The increase in surface area and the use of a very thin oxide (dielectric) lead to a significant increase in capacitance over that of a parallel plate capacitor. The anode here represents one capacitor; there is by necessity a second capacitor at the other electrode, and these capacitors are in series. Therefore, some aspects of the physics of the device are not well represented by the electrical symbol in Figure 11.1b.

To form the electrolytic capacitor, the metal anode is anodized by applying a constant potential to oxidize the surface (essentially a corrosion process, see Chapter 16). The magnitude of the potential applied may be several hundred volts and is greater than that for which the device is rated so that no appreciable oxidation takes place during later operation. A higher formation potential leads to a thicker metal oxide on the surface, which increases the operating voltage of the capacitor at the expense of a lower capacitance due to decreased surface area and the thicker oxide layer. For an aluminum foil, aluminum oxide forms on the anode, and the alumina serves as the dielectric that separates the surface of the conductor from the electrolyte. Other metals are also used in practice. In operation, the majority of the potential difference occurs across the oxide layer, rather than in the electrolyte. Consequently, very high voltages are possible. As mentioned previously, the capacitance is increased through the use of a very thin oxide, on the order of 100 nm, which is much thinner than the separation that can be achieved in a conventional capacitor without developing shorts. Growth of a thicker oxide layer (still quite thin) provides a higher voltage rating but tends to close up the smaller pores that were etched; therefore, a higher voltage rating also implies a smaller foil gain. Importantly, we note that the electrolytic capacitor (Figure 11.1b) has a polarity; that is, in contrast to a parallel plate capacitor, it must be connected so that the anodized metal is electrically positive relative to the other electrode. This polarity is also shown in the electrical symbol in the figure. The adverse consequences of reversing this polarity are explored in Problem 11.3. We also note that charge is supplied by the electrolyte on the other side of the oxide; this feature is similar to that found in the EDLC, which will be discussed in the next section.

Finally, we close this introductory section by reviewing how to calculate the total capacitance for multiple capacitors that are connected either in parallel or in series. This knowledge is important for understanding EDLCs. For capacitors that are electrically combined in parallel, the total capacitance is simply the sum of the individual capacitances:

$$C = C_1 + C_2 + C_3 + \cdots + C_n. \quad (11.6)$$

In contrast, for capacitors that are placed electrically in *series*, the total capacitance is

$$\frac{1}{C} = \frac{1}{C_1} + \frac{1}{C_2} + \frac{1}{C_3} + \cdots + \frac{1}{C_n}. \quad (11.7)$$

A useful way to help understand the physics behind these relationships is by considering charge storage at a given voltage. Capacitors in parallel simply add storage capacity to the system. In contrast, the charge stored on each capacitor must be the same for capacitors in series, and there is a voltage drop across each of the capacitors, which must sum to the total voltage difference. For two identical capacitors in series, the voltage drop across each capacitor will be the same and equal to half of the total voltage difference. Thus, the total capacitance will be half that of the individual capacitors. For two capacitors of unequal capacitance in series, a larger fraction of the voltage drop will be across the capacitor with the smallest capacitance. This imbalance is because it takes a greater voltage difference to store the same amount of charge if the capacitance is smaller. As a result, a capacitor with a very small capacitance in series with one or more larger capacitors will dominate the total capacitance as almost all of the voltage drop will be across this capacitor. These issues are explored in the following illustration.

ILLUSTRATION 11.1

Suppose that you have two capacitors in series, each with a capacitance of 1 farad. What is the total capacitance and what is the potential drop across each capacitor if the total voltage drop across the charged capacitors in series is 1 volt? Repeat this calculation for two capacitors where one of them has a capacitance of 1 F and the other a capacitance of 0.1 F.

Solution:

$$\frac{1}{C_{total}} = \frac{1}{C_1} + \frac{1}{C_2} = \frac{1}{1\,\text{F}} + \frac{1}{1\,\text{F}} = 2\,\text{F}^{-1} \Rightarrow C_{total} = 0.5\,\text{F}.$$

The charge on each capacitor is the same. Therefore, $Q_{total} = Q_1 = Q_2 = Q$.

$$Q = C_{total}V_{total} = (0.5\,\text{F})(1.0\,\text{V}) = 0.5\,\text{C}.$$

$$V_1 = \frac{Q}{C_1} = \frac{0.5}{1} = 0.5\,\text{V}.$$

$$V_2 = \frac{Q}{C_2} = \frac{0.5}{1} = 0.5\,\text{V}.$$

Now consider the situation with the two capacitors with different capacitances.

Solution:

$$\frac{1}{C_{total}} = \frac{1}{C_1} + \frac{1}{C_2} = \frac{1}{1\,\text{F}} + \frac{1}{0.1\,\text{F}} = 11\,\text{F}^{-1}$$

$$\Rightarrow C_{total} = 0.091\,\text{F}.$$

The charge on each capacitor is the same. Therefore, $Q_{total} = Q_1 = Q_2 = Q$.

$$Q = C_{total}V_{total} = (0.091\,\text{F})(1.0\,\text{V}) = 0.091\,\text{C}.$$

$$V_1 = \frac{Q}{C_1} = \frac{0.091\,\text{C}}{1\,\text{F}} = 0.091\,\text{V}.$$

$$V_2 = \frac{Q}{C_2} = \frac{0.091\,\text{C}}{0.1\,\text{F}} = 0.909\,\text{V}.$$

It is clear from these results that the smaller capacitor controls the behavior for the two capacitors in series, leading to a much lower charge capacity for a given voltage than what we calculated above. The majority of the change in voltage is across the smaller capacitor.

11.2 ELECTRICAL DOUBLE-LAYER CAPACITANCE

Recall from Chapter 3 that the interface between an electrode and the electrolyte is generally charged. There can be excess positive or negative charge in the metal that is balanced with an equal and opposite charge in the electrolyte adjacent to the surface. The counterbalancing

254 Electrochemical Engineering

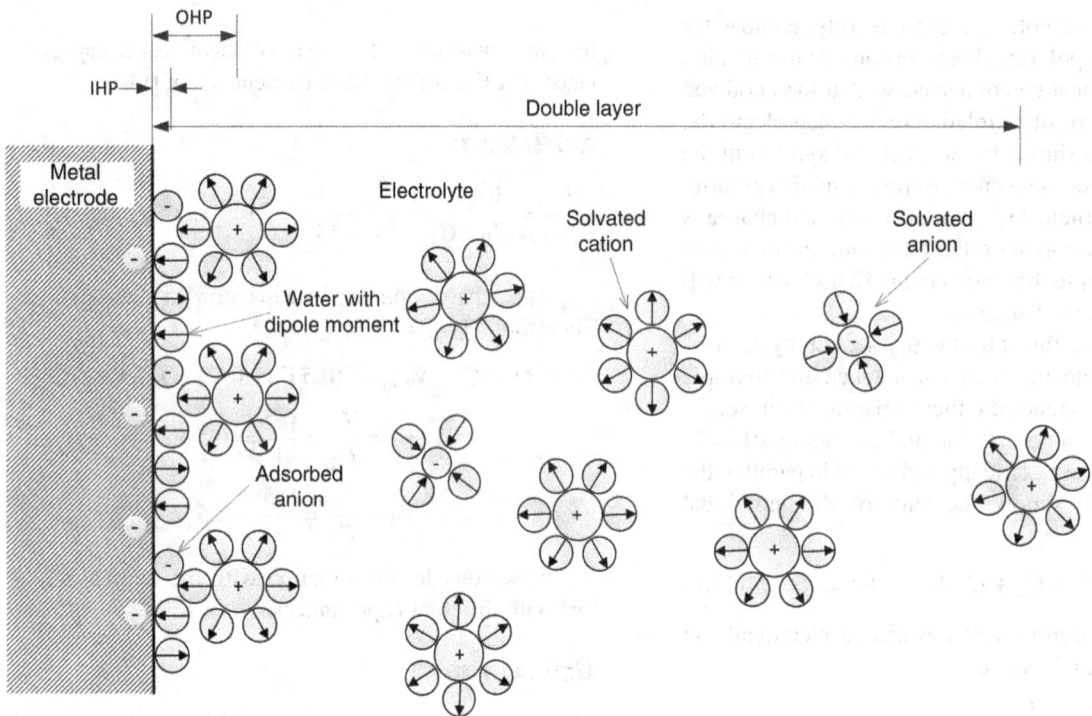

Figure 11.2 Structure of the electrical double layer.

charge may consist of adsorbed ions in the inner Helmholtz plane (IHP), solvated ions in the outer Helmholtz plane (OHP), and the ions in the diffused part of the double layer. This charge separation across the interface is effectively a capacitor and is capable of storing energy. As with other types of capacitors, work is required to achieve this charge separation, and work or energy can be extracted by eliminating this charge separation through the flow of current. Of course, there is a second electrode, not shown in Figure 11.2, that is required to create an EDLC as we will discuss shortly. Note that the separation of charge across the interface in an EDLC shown takes place without faradaic reaction (oxidation or reduction). In other words, there are no electrons transferred across the double layer and no change of oxidation state for the participating ions. In contrast, almost all of the current flow that we have discussed previously has been due to faradaic reactions. This distinction will be emphasized throughout the chapter, and an electrode that can be polarized (by applying a potential) without passing a faradaic current is called an *ideally polarizable electrode*.

The double layer does indeed provide charge separation and is capable of storing energy. But how can we achieve the charge separation depicted in Figure 11.2? We need a way to get charge on and off the electrodes. This is accomplished by adding a second electrode and by passing current between the two electrodes, which is equivalent to moving charge (electrons) from one electrode to the other. As electrons move through the external circuit, ions move in solution to balance the resulting charge on each electrode. This process is illustrated in Figure 11.3. Note that, as expected, a power supply is required for charging an EDLC. With a positive current, I, flowing through the leads as shown, electrons move in the opposite direction. An excess negative charge builds up in the left electrode, and an excess positive charge forms on the right electrode. To balance these charges,

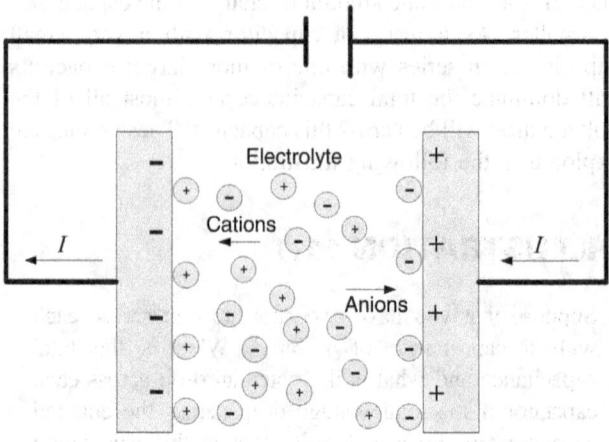

Figure 11.3 Charging of an EDLC.

cations from the electrolyte move toward the electrode on the left; simultaneously, anions move to the right side to compensate for the missing electrons from that electrode. The movement of ions constitutes a flow of current in the electrolyte equal to that in the leads. The net effect is to have two double-layer capacitors in series, and the capacitance of the device is calculated with Equation 11.7 for capacitors in series.

Classical Descriptions of the Electrical Double-Layer Capacitance

As noted previously, the charge in solution can be found at multiple locations; therefore, the capacitance of an EDLC is comprised of multiple parts. Restricting ourselves to a single electrode, its capacitance is divided into three parts: the diffuse layer (GC), ions in the compact layer near the electrode (OHP), and ions adsorbed on the surface (IHP). The contribution of each of these will be discussed separately, but it is important to recognize that the individual contributions are not really independent.

The capacitance per unit area associated with the diffuse layer is denoted as C_{GC}, where the subscript recognizes Gouy and Chapman for their independent contributions to the theory. The capacitance associated with the charges in the diffuse part of the double layer for a symmetric electrolyte is

$$C_{GC} = \frac{\varepsilon_r \varepsilon_o}{\lambda} \cosh\left\{\frac{zF\phi_2}{2RT}\right\}, \quad (11.8)$$

where λ is the Debye length. A symmetric electrolyte is one where a neutral salt dissolves into two ions of opposite charge, $z = z^+ = |z^-|$. For example, NaCl and ZnSO$_4$ are symmetrical electrolytes. Equation 11.8 was developed by Gouy and Chapman using reasoning similar to that used by Debye and Hückel for activity coefficients (see Chapter 2), although Gouy and Chapman's work predates DH by about a decade. The mathematical approximation used by DH for the spherical geometry is not needed here. Figure 11.4 shows the capacitance of a 1 : 1 symmetric electrolyte as a function of potential. The concentration of ions in solution is shown as a parameter in the figure. There are two points of note. First, C_{GC} is a strong function of potential. As the potential moves away from the point of zero charge (PZC), the capacitance increases exponentially. Second, we see that the capacitance increases with concentration. Recall that the Debye length is given by

$$\lambda = \sqrt{\frac{\varepsilon RT}{F^2 \sum_i z_i^2 c_i^\infty}}, \quad (2.39)$$

and provides a measure of the thickness of the diffuse layer. Physically, the amount of charge in the diffuse layer found close to the electrode increases as λ decreases, resulting in increased capacitance. Experimentally, this effect is shown in Figure 11.5 for a NaF solution using a Hg electrode. Observe that the hyperbolic cosine

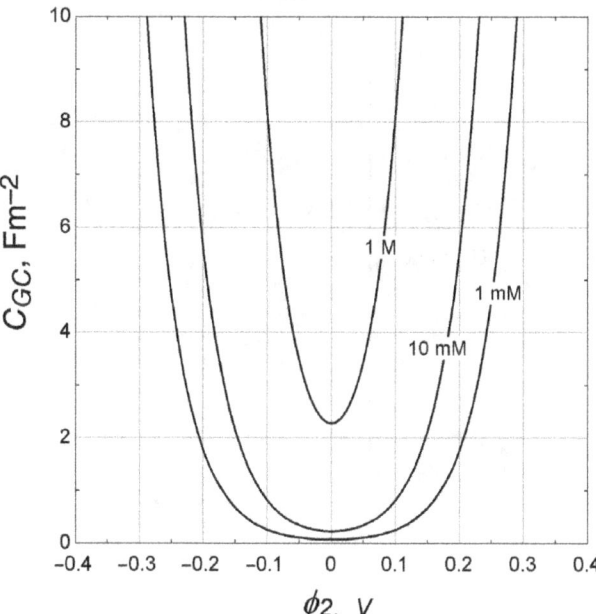

Figure 11.4 Double-layer capacitance based on Equation 11.8 for a 1 : 1 symmetric electrolyte.

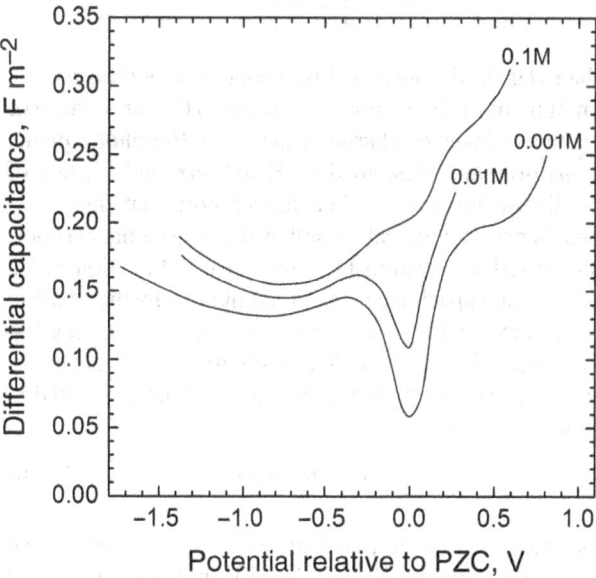

Figure 11.5 Experimental data for capacitance on mercury electrode. (Adapted from D.C. Graham (1947) *Chem. Rev.*, **41**(3), 441–501.)

behavior is present at lower ionic strengths. Sodium fluoride was used because F⁻ ions are only weakly adsorbed. A Hg electrode might seem an odd choice; indeed, if our objective were to make a practical device, we would not choose mercury. However, much of the fundamental work with capacitance uses liquid Hg because of its ability to provide a reproducible, polarizable surface. Hg is also a good electrical conductor at room temperature.

As will be discussed in more detail later, high electrolyte concentrations are desirable for EDLC devices in order to minimize resistance and ensure that there are sufficient ions in the electrolyte reservoir to balance the charges on the electrode (see Figure 11.3). At high concentrations, the contribution to the capacitance from the diffuse layer suggests a very large value for the capacitance, and one that grows exponentially with potential. Experimentally this is not observed—a typical value for the capacitance associated with the double layer per unit of real surface area is 0.1–0.5 F·m⁻², and it is relatively constant. Thus, the GC model alone does not adequately represent the double-layer capacitance, even though it does show the correct variation of capacitance with concentration near the PZC at low concentrations. This deficiency of the GC model was addressed by Stern, who recognized that charges have a finite radius and cannot get arbitrarily close to the electrode surface, in contrast to the assumptions of the Gouy Chapman model. Accounting for this, the capacitance of the double layer in the absence of specific adsorption is traditionally expressed as

$$\frac{1}{C_{DL}} = \frac{d}{\varepsilon_r \varepsilon_0} + \frac{\lambda}{\varepsilon_r \varepsilon_0 \cosh\left\{\frac{zF\phi_2}{2RT}\right\}} \equiv \frac{1}{C_{OHP}} + \frac{1}{C_{GC}}, \quad (11.9)$$

where d is the distance of closest approach for the ions. We can think of the two terms as defining (i) the capacitance of a compact layer of charge at the outer Helmholtz plane, C_{OHP}, originally described by Helmholtz, and (ii) that of the diffuse layer, C_{GC}. We further note that these two capacitances are typically combined as capacitors in series, as reflected in Equation 11.9 (refer also to Equation 11.7). The overall capacitance, C_{DL}, is dominated by the smallest capacitance in 11.9. At low concentrations, the smallest capacitance is C_{GC}, which controls the overall behavior. However, at the higher concentrations used in EDLC devices,

$$C_{GC} \gg C_{OHP}. \quad (11.10)$$

Therefore, for most practical systems, the capacitance associated with the diffuse layer is not important, and behavior is controlled by C_{OHP}.

We will now examine the capacitance associated with ions located at the OHP in a bit more detail. Once more, a physical picture of the structure of the double layer is key. Figure 11.6 presents a view of the double layer that focuses on the ions at the OHP. Here the surface of the electrode is envisioned to be completely covered with the solvent; for the moment, we are still not considering unsolvated ions that are directly adsorbed on the surface. The OHP is defined by solvated ions that are as close to the surface as possible while maintaining their waters of hydration. Clearly, even our simplified picture is far more complex than the normal parallel plate capacitor; nonetheless, we will see that Equation 11.9 gives a reasonable approximation for the capacitance. Two key parameters are d, the distance for charge separation, and the dielectric constant, ε_r. The dielectric constant of bulk water is about 80. The high relative permittivity of water is attributed to its polar nature and the ability of bulk water molecules to reorient freely under the influence of an electric field. This bulk value is no longer appropriate near the surface where movement is constrained. This is particularly true for the water molecules adjacent to the electrode surface (see Figure 11.6). We account for this by dividing the region between the OHP and the electrode surface into two parts with different dielectric constants, both of which are lower than the bulk value. C_{OHP} is approximated as (see Problem 11.5).

$$\frac{1}{C_{OHP}} = \frac{2r_w}{\varepsilon_0 \varepsilon_{rL}} + \frac{\sqrt{3} r_w + r_i}{\varepsilon_0 \varepsilon_{rH}}, \quad (11.11)$$

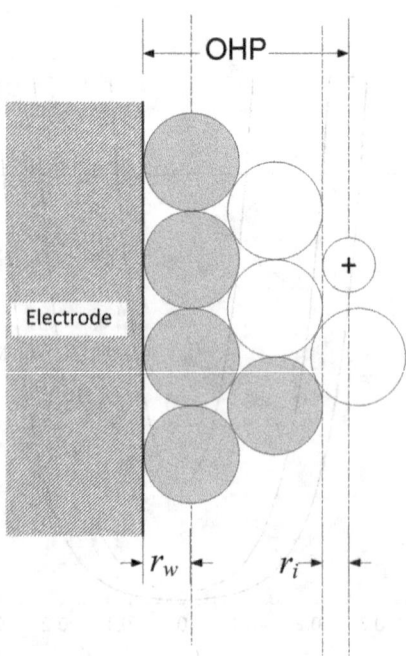

Figure 11.6 OHP contribution to capacitance. Large circles represent water molecules, white ones are hydrating cations. *Source:* Adapted from Bockris and Reddy 1970.

where r_w is the radius of the water, and r_i is the radius of the unsolvated ion. ε_{rL} is the low dielectric constant applicable for the water against the electrode surface, and ε_{rH} is a higher value appropriate for the intermediate region between the inner region and the OHP. Once again, these contributions are in series, and generally the first term dominates. Using a value of 6 for the low dielectric constant and 0.14 nm for the radius of water,

$$C_{OHP} \approx \frac{\varepsilon_0 \varepsilon_{rL}}{2 r_w} = 0.19 \, \text{F} \cdot \text{m}^{-2}. \quad (11.12)$$

The magnitude of this number is in the range of typical experimental values measured for the double-layer capacitance.

As noted previously, except for very dilute solutions near the PZC, C_{OHP} is much smaller than the value of C_{GC} calculated for the diffuse layer with Equation 11.8. Based on this, we would expect the double-layer capacitance to be more or less constant, and largely independent of the nature of the electrode surface and the type of ions in solution. The highest capacitance would result if the excess charge in solution, that which balances the charge on the electrode, were concentrated at the OHP. As the distance between the charge and the surface increases, we expect that the capacitance would decrease.

We will now consider the influence of ions that have shed their waters of hydration and are specifically adsorbed on the electrode surface. These ions define the IHP as described in Chapter 3. Development of the theory will not be attempted here; instead, key physical features that are important for understanding the influence of adsorbed ions on the capacitance are identified. To illustrate these features, we will use experimental results, which are often presented in the form of the differential capacitance defined previously (Equation 11.2).

In the end, differential capacitance curves can be quite complex. A representative, but idealized, curve for the differential capacitance of an electrode is shown in Figure 11.7. The magnitude of the capacitance is approximately what we would expect from Equation 11.12. However, as we move away from the PZC, the capacitance increases due to ion adsorption in the IHP. The increase is generally not symmetric about the PZC; rather, it increases more rapidly at positive potentials. The asymmetry is caused by the fact that negative ions are more likely to adsorb than positive ions.

What is the effect of ion adsorption? It is tempting (but wrong) to simply add the capacitance associated with the adsorbed ions, C_{IHP}, as another capacitance in series with the C_{GC} and C_{OHP}, similar to what was done in Equation 11.9. A simple electrical series addition of capacitance attributed to adsorption could only result in a decrease in capacitance, never an increase. Clearly, the

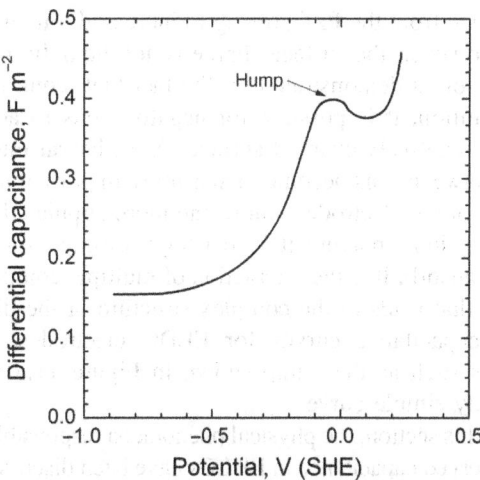

Figure 11.7 Representative experimental data for differential capacitance. Note the asymmetry. *Source:* Adapted from Bockris and Reddy 1970.

physical picture is more complex, and extending the above analogy is not fruitful.

Some insight into the problem can be gained from a charge balance. To maintain overall electroneutrality at the interface, the charge per unit area on the metal, q_m, is counterbalanced by the sum of the charges (i) in the adsorbed layer, q_{IHP}, (ii) at the outer Helmholtz plane, q_{OHP}, and (iii) residing in the diffuse layer, q_{GC}.

$$q_m + (q_{IHP} + q_{OHP} + q_{GC}) = 0. \quad (11.13)$$

As discussed in Chapter 3, ions (particularly anions) can adsorb directly onto the metal surface, forming the IHP. For a given charge on the metal, q_m, the presence of charge at the IHP means that there is less charge at the OHP and beyond. Consequently, charge storage at each location in the double layer is not independent. As the potential is increased in either the positive or negative direction from the PZC, the fraction of ions located at the IHP tends to increase. An increased fraction of ions at the IHP leads to increased capacitance since the ions that adsorb are very close to the electrode surface (i.e., the distance d is small).

The adsorption of ions at the IHP is influenced by several factors that include both chemical and coulombic interactions between ions, the solvent, and the surface. In general, anions are larger, form weaker interactions with the solvent, and are more likely to adsorb than cations. As expected, a positive q_m facilitates anion adsorption, while a negative value promotes the adsorption of cations. Because anions are more readily adsorbed, the differential capacitance curve tends to be asymmetric, with higher capacitance at positive potentials for a given potential difference from the PZC, owing to increased ion adsorption. However, the surface charge is not the only important factor as demonstrated by the fact that, contrary to our intuition, it is possible for negative ions to adsorb onto a negatively charged surface. Also, lateral interactions between ions become increasingly important as the surface of the electrode is more and more populated with adsorbed ions, making it increasingly difficult for new ions to adsorb. It is the interaction of multiple competing factors that leads to the complex structure of the differential capacitance curves for ELDC electrodes, with features such as the hump shown in Figure 11.7 for a relatively simple curve.

In this section, the physical phenomena responsible for the observed capacitance in EDLCs have been discussed in the context of classical double-layer theory. The resulting potential dependence of the differential double-layer capacitance is markedly different from that of a conventional plate capacitor. The magnitude of the double-layer capacitance has been estimated and justified in terms of the relevant physical processes. This information provides a background for understanding the performance of ELDCs.

Capacitance for Practical EDLC Analysis

Since we expect the potential of a practical device to be operated over the full range of stability allowed by the electrolyte, how important is it to include the complete potential-dependent differential capacitance for the EDLC? Such a treatment would greatly complicate our analysis. Therefore, for practical engineering calculations, we will assume that the C_{DL} is constant. While not perfect, this simplification allows us to reasonably approximate the behavior of the full EDLC. An EDLC has two electrodes, and charge is stored in the double layer of each of these electrodes. The total capacitance of the device is

$$\frac{1}{C_{device}} = \frac{1}{C_{elec1}} + \frac{1}{C_{elec2}}, \quad (11.14)$$

since the two double-layer capacitors are in series. One of the electrodes will have a positive charge, and the other a negative charge of equal magnitude. When the charge on the positive electrode increases, the charge on the negative electrode must also increase by an equal amount. Frequently, these two electrodes are identical, and we can use the same differential capacity curve for each of them. The asymmetry in the differential capacity curve implies that the electrode with the lower differential capacity will dominate. Thus, the fact that the two capacitances are in series reduces the variation of the overall capacitance with potential in the complete device. Also, the behavior of a device where the two electrodes are identical will be independent of the polarity to which it is charged. Finally, the practicing engineer is generally interested in the integral capacitance, where the variations in the differential curve are smoothed out. These behaviors are explored more fully in Problem 11.10.

As a final point related to practical devices, we note that a key advantage of the EDLC comes from using an electrode material with a high surface area. Carbon, for example, can have a surface area greater than $1000 \, m^2 \cdot g^{-1}$. A porous electrode, as described in Chapter 5, is a good way to achieve a large interfacial area between the electrode and electrolyte. Thus, in Figure 11.1c, we represent the two electrodes as porous electrodes. A typical value for the capacitance associated with the double layer per unit of real surface area is $0.1–0.5 \, F \cdot m^{-2}$. The capacitance of an electrode can be found by multiplying the electrode area by the capacitance per unit area. The total capacitance for the device would be approximately half of the value calculated for the single electrode as per Equation 11.7.

Nomenclature

We will restrict ourselves to using the terms *electrochemical double-layer capacitor* (EDLC) and *pseudocapacitor* (discussed in Section 11.8), where the names generally reflect the underlying physics of the devices. Within the scientific literature as well as product brochures, other commonly encountered terms are "supercapacitors" and "ultracapacitors," but there appears to be little consensus as to the precise features associated with these terms.

ILLUSTRATION 11.2

For a 1 F capacitor, calculate the area required for each type of capacitor:

1. Conventional capacitor where the plates are separated by an insulator ($\varepsilon_r = 100$) with a distance of 50 μm between plates.

$$A = \frac{Cd}{\varepsilon} = \frac{1F \, 50 \times 10^{-6} \, m}{(100) 8.85 \times 10^{-12} \, F \cdot m^{-1}} = 5.65 \times 10^4 \, m^2.$$

2. Electrolytic capacitor with foil gain of 25, dielectric constant of 10, and thickness of 225 nm.

$$A = \frac{Cd}{\varepsilon(\text{foil gain})} = \frac{1F \, 225 \times 10^{-9} \, m}{(10) \, 8.85 \times 10^{-12} \, F \cdot m^{-1}(25)} = 102 \, m^2.$$

The area calculated here is the superficial area of the electrode; the true surface area is 25 times larger. Note that geometric area is also commonly used in place of superficial area.

3. EDLC, 25 μm thick carbon electrode with a porosity of 0.5, $\rho_c = 2000$ kg·m^{-3}. The carbon has a specific surface area of 1000 m^2·g^{-1}, with a capacitance of 0.15 F·m^{-2} of carbon surface area.

$$\frac{1000 \text{ m}^2}{\text{g}} \frac{0.15 \text{ F}}{\text{m}^2} = 150 \text{ F} \cdot \text{g}^{-1}.$$

$$\frac{150 \text{ F}}{\text{g}} \frac{(2 \times 10^6 \text{g})(0.5)}{\text{m}^3} 25 \times 10^{-6} \text{ m} = 3750 \text{ F} \cdot \text{m}^{-2}.$$

Therefore, the superficial area needed for a 1 F, ECDL capacitor is 2.7 cm^2. This low area is mostly because the actual area per unit volume is so high.

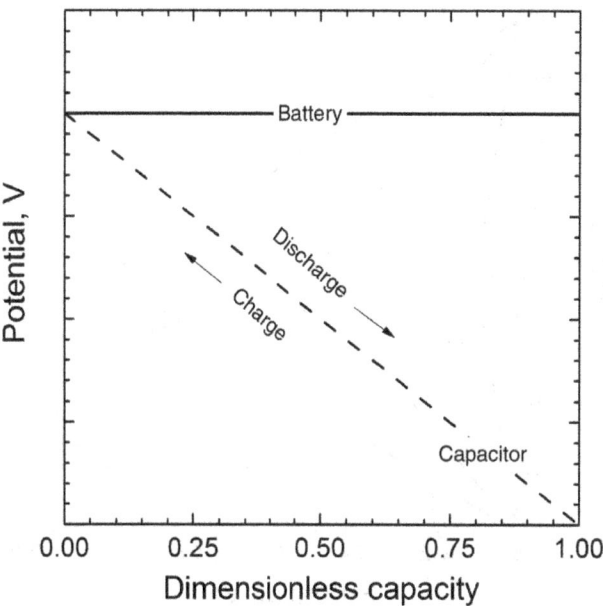

Figure 11.8 Comparison of ideal battery with an ideal capacitor.

11.3 CURRENT–VOLTAGE RELATIONSHIP FOR CAPACITORS

We are interested in how a capacitor behaves under a variety of circumstances. In particular, we seek out the current–voltage behavior of a capacitor under different conditions. This section explores these aspects. We begin with the expression for current from a capacitor (see Chapter 6):

$$I = C \frac{dV}{dt}. \tag{6.6}$$

This equation describes an ideal capacitor, where the capacitance, C, is a constant and there is no internal resistance. Suppose for a moment that we would like to charge or discharge the capacitor galvanostatically. According to Equation 6.6, a constant-current charge corresponds to a constant value of dV/dt. Therefore, the voltage would increase linearly with time as the capacitor was charged. Similarly, during constant-current discharge, the potential across the capacitor decreases linearly. This behavior highlights an important characteristic of capacitive energy storage, namely, a voltage that changes significantly with capacity. In Figure 11.8, we compare the behavior of an ideal battery to that of a capacitor for discharge at constant current. In most cases, a sloping potential with discharge capacity is undesirable. For instance, the changing potential at constant current leads to a steady decrease in the power available during discharge. On the positive side, one advantage of capacitors is that the state of charge is easy to determine since it is directly proportional to the voltage across the capacitor.

Next we will examine cyclic voltammetry, which is similar to the constant-current behavior just discussed. Recall from Chapter 6 that for cyclic voltammetry the potential is swept in a triangle wave, where ν is the sweep rate in [V·s^{-1}]. This process is illustrated in Figure 11.9, for which the sweep rate is a parameter. The sweep rate for each experiment is

$$\nu = \frac{dV}{dt} = \text{constant}. \tag{11.15}$$

We see immediately that under these conditions the current in the capacitor is constant and directly proportional to the sweep rate. Consequently, if the potential is swept at a constant rate to a specified value and then reversed to return at the same rate to the original potential, the current is as shown in Figure 11.9. On a plot of current versus potential, the response is a rectangle for the ideal capacitor. Note that the direction (sign) of the current changes immediately with the change in sweep direction. This response is different from the response of a faradaic reaction to a reverse in the sweep direction and is explored in greater detail in Problem 11.11. Once again, constant current charge or discharge of a capacitor is associated with a linear change in the potential across the capacitor with respect to time.

Another common operation is a step change in potential from an initial potential, V_i, to a final value, V_f. As seen in the definition of capacitance, the charge, Q, is proportional to the potential across the capacitor. In practice, we

260 Electrochemical Engineering

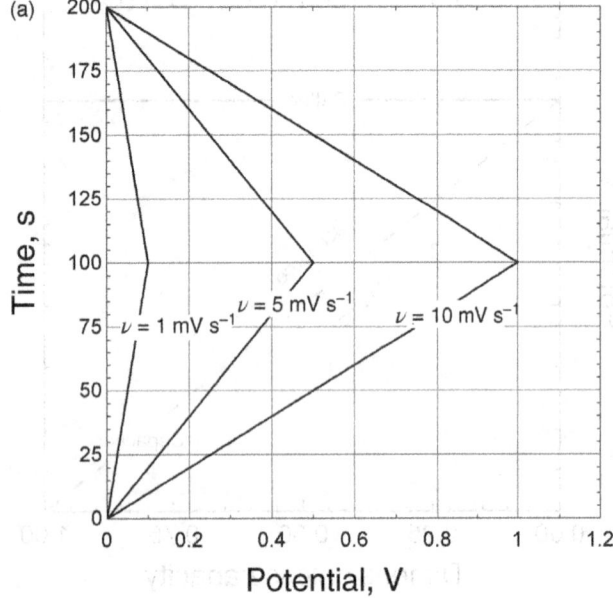

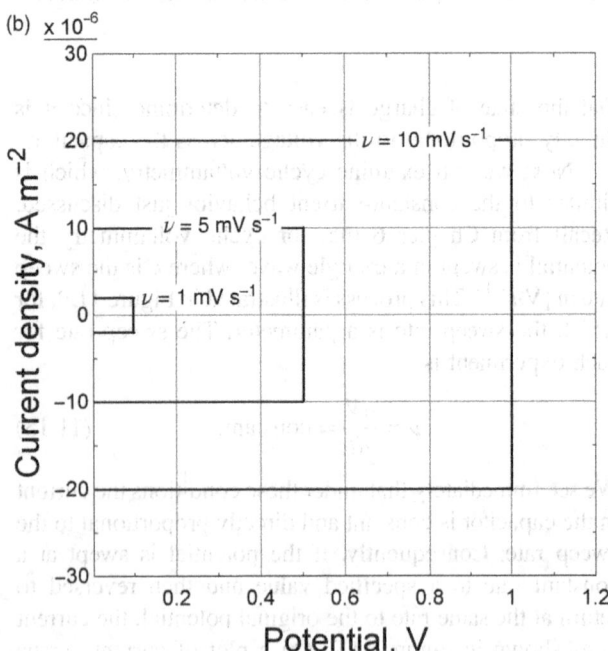

Figure 11.9 Potential and current for an ideal capacitor during a constant potential sweep.

cannot move the charges instantaneously; that is, there is some resistance to current flow. Therefore, we will introduce a resistor in series with our capacitor. As before, let the charge on the capacitor be Q, and let Q_f be the final value of the charge equal to CV_f, the charge at steady state. At time equal to zero, the potential applied to the circuit is changed to V_f. After some time, the capacitor will be fully charged or discharged, depending on whether the voltage is increased or decreased, and the current will be zero. The potential drop across the capacitor is simply Q/C. We can use Kirchhoff's voltage law to get

$$V_f = \frac{Q}{C} + IR_\Omega = \frac{Q}{C} + R_\Omega \frac{dQ}{dt}. \quad (11.16)$$

Equation 11.16 is integrated to give the following relationship for the charge on the capacitor as a function of time:

$$\frac{Q_f - Q(t)}{Q_f - Q_i} = e^{\frac{-t}{R_\Omega C}}, \quad (11.17)$$

where Q_i is the initial charge on the capacitor at $t = 0$, and Q_f is the final value (CV_f). The quantity $R_\Omega C$ has units of seconds and is a time constant related to the time required to charge or discharge the capacitor. Since the capacitance is the proportionality constant between charge and voltage, Equation 11.17 can also be written as

$$\frac{V_f - V(t)}{V_f - V_i} = e^{\frac{-t}{R_\Omega C}}. \quad (11.18)$$

The current is obtained by differentiation of Equation 11.17,

$$I(t) = \frac{dQ}{dt} = \frac{V_f - V_i}{R_\Omega} e^{\frac{-t}{R_\Omega C}}. \quad (11.19)$$

The sign of the current is positive for charge and negative for discharge. As evident from Equation 11.19, in response to a step change in potential, the current decreases exponentially to zero with a time constant $\tau = R_\Omega C$. Zero current is the endpoint associated with a constant voltage across the capacitor. A large resistance implies that the current is small and, therefore, the time required to charge or discharge the capacitor is large. Similarly, as the capacitance increases, it will also take longer to charge or discharge the capacitor.

ILLUSTRATION 11.3

A 25 cm² electrode has a capacitance of 900 F m⁻² (superficial area). Initially, the capacitor is completely discharged. A step increase in potential results in a 25 A current step. If the resistance is 4 mΩ, how long will it take the potential across the capacitor to reach 95% of its steady-state value?

First, calculate the time constant for an $R_\Omega C$ system:

$$\tau = R_\Omega C = (0.004\,\Omega)\,(25 \times 10^{-4}\,\text{m}^2)\left(\frac{900\,\text{F}}{\text{m}^2}\right) = 9\,\text{ms}.$$

Initially, all of the potential drop is across the resistor as the discharged capacitor offers no resistance to the flow of current. Therefore, $\Delta V = \Delta I R_\Omega = 100$ mV, which is the voltage drop across the resistor. At steady state, no current flows through the capacitor. From Equation 11.18, the variation of potential with time is

$$\frac{V_f - V}{V_f - V_i} = \frac{0.05 V_f}{V_f - 0} = 0.05 = e^{\frac{-t}{R_\Omega C}}.$$

Solving for t gives 27 ms. As can be seen from the solution, the time required does not depend on the magnitude of the voltage step.

11.4 POROUS EDLC ELECTRODES

As we noted in Chapter 5, a porous electrode is an effective means of increasing the surface area of an electrode. As with any porous electrode, the local resistance is a function of position in the electrode. An equivalent circuit diagram for an EDLC using a porous electrode is shown in Figure 11.10. The resistances along the top represent the electrolyte resistance through the thickness of the electrode. Along the bottom is another distributed resistance corresponding to that of the solid phase. The solid and electrolyte are connected by a capacitor associated with the double layer that is in parallel with a resistance, R_f, for faradaic reactions. For ionic current to travel to the back of the electrode, it must travel through the electrolyte. Current can also flow to charge the double layer along the thickness of the electrode. We now wish to examine the behavior of such a device and contrast it to a planar electrode. For our initial analysis we will assume that the effective conductivity of the solid phase is much larger than that of the electrolyte. $\phi = \phi_2 - \phi_1$, where ϕ_1 is a constant. We will then present more general results, but no attempt to derive them is made here. In the electrolyte we assume Ohm's law applies, which means that we are also neglecting the influence of any concentration differences.

$$i_2 = -\kappa \nabla \phi. \qquad (11.20)$$

It is understood that the conductivity in Equation 11.20 is an effective conductivity that accounts for porosity and tortuosity of the electrode. In Chapter 5, the charge balance in the porous electrode assumed that only faradaic reactions occurred; both adsorption and charging of the double layer were neglected. The resulting balance was

$$-\nabla \cdot i_2 + a i_n = 0. \qquad (5.5)$$

For the EDLC, we must modify this charge balance. The actual current density at the electrode surface, i_n, can either be due to a faradaic reaction (through R_f in the figure) or to double-layer charging. When double-layer charging is accounted for, the charge balance is

$$-\nabla \cdot i_2 + a \left(i_{n,faradaic} + C_{DL} \frac{\partial \phi}{\partial t} \right) = 0, \qquad (11.21)$$

where C_{DL} is the capacitance per unit area of electrode [F·m^{-2}], and a is the specific interfacial area [m^{-1}], as introduced previously. There are clearly cases where both faradaic and nonfaradaic reactions are important, for example, in batteries, fuel cells, and pseudo-capacitors, but for now we neglect faradaic reactions. Furthermore, we consider only one spatial dimension, where the potential and current density are functions of time and position. With these assumptions, Ohm's law and the charge balance reduce to

$$\frac{\partial \phi}{\partial z} + \frac{i_2}{\kappa} = 0 \qquad (11.22)$$

and

$$\frac{\partial i_2}{\partial z} + a C_{DL} \frac{\partial \phi}{\partial t} = 0. \qquad (11.23)$$

The current density can be eliminated from these equations to produce a single, second-order equation for the potential. To do this, Equation 11.22 is first differentiated with respect to z. We then multiply by $-\kappa$ and add the result to Equation 11.23 to yield

$$\frac{\partial \phi}{\partial t} = \frac{\kappa}{a C_{DL}} \frac{\partial^2 \phi}{\partial z^2}. \qquad (11.24)$$

Equation 11.24 may look familiar—it is similar in form to Fick's second law of diffusion, which we encountered in Chapter 4. The equation is also analogous to the equation

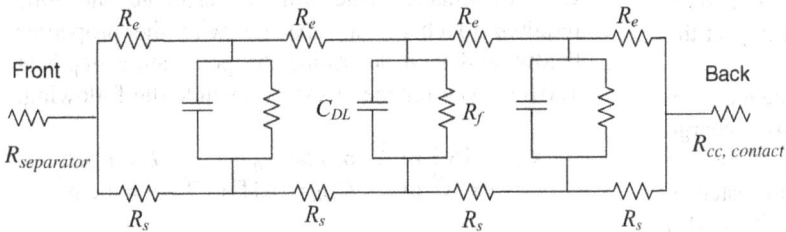

Figure 11.10 Equivalent circuit for one porous electrode of an EDLC.

for a transmission line familiar from electrical engineering. Once the initial and boundary conditions are specified, there are numerous techniques available for solving this linear, partial differential equation.

Our objective is to gain quantitative insight into the time scale and penetration depth that characterize a porous EDLC electrode. In other words, we want to know how quickly we can charge an electrode of a given size or, conversely, the thickness of the electrode that we can effectively use in a given time. To do this, let's consider a step change in potential at time equal to zero at the front of the electrode from ϕ_{ini} to a potential ϕ_f, for an electrode of thickness L. This process is equivalent to instantaneously charging (or discharging) the front of the electrode and then calculating the time it takes for the rest of the electrode to reach that same state of charge. Introducing the following dimensionless parameters:

$$\theta = \frac{\phi(t) - \phi_{ini}}{\phi_f - \phi_{ini}}, \quad x = \frac{z}{L}, \quad \tau = \frac{\kappa t}{aC_{DL}L^2},$$

where $\dfrac{aC_{DL}L^2}{\kappa}$ has units of time. Equation 11.24 becomes

$$\frac{\partial \theta}{\partial \tau} = \frac{\partial^2 \theta}{\partial x^2}. \quad (11.25)$$

The initial condition is at $\tau = 0$, $\theta = 0$, with boundary conditions at $x = 0$, $\theta = 1$ and at $x = 1$, $\frac{\partial \theta}{\partial x} = 0$. This last boundary condition is equivalent to setting the current density in the electrolyte to zero at the back of the porous electrode (i.e., at the current collector). The dimensionless potential from the solution of Equation 11.25 is plotted in Figure 11.11. From our result for the potential, we can calculate the current density at any position in the electrode from Ohm's law. At very short times, the gradient in potential is very steep at the front of the electrode and the current density there is high. In contrast, the current does not penetrate very far into the electrode and is essentially zero for a large fraction of the electrode. Thus, for these short times, the electrode appears infinite, and the penetration depth can be approximated as

$$\delta = \sqrt{\frac{4\kappa t}{aC_{DL}}}. \quad (11.26)$$

Equation 11.26 applies when semi-infinite conditions prevail, which is valid for $\tau \leq 1/16$, where the boundary condition at the back of the electrode does not impact the solution (see Figure 11.11).

Referring to the distributed capacitances in Figure 11.10, those near the front of the electrode charge or discharge more quickly because of lower resistance through the electrolyte due to a shorter transport distance. At short times, the current does not penetrate beyond a

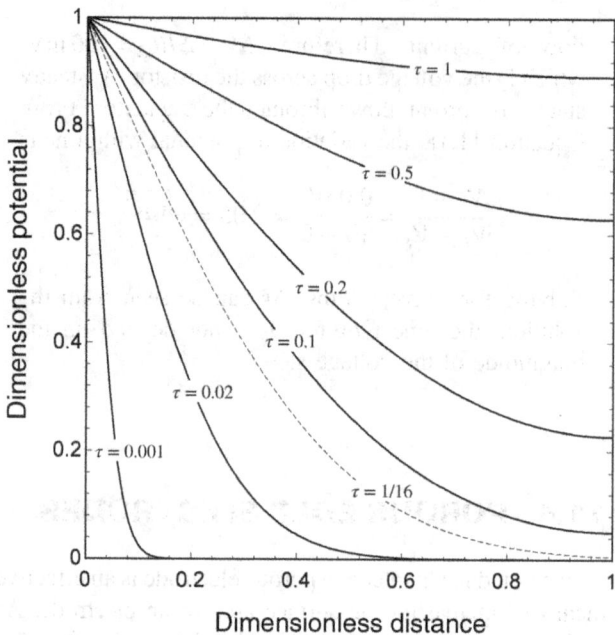

Figure 11.11 Solution to Equation 11.25 showing the change in potential across the electrode thickness with time as a parameter.

certain depth, and much of the electrode is left unused. Therefore, there is a limit to the thickness of the electrode that is effective for a particular application. Equation 11.26 can be used to determine the penetration depth corresponding to a given time associated with the application of interest, as shown in Illustration 11.4.

The group $\dfrac{aC_{DL}L^2}{\kappa}$ has units of time and represents a characteristic time for charging the porous EDLC electrode. This characteristic time comes naturally from the above analysis, and corresponds to the time at which $\tau = 1$. As seen in Figure 11.11, it provides a reasonable estimate of the time it takes to charge the capacitor. Its reciprocal, $\dfrac{\kappa}{aC_{DL}L^2}$, represents a characteristic frequency for the capacitor. The transient behavior of EDLCs will be explored in greater detail in the next section with use of impedance analysis, which builds upon the material that was introduced in Chapter 6.

ILLUSTRATION 11.4

Use the characteristic time to estimate the time required to charge a capacitor with the properties below, and then determine the penetration depth at 120 ms. Data for the electrode include the following:

$\kappa_{eff} = 15 \text{ S} \cdot \text{m}^{-1} \quad \sigma_{eff} \gg \kappa_{eff} \quad L = 1 \text{ mm}$
$a = 3 \times 10^8 \text{ m}^{-1} \quad C_{DL} = 0.3 \text{ F} \cdot \text{m}^{-2} \quad A = 1 \text{ cm}^2$

The time required to charge the capacitor is approximately $\frac{aC_{DL}L^2}{\kappa} = \frac{(3\times10^8)(0.3)(0.001^2)}{15} = 6$ seconds.

The penetration depth at 120 ms can be estimated from Equation 11.26, since the time is less than $\tau/16$.

$$\delta = \sqrt{\frac{4\kappa t}{aC_{DL}}} = \sqrt{\frac{4(15)(0.120)}{3\times 10^8(0.3)}} = 0.28 \text{ mm}.$$

11.5 IMPEDANCE ANALYSIS OF EDLCs

In this section, we apply impedance spectroscopy to electrochemical double-layer capacitors in order to gain insight into their transient behavior. Additionally, we use the impedance results as a basis for a simplified EDLC model that will facilitate our analysis of these devices.

Analysis for Highly Conductive Solid Phase

In order to use impedance to examine the transient behavior of EDLCs, we begin again with Equation 11.24. This partial differential equation describes the potential as a function of time and position in a porous electrode assuming that the solid matrix is infinitely conductive. To obtain an expression for the complex impedance, we apply an oscillating potential at $x = 0$, as demonstrated in Chapter 6. The details are not shown here as they are a bit tedious, but the results are quite useful. Specifically, we obtain the following expression for the complex impedance (in ohms):

$$Z(\omega) = \frac{L}{\kappa A} \frac{\coth\sqrt{j\omega^*}}{\sqrt{j\omega^*}}, \quad (11.27a)$$

where ω^* is a dimensionless frequency defined as the frequency, ω, divided by the characteristic frequency:

$$\omega^* = \frac{\omega}{\frac{\kappa}{aC_{DL}L^2}} = \frac{\omega aC_{DL}L^2}{\kappa}. \quad (11.27b)$$

The impedance described by Equation 11.27 is shown on a Nyquist plot, Figure 11.12, for three different electrode thicknesses. The parameters used to make this plot are included with the figure. First, we note that at high

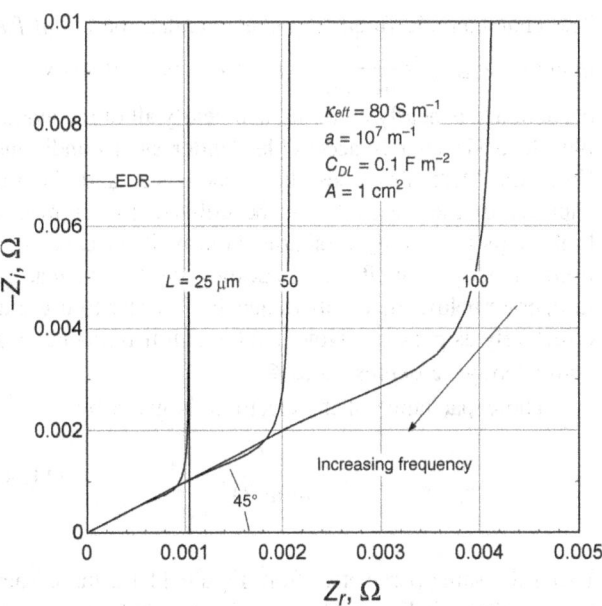

Figure 11.12 Nyquist plot for EDLCs of varying thickness using Equation 11.27a.

frequencies the impedance goes to zero. This is a consequence of assuming that the conductivity in the solid phase is infinite. At high frequencies, changes in potential do not penetrate significantly into the electrode and the effective resistance is zero. For a porous electrode, the slope of the impedance is 45° at high to moderate frequencies. As long as the time is short, the potential does not penetrate to the back of the electrode, and the electrode is effectively semi-infinite. As the frequency decreases, the potential penetrates farther and farther into the electrode. Eventually, the influence of the back of the electrode is felt. At frequencies sufficiently low that the capacitor has time to fully charge, the impedance becomes infinite as expected for an ideal capacitor. Note that the value for Z_r where the line becomes vertical depends on the thickness of the electrode. That makes sense since both the capacitance and the time required to charge fully the porous capacitor increase with the electrode thickness.

The capacitance at low frequencies is directly proportional to the electrode thickness since all parts of the electrode are accessible. The low-frequency capacitance in farads is simply

$$C(\omega \to 0) = aLC_{DL}A. \quad (11.28)$$

Actually, it is the relationship of the frequency to the characteristic frequency that is important. Therefore, a more precise statement of the criterion $\omega \to 0$ is

$$\omega < \frac{\kappa}{aC_{DL}L^2}. \quad (11.29)$$

The characteristic frequency is also called the *cutoff frequency*, $\omega_{cutoff} = \frac{\kappa}{aC_{DL}L^2}$. Below this frequency, the capacitance is nearly constant, and nearly all of the available electrode area is accessible. Under such conditions, Equation 11.28 describes the capacitance, and the full capacity of the electrode can be utilized. In contrast, at high frequencies, only a fraction of the available capacity is used. Generally, we'll want to design or select a capacitor to operate below the cutoff frequency in order to use it as effectively as possible. Note that the cutoff frequency can equivalently be expressed as $\frac{\kappa A}{LC}$.

The capacitance of the electrode is given by

$$C = \frac{-1}{\omega \text{Im}(Z)}. \qquad (11.30)$$

Using the same parameters from Figure 11.12, the capacitance is plotted in Figure 11.13 as a function of frequency. As noted above, the capacitance at low frequencies is constant and depends directly on the thickness of the electrode. At frequencies above the cutoff frequency, the capacitance drops quickly and goes to zero at high frequencies.

Referring back to Figure 11.12, we see that the impedance behavior roughly approximates that of an ideal capacitor (vertical line on Nyquist plot) in series with a resistance. This approximation provides the basis for a simplified series-RC circuit model as illustrated in Figure 11.14. The resistance is associated with the resistance to current flow in the electrolyte solution since we have assumed that the solid-phase resistance is zero (i.e., the solid is infinitely conductive). It is referred to as the *equivalent distributed resistance* (EDR). The value for the

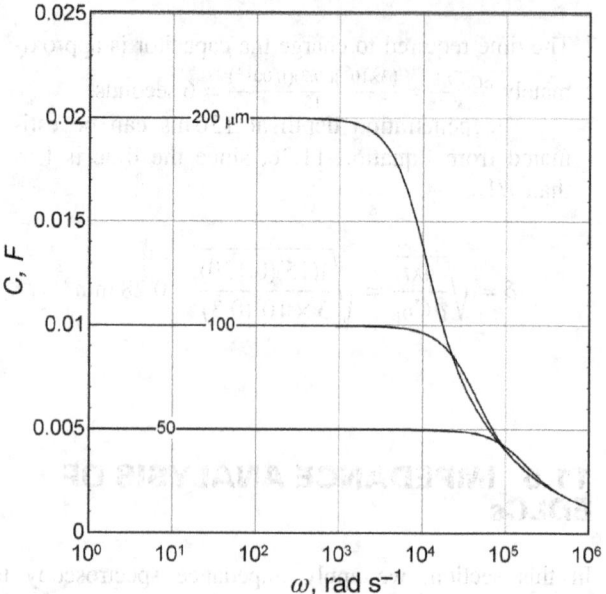

Figure 11.13 Capacitance of a 1 cm² porous electrode. The thickness of the electrode is a parameter.

EDR can be expressed as

$$\text{EDR} = \frac{1}{3}\frac{L}{\kappa A} \, [\Omega] \qquad (11.31)$$

and is the limit of the real portion of the complex impedance as the frequency goes to zero. Note that this resistance in ohms is directly proportional to the thickness of the electrode, L. This simplified model is applicable when the frequency is below the cutoff frequency.

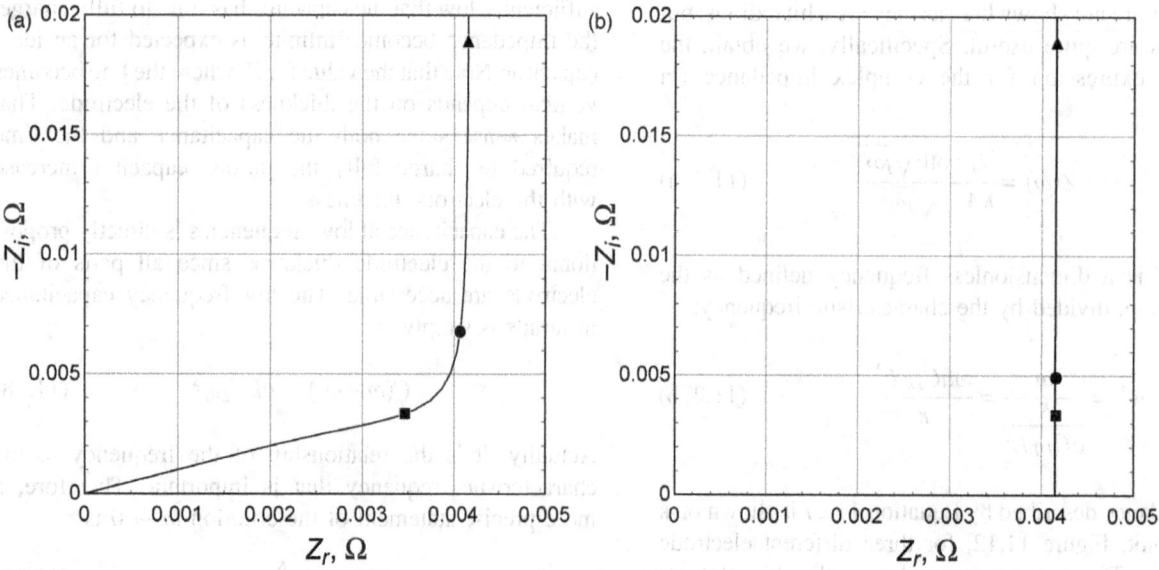

Figure 11.14 Comparison of porous electrode model (a) with simple model (b) at (1) $\omega = \omega^*$ (circle), (2) $\omega = \omega^*/3$ (triangle), and (3) $\omega = 3\omega^*$ (square).

How does this simplified series-RC model compare with the behavior of the actual electrode? At high frequencies, the simple model overestimates the electrode resistance, since the resistance in the simple model is constant and that of the porous electrode is very small at high frequencies and increases with decreasing frequency as the penetration into the electrode increases. Therefore, the simple model will tend to underestimate the maximum power. However, at lower frequencies where nearly all of the electrode capacity is used, the simplified series-RC circuit model does a good job of representing the actual behavior. As noted previously, we typically design the capacitor to use the available capacity. Under those conditions, the simple model does a good job of representing the actual behavior of the system. Use of this approximation is shown in Illustration 11.5.

ILLUSTRATION 11.5

For the electrode described in Illustration 11.4, determine a simplified series-RC circuit model consisting of the EDR and an ideal capacitor. At what frequencies does this model apply?

$$\text{EDR} = \frac{1}{3}\frac{L}{\kappa A} = \frac{1}{3}\frac{(0.001)}{15(10^{-4})} = 0.22 \ \Omega.$$

The capacitance, $C = aC_{DL}AL = (3 \times 10^8)(0.3)(10^{-4})(0.001) = 9 \text{ F}$.

Thus, the simplified model is a 9 F capacitor in series with a 0.22 Ω resistor.

The cutoff frequency can be estimated as follows:

$$\text{Cutoff frequency} \approx \frac{\kappa A}{LC} = \frac{\kappa}{aL^2 C_{DL}} = 0.167 \text{ s}^{-1}.$$

The frequency should be lower than this value for the full capacitance to be utilized. Now let's compare the charging time from the simple model with the characteristic time that we calculated from the more complex model. For the simple model,

$$\tau_{RC} = RC = (\text{EDR})(C) = (0.2)(9) = 2.0 \text{ s}.$$

For a capacitor in series with a resistor (Equation 11.14 with $Q_i = 0$) and a time of 6 seconds (see Illustration 11.4):

$$\frac{Q}{Q_f} = 1 - \exp\left(-\frac{t}{\tau_{RC}}\right) = 1 - \exp(-3) = 0.95$$

or 95% charged, which is consistent with the previous results, where charging required ≈6 seconds.

Analysis When Both Solid and Electrolyte Resistances Are Important

The results of the previous section are for the case where the conductivity of the solid phase is much larger than that of the electrolyte, a condition that is satisfied for many practical electrodes. If the resistances of both phases (solid and liquid) are important, the impedance can also be determined analytically:

$$Z(\omega) = \frac{L}{(\kappa + \sigma)A}\left(1 + \frac{2}{\sqrt{j\omega^*}\sinh\sqrt{j\omega^*}}\right)$$
$$+ \frac{L(\kappa^2 + \sigma^2)}{\kappa\sigma(\kappa + \sigma)}\frac{\coth\sqrt{j\omega^*}}{\sqrt{j\omega^*}}, \quad (11.32a)$$

where

$$\omega^* = \frac{\omega a C_{DL}L^2(\kappa + \sigma)}{\kappa\sigma}. \quad (11.32b)$$

Note that the dimensionless frequency, ω^*, has the same form as it did before, but now utilizes a composite conductivity that includes the contributions of both the electrolyte and the solid. Figure 11.15 shows the impedance behavior for a 100 μm thick electrode at different ratios of κ/σ as a parameter. All of the other parameters are the same as those used in the previous section to create Figure 11.12. Hence, the results can be compared directly.

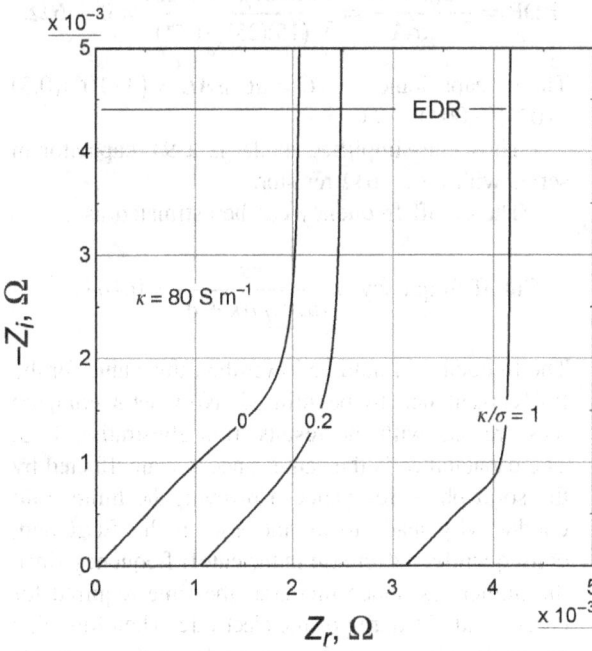

Figure 11.15 Nyquist plot for finite resistance in both the electrolyte and solid phases.

For an infinitely conductive solid phase ($\kappa/\sigma = 0$), the resistance for a porous electrode EDLC approaches zero at high frequencies, as we saw previously. These results are identical to those shown in Figure 11.12. As κ/σ increases, which is equivalent to a decrease in σ for these simulations since κ was kept constant, we see two marked differences in the impedance. First, Z_r does not approach zero at high frequencies when both σ and κ are finite. In fact, it approaches the following limit:

$$Z_r(\omega \to \infty) = \frac{L}{(\kappa + \sigma)A} \; [\Omega]. \quad (11.33)$$

The EDR is also impacted by a finite solid conductivity:

$$\text{EDR} \equiv \frac{1}{3}\frac{L(\kappa+\sigma)}{\kappa\sigma A} \; [\Omega], \quad (11.34)$$

reflecting the additional resistance that results from the finite solid conductivity. As expected, Equation 11.34 is equivalent to Equation 11.31 as $\sigma \to \infty$.

ILLUSTRATION 11.6

For the electrode described in Illustration 11.4, determine values for the simplified model consisting of the EDR and an ideal capacitor if the solid conductivity is finite and equal to $25 \, \text{S·m}^{-1}$. What is the cutoff frequency? How do these values compare with those from Illustration 11.5?

$$\text{EDR} = \frac{1}{3}\frac{L(\kappa+\sigma)}{\kappa\sigma A} = \frac{1}{3}\frac{(0.001)(15+25)}{(15)(25)(10^{-4})} = 0.356 \, \Omega.$$

The capacitance, $C = aC_{DL}AL = (3 \cdot 10^8)(0.3)(10^{-4})(0.001) = 9 \, \text{F}$.

Thus, the simplified model is a 9 F capacitor in series with a 0.356 Ω resistor.

The cutoff frequency can be estimated as

$$\text{Cutoff frequency} \approx \frac{\kappa\sigma}{aL^2 C_{DL}(\kappa+\sigma)} = 0.104 \, \text{s}^{-1}.$$

The frequency should be lower than this value for the full capacitance to be utilized. Now let's compare these results with the results from Illustration 11.5. The capacitance is the same, since it is unaffected by the solid phase resistance. However, the finite solid conductivity leads to an increase in the EDR and, consequently, a decrease in the cutoff frequency since the higher resistance increases the time required for charge and discharge of the electrode. Therefore, this electrode will not cycle as quickly as the electrode from Illustration 11.5.

11.6 FULL CELL EDLC ANALYSIS

With the analysis of porous electrodes presented above and the resulting simplified models, we can now address a full EDLC cell. Incorporating the simplified representation for each of the two electrodes yields two capacitors and two resistors in series. In addition, there are other resistances not directly associated with the porous electrodes themselves that have not yet been included. Hence, we add an equivalent series resistance (ESR) to account for these additional resistances. The most important resistance included in the ESR is that of the separator between the electrodes. Resistances such as contact losses are also included in the ESR. We can now combine all of the resistances into a single effective resistance for the cell:

$$\text{ESR}_{\textit{eff}} \equiv \text{ESR} + \text{EDR}_1 + \text{EDR}_2, \quad (11.35)$$

where the subscripts refer to the two electrodes. We can also use an equivalent capacitance to represent the capacitance of the two electrodes in series. We are now prepared to examine some interesting questions regarding EDLCs. For example, for a given electrode size (area), how thick must the electrodes be to achieve a particular capacity, and at what frequency can the device be operated and still maintain that capacity? Or, for a given capacity and required frequency of operation, what is the required area and thickness of the electrodes? Note that the specific area for capacitor materials is often expressed in area per mass. To use this number in the equations that we have developed, we also need to know the mass per volume of the material in an electrode. This and other concepts of interest are demonstrated in the following illustration.

ILLUSTRATION 11.7

An electrochemical double-layer capacitor is constructed from two porous carbon electrodes separated by a porous membrane. Please determine the capacity of the electrode, its effective resistance $\text{ESR}_{\textit{eff}}$, and the maximum frequency at which it can be operated and still use its full capacity. Assume that the contact resistance is 0.4 mΩ.

Thickness of separator	25 μm
Effective conductivity of electrolyte in separator	55 S·m^{-1}
Superficial (geometric) electrode area	30 cm^2
Porosity of separator, ε_s	0.5
Electrode thickness	50 μm
Capacitance per actual surface area	0.13 F·m^{-2}
Effective conductivity of electrolyte in electrode	35 S·m^{-1}

Specific surface area of electrode $600 \, \text{m}^2 \cdot \text{g}^{-1}$
Density of carbon $2050 \, \text{kg} \cdot \text{m}^{-3}$
Porosity of electrode, ε_e 0.4
Effective electronic conductivity $10^4 \, \text{S} \cdot \text{m}^{-1}$

The specific interfacial area, a, is $a = \dfrac{2050 \, \text{kg}}{\text{m}^3}$

$\dfrac{(1-0.4)}{\text{kg}} \dfrac{1000 \, \text{g}}{\text{kg}} \dfrac{600 \, \text{m}^2}{\text{g}} = 7.38 \times 10^8 \, \text{m}^{-1}$, where $1 - \varepsilon_e$ is the volume fraction of carbon in the electrode. Below the cutoff frequency, the capacitance of one electrode is

$$C = aC_{DL}LA = \dfrac{7.38 \times 10^8}{\text{m}} \dfrac{50 \times 10^{-6} \, \text{m}}{} \dfrac{0.13 \, \text{F}}{\text{m}^2}$$

$$\dfrac{30 \times 10^{-4} \, \text{m}^2}{} \; 14.4 \, \text{F}.$$

From Equation 11.7, the capacitance of the two electrodes in series is 7.2 F.

The conductivity of the solid phase is so high that we can neglect any contribution to the ESR from the solid carbon electrode and the resistance is simply

$$\text{ESR}_{eff} = R_c + 2\left(\dfrac{1}{3} \dfrac{L_e}{\kappa_e A}\right) + \dfrac{L_s}{\kappa_s A} = 0.0004$$

$$+ \dfrac{\frac{2}{3} 50 \times 10^{-6} \, \text{m} \, \Omega \cdot \text{m}}{30 \times 10^{-4} \, \text{m}^2 \; 35} + \dfrac{25 \times 10^{-6} \, \text{m} \, \Omega \cdot \text{m}}{30 \times 10^{-4} \, \text{m}^2 \; 55} = 0.87 \, \text{m}\Omega.$$

The maximum operating frequency for essentially full capacity will be lower than the cutoff frequency associated with the individual electrodes due to the influence of the additional resistance (separator and contact) that increases the time required to charge or discharge the electrode. To estimate the requested frequency, we use three times the RC time constant for the equivalent circuit, which is the time that it takes to reach 95% of the final value. The maximum frequency can be approximated as the reciprocal of this value.

$$3(\text{ESR}_{eff})(C_{eff}) = 3(8.7 \times 10^{-4} \, \Omega)(7.2 \, \text{F}) = 0.0188 \, \text{s}$$

$$\omega_{max} = 53 \, \text{s}^{-1}.$$

The maximum voltage at which an EDLC can be operated is typically determined by the stability of the electrolyte and would therefore be specified. However, our simple model also allows us, for example, to estimate the change in voltage or charge as a function of time for a step change in the current, similar to what we did in Section 11.3.

11.7 POWER AND ENERGY CAPABILITIES

Both the energy and power density are important characteristics of capacitors, and are considered in this section. The change in energy associated with a change in capacitor voltage is

$$dE = VdQ = CVdV. \qquad (11.36)$$

The total energy stored in the capacitor can be obtained by integration

$$E = \int dE = C \int VdV = \dfrac{CV^2}{2} = \dfrac{QV}{2}, \qquad (11.37)$$

where we have assumed that C is constant. Recall that capacitance has units of farads, which are equivalent to J V^{-2} or C V^{-1}. Using Equation 11.37, we can estimate the amount of energy stored in different types of capacitors, as demonstrated in the following illustration.

ILLUSTRATION 11.8

Calculate the energy stored in the three capacitors described in Illustration 11.2. Take the separator area as $10 \, \text{cm}^2$, and assume that the voltage for each is as follows:

1. Electrostatic, 20 V
2. Electrolytic, 100 V
3. EDLC, 3 V

1. For a 1 F capacitor, a surface area $5.65 \times 10^4 \, \text{m}^2$ was required. This corresponds to $1.77 \times 10^{-5} \, \text{F} \cdot \text{m}^{-2}$.

$$E = \dfrac{CV^2}{2} = \dfrac{1.77 \times 10^{-5} \, \text{F}}{\text{m}^2} \dfrac{0.001 \, \text{m}^2 (20 \, \text{V})^2}{2}$$

$$= 3.5 \times 10^{-6} \, \text{J},$$

2. $E = \dfrac{CV^2}{2} = \dfrac{0.0098 \, \text{F}}{\text{m}^2} \dfrac{0.001 \, \text{m}^2 (100 \, \text{V})^2}{2}$

$$= 0.049 \, \text{J},$$

3. $E = \dfrac{CV^2}{2} = \dfrac{1873 \, \text{F}}{\text{m}^2} \dfrac{0.001 \, \text{m}^2 (3 \, \text{V})^2}{2}$

$$= 8.4 \, \text{J}.$$

Note that the capacitance per unit area used for the EDLC calculation is half the value from Illustration 11.2 in order to account for the two capacitors in series that are inherent in the EDLC. The voltage is the total voltage across the EDLC.

The power is energy per unit time, and is perhaps most easily expressed as

$$P = IV. \quad (11.38)$$

The power will be a maximum for the initial discharge of a capacitor. The potential used in Equation 11.38 is the terminal potential, which is lower because of the resistive losses in the ESR. For an EDLC capacitor, one can show (Problem 11.19) that the maximum usable power that can be obtained is

$$P_{max} = \frac{V^2}{4 \times \text{ESR}_{eff}}, \quad (11.39)$$

where ESR_{eff} is the total effective series resistance defined by Equation 11.35, V is the terminal voltage across the EDLC, and Ohms law has been used to relate the current to the voltage. Equations 11.37 and 11.39 can be used to illustrate the trade-off between energy and power. We will explore this through Illustration 11.9.

ILLUSTRATION 11.9

For the capacitor from Illustration 11.7, calculate the maximum specific energy and specific power from this device assuming that the maximum potential is 1 V. All of the parameters are the same as those in Illustration 11.7. In addition,

Density of separator   1800 kg·m^{-3}
Density of electrolyte   1000 kg·m^{-3}

To determine the specific energy and power, we need the mass of the device. We will include the mass of all the EDLC components including the separator, two carbon electrodes, and the electrolyte. The mass of the casing is not included in this calculation.

$$m = m_s + 2m_c + m_e = A\{\rho_s(1 - \varepsilon_s)L_s + 2(1 - \varepsilon_e)\rho_c L_e$$
$$+ \rho_e[2\varepsilon_e L_e + L_s \varepsilon_s]\} = 0.594 \text{ g}.$$

$$\text{specific energy} = \frac{E}{\text{mass}} = \frac{CV^2}{2}\frac{1}{m} = \frac{(7.2)1^2}{2}\frac{1}{5.94 \times 10^{-4}}$$
$$= 6060 \text{ J/kg},$$

which when converted to W·h is 1.68 Wh·kg^{-1}.
The maximum specific power is

$$\frac{P_{max}}{\text{mass}} = \frac{V^2}{4 \times \text{ESR}_{eff} m} = \frac{(1\text{V})^2}{(4)(8.7 \times 10^{-4}\,\Omega)5.94 \times 10^{-4} \text{ kg}}$$
$$= 4.8 \times 10^5 \text{ W} \cdot \text{kg}^{-1},$$

where ESR_{eff} was determined as part of Illustration 11.7.

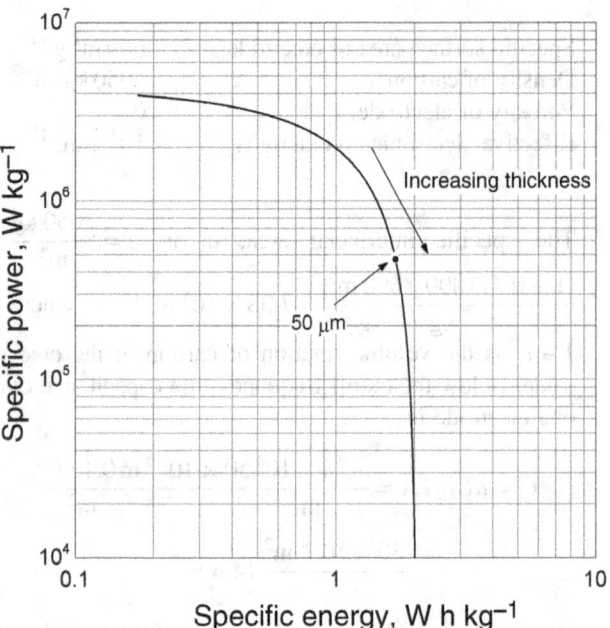

Figure 11.16 Comparison of maximum specific power and specific energy. The parameters are from Illustration 11.6 and the thickness of electrode varied.

The same procedure shown in Illustration 11.9 is used to generate Figure 11.16, which shows the specific energy and the maximum specific power for several different electrode thicknesses ranging from 1 μm to 1 mm. We note from Figure 11.16 that for thick electrodes, the rate capability is much lower and the specific energy approaches a constant value. The constant value of the specific energy is reached as the electrodes become sufficiently thick that the mass of the other cell components is no longer significant; we are therefore left with the specific energy of the electrodes themselves, which does not change with electrode thickness, even though the absolute energy stored in the electrodes does increase with thickness. Similarly, for thin electrodes, the maximum power approaches a constant as the resistance of the electrodes becomes less important as they get thinner, and the maximum power is limited only by ohmic losses in the separator and contact resistance. The specific energy, however, continues to drop as the electrodes are made thinner because the energy stored is directly proportional to the electrode thickness.

Finally, we note that heat is generated during both charging and discharging of an EDLC. As long as the device is not limited by the supply of ions in the electrolyte (see below), the power lost as heat can be estimated as

$$P_{loss} = I^2 \text{ESR}_{eff}. \quad (11.40)$$

The round-trip efficiency of an EDLC is defined as the ratio of the energy delivered by a capacitor to the energy that

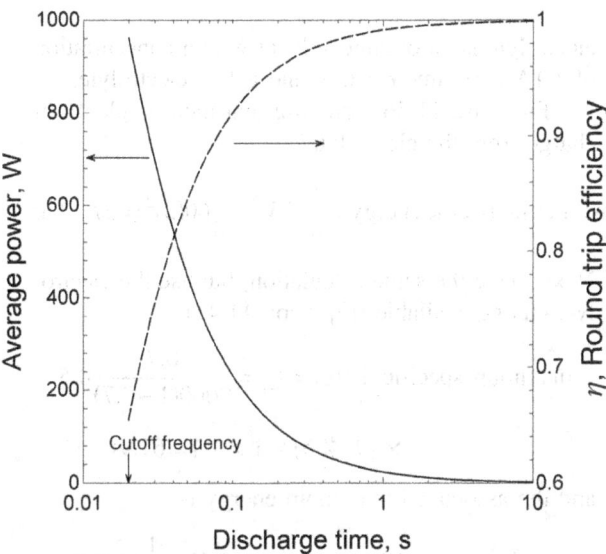

Figure 11.17 Effect of rate of charge and discharge on the efficiency of a capacitor modeled with a simplified series-RC circuit.

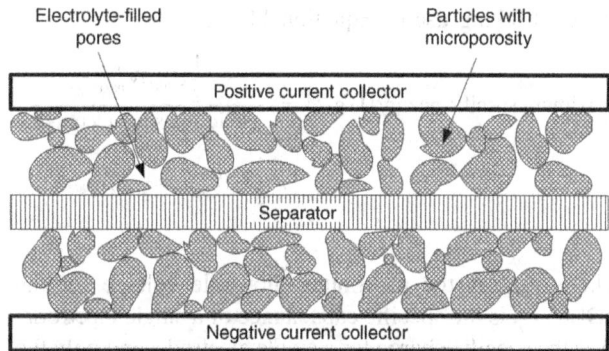

Figure 11.18 Construction of a typical EDLC. Notice the similarities of the cell sandwich with that of batteries and fuel cells.

was supplied to it during a specific discharge/charge cycle, provided that the beginning and ending SOC are the same. For a constant current charge and discharge below the cutoff frequency,

$$\eta = \frac{\text{energy delivered}}{\text{energy supplied}} = \frac{1 - \frac{4}{3}\frac{I\text{ESR}_{eff}}{V_{max}}}{1 + \frac{4}{3}\frac{I\text{ESR}_{eff}}{V_{max}}}. \quad (11.41)$$

Equation 11.41 assumes that the capacitor can be discharged only to half of V_{max}. Using the parameters in Illustration 11.7, the efficiency (right ordinate) and average power (left ordinate) are shown in Figure 11.17. Low currents correspond to long discharge times. The losses are resistive, and low currents therefore translate to high efficiency. For discharge times on the order of the $R_\Omega C$ or less, the efficiency falls off sharply because of the high resistive losses. In extreme cases, it is possible to damage the device by excessive heat generation.

11.8 CELL DESIGN, PRACTICAL OPERATION, AND ELECTROCHEMICAL CAPACITOR PERFORMANCE

Construction of a typical cell sandwich for an EDLC is shown in Figure 11.18. The differences in how energy is physically stored aside, these EDLCs have many similarities to batteries, two porous electrodes coated onto current collectors and separated by an electrolyte. The cell designs are similar too. Typical configurations for EDLCs include cylindrical, prismatic, button, or coin cells. In contrast to dielectric capacitors, EDLCs can operate over only a relatively small potential range, referred to as the *voltage window*, similar to batteries. Rather than dielectric breakdown, the electrochemical stability of the electrolyte determines the maximum potential difference that can be sustained. For an aqueous system, the stability of the solvent water is limited to about 1.2 V by the hydrogen and oxygen reactions. Organic electrolytes, similar to those used in lithium-ion batteries, are stable up to roughly 3 V. However, the electrical conductivities of aqueous electrolytes are generally much higher than those of organic electrolytes. Recall from the previous section that the energy stored is proportional to the potential squared; therefore, capacitors with organic electrolytes tend to store more energy. On the other hand, the power is inversely proportional to the ESR. Thus, the maximum power is roughly linear with electrical conductivity. As a result, capacitors with aqueous electrolytes tend to deliver much higher power. This trade-off is explored in Problem 11.13.

Since the potential of an individual cell is small, EDLCs are often connected electrically in series to build voltage. While the current through and the charge stored in each capacitor is the same, the voltage drop across nonidentical cells may be different. Therefore, it is possible for cells in the series string to exceed the allowable voltage unless means are taken to prevent this from occurring. Thus, either additional circuitry is required to balance cells or the practical voltage limits have to be reduced to ensure that the weakest cell is not overcharged. Moreover, it will also be rare to discharge a capacitor to near zero volts, since this practice would complicate the voltage regulation of the system. As a practical matter, voltage limits for these devices in operation may only be between 40–80% of what is considered the maximum potential rating of the device.

The specific capacity of the material, C_ρ [F·g^{-1} electrode], and the voltage stability window for the electrolyte are two critical factors that impact the specific energy of a device. Making use of Equation 11.37,

$$\text{maximum specific energy}[J \cdot g^{-1}] = \frac{\frac{1}{2}C_{total}V_{max}^2}{2 \text{ (solid mass per electrode)}}$$

$$= \frac{1}{8}C_\rho V_{max}^2 = \frac{1}{4}Q_\rho V_{max}, \quad (11.42)$$

where V_{max} is the maximum allowable voltage for the EDLC, C_ρ is the specific capacitance of a single electrode, Q_ρ is the specific charge for a single electrode, and only the combined mass of the solid portion of the electrodes has been considered in the specific energy. The additional factor of four arises because EDLCs must have two electrodes connected in series. The maximum voltage is determined by the stability window of the electrolyte. However, there is another factor that may limit the maximum energy as discussed in the next paragraph.

On the far right side of Equation 11.42, we see that the maximum specific energy is related to the charge density for the electrode. Let's think about the sources of this charge. In the solid, we simply move electrons from one electrode to the other so that one electrode has excess positive charge and the other excess negative charge. At each electrode, this excess charge in the solid phase is balanced by an equal but opposite charge of ions from the solution. However, there is not an unlimited supply of ions available in the electrolyte. Considering just the solution in the porous electrode, the maximum charge available is

$$\text{maximum specific charge } [C \cdot g^{-1}] = Q_\rho$$

$$= \frac{\varepsilon}{\rho(1-\varepsilon)}\alpha F \nu^+ z^+ c_0. \quad (11.43)$$

Here c_0 is the concentration of electrolyte, ε is the porosity of the electrode, ρ is the density of the solid material, and α is the degree of disassociation of the electrolyte. When the potential is raised, it is possible to deplete the electrolyte before the maximum voltage is reached. Thus, the stored energy can, under some conditions, be limited by the ions in the electrolyte. It is also worth noting that conductivity can also be a strong function of concentration. In cases where a significant fraction of the ions in solution are tied up at the electrodes, the conductivity of the solution may drop precipitously.

ILLUSTRATION 11.10

What is the maximum specific energy for a material with a capacitance per mass of 60 F·g^{-1}. Use $\rho = 2000$ kg·m^{-3}, $\varepsilon = 0.7$, and a maximum voltage of 3 V. The electrolyte is an organic solvent with a concentration of 1.1 M. Assume $\alpha = 0.5$, and a 1:1 electrolyte.

First, use 11.36 assuming that there is plenty of charge from the electrolyte:

$$\text{maximum specific energy} = \frac{1}{8}C_\rho V^2 = \frac{1}{8}60(3)^2 = 67.5 \, J \cdot g^{-1}$$

Next, make the same calculation, but use the electrolyte charge available (Equation 11.43):

$$\text{maximum specific charge} = Q_\rho = \frac{0.7}{2000(1-0.7)}0.5$$

$$\times (96485) \times 1 \times 1.1 = 61.9 \, C \cdot g^{-1}$$

and the associated maximum energy is

$$\text{maximum specific energy} = \frac{1}{4}Q_\rho V = \frac{1}{4}61.9(3)$$

$$= 46.4 \, J \cdot g^{-1}.$$

In this example, the charge available from the electrolyte is the limiting factor.

Often the volume of these devices is more important than their mass. For instance, in automotive applications space is at a premium, and achieving high-energy density [J·m^{-3}] is the key objective for developers of EDLCs. The above results are easily modified with the density of the different components to calculate the energy and power density.

Self-discharge occurs in EDLCs. Although the mechanisms can be complex, the path for leakage current is roughly represented as a simple resistance, R_p, in parallel with the capacitor—a so-called zero order model. If a capacitor is charged to a fixed potential and then held at open-circuit, that is, no net current at the terminals ($I_{app} = 0$), the potential will decrease slowly with time. From a current balance on the node circled in Figure 11.19,

$$I = I_{app} + I_{leakage}. \quad (11.44)$$

The leakage current is defined as the current that must be applied to keep the device at a fixed potential.

$$I_{leakage} = \frac{V}{R_p} = C\frac{dV}{dt}. \quad (11.45)$$

The leakage current is determined from the time rate of change of the voltage at open circuit or from the current that must be applied to maintain a constant voltage. As is apparent from this model, at open circuit the leakage current will decrease with time. Therefore, when leakage

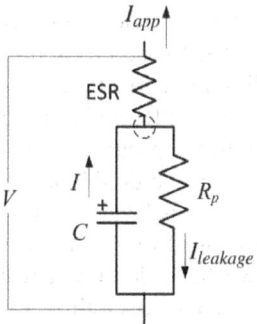

Figure 11.19 Equivalent circuit showing leakage current.

current is reported, the time must also be noted, for example, less than 5 μA at 72 hours.

ILLUSTRATION 11.11

To maintain a 1 F capacitor at 3.0 V, a current of 2.5 μA is required. Assuming a zero-order model for leakage, (a) calculate the parallel resistance, and (b) estimate the time constant.

$$R_p = \frac{V}{I_{leakage}} = \frac{3.0}{2.5 \times 10^{-6}} = 1.2 \text{ M}\Omega.$$

To approximate the time constant, we represent the self-discharge process as the discharge of the capacitor in series with R_p. The time constant for self-discharge is

$$\tau_{self\ discharge} = R_p C = 1.2 \times 10^6 (1) = 1.2 \times 10^6 \text{ s},$$

or about 2 weeks. This, of course, is quite short relative to the time required for self-discharge to completely discharge a battery and represents one of the key disadvantages of an EDLC.

The mechanisms for self-discharge are frequently similar to those seen in batteries and include electrical shorts, redox-shuttle mechanisms controlled by diffusion, and faradaic reactions of impurities. For well-made or commercial EDLCs, the zero-order model for leakage is too crude. In the absence of an internal short, the leakage current is quite small and often under activation control of a faradaic process. In this instance, Tafel kinetics is assumed:

$$I_{leakage} = A i_o \exp\left\{\frac{\alpha \eta F}{RT}\right\}. \quad (11.46)$$

Thus, the logarithm of the leakage current is proportional to the potential of the capacitor, so long as the overpotential remains large enough to remain in the Tafel regime. These two models provide markedly different behavior as

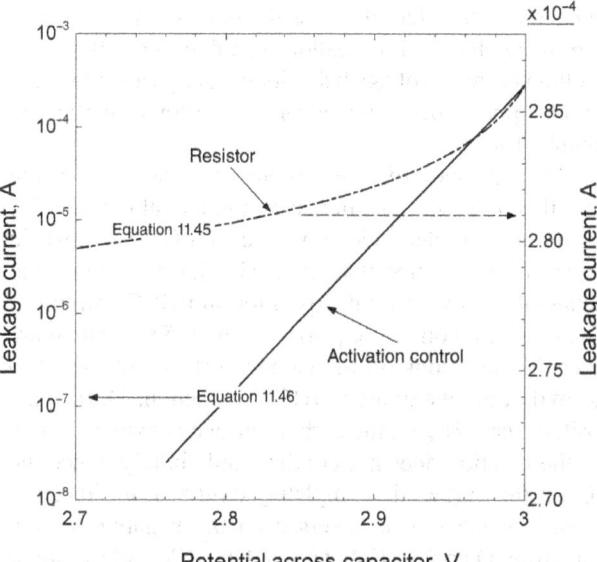

Figure 11.20 Comparison of two models for leakage current.

illustrated in Figure 11.20. Under activation control, the leakage current decreases rapidly as the potential of the capacitor decreases. A relatively small change in potential results in an order of magnitude reduction in leakage current. In contrast, the leakage current is directly proportional to the potential for the zero-order resistor model.

Because no intended faradaic reactions occur in EDLC, the principal mechanisms that limit the cycle life of batteries are not present in EDLCs, and cycling behavior can be exceptional by comparison. Lifetimes of 100,000–1,000,000 cycles are common. As noted previously, unintentional small faradaic reactions are often present. These reactions are frequently associated with impurities in the cell and affect not only the self-discharge of the capacitor but also the degradation of performance with cycling.

Finally, we note that similar to batteries, a state of charge is defined. In the broad sense, SOC is defined as the available charge as a percentage of nominal capacity. Rather than measuring the available charge, the SOC for an EDLC is usually determined from its open-circuit potential.

$$\text{SOC} = 100\% \frac{V_{ocp}}{V_{max}}. \quad (11.47)$$

As seen above, the voltage is linearly proportional to the charge when the capacitance is constant.

11.9 PSEUDO-CAPACITANCE

Other than self-discharge, our discussion of EDLCs has not involved faradaic reactions; that is, we have assumed that no charge is transferred across the double layer, and that there is no change in oxidation state due to reaction. The resulting current–voltage behavior is purely capacitive and would approximate the ideal box shown earlier in Figure 11.9.

The behavior of real systems can vary somewhat from this. For example, the current potential diagram for a carbon black electrode subjected to a CV is shown in Figure 11.21. Notice that the basic shape is somewhat rectangular, as we would expect for an EDLC. Additionally, there are noticeable peaks around 0.55 V. These are reversible reactions on the carbon surface, attributed to the hydroquinone/quinone (HQ/Q) reaction. During the positive scan, HQ on the carbon surface is oxidized to Q, but the reaction does not continue indefinitely. Once the HQ on the surface is completely oxidized, the reaction stops. The reaction is reversed on the negative sweep, converting Q back to HQ. Although the CV looks more or less like that of a capacitor, some of the current is due to charging of the electrical double layer and some is due to faradaic reaction of the HQ/Q redox couple. Is this a battery or capacitor? The answer is yes, it has characteristics of both a battery and a capacitor. We see that there isn't always a clear distinction in the behavior. However, for typical carbon EDLCs, no more than 5% of the current is due to the reactivity of surface oxides. Thus, these carbon materials are still treated as electrochemical double-layer capacitors.

Now imagine an electrode also undergoing a faradaic reaction, but where the potential changes linearly with charge. Physically, this behavior could occur from a redox reaction that is limited to the electrode surface, or from adsorption. Recalling Equation 11.2, the change in potential with charge defines a differential capacitance. Here, the physical origin of the capacitance does not arise from charge accumulation in the double layer; rather, it is the result of a faradaic reaction. We will use a new symbol, C_ϕ, to represent this so-called *pseudo-capacitance*. A device such as the one we have described would also have a capacitance associated with double-layer charging, but here $C_\phi \gg C_{DL}$. A device with both double layer and pseudo-capacitance may be represented by the equivalent circuit shown in Figure 11.22. Here these capacitances are in parallel, and the double-layer capacitance is typically much smaller than the pseudo-capacitance. A characteristic feature of pseudo-capacitors is that the faradaic reaction is more or less restricted to the surface. Thus, the charge-transfer process is limited by the surface area available.

A common example of pseudo-capacitance is ruthenium oxide. During charging and discharging,

$$\text{RuO}_x(\text{OH})_y + \text{H}^+ + e^- \leftrightarrow \text{RuO}_{x-1}(\text{OH})_{y+1} \quad (11.48)$$

Equation 11.48 is a faradaic reaction and involves the transfer of electrons to and from the metal oxide; however, because the reactions are limited to the surface, achieving a high surface area is critical for high energy storage. A typical CV for a pseudo capacitor is shown in Figure 11.22.

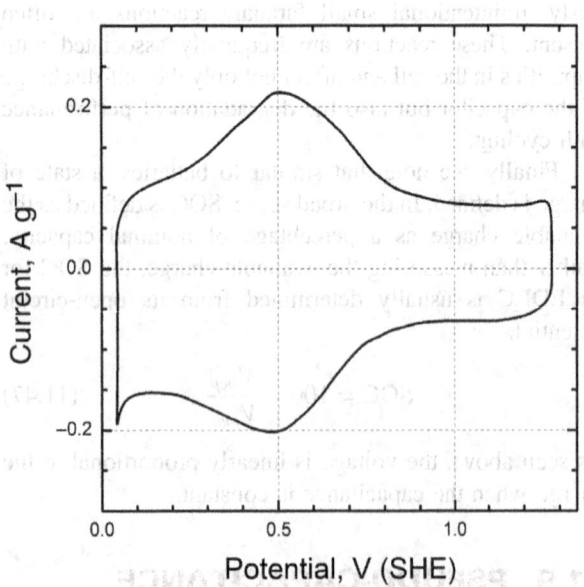

Figure 11.21 Cyclic voltammetry behavior of Vulcan XC-72 in phosphoric acid after oxidation. (Adapted from K. Kinoshita and J.A. S. Bett (1973) *Carbon*, **11**, 403.)

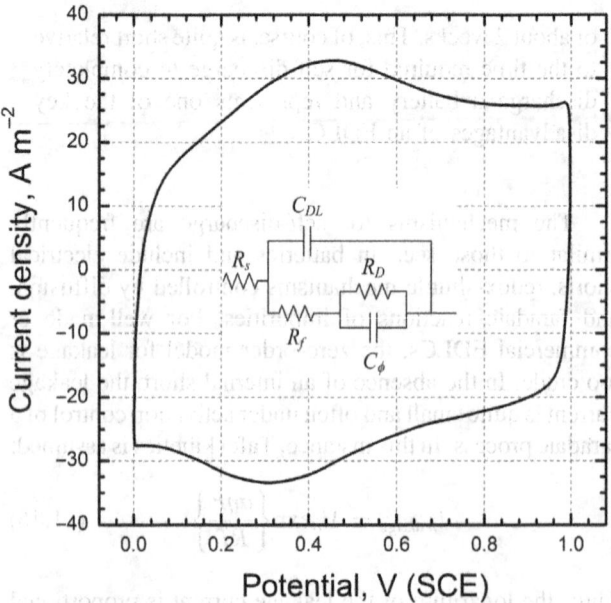

Figure 11.22 Cyclic voltammetry behavior of RuO$_2$. The scan rate is 20 mV·s^{-1}. Common equivalent circuit for pseudo-capacitor is also shown. *Source:* Adapted from Kim 2001.

The ideal rectangle is approximated and the capacitance is taken as

$$C_\phi = \frac{\frac{dq}{dt}}{\frac{dV}{dt}} \approx \left|\frac{i_{avg}}{\nu}\right|. \qquad (11.49)$$

Capacitance values for good pseudo-capacitor materials can be 5–10 F·m^{-2} of actual surface area, compared to typical C_{DL} values of approximately 0.25 F·m^{-2}. From Figure 11.22, the average current density (based on the superficial area) is ~25 A·m^{-2}. The data in Figure 11.22 correspond to a capacitance per superficial or geometric electrode area of about 420 F·m^{-2}. Typically, pseudo-capacitors have a greater capacitance than EDLCs, but tend to be much more expensive due to higher materials costs.

> ### *EDLCs and Pseudo-Capacitors*
>
> EDLCs store energy electrostatically in the electrical double layer through charge separation; no electron-transfer reactions are needed for EDLC energy storage. In contrast, with pseudo-capacitors, a highly reversible faradaic reaction occurs on the surface, and the energy is stored chemically as a result of electron transfer. Because the charge passed depends linearly on the potential, the behavior of these materials mimics that of a capacitor. Since there are faradaic reactions, they are called pseudo-capacitors.

CLOSURE

In this chapter, we introduced two types of electrochemical capacitors: EDLCs and pseudo-capacitors. Electrochemical double-layer capacitors do not rely on faradaic reactions for energy storage. Porous electrodes with high-surface areas are a key feature for these energy storage devices. The analysis of their behavior was examined using the same electrochemical principles that we applied to batteries and fuel cells. Notably, these devices are characterized by high-power density and relatively low-energy density. Pseudo-capacitors exhibit performance characteristics similar to EDLCs, but faradaic reactions are present.

FURTHER READING

Bequin, F. and Frackowiak, E. (2013) *Supercapacitors: Materials, Systems and Applications*, John Wiley & Sons, Inc.,

Conway, B.E. (1999) *Electrochemical Supercapacitors: Scientific Fundamentals and Technological Applications*, Springer.

O'M Bockris, J. and Reddy, A. K. N. (1970) *Modern Electrochemistry*, Vol. 2, Plenum Press, New York.

You, A., Chabot, V., and Zhang, J. (2013) *Electrochemical Supercapacitors for Energy Storage and Delivery: Fundamentals and Applications*, CRC Press.

PROBLEMS

11.1 In Illustration 11.1, the specific capacitance of carbon was calculated to be 150 F·g^{-1}. To fabricate an electrochemical double-layer capacitor, even if the separator, current collector, and packaging weights are ignored, the theoretical value for capacitance in F·g^{-1} must be reduced by a factor of exactly 4. Why?

11.2 Derive Equation 11.8 for a 1 : 1 electrolyte. *Hint*: Start with Poisson's Equation 2.32 and follow the development in Section 2.13. *Hint*: Use Cartesian coordinates and do not make the assumption of a small potential. Finally, the following transform is helpful:

$$\frac{1}{2}\frac{d}{d\phi}\left(\frac{d\phi}{dx}\right)^2 = \frac{d^2\phi}{dx^2}.$$

11.3 Electrolytic capacitors have a polarity and will be destroyed if subjected to even a modest voltage of more than about 1.5 V of the wrong polarity. Explain this behavior. What would be the effect of connecting two electrolytic capacitors with polarity in series?

11.4 Sketch the charge density and potential across a double layer that includes both charge in the compact layer near the electrode (OHP) and the diffuse layer (GC). Assume the metal is positively charged. Develop Equation 11.9 showing that the two capacitances combine in series.

11.5 Derive Equation 11.11 using the geometry of Figure 11.6. Assume that the region of low dielectric constant includes the first row of water on the electrode and a second region of high dielectric constant that extents from the ion at the OHP to the first row of water ($2r_w$).

11.6 A carbon porous electrode EDLC uses 0.05 M KOH at a temperature of 25 °C for the electrolyte. At the point of zero charge (PZC) and at a potential 0.1 V from the PZC, calculate the following capacitances per unit area [F·m^{-2}]:

   (a) Helmholtz contribution to the double-layer capacitance. Rather than the bulk dielectric constant of water, use a value of 11 for this inner region. The atomic diameter of K is 0.46 nm.

   (b) Contribution of the diffuse layer from the theory of Gouy and Chapman.

   (c) Combined value for the capacitance.

   (d) Typical experimental values for the double-layer capacitance in aqueous solution are 0.2–0.5 F·m^{-2}. Discuss importance of using the corrected dielectric constant for the Helmholtz region.

11.7 The table includes some data from Bockris and Reddy (Table 7.11) for 0.1 N aqueous chloride solutions. (a) for a 0.1 N solution, show that C_{OHP} has a relatively small effect on the differential capacitance. (b) What can be inferred about the adsorption of ions? (c) Using 0.14 nm for the radius of water, $\varepsilon_{rH} = 40$, and $\varepsilon_{rL} = 6$, does Equation 11.11 fit the data? Assume these are taken at room temperature.

Ion	r_i, Unsolvated radius [nm]	Measured capacitance [F·m^{-2}]
Li$^+$	0.060	0.162
K$^+$	0.133	0.170
Mg^{2+}	0.065	0.165
Al^{3+}	0.050	0.165
La^{3+}	0.115	0.171

11.8 Describe how the structure of electrode might be designed differently for aqueous and nonaqueous electrolytes.

11.9 If the space between two parallel oppositely charged, infinite plates is comprised of two regions of different permittivity, how is the capacitance expressed?

11.10 The differential capacitance of a single electrode [F·m^{-2}] is fitted with the following equation:

$$C_d = 0.2 + 0.1 \tanh(5\phi_2).$$

Find an expression for q_m. For a practical device, two of these electrodes are used, connected in series. In operation, opposite charges of equal magnitude are stored on each electrode. Plot the differential and integral capacitance of the device as a function of potential. Comment on the degree of variation between the single electrode and the device.

11.11 Compare and contrast differences in cyclic voltammograms for capacitors and redox reactions.

11.12 Calculate the round-trip efficiency of an EDLC that is discharged and charged at a constant current of 13 A. The capacitance is 150 F with an effective ESR of 14 mΩ. The maximum voltage is 2.7 V and the discharge is terminated at 1.35 V.

11.13 Create the equivalent of Figure 11.16 for a nonaqueous electrolyte. Except for the information provided below, use the data from Illustration 11.6 and assume a voltage of 2.3 V. Compare the curves and physically explain the difference between aqueous and nonaqueous capacitors.

Effective conductivity of electrolyte in separator	1.1 S·m^{-1}
Capacitance per actual surface area	0.100 F·m^{-2}
Effective conductivity of electrolyte in electrolyte	0.7 S·m^{-1}
Density of electrolyte	950 kg·m^{-3}

11.14 A EDLC with a maximum potential of 4.2 V is charged at constant current (CC) until the potential reaches 4.2 V. The charge is then continued at 4.2 V (CV) until the current becomes very small. Finally, the cell is discharged at a current of 0.1 A, starting at $t = 10$ seconds. The data are shown in the figure, the inset shows the step change in current to 0.1 A for the discharge. Calculate the nominal capacitance of the EDLC and its ESR$_{eff}$ from these data.

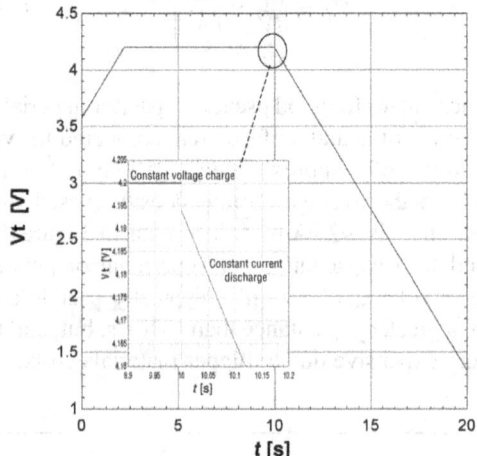

11.15 Below are Nyquist plots for two EDLC. The only difference is the loading of the electrode, which affects the thickness and capacitance of the device. What can be said about the conductivity of the solid compared to that of the electrolyte? What are the ESR$_{eff}$, ESR, and EDR for the device loaded to 11.3 mg cm^{-2}? Which electrode loading would have a higher cutoff frequency? (Data adapted from Taberna et al. (2003) *J. Electrochem. Soc.*, **150**, A292.)

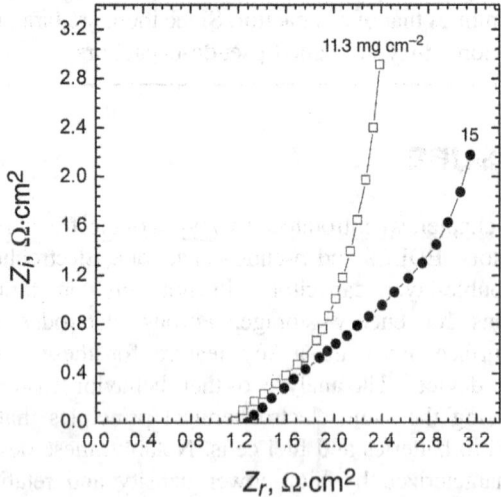

11.16 In Chapter 7, curves for the potential versus SOC for some common insertion electrodes used in Li-ion cells were presented. Lithium titanium disulfide is an insertion material where the potential varies linearly with the concentration of lithium between the titanium disulfide galleries. Considering the discussion in Section 11.8 on faradic reactions where potential changes with admitted charge, can you make the case that these insertion devices might equally well be called pseudo-capacitors?

11.17 In order to achieve a specific capacitance of $50\,\text{C}\cdot\text{g}^{-1}$, what minimum concentration of electrolyte is needed to avoid depletion? Assume a fully dissociated 1:1 electrolyte. The porosity is 0.65 and the density of the active material is $1800\,\text{kg}\cdot\text{m}^{-3}$.

11.18 One mechanism for self-discharge is caused by a faradaic reaction under kinetic control, perhaps from some impurity. If the faradaic reaction is controlled by Tafel kinetics, show that the leakage current depends on the logarithm of time. Assume the capacitance is constant.

11.19 Show that for an ideal capacitor in series with a resistance ESR, the maximum power is $V^2/4\text{ESR}$. For an ideal battery, the maximum power is $V^2/4R_{cell}$.

11.20 There are two common methods to measure leakage current. In the first, the potential of the device is measured at open-circuit over time. Explain how the $V(t)$ data can be converted to $I(t)$. Sketch how current would change over time. Why is it important to specify the time when leakage currents are reported?

11.21 Using the open-circuit data provided, what is the resistance associated with the leakage current for a 3 F capacitor? Estimate the 72-hour leakage current.

Time [s]	Potential [V]
0	2.800
600	2.799
1200	2.797
1800	2.796
2400	2.794
3000	2.793

FERDINAND PORSCHE

Ferdinand Porsche was born on September 3, 1875 in Maffersdorf, which is now called Liberec and is part of the Czech Republic. You surely know him as the founder of the Porsche Motor Company, but he was also a remarkable innovator in the development of the electric vehicle.

His father ran a metal-smithing business, and he was expected to apprentice for that trade. Ferdinand had other ideas—fascinated by electricity from an early age, Porsche traveled to Vienna at the age of 18 to work as a trainee for an electrical company, Vereinigte Elektrizitäts-AG (VEAG). During this period, Ferdinand attended lectures at the Technical University as a guest. There he gained some theoretical knowledge of electricity. After 4 years, he was appointed manager of the Testing Department at VEAG. It was at this point that he met Ludwig Lohner, a coachbuilder. Porsche collaborated with Lohner's company on several electric vehicle designs, and at the age of 24 he was lured away from VEAG and appointed Technical Director of Lohner's automobile manufacturing plant in Vienna. His first design was a front-wheel drive electric vehicle that used electric wheel-hub motors. This vehicle was the world's first front-wheel drive car. The battery consisted of 44 lead–acid cells connected in series (~80 V) with a capacity of 300 A·h. The battery weighed 410 kg and provided a range of 50 km. The massive weight of the battery caused the tires made from soft rubber to fail frequently; a more important challenge was that it was nearly impossible to recharge the battery in rural areas simply because electricity was unavailable at that time. To get around this issue, Porsche designed his first working gasoline battery hybrid-electric vehicle (Lohner–Porsche) in 1901. This design was a series hybrid, with the internal combustion engine (ICE) only serving to recharge the battery, thus easing the difficulty of recharging outside of metropolitan areas. This hybrid is thought to have been inspired by an exhibit at the 1899 International Motor Show in Belgium where series and hybrid designs were envisioned, but it isn't clear that these earlier vehicles ever worked.

The internal combustion engine improved dramatically and the automotive industry, along with Ferdinand Porsche, moved away from electric vehicles. Porsche had no formal training in engineering, but did receive an honorary doctorate in engineering from the Vienna Institute of Technology in 1916, and a second honorary degree from Stuttgart Technical University in 1924. Ferdinand Porsche died on January 30, 1951

Image Source: Courtesy of Porsche Museum, Stuttgart Germany.

Chapter 12

Energy Storage and Conversion for Hybrid and Electrical Vehicles

At the turn of the twentieth century, electric- and gasoline-powered vehicles were competing to dominate the growing vehicle market. Internal combustion engines (ICEs) were noisy, difficult to start, produced lots of smoke, and were not particularly dependable. The electric vehicle had similar reliability problems, but had none of the other drawbacks. In many major cities, there were fleets of hundreds of electric taxicabs that outnumbered the internal combustion competitors. Outside of major cities, the key challenges for electric vehicles were range and the availability of facilities to charge the batteries. At that initial stage, electrification was just beginning, and it was unlikely that one would find a means of recharging the battery in rural areas. On the other hand, range anxiety was much less important at that time because the roads connecting major urban areas were poor. Therefore, it was doubtful that drivers would even want to venture far from home. Nonetheless, range and charging were of significant concern then, and continue to be critical issues for electric vehicles.

From the need to address these concerns, the hybrid-electric vehicle was born—combining the advantages of the electric drive and allowing for recharging of the battery with an internal combustion engine. The first attempts at hybrids date from 1894; the first working petroleum-battery hybrid was in 1901. With the commercial introduction of the electric starter in 1912, the increasing availability of gasoline, and vast improvements in its efficiency, ability to burn fuel cleanly, and reliability, the internal combustion engine won out over electric and hybrid-electric vehicles. Today, automobiles powered by internal combustion engines are relatively clean, highly reliable, and manufactured at a low cost. Yet, some 100 years after the first hybrids, the rivalry between the internal combustion engine and electrically driven vehicles has reignited. The newest battle, however, is being conducted in a different setting, a landscape where energy security and environmental concerns have emerged as critical factors in a new high-stakes competition.

12.1 WHY ELECTRIC AND HYBRID-ELECTRIC SYSTEMS?

A key feature of systems that include energy storage is that excess energy can be accumulated for later use. Our focus will be on vehicles, but applications for hybrid power systems are common. For example, a renewable energy system based on wind power alone will struggle to match electrical power production and demand. What do you do when electrical demand is less than the amount being produced by the wind turbine, or when the wind isn't blowing at all? One option is to store energy when the supply is greater than the demand. This storage could be accomplished with a battery or similar rechargeable energy storage system (RESS). The stored energy can then be used later when the demand is greater than the supply. Similarly, with conventional base-load power generation devices, a nuclear power plant for instance, it is desirable to keep electrical power production constant. Since demand for electricity varies over the day, a method of efficiently storing energy during off-peak times is valuable. Electrolysis of water to produce hydrogen has been proposed to be paired with a nuclear power plant as one means to deal with this issue. The hydrogen would then be recombined with oxygen in a fuel cell (FC) when the electrical demand was high.

The situation in a vehicle is not completely analogous, but there is a clear benefit of having the ability to store energy. Onboard vehicle energy storage will be the primary emphasis of this chapter. The main incentive for electric and hybrid-electric vehicles is to improve their efficiency, which leads to reduced use of fossil fuels and lower emissions of pollutants. Let's explore this

---

Electrochemical Engineering, First Edition. Thomas F. Fuller and John N. Harb.
© 2018 Thomas F. Fuller and John N. Harb. Published 2018 by John Wiley & Sons, Inc.
Companion Website: www.wiley.com/go/fuller/electrochemicalengineering

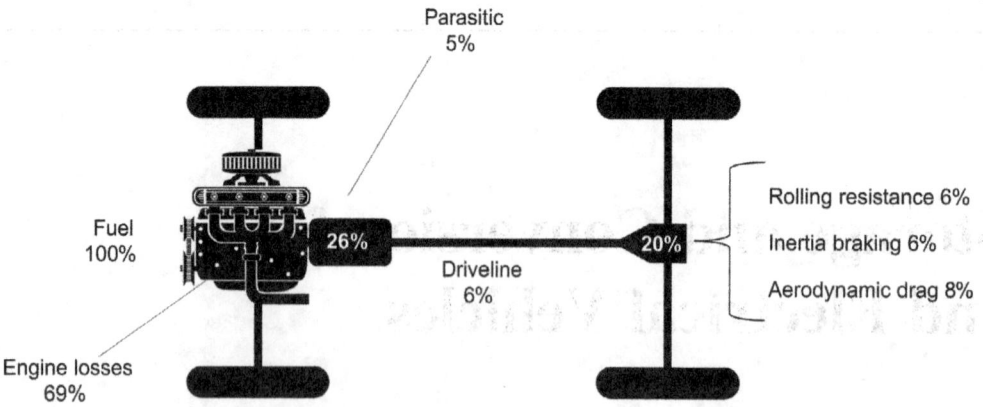

Figure 12.1 Representation of where the energy of the fuel goes for a typical vehicle driving a combination of city and highway. *Source:* Data taken from http://www.fueleconomy.gov/feg/atv.shtml

motivation in more detail. The starting point is a conventional vehicle powered by an internal combustion engine.

> The motivation and goals for electric and hybrid-electric vehicles are as follows:
> - Reduced petroleum use
> - Lower releases of greenhouse gases
> - Decreased emissions of criteria pollutants
> - Increased energy efficiency

Figure 12.1 provides an example of how the chemical energy from the fuel is used in an ICE vehicle. Chemical energy is released by combustion and converted into mechanical energy in the engine. This process is a thermal one and limited by the Carnot efficiency. In this example, which combines city and highway driving, only 26% of the available chemical energy reaches the driveline. A further 6% is lost there. As a result about 20% of the chemical energy in the fuel applies torque to the wheels. Of this, 6% is used to accelerate the vehicle after braking. Eight percent is needed to overcome aerodynamic drag on the vehicle, and another 6% to overcome rolling resistance of the tires. Clearly, the situation depicted in Figure 12.1 is not an efficient use of the energy from the fuel.

Understanding a little about ICEs and reflecting on how a typical vehicle is driven can help to explain these results. First, the fuel efficiency of an ICE is not constant with power. Typical of thermal devices, there is a point where the efficiency is maximized, but as the power level decreases from this value, the efficiency also decreases. Therefore, a key consideration is where does this maximum occur and at what point is the engine being operated most frequently. As you know, most driving involves frequent starts and stops; on average, the power demanded by the driver is well below the rated power. In fact, this maximum-rated power is needed infrequently. If the engine is sized to provide high power, but operates almost all the time at low power, its efficiency will be low. Consequently, the largest loss in Figure 12.1 (69%) is attributed to engine losses. Ideally, we would operate the engine near the point of maximum efficiency. How can we do that? One answer is to use a hybrid propulsion system. The ICE could provide the average traction power required and be sized so that its efficiency was high at this average power. A second power source would supplement the ICE when high power is needed. Another opportunity to improve efficiency is to recover energy during braking. The kinetic energy of the vehicle is reduced by friction of the brakes, generating heat. This energy is lost and additional fuel must be burned to accelerate the vehicle back up to speed. Lost energy due to braking accounts for about 6% of the total as shown in Figure 12.1. Efficiency would be improved if we could store this kinetic energy and reuse it later. There are other important approaches to improve use of the chemical energy of the fuel: lowering the aerodynamic drag, reducing the weight of the vehicle, and minimizing rolling resistance, for instance. These topics are beyond the scope of this chapter, which explores the use of energy storage and hybridization to increase efficiency.

An onboard RESS coupled with a means of proving torque to the wheels separate from the ICE addresses both of the two ideas described above. In addition to batteries and electrochemical capacitors, there are mechanical ways to store energy: pumped hydroelectric, compressed air systems, hydraulic energy storage, and flywheels. Some of these are used in vehicles, but our focus will be on those systems that use electrochemical energy storage. Let's briefly consider a few aspects of a hybrid-electric vehicle as illustrated in Figure 12.2—These will be examined in more detail later in this chapter. First, both electrical energy from the RESS and the engine can be used to propel the vehicle. This parallel architecture decouples the torque to

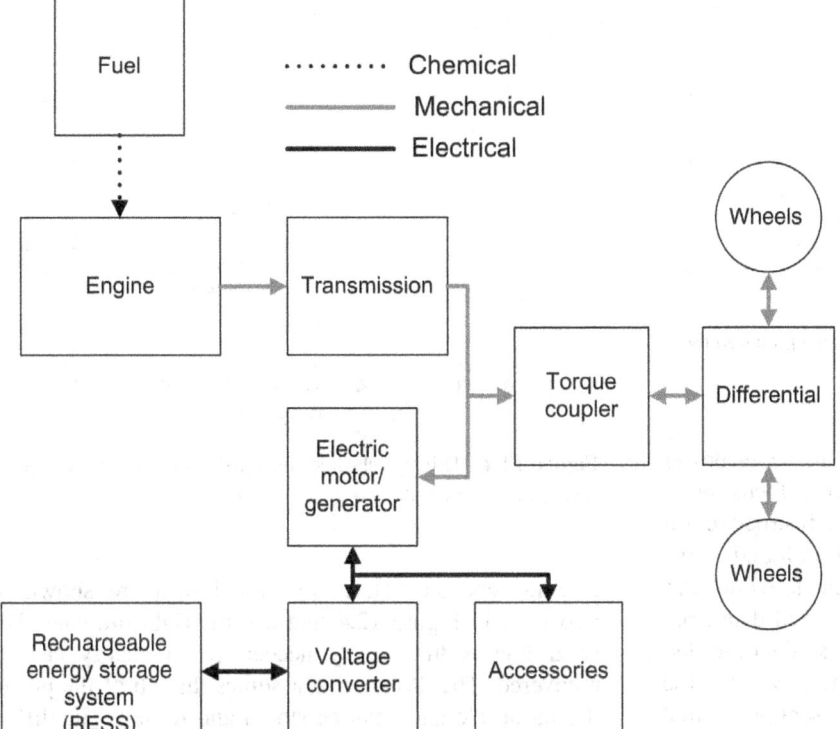

Figure 12.2 Parallel-hybrid architecture, one implementation of a hybrid vehicle.

the wheels from the torque provided by the engine. Because of this broken link, the engine can be operated selectively under conditions of higher efficiency, which is an important advantage of a hybrid system. Furthermore, electric-only operation is also possible if the energy storage system is large enough. In addition, the hybrid architecture enables the capture and storage of kinetic energy when stopping the vehicle. This energy would otherwise have been lost (dissipated) as heat in friction brakes. To recover this energy, the electric motor also functions as a generator, and the energy is stored in a rechargeable storage system for later use.

It is important to note that there is a price to be paid for these improvements in efficiency. The added flexibility with hybrid systems is achieved by adding components. The RESS, electric motor/generator, and voltage converter increase the mass of the vehicle, and add cost and complexity. The additional complexity requires more elaborate controls. It is the task of the design engineer to evaluate the trade-offs between efficiency, complexity, and cost in order to provide an optimal vehicle system.

To summarize, the key concepts used in hybrid systems to increase the fuel efficiency of vehicles are (i) to operate the engine at speeds where it is most efficient, and (ii) to recover the kinetic energy wasted during friction braking. Efficient operation of the engine may include turning it off to lower standby/idle losses when the vehicle is stopped for short periods, or to permit all-electric operation if the RESS is of sufficient size. Each of these improvements requires a means of energy storage onboard the vehicle, and hence motivates our interest in the topic. Batteries (Chapters 7 and 8) and electrochemical double-layer capacitors (Chapter 11) are energy storage devices, and their role in hybrid vehicles will be highlighted. Finally, we also consider hybrids where electrochemical energy storage is combined with a fuel cell (Chapters 9 and 10) operating on hydrogen rather than with a petroleum-based ICE.

12.2 DRIVING SCHEDULES AND POWER DEMAND IN VEHICLES

In evaluating powertrain systems, the vehicle speed is often prescribed as a function of time. This relationship is generalized with a *driving schedule*. Figure 12.3 provides such a relationship for an urban dynamometer driving schedule, one of many standardized schedules that are available. This plot shows speed versus time for the vehicle driven in an urban area. The distance traveled is 12 km with an average speed of about 40 km·h^{-1}. This schedule will serve as an example for this section.

The speeds from the driving schedule can be converted to the power required for traction or, more specifically, to the torque to the wheels. The key parameters needed to make the calculation are specific to the vehicle: the mass of the vehicle, rolling resistance of the tires, and the aerodynamic drag. With a *vehicle model*, the driving

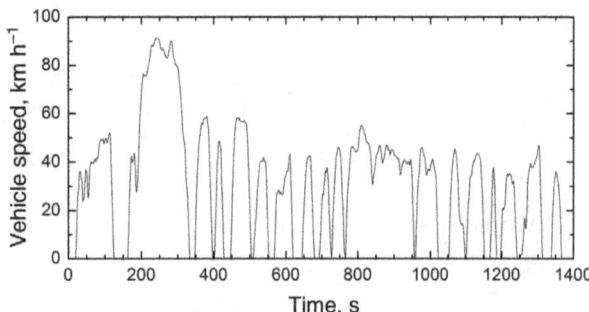

Figure 12.3 Urban dynamometer driving schedule produced by the EPA.

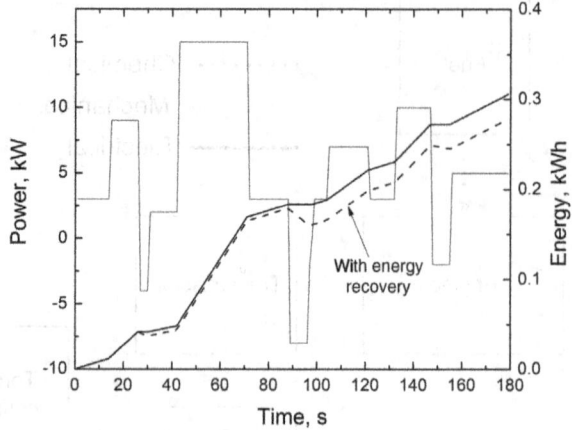

Figure 12.4 Driving schedule converted to power (left axis) and energy in kWh on right axis.

schedule can be transformed into the instantaneous power required. The vehicle model is not our main focus, but a short discussion of vehicle dynamics is instructive for those who would like a review, and can be found in the appendix at the end of this chapter. A vehicle model was used to convert a portion of the vehicle speed data from Figure 12.3 to instantaneous power. These data are displayed in part in Figure 12.4, which shows power (on the left axis) versus time. Positive values represent power that is needed from a combination of either an internal combustion engine or fuel cell and the rechargeable energy storage system. Negative values correspond to deceleration of the vehicle and represent opportunities for energy recovery and storage. The power required to move the vehicle does not represent the entire power requirement as some power is needed for accessories. For instance, power may be needed to run vehicle controls, lights, and the air conditioner regardless of the vehicle speed. Data such as those shown in Figure 12.4 link together the desired speed profile (Figure 12.3), the characteristics of the vehicle, and power consumption or generation.

We can also integrate the instantaneous power with time to determine the energy needed to complete the driving schedule. These integrated data are shown as two lines in Figure 12.4 that use the right ordinate. The solid line is the energy needed if no energy can be recovered. The dashed line assumes that all of the power during deceleration can be stored and reused. The difference between the two lines illustrates one of the advantages of a hybrid system with a RESS. This example driving schedule covers a time period of only a little more than 20 minutes, but the schedule can be repeated or combined with a highway or another schedule to represent use of the vehicle over a longer period of time. Table 12.1 shows the speed and energy consumption for a few different driving schedules. Of most interest is the total energy required to follow the driving schedule and the energy associated with braking. Note that the braking energy is close to half of the total energy for some city driving (10.47 versus 4.52 kWh per 100 km).

Our interest is in powertrain systems that use a rechargeable energy storage system, which for our

Table 12.1 Energy Associated with Different Driving Schedules for a 1500 kg Vehicle

	FTP-75, city driving, frequent stops, idling	FTP-75, highway driving, no stops	US06, high speed, aggressive driving
Average speed [km·h^{-1}]	27.9	79.3	77.5
Maximum speed [km·h^{-1}]	86.4	97.7	128.05
Traction energy [kWh·100 km^{-1}]	10.47	10.45	17.03
Traction energy efficiency [km·kWh^{-1}]	9.6	9.6	5.9
Braking energy, [kWh·100 km^{-1}]	4.52	0.98	5.30

The traction energy is at the wheels.
Source: Adapted from Ehsani 2009.

purposes is either a battery or an electrochemical double-layer capacitor. Of particular interest is the size of the RESS. There are three aspects that are important for RESS sizing: its power, total useable energy, and life—a theme that will be repeated throughout this chapter. We first consider sizing of the RESS for power. The designer is free to choose how power is divided between the RESS and the engine (either ICE or fuel cell). Similarly, the designer can select the maximum rate that energy can be absorbed into the RESS from regenerative braking. The power ratings (kW) for accepting and delivering energy are frequently similar, since both processes are often limited by the internal resistance and heat removal capability of the RESS. However, the ultimate power rating of the RESS depends strongly on the design objectives and system architecture and can vary dramatically from system to system.

As you would expect, energy requirements also have a large impact on the size of the RESS. The key factor in sizing for energy is the desired range of the vehicle on the battery alone or, in the case where the RESS doesn't provide traction power, the desired stopped time without assistance from the ICE.

Finally, the lifetime of the RESS is critical in the design of hybrid systems. One of the key factors that influences battery life is the SOC range over which the battery is designed to operate. Battery requirements vary significantly across the range of hybrid vehicles that are available and have a significant impact on RESS design with respect to lifetime.

As we explore sizing of the rechargeable energy storage system in more detail, we will examine four types of vehicles: all electric, start–stop hybrid-electric, full hybrid-electric, and a fuel-cell hybrid vehicle. In the discussions that follow, we also adopt the simplistic view that the only aspect of the energy storage system that can be altered is its size. As we have seen from Chapters 8 and 11, other aspects such as electrode design can also affect performance. These facets are, nonetheless, ignored in this chapter.

12.3 REGENERATIVE BRAKING

Before turning to specific vehicle architectures, we pause to consider regenerative braking, which plays an important role in increasing hybrid vehicle efficiency. A key benefit of the vehicle strategies considered in this chapter is that kinetic and potential energy can be recovered during braking. Clearly, energy is required to accelerate a vehicle to a higher speed or to drive up a hill. It makes sense to try to recover and store this energy when stopping the vehicle or when traveling downhill. For highway driving, the braking energy may be a small portion of the total traction energy. However, for urban driving that includes frequent starts and stops, the energy dissipated in braking may be more than 50% of the total traction energy. This is consistent with the data in Figure 12.3 that shows many periods of deceleration. Thus, there is a strong incentive to recover this energy. The kinetic energy of the vehicle is converted to electrical energy with a generator (the same motor that is used to propel the vehicle can be controlled to operate as a generator) and this energy is stored in a battery or capacitor.

ILLUSTRATION 12.1

A 1600 kg passenger hybrid-electric vehicle is moving at $60 \, \text{km} \cdot \text{h}^{-1}$. The battery has a nominal voltage of 300 V and a capacity of 8 kWh.

a. To stop the vehicle in 3 seconds, determine the power available in principle from regenerative braking. Assume that the power is constant and neglect rolling resistance and aerodynamic drag during braking.

   The kinetic energy is $\frac{1}{2} M_v v^2 = \frac{(1600 \, \text{kg})}{2} \left(\frac{60 \, \text{m}}{3.6 \, \text{s}}\right)^2 = 0.222 \, \text{MJ} = 222 \, \text{kJ}$.

   This energy must be dissipated or captured in the time that it takes to stop the vehicle (3 seconds). Therefore, the power is 222 kJ/3 seconds = 74 kW.

b. What is the current through the generator? Express the current in terms of both amperes and the battery C-rate (see Chapter 8).

   The current is simply the power divided by the battery voltage, $I = \frac{P}{V} = \frac{74000 \, \text{W}}{300 \, \text{V}} = 247 \, \text{A}$.

   The capacity in A·h of the battery is $\frac{8000 \, \text{Wh}}{300 \, \text{V}} = 26.7 \, \text{A} \cdot \text{h}$. Therefore, by definition, the 1C rate is 26.7 A. The C-rate through the generator is $\frac{247 \, \text{A}}{26.7 \, \text{A}} = 9.3$.

c. What is the change in the battery SOC?

$$\Delta SOC = \frac{247 \, \text{A}}{} \left|\frac{\text{hour}}{3600 \, \text{seconds}}\right| \frac{3 \, \text{seconds}}{} \left|\frac{}{26.7 \, \text{A} \cdot \text{h}}\right| \times 100 = 0.8\%.$$

Illustration 12.1 provides a simple calculation assuming a constant rate of energy recovery. Alternatively, we could maintain a constant rate of deceleration. Regardless, we can make some general observations. First, note that the power generated can be large during periods of hard braking. Stopping in 3 seconds from $60 \, \text{km} \cdot \text{h}^{-1}$ is not particularly aggressive; nonetheless, the power that must be absorbed was large compared to the size of the battery—the C-rate is nearly 10. This power must pass through the electric motor/generator. Often, the motor torque from braking is larger than the maximum torque allowed on the motor/generator, and the extra cost, weight, and volume of the larger motor needed to recover all of the energy may not be warranted from a system perspective. Similarly, the current may be larger than that which the battery can accept without exceeding the maximum charge voltage for the cells. This possibility is evident from the relatively high C-rate at which the battery must be charged in order to store the energy from braking. Finally, we can note that in most cases the energy from a single braking event is small compared to the total energy stored in the battery.

12.4 BATTERY ELECTRICAL VEHICLE

We begin our discussion of vehicles by first exploring the use of a battery to store energy for an all-electric vehicle. In a battery electric vehicle (BEV), the battery provides all of the power and energy needs. As mentioned earlier in the chapter, here are three key aspects to sizing the battery: energy capacity, power, and life. Each of these is influenced by the operating voltage of the cells. Figure 12.5 shows three types of voltages discussed for any electrochemical device: maximum, nominal, and minimum. The nominal voltage is a typical or average value seen during normal operation. Of course, the potential changes with current, state of charge (SOC), temperature, and age. Nonetheless, characterization by a nominal voltage is still a useful concept for almost any electrochemical device. The maximum and minimum voltages represent limits on the operating potential and define the operating range. There are many reasons for placing maximum and minimum limits on the potential of the electrochemical storage system, such as preventing damage to the cells or allowing for proper operation of the power electronics.

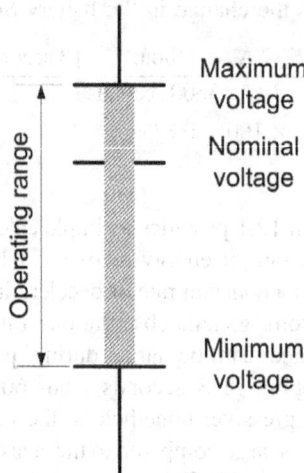

Figure 12.5 Range of operating voltage for electrochemical device.

Since the battery is the only source of energy for vehicle, its *energy capacity*, typically expressed in kWh, must be sufficient to achieve the desired range. This energy is used to overcome the rolling resistance of the wheels, the aerodynamic drag, and losses in the drive train and in power conversion, as well as to provide power to the accessories. As discussed previously, the energy needed is obtained from a representative driving schedule and the specific vehicle design. For a passenger vehicle, a good rule of thumb is that it is possible to travel $6\,\mathrm{km\cdot kWh^{-1}}$ on the battery alone. You'll note from the last row in Table 12.1 that, under the right conditions, a greater distance can be covered per kWh. It is important to remember, however, that the values in Table 12.1 are reference energy at the wheels and do not account for losses in the powertrain.

As the energy of the battery is stored chemically in the active material, battery capacity is proportional to the mass of active material as we saw in Chapter 7. Thus, to double the range of the vehicle, the mass of active material must be doubled. This would double the rating, expressed in either A·h or kWh, for the battery. In determining the required mass, an important consideration is the useable SOC window of the battery. Rather than operating the battery from fully charged to fully discharged, it is common to use less than the rated capacity or, in other words, to restrict the window for the variation in state of charge. The most important reason for shrinking the SOC window is to improve cycle life (discussed below). As the SOC window is reduced, a larger battery is needed to achieve the same range, but the useable lifetime of the battery is increased.

The battery must also supply all of the *power* needed to meet the driving requirements of the vehicle. This power depends on the performance targets for the vehicle (acceleration, speed) and, of course, is highly dependent on the size of the vehicle. The power requirement for the vehicle might be estimated as the power needed to complete a UDDS driving schedule or, alternatively, the power to maintain $90\,\mathrm{km\cdot h^{-1}}$ speed on a 6% grade at 2/3 of maximum power. A typical value of the power required for a passenger vehicle is 50 kW, but again this value is highly dependent on the performance targets for the vehicle.

For our purposes in evaluating its power capability, we use a simple resistance model to describe the battery. Specifically, we assume that the potential of the cell is given by

$$V = V_{ocv} - IR_\Omega. \tag{12.1}$$

We seek an expression for the maximum power available from the battery. Recalling that power is the product of current and voltage, we can use Equation 12.1 to express the voltage, V, as a function of current. We then multiply V, which is now expressed as a function of current, by I to get an expression for power as a function of current. That expression for the power can be differentiated with respect to current to find the maximum, followed by rearrangement to yield:

$$P_{max} = \frac{V_{ocv}^2}{4R_\Omega}. \qquad (12.2)$$

Since R_Ω is inversely proportional to the area, it follows from Equation 12.2 that, for a fixed cell capacity, the maximum power produced from a battery is roughly proportional to the area of the separator.

> **Run-Time Is the Key Design Criterion for Battery Design**
>
> The energy of the battery scales with the mass of active material (kWh).
>
> The power of the battery scales with the separator area (kW).
>
> The energy to power ratio is the run-time of the battery.
>
> $$\text{run-time} = \frac{\text{energy}}{\text{power}}.$$

According to Equation 12.2, the maximum power corresponds to a cell voltage equal to half of the open-circuit potential. That voltage, however, may be unacceptably low. In practice, as we discussed before, there will be a minimum limit on the potential, referred to as the cutoff voltage. If the cutoff voltage is greater than half of the open-circuit potential, then the maximum power is limited to

$$P_{max} = V_{cutoff} \frac{V_{ocv} - V_{cutoff}}{R_\Omega}. \qquad (12.3)$$

Both the energy and power requirements must be met simultaneously. The ratio of the two requirements has units of time and is called the *run-time* of the battery. This value is perhaps the key criterion for battery selection and design. As you might have guessed, the battery design used to provide power for a few seconds is quite different from one that needs to supply power for hours. The difference in battery design, discussed in Chapter 8, is largely reflected in changes in the thickness of the electrodes.

A third factor in sizing the battery is its useable *lifetime*. From our own experience with batteries in consumer devices, we know that the capacity of any battery decreases over its lifetime. Predicting life of a battery is complex, and here we adopt a primitive approach. Namely, we assume that the battery is degraded in proportion to the number of coulombs passed in the battery or what is called *capacity turnover*. The cell can only charge and discharge a certain number of coulombs before it no longer has an acceptable capacity. Clearly, however, a large

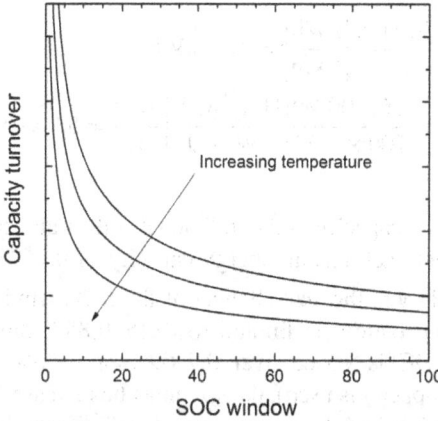

Figure 12.6 Capacity turnover for a hypothetical rechargeable battery.

battery is capable of passing more charge than a small battery. Consequently, the capacity turnover, which is dimensionless, is defined as the coulombs passed before the capacity is no longer acceptable divided by the nominal capacity of the battery. In a sense, the capacity turnover represents the number of times that the nominal capacity of the battery can be used before the per-cycle capacity is no longer adequate. Figure 12.6 shows a generic set of curves. Capacity turnover is highly dependent on battery chemistry, but generally decreases as the SOC window expands and as the temperature increases.

Cycle life is the number of cycles possible from a battery over its lifetime, and is related to the capacity turnover as follows:

$$\text{cycle life} = \text{rated capacity}[A \cdot h] \times \frac{\text{cycle}}{A \cdot h} \qquad (12.4)$$
$$\times \text{ capacity turnover},$$

where the middle term provides a measure of the charge passed per cycle. These three factors are needed to size a battery for an electrical vehicle, and their use is shown in Illustration 12.2.

ILLUSTRATION 12.2

The building block for the energy storage device is an electrochemical cell with a nominal voltage of 3.2 V. Of these cells, 100 are connected in series to form the battery.

a. Using a value of $6 \text{ km} \cdot \text{kWh}^{-1}$, estimate the size of the battery required to achieve a range of 100 km. Express the size in kWh and A·h. The capacity of the battery in kWh is determined directly.

$$\frac{100 \text{ km}}{\cdot} \left|\frac{\text{kWh}}{6 \text{ km}}\right| = 16.7 \text{ kWh},$$

$$\frac{16,700 \text{ Wh}}{100 \times 3.2 \text{ V}} \left|\frac{\text{J} \cdot \text{s}^{-1}}{\text{W}}\right| \frac{\text{C V}}{\text{J}} \left|\frac{\text{A} \cdot \text{s}}{\text{C}}\right| = 52 \text{ A} \cdot \text{h}.$$

The capacity of the individual cells is also 52 A·h, but each has an energy capacity of 0.167 kWh.

b. How is the size changed if the SOC window for the battery is limited to (0.15–0.85)? Since the SOC is varied over 0.7 (i.e., only 70% of the capacity is used), the size must be increased to 52/0.7 = 74 A·h per cell and 24 kWh for the battery.

c. What value for battery resistance is required to achieve a maximum power of 75 kW from the battery? Assume the cutoff potential for the cell is 2.5 V. Since the cutoff potential is more than half of the nominal potential, use Equation 12.3:

$$R_\Omega = \frac{V_{cutoff}(V_{ocv} - V_{cutoff})}{P_{max}} = \frac{250 \times (320 - 250)}{75000}$$

$$= 0.23 \text{ }\Omega.$$

d. If the capacity turnover is 1300 for the 70% SOC window, and assuming that on average the vehicle travels 50 km·day^{-1}, what is the useable life for the battery?

The total charge available over the lifetime of the battery can be obtained from the capacity turnover and the nominal capacity. We also know the energy required to travel 6 km, and the kWh·A^{-1}·h^{-1} for the battery.

$$\frac{6 \text{ km}}{\text{kWh}} \left|\frac{\text{day}}{50 \text{ km}}\right| \frac{24 \text{ kWh}}{74 \text{ A} \cdot \text{h}} \left|\frac{74 \text{ A} \cdot \text{h} \times 1300}{\cdot}\right| \frac{1 \text{ year}}{365 \text{ days}}$$

$$= 10.3 \text{ years}.$$

The round-trip efficiency of a battery is typically quite high. Why, then, isn't the BEV the right solution for all applications? The principal reasons are range and cost. This challenge is illustrated with the Ragone plot, Section 7.5. The energy density (and specific energy) of the typical battery is too low to achieve a range of say 400 km in a practical vehicle. Of course, the battery can be made larger, but takes up more space, adds mass to the vehicle, and increases the cost for the battery. In addition, the efficiency of the vehicle operated over a fixed driving schedule decreases as the mass of the vehicle increases. More advanced batteries offer better specific energy, but their cost is high.

12.5 HYBRID VEHICLE ARCHITECTURES

There are many architectures used in hybrid systems. We will not attempt to cover them extensively; rather, our objective is to review some typical architectures and to provide a broad overview of terminology. In subsequent sections, we will explore in more detail electrochemical devices for energy storage for specific hybrid architectures. Because there are many different applications and usage patterns, our treatment in this chapter provides just a small sample of the possibilities.

The simplest form of a hybrid is the so-called start–stop[1] hybrid (Figure 12.7). Here, the IC engine is turned off when the vehicle is stopped. While the engine is off, electrical power is provided by the energy storage system to handle the accessory load (e.g., air-conditioning). The same RESS is subsequently used to restart the engine. The RESS is then recharged by the engine. For the start–stop hybrid, all traction power is provided by the engine. Nonetheless, improvement in the range of 3–8% in fuel economy is possible with this simple system. This increase in efficiency comes about because of the low efficiency of the ICE during idling. The start–stop hybrid has the added advantage of reducing pollution by avoiding idling of the engine.

A block diagram for a conventional vehicle would look the same as that shown in Figure 12.7. The only modifications made for a start–stop hybrid is to increase the size of the starter/alternator and the RESS. These modifications allow for very rapid starting of the engine and

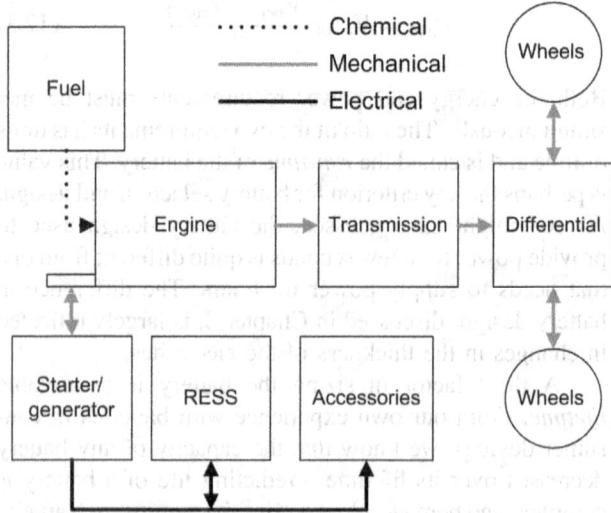

Figure 12.7 Start–stop hybrid.

[1] Sometimes the start-stop hybrid is referred to as a micro-hybrid. The micro-hybrid is differentiated by allowing recovering of energy from braking, where a start-stop hybrid does not include energy recovery.

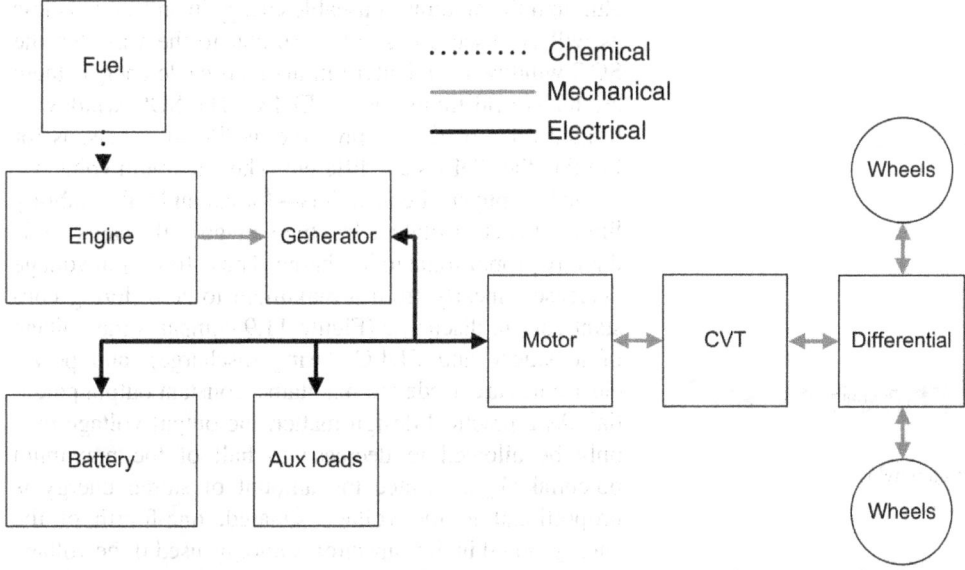

Figure 12.8 Series hybrid system.

provide sufficient energy to run the accessories while the vehicle is stopped briefly.

Next, we consider series- and parallel-hybrid architectures, which are both full-hybrid architectures. Two features distinguish these architectures from that of the start–stop hybrid. First, energy from braking can be recovered. Second, energy from the RESS can be used to provide traction to the wheels. Energy storage is a major component of these hybrid systems, and the large amount of energy needed makes a battery the only practical electrochemical energy storage system. For most hybrid designs, all of the power ultimately comes from the conversion of the fuel, although some of the energy from the fuel is temporarily stored in the RESS for later use in order to optimize total system performance.

The *series-hybrid* design is shown in Figure 12.8. Solid, black lines depict the flow of electrical energy; the gray lines show transfer of mechanical energy. In contrast to the stop–start and microhybrids, the RESS is used for vehicle propulsion. In the series design, the electric motor is the only means of providing torque to the wheels. The engine, typically an ICE, powers an electrical generator, which converts all of the power generated from the engine into electrical power. This electrical power is either used to charge the battery or supplied to the motor for propulsion. Since all power to the wheels must go through the motor, the motor is sized to meet the maximum power demand. Because the combustion engine is not directly connected to the drive train, it can be run at optimal conditions, which increases efficiency and decreases emissions. The advantages of this architecture are most significant for driving that involves frequent starts and stops. The result is a significant reduction in fuel consumption relative to that of a conventional vehicle. The advantages of the series-hybrid are less significant for long periods of continuous driving, on the highway for instance, because the power from the engine must be converted to electrical energy and then back to mechanical energy since there is no direct coupling of the ICE to the wheels.

You were already introduced to a parallel design, shown in Figure 12.2. This design is an alternative architecture that addresses some of the challenges with the series configuration. Power to the wheels can be delivered either by the internal combustion engine through mechanical coupling or from the battery through an electrical motor. The battery is charged from the engine and during braking. A motor-generator is required to convert between mechanical and electrical energy.

Compared to the internal combustion engine, which uses a hydrocarbon fuel with a very high energy density, battery and fuel-cell systems have low specific energy. For a fixed vehicle mass, low energy density or low specific energy translates to reduced range. The key to the success of hybrid vehicles is that these electrochemical devices are able to store and use energy efficiently; that is, the round-trip efficiency can be more than 80%.

Now that we have briefly examined hybrid architectures, we will consider the RESS for several specific cases, beginning with the start–stop hybrid.

12.6 START–STOP HYBRID

As we consider the energy storage requirements for the start–stop hybrid, it is important to know the power

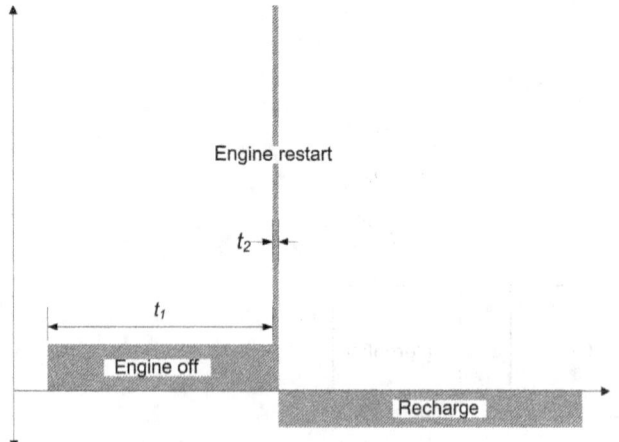

Figure 12.9 Power cycle for a start–stop hybrid.

required as a function of time. The intended driving schedule and many of the specifics for the vehicle are not needed for design of the energy storage system for this type of hybrid, since the RESS does not provide power to move the vehicle. The needed power and energy requirements can be approximated from a simple cycle with just three features that are repeated: engine off, engine restart, and recharge. The three states are shown in Figure 12.9, where positive values represent energy being supplied by the energy storage device and negative values correspond to recharge. We see that, in addition to relatively long periods (on the order of a minute) of charging and discharging, a high pulse of power is needed for restarting the engine. Sizing an energy storage device for the power and energy needs dictated by the cycle shown in Figure 12.9 is straightforward as shown in Illustration 12.3.

Here we contemplate two options for energy storage: electrochemical double-layer capacitors (EDLC) and batteries. Each must meet both the energy and power needs of the cycle. Although the fundamentals of these devices have been covered in earlier chapters, we introduce a couple of additional considerations. For batteries, we include cycle life and capacity turnover (Section 12.3). Examining the energy required to complete a typical cycle as shown in Figure 12.9, we observe that the capacity of the battery needed is quite small. A typical starting–lighting–ignition (SLI) battery (Chapter 8) for a conventional vehicle has a rated capacity on the order of 60 A·h. Why is the capacity so much larger than that required to start the vehicle? Part of the answer has to do with providing the requisite power, but low-temperature performance and the useable life of the battery are also important.

By contrast, EDLC performance degrades only minimally over tens of thousands of cycles; therefore, cycle life is generally much less of an issue. We learned in Chapter 11 that the energy stored in a capacitor is ½ CV^2,

although the amount of useable energy in an EDLC is less as will be discussed shortly. Similar to the way that the SOC window for a battery limits its useable energy, there are real-world limits for the EDLC. The SOC window is limited for a battery to preserve its life—the reasons for limiting the EDLCs are different. The maximum voltage is limited by physical constraints—for example, the stability limit of the electrolyte. Also, the voltage of the capacitor is directly proportional to its charge. Thus, the output voltage decreases linearly from a maximum to zero during constant-current discharge (Figure 11.9 compares the voltage of a battery and EDLC during discharge), and power electronics are needed to maintain a constant output potential. As a practical design matter, the output voltage may only be allowed to decrease to half of the maximum potential (V_{max}). Since the amount of stored energy is proportional to the voltage squared, one-fourth of the energy stored in the capacitor cannot be used if the voltage is limited in this fashion, as illustrated in Figure 12.10. The usable energy is therefore equal to three-fourths of the maximum energy when the minimum voltage is equal to $V_{max}/2$.

$$\text{Useable energy} = \left(\frac{3}{4}\right)\left(\frac{1}{2}CV_{max}^2\right) = \frac{3}{8}CV_{max}^2. \quad (12.5)$$

When regenerative braking is used, it is desirable to keep some capacity available at all times to allow energy recovery. Thus, the charge may be stopped before the maximum voltage, V_{max}, of the EDLC is reached. The difference between the nominal voltage and maximum voltage (refer back to Figure 12.5) will determine the amount of energy that may be unused on the high end.

On the other hand, compared to a typical secondary battery, EDLCs self-discharge much more quickly. Thus,

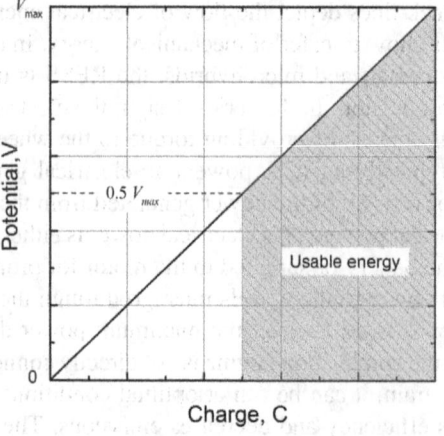

Figure 12.10 Energy for EDLC is equal to the area under the curve. Limiting voltage to half of V_{max} results in 25% of energy unavailable.

the EDLC must be made larger to make up for the self-discharge, or combined with a battery to form a hybrid RESS.

ILLUSTRATION 12.3

Size a RESS for a start–stop hybrid using the profile shown in the table. Assume 12,500 stops per year and a 5-year life. Consider both a lead–acid battery and an electrochemical double-layer capacitor. The battery has a capacity turnover of 800, a nominal voltage of 12 V, and a specific energy of 42 W·h·kg^{-1}. EDLC modules (15 V, 230 F) are available with 3 mΩ resistance (ESR). Each module has a mass of 1.7 kg.

	Power [W]	Time [s]
Engine off, P_{acc}	600	$t_1 = 45$
Restart, P_s	5000	$t_2 = 0.4$

Solution:

First, let's size the lead–acid battery. Calculate the A·h associated with one cycle:

$$\frac{(600 \text{ J} \cdot \text{s}^{-1})45 \text{ s} + (5000 \text{ J} \cdot \text{s}^{-1})0.4 \text{ s}}{} \left| \frac{\text{Wh}}{3600 \text{ J}} \right| \frac{1}{12 \text{ V}}$$

$$= 0.671 \text{ A} \cdot \text{h}.$$

Now, calculate the charge capacity needed for the specified 5-year lifetime (see Equation 12.4).

$$\frac{12,500 \text{ cycles}}{\text{year}} \left| \frac{5 \text{ years}}{} \right| \frac{0.671 \text{ A} \cdot \text{h} - \text{passed}}{\text{cycle}}$$

$$\left| \frac{1 \text{ A} \cdot \text{h cap}}{800 \text{ A} \cdot \text{h} - \text{passed}} = 52.4 \text{ A} \cdot \text{h}.$$

The last term on the left-hand side of the equation follows directly from the definition of capacity turnover. The calculated capacity corresponds to $(52.4 \text{ A} \cdot \text{h})(12 \text{ V}) \frac{\text{kg}}{42 \text{ Wh}} = 15.0 \text{ kg}$.

Next, we calculate the size of the EDLC. To do so, we assume that cycle life is not limiting. The EDLC is consequently sized to meet the energy and power needs of the cycle. Given the energy required per cycle and assuming 15 V for the capacitor (as specified above), Equation 12.5 provides the capacitance, which is a measure of size for the capacitor. Note that there is no need for additional capacity at the "top end" since the start–stop hybrid does not include regenerative braking. We have not included allowance for self-discharge in our design.

$$\frac{(600 \text{ J} \cdot \text{s}^{-1})45 \text{ s} + (5000 \text{ J} \cdot \text{s}^{-1})0.4 \text{ s}}{} \left| \frac{8}{3} \right| \frac{1}{(15 \text{ V})^2} = 344 \text{ F}.$$

Given the above module size (230 F), a minimum of two modules connected in parallel are needed. The power that the capacitor can deliver must also be considered. The maximum power from a capacitor is given by (see Chapter 11)

$$P = \frac{V^2}{4ESR}.$$

For the start–stop hybrid, it is critical that we have sufficient power to start the engine after running the accessories during the time that the vehicle is stopped. Therefore, the potential, V, that is used in this equation is the voltage of the capacitor after 45 seconds of supplying accessory power. In addition, to ensure that the power is available throughout the start portion of the cycle, we include the energy consumed during engine start in our calculation of the voltage drop. Assuming that the capacitor starts with a voltage of 15 V = V_{max}, the potential at the end of the start cycle, determined by calculating the voltage drop associated with the energy consumed, is

$$V = \sqrt{V_{max}^2 - \frac{2(P_{acc}t_1 + P_s t_2)}{mC}},$$

where m is the number of modules and C is the module capacitance. Therefore, the power available at the end of the start cycle is

$$P_{start} = \frac{V_{max}^2 - \frac{2(P_{acc}t_1 + P_s t_2)}{mC}}{4ESR}$$

For $m = 2$, there is about 8 kW of power still available. Thus, there is more than sufficient power to start the engine. We conclude that two modules are needed; the mass is 3.4 kg. Clearly, this mass is much lower than the battery mass estimated above. Remember, however, that capacitors do not hold a charge for long periods, and a battery would be needed to start the engine in the event that the capacitors had discharged due to lack of use of the vehicle.

12.7 BATTERIES FOR FULL-HYBRID ELECTRIC VEHICLES

In this section, we are interested in hybrids where energy can be recovered during braking and where energy from the RESS can be used to propel the vehicle (so-called *full hybrids*). Full-hybrid vehicles span a range of architectures, including both the parallel and series architectures discussed previously. They have many advantages over a start–stop hybrid, but the systems themselves are also more complex and costly. The starting point for our analysis is still the driving schedule (vehicle speed versus time) combined with a vehicle model to convert these data to power versus time. With a battery electric vehicle, the battery provides all of the power as discussed above; in contrast, the battery provides no traction power for the start–stop hybrid. Now with the full hybrid, the designer has the flexibility to decide how much of the required power at each point during the driving schedule comes from the engine and how much from the battery.

The motivation for the full hybrid can be brought into focus with an engine map. The efficiency of the ICE is inversely proportional to the specific fuel consumption (SFC), g (kWh)$^{-1}$, which is shown as a function of engine speed with the solid contour lines in Figure 12.11. Shown with the dashed lines are curves of constant power. As the torque increases at a fixed speed, the efficiency rises and then falls. The engine power is set by the requirements of the prescribed driving schedule. In addition, the speed of the ICE engine is fixed because the engine is connected to the wheels through a mechanical link. Since the engine must be sized to meet the maximum power requirements and because the speed of the engine is connected to the driving conditions, ICEs tend to operate at a relatively low efficiency under most conditions (refer to Figure 12.1). Even if it were possible to follow the optimum efficiency curve, we see that from 40 to 5 kW, the minimum SFC increases from about 225 to 300 g·kWh^{-1}. Given that a typical driving schedule has an average power much lower than the peak power, most of the time the engine will be operating at a fraction of the maximum power where the efficiency is low.

How do full hybrids further improve efficiency beyond that possible with energy recovery during braking? As alluded to in the previous paragraph, the answer to this question is directly related to how the engine is operated. For all hybrids except for plug-in hybrids, which will be discussed shortly, all of the energy used to operate the vehicle comes from the fuel. The hybrid architecture allows the engine speed to be decoupled from the vehicle speed and permits the ICE to be operated more efficiently. If the engine is operated at higher efficiency and turned off at other times, the overall efficiency of the vehicle can be improved significantly. This is possible because, in contrast to the ICE, batteries, capacitors, and electric motors typically operate at much higher efficiencies.

Let's examine how this works for simple parallel and series hybrids. For a parallel hybrid, there is a mechanical link between the engine and the wheels; therefore, the engine cannot be completely decoupled from the driving conditions. As a result, ICE operation cannot be completely restricted to a narrow or constant RPM range, and efficiency is reduced when the engine operates, for example, at low rotation rates. The engine is typically off when the vehicle is stopped, which increases efficiency by a small amount as we saw for the start–stop hybrid. Efficiency is also increased by reducing the power required from the engine. This is done by "power assist" where the RESS is used in parallel with the ICE to provide additional power when needed. An example of the operation of such a hybrid is shown in Figure 12.12. A maximum power of approximately 10 kW is available from the engine. When the total power required exceeds 10 kW, additional power is supplied by the RESS. When the total power required is less than the capacity of the

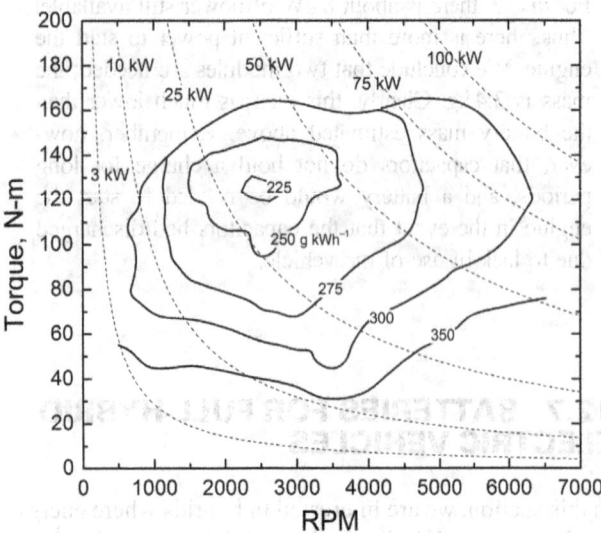

Figure 12.11 Engine map for a 1.9 L spark ignition engine.

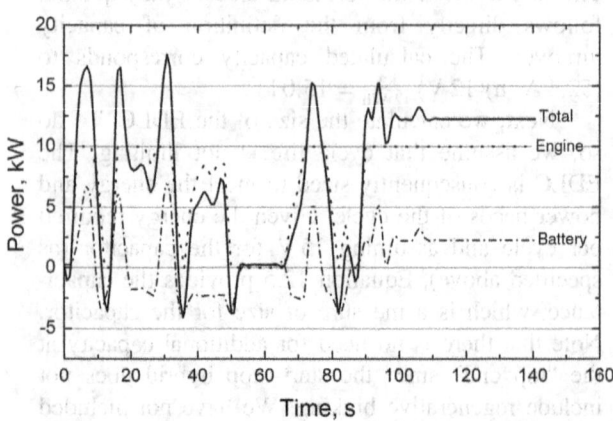

Figure 12.12 Example of power usage for parallel hybrid. Power of RESS (battery) is positive for discharging and negative for charging.

engine, the engine can be used to recharge the battery. This is shown in Figure 12.12 in the areas where the engine power is greater than the total power. The parallel hybrid effectively reduces the power required of the engine and allows it to run more efficiently. Thus, a smaller, more efficient engine is sufficient to meet the power needs of the same driving cycle. Because there is a mechanical link with the wheels, the combustion engine in a parallel hybrid cannot be used to charge the RESS when the vehicle is stopped.

The decoupling of the ICE from the drive cycle is even more significant with a series hybrid where there is no mechanical link to the wheels and the only role of the engine is to recharge the RESS. In that case, the engine speed is completely decoupled from the drive cycle and the engine can be run as efficiently as possible to power the generator. Series hybrids are the most efficient hybrid for stop-and-go city driving. However, for long distance highway driving, they are less efficient than a parallel hybrid since, beginning with the fuel, energy must be converted multiple times before providing power to the wheels because there is no direct mechanical link between the ICE and the wheels.

There are two basic modes of operating the battery in a hybrid-electric vehicle: *charge-sustaining* and *charge-depleting*. In the charge-sustaining mode, the battery is available for power assist, providing accessory power when the engine is off, capturing energy during braking, and providing very limited electric-only operation. In charge-sustaining mode, the SOC may fluctuate, but as the vehicle is driven over time, the SOC remains in a small window. In contrast, in charge-depleting mode, the battery is also used for electric-only propulsion over extended periods. In charge-depleting mode, the SOC may fluctuate, but over time there is a net decrease in the SOC of the RESS. The depletion can be due to all electric or blended operation.

Charge-Sustaining

The data in Figure 12.12, which were examined earlier, represent a parallel hybrid operating in a charge-sustaining mode. In this case, the engine was limited to 10 kW and any power demand in excess of 10 kW was supplemented by the battery. The RESS system must be large enough to make up the difference between required and engine power. Sizing of the RESS must also consider the power coming back into the battery during periods of braking. Either the RESS is sized to accommodate all of this power, or some of the energy from braking is dissipated as heat.

Of course, the SOC of the battery for this charge-sustaining system is not strictly constant; it varies, but over a small range. For the data from Figure 12.12, the changes

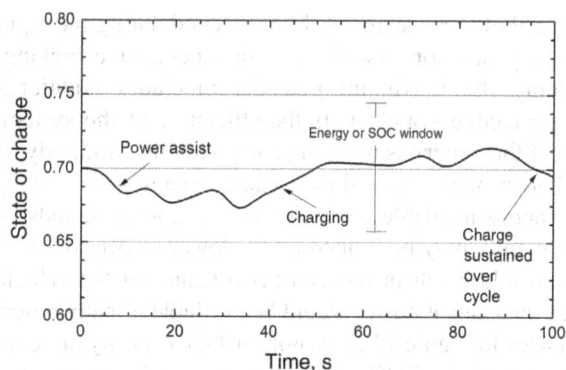

Figure 12.13 Charge-sustaining details. Battery SOC is maintained in a small window.

in energy for the battery in kWh are shown in Figure 12.13. During periods of electric-only launch, power assist, or when the vehicle is stopped, the SOC of the battery decreases. However, this energy is quickly replaced so that the SOC stays in a narrow window. Both the size of the SOC window and number of cycles required of the battery must be known in order to size the battery.

Clearly, the size of the battery must increase as the engine is made smaller since they both combine to meet the peak power needs of the vehicle. In addition, the energy-window of the battery will grow as the size of the engine shrinks. One way to quantify the relative size of the battery is with the degree of hybridization:

$$\text{degree of hybridization} = DOH$$
$$= \frac{\text{battery power}}{\text{engine power} + \text{battery power}} \quad (12.6)$$

For a charge-sustaining hybrid, the *DOH* should not be too high or too low. If the battery is too small, which corresponds to *DOH* of about 25% or less, then the advantages of energy recovery during braking cannot be fully realized—There simply is not enough capacity or power capability in the battery to absorb the energy. A high *DOH* can also create problems. A vehicle must also be able to sustain speed on a 6.5% grade, for example, over an extended distance. A 1% grade means that the elevation increases 1 m for every 100 m of lateral distance, that is, rise/run. Under these conditions, if the engine power is too small, then the battery charge cannot be sustained. The upper limit for *DOH* is about 60%.

There are several reasons for limiting the SOC window. First, in a hybrid vehicle, it would be undesirable to have the battery either fully charged or fully discharged. For instance, if the battery is fully charged, then it would not be possible to recover energy during braking. Also, a reduction in the SOC reduces the potential of the battery and makes it less likely that the maximum charging potential of the system would be reached during the rapid energy generation associated with regenerative braking; reaching the maximum potential precludes additional energy recovery and limits the efficiency of the system, even if the battery is not completely charged. Similarly, if the battery were allowed to discharge to a near-zero SOC, the energy available from the battery during periods of acceleration may be limited by the lower cutoff potential. This would result in reduced performance of the vehicle since inadequate power would be available. Finally, referring back to Figure 12.6, we note that the capacity turnover decreases as the SOC window is expanded. Limiting the SOC window increases the lifetime of the battery.

ILLUSTRATION 12.4

Evaluate the suitability of existing Ni-MH cells for constructing a battery for a power-assist hybrid operating in the charge-sustaining mode. The battery must provide 25 kW of peak power and 260 Wh of energy, and a potential of about 200 V. Individual cells have a nominal potential of 1.2 V, an open-circuit potential of 1.3 V, and a cutoff potential of 1.0 V. Six cells are assembled together and connected in series in modules. The capacity of the individual cell is 6.6 A·h; the resistance of a single module is 20 mΩ. The SOC window is limited to 20% to achieve the desired life.

1. How many cells are needed in series?

$$\frac{200\,\text{V}}{1.2\,\text{V/cell}} = 167\,\text{cells}.$$

Convert to modules, $\frac{167\,\text{cells}}{6\,\text{cells/module}} = 27.8$ modules, round to 28.

2. What capacity is required for each cell in A·h?

$$\frac{260\,\text{Wh}}{28 \times 6 \times 1.2\,\text{V/cell}} = 1.29\,\text{A}\cdot\text{h}.$$

This value assumes that the cells can be fully discharged. If the SOC window is limited to 20%, the A·h capacity of each cell is

$$\frac{1.29\,\text{A}\cdot\text{h}}{0.2} = 6.45\,\text{A}\cdot\text{h}.$$

This value is close to the rated capacity of the cells and is acceptable.

3. Since the cutoff voltage is greater than half of the OCV, Equation 12.3 applies, which is rearranged to

$$R_\Omega \approx N_m V_{cutoff} \frac{N_m(V_{oc} - V_{cutoff})}{\frac{P}{(\text{mod})}}$$

$$= \frac{6^2(1.0)(1.3 - 1.0)}{\frac{25{,}000}{28}} = 12.1\,\text{m}\Omega,$$

where N_m is the number of cells in a module. The calculated resistance value is smaller than the actual resistance (20 mΩ) of the module. Therefore, the module resistance is too high and the cells will not deliver the required power and are unacceptable for this design. Finally, there will be some additional resistance that comes from connecting the modules together as discussed in Chapter 8. Although these cells can meet the energy and voltage requirements, their resistance makes them unsuitable for this application.

Charge-Depleting

A second common mode of operation for a full hybrid is one where propulsion is supplied by the battery alone for extended periods. In this mode, the SOC of the battery decreases with time. When the SOC reaches a lower limit, then the mode is switched from charge-depleting to charge-sustaining as shown in Figure 12.14. This feature differentiates the charge-depleting hybrid from the all-electrical vehicle.

An important design objective is the all-electrical range, which directly impacts the sizing of the battery. The greater the all-electrical range, the larger the size of the battery. Therefore, the two main differences in sizing are (i) the battery must be large enough to supply all of the power requirements, and (ii) the window for SOC is

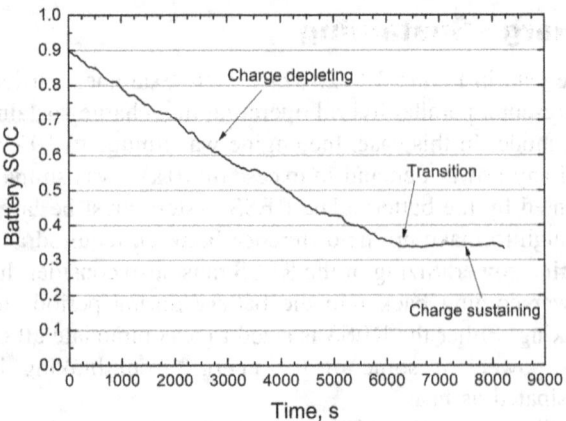

Figure 12.14 Window over which state of charge is varied for a full-hybrid operating in charge depleting mode.

Table 12.2 Comparison of Batteries for Hybrid and All-Electric Vehicles

	Mild hybrid	Strong hybrid	All electric vehicle
Average power	5 kW	20 kW	20 kW
Energy	0.5 kWh	8 kWh	25 kWh
Run-time	0.1 hours	0.4 hours	1.2 hours

much larger than for charge-sustaining designs. Table 12.2 shows examples of battery capacity and power for different types of hybrid and electric vehicles.

In addition to the improved fuel economy, full hybrids that are designed to operate in charge-depleting mode provide flexibility in the source of energy and hence greater energy security. Specifically, vehicles can be designed so that the depleted battery can be charged from an electrical outlet, allowing the vehicle to be driven over a limited range without ever using the engine. This concept is the basis of the so-called plug-in hybrid vehicle (PHEV). Operation in charge-depleting mode closely approximates an all-electric vehicle; consequently, vehicles designed to operate in this mode tend to favor a series architecture where the internal combustion engine acts only to recharge the battery system. In practice, combined hybrids that incorporate the advantages of both the series and parallel architectures are also available and represent an important part of the hybrid market. The advantages of these combined vehicles come at the price of significantly increased complexity of both hardware (requires both mechanical and electrical connection of ICE with drive axle) and software (sophisticated control algorithms are required).

Summary of Hybrid Designs

Figure 12.15 summarizes the functionality and nomenclature for hybrid electric vehicles. Full hybrids are further characterized by the degree of hybridization. All full hybrids use the electric motor and battery as an assist to the internal combustion engine, and have regenerative braking. The movement from *mild* to *strong* indicates a higher degree of hybridization, which corresponds to increasing amounts of electrical power to the drive train and increased levels of energy recovery through regenerative braking. We also see that as the capacity of the battery increases, so does the voltage of the battery pack. A start–stop hybrid has a battery voltage of 12–42 V; by comparison, full hybrids and all-electric vehicles use batteries of 300–400 V.

12.8 FUEL-CELL HYBRID SYSTEMS FOR VEHICLES

An alternative to the ICE is a fuel-cell power source, typically operating with hydrogen as the fuel. A key advantage is that emissions of carbon are eliminated as well as criteria pollutants. There are a few important distinctions to be made for fuel-cell hybrids. First, the fuel cell, like the battery, generates DC electrical power; therefore, the fuel-cell hybrid is an all-electrical vehicle. The architecture is invariably a series hybrid as shown in Figure 12.16, where all of the traction power passes through the electrical motor. Electrical power from the fuel cell can either be used to drive the electrical motor or to charge the battery. As with any series configuration, the electric motor must be sized large enough to deliver the maximum power to the wheels.

The efficiency characteristics of a fuel-cell system are markedly different from those of an ICE. Recall from Chapters 9 and 10 that the efficiency of a fuel cell is roughly proportional to its operating potential, with higher efficiencies at high voltages and low levels of power. This trait is easy to comprehend in terms of cell polarizations. The current increases with increasing power, resulting in increases in the ohmic, kinetic, and mass-transfer polarizations. Each of these polarizations represents a loss in

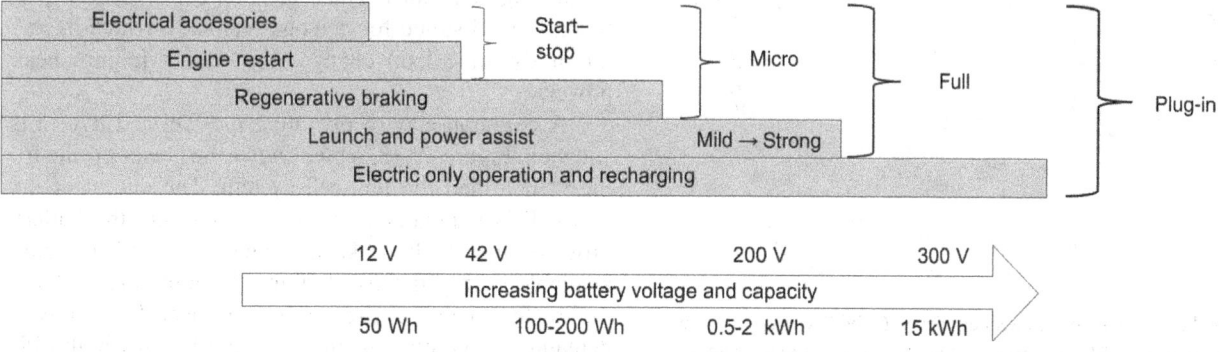

Figure 12.15 Summary of hybrid-electric vehicles. As the degree of hybridization increases, the voltage of the battery and its capacity increase.

292 Electrochemical Engineering

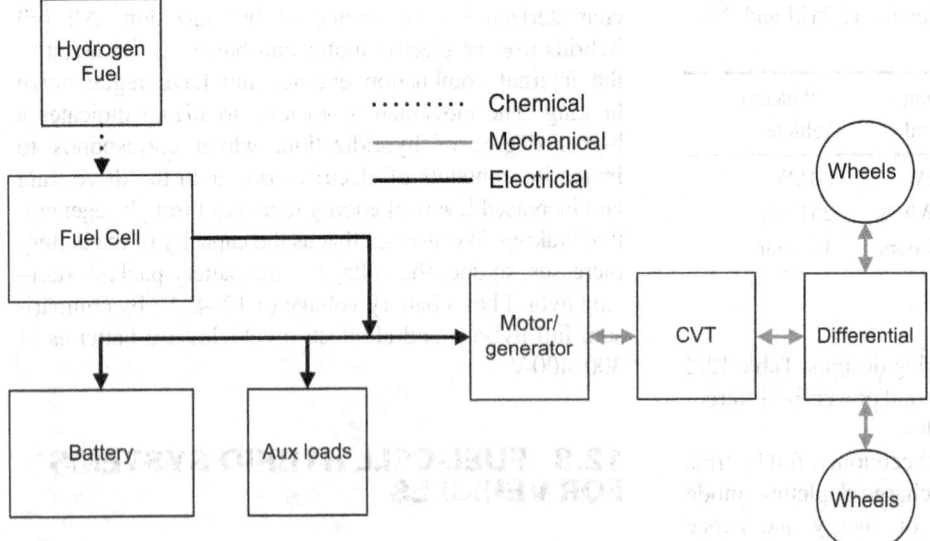

Figure 12.16 Typical architecture for a fuel-cell hybrid system.

efficiency loss; consequently, the efficiency of the fuel cell itself increases as the power decreases.

The efficiency of a fuel-cell *system*, however, does not continue to increase indefinitely with decreasing power. Instead, as shown in Figure 12.17, there is a maximum in the system efficiency. We can understand this maximum by remembering that these data represent system efficiency given by

$$\eta_{sys} = \frac{IV - \text{ancillary power} - \text{electrical losses}}{\text{availability of the fuel}}. \quad (12.7)$$

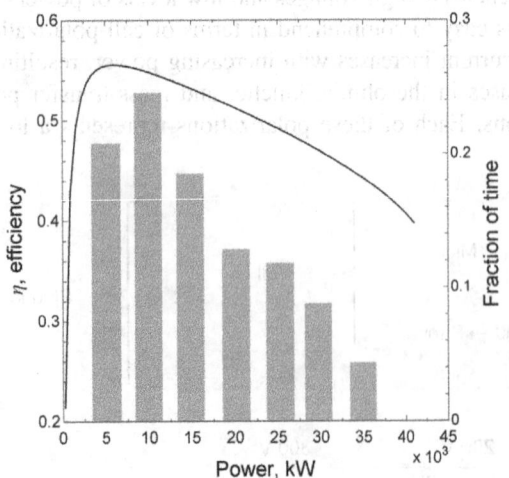

Figure 12.17 Fuel-cell system efficiency (solid line) as a function of fuel-cell power. The bars represent the frequency of time spent at each power level for an example driving schedule.

Some ancillary power is always needed to operate the fuel-cell system, irrespective of power level. For instance, a blower may be needed to provide air to the fuel cell. This ancillary power, as well as any electrical losses, must be subtracted from the gross electrical output of the fuel-cell stack (IV) to get the net power. The losses attributed to these ancillary devices dominate at low power and the efficiency of the fuel-cell system goes through a maximum.

Notably, this maximum in efficiency occurs at low power levels for a properly designed fuel-cell system. Compare this behavior with the engine map shown in Figure 12.11, which shows decreased efficiency at low power. Also shown in Figure 12.17 is a Pareto chart for the required traction power corresponding to a typical driving schedule. For the vast majority of driving, the vehicle is operating at part power; furthermore, the most frequent power level is well below half of the maximum power. This outcome is characteristic of nearly all driving schedules. Thus, the fuel-cell hybrid is particularly well suited for personal vehicles, which have a lot of dynamic load changes, but rarely require peak power.

A common way of operating the fuel-cell hybrid is shown in Figure 12.18, which shows the power profile for the vehicle and the fuel-cell system. The output of the fuel cell is maintained nearly constant, and the battery provides the small peaks in demand as well as being available for regenerative braking. Both the fuel cell and the battery can be used fully to meet large power demands. In contrast to the ICE, the FC has high efficiency at part power.

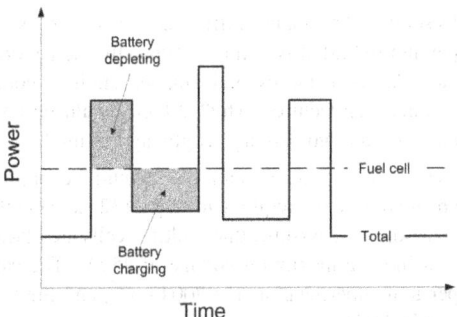

Figure 12.18 Typical method of operating a fuel-cell hybrid. Fuel-cell power is constant.

ILLUSTRATION 12.5

Compare the power requirements for of the engine, fuel cell, and battery for three systems: (i) ICE only, (ii) ICE battery hybrid, and (iii) fuel-cell battery hybrid. The hybrids are charge-sustaining. The average power to complete the driving schedule is 10 kW, and a maximum power of 90 kW is required. For the hybrids, size the ICE and the fuel cell to achieve maximum efficiency while supplying the average 10 kW. Use data in Figure 12.11 for the ICE, and the plot below for the fuel cell.

1. In systems where the ICE is the only source of power, the engine must be sized to meet the maximum power required. Energy storage is not an issue since it comes in the form of a hydrocarbon fuel with very high energy density. In this case, the engine must provide 90 kW, which corresponds roughly to the maximum value in Figure 12.11. Its maximum efficiency is near 35 kW, and at 5 kW, specific fuel consumption (300 g·Wh^{-1}) compared to 225 g·Wh^{-1} at 35 kW, corresponding to a significantly lower efficiency.

2. For the ICE/battery hybrid, the engine can be reduced in size to provide the average power at maximum efficiency. To do this, the engine is scaled from the one shown in Figure 12.11 so that the maximum efficiency is at 10 kW. Assuming linear scaling, the maximum power of this ICE is

$$P_{max,ice} = \frac{90}{35} \times 10 = 26 \text{ kW},$$

which is not sufficient to meet the maximum power requirement. Since, the combined power of the battery and engine must still be 90 kW,

$$P_{batt,ice} = 90 - 26 = 64 \text{ kW}.$$

3. A fuel-cell system is designed to achieve a maximum in system efficiency at roughly 10 kW as shown in Figure 12.17. The net power of the same system is shown below. The maximum power of this fuel-cell system is about 43 kW. Thus, the required battery power is

$$P_{batt,fc} = 90 - 43 = 47 \text{ kW}.$$

While greatly simplified, this example illustrates trade-offs between the different types of vehicles.

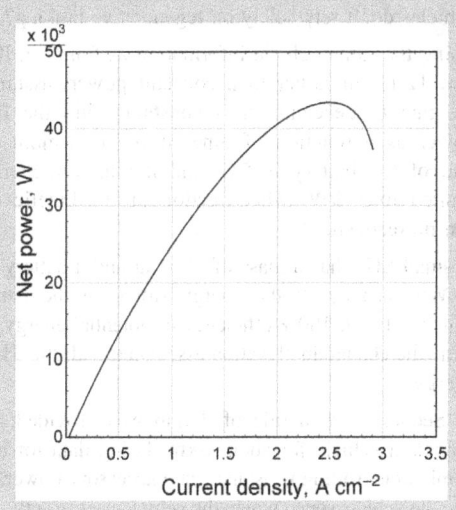

CLOSURE

In this chapter, we considered hybrid vehicles, with particular emphasis on vehicles that use batteries as a rechargeable energy storage system. Efficiency gains made possible through the use of hybrids were discussed for several different hybrid strategies and architectures. The connection between the driving schedule, vehicle characteristics, and power requirements was also examined in the context of a driving schedule and a vehicle model. Hybrid vehicles that utilize combustion engines offer significant efficiency advantages as a result of energy recovery through regenerative braking and increased engine efficiency through decoupling of the engine speed from that of the vehicle. The use of electrochemical capacitors and fuel cells in hybrid vehicles was also examined. Finally, the power and energy requirements of the energy storage system were determined for different levels of hybridization. Because of the increased efficiency of hybrid vehicles, they will likely play an important role in future transportation.

FURTHER READING

Ehsani, M., Gao, Y., and Emadi, A. (2009) *Modern Electric, Hybrid Electric, and Fuel Cell Vehicles*, CRC Press, Boca Raton.

Husain, I. (2011) *Electric and Hybrid Vehicles: Design Fundamentals*, CRC Press, Boca Raton.

Scrosati, B. Garche, J. and Tillmet, W., eds., (2015) *Advances in Battery Technologies for Electric Vehicles*, Woodhead Publishing.

PROBLEMS

12.1 What are some of the reasons that hybrid and all-electric vehicles don't rely solely on regenerative braking?

12.2 Using the same vehicle information as found in Illustration 12.1, but rather than constant power, assume that the rate of deceleration is constant. Find the braking power as a function of time. If the maximum charge rate of the battery is 5 C, and the motor/generator is limited to 50 kW, what fraction of the kinetic energy can be recovered?

12.3 A small SUV has a mass of 1750 kg and a battery with a 2 kWh capacity. If the energy stored in the battery is converted with 100% efficiency to potential energy, determine the change in elevation associated with the 2 kWh of energy.

12.4 In Section 12.4, a rule of thumb was provided: 1 kWh propels a vehicle for about 6 km. Using data for aerodynamic drag, rolling resistance, and accessory power, calculate the vehicle speed where the vehicle uses exactly 1 kWh of energy to travel 6 km. What other factors will affect this value?

$m = 1950$ kg, $A_f = 2.8$ m^2, $C_d = 0.39$,

$f_r = 0.015$, rolling resistance; air density $\rho_a = 1.2$ kg·m^{-3},

$P_{acc} = 400$ W

12.5 Use the rule-of-thumb of 6 km·kWh^{-1}, configure a battery for an all-electric vehicle with a range of 150 km. The design voltage of the battery is 300 V, which is made from 2S-4P modules. Each of the cells has a nominal voltage of 3.8 V and a cutoff potential of 3.2 V. How many modules are required? What is the required capacity of each cell? If the peak power is 75 kW, what is the maximum resistance of each cell in mΩ?

12.6 A battery for a power-assist HEV is required to deliver 25 kW and 300 Wh. The characteristics of the cell are a capacity of 12.2 A·h·m^{-2}, a resistance of 2 mΩ·m^2, and an open-circuit potential given by $U = 3.8 - B$ (1-SOC). Determine the separator area required by the battery if $B = 0.2$ V and the allowable change in SOC is 0.3. What if the permitted change in SOC is only 0.2? Explain the results.

12.7 Describe the main differences between series- and parallel-hybrid drive trains. Typically, series-drive trains have larger batteries and smaller engines compared to parallel architectures, why? Which would be preferred in city stop-and-go driving? Highway driving?

12.8 Design an EDLC system that would be appropriate to absorb the energy required to stop a 1325 kg vehicle moving at 50 km·h^{-1}? Assume that multiple cells are place in series to achieve a maximum voltage of 450 V. The capacitance per unit superficial area is 800 F·m^{-2}, and the voltage per cell is 3.0 V.

12.9 Using a single EDLC module described in Illustration 12.3, estimate the standby time permissible if there is a self-discharge current of 3 mA. At the beginning of the standby period, the potential was 13 V and the starting power required is 5 kW.

12.10 A key difference between a start–stop hybrid and a micro-hybrid is that energy is recovered. How much kinetic energy does a vehicle with a mass of 1520 kg traveling at 45 km·h^{-1} have? Express your answer in Wh. The RESS is a number of EDLC modules of 230 F arranged in parallel, with a maximum voltage of 15 V. The braking event takes 10 seconds and at the start the potential of the EDLC is 8.8 V. If the regenerated power is constant, what is the maximum current? How many modules are needed to recover this energy?

12.11 In Illustration 12.3, the lead–acid battery is replaced with a lithium-ion battery that has a voltage of 42 V, a turnover capacity of 400, and a specific energy of 130 W·h·kg^{-1}. What is the size of the battery required in kg?

12.12 Identify three advantages and three disadvantages of full-hybrid vehicles. How would improvements in energy storage technology mitigate these disadvantages?

12.13 Cells have an average potential of 3.8 V, a loading of 31 A·h·m^{-2}, and a resistance of 40 mΩ·m^2. The maximum battery voltage is 300 V, and the cutoff potential for the cells is 3.1 V. Finally, the cell size is limited to 30 A·h. Configure a battery for a vehicle with an all-electric range of 100 km. What is the maximum power of the battery? Would there be an advantage to increasing the maximum battery voltage?

12.14 A fuel-cell hybrid vehicle is required to complete a specified driving schedule. A simplified representation of the power needed for traction and parasitic power is shown in the table. Assume that the system is operating in charge-sustaining mode, and that the cycle is repeated many times. The battery is limited to being discharged at 5 C, and the rate of charging is limited to 2 C. Suggest how to size the battery and fuel cell to accomplish this profile. Comment on the effectiveness of the control strategy outlined in Figure 12.18. There is not a "right" answer, but you need to justify your choices.

Time [s]	Power [kW]
30	3
10	40
58	25
14	−10
34	8
16	60
40	20
18	−5
38	15
14	30
26	10

12.15 Using data from the engine map (Figure 12.11), sketch the efficiency of the engine as a function of power. Use a heating value of the fuel of 44 MJ·kg^{-1} Compare this efficiency versus power with that for a fuel-cell system (Figure 12.17). How might hybrid designs and operation differ?

12.16 Calculate the additional power needed to sustain 90 km·h^{-1} up a 6% grade compared to level ground. Neglect the change in rolling resistance. The vehicle mass is 1700 kg.

12.17 A vehicle requires 12 kW of average power, and 70 kW maximum power to complete a typical driving schedule. If 22 kW additional power is required to sustain 90 km·h^{-1} up a 6% grade, what is the maximum degree of hybridization?

12.18 Using the start–stop information from Illustration 12.3, for two EDLC modules how many attempts to start the vehicle can be made after a period of stopping for 45 seconds?

12.19 Using data from the table, determine power to keep vehicle moving at different speeds. Express these answers in terms of km·kWh^{-1} of battery power assuming 80% efficiency. The frontal area is 0.6 m^2, $C_d = 0.50$, and $f_r = 0.016$. Assume parasitic power is 1 kW.

Speed, km·h^{-1}	60	60	90
Mass, kg	1300	1400	1400

12.20 How much power is needed to accelerate a 1500 kg vehicle from 0 to 90 km·h^{-1} in 10 seconds? What about power to stop the vehicle in 4 seconds.

12.21 A battery for a hybrid electric vehicle has 80 cells connected in series. Treat the battery as ohmically limited with $U = 3.8$ V, and $V_{max} = 4.2$ V. What is the maximum value of resistance that would still allow recovery of 80 kW during braking? What is the corresponding value of joule heating during this braking?

12.22 A large hybrid-electric transit bus is being designed. It is desired to recover the kinetic energy during stopping. The fully loaded mass is 20,000 kg, and the bus is estimated to make 200 stops per day from 50 km·h^{-1}, each in 10 seconds. Using information provided below, first size a battery and then an EDLC for 1 year of operation, and complete the table below. The maximum allowable voltage is 600 V.

	Battery	EDLC
Individual cell voltage	3.5 V (constant with SOC)	3.0 V (maximum) 1.5 V (minimum)
Specific energy	125 Wh·kg^{-1}	5 Wh·kg^{-1}
Energy density	275 Wh·dm^{-3}	7 Wh·dm^{-3}
Number of cells		
C-rate during braking		NA
Maximum current		
Capacity of RESS, kWh		
Volume of RESS		
Mass of RESS		

For the battery, the capacity turnover $= 500 + 2000(1 - \Delta_{SOC})^4$. For the EDLC, assume that the size must be increased by 20% to meet life requirement.

APPENDIX: PRIMER ON VEHICLE DYNAMICS

The fundamental equation is Newton's second law of motion:

$$F = M_v a = M_v \frac{dv}{dt}. \quad (A.1)$$

F is the net force acting on the vehicle of mass M_v. The forces acting on the vehicle include the tractive force, road resistance (including the effect of the grade), and aerodynamic drag. The tractive force F_t is related to the torque produced by the internal combustion engine or the electric motor:

$$F_t = \frac{T_p i_g \eta_t}{r_d}. \quad (A.2)$$

T_p is the torque, i_g is the gear ratio, η_t is the mechanical efficiency of the drivetrain, and r_d is the effective radius of the wheels. At steady speed, this tractive force is balanced against the frictional force associated with rolling the wheels, gravity, and aerodynamic drag. Both the forces of rolling resistance, F_f, and gravity, F_g, depend on the weight of the vehicle and are conveniently combined. For a vehicle of mass, M_v, climbing a grade with an angle α, the road resistance is

$$F_{rd} = F_f + F_g = M_v g(f_r \cos \alpha + \sin \alpha). \quad (A.3)$$

f_r is the coefficient of rolling resistance and g is the acceleration of gravity. The aerodynamic drag is

$$F_w = \frac{1}{2}\rho_a C_D A_f v^2. \quad (A.4)$$

ρ_a is the density of air, C_D is the drag coefficient, and A_f is the projected area of the vehicle.

The sum of these forces are used with Equation A.1:

$$M_v \frac{dv}{dt} = \sum F = F_t - (F_f + F_g) - F_w. \quad (A.5)$$

Equation A.5 forms the basis of a dynamic model that can be used to simulate a driving schedule. When the velocity is constant, the net force is zero, and the required torque can be calculated:

$$\frac{T_p i_g \eta_t}{r_d} = M_v g(f_r \cos \alpha + \sin \alpha) + \frac{1}{2}\rho_a C_D A_f v^2. \quad (A.6)$$

From Equation A.5 we can calculate the time, energy, and instantaneous power for a prescribed vehicle velocity versus time. Below we consider a highly simplified case. The time required to accelerate uniformly from speed 1 to 2. Since the force is constant, Equation A.5 is integrated.

$$t = M_v \int_{v_1}^{v_2} \frac{dv}{\frac{T_p i_g \eta_t}{r_d} - M_v g(f_r \cos \alpha + \sin \alpha) - \frac{1}{2}\rho_a C_D A_f v^2}. \quad (A.7)$$

The distance travel during this period of constant acceleration is

$$S = M_v \int_{v_1}^{v_2} \frac{v\,dv}{\frac{T_p i_g \eta_t}{r_d} - M_v g(f_r \cos \alpha + \sin \alpha) - \frac{1}{2}\rho_a C_D A_f v^2}. \quad (A.8)$$

Thus, if the distance S, the two speeds (v_1 and v_2), and the grade (α) are specified, the required torque of the ICE or electric motor can be determined. Since the torque is a function of engine or motor speed, which is related to vehicle speed and gear ratio, the integration of Equation A.8 must be done numerically.

ILLUSTRATION A.1

A 1350 kg passenger vehicle is moving at 90 km·h^{-1} on a level grade.

a. Determine the power needed to overcome aerodynamic drag and rolling resistance. The following data are provided: $A_f = 0.58\,\text{m}^2$, $\rho_a = 1.188\,\text{kg·m}^{-3}$, $C_D = 0.25$, and $f_r = 0.015$. The speed of 90 km·h^{-1} is 25 m s^{-1}.

Using Equation A.4,

$$F_w = \frac{1}{2} \left| \frac{0.25}{} \right| \frac{0.58\,\text{m}^2}{} \left| \frac{1.188\,\text{kg}}{\text{m}^3} \right| \frac{(25)^2 \text{m}^2}{\text{s}^2} = 53.8\,\text{N}.$$

The power to overcome aerodynamic drag is $F_w v = 1.35\,\text{kW}$. The force due the road is given by Equation A.3.

$$F_{rd} = \frac{1350\,\text{kg}}{} \left| \frac{9.81\,\text{m}}{\text{s}^2} \right| \frac{0.015}{} \left| \frac{\cos(0)}{} \right| = 198.6\,\text{N}.$$

The associated power is $F_{rd}v = 4.96\,\text{kW}$.

b. How much less power is needed to maintain this speed going down a grade of 6%?

$$\tan \alpha = \frac{-6}{100}, \quad \text{solving for the angle } \alpha$$

$$= -3.434\,\text{degrees}.$$

Again, using Equation A.3

$$F_{rd} = \frac{1350\,\text{kg}}{} \left| \frac{9.81\,\text{m}}{\text{s}^2} \right| \frac{[0.015 \cdot \cos(-3.434) + \sin(-3.434)]}{}$$

$$= -592.5\,\text{N}.$$

The difference is $\Delta P = v \times \Delta F = \frac{25\,\text{m}}{\text{s}} |(-595 - 198.6)\text{N}| = -19.8\,\text{kW}.$

In this case, the force from the acceleration of gravity is greater than the forces of rolling resistance and aerodynamic drag. You would need to brake (and recover energy with a hybrid) to maintain a speed of 90 km·h^{-1}. Multiplying the right side of Equation A.6 by the speed will give the power available to be stored.

$$P_{storage} = v\left(M_v g(f_r \cos\alpha + \sin\alpha) + \frac{1}{2}\rho_a C_D A_f v^2\right)$$
$$= 13.5 \text{ kW}.$$

RICHARD C. ALKIRE

Richard C. Alkire was born in Easton, Pennsylvania, in 1941. He grew up around his parents' music business and realized early on that he wanted to teach and to play the piano, both of which he has done throughout his life. Inspired by an older brother's career choice, he studied Chemical Engineering at Lafayette College, located in his home town of Easton. While in college, he performed piano at some 300 musical events throughout the East. During that time, Professor Zbigniew Jastrzebski invited him to carry out research on the corrosion of stainless steel, and Alkire has been investigating various electrochemical systems ever since. In 1963, he entered the University of California at Berkeley to study electrochemical engineering under Professor Charles Tobias. His M.S. (1965) and Ph.D. (1968) theses were on experimental and modeling investigations of shape evolution in porous electrodes. Subsequently, under the direction of Professor Carl Wagner at the Max Planck Institute for Physical Chemistry in Göttingen, he carried out high-temperature experiments to measure thermodynamic properties of solid-state fluoride compounds.

In 1969, he joined the Chemical Engineering faculty at the University of Illinois in Urbana. He advanced to full Professor in 1977, and served as department Head, 1986–1994, and as Vice Chancellor for Research and Dean of the Graduate College, 1994–1999. In 1999, he was appointed to the Charles J. and Dorothy G. Prizer Chair, which he held until retiring in 2008.

Alkire is particularly known for bridging the gap between fundamental scientific understanding and the engineering ability to design, scale over, and optimize complex, technologically significant processes. His works contributed to the physical understanding and the numerical simulation of electrochemical phenomena at the solid–liquid interface over many length scales. By taking advantage of unprecedented increases in computational power and experimental resolution, Alkire extended traditional continuum electrochemical engineering methods to the molecular scale associated with nucleation, additives, and evolution of surface morphology. His works laid the foundation for significant technological applications involving precision fabrication, surface modification, and corrosion prevention. These accomplishments were reported in the theses of 58 Ph.D. and 86 M.S. students and are described in over 200 publications.

Professor Alkire has been active in the Electrochemical Society since 1970. He served as a Divisional Editor for the Industrial Electrolytic Division from 1973–1990. He was also Chairman of the Industrial Electrolytic Division and was the Society representative for the American Association for the Advancement of Science. Professor Alkire has received many awards from the Electrochemical Society: the Research Award of the Electrodeposition Division in 1983, the Carl Wagner Award in 1985, Honorary Member of the Society in 1991, and Fellow of The Electrochemical Society in 1992. Professor Alkire also served as President of The Electrochemical Society from 1985 to 1986. Since then he has received two more major awards from the Electrochemical Society: the Edward Goodrich Acheson Award in 1996 and the Vittorio de Nora Award in 2005.

Dr. Alkire has also received major awards from the American Institute of Chemical Engineers, the Federation of Materials Societies, and the U.S. National Academies; most notable is the E.V. Murphree Award in Industrial and Engineering Chemistry from the American Chemical Society in 1991. These awards span activities in research, industrial impact, education, and service to the broader technical community. He was elected to the National Academy of Engineering in 1988. Since 1996, he has been coeditor of *Advances in Electrochemical Science and Technology*, a review series begun by Charles Tobias in 1961. Still an avid musician, Dr. Alkire maintains a broad keyboard repertoire on piano and harpsichord.

Image Source: Reproduced with kind permission of Richard C. Alkire.

Chapter 13

Electrodeposition

Electrodeposition is the reduction and deposition of an electronically conductive species at the cathode of an electrochemical cell. It is most frequently used for depositing metals on surfaces (electroplating) or for fabrication of devices from metals, although it is possible to electrodeposit nonmetals such as semiconductors or conducting polymers. Surface finishing of a variety of materials is performed by electrodeposition for corrosion protection and for aesthetic purposes, among others. Another important application is fabrication of the nanoscale metallic interconnects that enable state-of-the-art computing.

In this chapter, we examine the fundamental principles that govern electrodeposition and the application of those fundamentals to several systems of interest. In doing so, we will apply many of the concepts that you learned in the first few chapters of this book. In addition to the fundamentals that are the focus of this chapter, there are several practical aspects that are critical for the successful application of electrodeposition in industrial settings. We refer you to sources beyond this text for that important information.

13.1 OVERVIEW

Electrodeposition is a shortened form of "electrolytic deposition," which is the use of an electrolytic cell to deposit metal(s) or other material(s) on the target electrode. Therefore, like other electrolytic processes, electrodeposition consumes power to produce the desired product. For the moment, we focus our discussion on the electroplating of metals from aqueous solutions. Figure 13.1 illustrates the basic components of an electroplating cell. The metal of interest is deposited on the cathode. Consequently, the cathode must be conductive in order to transfer the electrons needed to reduce the metal over the entire surface of the cathode. Variation of the potential across the metal surface will affect the local deposition rate as we will see later in this chapter. To plate on nonconductive materials, such as plastic, a conductive seed layer must be deposited prior to electrodeposition of the metal.

The electrolyte contains the soluble form of the metal to be deposited. This may simply be the metal ion in solution, or may be a complex that contains that metal. A salt is frequently added to enhance conductivity. Buffers can also be added to stabilize the pH of the plating solution. What's more, the electrolyte typically contains a variety of additives, often proprietary, that influence factors such as the morphology and local deposition rate of the metal on the cathode surface.

The anode may be either consumable (soluble) or inert (insoluble), depending on the system. Ideally, a soluble anode that is made up of the same metal that is being deposited at the cathode can be used. If this is possible, the metal of interest would enter the solution due to oxidation at the anode and then deposit at the cathode. Under these ideal conditions, the average composition of the electrolyte bath would remain constant with time. The cell voltage would also be minimized since the equilibrium potential of the cell would be zero for two electrodes of the same type in the same solution. Unfortunately, these ideal conditions are rarely present in real systems. For example, the current efficiency at one or both of the electrodes is frequently less than 100%, disrupting the balance between the amount of metal added to and plated from the bath. The anode may contain impurities that remain in the bath or plate onto the cathode. Dissolution of the anode may not readily occur in the plating solution due to passivation.

An insoluble or nonconsumable inert anode may also be used. In such cases, a reaction other than dissolution of the metal to be deposited occurs, the most common of which is oxygen evolution. The metal ions in the electrolyte bath need to be regularly replenished under these conditions. The pH at the depositing electrode will also change with time unless the electrolyte is adequately buffered or unless a membrane is used to isolate the

---

Electrochemical Engineering, First Edition. Thomas F. Fuller and John N. Harb.
© 2018 Thomas F. Fuller and John N. Harb. Published 2018 by John Wiley & Sons, Inc.
Companion Website: www.wiley.com/go/fuller/electrochemicalengineering

300 Electrochemical Engineering

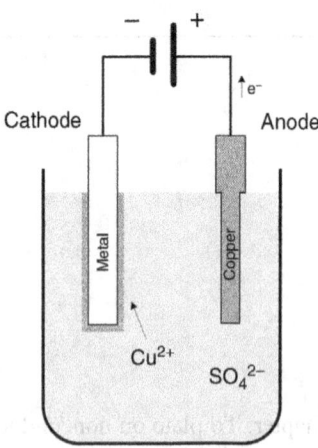

Figure 13.1 Illustration of an electroplating cell with a consumable anode.

anode compartment. Contamination of the bath can also be of concern when using this type of anode, as it is with soluble anodes.

13.2 FARADAY'S LAW AND DEPOSIT THICKNESS

An important aspect of electrodeposition is that the total amount deposited is directly related to the amount of charge passed in carrying out the desired reaction, and hence to the thickness of the plated layer. The mass of the deposit is

$$m_i = \frac{M_i Q}{nF}, \quad (13.1)$$

where Q is the charge passed due to the plating reaction. Assuming that the current density is uniform, the charge, Q, is

$$Q = \int_0^t \eta_f \, iA \, dt, \quad (13.2)$$

where i is the total current density (all reactions, not just the desired reaction) and η_f is the faradaic efficiency for the deposition reaction at the cathode. If the current density, area, and faradaic efficiency do not change with time,

$$Q = \eta_f iAt = \eta_f It, \quad (13.3)$$

where I is the current, and t is the time since the plating began. Under these conditions, the thickness of the deposited layer, L, can be calculated from the mass

$$L = \frac{m_i}{\rho_i A} = \frac{M_i \eta_f It}{\rho_i AnF}. \quad (13.4)$$

The average deposition rate is L/t.

ILLUSTRATION 13.1

A continuous sheet of copper is made by electrodeposition on a rotating drum of lead. For the conditions given below, what should be the rotation speed of the drum in revolutions per hour?

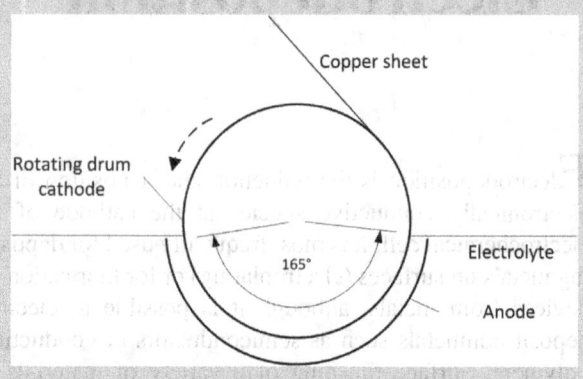

Cathode current density = 1000 [A·m^{-2}]
Faradaic efficiency = 95%
Desired thickness = 10 μm
Copper density = 8900 kg·m^{-3}
Angle of cathode immersion = 165°

In order to find the revolution speed, we need to know the length of time required to produce the deposit of the desired thickness. From Equation 13.4,

$$L = \frac{M_i \eta_f It}{\rho_i AnF} = \frac{M_i \eta_f it}{\rho_i nF}.$$

Solving this equation for t yields

$$t = \frac{L \rho_i nF}{M_i \eta_f i} = \frac{(10 \times 10^{-6})(8900)(2)(96485)}{(0.063546)(0.95)(1000)}$$

$$= 2845 \text{ seconds}$$

$$t = (3600)\left(\frac{165}{360}\right)\left(\frac{1}{\text{rotation rate}}\right)$$

$$\text{rotation rate} = (3600)\left(\frac{165}{360}\right)\left(\frac{1}{2845}\right) = 0.580 \text{ rev h}^{-1}.$$

13.3 ELECTRODEPOSITION FUNDAMENTALS

To this point, we have treated electrodeposition in a macroscopic way. However, deposit morphology and properties vary dramatically with the conditions under which the deposition takes place as illustrated in Figure 13.2, where current density is varied. In order to understand these different morphologies, we need to explore the fundamental processes that take place during

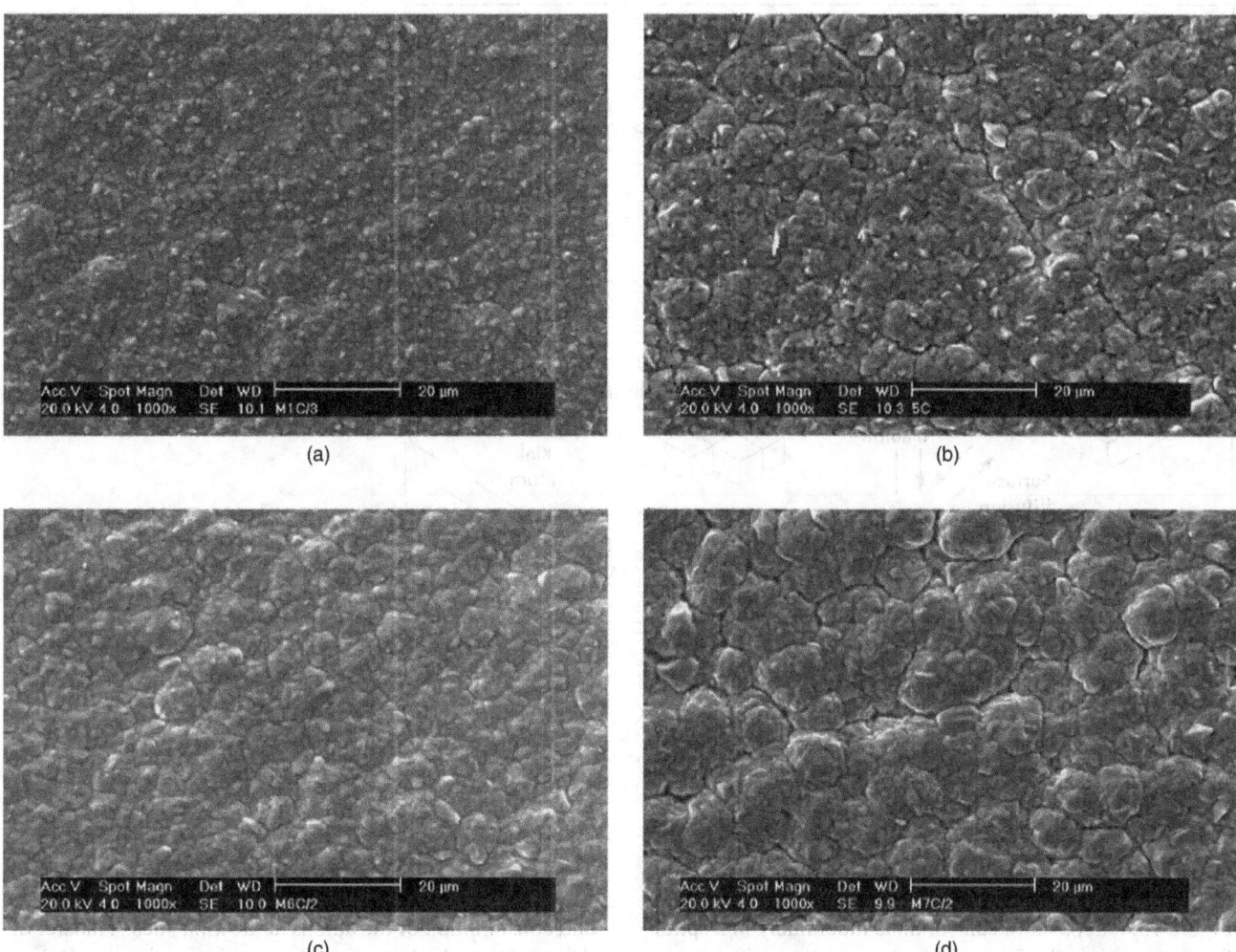

Figure 13.2 Examples of morphologies of copper deposits grown at (a) 60, (b) 1200, (c) 2400, and (d) 5000 A·m^{-2}. (From A. Ibañez and E. Fatás (2005) *Surface & Coatings Technology*, **191**, 7–16. Reprinted with permission from Elsevier.)

deposition. The purpose of this section is to explore those fundamentals at an introductory level. Electrodeposition is a rich and interesting field about which much has been written, and we regret that we are limited by space to provide only a brief introduction. Please refer to the Further Reading section at the end of this chapter and to the literature for additional information.

The charge transfer that takes place during electrodeposition follows the same processes that we have discussed previously in other chapters. Consider a hydrated cation that approaches a surface. The cation must get sufficiently close to the surface to permit electron transfer from the surface in order to be reduced. However, electrodeposition offers an additional complexity that we have not yet considered—the formation of a new phase. After reduction, the metal atom must be incorporated into a metal lattice on the surface. In this section, we consider the physical processes by which this takes place and how these processes are influenced by the deposition conditions.

As we begin our discussion, imagine a perfect metal crystal, atomically flat with no defects. Since the lattice is complete, there is no room for an additional metal atom to be incorporated. Hence, a metal ion reduced on this surface would form a surface adion that is in a higher energy state than the metal atoms that are already incorporated into the lattice. We refer to the newly adsorbed species as an *adion* because evidence indicates that it retains a partial charge and some of the hydration water molecules rather than directly forming a charge-free surface atom. If there are other adions in close proximity and sufficient energy has been supplied, it is possible to form a stable nucleus from which deposition can continue. Nucleation is an important aspect of electrodeposition and is discussed below. However, before doing that, let's consider the impact of having a real surface rather than the perfect, defect-free surface imagined above.

Real surfaces have a variety of defects, as illustrated schematically in Figure 13.3. These defects greatly reduce

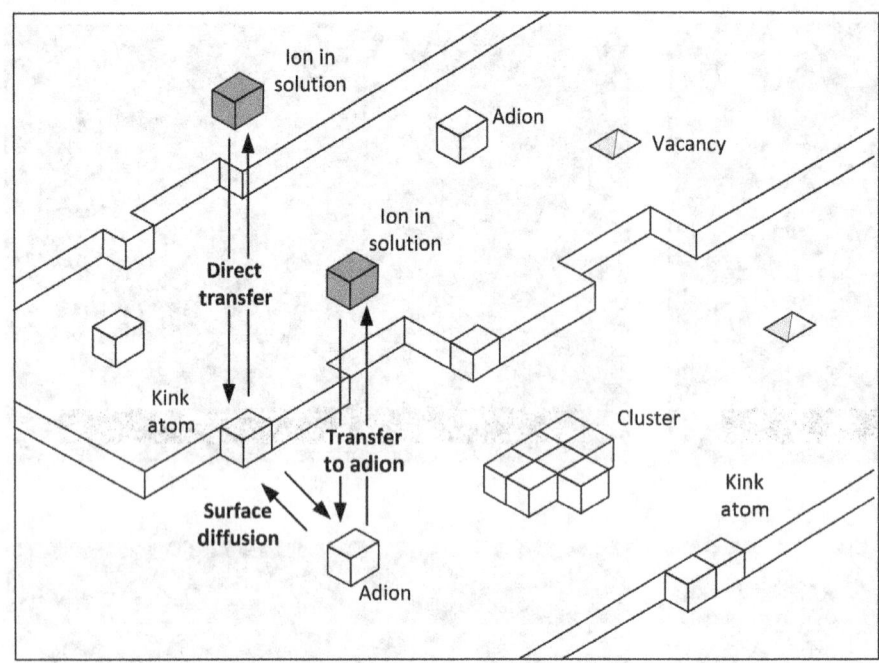

Figure 13.3 Schematic diagram of the structure of a crystal face with a simple cubic lattice. *Source:* Adapted from Budevski 1996.

the energy required for incorporation of a new atom. Integration of a metal adion into the lattice occurs primarily at kink sites (see Figure 13.3) where half of metal atom is bound to the lattice. One of the important aspects of kink sites is that another kink site is produced as a result of lattice incorporation. Therefore, kink sites are not consumed until the edge of the surface is reached. In fact, for dislocation defects, such as the screw dislocation illustrated in Figure 13.4, deposition can continue indefinitely in a spiral form as shown in the figure. In contrast, vacancy sites are no longer available for adion incorporation once filled.

Because of the availability of kink sites and other low-energy sites on real surfaces, deposition at low overpotentials takes place almost exclusively at those sites, and the nucleation of new sites does not play a role. Electron exchange takes place primarily on the flat portion of the surface, which represents the largest fraction of the total surface area and the location on the surface where less distortion of hydrated cation is required in order to get sufficiently close for electron transfer to occur. Adions then diffuse to step sites and eventually to kink sites where they are incorporated into the lattice as shown in Figure 13.3. The driving force for transport is the concentration gradient formed by the addition of adions away from the kink sites and consumption of the adions at the kink sites. Transfer of the adions to kink sites is the limiting resistance at low overpotentials where the concentration of adions on the surface is low. An increase in the overpotential leads to an increase in the adion concentration and a reduction in the transport resistance that must be overcome for incorporation into the lattice. At sufficiently high overpotentials, the process is no longer limited by adion diffusion.

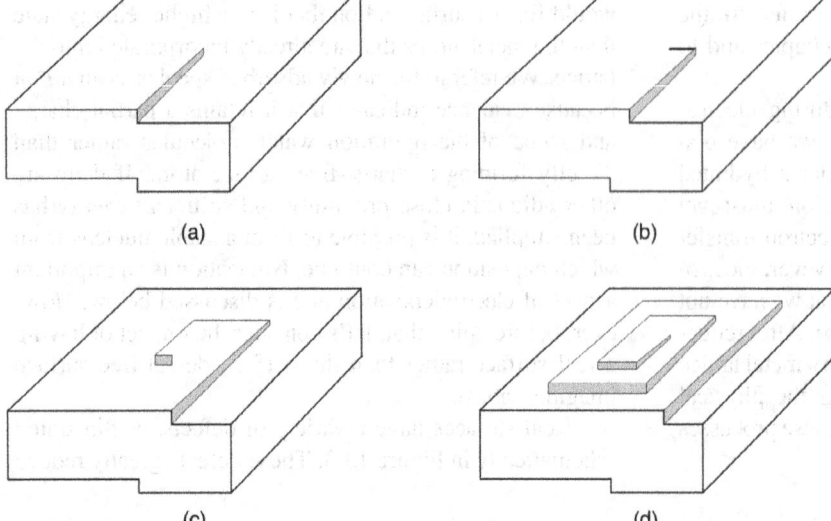

Figure 13.4 Spiral growth around a screw dislocation during deposition of a metal.

Figure 13.5 Crystallographic copper deposit showing macro-spiral growth. *Source:* Reproduced with permission of German Bunsen Society.

What does this mean from a practical perspective? At low overpotentials growth occurs only at existing sites, and the result is a crystallographic deposit such as that shown in Figure 13.5. Growth can continue until all of the available sites on the surface are consumed. Any subsequent growth requires an overpotential that is large enough for nucleation to occur.

13.4 FORMATION OF STABLE NUCLEI

The formation of new sites or nuclei from which deposition may occur is called nucleation, and it is a critical aspect of electrodeposition. The number and type of nuclei strongly influence the morphology of the deposit. Both the growth rate of the deposit and the rate of nucleation increase with increasing overpotential. If the nucleation rate is low relative to the growth rate, deposit growth will take place from relatively few nuclei and deposits with larger grains will be formed. In contrast, a high nucleation rate will lead to finer crystal grains and highly granular deposits. The manner in which the crystals grow also influences the appearance and structure of the deposit as discussed later in this chapter. The nuclei formed can be either 2D or 3D. The dimensionality depends to a large degree on the strength of the interaction between the substrate (the surface upon which the deposition is taking place) and the depositing metal, as well as the degree to which the crystal lattice of the substrate matches that of the depositing metal.

What are the factors that influence nucleation on a metal surface? The formation of a nucleus involves, of course, the transfer of metal ions from the solution to the surface. If just a single ion is transferred, it is in a higher energy state than the atoms incorporated into the metal lattice and does not remain on the surface long. In order for the metal to remain on the surface, it must move to a kink site, as discussed previously, or must combine with other adions to form a stable nucleus (the subject of this section).

The new nucleus (crystal) formed will be stable and continue to grow only if it reaches a critical dimension. To determine that critical size and the factors that influence that size, we follow the work of Budevski *et al.* (1996) and begin with the associated Gibbs energy change, which consists of two principal parts:

$$\Delta G(N) = -Nzq|\eta_s| + \varphi(N), \quad (13.5)$$

where $\Delta G(N)$ is the Gibbs energy of cluster formation for a cluster of N atoms. The first term on the right describes the change in the electron energy due to the difference in potential between the metal and the solution. It is always negative under conditions where deposition can occur. The second term, $\varphi(N)$, is the excess energy associated with the formation of the cluster. It is always positive and represents the energy that is consumed as a new phase is formed with interface boundaries between (i) the cluster and the solution and (ii) the cluster and the substrate.

Three-Dimensional Nucleation

Let's now consider the influence of the two terms in Equation 13.5 for 3D nucleation. In order to do this, we need an expression for $\varphi(N)$. We begin with the following approximation:

$$\varphi(N) = A_s \gamma. \quad (13.6)$$

A_s is the surface area of the 3D crystal and γ is the average specific surface energy [J·m^{-2}]. The surface area depends on the size of the cluster and hence on N, the number of atoms in the 3D cluster; in contrast, γ is independent of N. Our next task is to express A_s as a function of N. The volume of a cluster is directly related to the number of atoms in that cluster $\mathbb{V} = V_m N$, where V_m is the volume of one atom in the lattice. The surface area is also related to the volume. For a known 3D geometry, the relationship between A_s (proportional to L^2) and $\mathbb{V}$ (proportional to L^3) can be written as $A_s^3 = B\mathbb{V}^2$. An example of B is shown in the following illustration for a spherically shaped cluster.

ILLUSTRATION 13.2

Please determine B for a spherically shaped cluster.

Solution:

From above, $A_s^3 = B\mathbb{V}^2$. Applying this to a sphere,

$$A_s = 4\pi r^2,$$

$$\mathbb{V} = \frac{4}{3}\pi r^3,$$

$$B = \frac{(4\pi r^2)^3}{\left(\frac{4}{3}\pi r^3\right)^2} = \frac{(64)(9)\pi}{16} = 36\pi.$$

Using the relationship between A_s and $\mathbb{V}$, we can now write the desired relationship for $\varphi(N)$:

$$\varphi(N) = A\gamma = B^{1/3}\mathbb{V}^{2/3}\gamma = \left(B^{1/3}V_m^{2/3}\gamma\right)N^{2/3}. \quad (13.7)$$

With this, Equation 13.5 for 3D nucleation becomes

$$\Delta G(N) = -Nzq|\eta_s| + \left(B^{1/3}V_m^{2/3}\gamma\right)N^{2/3}. \quad (13.8)$$

We can now plot the Gibbs energy of cluster formation as a function of the number of atoms in a cluster, Figure 13.6. This plot shows some interesting and important results. First of all, the surface energy term (Equation 13.7) is positive and dominates at low values of N, while the negative term associated with the difference in overpotential dominates at high values of N. This means that $\Delta G(N)$ goes through a maximum at a particular value of N, which we will call N_{crit}. For cluster sizes below N_{crit}, an increase in the cluster size is accompanied by an increase in $\Delta G(N)$. The opposite is true for N values above N_{crit}, where an increase in the cluster size reduces the free energy. Therefore, spontaneous growth is energetically favorable for cluster sizes greater than N_{crit}, but not for cluster sizes smaller than the critical value. The two curves in Figure 13.6 correspond to two different values of the overpotential. As seen in the figure, at higher values of η_s the cluster energy curve shifts down and to the left, which corresponds to a lower value of N_{crit}. In other words, the critical cluster size for stable growth decreases with increasing (negative) overpotential, and deposits can undergo stable growth from smaller seed clusters. Succinctly stated, nucleation is favored at higher overpotentials.

An expression for N_{crit} can be derived by differentiating Equation 13.8, setting the derivative equal to zero, and solving for N_{crit}:

$$N_{crit,3D} = \frac{8BV_m^2\gamma^3}{27(zq|\eta_s|)^3}. \quad (13.9)$$

For the parameters used in Figure 13.6, application of this equation yields N_{crit} values of 83 and 10 atoms, respectively, consistent with the maximum value for each of the curves in the figure. The Gibbs energy of cluster formation at the critical cluster size is

$$\Delta G_{crit,3D} = \frac{4BV_m^2\gamma^3}{27(zq|\eta_s|)^2} = \frac{N_{crit}zq|\eta_s|}{2}. \quad (13.10)$$

This quantity plays an important role in determining the nucleation rate as we shall see below.

Two-Dimensional Nucleation

Let's now consider the influence of the two terms in Equation 13.5 for 2D nucleation. In contrast to 3D nucleation where growth occurs in all directions, 2D growth takes place only on the edges of the nucleus so that the growth is outward and not upward. Therefore, the expression for $\varphi(N)$ involves the perimeter rather than the surface area as follows:

$$\varphi(N) = P\varepsilon. \quad (13.11)$$

P is the perimeter of the 2D crystal and ε is the average specific edge energy. As ε is independent of N, our next task, analogous to the procedure followed above, is to express P as a function of N, the number of atoms in the 2D cluster. The area of the cluster is directly related to the number of atoms in the cluster $A_s = \Omega N$, where Ω is the area of one atom on the surface. For a known 2D geometry, the relationship between P (proportional to L) and A_s (proportional to L^2) can be written as $P^2 = 4bA_s$, where b is the constant of proportionality. From this it follows that

$$\varphi(N) = P\varepsilon = 2\varepsilon\sqrt{b\Omega N}. \quad (13.12)$$

With this, Equation 13.5 for 2D nucleation becomes

$$\Delta G(N) = -Nzq|\eta_s| + 2\varepsilon\sqrt{b\Omega N}. \quad (13.13)$$

The corresponding 2D expressions for N_{crit} and ΔG_{crit} are as follows:

$$N_{crit,2D} = \frac{b\Omega\varepsilon^2}{(zq|\eta_s|)^2} \quad (13.14)$$

$$\Delta G_{crit,2D} = \frac{b\Omega\varepsilon^2}{zq|\eta_s|} = N_{crit}zq|\eta_s|. \quad (13.15)$$

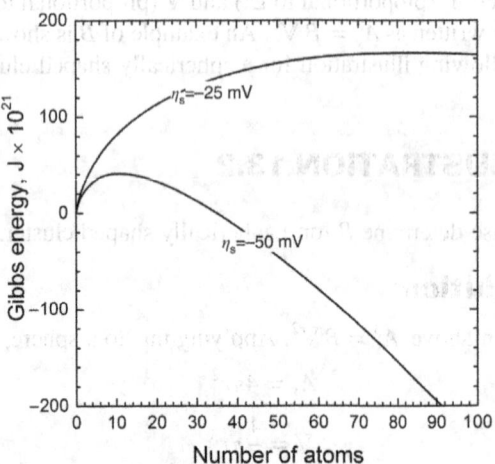

Figure 13.6 Gibbs energy of cluster formation as a function of the number of atoms in the cluster. *Parameters:* Ag atom (0.1444 nm radius), $\gamma = 0.1$ J·m^{-2}, $z = 1$, atoms and clusters assumed spherical.

The behavior for two-dimensional nucleation is qualitatively similar to that shown above for 3D clusters. In two-dimensional growth, metal deposition takes place in the form of 2D monoatomic layers. This type of nucleation is important for deposition onto foreign metal substrates that have a strong metal–substrate interaction, and it can lead to underpotential deposition (UPD), which is 2D deposition at potentials below that needed for sustainable deposition onto a native metal substrate. UPD is also important in the absence of growth sites on single crystal faces of the native metal.

ILLUSTRATION 13.3

The calculations in Figure 13.6 assumed 3D spherical deposition of Ag. For the same conditions shown in Figure 13.6, $r = 0.1444$ nm, $\gamma = 0.1$ J·m^{-2}, $z = 1$, what value of the overpotential would be required to reduce the critical cluster size to five silver atoms?

Solution:

From Equation 13.9,

$$N_{crit,3D} = \frac{8BV_m^2 \gamma^3}{27(zq|\eta_s|)^3},$$

$$N_{crit,3D} = 5, \quad B = 36\pi, \quad q = 1.602 \times 10^{-19} \text{ C}$$

$$V_m = \frac{4\pi}{3} r^3 = \frac{4\pi}{3}(1.444 \times 10^{-10})^3 = 1.261 \times 10^{-29} \text{ m}^3,$$

$$\eta_s = 0.064 \text{ V}.$$

In this section, we explored the energy change associated with the formation of nuclei during electrodeposition. We have seen that a critical cluster or nucleus size is required for stability. That critical size is strongly dependent on the applied overpotential, where higher values of the overpotential reduce the cluster size required for stable growth. Having examined the conditions required for stability, we now explore to the rate at which nuclei are formed.

13.5 NUCLEATION RATES

Above we treated the thermodynamics associated with stable cluster formation. We now turn our attention to the rate at which nuclei are formed. In doing this, we restrict ourselves to the classical expressions for the nucleation rate attributed to Volmer and Weber, noting that similar expressions can be derived from atomistic theory. Please refer to the references at the end of this chapter for additional information.

Nucleation is a probabilistic process that can be expressed as a function of the critical energy of formation for the cluster as follows:

$$J = A_0 \exp\left(-\frac{\Delta G_{crit}}{kT}\right), \qquad (13.16)$$

where k is Boltzmann's constant, and the pre-exponential factor, A_0, is approximately constant. The nucleation rate, J, has units of cm^{-2}·s^{-1} (in practice, [cm^{-2}·s^{-1}] are used rather than [m^{-2}·s^{-1}]). It represents the rate per area of nuclei formed on the surface of interest. Substituting the expression for ΔG_{crit} from Equation 13.10, the nucleation rate becomes

$$J = A_{3D} \exp\left(-\frac{4BV_m^2 \gamma^3}{27(zq|\eta_s|)^2 kT}\right). \qquad (13.17)$$

The equivalent expression for 2D nucleation is

$$J = A_{2D} \exp\left(-\frac{b\Omega\varepsilon^2}{zq|\eta_s|kT}\right). \qquad (13.18)$$

The dependence of the nucleation rate on the overpotential is evident from Equations 13.17 and 13.18, and it is different for 3D and 2D nucleation. In both cases, the nucleation rate is independent of time. This will be the case in practice as long as the nuclei are sufficiently separated that they do not influence the formation of additional nuclei. Once a sufficient number of nuclei are formed, growth at the existing nuclei will be favored over the formation of new nuclei from adions on the surface.

The strong impact of the overpotential is readily apparent from a plot of the nucleation rate as a function of η_s. As shown in the Figure 13.7, the nucleation rate increases rapidly over a relatively narrow range of overpotentials (for this case, -0.080 to -0.105 V). In contrast, the nucleation rate at low overpotentials is essentially zero, and growth takes place only at existing sites as discussed

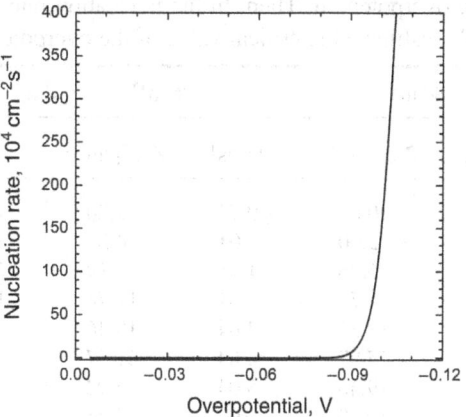

Figure 13.7 The rate of nucleation as a function of overpotential showing the strong dependence on the overpotential.

306 Electrochemical Engineering

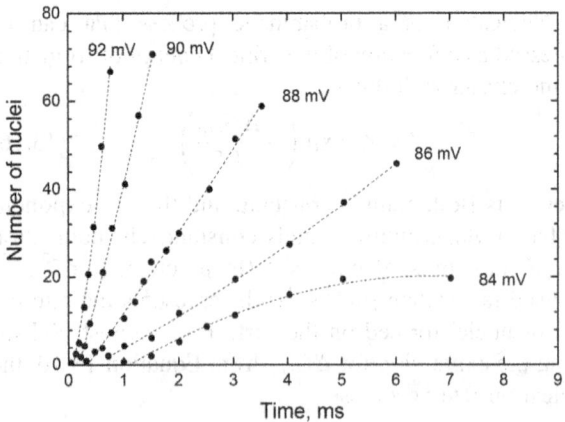

Figure 13.8 The number of nuclei versus time for electrodeposition of mercury on Pt at different overpotentials. *Source:* Adapted from Toshev 1969.

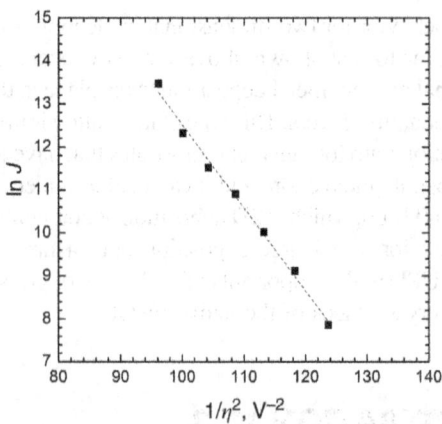

Figure 13.9 Determination of ΔG_{crit} from a plot of $\ln J$ versus $1/\eta^2$.

previously. In practice, the value of the overpotential at which the nucleation rate rises sharply is sometimes referred to as $|\eta_{crit}|$. For our purposes, we define η_{crit} to be the value of the overpotential at which the nucleation rate is equal to 1 ($J = 1\,\text{cm}^{-2}\cdot\text{s}^{-1}$ or $10^4\,\text{m}^{-2}\cdot\text{s}^{-1}$). It can be determined from Equation 13.17 or 13.18 if the parameters are known, or from experimental data.

Given that J, the rate of nucleation, does not change with time, the number of nuclei formed increases linearly with time. If Z_{nucl} is the number of nuclei per area, then

$$\frac{dZ_{nucl}}{dt} = J. \qquad (13.19)$$

Therefore, J can be measured experimentally by counting the number of nuclei formed in a given period of time. One method of doing this is to apply an overpotential above $|\eta_{crit}|$ for a specific period of time and then drop the potential below $|\eta_{crit}|$, where the nucleation rate is essentially zero. The nuclei can then be grown to a size where they can be more easily seen and counted. This process can be repeated for various times at each of several values of the overpotential to yield curves such as those shown in Figure 13.8. Note that there is a short incubation period before the expected linear growth of the number of nuclei with time is observed.

With these data, assuming 3D nucleation, it is now possible to determine ΔG_{crit} as a function of the overpotential from a plot of $\ln J$ versus $1/\eta_s^2$ as shown in Figure 13.9 and demonstrated in Illustration 13.4. A similar procedure can be done for systems that exhibit

ILLUSTRATION 13.4

The following data present the number of nuclei present as a function of time for mercury on a platinum substrate (*Ber. Bunsenges. Phys. Chem.* **73**, 184 (1969).) Please use these data to determine the nucleation rate at each value of the overpotential. Then, fit the nucleation rate data to the expression for the 3D nucleation rate (Equation 13.17). Finally, estimate the critical value of the overpotential.

84 mV		86 mV		88 mV		90 mV		92 mV	
t [ms]	Z_{nuc} [cm^{-2}]	t [ms]	Z_{nuc} [cm^{-2}]	t [ms]	Z_{nuc} [cm^{-2}]	t [ms]	Z_{nuc} [cm^{-2}]	t [ms]	Z_{nuc} [cm^{-2}]
1.00	0.61	0.72	2.00	1.21	14.34	0.22	1.77	0.05	0.37
1.50	2.00	1.01	4.33	1.37	18.99	0.29	4.33	0.13	2.47
2.01	5.26	1.51	6.19	1.50	23.42	0.38	9.45	0.19	5.03
2.52	8.75	2.01	11.78	1.77	25.98	0.62	21.09	0.29	13.18
3.03	11.31	3.04	19.46	2.01	29.93	0.79	31.10	0.36	20.62
4.04	15.97	4.04	27.37	2.57	39.94	1.03	41.11	0.45	31.33
5.02	19.46	5.04	36.92	3.04	51.35	1.28	56.70	0.60	49.72
7.02	19.69	6.02	45.76	3.53	58.80	1.54	70.67	0.77	66.71

Solution:

The rate of nucleation at each overpotential is obtained by fitting each dataset to a line. The nucleation rate, J, is the slope of the line (cm^{-2}·s^{-1}). These are the data from Figure 13.8. The corresponding J values are shown in the table. Note that time was converted from "ms" to "s" to get the proper units on J. Also note the large magnitude of the nucleation rate.

η_s [V]	J [cm^{-2}·s^{-1}]
0.084	3,532
0.086	8,263
0.088	18,735
0.090	52,284
0.092	97,477

To fit the data to Equation 13.17, we plot $\ln J$ versus $1/\eta_s^2$. The slope of this line is -0.1439, and the intercept is 28.512. This yields a value for A_{3D} of 2.41×10^{12}. Therefore, the expression for J is

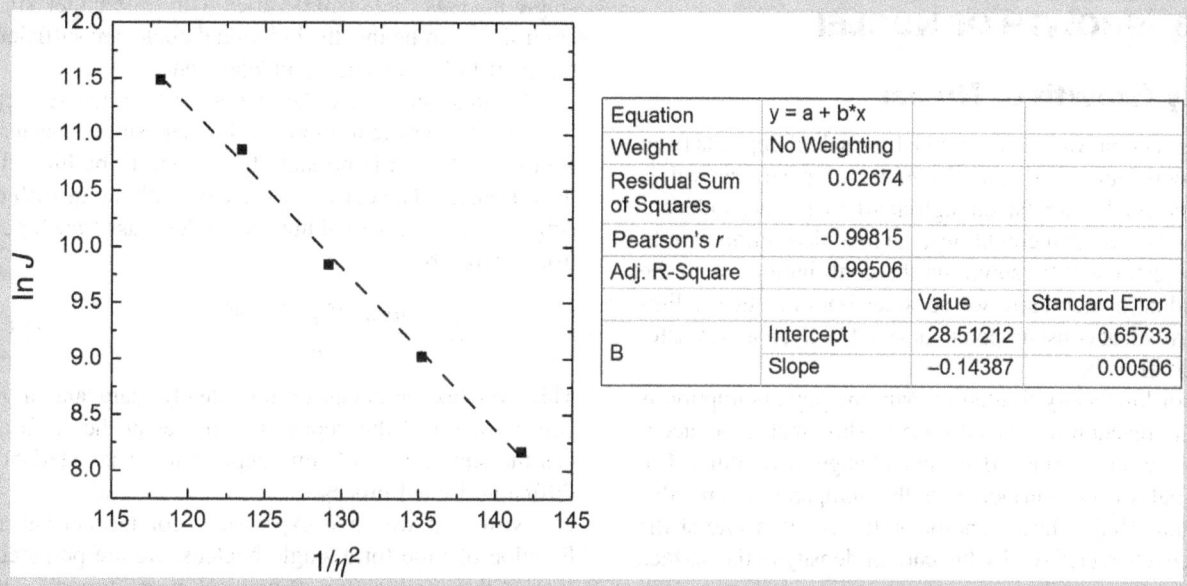

$$J = 2.41 \times 10^{12} \exp\left(\frac{-0.1439}{\eta_s^2}\right).$$

Finally, η_{crit} can be found by taking the natural logarithm of both sides, setting J equal to 1, and solving for η_s.

$$\ln J = \ln(2.41 \times 10^{12}) - \frac{0.1439}{\eta_s^2},$$

$$\eta_{crit} = \sqrt{\frac{0.1439}{28.512}} = 0.071 \text{ V}.$$

2D nucleation. In addition, N_{crit} can be estimated from the following equation:

$$\frac{d \ln J}{d|\eta|} = -\frac{1}{kT}\frac{d \Delta G_{crit}}{d|\eta|} = N_{crit}\frac{zq}{kT} = N_{crit}\frac{zF}{RT}. \quad (13.20)$$

Physically, this represents the tangent to the curve at the point of interest since a plot of $\ln J$ versus $|\eta_s|$ is not linear. While this equation "stretches" the continuum approximation, since it does not account for the atomistic nature of the clusters that becomes increasingly important with decreasing size, it has been shown to give estimates of N_{crit} in good agreement with those from atomistic models.

In this section, we learned that the rate at which nuclei are formed is also strongly dependent on the overpotential. Above a certain critical overpotential, the nucleation rate increases very rapidly. This will be important as we examine deposition on a more macroscale. First, let's investigate the early stage growth of nuclei.

13.6 GROWTH OF NUCLEI

Early Growth of Nuclei

In this section, we examine initial growth of deposits from distinct nuclei. The distinguishing factor in early growth is that the nuclei are far enough apart to behave independently. Under such conditions, the total deposition current can be estimated by summing the contributions from the individual nuclei. Here we consider both kinetically limited growth at constant potential and, briefly, mass-transfer limited growth.

For kinetically limited growth, the key assumption is that the reaction rate is sufficiently slow that the concentration at the surface does not change with time. The potential of the surface and the temperature are also constant. Under these conditions, the reaction rate at the surface, characterized by the current density at the surface (i_{surf}), is constant. The total current, however, changes with time as the nuclei grow. Considering a 2D single nucleus that is cylindrical in shape with a height h and a radius r, only the sides of the cylinder are active. We begin with a mass balance where we express the change in mass of the seed particle in terms of the current density at the surface. The 2D growth is characterized by an increase of the radius with time.

$$\frac{dm_i}{dt} = \rho_i \frac{d\mathbb{V}}{dt} = \rho_i(2\pi hr)\frac{dr}{dt} = \frac{M_i i_{surf}}{nF}(2\pi hr). \quad (13.21)$$

We can use this balance to solve for the rate of change of the radius:

$$\frac{dr}{dt} = \frac{M_i i_{surf}}{\rho_i nF}. \quad (13.22)$$

Assuming that $r \approx 0$ at $t = 0$, this expression can be integrated to yield

$$r = \frac{M_i i_{surf}}{\rho_i nF}t. \quad (13.23)$$

The current, I_s, from a single nucleus is simply the current density at the surface multiplied by the surface area of the growing nucleus (in this case a cylinder growing in the radial direction):

$$I_s = i_{surf}(2\pi rh) = \frac{2\pi h M_i i_{surf}^2}{\rho_i nF}t. \quad (13.24)$$

Following a similar procedure, the analogous expression for a 3D hemispherical nucleus growing under kinetic control can be derived:

$$I_s = i_{surf}(2\pi r^2) = 2\pi i_{surf}^3 \left(\frac{M_i}{\rho_i nF}\right)^2 t^2. \quad (13.25)$$

Equations 13.24 and 13.25 show how the current from a single nucleus (2D or 3D) varies with time under kinetic control, assuming that the individual nuclei are sufficiently far apart to be considered independent.

In situations where the reaction rate at the surface is fast and the concentration of the depositing species in solution is low, it is possible for growth to be limited by mass transfer. The current associated with the growth of a single 3D hemispherical nucleus under mass-transfer control is given by

$$I_s = \frac{nF\pi(2D_i c_i^\infty)^{1.5}M_i^{0.5}}{\rho_i^{0.5}}t^{0.5}. \quad (13.26)$$

This equation assumes pseudo-steady state and a zero concentration of the depositing species at the surface. It has the square root of time dependence characteristic of diffusion-limited processes.

Now that we have expressions for the current as a function of time for a single nucleus, we are prepared to look at the combination of nucleation and growth. There are two important cases to consider.

For *instantaneous nucleation*, nucleation takes place much faster than growth. Under such conditions, the current from the entire surface is simply the product of the single-nucleus current (Equations 13.24–13.26) and the number of nuclei instantaneously formed, assuming, once again, that the nuclei are sufficiently far apart that they do not interact.

For *progressive nucleation*, nucleation and growth take place on a similar timescale. Therefore, the current due to independent nuclei increases as a function of time due to the increase in surface area of existing nuclei and the formation of new nuclei. For a known constant nucleation rate J, the relevant expressions follow. For 2D kinetically controlled growth and progressive nucleation,

$$I = A \cdot \frac{J\pi h M_i \, i_{surf}^2}{\rho_i n F} t^2. \quad (13.27)$$

Similarly, for a 3D hemispherical nucleus growing under kinetic control with progressive nucleation,

$$I = A \cdot \frac{2J\pi \, i_{surf}^3}{3} \left(\frac{M_i}{\rho_i n F}\right)^2 t^3. \quad (13.28)$$

Finally, for mass-transfer limited growth with progressive nucleation,

$$I = A \cdot \frac{2J\pi n F (2 D_i c_i^\infty)^{1.5} M_i^{0.5}}{3 \rho_i^{0.5}} t^{1.5}. \quad (13.29)$$

Equations 13.27–13.29 are for the total current on the surface with a geometric area A and include the combined influence of nucleation and growth. The illustration below considers the difference between instantaneous nucleation and progressive nucleation for a situation where the total number of nuclei at the end of the time period of interest is the same. Although this is an artificial constraint, it illustrates the difference between the two situations.

Finally, we end this section by noting that the above expressions apply during the early stages of electrodeposition, where a combination of the observed time dependence of the growth and microscopic examination can be used in tandem to identify and quantify the processes that control deposit growth.

ILLUSTRATION 13.5

In this illustration, we compare the current that results from instantaneous and progressive nucleation for the growth of 3D hemispherical nuclei where the number of nuclei at the end of the growth period is the same. Assume that the nucleation rate, J, is $10{,}000 \, cm^{-2} \cdot s^{-1}$, and that growth takes place for 15 seconds. Assume that the geometric area (superficial surface area) is $1 \, cm^2$. The following additional parameters are known:

$$M_i = 107.87 \, g \cdot mol^{-1};$$
$$\rho_i = 10.49 \, g \cdot cm^{-3};$$
$$n = 1, \, i_{surf} = 0.02 \, A \cdot cm^{-2}.$$

Solution:

First, we need to determine the total number of nuclei at the end of the growth period so that we can use this same number of nuclei for instantaneous nucleation.

For a nucleation rate of $10{,}000 \, cm^{-2} \cdot s^{-1}$, an area, A, of $1 \, cm^2$ and a growth time of 15 seconds, the number of nuclei at the end of the growth period assuming progressive nucleation is $AJt = (1)(10{,}000)(15) = 150{,}000$. We will use this same number of nuclei for the instantaneous calculation as follows:

$$I = N_0 I_s = N_0 \left[2\pi \, i_{surf}^3 \left(\frac{M_i}{\rho_i n F}\right)^2 t^2 \right],$$

where N_0 is the number of nuclei formed instantaneously. For progressive nucleation, we use Equation 13.28, where nucleation takes place simultaneously during growth. The resulting currents are shown in the plot.

Even though the number of nuclei is the same after 15 seconds, the current is less for progressive nucleation because the nuclei are at all stages of growth, in contrast to the instantaneous situation where all of the nuclei have been growing for the entire 15-second period.

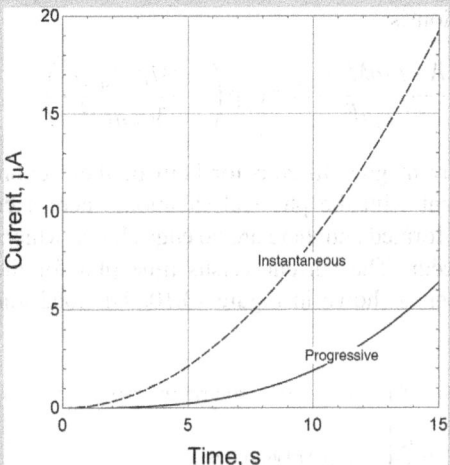

A quick check of the assumption of independent nuclei is appropriate. If we assume that the nuclei are evenly spaced over the $1 \, cm^2$ area, the spacing between nuclei is about $1/\sqrt{150{,}000} = 0.0026 \, cm$. A nucleus that grew for the entire 15 seconds would have a radius of

$$r = \frac{M i_{surf}}{\rho_i n F} t = \frac{(107.87)(0.02)}{(10.49)(1)(96485)} (15) = 3.2 \times 10^{-5} \, cm,$$

which is much less than the average spacing between nuclei. Independent growth appears to be a reasonable assumption.

Interaction between Growing Nuclei

All of the above equations that describe current as a function of time for early deposit growth predict a continuing increase in the current due to an increase in the size of the nuclei and, for progressive nucleation, the addition of more nuclei. Such an increase, of course, is not sustainable since nuclei eventually overlap and coalesce. We illustrate the concept of overlap here by providing an expression for 2D deposit growth under conditions of both instantaneous and progressive nucleation. As you will remember, 2D growth occurs in a layer-by-layer fashion due to growth at the edges of the nuclei only (growth out, but not up); hence, growth stops once the layer is complete. Paunovic and Schlesinger (2006) provide expressions for 2D growth with overlap. For instantaneous nucleation,

$$I = \frac{2\pi h M_i N_0 \, i_{surf}^2}{\rho_i n F} t \exp\left\{\frac{-\pi M_i^2 (N_0/A) i_{surf}^2 t^2}{(\rho_i n F)^2}\right\}. \quad (13.30)$$

Note that this is the total current since it has been multiplied by N_0. The analogous expression for progressive nucleation is

$$I = \frac{A \cdot J \pi h M_i \, i_{surf}^2}{\rho_i n F} t^2 \exp\left(-\frac{\pi M_i^2 J i_{surf}^2 t^3}{3(\rho_i n F)^2}\right). \quad (13.31)$$

The current goes to zero for both of these expressions, consistent with the physical situation where a complete layer is formed and there are no edges left at which growth may occur. The current versus time plot for these two situations is shown in Figure 13.10. The total number of nuclei was artificially constrained to be equal at the end of 80 seconds in order to illustrate the difference between the two types of nucleation and growth with overlap. The nucleation rate was taken to be $100{,}000\,\text{cm}^{-2}\cdot\text{s}^{-1}$. The other parameters are the same for both cases and correspond to Ag metal. The current initially varies linearly for instantaneous nucleation, as described by Equation 13.24. As before, progressive nucleation lags behind as expected. It peaks later and persists longer, but the magnitude of its peak current remains lower than that resulting from instantaneous nucleation.

Qualitatively similar behavior is expected for 3D nucleation, although prediction is much more difficult since the rates of nucleation and of growth in each of the coordinate directions are influenced by the different crystal planes and the relative growth rates of those planes, which make a variety of different morphologies or "textures" possible. Expressions that account for overlap under mass-transfer limited conditions are available in the literature, but are less important since electrodeposition is not typically performed under mass-transfer control.

13.7 DEPOSIT MORPHOLOGY

Description of the deposition process rapidly becomes more complex as we move from the isolated nuclei of initial growth to interacting growth sites and, finally, coalescence and layer growth. The fundamental processes described above, however, provide a basis for understanding the deposit morphologies that are observed in practice. They also help us to understand ways in which additives may influence the deposition process as we will see later in this chapter.

The role of the fundamental processes in determining the deposit morphology is a strong function of the surface overpotential and, therefore, current density. In this section, we make some generalizations regarding that relationship. In doing so, we note that there are a number of subtleties not factored into these generalizations that can have an important impact on the morphology and other physical properties of the deposit. For example, growth rate differences or differences in the reaction mechanisms on individual crystal planes can make a huge difference on the size and orientation of crystal grains in the resulting deposit. Also, our generalizations do not explicitly consider the impact of additives in the plating bath, which are discussed later in the chapter.

At low overpotentials, deposition is limited by surface diffusion of adions to preferred sites as there is insufficient energy for nucleation. The adion concentration on the surface is low, as is the concentration gradient for surface diffusion. Growth is primarily at kink sites, and the

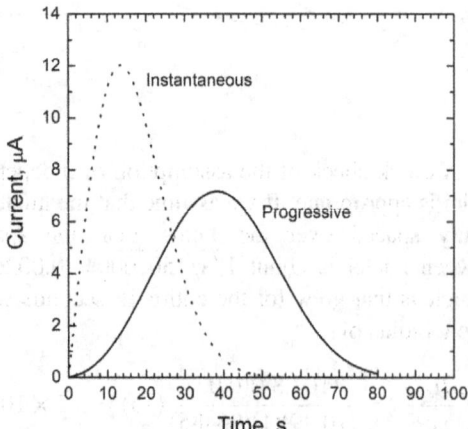

Figure 13.10 The current as a function of time for 2D growth with overlap under the assumption of both instantaneous and progressive nucleation. $J = 100{,}000\,\text{cm}^{-2}\cdot\text{s}^{-1}$ for progressive nucleation. The total numbers of nuclei are equal and the area is 1 cm^2. Other parameters correspond to Ag.

structure of the imperfect crystal surface is critical in providing those sites. Deposits are almost always polycrystalline, although crystal sizes can be large; crystal structures are well-formed.

As the overpotential is increased, the adion concentration increases and growth may occur at less energetically favorable sites (e.g., step sites). Growth is still crystallographic, and may result in bunched layers and ridges. As the overpotential is increased further, nucleation becomes important and deposit growth is no longer dependent on existing growth sites. As described previously, nucleation increases sharply with overpotential leading to deposits with large numbers of crystal grains because of the additional sites available for deposit growth. Nucleation occurs at multiple sites and the resulting crystallites coalesce to form the deposit. Defects are present in the crystallographic structure of the deposit due to coalescence of the distinct crystallites or grains. The relative rates of nucleation and growth determine the number of crystals and hence the granularity of the deposit. The large number of crystal grains may persist as the deposit continues to grow, leading to a hard, rigid polycrystalline deposit. In contrast, nonuniformities in the growth rate of different crystal planes can result in the formation of larger crystals as the deposit thickens, and may even lead to columnar deposits. Deposits with fewer large grains tend to be more ductile. Coalescence to form a complete layer is typically complete by the time a thickness of about 10 nm has been reached.

Mass transfer of the reactant species to the surface from the bulk eventually becomes important as the overpotential, and hence the rate of the reaction at the surface, continues to increase. Deposit growth under mass-transfer control leads to roughened deposits that are often powdery, and thus, unsatisfactory from a metal finishing perspective. For this reason, electrodeposition is typically performed at only a fraction of the limiting current. Dendritic growth is also possible under mass-transfer limiting conditions where deposition occurs preferentially at protrusions due to higher rates of mass transfer to those locations.

The above generalizations are illustrated in Figure 13.11. As shown, side reactions such as hydrogen evolution also become important as the cathodic overpotential is increased. While the figure shows hydrogen evolution following the mass-transfer plateau, this is certainly not always the case. For example, improvement of mass transfer in the cell due to the addition or increase of convection may raise the mass-transfer limit so that hydrogen evolution occurs prior to reaching that limit. Typical faradaic efficiencies for deposition approach 90%, although some processes, such as chromium deposition, have much lower efficiencies (20% or less). Hydrogen evolution is the principal side reaction that occurs during the electrodeposition of metals from aqueous solutions.

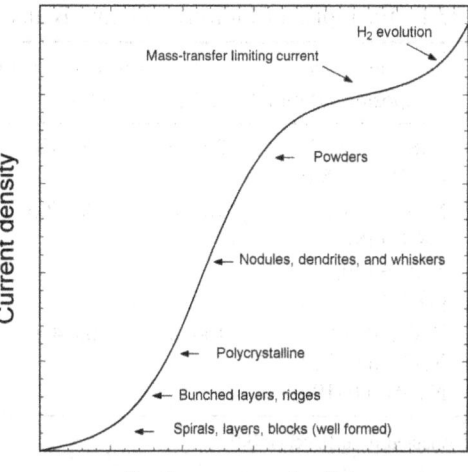

Figure 13.11 Generalization of deposit morphology as a function of overpotential and current. *Source:* Adapted from Pletcher and Walsh.

According to Pletcher, industrially electrodeposited layers range, in general, between 0.01 and 100 μm, although most are 1–10 μm thick. Adhesion and mechanical properties are also important so that the best deposits are not necessarily those formed at the lowest current density. For example, the adhesion of small-grained polycrystalline deposits is frequently better than that of large-grained deposits. For electrodeposition, average current densities range from 100 to 1000 $A \cdot m^{-2}$, although higher rates are possible with some baths. Most electrodeposition is performed with a DC power supply at constant current. Pulse plating, however, is important for special applications, but is beyond the scope of this chapter. Since good quality deposits are formed at current densities well below the mass-transfer limit, convection is an important aspect of electrodeposition cell design. Note that the methods described in Chapter 4 can be used to estimate the mass-transfer limiting current.

13.8 ADDITIVES

Additives have been and continue to be a critical component of industrial plating baths. Most additive work has been done empirically, although recent years have seen increased fundamental understanding of these complex systems. A detailed treatment of additive effects is left to more specialized texts. Instead, we provide just a brief introduction to this important topic.

Additives act in a variety of ways to influence the deposition process. For example, additives may complex with ions in solution and influence solution stability,

Table 13.1 Examples of Industrial Plating Baths and Conditions

Metal	Electrolyte composition [kg·m^3]	T [°C]	Current density [A·m^{-2}]	Additives	Anode	Current efficiency (%)
Cu	CuSO$_4$ (200–250) H$_2$SO$_4$ (25–50)	20–40	200–500	Dextrin, gelatin, S-containing brighteners, sulfonic acids	P-containing rolled Cu	95–99
Ni	NiSO$_4$ (250) NiCl$_2$ (45) H$_3$BO$_3$ (30) pH 4–5	40–70	200–500	Coumarin, saccharin, benzenesulfonamide, acetylene derivatives	Ni pellets or pieces	95
Zn	ZnO (20–40) NaCN (60–120) NaOH (60–100)	15–30	100–400	Glycerin, organic additives	Zn	70–90

Source: Adapted from Pletcher 1993.

deposition potentials and rates, and reaction mechanisms. Additives can also affect the stability of surface layers and prevent, for example, the formation of passive layers on anodes.

Organic additives act by adsorbing onto the deposit surface and can impact deposition by changing, for example, the concentration of available growth sites, the concentration of adions, and the rate of surface diffusion of adions. Such changes may lead to an increase in the local adion concentration, which may enhance nucleation and significantly change the structure and properties of the deposit. Preferential deposition of additives at specific sites may dramatically alter the growth mechanism (e.g., by blocking preferred growth sites) and the resulting deposit morphology. Industrially, organic additives can be classified as follows:

Brighteners: These are additives that lead to bright deposits by reducing the deposit roughness so that light is reflected rather than scattered. One way to reduce roughness is to enhance nucleation and randomize deposition in order to form small grains uniformly over the entire surface.

Levelers: Levelers control deposit roughness on a more "macroscale." For example, some levelers act by preferentially adsorbing onto areas of high deposition, such as areas favored by diffusion, leading to a reduced rate of deposition at those areas.

Structure Modifiers: These additives can change the crystal orientation of the deposit or otherwise impact the structure and stresses in the deposit.

Wetting Agents: Wetting agents function by reducing the surface tension in order to prevent hydrogen bubbles from adhering to the surface and negatively impacting the deposit by the formation of pits under bubbles or by the entrapment of bubbles and possible hydrogen embrittlement. The wetting agent accelerates the release of hydrogen from the surface before bubbles of sufficient size to cause damage can be formed.

The above classification is not used universally. For example, in the semiconductor industry, the plating of copper to fill high-aspect ratio trenches is enabled by plating additives that are typically classified as *accelerators*, *suppressors*, and *levelers*. Independent of how they are classified, additives play a critical role in electrodeposition by interacting with the fundamental processes, either in the bath or on the deposit surface, that determine deposit growth and morphology. Table 13.1 provides an example of plating baths used industrially.

13.9 IMPACT OF CURRENT DISTRIBUTION

In the above discussion, we have implicitly characterized deposition by a single overpotential and rate for the surface of interest. In practice, however, the rate varies over the surface of the piece being plated, as illustrated schematically in Figure 13.12. To avoid the problems associated with deposition under mass-transport control, practical systems typically operate at a fraction of the limiting current. Because we are operating well below the limiting current, local deposition rates are governed by the potential field and described by the secondary current distribution.

Current Distribution and Wagner Number

For the Tafel region most relevant to industrial electrodeposition, the uniformity of the current distribution is characterized by the Wagner number (Wa) defined in

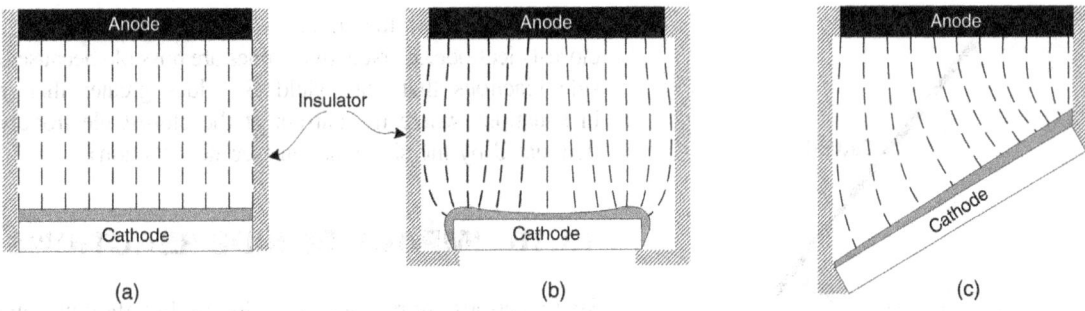

Figure 13.12 Schematic illustration of the impact of the current distribution on plating uniformity.

Chapter 4 as

$$\text{Wa} = \frac{RT\kappa}{FL_c} \frac{1}{|i_{avg}|\alpha_c}. \qquad (4.65)$$

This dimensionless group represents the kinetic resistance divided by the ohmic resistance. Higher values of Wa correspond to a more uniform current distribution. The factors that influence the uniformity of the current distribution are those reflected on the right side.

The conductivity of the solution is one of the primary factors that influence the uniformity of the current distribution. The conductivity, in turn, is influenced by the concentration of ions in solution, the temperature of the electrolyte, and the degree of complexation of the electrolyte. Concentrations of metal ion are often limited by solubility, and salts are added to enhance solution conductivity. The cost of precious metals leads to the use of reduced concentrations of these ions and lower solution volumes than what otherwise might be used. Effluent considerations (e.g., disposal or treatment) may also influence metal ion concentrations, as well as the types of ions used in solution. Complexation reduces the charge and increases the size of the ions being transferred, and thus has a substantial influence on conductivity. Although temperature influences the conductivity, the impact of temperature on the rate of the desired reaction and side reaction(s) is probably more important.

As discussed in Chapter 4, the length over which the current must travel in solution to different locations on the work piece is critical as even a small difference in the overpotential can cause a substantial difference in the local reaction rate (see Figure 13.12). Strategies such as the use of multiple anodes can be used to make the current density more uniform.

Deposit uniformity decreases as the average current density is increased. This happens because the ohmic resistance becomes a larger portion of the total resistance at higher current densities. Both the conductivity and exchange current density increase with increasing temperature, leading to a current density that increases with temperature at a given cell potential. Since the i_{avg} is a stronger function of temperature than the conductivity, an increase in temperature at a constant cell potential tends to decrease the Wa and make the current distribution less uniform. In contrast, increasing the temperature while maintaining a constant current density will decrease the cell potential and increase uniformity since the current is constant, but the conductivity is higher at the higher temperature. Finally, we note that the rate of side reactions also increases with increasing temperature and, depending on the activation energy, may lead to increased relative importance of a given side reaction and, therefore, to greater loss of efficiency. The side reaction may, however, improve the uniformity of the plating as discussed below.

The final parameter in the Wa is the transfer coefficient or, equivalently, the Tafel slope. Small values of α_c correspond physically to reactions that are less sensitive to potential. Consequently, the current distribution that corresponds to a given potential distribution will be more uniform for a reaction with a small transfer coefficient. In practice, additives that complex with the metal ions can be used to increase the Tafel slope (decrease α_c), by changing the reaction mechanism.

Experimental Devices for Examining Current Distribution

Two different types of experimental cells are used routinely in industry for plating diagnosis and optimization. The first is the Hull Cell shown in Figure 13.13a. This type of cell comes in a standard size and is designed to explore a range of current densities in a single test, a task made possible by the slanted cathode. It is an excellent tool for comparing bath compositions and for determining the approximate current density at which the desired plating characteristics can be achieved.

The second type of experimental cell is the Haring–Blum cell shown in Figure 13.13b. It is used to quantify the *throwing power* of a bath. A bath with a high throwing power yields more uniform deposition. The cell consists of an

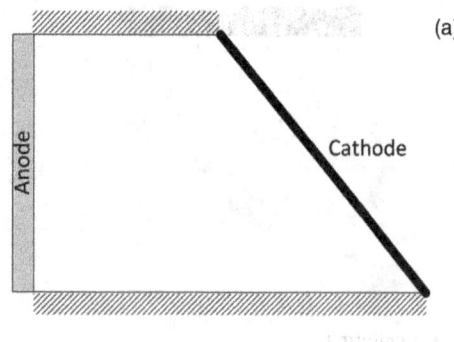

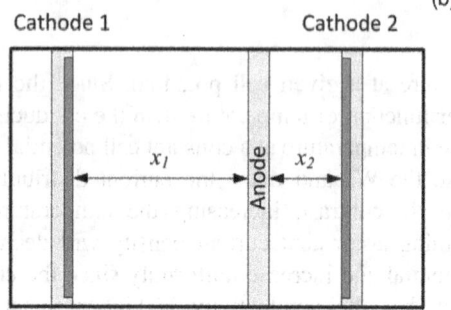

Figure 13.13 Cells for experimental examination of current distribution. (a) Hull cell. (b) Haring–Blum cell.

anode with two cathodes, one on either side of the anode. The backsides of the cathodes are insulated so that only the surface facing the anode is active. The two cathodes are placed at different distances from the anode, x_1 and x_2. Usually, $x_1/x_2 \approx 5$. The different distances correspond to different values of the electrolyte resistance, which is larger for the cathode furthest away from the anode. For $Wa \gg 1$, the electrolyte resistance is not significant and the weight of the deposits on the two cathodes will be the same. For $Wa \ll 1$, the weight of the deposit on the electrode closest to the anode will be greater, consistent with a nonuniform current distribution. The results from the Haring–Blum cell are most frequently expressed in terms of the throwing power, which is defined as follows:

$$\text{Throwing power (\%)} = \frac{100(K - B)}{K + B - 2}, \quad (13.32)$$

where $K = x_1/x_2$ and $B = w_2/w_1$ (weight ratio of the deposits on the two cathodes). The equation is designed to give throwing powers between -100 (very poor) and $+100$ (very good). The best possible uniformity would correspond to equal weights of the deposits at the two electrodes ($B = 1$), which corresponds to a throwing power of 100%. A throwing power of 0% corresponds to the situation where K and B are equal and deposition is inversely proportional to the solution resistance. To illustrate, if $K = 5$ and the throwing power $= 0\%$, the mass of the deposit at x_1, the electrode farthest away from the anode, would be 1/5 of that on the closest electrode, x_2. Negative values are possible because of side reactions that may yield B values greater than K in situations where the current at the closest electrode is dominated by the side reaction (see next section).

13.10 IMPACT OF SIDE REACTIONS

Side reactions are reactions that do not directly contribute to the formation of the desired deposit. For the deposition of metals from aqueous solutions, the most common side reaction is electrolysis of water or reduction of hydrogen ions, which results in the evolution of hydrogen gas at the cathodic potentials needed for deposition. The portion of the applied cathodic current that results in hydrogen evolution is current that does not result in metal deposition and, therefore, represents a loss of efficiency. Loss of efficiency is one of the primary impacts of side reactions. The loss of efficiency can have a significant impact on the economic viability of a plating process, and is an important factor in determining the conditions under which feasible deposition can be performed.

Side reactions may also have a beneficial effect on deposit uniformity and hence on the throwing power of the bath. Because such reactions occur preferentially where the cathodic overpotential is largest, the side reaction can, under certain conditions, "scavenge" the current at locations where the deposition rate of the metal would otherwise have been much higher. This scavenging occurs because the side reaction lowers the cathodic overpotential at those locations by contributing to the iR drop (by increasing the current) without adding to the metal deposit. A necessary condition for this beneficial effect is that the side reaction takes place to a varying extent over the surface upon which deposition is taking place. An example of how this might occur is illustrated by the curves shown in Figure 13.14, which identifies a plating region over which hydrogen evolution varies significantly. In the situation illustrated, hydrogen constitutes a higher fraction of the total current at high (cathodic) potentials, and would tend to reduce nonuniformities. Illustration 13.6 further explores efficiency and its impact on the local deposition rate.

Finally, gas evolution is frequently the result of side reactions. The gas bubbles themselves may impact deposition by influencing the conductivity of the solution. Specifically, the cell potential and local nonuniformity may increase to some degree due to a decrease in both the overall and local conductivity of the plating bath as a result of bubble formation. Bubbling can have the positive effect of mixing the solution to minimize concentration gradients.

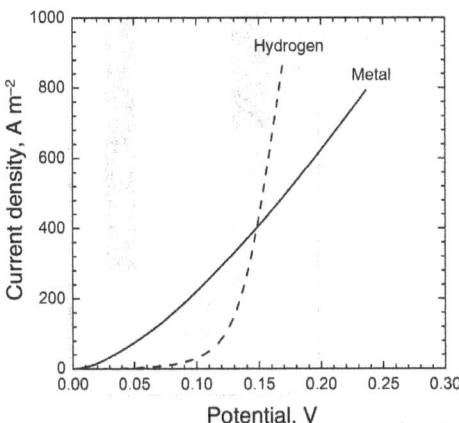

Figure 13.14 Impact of side reaction on deposition showing how both metal deposition and hydrogen evolution can contribute to the total current.

ILLUSTRATION 13.6

Nickel is electrodeposited from a bath that is 1 M in Ni^{2+} at a pH of 4.5. The anode is also nickel, and the potential applied across the cell is −1.3 V. Due to the larger size and surface area of the anode, the anodic surface overpotential can be neglected for this problem (not true in general); concentration gradients can also be neglected. However, iR losses in solution are important. Because of this, the current density is not uniform. We are interested in the relative deposition rate at two specific points on the surface, and the impact of the side reaction (hydrogen evolution) on that relative rate, as well as on the current efficiency. The solution resistance from the anode to the cathode is $0.002\ \Omega\cdot m^2$ at the first point of interest, and $0.003\ \Omega\cdot m^2$ at the second. Please determine the value of the current density at each of the two points, as well as the relative rate of deposition. Next, include the hydrogen reaction and repeat the calculation. In this example, how did H_2 evolution impact the absolute and relative rates of deposition? The following parameters are known:

Nickel reaction: Tafel slope = −0.06 V per decade; $i_0 = 0.1\ A\cdot m^{-2}$

Hydrogen evolution reaction: Tafel slope: −0.11 V per decade; $i_0 = 1.0\ A\cdot m^{-2}$

Solution:

1. In the absence of hydrogen evolution, the total voltage drop is due to the sum of the kinetic overpotential and the iR drop:

$$V_{total} = b\log\left(\frac{i}{i_0}\right) + U - iR,$$

where b is the Tafel slope and R is the resistance that corresponds to the point of interest. U is equal to zero since the potential is referred to the nickel anode. This equation can be solved for i at each point. The answers are shown in the table. The relative deposition rate is 1.49.

	Resistance	i (A·m^{-2})
Point 1	0.002	538
Point 2	0.003	362

2. With the hydrogen reaction, we now have two equations to solve:

$$V_{total} = b\log\left(\frac{i}{i_0}\right) + U - (i + i_H)R.$$

$$b\log\left(\frac{i}{i_0}\right) + U = b_H\log\left(\frac{i_H}{i_0}\right) + U_H.$$

The first is the same as the equation solved earlier, except that the hydrogen current has been added to the iR term. The second equation states that the voltage drop (not the overpotential) across the cathode interface is the same for the two reactions, which of course must be true since there is only one interface at which the reactions are occurring. U_H is the difference between the potential of a hydrogen electrode and a nickel electrode at equilibrium. Since the concentration of the nickel ions is 1 M, the equilibrium potential of the nickel electrode versus SHE (neglecting activity coefficients) is just the standard potential or −0.257 V. At a pH of 4.5, the equilibrium potential of the hydrogen electrode is −0.2664 V. Therefore, $U_H = -0.0094$ V. Solving yields

	Resistance	i [A·m^{-2}]	i_H [A·m^{-2}]
Point 1	0.002	459	81.6
Point 2	0.003	299	64.6

The relative deposition rate (high/low) is 1.53, which is close to that calculated without hydrogen evolution. In this case, the fraction of hydrogen did not change significantly between the two points of interest and, therefore, the relative rate was nearly the same (very slightly worse). Hydrogen evolution did impact the magnitude of the deposition rate, however, as more than 15% of the current went to gas evolution and did not contribute to the deposition rate.

13.11 RESISTIVE SUBSTRATES

Up to this point we have assumed that the material being plated is highly conductive and essentially at a single potential; in other words, we have assumed that the conductivity of the material is sufficiently high that the potential gradient accompanying current flow in the material is very small and can be neglected. There are important practical situations where this assumption is not true. Resistive materials and even insulators can be electroplated. In the case of insulators, a thin conductive layer is first deposited by techniques such as chemical vapor deposition, physical vapor deposition, or electroless deposition. Electrodeposition, which is driven by an applied potential or current, is then used to provide a layer with the desired thickness and properties. A couple of important examples that involve resistive substrates are copper electrodeposition to form interconnects on semiconductor chips and chrome plating on plastics. The key challenge with resistive substrates is that *the current distribution is influenced by the potential drop in the solid substrate* as well as by the potential drop in the electrolyte solution.

Figure 13.15 illustrates a simple electrochemical cell with a resistive substrate. It looks similar to the electrochemical cell shown in Figure 13.1, except that the electrical resistance of the cathode is now important. The scale of the thin conductive layer in the figure is not necessarily accurate in proportion to the rest of the dimensions. For example, a semiconductor wafer 300 mm in diameter may have an initial layer on the order of 10 nm thick, upon which electrodeposition would be performed. While intentionally simple, Figure 13.15 also illustrates an important aspect of behavior with resistive substrates, namely, that the general flow of current in the substrate is orthogonal to the flow of current in solution. Hence, the problem is inherently multidimensional and does not, in general, lend itself to simple numerical analysis, even when concentration effects are unimportant.

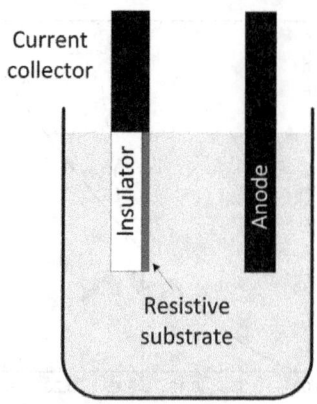

Figure 13.15 Schematic diagram of deposition cell for deposition on a resistive substrate where the potential drop in the substrate is important.

However, if we assume that the kinetic resistance is large relative to the ohmic resistance in the solution (Wa > 10), then the potential in the electrolyte solution at the surface of the working electrode is essentially constant along the length of the electrode, and the current distribution will be controlled by the potential drop in the resistive substrate. The situation is illustrated in Figure 13.16. This assumption is essentially the opposite of that which we have made in previous chapters of this text. Initially, we assumed that the metal substrate was much more conductive than the solution, so the potential drop in the metal was not significant and the metal was at a constant potential. Now we assume that the solution is much more conductive than the resistive substrate, so the solution is at a constant potential and the behavior is controlled by the potential drop in the resistive substrate. This assumption will allow us to develop a dimensionless number analogous to the Wagner number to describe the ratio of the kinetic resistance to the substrate resistance in order to determine whether the substrate resistance is important.

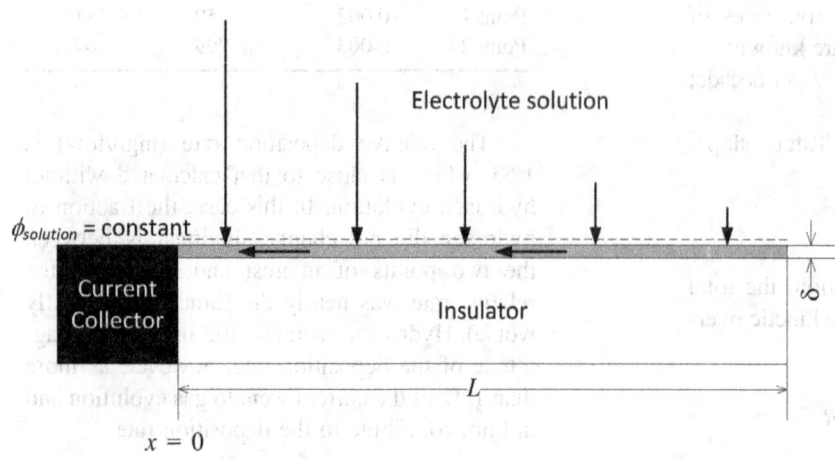

Figure 13.16 Schematic diagram of the physical situation described by the model below. ϕ_{metal} decreases from $x = L$ to $x = 0$ (not shown). Current flow shown qualitatively by filled arrows.

Consistent with the physical situation shown in Figure 13.16, we model one-dimensional current flow in a thin layer that is connected to the current collector at $x=0$, and continues for a length L. We assume a rectangular electrode of width W with an area $L \times W$ exposed to the electrolyte. We begin by writing a charge balance for current flow in the resistive substrate (shown in gray in the diagram)

$$\frac{di_1}{dx} = -ai_n, \qquad (13.33)$$

where i_1 is the current density in the thin metal layer, a is the area of the metal electrode exposed to the electrolyte per volume of metal, and i_n is the current flow between the solution and metal. Note that for deposition, which is a cathodic process, i_n is negative and the term on the right-hand side of Equation 13.33 is positive since current flows from the solution to the metal. The boundary conditions are as follows:

$$i_1 = -I/A_{rs}, \quad \text{at} \quad x=0,$$
$$i_1 = 0, \quad \text{at} \quad x=L,$$

where I is a positive number and represents the current in amperes, and $A_{rs} = W\delta$ is cross-sectional area of the conductive layer through which that current flows. The specified boundary condition is negative at $x=0$ since the current flow is out of the electrode (negative x direction) for a cathodic reaction. The current flow through the metal layer is related to the potential drop:

$$i_1 = -\sigma \frac{d\phi_1}{dx} = -\sigma \frac{d\eta}{dx}, \qquad (13.34)$$

where ϕ_1 is the potential in the metal and η is the solution overpotential. The two derivatives are equal since the solution potential is constant. Assuming Tafel kinetics, Equation 13.33 becomes

$$\frac{di_1}{dx} = ai_0 \exp\left(-\frac{\alpha_c F}{RT}\eta\right). \qquad (13.35)$$

Differentiating this equation with respect to x yields

$$\frac{d^2 i_1}{dx^2} = \left(\frac{-\alpha_c F}{RT}\right)\left[ai_0 \exp\left(-\frac{\alpha_c F}{RT}\eta\right)\right]\frac{d\eta}{dx} = \frac{\alpha_c F}{RT} \cdot \frac{di_1}{dx} \cdot \frac{i_1}{\sigma}. \qquad (13.36)$$

This second-order differential equation can be solved analytically for i_1 as a function of x. Before doing so, it is instructive to make the equation dimensionless. To do so, we define the following:

$$i^* \equiv \frac{i_1}{I/A_{rs}} \quad \text{and} \quad x^* \equiv \frac{x}{L}.$$

The resulting dimensionless equation is

$$\frac{d^2 i^*}{dx^{*2}} = \frac{IL\alpha_c F}{W\delta RT\sigma} \cdot \frac{di^*}{dx^*} \cdot i^*. \qquad (13.37)$$

The dimensionless group $\frac{IL\alpha_c F}{W\delta RT\sigma}$ represents the ratio of the substrate resistance to the kinetic resistance. It is expressed in terms of I, the current density in the metal at the current collector. At small values of this ratio, equation approaches

$$\frac{d^2 i^*}{dx^{*2}} \approx 0 \quad \text{or} \quad \frac{di^*}{dx^*} \approx \text{constant}. \qquad (13.38)$$

This implies a uniform current distribution (you should understand and be able to explain why!).

It is common to address these types of deposition problems in terms of the average current density in the cell, which is based on the area WL. The average current density is related to the current I by

$$I = |i_{avg}|WL. \qquad (13.39)$$

Using this relationship, Equation 13.37 becomes

$$\frac{d^2 i^*}{dx^{*2}} = \frac{|i_{avg}|L^2 \alpha_c F}{RT\sigma\delta} \cdot \frac{di^*}{dx^*} \cdot i^* = \frac{1}{\chi} \cdot \frac{di^*}{dx^*} \cdot i^*, \qquad (13.40)$$

where

$$\chi \equiv \frac{R_{ct}}{R_{sub}} = \frac{RT\sigma\delta}{|i_{avg}|L^2 \alpha_c F}. \qquad (13.41)$$

The dimensionless number χ is analogous to Wa and represents the ratio of the charge-transfer resistance to the electrical resistance of the substrate. Large values of χ are desired for a uniform current density. It is interesting to note the role of L, which is squared in this ratio. This reflects its dual role as both the distance that the current must travel in the substrate and the magnitude of the overall current for a given value of i_{avg} scale with L. It is necessary for both Wa $\gg 1$ (unless the primary current density is uniform) and $\chi \gg 1$ for the current density to be uniform when electrodepositing on a resistive substrate. Since the ohmic resistance in solution and in the substrate are essentially in series, one must also consider the ratio of the two resistances in situations where Wa and χ indicate that both of these resistances are important relative to the charge transfer resistance, as it is possible that one or the other of the resistances dominates. The relevant ratio is

$$\frac{\chi}{\text{Wa}} = \frac{R_\Omega}{R_{sub}} = \frac{L_{\text{Wa}}\sigma\delta}{L_\chi^2 \kappa}. \qquad (13.42a)$$

Subscripts have been added to distinguish the characteristic lengths for the solution (L_{Wa}) and the substrate (L_χ). In cases where the characteristic lengths are similar, this ratio

becomes

$$\frac{\chi}{\text{Wa}} \approx \frac{\sigma\delta}{L\kappa}. \qquad (13.42b)$$

To summarize the use of these dimensionless numbers:

1. $\text{Wa} \gg 1$ and $\chi \gg 1$: current density is uniform, charge-transfer resistance dominates.
2. $\text{Wa} \leq 1$ and $\chi \gg 1$: resistive substrate is not important, current distribution determined by Wa.
3. $\text{Wa} \gg 1$ and $\chi \leq 1$: solution resistance is not important, current distribution determined by χ.
4. $\text{Wa} \leq 1$ and $\chi \leq 1$: both substrate and solution resistances potentially important; need Wa, χ, and ratio (Equation 13.42).

When we introduced the current distribution in Chapter 4, we assumed that $\chi \gg 1$, which is common for highly conductive substrates. Hence, we considered only the first two possibilities in the above list. Items 3 and 4 allow us to consider the impact of the substrate. We now return to the solution of Equation 13.40. The analytical solution is

$$i^*(x^*) = 2\theta\chi \tan(\theta x^* + \psi), \qquad (13.43)$$

where θ and ψ are integration constants. Application of the boundary conditions yields two equations that can be solved for the two constants:

$$\tan \psi = -\frac{1}{2\theta\chi}. \qquad (13.44a)$$

$$\tan \theta = \frac{1}{2\theta\chi}. \qquad (13.44b)$$

We can take advantage of the fact that $\tan \psi = -\tan \theta$, which comes from the boundary conditions, to simplify Equation 13.43:

$$i^*(x^*) = 2\theta\chi \tan(\theta(x^* - 1)). \qquad (13.45)$$

Thus, only the second relationship of 13.44 is needed. Note that θ is bound between 0 and $\pi/2$. Equation 13.45 can be differentiated to yield the local current density across the electrode–solution interface (see Equation 13.33):

$$\frac{di^*}{dx^*} = \frac{i_n(x^*)}{i_{avg}} = 2\theta^2\chi \sec^2[\theta(x^* - 1)]. \qquad (13.46)$$

Remember that implicit to this solution is the condition that $\text{Wa} \gg 1$, which was used as an assumption in the derivation. Solution of a problem with these equations will typically involve the use of either Equation 13.45 or 13.46 with Equation 13.44b, and may require iterative solution of 13.44b to find θ. Figure 13.17 shows the local current distribution as a function of dimensionless distance for several different values of χ.

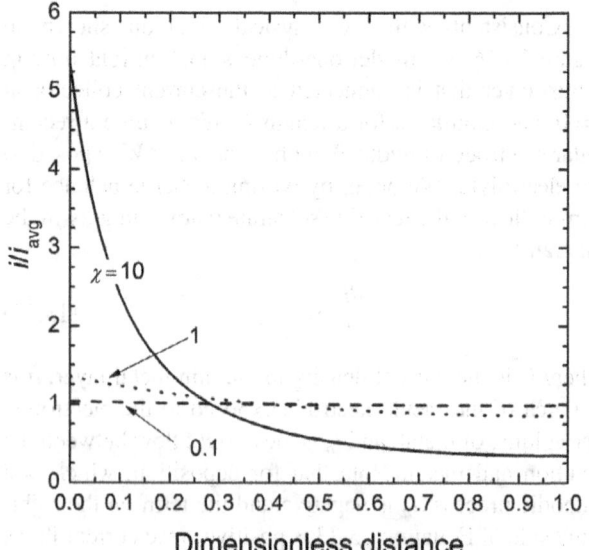

Figure 13.17 Influence of the dimensionless ratio χ on the uniformity of the current distribution for a resistive substrate.

ILLUSTRATION 13.7

How thick of a copper layer must be deposited on a rectangular electrode 15 cm long and 5 cm wide connected to the current collector on one side in order that the current density across the electrode does not vary by more than 15%? Assume that the conductivity of the copper is equal to its bulk value, and $\alpha_c = 0.5$. Also, assume that the initial copper layer on the surface is of uniform thickness. The average current density is 100 mA·cm^{-2}.

Solution:

Application of Equation 13.46 at the two ends of the resistive substrate ($x = 0$ and $x = L$) must yield current densities that differ by no more than 15%. Therefore,

$$\frac{i_n(0)}{i_n(1)} = \frac{2\theta^2\chi \sec^2(\theta(0-1))}{2\theta^2\chi \sec^2(\theta(1-1))} = \frac{\sec^2(-\theta)}{\sec^2(0)} = \sec^2(\theta) = 1.15.$$

Solving this expression for θ yields $\theta = 0.3695$. We can now use this value of θ in Equation 13.44b to find χ.

$$\chi = \frac{1}{2\theta \tan \theta} = 3.494.$$

From the definition of χ,

$$\chi = \frac{RT\sigma\delta}{|i_{avg}|L^2\alpha_c F},$$

$$\delta = \frac{\chi|i_{avg}|L^2\alpha_c F}{RT\sigma} = \frac{(3.494)(0.100)(15^2)(0.5)(96485)}{(8.314)(298.15)(5.96 \times 10^5)}$$

$$= 0.00257 \text{ cm} = 26 \text{ μm}.$$

CLOSURE

In this chapter, we examined several aspects of electrodeposition that range from the fundamental processes that contribute to the formation of new phases to macroscopic issues such as the role of side reactions and of the current distribution. Importantly, the connection between the fundamentals and the characteristics of electrodeposited materials was discussed. Finally, the influence of the substrate resistance was investigated and quantified in terms of a dimensionless number similar to the Wagner number.

FURTHER READING

Bockris, J.O'M. and Reddy, A.K.N. (1970) *Modern Electrochemistry*, vol. 2, Plenum Press, New York.

Budevski, E.B., Staikov, G.T., and Lorenz, W.F. (1996) *Electrochemical Phase Formation and Growth*, Wiley-VCH Verlag GmbH, Weinheim, Germany.

Paunovic, M. and Schlesinger, M. (2006) *Fundamentals of Electrochemical Deposition*, 2nd edition, John Wiley & Sons, Inc., Hoboken, NJ.

Pletcher, D. and Walsh, F.C. (1993) *Industrial Electrochemistry*, 2nd edition, Springer Science+Business Media, LLC.

Schlesinger, M. and Paunovic, M. (2010) *Modern Electroplating*, 5th edition, John Wiley & Sons, Inc., Hoboken, NJ.

West, A.C. (2012) *Electrochemistry and Electrochemical Engineering: An Introduction*, Independent Publishing Platform.

PROBLEMS

13.1 Corrosion protection is frequently provided by coating a steel structure with a thin coating of zinc. Please determine the time required to electroplate a 25 μm layer. The density of zinc is 7.14 g·cm^{-3}. The current density is 250 A·m^{-2}, and the faradaic efficiency is 80%.

13.2 Nickel is plated from a sulfate bath at 95% faradaic efficiency onto a surface with a total area of 0.6 m^2. The density of nickel is 8910 kg·m^{-3}, and plating is performed at a current of 300 A. What is the thickness of the plated nickel after 30 minutes? What is the average rate of deposition?

13.3 Electrorefining of copper involves the plating of copper onto large electrodes in tank-type cells. In one design, electrodes are plated with 5 mm of copper before they are mechanically stripped. If the electrode area is 2 m^2 and plating takes place at 200 A·m^{-2} at a faradaic efficiency of 96%, how long does it take to deposit a layer of the desired thickness? What is the mass of the plated copper? What is the total current for a production cell that contains 50 cathodes? The density of copper is 8960 kg·m^{-3}.

13.4 Chromium is plated from a hexavalent bath at a faradaic efficiency of about 20%. A metal piece is to be plated to a thickness of 0.5 μm. The current density is 500 A·m^{-2}.
  (a) How long will it take to deposit the desired amount of Cr?
  (b) How does the low faradaic efficiency impact the time and energy required to plate Cr?
  (c) There is a movement away from the use of hexavalent Cr. Why?

13.5 Please determine the B value for a 3D cubic cluster of atoms.

13.6 Show that the perimeter $P = 2\sqrt{b\Omega N}$ for a 2D cluster. What is Ω in this expression? Why do we need P as a function of N? How does the variation of P with N affect the size of the critical size of the nucleus? The critical number of atoms in a nucleus changes with overpotential. Does that mean that the relationship between P and N changes? Please explain.

13.7 Please determine the critical cluster size for the system shown in Figure 13.6 at an overpotential of -40 mV. Also, plot the Gibbs energy of cluster formation as a function of cluster size for this value of the overpotential, similar to Figure 13.6. Why does the critical cluster size change with overpotential? How does this impact nucleation?

13.8 The following data were taken for the number of nuclei as a function of time at the overpotentials indicated for the same system considered in Illustration 13.4, but over a different range of overpotentials. Please use these data to determine the nucleation rate at each of the two overpotentials given. Then, fit the nucleation rate data from these two points together *with* the data from the illustration to the expression for the 3D nucleation rate. Finally, estimate the critical value of the overpotential. How do your results compare to those from the illustration?

94 mV		98 mV	
t [ms]	Z_{nuc} [cm^{-2}]	t [ms]	Z_{nuc} [cm^{-2}]
0.112	6.0	0.057	4.6
0.145	10.5	0.072	13.6
0.189	20.5	0.089	24.6
0.231	33.2	0.110	38.4
0.289	49.7	0.129	54.9
0.346	65.6	0.156	77.6

13.9 Please derive the expression for r as a function of t for a growing hemispherical nucleus beginning with a mass balance similar to that given in Equation 13.21. Compare the resulting relationship to that given for a 2D cylindrical nucleus in Equation 13.23. Comment on similarities and differences between the two equations. What does the equation represent physically and what was a key assumption in its derivation?

13.10 Figure 13.10 shows current as a function of time for 2D layer growth under the assumptions of instantaneous and progressive nucleation.

(a) In both cases, the current drops to zero at "long" times. Why?

(b) Please sketch analogous curves for 3D nucleation and growth. How are they different? Why?

13.11 In this problem, we examine instantaneous nucleation both with and without overlap.

(a) Calculate the current as a function of time for 2D instantaneous nucleation without overlap and plot the results ($t \sim 10$ seconds). Assume a nucleation density of 4×10^{10} cm^{-2}. Assuming equally spaced nuclei, estimate the time at which you would expect overlap to occur.

(b) Calculate the current as a function of time assuming overlap. Plot the results on the same figure as the data from part (a). Assume the same nucleation density. Based on the results, comment on the accuracy of the estimate made in part (a).

The following parameters are known (Ag): $n = 1$; $M_{Ag} = 107.87$ g·mol^{-1}; $\rho = 10.49$ g·cm^{-3}; $i_{surf} = 0.005$ A·cm^{-2}; $h = 0.288$ nm.

13.12 In the section on deposit morphology, we discussed the morphological development of deposits in terms of the fundamental processes that occur as discussed in previous sections.

(a) Please reread that section and comment briefly on the role of the overpotential in at least one aspect of deposit growth.

(b) Assume that an additive is added to the solution that preferentially adsorbs and inhibits growth at step sites and kink sites. How might this affect deposit growth?

13.13 Suppose that you are the engineer put in charge of implementing an electroplating plating process for your company. After considerable effort, you get the process running only to discover that the uniformity of the plating layer is not acceptable. Using Wa as a guide, and assuming that the solid phase is very conductive, describe three or four changes that you might try to improve plating uniformity. Please justify each recommended change.

13.14 Suppose that you are conducting tests on a plating bath with use of a Hull cell.

(a) Assume that the bath does not plate uniformly. Where and under what conditions would you expect to see the highest deposition rate?

(b) In measuring the local deposition rate, you find that the deposition rate on the surface closest to the anode is less than observed in the middle of the cathode. Please provide a possible explanation.

13.15 A Haring–Blum cell is used to measure the throwing power of a plating bath. The distance ratio, x_1/x_2 is 5, and the conductivity of the solution is 20 S·m^{-1}. The measured deposit loading at x_1 is 0.8 kg·m^{-2}, and that at x_2 is 1.4 kg·m^{-2}. Both electrodes have the same surface area. What is the throwing power?

13.16 Iron is electrodeposited from a bath that is 0.5 M in Fe^{2+} at a pH of 5. The anode is also iron, and the potential applied across the cell is −3.0 V. Due to the larger size and surface area of the anode, the anodic surface overpotential can be neglected for this problem, as it was in the illustration (not true in general); concentration gradients can also be neglected. However, iR losses in solution are important. Because of this, the current density is not uniform. We are interested in the relative deposition rate at two specific points on the surface, and the impact of the side reaction (hydrogen evolution) on that relative rate, as well as on the current efficiency. The solution resistance from the anode to the cathode is 0.005 Ω·m^2 at the first point of interest, and 0.006 Ω·m^2 at the second. Please determine the value of the current density at each of the two points, as well as the relative rate of deposition. Next, include the hydrogen reaction and repeat the calculation. In this example, how did H$_2$ evolution impact the absolute and relative rates of deposition? The following parameters are known:

Iron reaction: Tafel slope $\approx$ −0.1 V; $i_0 = 0.8$ A·m^{-2}.

Hydrogen evolution reaction: Tafel slope: −0.11 V; $i_0 = 10^{-2.3}$ A·m^{-2}.

13.17 Electrodeposition is performed on a thin chrome layer that has been deposited on an insulating substrate. The layer is 0.5 μm thick and of uniform thickness. About how far from the point where electrical connection is made can electrodeposition be performed (distance L) if the current density across the surface must not vary by more than 20%? The solution phase is not limiting. Assume that the conductivity of the chrome is equal to its bulk value and $\alpha_c = 0.5$. The average current density is 200 A·m^{-2}.

13.18 Assuming Wa $\gg$ 1, please use the appropriate equations from the chapter to generate Figure 13.17. Plot for χ values of 10, 1, and 0.05 (in other words, replace 0.1 with 0.05). *Hint:* You will need the θ value that corresponds to each value of χ.

13.19 Copper is to be electrodeposited onto a thin copper seed layer, 20 nm thick, which has been deposited onto a silicon oxide insulator. If the solution-phase conductivity is high, how much would you expect the deposition current to vary over a 150 mm distance if the average current is 10 A·m^{-2}? Assume that the conductivity of the copper is the same as its bulk conductivity. Thin layers often have conductivities that are significantly lower than the bulk value. Comment on the impact of the conductivity on the results reported above.

13.20 Electrodeposition is being performed onto a thin metal layer ($\sigma = 6 \times 10^6$ S·m^{-1}) that has been deposited onto an insulating substrate. The length scale associated with solid phase (distance from connection point) is 0.5 m, and the thickness of the solid layer is 100 nm. The conductivity of the solution is 20 S·m^{-1}, and the length scale associated with the solution transport is 0.1 m. The average current density is 500 A·m^{-2}. Would you expect the current distribution to be uniform? If not, which phase (solution, solid, or both) would determine the nonuniform current distribution? In which direction would the thickness of the metal layer need to change in order to change the limiting process?

Fumio Hine

Fumio Hine was born on November 23, 1927 in Osaka Japan. His father was a committed businessman, but Fumio was a bit of a romantic. Nevertheless, after taking some electrical engineering classes in Osaka, he began his studies of chemistry under the direction of Professor Shinzo Okada in the Department of Industrial Chemistry in 1948. Hine received his Ph.D. in 1960 from Kyoto University. He was a Fulbright Scholar, and without much foreknowledge of what to expect, he was sent to work with Earnest Yeager at Case Western Reserve University in Cleveland. During this period he began a close and long collaboration with Robert B. MacMullin. Working in Niagara, MacMullin was well known for his work on the development of the chlor-alkali process and had a great influence on the young Hine. Given that electricity is the major input to most electrolytic processes, such processes are often located close to low-cost electrical power. This was the case in Niagara, New York, which supported a flourishing electrolytic industry beginning in the early 1900s because of Niagara Falls. Cleveland is not too far away, and Fumio recounts visiting MacMullin's home several times in the 1960s, where he discussed concepts of industrial electrochemical processes and explored the natural beauty of the Niagara gorge. In addition to picking up MacMullin's electrochemical engineering philosophy, together they came up with a new phrase, "engineering concepts on industrial processing and electrolyzers." Probably more than anyone, Fumio developed a view of the electrochemical industry as a chemical process industry, using electrical energy to drive electrochemical reactions.

In 1965, he became a Professor at Kyoto, and in 1969 he moved to the Nagoya Institute of Technology. He is best known for two books: *Electrode Processes and Electrochemical Engineering*, published in 1985, and *Handbook of Chlor-Alkali Technology*. The handbook is a comprehensive five-volume set that covers fundamentals, cell design, facility design, and plant commissioning; it was published in 2005. In addition to his academic research, Professor Hine has advised many industries and organizations, particularly on chlor-alkali technology. He has received many awards over his long career, including awards from the Kinki Chemical Society and the Osaka and Japan Soda Industry Association. He was actively involved in and dedicated to the Electrochemical Society, particularly the Industrial and Electrochemical Engineering Division—in 1998 he was named a Fellow of the Society. He has also been active in the American Institute of Chemical Engineering, the Chemical Society of Japan, and the National Association of Corrosion Engineers, where he was a member for more than 50 years.

Professor Hine retired from the Nagoya Institute of Technology in 1991. Fumio Hine, Carl Wagner, Charles Tobias, and Norbert Ibl have been aptly described as the pioneers of Electrochemical Engineering. At the time of the writing of this book, Fumio Hine was living a comfortable life in rural Japan with a large collection of books—about half on electrochemical engineering and half on poetry. Now, Dr. Hine spends much of his time reading and writing Waka poetry. Waka means Japanese poem, and this classical form predates the better known Haiku by more than a thousand years. One of his favorites is Yononaka-ha Nanika Tsunenaru Asukagawa Kinou-no Fuchi-zo Kyou-ha Seni-naru from Kokin Wakashū.

Image Source: Courtesy of Fumio Hine.

Chapter 14

Industrial Electrolysis, Electrochemical Reactors, and Redox-Flow Batteries

The objective of this chapter is to explore a few processes and concepts related to the industrial use of electrochemical systems. In doing so, we will consider mature commercial processes, semicommercial processes, and a process or two that still need development before commercial application is viable. Of course, the economics of these processes will determine their commercial importance. As we examine these industrial processes, we will apply the principles learned in earlier chapters covering thermodynamics, kinetics, and transport.

> Industrial electrolytic processes consume about 6% of U.S. electricity supply. Late in the nineteenth century, the price of aluminum was the same as silver. The advent of the electrical generator and associated low-cost electricity led to the development of industrial electrochemical processes. The industry flourished once low-cost, DC power became available.

14.1 OVERVIEW OF INDUSTRIAL ELECTROLYSIS

The purpose of industrial electrolysis is to use electrical energy to convert raw materials into desired products. This conversion of raw materials takes place in an electrochemical reactor. Since energy is added, the electrochemical cells used are electrolytic rather than galvanic. Hence, the term *industrial electrolysis* is used to describe these processes. Just two electrolytic processes, the production of aluminum and the chlor-alkali process, consume about 90% of the electricity used in all electrolytic processes. These examples merit special attention in this chapter.

An important difference between industrial electrolytic processes and many of the electrochemical processes that we have considered to this point is the use of flow. Industrial processes frequently operate with a continuous flow of reactants and products. Semicontinuous and batch processes are also used. Reaction rates and, in particular, reaction rates per volume are critical, and can frequently be raised by increasing the transport rate of reactants and products to and from the electrodes. A key parameter in the design of electrochemical reactors is the current density, which determines the area needed and, hence, the number and size of cells required to achieve a desired rate of production. Voltage losses in the cell are also important and determine the energy necessary for production, as well as the energy efficiency of the process. Finally, several important processes involve products in the form of evolved gases.

We begin with an example of an industrial electrolytic process—chlorine production by what is known as the *chlor-alkali* process. The reactant stream for this process is a purified, saturated brine of NaCl, which flows continuously into the reactor. In addition to chlorine, sodium hydroxide and hydrogen gas are produced simultaneously in the reactor. The two electrochemical reactions are

$$2Cl^- \rightarrow Cl_2 + 2e^- \quad (U^\theta = 1.3595 \text{ V})$$
$$2H_2O + 2e^- \rightarrow H_2 + 2OH^- \quad (U^\theta = -0.828 \text{ V})$$

The standard potential for full cell is 2.188 V. Figure 14.1 illustrates a diaphragm cell, which has been the dominant type of chlor-alkali cell used in the United States for many

Electrochemical Engineering, First Edition. Thomas F. Fuller and John N. Harb.
© 2018 Thomas F. Fuller and John N. Harb. Published 2018 by John Wiley & Sons, Inc.
Companion Website: www.wiley.com/go/fuller/electrochemicalengineering

324 Electrochemical Engineering

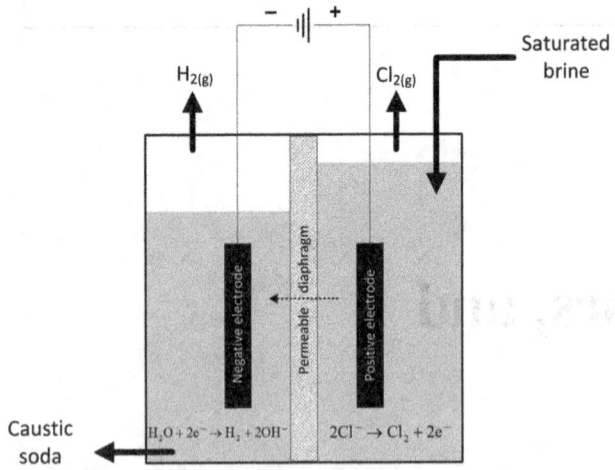

Figure 14.1 Diaphragm cell for the production of chlorine and caustic soda.

years. A newer membrane cell is replacing the diaphragm cell in applications across the globe and will be discussed later in the chapter. Early diaphragm cells consisted of vertical anodes of graphite, steel mesh cathodes, and an asbestos separator (diaphragm). Chlorine is evolved at the positive electrode (anode) and the brine flowing in the compartment is called the *anolyte*. At the cathode, hydrogen is evolved and hydroxide ions are produced. The electrolyte is called the *catholyte*. A diaphragm keeps the product gases from mixing and is an effective barrier for separating the chlorine and hydrogen. Importantly, however, the diaphragm is permeable to the liquid solution, thus allowing for transport of ions between electrodes. Figure 14.2 shows transport in the diaphragm in more detail. In this design, the solution flows from the anode through the diaphragm to the cathode. This flow helps prevent NaOH from back-diffusing into the anode side of the diaphragm. Within the diaphragm, bulk fluid flow, migration, and molecular diffusion are all present. On the cathode side, the catholyte is removed as one of the products, containing NaOH with some chloride ion contamination. The NaOH is recovered in a subsequent process. Assuming that all of the electrical current goes to support the above reactions, a simple material balance can

establish the exit composition if the inlet flow rate and cell current are known. Both Cl_2 and H_2 are removed from the cell as gaseous products. The composition of NaOH at the exit for this type of cell is ~12 wt%, and the purity of the chlorine gas produced is 98%. Since chloride ions are consumed at the positive electrode (anode) and hydroxyl ions are produced at the negative electrode (cathode), ions must move to balance the charge; that is, there is an electrical current flow in the solution. Ideally, only Na^+ would move from anode to the cathode. This best case is not achieved completely with this design; it is important, nonetheless, to minimize the amount of hydroxide ions that reach the anode and the concentration of Cl^- ions that enter the cathode chamber. As you have probably already noted, the diaphragm cell illustrates many important aspects of industrial electrolysis.

About 44 million metric tons of chlorine are produced annually. Assuming that all of this production takes place in diaphragm cells, we estimate the total power required for this production in Illustration 14.1.

We next consider several important performance measures for industrial electrochemical reactors.

14.2 PERFORMANCE MEASURES

This section describes three performance measures that are useful for industrial electrolytic systems. We begin with the *faradaic efficiency*, which is the ratio of the product mass to the amount that could be obtained based on the current and Faraday's law, as was introduced previously in Chapter 1. It can be written as

$$\eta_f = \frac{\text{mass of desired product recovered}}{\text{theoretical mass from Faraday's law}}$$
$$= \frac{m_i}{QM_i/nF} = \frac{\dot{m}_i}{IM_i/nF}, \quad (14.1)$$

where m_i is the mass of the product, Q is the total charge passed during electrolysis, $\dot{m}_i$ is the mass flow rate of the desired product, and I is the total current (assumed constant). In cases where multiple products are formed,

(a)
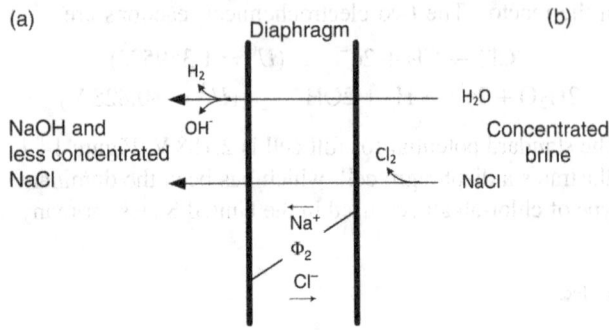

(b)
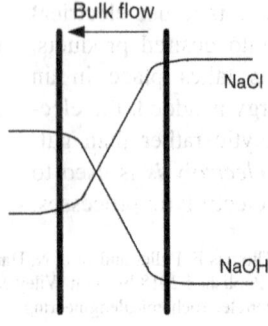

Figure 14.2 Transport in diaphragm separator of a chlor-alkali cell. (a) Movement of species and potential gradient. (b) Concentration profiles.

ILLUSTRATION 14.1

Calculate the electrical power needed to produce the global supply of chlorine with diaphragm cells. Use an annual production of 44 million metric tons Cl$_2$ per year. The operating voltage of a diaphragm cell is about 3.45 V, considerably higher than the equilibrium voltage.

a. Convert the rate of production to kg·s^{-1} of chlorine

$$\frac{44 \times 10^6 \text{ tons}}{\text{year}} \left|\frac{\text{year}}{365 \times 24 \times 60 \times 60 \text{ s}}\right| \frac{1000 \text{ kg}}{\text{ton}} = 1395 \text{ kg} \cdot \text{s}^{-1}.$$

b. Use Faraday's law and the cell potential, 3.45 V

$$\frac{1395 \text{ kg Cl}_2}{\text{s}} \left|\frac{\text{mol Cl}_2}{0.07091 \text{ kg Cl}_2}\right| \frac{2 \text{ equiv}}{\text{mol Cl}_2} \left|\frac{96,485 \text{ C}}{\text{equiv}}\right| 3.45 \text{ V} \left|\frac{\text{J} \cdot \text{C}^{-1}}{\text{V}}\right| = 13.1 \text{ GW}.$$

---

the faradaic efficiency for each product may be different. Note that the faradaic efficiency is dimensionless and has the same value if based on mass, moles, or amperes since it is calculated for a single chemical species. In Illustration 14.2, we rework the problem from Illustration 14.1 while accounting for the faradaic efficiency of the chlorine reaction, which for a diaphragm cell is ~0.96 or 96%. It turns out that the faradaic efficiency for the Cl$_2$ reaction in a chlor-alkali cell is quite high. However, that is not the case for reactions in general, and faradaic efficiencies much less than 1 are frequently encountered.

There are a number of reasons why the faradaic efficiency may be less than 1. One of the most important reasons is the presence of side reactions, which are reactions driven by the current flow that do not produce the desired product. We saw previously in Chapter 3 how a current efficiency can be used to account for such reactions. In a chlor-alkali cell, oxygen evolution at the positive electrode is an example of a parasitic (current consuming) side reaction.

$$2H_2O \rightarrow O_2 + 4H^+ + 4e^- \quad (U^\theta = 1.229 \text{ V})$$

The faradaic efficiency is generally different for the anode and the cathode. For example, the evolution of oxygen, an anodic reaction, reduces the amount of chlorine evolved, and thus, lowers the faradaic efficiency of the anode. However, oxygen evolution does not affect the efficiency of the cathode, since it does not influence the amount of caustic or hydrogen produced at that electrode.

Note that the faradaic efficiency as defined in Equation 14.1 is slightly different from the current efficiency introduced in Chapter 3. You should carefully compare the definitions and note the difference. The faradaic efficiency is focused on the final rate of product formation and not just on the electron transfer reaction(s) at the electrode. Therefore, while the current efficiency constitutes an important part of the faradaic efficiency, there are processes that affect the faradaic efficiency, but do not affect the current efficiency. One such process that causes a diminished faradaic efficiency involves the transport of material across the cell. For instance, current is consumed to produce chlorine gas at the positive electrode of a diaphragm cell. This chlorine has a small but significant solubility in the anolyte. The dissolved chlorine can be transported across the diaphragm to the catholyte. In the catholyte, chlorine reacts with NaOH and is not recovered. The chlorine lost to reaction in the catholyte represents a reduction in the faradaic efficiency, even though the current efficiency has not changed. Similarly, hydrogen and caustic can diffuse from the catholyte to the anolyte. This diffusion process lowers the faradaic efficiency and may also contaminate the products.

Contaminants that react with the desired products represent another possible source of lower faradaic efficiency. For example, sodium carbonate in the brine feed reacts with chlorine to reduce the amount of chlorine produced in the process per coulomb passed in the cell,

$$Na_2CO_3 + 2Cl_2 + H_2O \rightarrow 2HOCl + 2NaCl + CO_2$$

Finally, product recovery can affect the faradaic efficiency. For example, because of its solubility, a small fraction of

> **ILLUSTRATION 14.2**
>
> Calculate the electrical power needed to produce the global supply of chlorine using diaphragm cells. Use an annual production of 44 million metric tons Cl$_2$ per year, and a faradaic efficiency of 96%.
>
> **a.** As before, we convert the rate of production to kg·s^{-1} of chlorine:
>
> $$\frac{44 \times 10^6 \text{ tons}}{\text{year}} \left| \frac{\text{year}}{365 \times 24 \times 60 \times 60 \text{ s}} \right| \frac{1000 \text{ kg}}{\text{ton}} = 1395 \text{ kg} \cdot \text{s}^{-1} = \dot{m}_i.$$
>
> **b.** With a known mass flow rate and a known efficiency, we can use Equation 14.1 to determine the theoretical mass flow rate and the total required current.
>
> $$\frac{\dot{m}_i}{\eta_f} = \text{theoretical mass flow rate} = \frac{I_{total} M_i}{nF} \quad \therefore I_{total} = \frac{\dot{m}_i n F}{\eta_f M_i}$$
>
> $$\frac{1395 \text{ kg Cl}_2}{\text{s}} \left| \frac{\text{mole Cl}_2}{0.07091 \text{ kg Cl}_2} \right| \frac{2 \text{ eq}}{\text{mol Cl}_2} \left| \frac{96,485 \text{ C}}{\text{equiv}} \right| \frac{1}{0.96} = 3.955 \times 10^9 \text{ A}.$$
>
> **c.** To calculate the power, we simply multiply the total current by the operating voltage:
>
> $$(3.955 \times 10^9 \text{ A})(3.45 \text{ V}) = 13.6 \text{ GW}.$$
>
> Another way to approach this problem is as follows:
>
> **a.** Determine $I_{reaction}$, which is the current associated with just the desired product.
>
> $$\frac{44 \times 10^6 \text{ tons}}{\text{year}} \left| \frac{\text{year}}{365 \times 24 \times 60 \times 60 \text{ s}} \right| \frac{1000 \text{ kg}}{\text{ton}} \left| \frac{\text{mol Cl}_2}{0.07091 \text{ kg Cl}_2} \right| \frac{2 \text{ equiv}}{\text{mol Cl}_2} \left| \frac{96,485 \text{ C}}{\text{equiv}} \right| = 3.797 \times 10^9 \text{ A} = I_{reaction}.$$
>
> **b.** Determine $I_{total} = I_{reaction}/\eta_f = 3.955 \times 10^9$ A.
> The power can then be determined as above.

the chlorine evolved in a diaphragm cell will be removed with the flow of the anolyte. This chlorine is lost and not recovered as product.

Even though it is not the only contributing factor, the current efficiency remains a critical component of the faradaic efficiency. The current efficiency for multiple electrochemical reactions that take place on a single electrode can be calculated as a function of potential and concentration if the kinetics of the reactions are known as a function of those variables. Such a calculation, useful for process optimization, is given in Illustration 14.2.

The next performance measure of interest is the *space–time yield*, which is the rate of production per volume of reactor. It is essentially a measure of reactor efficiency and is defined as

$$Y = \frac{\dot{m}_i}{\mathbb{V}_R} = \eta_f \times \frac{I M_i / nF}{\mathbb{V}_R} = \eta_f \times \frac{i a_r M_i}{nF}, \quad (14.2)$$

where a_r, the specific area for the reactor, is the area of the electrode at which the production takes place divided by the volume of the reactor. It is similar to the specific area, a, that was defined in Chapter 5 for porous electrodes. The difference is that the volume used for a_r is the total reactor volume rather than just the electrode volume used previously for a. M_i is the molecular weight, and $\mathbb{V}_R$ is the reactor volume. Y has units of kg s^{-1}·m^{-3}. The current, I, and the current density, i, both correspond only to the portion of the current associated with the product of interest. In a situation where there are multiple products, for example, one product at the anode and another at the cathode, a space–time yield can be specified for each product with use of the area of the corresponding electrode. The quantity ia_r represents the current per unit volume of reactor. Economic analysis is at the heart of industrial processes—both capital and operating costs must be considered. The space–time yield is a parameter that includes the reactor volume, which will directly impact the capital cost. Our main tool to minimize the volume of the reactor is to raise the operating current density. The interplay of the current and volume is apparent from the space–time yield. For a fixed rate of production, $\dot{m}_i$, the volume is inversely proportional to the current. Thus, high currents or high current densities lead to smaller

ILLUSTRATION 14.3

In a chlor-alkali cell, it is possible to have oxygen evolution in addition to chlorine evolution at the anode. The evolution of unwanted oxygen is, of course, a side reaction that reduces the faradaic efficiency. Determine the reaction rates for each of the two reactions for a chlorine overpotential of 0.08 V. Also calculate the faradaic efficiency at the anode, considering just the relative rate of these two reactions. Both reactions can be represented reasonably well with a Tafel expression. The pH = 4 on the anodic side of the cell, and the temperature is 60 °C. The equilibrium potentials versus SHE are given below at the conditions in the reactor.

$$\text{Chlorine}: \quad \text{Tafel slope} = 30 \text{ mV}; \quad i_0 = 10 \text{ A} \cdot \text{m}^{-2} \quad U_{Cl_2} = 1.31$$

$$\text{Oxygen}: \quad \text{Tafel slope} = 40 \text{ mV}; \quad i_0 = 10^{-9} \text{ A} \cdot \text{m}^{-2} \quad U_{O_2} = 0.99 \text{ V}$$

Let's first determine the overpotentials for the two reactions.

$$\eta_{chlorine} = 0.08 \text{ V(given)} = V - U. \quad \text{Therefore, } V = 0.08 + 1.31 = 1.39 \text{ V}.$$

$$\eta_{oxygen} = V - U = 1.39 - 0.99 = 0.4 \text{ V}.$$

Now, we can calculate the current corresponding to each reaction. Please review Chapter 3 for the definition of the Tafel slope if necessary.

$$i_{chlorine} = i_0 \exp\left(\frac{2.303}{\text{Tafel slope}} \cdot \eta_{chlorine}\right) = 10 \text{ A} \cdot \text{m}^{-2} \exp\left(\frac{2.303}{0.03} \cdot 0.08\right) = 4647 \text{ A} \cdot \text{m}^{-2}.$$

$$i_{oxygen} = i_0 \exp\left(\frac{2.303}{\text{Tafel slope}} \cdot \eta_{oxygen}\right) = 10^{-9} \text{ A} \cdot \text{m}^{-2} \exp\left(\frac{2.303}{0.04} \cdot 0.04\right) = 10 \text{ A} \cdot \text{m}^{-2}.$$

Based on the relative rates of these reactions alone (ignoring other contributions to the faradaic efficiency),

$$\eta_{f,chlorine} = \frac{4647}{4647 + 10} = 0.998.$$

volumes and lower capital cost. On the other hand, high current densities result in high cell potentials, and high operating costs. Operating efficiency is addressed below with the third performance measure.

The magnitude of the current density also influences the type of reactor that can be used economically. When the current density is sufficiently high (>100 A·m^{-2}), then simple two-dimensional electrodes can be used. However, high current densities are not possible for some systems due, for example, to mass-transfer limitations or side reactions. Processes with very low current densities (≤ 10 A·m^{-2}) may require three-dimensional electrodes, such as the porous electrodes considered in Chapter 5, in order to be economically feasible; such electrodes can have high specific areas of ~ 5000 m^2·m^{-3}. The reactor configurations considered in this chapter are quite simple. In contrast, a large variety of configurations of varying complexity are used in practice (see Further Reading section for more details).

The third performance measure we consider in this section is the *energy efficiency* defined as

$$\eta_{energy} = \eta_f \times \eta_V = \eta_f \times \frac{U}{V_{cell}}. \tag{14.3}$$

This parameter is the product of the faradaic efficiency and the ratio of the equilibrium potential to the cell potential. The last term on the right side is the voltage efficiency defined for an electrolytic cell. Physically, the energy efficiency is the theoretical power required to complete the chemical conversion ($I_{Rx}U$) divided by the actual power used (IV_{cell}). A key element of this efficiency is calculation of the cell potential, V_{cell}, which is discussed in the next section.

ILLUSTRATION 14.4

If the equilibrium voltage of the diaphragm cell considered above is 2.25 V, determine the energy efficiency of the cell if the operating voltage is 3.45 V.

$$\eta_{energy} = \frac{U}{V}\eta_f = \left(\frac{2.25\text{ V}}{3.45\text{ V}}\right)(0.96) = 0.65.$$

Thus, only two-thirds of the energy added goes into the reaction itself. The balance of the energy ends up as heat or drives unwanted reactions.

14.3 VOLTAGE LOSSES AND THE POLARIZATION CURVE

In this section, we examine the different polarizations or voltage losses present in typical electrolytic cells. The analysis is largely the same as for any electrochemical cell. Again, the objective is to establish the relationship between the potential of the cell and the current density. Recall from Chapter 4 for an electrolytic cell,

$$V_{cell} = U_{cell} + |\eta_{s,anode}| + |\eta_{s,cathode}| + |\eta_{conc,anode}| + |\eta_{conc,cathode}| + |\eta_{ohmic}|. \quad (4.58b)$$

We again use a diaphragm cell as an example. Let's start by examining ohmic losses, which are particularly important for industrial electrolysis cells. The potential drop across a gap of distance h due to current flow between two parallel electrodes is

$$\eta_{ohmic} = \frac{ih}{\kappa} = iR_\Omega. \quad (14.4)$$

For the diaphragm cell, however, the situation is a bit more complex as shown in Figure 14.3. Separating the two electrodes is a diaphragm, which is needed to prevent the product gases from mixing. The diaphragm of thickness h_d is a porous sheet with an effective conductivity. We can estimate the effective conductivity by modifying the conductivity of the electrolyte to account for the porosity and tortuosity of the diaphragm, similar to what was done in Chapter 5. Additionally, there is a gap between the diaphragm and each of the two electrodes, h_a and h_c. Finally, there is resistance associated with current flow through and connections to the electrical leads. Thus, the total ohmic resistance of the diaphragm cell is

$$R_\Omega[\Omega \cdot m^2] = \frac{h_a}{\kappa} + \frac{h_d\tau}{\kappa\varepsilon_d} + \frac{h_c}{\kappa} + R_{leads}. \quad (14.5)$$

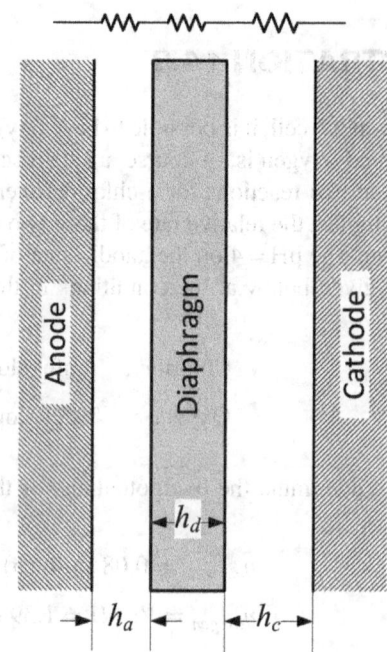

Figure 14.3 Cell resistances.

The first three terms on the right side of Equation 14.5 can be readily evaluated. R_{leads} depends on the specific connections and bus bars used in the system. For our purposes, you will be given the value for this resistance. Some systems may also have an additional term in Equation 14.5 to account for ohmic losses across imperfect interfaces, frequently referred to as contact resistance.

Gas evolution occurs in many industrial electrolysis cells and can be associated with a product (e.g., Cl_2 in the diaphragm cell) or with side reactions. In Chapter 4, we saw how gas evolution increases the rate of mass transfer in a cell. Gas evolution can also have a negative effect by increasing ohmic losses. As shown in Figure 14.4, gas that evolves at an electrode produces bubbles that rise along the length of the electrode due to buoyancy forces. Gas bubbles displace the electrolyte and cannot carry current. These bubbles reduce the effective conductivity of the solution in a fashion similar to the porous membrane discussed in Chapter 5. One simple expression for this effect, useful for gas fractions up to about 40%, is

$$\kappa_{eff} = \kappa(1 - \varepsilon_g)^{3/2}, \quad (14.6)$$

where ε_g is the volume fraction of gas in the gap. As expected, the conductivity decreases as the volume fraction of bubbles increases, providing greater resistance to current flow. A simple way to use this expression is to assume that the distribution of bubbles in the electrolyte is uniform and that the bubbles occupy a certain fraction of the volume, which can be estimated from the height change

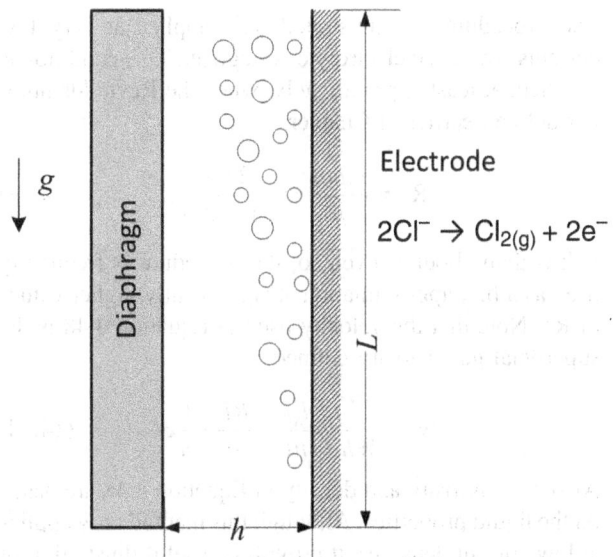

Figure 14.4 Evolution of gas on an electrode.

that occurs in the level of the electrolyte as a result of the gas evolution.

To more accurately account for the volume fraction of bubbles as a function of height and to connect the local volume fraction explicitly to the local current density, we consider some early work in this area by Charles Tobias that applies at low current densities where the bubble formation is not sufficient to cause circulation of the electrolyte. Specifically, the analysis assumes a stagnant electrolyte, no interaction between bubbles, and a single, constant bubble velocity for all bubbles. The ideal gas law is also assumed to apply. We make the additional assumption that kinetic overpotentials are not significant and that the current is controlled by ohmic losses at these low current densities. With these assumptions, the following expressions result:

$$i_x = \frac{8\Delta V \kappa}{h(2 + Kx/L)^3}, \quad (14.7)$$

$$i_{avg} = \frac{\Delta V \kappa (4 + K)}{h(2 + K)^2}, \quad (14.8)$$

where K is the *gas effect parameter*,

$$K = \frac{RT\kappa L \Delta V}{nFph^2 v_{bubble}}. \quad (14.9)$$

In these equations, i_x is the local current density beginning at the bottom of the electrode, where $x = 0$, and i_{avg} is obtained by integrating the local current density over the electrode surface. The local current density decreases with increasing height due to the presence of more bubbles (see Figure 14.4). Also, as before, h is the gap between electrodes and L is the vertical height of the electrode.

With the assumptions described above, we can also write

$$i_{avg} = \frac{\Delta V \kappa_{eff}}{h}. \quad (14.10)$$

Combining Equations 14.8 and 14.10 yields

$$\kappa_{eff} = \frac{\kappa(4 + K)}{(2 + K)^2} \quad (14.11)$$

and

$$\varepsilon_{avg} = \frac{0.5K}{1 + 0.5K}. \quad (14.12)$$

Our goal is to find the potential of the cell that corresponds to a particular current density, which of course is related to the production rate. To do this, we substitute the known current density into i_{avg} in Equation 14.8 and solve for ΔV. Note that K is a function of ΔV. K also includes the bubble velocity, which depends on the size of the bubble, the difference in density, and viscosity of the liquid. For single, small (less than 0.7 mm in diameter), spherical bubbles at low Re, Stokes flow gives

$$v_{bubble} = \frac{g d_b^2 \Delta \rho}{18\mu}. \quad (14.13)$$

Bubble diameters (d_b) of 0.05–0.1 mm are typical. Use of these equations to calculate ΔV is shown in Illustration 14.5.

Dimensionally Stable Anode (DSA)

One of the most significant advancements in industrial electrochemistry is the development of stable metal oxide electrodes or DSA. The first anodes for chlor-alkali were carbon and then graphite, both of which were consumed over time, increasing the electrode gap and requiring costly maintenance. Titanium is stable in the harsh environment of the chlor-alkali cell, but forms a nonconducting oxide. The key innovation was the application of conductive oxides on Ti to produce a "nonconsumable," dimensionally stable electrode. Also, with titanium as a support, the electrode could be constructed in the form of meshes and expanded metals. The open structure allowed gases to be removed more easily, thus allowing the gap to be reduced.

ILLUSTRATION 14.5

Estimation of Ohmic Resistance in a Cell Gap with Gas Evolution

A two-electron reaction with a gas-phase product takes place in an undivided cell (no separator) at an average current of 20 A. The pressure is 250 kPa and the temperature is 300 K. The gap between the electrodes is 4 mm, the length (vertical) of the electrode is 0.5 m, and its width is 0.5 m. The conductivity of the solution without bubbles is 5 S m^{-1}. The density and viscosity of the liquid are, respectively, 1100 kg·m^{-3} and 0.00105 Pa·s. The gas density is approximately 3.2 kg·m^{-3} at the stated pressure.

Solution:

All the quantities needed for the gas-effect parameter, K, are known except the bubble velocity and ΔV. Therefore, we will first determine the bubble velocity. Once that is known, we can express K as a function of ΔV and then substitute that function into Equation 14.9 in order to solve for ΔV.

$$v_{bubble} = \frac{g d_b^2 \Delta \rho}{18\mu} = \frac{(9.81)(0.0001)^2(1100-3.2)}{18(0.00105)}$$
$$= 0.00569 \text{ m}\cdot\text{s}^{-1}.$$

We can now calculate K as function of ΔV. Substituting the appropriate values into Equation 14.9 yields $K = 1.419 \Delta V$.

Next we substitute this expression for K into Equation 14.10 for the average current density. The average current density is

$$i_{avg} = \frac{20 \text{ A}}{(0.5 \text{ m})(0.5 \text{ m})} = 80 \text{ A}\cdot\text{m}^{-2}.$$

Therefore,

$$i_{avg} = 80 \text{ Am}^{-2} = \frac{\Delta V \kappa (4+K)}{h(2+K)^2}$$

$$= \frac{\Delta V (5 \text{ S}\cdot\text{m}^{-1})(4+1.419\Delta V)}{(0.004 \text{ m})(2+1.419\Delta V)^2}.$$

Solving this expression for ΔV gives $\Delta V = 0.0687$ V.

The voltage drop is not large due to the low current density, which is a requirement for use of the simplified analysis for the effect of bubbles. The impact of the bubbles on the effective conductivity is

$$\kappa_{eff} = \frac{\kappa(4+K)}{(2+K)^2} = 4.66 \text{ S}\cdot\text{m}^{-1}.$$

This value is 93% of the original value without bubbles.

The procedure just illustrated only applies at very low currents where the electrolyte is stagnant. This condition is satisfied, at least approximately, when the Reynolds number defined earlier in Chapter 4,

$$\text{Re} = \frac{v_s \rho d_h}{\mu} = \frac{2\dot{V}\rho}{(W+h)\mu}, \quad (4.48)$$

is less than about 3; even so, the procedure is frequently used as a first approximation at significantly higher values of Re. Note that the velocity used in Equation 4.48 is the superficial gas velocity defined as

$$v_s = \frac{\dot{V}}{Wh} = \frac{i_{gas}}{nF} \cdot \frac{RT}{p} \cdot \frac{L}{h}. \quad (14.14)$$

Also, the viscosity and density in Equation 4.48 are based on the liquid properties. Although this method only applies at low current densities, it provides a useful illustration of the effect of bubbles on the ohmic drop. In general, the flow is more complex as bubbles induce electrolyte flow and interact with each other. Some of the factors that are important include coalescence of bubbles, turbulence, the nonspherical shape of the bubbles, and the effect of the walls on the flow. Regardless of the complexity, there is a relationship between the gas evolution rate and the void volume, and that the void volume impacts the ohmic losses in solution.

The impact of gas evolution can be reduced by lowering the current density, increasing the pressure, and increasing the gap distance. At the same time, there are strong incentives to avoid these remedies in order to keep the overall size and cost low, and the energy efficiency high. There is clearly a need for system optimization. One notable engineering solution that has significantly reduced the ohmic resistance associated with bubbles is the use of perforated electrodes in some types of cells that allow gas to be removed from the backside of the electrode out of the current path, specifically the DSA used in the chlor-alkali industry. Another important strategy is the use of convective flow through the gap between electrodes to limit gas buildup by sweeping the gas out of the cell.

In addition to the ohmic losses described previously, kinetic losses or surface overpotentials can also be important. Industrial processes can most often be described with Tafel kinetics, and the corresponding overpotential can be calculated with use of that expression. This is true for a chlor-alkali cell where both electrode reactions are a bit sluggish. For chlorine evolution at the anode,

$$i = i_0 \exp\left\{\frac{\alpha_a F \eta_a}{RT}\right\} = i_{0,ref} \frac{c_{Cl^-}}{c_{Cl^-}^{ref}} \exp\left\{\frac{\alpha_a F \eta_a}{RT}\right\}. \quad (14.15)$$

With α_a equal to 2, the Tafel slope is roughly 30 mV per decade. Using $i_0 = 10 \text{ A}\cdot\text{m}^{-2}$ at 60 °C, the anode overpotential at a current density of 1940 A·m^{-2} is about 75 mV. Similarly, for the cathodic hydrogen reduction

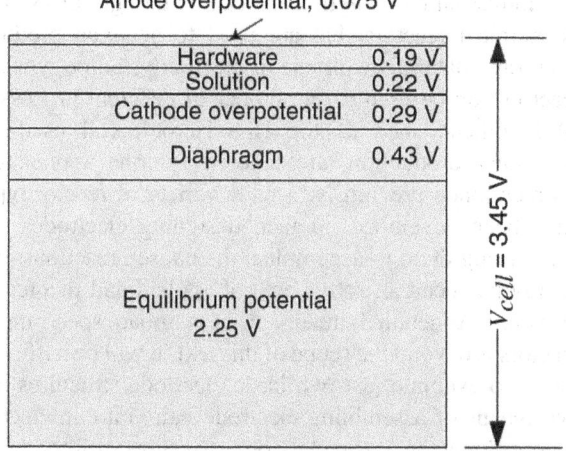

Figure 14.5 Polarization in an operating diaphragm chlor-alkali cell.

process,

$$i = i_0 \exp\left\{\frac{\alpha_c F \eta_c}{RT}\right\}. \quad (14.16)$$

With $\alpha_c = 1$ and $i_0 = 0.07\,\text{A·m}^{-2}$, the overpotential at $1940\,\text{A·m}^{-2}$ is about 0.29 V. The various polarizations or voltage losses at this current density are shown in Figure 14.5. Now that the voltage losses are known, we can estimate the cell voltage with Equation 4.58b to be 3.45 V. Concentration overpotential, which is not likely to be large, has been ignored. For a diaphragm cell operating at 60 °C, the equilibrium potential is 2.25 V. This is the same cell voltage used in Illustration 14.4 to determine the energy efficiency.

This methodology can be extended to develop a full polarization curve that relates the cell potential to the operating current density. Just as in other systems, this relationship is essential in designing electrolytic systems.

14.4 DESIGN OF ELECTROCHEMICAL REACTORS FOR INDUSTRIAL APPLICATIONS

The process of designing an industrial electrochemical system is multifaceted, typically specialized to the application, and iterative. Our description is limited and focuses on the key trade-off between size and efficiency. This balance, illustrated in Figure 14.6, dominates the design process. Economic considerations are at the heart of the design of industrial electrolytic processes. As we'll explore in more detail, low current densities correspond to high efficiencies and low operating costs for electricity. On the other hand, lower current densities require larger electrode areas, and the increased size leads to greater capital costs.

As noted earlier, there are three important performance measures for electrolytic processes: the faradaic efficiency (η_f), space–time yield (Y), and the energy efficiency (η_{energy}). These are design variables—the engineer has a hand in selecting these to meet the needs of the application. In contrast, quantities such as exchange-current densities and electrical conductivity are considered physical properties. Additional variables that underlie the performance measures are shown in Table 14.1. A starting point for reactor selection and sizing is the desired production rate of the material, $\dot{m}$, typically expressed on an annual basis. The

Figure 14.6 Trade-off between size and efficiency is essential part of design.

Table 14.1 Eight Key Variables for an Electrochemical Reactor

Variable	Units	Description	Comments
$\dot{m}_i$	kg·s^{-1} metric tons·yr^{-1}	Rate of production of desired material	Size scales with production rate
i	A·m^{-2}	Current density	These three along with the *configuration of the cell* would be optimized simultaneously
V_{cell}	V	Potential of an individual cell	
η_{energy}	–	Energy efficiency	
A	m^2	Total electrode area, sometimes referred to as the separator area	Follows directly from Faraday's law once i is established
V_s	V	Voltage of the DC power system	
m	–	Number of cells that are connected in series	Follows directly, $m = \frac{V_s}{V_{cell}}$
A_c	m^2	Area of individual electrodes	Cells may be placed in parallel

production rate is determined by the opportunity in the market and will include a target price that can be used for profitability analysis. As we know, Faraday's law is used to convert the production rate to a current for use in reactor sizing. For the purposes of this text, the production rate is provided as an input. Detailed costing and profitability analysis is beyond the scope of this text and, if this is your objective, you should refer to a design text that covers economic analyses of processes. Both the current density and cell potential appear in Table 14.1. Of course, the current density and potential are coupled through the polarization curve. This relationship is central to the analysis of electrolytic systems.

When designing an industrial electrochemical process, there are many questions to be answered. At what current density should the system operate? Is it preferred to have a few very large cells, or would more, smaller cells be better? What shape should the electrodes take? How should these individual electrodes and cells be arranged? How will flow of reactants be distributed within each cell and between multiple cells? We will discuss these topics in the sections that follow; you should remember, however, that the topics are all interrelated. Before going on, let's look at a quick example where the current density and production rate are given.

ILLUSTRATION 14.6

Design an electrolysis reactor by determining the total cell area and number of cells required to produce 6800 metric tons of chlorine per year. The electrolyzer is to operate 360 days per year at a current density of 4000 A·m^{-2}. The electrodes are 1 m × 2 m in size.

Solution:

1. We first determine the total current required to produce the desired amount of chlorine. The faradaic efficiency is assumed to be one.

$$\frac{6800 \text{ tons}}{\text{year}} \cdot \frac{1000 \text{ kg}}{\text{ton}} \cdot \frac{\text{mol}}{0.0709 \text{ kg}} \cdot \frac{\text{year}}{360 \text{ days}} \cdot \frac{\text{day}}{24 \text{ hours}}$$

$$\cdot \frac{\text{hour}}{3600 \text{ s}} \cdot \frac{96485 \text{ C}}{\text{equiv}} \cdot \frac{2 \text{ equiv}}{\text{mol}} = 5.95 \times 10^5 \text{ A}.$$

2. Next, we calculate the total area required (anode area or cathode area).

$$\frac{5.95 \times 10^5 \text{ A}}{4000 \text{ Am}^{-2}} \approx 150 \text{ m}^2.$$

3. Finally, the number of cells:

$$\frac{150 \text{ m}^2}{2 \text{ m}^2 \text{(anode or cathode)/cell}} = 75 \text{ cells}.$$

Industrial electrolytic cells can be thought of as electrochemical reactors that are used to generate products from raw materials with use of electricity. Some types of reactors can be used for a number of different processes, while others, such as the Hall–Héroult cell used for aluminum production, are tailored for one application. As mentioned previously, a wide variety of reactor types have been developed; in fact, designing electrodes and configuring them to accomplish the chosen reactions efficiently has been an active area of applied and theoretical research. A detailed treatment of a broad spectrum of reactors is beyond the scope of this text. It will be sufficient for us to examine just two basic electrode structures and two means of assembling electrode pairs into an electrochemical reactor. Basic electrode structures of importance here are planar (2D) and porous (3D) electrodes. Configurations will be restricted to (i) assemblies with parallel plate electrodes, which may be 2D or 3D, and (ii) plug-flow reactors with porous electrodes. In the first instance, anode and cathode plates are placed parallel to each other and separated by a fixed distance that is filled with electrolyte as shown in Figure 14.3. When the exchange-current density for the reaction is low, these electrodes are often made porous to increase the specific interfacial area. Almost all the applications that we examine will use these assemblies of parallel plate electrodes. The second approach is limited to porous electrodes or packed beds where reactant streams flow through the electrodes.

Establishment of Operating Current Density

Understanding the relationship between current density and voltage is essential to setting the design current density. There are several factors that influence this choice. One strategy would be to operate at the highest current density possible, which would be at the limiting-current density. As seen from Equation 14.2 for the space–time yield, higher current densities increase Y and result in a smaller reactor volume for a fixed rate of production. A smaller reactor (smaller electrodes and less separator area) corresponds to lower initial costs. However, operation near the limiting current has its drawbacks.

1. As the limiting current is approached, V_{cell} increases rapidly. Both the voltage efficiency and the energy efficiency decrease with increasing current density, see Equation 14.5.
2. Operation at the limiting current may lower the current efficiency, especially for multistep reactions. Thus, the faradaic efficiency, η_f, is reduced. A lower faradaic efficiency reduces both the energy efficiency and the space–time yield. This situation is therefore particularly bad—a larger reactor that is less efficient.

3. The limiting-current density may exceed the capability of the available electrode materials or separator membranes and may therefore unacceptably shorten the lifetime of these materials. For example, high current operation may lead to excessive cell temperatures that damage the physical components of the cell.
4. Constant current operation at the limiting current may not be robust from an operational standpoint, since a change in the inlet conditions or in the operating conditions may reduce the limiting current and lead to undesirable side reactions if operation at the same value of the current is continued.

In the end, the current density will be selected so that the profitability of the electrolytic process is maximized. The choice of current density is both important and complex. Data on reaction rates, mass-transport rates, ohmic losses, current efficiencies, and heat removal are needed for design purposes. For the problems that we will consider in this chapter, you will either be given the current density, instructed to operate at the mass-transfer limit, or be given sufficient information to establish the relationship between the cell voltage and current in order to determine operation below the limiting current. In all cases, we will assume a uniform current density for simplification purposes.

While simplification is necessary for our initial treatment of the topic, let's not forget that we, the electrochemical engineers, have many tools at our disposal to alter the polarization curve. The flow rate of reactants can be increased. Higher flows lead to increased rates of mass transfer and greater limiting-current densities. The gap between electrodes can be reduced, resulting in a lower ohmic drop and better energy efficiency. We may be able to change the concentration of reactants, which directly affects the limiting-current density. Catalysts can be added to reduce kinetic polarizations and improve faradaic efficiency. Porous electrodes may be used to increase the specific surface area of the electrodes, and the temperature of the process may be changed.

Electrical Configurations

Once the current density is fixed, the electrode area follows directly from Faraday's law and the production rate. From our experience with electrochemical systems, we know that the potential of individual cells will be on the order of a few volts, and certainly less than 10 volts. The scale of industrial processes requires enormous amounts of electrical power. Most often the electrical power to drive the electrolytic process comes from high voltage alternating current (AC) that is then rectified. It's not practical to supply vast quantities of direct current (DC) at a potential of just a few volts. The solution is to place cells electrically in series to build voltage. The same approach is used for batteries, double layer capacitors, and fuel cells for high-power applications. The number of series connections is established from the system voltage and potential of an individual cell.

$$\text{number of cells in series} = m = \frac{V_S}{V_{cell}}. \quad (14.17)$$

ILLUSTRATION 14.7

Aluminum is produced in an electrolytic process. The rectified electrical power supplied is 1200 VDC. If the individual cells operate at 4.2 V, how many cells are connected in series?

$$\text{number of cells in series} = m = \frac{V_S}{V_{cell}} = \frac{1200}{4.2} = 286.$$

Determining the electrode area and the number of cells connected in series is not the end of the story. When a system contains more than a single anode and cathode pair in series, there are two general methods for making electrical connections: monopolar and bipolar. These terms have the same meanings as they did for batteries and fuel cells. In the monopolar configuration (Figure 14.7a), a separate electrical connection is made to each electrode. The current through the cell is divided among the electrodes electrically connected in parallel, $n_\parallel$. Many electrode pairs can be combined in a single cell. In the individual cell, all of the anodes in a cell are at the same potential, as are all of the cathodes. The voltage between each anode and cathode pair is the same and equal to the cell voltage. Both surfaces of each electrode are active. These cells can then be connected in series as needed to add voltage.

The second means of connecting multiple electrode pairs is the bipolar stack. Here, assemblies of electrode pairs separated with a solid conductive plate are stacked like a deck of cards. Current in a bipolar stack flows straight through the stack and eliminates the need for connections to each internal electrode; the current distribution tends to be more uniform in the bipolar arrangement. The use of narrow-gap cells is considerably easier in bipolar stacks. There are, however, a couple of important disadvantages. Because the current flows from cell to cell through a bipolar stack, failure of one cell results in failure of the entire stack. In contrast, electrodes in a monopolar arrangement function independently. Also, because the difference in potential from one end of the bipolar stack to the other is large, it is possible for a portion of the current to skip one or more cells and flow directly to another cell downstream (see Figure 14.7b). This phenomenon is referred to as a *bypass* or *shunt current*. Bypass currents reduce the faradaic efficiency. Electron-transfer reactions are still needed for the bypass

334 Electrochemical Engineering

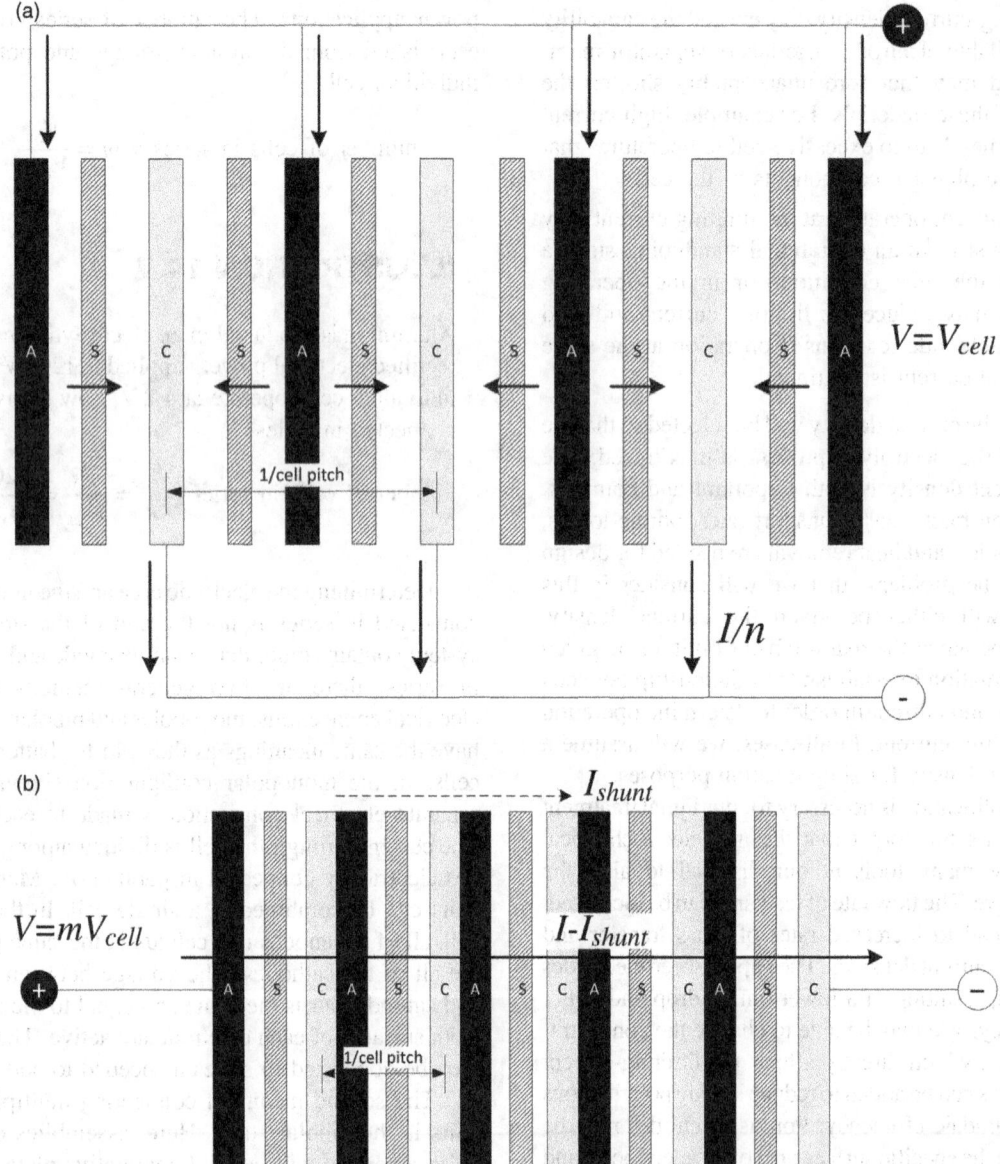

Figure 14.7 Monopolar (a) shown with a separator and bipolar (b) configurations. Current flow is shown for electrolysis. Cell pitch is the number of electrode pairs per unit length.

currents, and these reactions may be undesired or destructive to the cell, as is the case for corrosion reactions. Often this damage is more important than the small loss of efficiency. Bypass currents can be reduced by eliminating bypass pathways, but this can be difficult to do in an industrial cell where, for example, electrolyte from different cells flows into a common manifold. Because of these characteristics, bipolar stacks are standard in fuel cells and redox-flow batteries, but used less frequently in industrial electrolysis.

Another key reactor characteristic that needs to be determined is whether or not a divided cell should be used. A divided cell uses a separator to create distinct anolyte and catholyte solutions. If possible, we prefer not to have a separator since it represents an extra resistance to current flow between the electrodes. However, as we have noted previously, the anode is at a higher potential than the cathode in an electrolytic cell. Therefore, a product or by-product produced at the anode can be reduced spontaneously at the cathode. Similarly, a product or by-product produced through reduction at the cathode can be oxidized at the anode. In addition, soluble products may react with each other in solution. Thus, a principal purpose of the separator is to prevent loss of faradaic efficiency by minimizing or eliminating transport of reaction products in order to prevent undesirable reactions. For example, the diaphragm in a chlor-alkali cell helps to keep Cl_2 that is dissolved in the electrolyte from reaching the cathode where it would react. Consequently, the faradaic efficiency is increased by preventing Cl_2 reduction to Cl^- at the cathode. A separator can also maintain purity of the anolyte and catholyte solutions. For example, the diaphragm in a chlor-alkali cell helps to reduce the amount of Cl^- in the catholyte, which increases the value of the liquid NaOH product. Finally, separators can prevent the formation of explosive mixtures such as H_2/Cl_2. The following questions may be useful in considering whether or not to use a separator:

1. To what extent is the desired product likely to react at the opposite electrode?
2. Are there undesirable solution phase reactions that may be prevented through the use of a separator?
3. Are there safety issues that can be addressed through the use of a separator?
4. Will use of a separator to create distinct anolyte and catholyte solutions enhance the value of product streams or avoid an expensive downstream separation process?

Flow Configurations

Industrial electrochemical reactors are usually flow reactors. Streams of reactants into and out of the reactor are an essential aspect of both continuous and semicontinuous operation. Reactors can also incorporate internal convection to improve rates of transport, as well as to improve the concentration and temperature distributions. Flow can be important for the removal of gases evolved in the reactor in order to minimize the resistance as discussed in Section 14.3. For assemblies of parallel plate electrodes, the flow is principally coplanar across the electrode surface; whereas with plug-flow reactors, the flow is in void spaces of the porous electrode.

Flow patterns are often unique to the application and too numerous to categorize succinctly. Nonetheless, we will identify the basic flow arrangements for an assembly of cells. For multiple electrode pairs that are housed together in one assembly (a cell for monopolar or a stack for bipolar), there are two principal flow arrangements: *parallel flow* and *series flow*. Figure 14.8 illustrates the

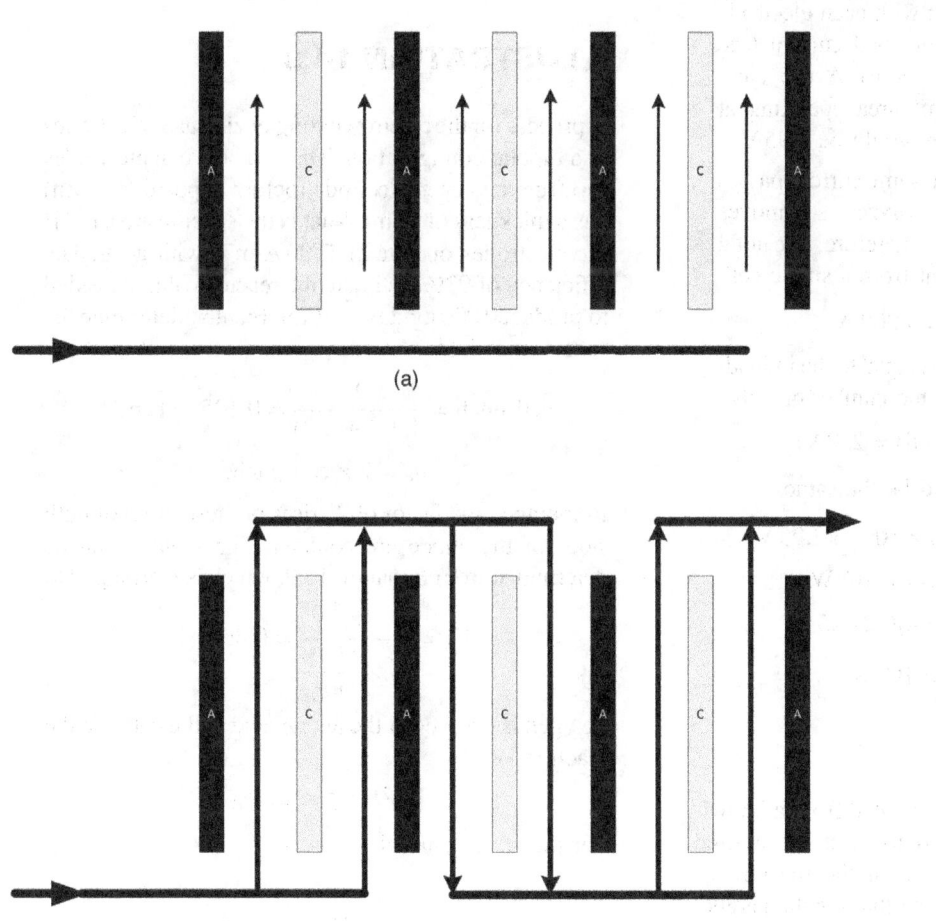

Figure 14.8 Basic flow arrangements. (a) Parallel flow. (b) Series flow.

difference between series and parallel flow for a monopolar design. It is clear that parallel flow will have a lower pressure drop. Series flow enables a greater fraction of the reactants to be converted in a single pass through the assembly, at the expense of a larger pressure drop. Problems associated with gas evolution can be exacerbated with series flow as bubbles accumulate along the flow path. Hybrids of parallel and series are also possible. Industrial practice favors parallel flow where conversion can be increased by placing cell stacks in series with respect to flow, or by separation and recycle of reactants.

ILLUSTRATION 14.8

We calculated the total current, electrode area, and number of individual cells that an electrolysis reactor would need to have to produce 7500 metric tons of chlorine per year in Illustration 14.6. Here we assume membrane-type cells with a cell voltage of 2.95 V. What would be the electrolyzer current and voltage for a completely monopolar and a bipolar configuration of the electrodes?

1. In a monopolar configuration with each electrode pair connected in parallel, the total current (see Illustration 14.6) would be 6×10^5 A (75 electrode pairs, each with a 2 m^2 area operating at 4000 A·m^{-2}) and the voltage would be 2.95 V.

2. In a bipolar configuration, the same current passes through each cell in the electrolyzer as it moves from one end to the other. Therefore, the total current is equal to the current from a single cell:
$$(4000 \text{ A} \cdot \text{m}^{-2})(2 \text{ m}^2) = 8000 \text{ A}.$$
The total electrolyzer voltage is equal to the individual cell voltage multiplied by the number of cells:
$$(2.95 \text{ V per cell})(75 \text{ cell}) = 220 \text{ V}.$$

3. The power, of course, should be the same.
Monopolar: Power $= IV = (6 \times 10^5 \text{ A})(2.95 \text{ V})$
$$= 1.77 \times 10^6 \text{ W}$$
Bipolar: Power $= IV = (8000 \text{ A})(221.25 \text{ V})$
$$= 1.77 \times 10^6 \text{ W}.$$

Reactor Volume

The volume of the reactor can be estimated from a knowledge of the electrode area required to meet the desired production rate and the specific area of the reactor, a_r. Referring back to Figure 14.7, we see that for the parallel plate construction, we can estimate the specific area from the cell dimensions, specifically the thicknesses of the electrodes and the width of the gap between electrodes. The calculation is straightforward, and the result can be summarized by a quantity called *cell pitch* (see Section 10.4). This parameter is simply the number of electrode pairs per unit length when the repeating units are stacked together.

$$\text{cell pitch} = \frac{1}{\text{center to center distance between electrodes of the same type}}.$$
(14.18a)

$$a_r = \frac{\text{electrode area}}{\text{total volume}} = \frac{2(\text{length})(\text{width})}{\frac{(\text{length})(\text{width})}{\text{cell pitch}}} = 2 \times \text{cell pitch},$$
(14.18b)

where the factor of 2 in Equation 14.18b accounts for the two active faces of the electrode. With porous electrodes, there is a lot of internal surface area inside the electrode, represented by a, the specific interfacial area of the electrode. However, even for porous electrodes, we often speak of the superficial current density rather than the true current density based on the internal area. When using the superficial current density with a porous electrode, it is not necessary to include the internal surface area in our analysis.

ILLUSTRATION 14.9

A process for the electrowinning of zinc uses electrodes in a bipolar configuration. The gap between electrodes is 3.0 cm and each electrode (before deposition of Zn) has a thickness of 1 cm. What is the specific area, a_r? If the electrodes operate at 1700 A·m^{-2} with a faradaic efficiency of 93%, estimate the reactor volume needed to produce 1000 kg·day^{-1} of Zn. Finally, determine the space time yield.

$$\text{cell pitch} = \frac{1}{1+1+2(3)} = 0.125 \text{ cell cm}^{-1}$$

$$a_r = 2 \times \text{cell pitch}.$$

Remember, the factor of 2 arises because areas of both sides of the anode are counted. The volume can be determined from Equation 14.2, which is rearranged to

$$V_R = \frac{\dot{m}_i}{\eta_f \times \frac{i a_r M_i}{nF}} = 0.86 \text{ m}^3.$$

Oxygen is evolved at the anode, and at the cathode the reaction is

$$\text{Zn}^{2+} + 2e^- \rightarrow \text{Zn}$$

For the space time yield

$$Y = \eta_f \times \frac{i a_r M_i}{nF} = 48 \ \frac{\text{kg}}{\text{m}^3 \cdot \text{h}}.$$

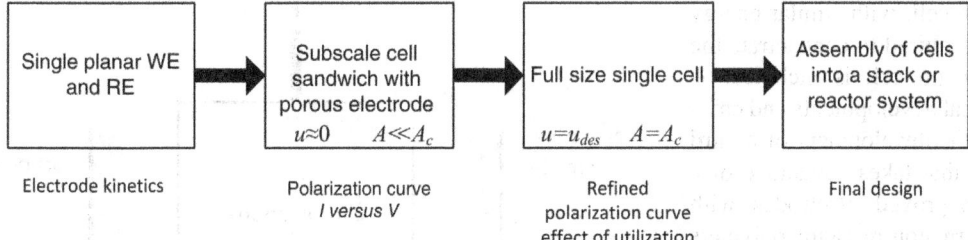

Figure 14.9 One possible scale-up process from fundamental electrochemistry to prototype reactor.

Scale-Up

The process of designing a system for industrial electrolysis is sequential but with some iteration as noted previously. This characteristic is best illustrated through an envisioned process to scale-up a reactor. Referring to Figure 14.9, we might start with fundamental electrode studies as described in Chapter 6. Basic kinetic data are obtained and side reactions identified. The effects of temperature and reactant concentration are often examined at this stage. The second stage is a complete system of anode, cathode, and electrolyte, but at a subscale—a small single cell where the reactants are supplied in large excess to each electrode. The electrode area of this subscale cell might be a factor of 10 or more less than A_c. Of course, at this point, the area for an individual electrode and the total area are only estimates. These estimates will be refined at each stage. For this subscale cell, uniform current density is assumed and often there are no mass-transfer limitations.

The next step is to increase the area of the cell to its full size; that is, A_c. At this stage, the flow configuration is set and the design includes the effect of finite utilization (conversion) of reactants, u:

$$u = \frac{\text{amount reacted}}{\text{amount supplied}}. \qquad (14.19)$$

Utilization can be defined for an electrode, a cell, or a cell stack. Finally, these individual electrode pairs are connected together to form a system. As noted previously, electrodes are almost invariably connected together electrically (series–parallel combination) to build voltage. The cell may also be combined to form one or more mechanical assemblies. The flow rates of reactants and products are critical elements of the final cell design.

Finally, we note that there is off-the-shelf hardware available for initial evaluation of a process and prototype development. One example is the so-called *plate-and-frame* assembly shown in Figure 14.10. An off-the-shelf reactor such as this facilitates the development of new electrochemical processes.

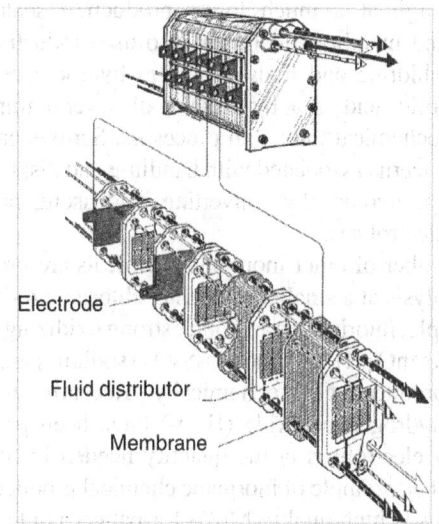

Figure 14.10 Plate-and-frame system that is commercially available for process evaluation. Image provided by ElectroCell A/S.

14.5 EXAMPLES OF INDUSTRIAL ELECTROLYTIC PROCESSES

Industrial electrolytic processes consume about 6% of the total electrical generating capacity of the United States, and represent the principal or only method for producing several important products. This section provides a brief summary of some important industrial applications.

Synthesis of Inorganic Chemicals

Electrolytic production of chlorine and sodium hydroxide, introduced earlier in this chapter, represents the largest electrolytic industry. The process produces chlorine, sodium hydroxide, and hydrogen from a salt solution. Production takes place at 60–95 °C and 0.1–1 MPa. Two types of cells dominated this industry for many years: the diaphragm cell described at the beginning of this chapter and a mercury cell. The mercury cell permitted operation at higher current densities and resulted in products of higher purity relative to the diaphragm cell, with similar energy requirements when the energy required to concentrate the dilute NaOH from the diaphragm cell is included. A combination of new technological developments and environmental concerns has led to the development of a third type of cell, a membrane cell, that takes advantage of a cation-exchange membrane, improved electrodes with reduced overpotentials, and corrosion-resistant polymers for cell construction to produce higher purity products than the diaphragm cell at lower energy consumption rates than the mercury cell. Therefore, most new chlor-alkali cells are of the membrane type.

Although at a much lower production scale, diaphragm and membrane cells are also used industrially to produce chlorine and hydrogen from hydrochloric acid. Hydrochloric acid is a by-product of several important nonelectrochemical industrial processes. Serious environmental concerns associated with handling and disposing of HCl can be avoided by converting it to useful products through electrolysis.

A number of other inorganic chemicals are produced by electrolysis at a smaller scale than chlorine production. For example, fluorine gas and other strong oxidizing agents such as $KMnO_4$, H_2O_2, and $Na_2S_2O_8$ (sodium persulfate) can be produced electrochemically. Recently, reagents such as hydrogen peroxide (H_2O_2) have been produced *in situ* by electrolysis at the quantity needed for optimal use. Another example of inorganic chemical production by electrolysis is high quality MnO_2 for battery applications. In addition to these and other contemporary products, there are many other products that can be made electrochemically, but are not currently manufactured that way because of cost. For example, water electrolysis can be used to produce hydrogen and oxygen gas at high purity; however, except for some specialized applications, other methods of producing hydrogen and oxygen are currently more economical.

Electrowinning of Metals

Electrowinning is the production of metals from ores by electrodeposition from a melt or solution. The most important industrial electrowinning process is the production of aluminum using the Hall–Héroult process. In fact, more electrical power is consumed in aluminum production than in any other electrolytic process. The key innovation, made simultaneously and independently in 1886 by Hall in the United States and Héroult in France, was the discovery that alumina (Al_2O_3) is soluble in cryolite (sodium hexafluoroaluminate) at about 1000 °C, resulting in a conductive solution. The overall reaction for the production of aluminum is

$$2Al_2O_3 + 3C \rightarrow 4Al + 3CO_2$$

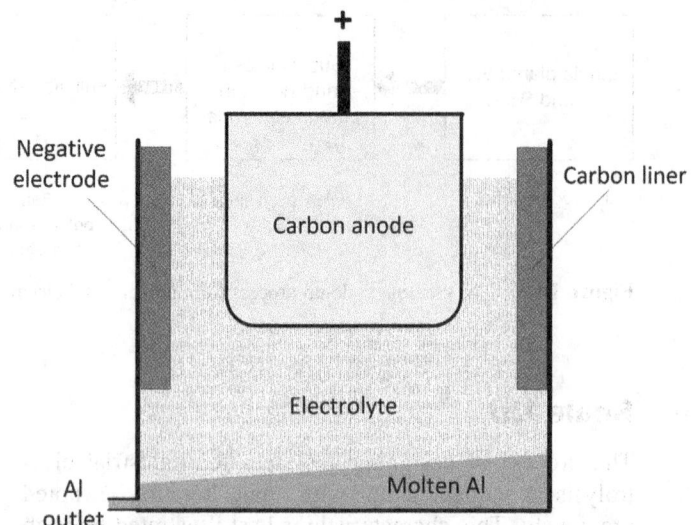

Figure 14.11 Hall–Héroult process for the production of aluminum.

The precise details of the chemistry are not fully understood; consequently, it is difficult to write a complete set of electrochemical reactions. However, the cathodic reaction is clearly the reduction of aluminum. Molten aluminum is denser than the cryolite solution and falls to the bottom of the crucible where it forms the cell cathode; it is periodically siphoned off as the desired product as shown in Figure 14.11. The carbon anode is consumed in the reaction and is lowered gradually into the cell at a rate of about 2 cm per day to maintain the desired cell gap. The other principal reactant, alumina, is added periodically to the melt through a hopper. Typical faradaic efficiency is near 90%, but energy efficiency is low, on the order of 25%.

Other reactive metals produced by electrowinning from a molten salt include lithium, magnesium, and sodium, where chloride-based salts are typically used.

Copper and zinc are the principal metals recovered by electrowinning from aqueous solutions. The hydrometallurgical process used to do this includes acid leaching followed by extraction and then electrowinning. Historically, most copper has been made by smelting, a competing process. Electrowinning is performed in lined concrete tanks into which alternate rows of anodes and cathodes are placed. The spacing between electrodes is about 5 cm. The operating current density for copper ranges from 150 to 1500 $A \cdot m^{-2}$, although maximum values of 350–400 $A \cdot m^{-2}$ are more common. The cathodic reaction is the reduction of the metal, which is plated onto the cathode. The anodic reaction is oxygen evolution on, for example, Pb electrodes. Electrolyte temperatures range from 40 to 60 °C, cell voltages from 1.9 to 2.5 V, and current efficiencies from 80 to 95%. Air sparging, electrolyte circulation, or ultrasonic agitation can be used to increase mass transport and, consequently, the maximum current density. The purity of the copper produced by electrowinning can be quite high (99.999%) and is typically ready for market. In contrast to copper, most zinc is now produced by electrowinning, where the process used is similar to that used for copper. The cell voltage for zinc electrowinning, however, is somewhat higher at 3.3 V.

ILLUSTRATION 14.10

How much carbon is consumed for each kg of Al produced? Assume a faradaic efficiency of 100%.

$$1\,\text{kg Al} \left| \frac{\text{mol Al}}{0.02698\,\text{kg Al}} \right| \frac{3\,\text{mol C}}{4\,\text{mol Al}} \left| \frac{0.012\,\text{kg C}}{\text{mol C}} \right. = 0.33\,\text{kg}.$$

Estimate time before carbon electrode must be replaced. The carbon electrodes are $1.5\,\text{m} \times 0.7\,\text{m} \times 0.7\,\text{m}$ with a bulk density of $1500\,\text{kg·m}^{-3}$. Assume a rate production of 1500 kg aluminum per day.

$$1.5 \times 0.7 \times 0.7\,\text{m}^3 \left| \frac{1500\,\text{kg C}}{\text{m}^3} \right| \frac{1\,\text{kg Al}}{0.33\,\text{kg C}} \left| \frac{\text{day}}{1500\,\text{kg Al}} \right.$$
$$= 2.2\,\text{days}.$$

Electrorefining

In contrast to electrowinning, the purpose of electrorefining is to purify rather than to recover the metal. Aspects of copper electrorefining were used to illustrate several concepts in Chapter 4—you may want to review those parts. Metallic copper, often from a smelting process (approximately 99.5% Cu), is used as the anode. During the refining process, the copper anode is dissolved and copper is plated at the cathode. Any impurities that are more noble than copper stay with the anode and are not dissolved. Impurities that are more active than copper dissolve with the copper into the electrolyte. These active impurities, however, remain in the electrolyte and do not plate out with the copper at the cathode; they are later precipitated out or otherwise removed or recovered from the electrolyte. The net result is the electrodeposition of high-purity copper at the cathode (e.g., 99.999%).

A variety of metals can be purified by electrorefining. For example, nickel, cobalt, lead, and tin can all be refined electrochemically in aqueous solution. Active metals such as aluminum can also be purified in this manner with use of a molten salt electrolyte. The cell voltage for electrorefining tends to be lower than that used for electrowinning as the equilibrium potential is essentially zero for the electrodes of nearly the same composition. Current densities are also modest in order to maintain high purity product and avoid anode passivity (where applicable). As a result, the operating cell voltage for copper electrorefining is only about 0.25 V.

Electrosynthesis of Organic Compounds

A large number and variety of organic reactions can be carried out electrochemically. In fact, organic electrochemistry is considered to be a mature branch of organic synthesis, and most organic reactions that involve electron transfer can be performed by electrochemistry. Types of reactions include oxidation and reduction of functional groups, cleavage, substitutions (e.g., halogenation), additions (e.g., hydrogenation), coupling (e.g., dimerization), and rearrangement. See Further Reading at the end of this chapter for examples of specific reactions.

Organic electrosynthesis reactions may be performed directly or indirectly. *Direct* synthesis reactions are heterogeneous reactions that take place directly on the surface of the electrode. In most cases, the electrochemical reaction forms a reactive intermediate or radical that undergoes further reaction in close proximity to the electrode surface to produce the desired product. *Indirect* electrosynthesis reactions take place via a mediator, which in turn reacts homogeneously in solution with the organic reactant to produce the desired product. The mediator is regenerated electrochemically once it has reacted to affect the desired synthesis. Therefore, there are no waste or disposal concerns since the mediator is recycled and not consumed. Most mediators (catalysts) for indirect synthesis are inorganic redox couples such as the following:

Reductions: Sn^{4+}/Sn^{2+}, Cr^{3+}/Cr^{2+}, Ti^{4+}/Ti^{3+}, Zn^{2+}/Zn, $Na^+/NaHg$

Oxidations: Ce^{3+}/Ce^{4+}, Cr^{3+}/Cr^{6+}, Mn^{2+}/Mn^{3+}, Mn^{2+}/Mn^{4+}, $Ni(OH)_2/NiOOH$, I^-/I_2, Br^-/Br_2, Cl^-/ClO^-

Indirect reactions may be advantageous when they can be used to replace organic reactions that have a high overpotential and sluggish kinetics or that tend to passivate the electrodes. Indirect reactions are also favored when the redox catalysts can provide enhanced selectivity. The catalyst regeneration and the chemical reaction steps can take place in the same reactor (in-cell) or in different reactors (ex-cell). An ex-cell strategy, made possible through the use of indirect reactions, permits separate optimization of the catalyst and organic reactions. Multiphase reactions are also possible with the catalyst regeneration in the aqueous phase and the organic reaction in a separate organic phase. Use of multiple phases can facilitate product separation and enhance the commercial viability of a process. As always, however, there are trade-offs between the simplicity of a direct process and the enhanced flexibility of an indirect process that must be considered carefully in the design process.

In spite of the many possibilities that exist, relatively few organic electrosynthesis reactions have been successful industrially. Even some of the early successes are no longer performed commercially. Steckhan (2012) estimated that approximately 200 reactions have been performed at the pilot scale with more than 100 commercially available. It is difficult to get a precise number because the details of many of industrial processes are often kept confidential.

The most significant industrial process is the production of adiponitrile, an intermediate in the production of Nylon®. It is the only organic electrosynthesis process where the volume of production is consistent with that of a commodity chemical (300,000 metric tons·yr^{-1}). The reactions are shown below. The cathode reaction is the electro-hydro-dimerization of acrylonitrile, and oxygen is evolved at the anode.

$$2CH_2CHCN + 2H^+ + 2e^- \rightarrow NC(CH_2)_4CN \quad (14.20)$$

$$H_2O \rightarrow 2H^+ + 0.5O_2 + 2e^-$$

The overall reaction is

$$2CH_2CHCN + H_2O \rightarrow 0.5O_2 + NC(CH_2)_4CN \quad (14.21)$$

These reactions occur in an undivided bipolar stack using aqueous sulfuric acid as the electrolyte. Another process, currently under development, is the electrochemical synthesis of ethylene glycol, which has the potential to become another high-volume process.

There are many potential advantages to the electrochemical synthesis of organic compounds. The inherent advantage is that electrons serve as the oxidizing and reducing agents. These electrons are, in general, inexpensive and clean relative to chemical agents. The rate of reaction is activated with potential rather than temperature. Thus, the mild conditions characteristic of electrochemical synthesis are well suited for chemicals that are heat sensitive. Also, since the current is directly proportional to the reaction rate, these reactions are inherently easier to control. Closely connected to the ability to control the process is the potential for high selectivity from electrochemical processes. Selectivity is particularly important for high-value specialty products. In spite of these advantages, the number of commercial processes is small, as are the volumes produced, as mentioned previously.

Given the advantages of organic electrosynthesis, why are there not more successful industrial processes? What are the key factors that contribute to a successful process? It turns out that energy costs and initial capital costs associated with the electrochemical cells are not typically the problem. At the risk of overgeneralizing, the factors that are often most important are the availability and cost of the reactants, the reaction yield that can be obtained, the ability to inexpensively separate the product(s) from reactant(s), the availability of a suitable, stable electrolyte, and the ability to achieve an acceptable production rate. Because of the low conductivity of organic solvents, it is necessary to add a supporting electrolyte, which must then be separated from the product downstream. Many commercially successful processes involve water soluble reactants and products and utilize sulfuric acid as the electrolyte. Alcohols and acetic acid are also used industrially with some frequency. Cosolvents can be used to enhance solubility if needed. Separation can be facilitated by phase separation where feasible.

Another important factor that may easily be overlooked is the need to consider organic electrosynthesis as a design alternative early in process development. It is difficult and expensive to consider such options at an advanced stage of design. Consequently, it is important for commercial success that a company has the expertise needed to consider electrochemical options as part of their normal design process. This factor is likely to become more important in the future as society shifts to solar power as the primary energy source from which electricity can be generated directly rather than from fuels as is currently the case.

ILLUSTRATION 14.11

Naphthoquinone can be formed from naphthalene by the following reaction, where cerium acts as the mediator.

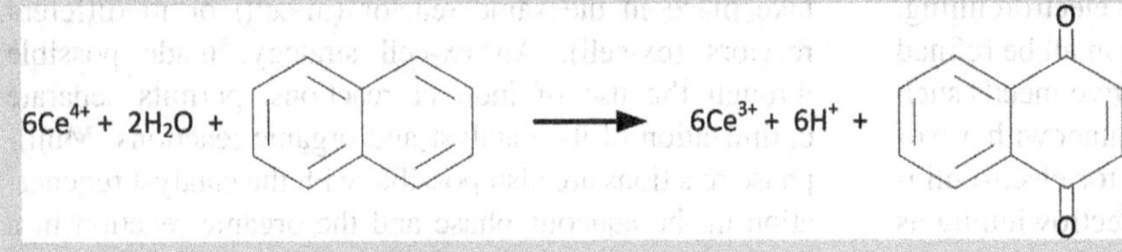

$$6Ce^{4+} + 2H_2O + \text{(naphthalene)} \longrightarrow 6Ce^{3+} + 6H^+ + \text{(naphthoquinone)}$$

Regeneration of the mediator takes place in a separate reactor. What is the desired reaction in the regeneration reactor? If regeneration is done in aqueous solution, what is the most likely cathodic reaction? The faradaic efficiency can be greater than 90%. What is the most likely side reaction? The standard potential of the Ce^{3+}/Ce^{4+} is 1.72 V.

Solution:

Since cerium is reduced in order to make naphthoquinone, it must be reoxidized in the regeneration reactor. The most likely cathodic reaction in an aqueous system is hydrogen evolution. Since the standard potential of the cerium reactions is well above the oxygen potential, it is most likely that oxygen is evolved during the oxidation process. Oxygen evolution can be reduced by choosing an electrode surface with a high O_2 overpotential.

14.6 THERMAL MANAGEMENT AND CELL OPERATION

As you may have noted, some of the industrial processes discussed in this chapter operate at high temperatures. The most extreme example considered is the electrowinning of aluminum, which takes place in molten salt at temperatures of almost 1000 °C. How much heat is required to maintain the required temperature and how is this heat supplied? In this section, we consider heating and cooling of electrochemical systems since temperature control is a critical part of any industrial process.

Let's begin with an overall energy balance that applies to a system in which multiple reactions take place. For open systems, the following energy balance applies:

$$mC_p \frac{dT}{dt} = \sum_m \dot{n}_m H_{in,m} - \sum_p \dot{n}_p H_{out,p} + \dot{q} - \dot{W}$$
$$- \sum_j r_i \Delta H_{Rx,j}, \qquad (14.22)$$

where

m = mass of the system, assumed constant [kg]

C_p = average heat capacity of the system [J·kg^{-1}·K^{-1}]

$H_{out,p}$ = enthalpy of outlet stream p [J·mol^{-1}]

$H_{in,m}$ = enthalpy of inlet stream m [J·mol^{-1}]

$\dot{n}_m$ = molar flowrate of inlet stream m, [mol·s^{-1}]

$\dot{n}_p$ = molar flowrate of outlet stream p, [mol·s^{-1}]

$\dot{q}$ = heat transferred to the system from the environment [W]

$\dot{W}$ = Rate of work done by the system on the environment [W]

r_i = Rate of reaction of species i [mol·s^{-1}]

$\Delta H_{Rx,j}$ = Heat of reaction j per mole of species i [J·mol^{-1}]

For the electrolytic systems considered in this chapter, $-\dot{W}$ is positive and equal to the power added to the cell in order to carry out the reaction, IV_{cell}. The heats of reaction apply to full reactions rather than to half-cell reactions, and are determined as described in Chapter 2. When operating at steady state, the term on the left side of Equation 14.22 is zero. Consequently, one important use of the energy balance is to determine the rate of heat ($\dot{q}$) that must be added to or removed from the system in order to maintain a steady temperature. Let's illustrate the use of the balance by applying it to aluminum electrowinning (see Illustration 14.12).

As the illustration demonstrates, heat must be removed from cells used to produce aluminum. This situation is typical for an industrial electrolytic process. Therefore, the focus is on rejecting heat in order to maintain the desired temperature. The high operating temperature for aluminum production facilitates heat transfer from the cell to the environment, which directly provides the necessary cooling. The formation of a solidified molten salt insulating layer on top of the melt permits the system to flexibly maintain the needed temperature. However, these characteristics are not typical of industrial electrolytic cells. For operation closer to room temperature, heat is removed by placing heat exchangers in the anolyte and catholyte loops. In other words, the cooling takes place outside of the reactor with use of heat exchangers placed in the flow loops as illustrated in Figure 14.12. Evaporative cooling towers are frequently used to provide a heat sink for large systems. If operation is assumed to be steady, Equation 14.22 can be used to calculate the heat that would need to be removed to maintain a constant temperature. Similarly, Equation 14.22 can be used to approximate the increase in electrolyte temperature in the electrochemical reactor by assuming adiabatic operation or a known finite heat loss and then calculating the outlet temperature.

342 Electrochemical Engineering

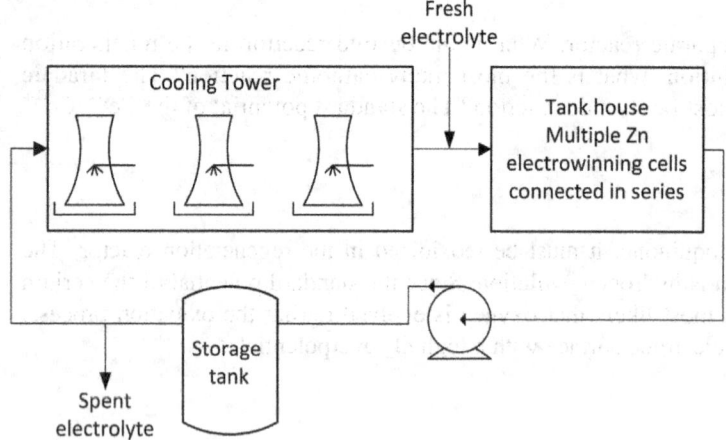

Figure 14.12 Process diagram for Zn electrowinning that emphasizes cooling of the electrolyte.

ILLUSTRATION 14.12

Energy Balance for Hall–Héroult Cell. An aluminum cell is operating at 4.2 V and 200 kA. The faradaic efficiency is 95%. Assume that the system is at steady state and that the reaction occurs as follows:

$$2Al_2O_3 + 3C \rightarrow 4Al + 3CO_2$$

Determine the amount of heat that would need to be added to the reactor in order to maintain the operating temperature of 970 °C. The Al_2O_3 and C added to the reactor are at room temperature. The carbon dioxide and molten aluminum leave at the operating temperature of the reactor. The following physical data are known:

Heat capacities [J·mol^{-1}·K^{-1}]:

$$Al(s): \quad 20.38 + 1.29 \times 10^{-2}\,T$$
$$Al(l): \quad 31.75$$
$$CO_2: \quad 32.2 + 2.22 \times 10^{-2}\,T - 3.47 \times 10^{-6}\,T^2$$

For aluminum: $T_{melt} = 933.47$ K, and $\Delta H_{fus} = 10{,}700$ J·mol^{-1}.

First, we determine the molar flow rates. Noting that six electrons are transferred for every mole of Al_2O_3 that reacts,

$$\dot{n}_{Al_2O_3} = \frac{\eta_f I}{6F} = 0.328 \text{ mol}\cdot\text{s}^{-1}, \quad \dot{n}_C = \dot{n}_{Al_2O_3} \cdot \frac{3}{2} = 0.492 \text{ mol}\cdot\text{s}^{-1},$$

$$\dot{n}_{Al} = \dot{n}_{Al_2O_3} \cdot 2 = 0.656 \text{ mol}\cdot\text{s}^{-1}, \quad \dot{n}_{CO_2} = \dot{n}_C = 0.492 \text{ mol}\cdot\text{s}^{-1}.$$

We now calculate the heat of reaction at 25 °C with use of heat of formation data from standard tables.

$$\Delta H_{Rx} = \frac{3}{2}\Delta H_{f,CO_2} - \Delta H_{f,Al_2O_3} = \frac{3}{2}\left(-393.52\,\frac{\text{kJ}}{\text{mol}}\right) - \left(-1675.7\,\frac{\text{kJ}}{\text{mol}}\right) = 1.085 \times 10^3 \,\frac{\text{kJ}}{\text{mol } Al_2O_3}.$$

For this problem we choose a pathway where the reactants enter and react at 25 °C, and then the products are heated to the outlet temperature of 1243 °C. All enthalpies are referenced to 25 °C. Therefore, the inlet enthalpies are zero since they are at the reference temperature. The exit enthalpies are calculated as follows:

$$H_{CO_2} = \int_{298}^{1243} C_p(T)\,dT = 44.403 \text{ kJ}\cdot\text{mol}^{-1},$$

where the expression for the heat capacity was inserted into the integral and evaluated.

For aluminum, we must account for the change in phase and the heat of fusion.

$$H_{Al} = \int_{298}^{933.47} C_p(T)dT + \int_{933.47}^{1243} C_p(T)dT + \Delta H_{fus} = 38.526 \text{ kJ}\cdot\text{mol}^{-1}$$

We can now apply the full energy balance, noting that the reaction rate of Al_2O_3 is equal to $\dot{n}_{Al_2O_3}$. Solving for $\dot{q}$

$$\dot{q} = \dot{W}_{elect} + \dot{n}_{CO_2}H_{CO_2} + \dot{n}_{Al}H_{Al} + \dot{n}_{Al_2O_3}\Delta H_{Rx}$$

$$\dot{q} = -(200 \text{ kA})(4.2 \text{ V}) + \left(0.492 \frac{\text{mol}}{\text{s}}\right)\left(44.402 \frac{\text{kJ}}{\text{mol}}\right) + \left(0.656 \frac{\text{mol}}{\text{s}}\right)\left(38.526 \frac{\text{kJ}}{\text{mol}}\right)$$
$$+ \left(0.328 \frac{\text{mol}}{\text{s}}\right)\left(1.085 \times 10^3 \frac{\text{kJ}}{\text{mol}}\right) = -437 \text{ kW}.$$

The work term in the equation is negative because it is done on the system rather than by the system. Importantly, we note that $\dot{q}$ is negative. Therefore, heat must be removed from the system rather than added to the system, even for this high-temperature reaction. This is the typical case for electrochemical systems.

14.7 ELECTROLYTIC PROCESSES FOR A SUSTAINABLE FUTURE

Electrolytic Fuel Generation

As we look to the future, it seems clear that the sun will be our primary source of energy. In addition to solar thermal methods, solar energy can be captured in the form of energetic electrons and holes, inherently an electrochemical process. Also, since the availability of solar energy is cyclic, electrochemical processes can be used for energy storage in order to provide the energy needed in off cycles.

Perhaps more important, the transition to solar energy as the primary source of energy for society will likely be accompanied by a shift from fuel-based energy use to direct use of electricity. Electrochemical devices will undoubtedly play a critical role in this shift. Electricity has traditionally been the "high-end" form of energy since it has most frequently been generated using fuels. Consequently, it has historically been more efficient to use fuels directly, where possible, rather than to use fuels to generate electricity, which is subsequently used for the intended application. This will change as electricity is generated directly from renewable sources; hence, electrochemical processes that use electricity directly will have an added economic advantage. There will also likely be a shift from centralized generation of electricity to a distributed solar-based system, which will again impact the use and scale of electrochemical devices. The same type of shift is already occurring in the transportation industry, driven by environmental concerns, where portability will continue to depend on electrochemical devices.

What role, if any, will electrolytic processes play with respect to fuels in a solar-based system? In applications where direct use of electricity is not viable, electric power can be used to generate *solar fuels* such as hydrogen. This process is a reversal of the previous paradigm where fuels were used to generate electricity. Electrolysis of water to produce hydrogen and oxygen is perhaps the electrolytic process first considered for fuel generation.

Water Electrolysis

The electrolysis of water is essentially a fuel cell in reverse, where electricity is used to create hydrogen and oxygen from water. Therefore, water electrolyzers reflect the types of technologies that we considered for fuel cells in Chapter 9. The three principal types of electrolyzers are alkaline, PEM, and solid oxide.

Most commercial water electrolysis is performed with alkaline electrolyzers. The electrode reactions are as follows:

$$\text{Cathode}: \quad 2H_2O + 2e^- \rightarrow H_2 + 2OH^-$$

and

$$\text{Anode}: \quad 2OH^- \rightarrow \frac{1}{2}O_2 + H_2O + 2e^-$$

The net reaction is, of course, just the splitting of a water molecule to produce hydrogen and oxygen. Figure 14.13 shows the equilibrium potential and thermoneutral potential ($-\Delta H_{Rx}/nF$) for water electrolysis as a function of temperature. These data assume the water is a vapor. Near room temperature, where water is a liquid, the two values are close to one another. At high temperatures, there is large difference—which is important for high-temperature electrolysis. The equilibrium potential, which is proportional to the free energy change for the reaction and,

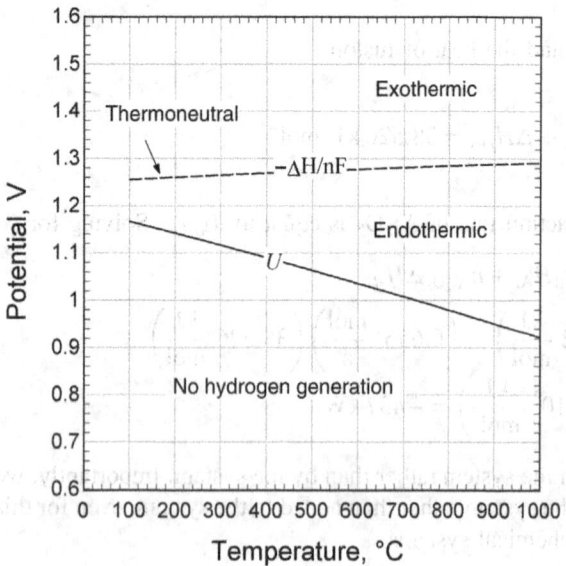

Figure 14.13 Equilibrium and thermoneutral potentials assuming gaseous water.

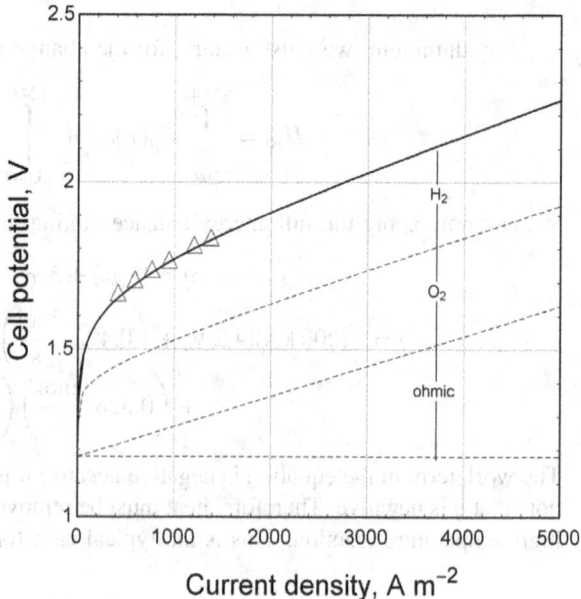

Figure 14.14 Electrolysis of water at 40 °C.
Source: Adapted from Ulleberg 2003.

therefore, the minimum potential needed to split the water, decreases with increasing temperature. In contrast, the thermoneutral potential does not change significantly with temperature. Operation at a cell voltage greater than the equilibrium potential but less than the thermoneutral potential would lead to net cooling due to the reversible heat term ($T\Delta S$) for water electrolysis, and would require the addition of heat to the reactor. Above the thermoneutral potential, the irreversible losses in the cell are sufficient to generate the heat needed for the reaction. Note that, while convenient, analysis with the thermoneutral potential is approximate for an open system such as an electrolyzer since it does not account for the enthalpy of the inlet and outlet streams. Therefore, where possible, use of the full energy balance, Equation 14.22, is preferred.

Voltage losses as a function of current density in a typical alkaline electrolyzer are shown in Figure 14.14. The overpotentials for both the anodic and cathodic reactions are significant. As expected, ohmic losses become more important at the higher current densities. Since the energy efficiency is a direct function of the operating voltage, a trade-off exists between the absolute amount of hydrogen that can be produced in a given electrolyzer and the energy efficiency at which it can be produced. Commercial electrolyzers typically operate at cell potentials below 2 V and current densities between 1000 and 3000 A·m^{-2}. Note that at 25 °C, $U^\theta = 1.229$ V and the thermoneutral potential is 1.481 V (assuming liquid water). Given the relatively sluggish kinetics at these temperatures, operation below the thermoneutral potential is not practical.

Both monopolar and bipolar alkaline electrolyzers are available, although the bipolar configuration is more common. The electrolyte consists of 25–30 wt% KOH, which has a relatively high electrical conductivity. The decision to operate at high rather than low pH is based on material's cost and stability. Specifically, corrosion problems are less severe at high pH than under acidic conditions. Cells traditionally operate at temperatures ranging from 65 to 100 °C. Input or makeup water must be relatively pure ($\kappa < 500\ \mu$S·m^{-1}) in order to avoid the buildup of impurities in the cell. Alkaline electrolyzer technology is considered to be mature with a life expectancy of up to 15 years. High-temperature alkaline cells have also been developed; these cells operate at temperatures up to 150 °C in order to improve conductivity and reaction kinetics, although water management is an issue at the high temperatures.

ILLUSTRATION 14.13

Efficiency of an Alkaline Water Electrolyzer

A bipolar alkaline water electrolyzer is operated at 80 °C and consumes 100 kW during operation. The voltage between each anode and cathode is 1.85 V, and the corresponding equilibrium voltage at the operating temperature is 1.18 V. The current density is 1500 A·m^{-2} (assume uniform) and the faradaic efficiency is 98.5%. The electrodes are 1 m^2. Please determine the operating current and voltage of the stack, the number of cells in the stack, the rate of hydrogen production, the energy efficiency of the electrolyzer, and the specific energy for hydrogen production (kWh·Nm^{-3}, where Nm^3 is a "normal" m^3 at 0 °C and 100 kPa).

Operating current: $(1500 \text{ A·m}^{-2})(1 \text{ m}^2) = 1500 \text{ A}$.

Stack voltage: $(100 \text{ kW})\left(\frac{100,000 \text{ W}}{\text{kW}}\right)\left(\frac{1}{1500 \text{ A}}\right) = 66.7 \text{ V}$.

Number of cells: $66.7 \text{ V}\left(\frac{1 \text{ cell}}{1.85 \text{ V}}\right) = 36$ cells.

Rate of H$_2$ production:

$0.985\left(\frac{1500 \text{ A}}{\text{cell}}\right)(36 \text{ cells})\left(\frac{1}{2(96,485) \text{ C·mol}^{-1}}\right) =$

$0.276 \text{ mol·s}^{-1} = 993 \text{ mol·h}^{-1}$.

Use the ideal gas law to determine the volumetric flow rate of hydrogen.

Use normal volume of hydrogen (Nm^3): $= 22.5$ Nm^3·h^{-1}.

Specific energy consumption: $\frac{100 \text{ kW}}{22.5} = 4.44$ kW·Nm^{-3}h^{-1}.

Energy efficiency (percent of energy consumed that results in hydrogen production): $0.985\left(\frac{1.18}{1.85}\right)100 = 63\%$.

An efficiency based on the HHV of hydrogen gas (3.54 kW h/Nm^3) is also sometimes reported,

$$\frac{3.54 \text{ kW h/Nm}^3}{4.44 \text{ kW h/Nm}^3}(100) = 80\%.$$

Water electrolyzers based on proton exchange membranes (PEM) are also manufactured. These electrolyzers first appeared during the Space Race in the 1960s, and use technology similar to that of PEM fuel cells. The electrode reactions are as follows:

$$\text{Cathode}: \quad 2\text{H}^+ + 2\text{e}^- \rightarrow \text{H}_2$$

$$\text{Anode}: \quad \text{H}_2\text{O} \rightarrow \frac{1}{2}\text{O}_2 + 2\text{H}^+ + 2\text{e}^-$$

Hydrogen production rates in PEM electrolyzers are low relative to alkaline cells, in spite of the fact that the current densities are higher. The low production rate is due to the smaller electrode surface area (superficial) in the cell. Use of an ion-exchange membrane provides enhanced safety and increased purity due to the low gas permeability of the membrane. PEM electrolyzers operate well at partial load and have good dynamic performance. The major disadvantages of these electrolyzers are higher initial cost due to the membrane and catalysts, shorter lifetimes, and lower hydrogen production capacity.

A promising technology for water electrolysis that is still at the research stage is that of solid oxide electrolyzers (SOEs). Analogous to solid oxide fuel cells, these devices operate at high temperature and utilize a solid, ceramic O^{2-} conducting electrolyte. The reactions are as follows:

$$\text{Cathode}: \quad \text{H}_2\text{O} + 2\text{e}^- \rightarrow \text{H}_2 + \text{O}^{2-}$$

$$\text{Anode}: \quad \text{O}^{2-} \rightarrow \frac{1}{2}\text{O}_2 + 2\text{e}^-$$

Operation at temperatures up to 1000 °C provides some important advantages. Reaction rates are much faster at high temperature, which circumvents the need for expensive catalysts. Also, the equilibrium voltage decreases with increasing temperature—at 1000 °C $U = 0.922$ V. That means that less electrical power is needed to drive the reaction in the desired direction to produce hydrogen and oxygen. As seen in Figure 14.13, at 1000 °C, the thermoneutral potential is 1.291 V. It is easy to conceive of a condition where the cell potential is below the thermoneutral potential. This means that, if available, high-temperature heat rather than electricity could be used to provide some of the energy needed to carry out the reaction. Consequently, these SOEs are viewed as a candidate for coupling with high-temperature gas-cooled nuclear reactors to provide both the electrical energy and high-temperature heat for optimal operation. SOEs also provide the advantage of fuel flexibility since, for example, they are capable of directly reducing CO$_2$ to CO or a combination of H$_2$O and CO$_2$ to syngas (H$_2$ and CO) if desired. The primary concern with SOEs is the lack of suitable materials to provide the lifetime needed for industrial applications.

In spite of recent and continuing developments, water hydrolysis is still more expensive, in general, than hydrogen generation from hydrocarbon sources. Consequently, only about 4% of hydrogen is currently produced from water. The largest electrolytic hydrogen plants worldwide are located in proximity to hydroelectric generation facilities in order to take advantage of inexpensive off-peak power.

Other Electrolysis Processes

There are several other electrolytic processes that may offer advantages over water electrolysis. Two key advantages include: (1) use of a waste stream to produce the desired fuel and perhaps other valuable products while simultaneously cleaning up the stream and (2) reduction of the voltage, and hence the energy, required for hydrogen production. It may also be possible to design or modify processes that have traditionally treated hydrogen as an undesirable by-product to produce hydrogen as one of the intended products. Two examples of alternative processes for hydrogen generation include the electrolysis of HCl and the electrolysis of NH$_3$.

Waste streams rich in ammonia can be treated by electrolysis to produce hydrogen gas while simultaneously cleaning up the stream. The equilibrium potential for the electrolysis cell is 0.06 V, much lower than that for water electrolysis (1.229 V). Thus, hydrogen production from ammonia waste streams would require much less energy than that required for water electrolysis. In a similar fashion, urea-contaminated wastewater can also be treated by electrolysis ($U = 0.37$ V). In addition, both of these processes contribute in a positive way to the reduction of the amount of fixed nitrogen, which has increased significantly as a result of human activity. Finally, another electrochemical process for hydrogen production involves the use of a photoelectrochemical cell. This process will be treated separately in Chapter 15.

Wastewater Treatment

Several of the processes just mentioned involve the use of electrochemical reactors in environment friendly ways to produce useful products while simultaneously cleaning up waste streams. Another use of electrochemical technologies is for the cleanup of industrial effluent streams containing dilute concentrations of toxic materials. To illustrate, electrochemical methods can be used to remove toxic metal ions and have been the subject of renewed interest as regulations have been tightened. For many metals, it is no longer feasible to meet the specified limits for effluent discharge with use of conventional hydroxide precipitation methods. Also, the cost of disposing of the precipitation sludge has increased dramatically as such disposal must ensure that ground contamination by leaching out of the metal ions does not occur. In addition, remediation of water pollution caused by low concentrations of pharmaceutical residues is of significant recent interest and can be done electrochemically. In the treatment below we assume the following:

- The concentration of the contaminant is dilute. Therefore, removal of the contaminant does not change the liquid flow rate to any appreciable extent.
- The removal or cleanup of the contaminant is mass-transfer limited.
- The inlet flow rate and concentration are known.
- The outlet target concentration is known (typically set by regulation).
- Operation is continuous and steady state.
- The reactor cross section is constant and variations only occur in one dimension along the length of the reactor.
- Axial diffusion is not significant.
- Excess supporting electrolyte is used to enhance the conductivity and reduce the potential drop in solution.

The situation that we consider here involves a three-dimensional porous electrode. Our objective is to estimate the size of reactor needed to clean up the stream to the desired level. The *flow-through* and *flow-by* configurations, as well as the required material and charge balances, were presented in Chapter 5. Key results are repeated here for convenience. The concentration distribution under mass-transfer control is a function of x only:

$$c_A = c_{A,in}\, e^{-\alpha x}, \quad \text{where} \quad \alpha = \frac{ak_c}{\varepsilon v_x}, \qquad (5.57)$$

where v_x is the actual velocity of the fluid in the pores in the direction of interest (εv_x would be the superficial velocity). Even though the local rate of reaction is controlled by mass transfer, current must still flow in solution between the upstream counter electrode and the three-dimensional electrode of interest. Again, as we saw in Chapter 5, the current in solution is

$$i_2 = nF\varepsilon v_x c_{A,in}\left(e^{-\alpha x} - e^{-\alpha L}\right). \qquad (5.60)$$

Finally, there is a potential drop in solution associated with the current flow. We consider the most common situation where $\sigma \gg \kappa$. Under such conditions, the change in overpotential is equal to the change in the potential in solution across the thickness of the electrode.

$$\Delta\phi_2 = \frac{\beta}{\alpha}\left(1 - e^{-\alpha L}\right) - \beta L e^{-\alpha L}, \qquad (5.62)$$

where $\beta = \dfrac{nF\varepsilon v_x c_{A,in}}{\kappa_{eff}}$.

You may be wondering why we care about the potential drop and change in overpotential if the system is at limiting current and controlled by mass transfer. To illustrate, consider Figure 14.14, which shows the current density as a function of overpotential. At limiting current, the current does not change with changing overpotential. However, if the overpotential is increased too much, the current will again increase due to side reaction(s). Side reactions should be avoided since they consume power and may have significant additional negative impacts on the process. For the process under consideration, side reactions are avoided by limiting the potential drop in solution, which limits the range of overpotentials in the system. The maximum $\Delta\phi_2$ depends on the specific chemical system, but is usually between 100–300 mV. Equation 5.62 can be rewritten as

$$\Delta\phi_2 = \phi_2|_{x=0} - \phi_2|_{x=L} = \frac{nFv_s^2}{\kappa_{eff}\, k_c a}\left(c_{A,in} - c_{A,out} - \frac{k_c a}{v_s}L c_{A,out}\right), \qquad (14.23)$$

where the superficial velocity, $v_s = \dot{V}/A_c$. Equation 5.57 can be used to relate the concentration at the outlet to the thickness L of the electrode as follows:

$$L = \frac{v_s}{ak_c}\ln\frac{c_{in}}{c_{out}}. \qquad (14.24)$$

In writing the equation in this form, we recognize that k_{cm} is a function of the velocity through the bed, v_x, which will vary with both $\dot{V}$ and A_c. The following correlation from Wilson and Geankoplis for packed beds at low Re can be used:

$$\frac{k_c \varepsilon}{v_s}(\text{Sc})^{2/3} = 1.09(\text{Re})^{-2/3}. \tag{14.25}$$

This correlation applies to a packed bed of spherical particles: $0.0016 < \text{Re} < 55$, $168 < \text{Sc} < 70{,}600$, and $0.35 < \varepsilon < 0.75$. Mass-transfer coefficients in electrochemical reactors typically vary between 10^{-6} and $10^{-4}\,\text{m·s}^{-1}$.

Our goal is to perform a preliminary calculation of the size of the flow-through reactor. In doing so, we note that there are multiple combinations of cross-sectional area and length that will provide the desired outlet concentration. Also, the simple model used here seems to indicate that the length of the porous bed should be as short as possible since a short length would minimize the potential drop and the pressure drop through the bed (Figure 14.15). However, practical issues such as the nonuniformity of current across a large area and the difficulty of fabricating and obtaining uniform flow over a thin bed with a large cross section become important. Where possible, we recommend that a cross-sectional area consistent with a commercially available electrolyzer be chosen in order to avoid the design and fabrication of a custom system. In practice, the final design decisions will be based on an optimal return on investment.

The preliminary design calculation can be performed as follows:

1. Choose an initial cross-sectional area for the reactor (best if aligns with that of a commercial available reactor).
2. From the total flow rate, cross-sectional area, and void fraction, calculate v_s and v_x.
3. Use v_x to estimate k_c for the bed with use of a correlation or from experimental data.
4. Calculate the length L of the bed with use of Equation 14.24.
5. Repeat the above steps for several possible values of the cross-sectional area until a value for L that is comparable to that of the height and width of the unit is obtained.
6. Use Equation 14.23 to check if the maximum potential drop has been exceeded for the chosen value of L.
7. If the absolute value of the potential drop is lower than the specified maximum, the initial sizing of the electrolyzer is complete. If the potential drop is too high, increase the cross-sectional area until the potential drop no longer exceeds the specified maximum.

Illustration 14.14 demonstrates this general procedure.

ILLUSTRATION 14.14

Removal of Hg from a dilute product waste stream. A stream containing 4 ppm by weight of Hg is to be cleaned to an outlet concentration of 0.05 ppm with use of a particulate flow-through electrode. The flow rate of the stream is $20\,\text{m}^3\cdot\text{h}^{-1}$. The packed bed is 45% porous and is made from particles that are 1 mm in diameter. The effective conductivity of the electrolyte is $10\,\text{S·m}^{-1}$, significantly lower than the electrical conductivity of the solid bed. The removal reaction is cathodic, and the anode is located upstream from the porous electrode. Please determine the size of the porous bed needed to carry out the desired reduction in Hg concentration. The maximum potential drop across the 3D electrode is 200 mV. The diffusivity of the soluble Hg species is $0.7 \times 10^{-9}\,\text{m}^2\cdot\text{s}^{-1}$. Assume a two-electron process.

Solution:

Initially, we choose $A_c = 1\,\text{m}^2$ as this size is frequently used in practice.

The superficial velocity is: $v_s = \dot{V}/A_c = (20\,\text{m}^3\cdot\text{h}^{-1})/(1\,\text{m}^2) = 20\,\text{m·h}^{-1} = 0.0056\,\text{m·s}^{-1}$

Using $\rho = 997\,\text{kg·m}^{-3}$, $\mu = 0.89\,\text{mPa·s}$, we find that $\text{Sc} = 1276$, and $\text{Re} = 6.22$; therefore, from

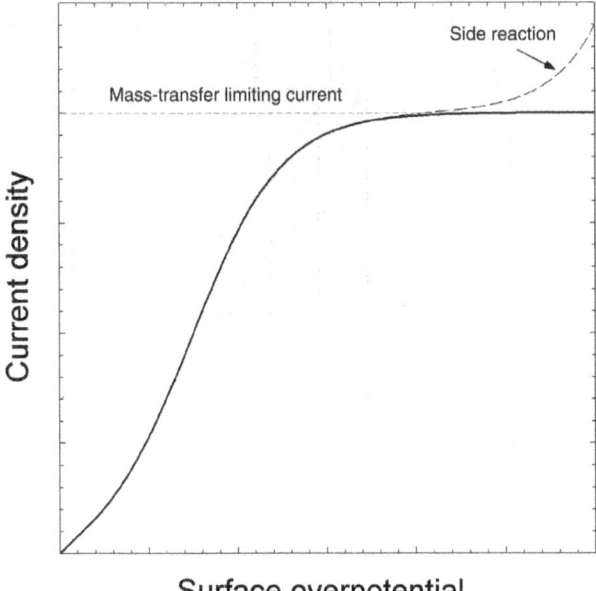

Figure 14.15 Current potential relationship with a side reaction.

Equation 14.25,

$$k_c = 3.4 \times 10^{-5} \text{ m} \cdot \text{s}^{-1}.$$

The specific area of the bed can be approximated from the particle size assuming spherical particles:

$$a = \frac{6}{d_p}(1-\varepsilon) = 3300 \text{ m}^{-1}$$

We can now calculate the length of the bed:

$$L = \frac{v_s}{ak_c} \ln \frac{c_{in}}{c_{out}} = \frac{0.0056 \text{ m} \cdot \text{s}^{-1}}{(3300 \text{ m}^{-1})(3.4 \times 10^{-5} \text{ m} \cdot \text{s}^{-1})}$$

$$\ln\left(\frac{4}{0.05}\right) = 0.22 \text{ m}.$$

Note that the units on the concentration do not matter for this calculation as long as the ratio is accurate. However, concentration in mol·m^{-3} is needed to check the potential drop. The values of the concentration are

$$c_{in} = \frac{0.0199 \text{ mol}}{\text{m}^3}; \quad c_{out} = \frac{0.000249 \text{ mol}}{\text{m}^3}.$$

We can now check the potential drop in solution:

$$\phi_2|_{x=0} - \phi_2|_{x=L} = \frac{nFv_s^2}{\kappa_{eff}\, k_c a}$$

$$\left(c_{A,in} - c_{A,out} - \frac{k_c a}{v_s} L c_{A,out}\right) = 0.10 \text{ V}.$$

This value is smaller than the specified maximum. Therefore, these dimensions are acceptable.

A_c [m²]	L [m]	$\Delta\phi$ [V]
1	0.22	0.10
0.75	0.27	0.16
0.5	0.35	0.32

Noting that L is a bit smaller than the height and width (assuming a square cross section), it should be possible to reduce the cross-sectional area and increase the thickness. The table provides a summary of values considered.

Examination of these values shows that L cannot be increased substantially without reaching the voltage limit of 0.2 V. In this case, one would look for a commercial reactor whose cross section is about 0.75 m².

14.8 REDOX-FLOW BATTERIES

The *redox-flow battery* is a battery in the sense that it is used to store and release energy. However, it operates much like a combination of a fuel cell (discharge) and an electrolyzer (charge). In contrast to typical secondary batteries where the reactants are part of the electrode, the reactants and products in a flow battery are contained within the electrolyte, which circulates through the cells. As we'll see in a moment, this situation allows for decoupling of the power and energy requirements for the system. There are numerous possible redox couples; here we will focus on the vanadium system as an example.

The vanadium redox-flow battery (VRB) employs two vanadium redox couples for the negative (V^{3+}/V^{2+}) and positive electrodes (VO_2^+/VO^{2+}) as follows:

$$V^{3+} + e^- = V^{2+} \quad (U^\theta = -0.26 \text{ V}) \quad (14.26)$$

$$VO_2^+ + 2H^+ + e^- = VO^{2+} + H_2O \quad (U^\theta = 1.00 \text{ V}) \quad (14.27)$$

Figure 14.16 illustrates how the system operates. A solution (anolyte) circulates through the negative electrode. A separate solution circulates through the positive electrode, the catholyte. Both solutions are acidic, and a proton exchange membrane is used to keep the solutions separate. During discharge, V(II) is oxidized to V(III) at the negative electrode. Electrons travel through the external circuit as usual, and the current in solution is carried primarily by protons. One mole of protons moves across the ion-

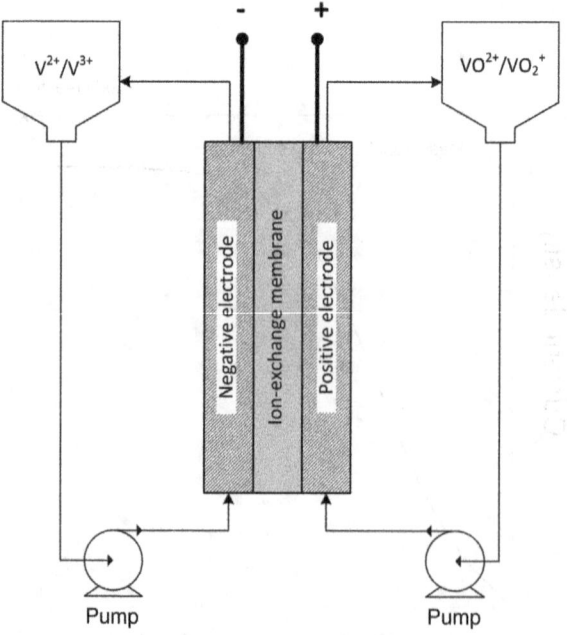

Figure 14.16 Vanadium redox-flow battery. Reactions and transport are shown for discharging.

exchange membrane for each mole of vanadium that is oxidized. These protons react with VO_2^+ and electrons from the external circuit to form VO^{2+} and water according to Equation 14.27.

As just mentioned, the negative and positive electrodes are separated by an ion-exchange membrane (see also PEM, Chapter 9) that prevents the mixing of electrolytes and allows for transport of protons between the electrodes. The ideal characteristics of this separator are low permeability to vanadium cations, high proton conductivity, as well as mechanical and chemical stability.

The electrochemical cell is sized to meet the maximum power [W] and voltage requirements. Independently, the energy requirement [W·h] can be accomplished simply by adding a larger storage tank for reactant-containing electrolyte. This decoupling is shown in Illustration 14.15.

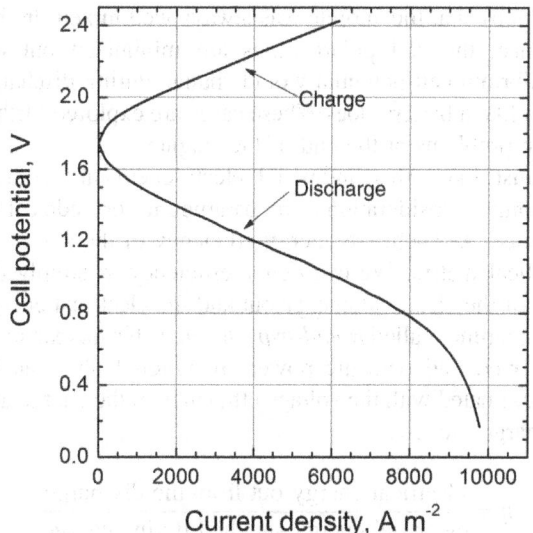

Figure 14.17 Example of polarization curve for the vanadium redox-flow battery at a high state of charge.

ILLUSTRATION 14.15

A VRB energy storage system is to provide 1 MW of power for a period of 4 hours at 400 VDC. If a cell operates at $4\,kA\cdot m^{-2}$ and 1.23 V, determine the total cell area and number of cells in the cell stack. Next, calculate the size of the storage tanks. Assume a one-electron reaction and 1 M concentration for both electrodes.

Cells are sized for power. First, estimate total cell area needed to satisfy power requirements:

$$A = \frac{P}{iV_{cell}} = \frac{1 \times 10^6\,W}{(4000\,A\cdot m^{-2})1.23\,V} = 203\,m^2.$$

The number of cells and individual cell size are

$$m = \frac{V_s}{V_{cell}} = \frac{400\,V}{1.23\,V} = 325; \quad A_{cell} = \frac{A}{m} = \frac{203}{325} = 0.625\,m^2.$$

Tanks are sized to meet the energy requirements. The volume of each tank needed for a 4 hour capacity is

$$\mathbb{V} = \frac{Ait}{Fc} = \frac{Pt}{FcV_{cell}}$$

$$= \frac{(1 \times 10^6\,W)(4)3600\,s}{96{,}485\,C\cdot mol^{-1}\,1000\,mol\cdot m^{-3}\,1.23\,V} = 121\,m^3.$$

Note that the discharge time does not enter into the determination of cell area. Similarly, the volume depends on the product of power and time (energy) and is independent of the rate.

The analysis to find the relationship between current density and potential closely follows the developments made in earlier chapters for electrochemical cells. During discharge, the potential of the cell is the equilibrium potential minus the ohmic, kinetic, and concentration polarizations. Example charge and discharge curves are shown in Figure 14.17.

There are many ways to operate the system. It is instructive to write the equilibrium potential for the VRB. With activity coefficients neglected, the change in cell potential with reactant and product concentrations is clear.

$$U \approx U^\theta + \frac{RT}{2F} \ln \frac{c_{V^{2+}}^2 c_{VO_2^+}}{c_{V^{3+}}^2 c_{VO^{2+}}}. \quad (14.28)$$

To minimize the volume of anolyte needed, we would like to convert nearly all of the V(II) to V(III). There are a couple of ways to effect this conversion. First, we could have the anolyte pass through the cells just once. Achieving full conversion would mean that the concentration of our reactant, V(II), would go toward zero at the exit of the cell. At that point, the equilibrium potential would drop and kinetic and mass-transfer polarizations would become very large. The negative effects associated with this approach could be mitigated to some degree by lowering the operating current density, but that would increase the size of the cell. A second option is to recirculate the anolyte as shown in the figure so that the reactants make many passes through the cell. The conversion during each pass is not high, but after many passes a high overall conversion is achieved. In the extreme, let's imagine that we circulated so rapidly that the anolyte was always well mixed. In this instance, the cell polarizations are minimized but the equilibrium cell potential would change during discharge much like a battery does. These cases are explored further in the problems at the end of the chapter.

Just like other industrial electrochemical systems, economic considerations are paramount for redox-flow batteries. As such, the energy efficiency of the system is a critical metric. We can define efficiency as simply the ratio of the electrical energy out and the electrical energy in, sometimes called *round-trip efficiency*. If side reactions, crossover, and parasitic power are ignored, this can be approximated with the voltage efficiency of the charge and discharge process,

$$\eta = \frac{\text{electrical energy out from the discharge}}{\text{electrical energy supplied during charge}}$$

$$\approx \frac{\eta_V^{\text{discharge}}}{\eta_V^{\text{charge}}} = \frac{V_{\text{cell}}^{\text{discharge}}}{V_{\text{cell}}^{\text{charge}}} \tag{14.29}$$

Using the data from Figure 14.17, we can estimate the current density required to achieve a round-trip efficiency of 65% to be about 2400 A·m^{-2} (1.35/2.09 = 0.65).

Redox couples for flow batteries

The vanadium couple is only one of many that are possible for the redox-flow batteries. The key factor in commercialization of flow cells is, of course, economics, and here the cost of vanadium is quite high. The ideal couple would have

- Highly reversible reactions
- Earth abundant materials
- Low-toxicity
- Noncorrosive
- High solubility
- High equilibrium potential

CLOSURE

Industrial electrolysis uses electrical energy to create desirable products via electrochemical processes. In this chapter, we considered several industrial processes of economic importance, including the production of chlorine and aluminum. We also applied the fundamental principles learned previously to help us understand the operation of electrolysis cells. Design criteria for electrochemical reactors were discussed and applied, and methods for approximating reactor size were demonstrated. The thermal management of industrial cells was also considered, with the help of an overall energy balance. Finally, we examined several ways in which electrochemical processes can be used to help achieve a sustainable future. As with any commercial product, economic considerations are critical and the cost of electricity often drives the feasibility of large industrial processes.

FURTHER READING

Goodridge, F. and Scott, K. (1995) *Electrochemical Process Engineering*, Plenum, New York.

Hine, F. (1985) *Electrode Processes and Electrochemical Engineering*, Plenum Press, New York.

Muthuvel, M. and Botte, G. (2009) *Trends in Ammonia Electrolysis: Modern Aspects of Electrochemistry*. Springer, 45, 207–243.

Newman, J. and Thomas-Alyea, K.E. (2004) *Electrochemical Systems*, John Wiley & Sons, Inc., Hoboken, NJ.

Pletcher, D. and Walsh, F.C. (1993) *Industrial Electrochemistry*, 2nd edn, Springer Science+Business Media, LLC.

Rajeshwar, K. and Ibanez, J.G. (1997) *Environmental Electrochemistry: Fundamentals and Applications in Pollution Sensors and Abatement*, Academic Press, San Diego.

Steckhan, E. (2012) Electrochemistry, 3. Organic Electrochemistry, *Ullmann's Encyclopedia of Industrial Chemistry*, Wiley-VCH Verlag GmbH, Weinheim, Germany.

Tobias, C.W. (1959) Effect of gas evolution on current distribution and ohmic resistance in electrolyzers. *J. Electrochem. Soc.*, **106**, 833.

PROBLEMS

14.1 What is the minimum energy required to produce a kg of NaOH from a brine of NaCl? Compare this number with typical values reported industrially of 2100–2500 Wh·kg^{-1}. What are some factors that account for the difference? A rule of thumb for the chlor-alkali industry is that two-thirds of the production costs are electrical energy. If the cost of electricity is $0.06/kWh, estimate the cost to produce a kg of Cl$_2$.

14.2 The third and most modern design for the chlor-alkali process uses an ion-exchange membrane instead of the porous diaphragm. The membrane allows cations to permeate through but is an effective barrier for anions and water. How would the cell design change for this approach? Identify possible advantages and disadvantages to the membrane design.

14.3 An alternative chlor-alkali process has been proposed. Rather than evolving hydrogen, the cathode for a membrane cell design is replaced with an oxygen electrode.

Write the cathodic reaction at the oxygen electrode. Compare the equilibrium potential and theoretical specific energy for Cl_2 production with that for the conventional cell. What are some advantages and challenges with this concept?

14.4 HCl (g) is a waste product in a number of chemical processes. Ideally, the HCl could be converted to $Cl_2(g)$ economically. There are several paths to do this conversion including the Deacon process, which is not electrochemical. You are investigating the direct conversion of anhydrous HCl in an electrochemical cell analogous to a proton exchange membrane fuel cell. A proton-conducting membrane serves as the separator and electrolyte. Write the overall and electrode reactions for this process. What is the equilibrium potential?

14.5 Calculate the theoretical specific energy ($W \cdot h \cdot kg^{-1}$) to produce Al. Use the reaction shown in Section 14.5 at standard conditions. The U.S. Geological Survey provides the commodity price of Al, http://pubs.usgs.gov/sir/2012/5188/. If the energy efficiency of the process is 30%, what is the maximum price of electricity needed to make the process feasible?

14.6 An aluminum electrowinning process operates at 960 °C. At these conditions, the equilibrium potential is 1.22 V. If the cell is operating at 4.19 V with a faradaic efficiency of 92%, what is the energy efficiency of the process?

14.7 How much electrode area is needed to supply the world with Al? Use a production rate of 20 million metric tons per year. Assume the current density is $900 \text{ A} \cdot m^{-2}$ and a faradaic efficiency of 0.97. What is the annual release of carbon dioxide to the atmosphere? For comparison, in the United States, about 1.5 billion metric tons per year of carbon dioxide are emitted by the transportation sector.

14.8 During electrowinning of aluminum, some of the deposited material dissolves back into the melt. If the rate of dissolution is constant ($x \text{ g} \cdot s^{-1}$), develop a relationship for the faradaic efficiency of the process as a function of current density, i. Do the experimental data in the table support this model?

$i \text{ [A} \cdot m^{-2}]$	η_f
441	32.8
803	59.0
1190	71.9
5010	93.5
7520	94.6
9990	94.4

14.9 Use the parameters from Section 14.3 for the chlor-alkali cell to develop a polarization curve. If the total cost (installation and operating) is given by $0.01 A_c + 5 V_{cell}$, at what current density are costs minimized?

14.10 A key advantage of the DSA for the chlor-alkali process is the removal of gas from the electrode gap. Consider an older design where the gas bubbles remain in the gap. Estimate the optimum electrode gap (minimum ohmic loss) for a 1 m long cell operating at 65 °C, 150 kPa, and $2500 \text{ A} \cdot m^{-2}$. If it were desired to keep the ohmic loss to less than 200 mV to achieve high-energy efficiency, at what current density would the cell need to operate? Use 0.5 mm for the bubble diameter.

14.11 A chlor-alkali plant has a capacity of 35,000 metric tons chlorine per year. If electricity costs are 0.06 [\$·kWh^{-1}], what are the annual savings and reduction in production costs per ton of chlorine for each mV of reduction in overpotential? Assume a faradaic efficiency of 0.95. The dimensionally stabilized anode (DSA) is able to reduce the overpotential by about 1 V at $10 \text{ kA} \cdot m^{-2}$. Does describing the DSA innovation as "one of the greatest technological breakthrough of the past 50 years of electrochemistry" seem justified?

14.12 For the diaphragm process shown in Figure 14.1, develop equations for material balances around the cell. Show how to relate the composition of the cathode liquor (caustic) to design parameters. Assume a production rate of $20 \text{ kg} \cdot h^{-1}$ chlorine per hour and that the feed is a saturated brine of NaCl. Include parameters for current efficiency, solubility of chlorine in the brine, and back diffusion of OH^-.

14.13 A gas-evolving electrode operates with a gap of 4 mm and the height of the electrode is 0.5 m. The two-electron reaction with a gas-phase product takes place in an undivided cell (no separator) at an average current density of $200 \text{ A} \cdot m^{-2}$. The pressure is 150 kPa and the temperature is 35 °C. Use a bubble velocity of $6 \text{ mm} \cdot s^{-1}$. The conductivity of the solution is $5 \text{ S} \cdot m^{-1}$. It is proposed to reduce the gap to 2 mm to reduce ohmic losses. Is this a good idea? Explain your answer and include a sketch of the current distribution for the two cases.

14.14 For the following processes, indicate whether a divided cell is needed and why. Also indicate whether a bipolar configuration is feasible: (a) electrorefining of tin, (b) production of naphthoquinone (Illustration 14.10), (c) redox-flow battery, and (d) production of adiponitrile.

14.15 A monopolar cell for electrowinning of copper is an open tank that contains multiple parallel plate electrode pairs electrically connected in parallel. If the distance from cathode to cathode is 75 mm, the current density is $280 \text{ A} \cdot m^{-2}$, and the faradaic efficiency is 0.84, calculate the space–time yield. Why is there one more anode than cathode in the cell? Discuss any advantages that might result from making the anodes and cathodes slightly different sizes.

14.16 The cell for electrowinning of zinc is made up of 100 electrode pairs in a monopolar arrangement. Each electrode is 1.4×0.7 m. Given an electrode gap of 30 mm, and the $i = 3000$ A·m^{-2}. The rate of flow of electrolyte is 120 m^3·h^{-1}, and enters the cell at a concentration of 100 kg Zn^{2+} m^{-3}. How would you propose to flow electrolyte to the cells. Sketch the current density on the electrode. Why wouldn't bipolar stack be practical for this application?

14.17 You are setting up a tank house for the electrowinning of Cu. The electrical supply is 180 VDC and 40 kA. For a 1 m^2 cell operating at 300 A·m^{-2}, the potential of the cell is 2.0 V and the faradaic efficiency is 85%. Using these conditions, how would you configure the tank house? If the equilibrium potential is 1.45 V, what is the energy efficiency of the process? The cathode cycle is 4 days, during this time by how much does the mass of the cathode increase? Assuming that the tank house operates 340 days per year, what is the annual production rate of copper?

14.18 A large tank house for the electrowinning of zinc contains 200 tanks, each containing 100 electrodes with an area of 1.3 m^2 (both sides) in a bipolar arrangement. The electrolyte is acidic. If the operating current density is 490 A·m^{-2}, and the cell potential is 3 V, what is the steady-state cooling requirement for the tank house? If the rate of flow of electrolyte through the cells is 4800 m^3·h^{-1}, estimate the change in temperature of the electrolyte through the tank house. Use the heat capacity and density of pure water for this calculation.

14.19 An alternative to the carbon anode in the electrowinning of aluminum is the so-called inert anode. The cathode reaction is unchanged, but here oxygen is evolved instead of consumption of carbon. Write the overall reaction for the inert process. Compare the standard potential for the reaction with the reaction from Equation 14.13. The Hall–Héroult process is already notoriously inefficient. What then are the possible advantages of the inert-anode process?

14.20 Consider the wastewater cleanup with porous electrodes shown in Illustration 14.14. If the particle size were increased to 2 mm, to what value would the effective conductivity need to be increased in order to keep the Hg within the limits. Use an area of 0.75 m^2.

14.21 Lithium metal is produced using an electrolytic process. The electrolyte is LiCl-KCl eutectic melt at 427 °C. At these operating conditions, the equilibrium potential is 3.6 V. What is the cathodic reaction? The faradaic efficiency for lithium is reported to be 95%, and the energy consumption is 35 kWh·kg^{-1}. At what potential is the cell operating? What is the energy efficiency of the cell?

14.22 The growth in the lithium-ion battery market has raised demand for lithium. Rechargeable batteries typically use lithiated metal oxides, and the precursor is LiOH not lithium metal. Describe a method to produce LiOH by a process similar to that used for caustic soda using an ion-exchange membrane.

14.23 Calculate the power needed to produce 100 million kg·yr^{-1} of adiponitrile. Assume the faradaic efficiency is 95% and that the cells operate at a potential of 4.6 V.

14.24 A plant produces 20,000 metric tons of adiponitrile per year. If the power requires is 50 MW, calculate the specific energy (kWh·kg^{-1}) and energy efficiency of the process assuming a faradaic efficiency of 90%. Use 3.08 V for the equilibrium potential of the reaction. How might hydrogen evolution be avoided?

14.25 Compare the energy efficiency of the alkaline electrolyzer from Illustration 14.13 with the efficiency that you would achieve if the current density were doubled and the voltage of the cell increased to 2.25 V. Look at energy efficiency and space–time yield.

14.26 Electrolyzers and fuel cells are envisioned to be a part of an energy storage system for the electrical grid. When supply exceeds demand, hydrogen is generated and stored. When demand exceeds supply, this stored hydrogen is used in a fuel cell to generate electricity. Using the efficiency for electrolysis from Illustration 14.13 and assuming a fuel-cell system efficiency of 60%, what is the round-trip efficiency?

14.27 Calculate change in equilibrium potential as a function of SOC for vanadium redox-flow battery. Assume that the junction region is an ion-exchange membrane that only allows transport of water and protons and completely excludes anion and vanadium ions.

α \quad\quad\quad α'

Pt(s), V^{2+}, V^{3+} H$_2$SO$_4$(aq) \quad\quad\quad Pt(s), VO^{2+}, VO$_2^+$ H$_2$SO$_4$(aq)

14.28 An iron–chromium redox-flow battery energy storage system is to provide 8 MW of power for a period of 3 hours at a minimum of 650 VDC. The two reactions are

$$Fe^{3+} + e^- \leftrightarrow Fe^{2+}$$
$$Cr^{3+} + e^- \leftrightarrow Cr^{2+}$$

(a) What is the standard potential of the full cell?

(b) Assume that the potential of the cell can be treated as ohmically limited. Using the equilibrium potential calculated in part (a) and a resistance of 0.05 mΩ·m^2, what cell area is required to provide the desired power while maintaining a round-trip efficiency of 80%? Assume that the current density is the same for charging and discharging.

(c) If it is impractical to have single cells with an area of greater than 1 m², what series–parallel configuration would you recommend?

(d) What is the total volume of electrolyte needed if the solubility of reactants is 0.7 M?

14.29 For the system shown in Illustration 14.15, calculate the rate of heat removal required to maintain a constant temperature. Use a utilization of 0.8 for both the anolyte and the catholyte. The heat capacity can be approximated as that of water. Equation 7.20 can be used to estimate the rate of heat generation.

Heinz Gerischer

Heinz Gerischer was born on March 31, 1919 in Wittenberg, Germany. He studied chemistry nearby at the University of Leipzig. Gerischer endured tragedy as a young man. His studies were interrupted for 2 years by military service during World War II. This ended with expulsion from the army in 1942 because his mother was born Jewish. She committed suicide in 1943, and his one sister escaped a Gestapo prison only to be killed later in an air raid in 1944.

Some of the details of Gerischer's early life are unclear, but we know that he was able to complete his studies. He received his doctorate in 1946 from the University of Leipzig. His dissertation examined periodic oscillations on an electrode surface. Gerischer spent the next several years in Göttingen and then Berlin. His early work examined kinetics at metal electrode surfaces. He played an important role in the development of electroanalytical techniques and the potentiostat, a mainstay in any electrochemical laboratory today (*Z. Electrochem.*, **61**, 789 (1957)). He received his habilitation from the University of Stuttgart in 1955. It was here in the mid-1950s that Gerischer was inspired by work at AT&T Bell Laboratories on semiconductors. On hearing a report in 1956 of the work of Brattain and Garrett on semiconductors in contact with electrolytes, Gerischer said, "I realized that a new type of electrochemical process occurred at these interfaces." This phenomenon resulted from the two types of reactions that could occur, either using electrons from the conduction band or from the valence band, and this is the main topic of this chapter. Gerischer's intense efforts, passion, and keen insight made him a pioneer in the electrochemistry of semiconductors. He published seminal papers on the topic in the late 1950s and early 1960s in *Zeitschrift für Physikalische Chemie* focusing on the theory of redox reactions at semiconductor surfaces.

In recognition of his growing contributions, Gerischer moved to Technical University of Munich in 1962 to head the electrochemistry department. In 1964, he became head of the physical chemistry department. In 1967–1968, he spent a year sabbatical with Professor Charles Tobias (Chapter 1) at University of California, Berkeley. It wasn't long before he was offered a new and better position in Berlin. From 1969 to 1987, he was Director of the Fritz Haber Institute of the Max Planck Society in Berlin. For many years, he was the foremost electrochemist in Europe. Gerischer retired in 1987 and then spent 2 years as a visiting scholar at Berkeley. Here he offered a course in semiconductor electrochemistry, which one of us (TF) audited as a first-year graduate student.

He published more than 300 papers in the areas of electrocatalysis, photoelectrochemistry, electrochemistry of semiconductors, and corrosion. Over his long and productive career, Gerischer received numerous awards. He served as President of the International Society of Electrochemistry (1971–72) and was awarded the Palladium Metal of the Electrochemical Society in 1977. He died unexpectedly on September 14, 1994. In 2001, the Electrochemical Society founded the Heinz Gerischer Award of the European Section of the Electrochemical Society. This biennial award recognizes the scientific leadership of Professor Gerischer and honors his seminal contributions to the science of the electrochemistry of semiconductors.

Image Source: Reproduced with kind permission of Archives of the Max Planck Society, Berlin.

Chapter 15

Semiconductor Electrodes and Photoelectrochemical Cells

Up to this point in the book, we have only considered electrochemical reactions at metal–electrolyte interfaces. In Chapter 3, we presented a model for reactions at those interfaces and have used that model throughout the text to describe a variety of electrochemical systems. In this chapter, we examine the behavior of a semiconductor in an electrolyte solution. As we will see, there are some important differences between the semiconductor–electrolyte interface and the interface that exists between a metal and the electrolyte that lead to different electrochemical behavior. A quantitative description of that behavior will be developed for simplified systems as we introduce this important topic.

Why semiconductors? Metals have excellent properties and are used pervasively in electrochemical systems. In general, semiconductors are less conductive, more expensive, more difficult to fabricate, and inferior in performance to conducting materials for electrochemical applications. Given that, what motivates us to consider semiconductors? In a word, light. Semiconductors are photosensitive and, therefore, have the potential to use light to generate power and fuels electrochemically in photoelectrochemical cells. Photoelectrochemical processes can also be used for cleanup of contaminated process streams.

15.1 SEMICONDUCTOR BASICS

Solid materials can be classified by their ability to conduct electrical current as metals, semiconductors, and insulators. The ability to conduct is directly related to the availability of and mobility of charge carriers (e.g., electrons). As you may recall from your basic chemistry course, electrons can only occupy discrete energy levels in atoms. Hence, you will remember filling orbitals with electrons in unique states as you learned about the periodic table. For atoms in a crystalline solid, discrete atomic levels with similar energy blend together to form energy bands as illustrated in Figure 15.1 for a material with an energy gap between the two bands. The valence band is the highest energy band that contains electrons, and the conduction band is the lowest energy band that is unoccupied. Bands that are completely filled with electrons or are empty do not conduct. The energy band situation for solid materials is illustrated in Figure 15.2. Electrons in a metal reside in a band that is only partially filled. Hence, a large number of electrons can easily become excited from the occupied states to the unoccupied states where they are free to move between atoms and provide electrical conduction. In a semiconductor, the valence band is completely full (0 K), and there is an energy gap between the valence band and the conduction band. This difference in energy between the valence band maximum (E_V) and conductivity band minimum (E_C) is known as the band gap (E_g). In order for conduction to occur, electrons must be excited from the valence band to the conduction band. Fortunately, the energy gap is sufficiently small that thermal energy is able to excite a certain number of electrons into the conduction band, leading to conductivity. Importantly, the number of charge carriers is critical for semiconductor conductivity. With insulators, the energy gap is too large for a significant number of charge carriers to be excited into the conduction band, leading to very low electrical conductivity.

Figure 15.3 illustrates thermal excitation of an electron from the valence band to the conduction band, leaving in its place an electron vacancy that is known as a *hole*. This results in the ability to move charge in both bands and gives rise to conductivity. In a pure semiconductor devoid of impurities or dopants, the number of free electrons in the conduction band and holes in the valence band are equal, and the resulting conductivity is known as the *intrinsic*

Electrochemical Engineering, First Edition. Thomas F. Fuller and John N. Harb.
© 2018 Thomas F. Fuller and John N. Harb. Published 2018 by John Wiley & Sons, Inc.
Companion Website: www.wiley.com/go/fuller/electrochemicalengineering

356 Electrochemical Engineering

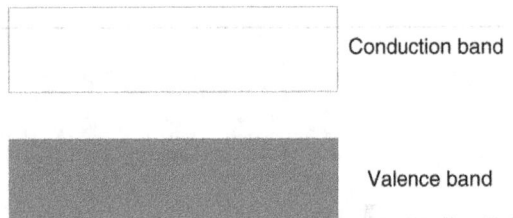

Figure 15.1 Illustration of energy bands formed in a crystalline solid.

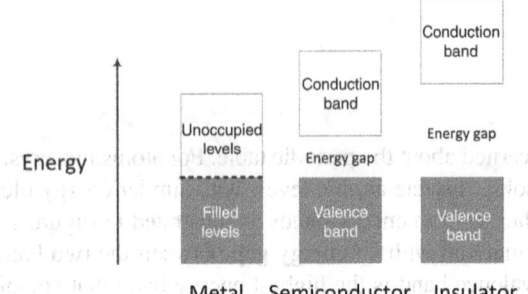

Figure 15.2 Energy bands and energy (band) gaps for different types of solid materials.

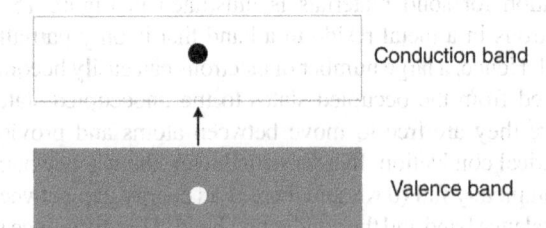

Figure 15.3 Excitation of an electron from the valence band to the conduction band of a semiconductor.

conductivity. The charge carriers in the conduction band are electrons, while the holes (missing electrons) are the charge carriers in the valence band. These holes move like positive charges, and are a convenient way to represent charge movement in the valence band. In reality, it is electrons that move by breaking and reforming bonds in the valence band. Because of this, electrons in the conduction band have a higher mobility, meaning that they move more easily in response to a potential gradient than do holes since their movement does not require the making and breaking of chemical bonds. In intrinsic (pure) semiconductors, the number of charge carriers, and thus the intrinsic conductivity, is strongly influenced by the crystallographic quality of the semiconductor because defects in the crystal structure increase the recombination rate of the electron–hole pairs. The rate of recombination is closely tied to a parameter called carrier lifetime, which is the average amount of time that a charge carrier can exist in an excited state before recombination occurs. Carrier lifetimes are a few microseconds for high-quality silicon, but can be much shorter for defect-riddled oxide semiconductors. The number of carriers depends on temperature, as you would expect, with more carriers and higher conductivity at higher temperatures. For intrinsic silicon, the number of carriers at room temperature is 9.65×10^9 carriers per cubic centimeter, and the room temperature conductivity is $3.1 \times 10^{-6}\,\text{S·cm}^{-1}$.

If intrinsic conductivity were the end of the story, semiconductors would not be particularly interesting. One of the key advantages of semiconductors is that their conductivity can be changed by *doping* with impurity atoms. For silicon, which is the most commonly used semiconductor material in photovoltaic cells, each Si atom in the pure material shares an electron with the four adjacent Si atoms in order to provide a stable outer shell with eight electrons as illustrated in Figure 15.4a. Dopants for silicon typically come from the elements in the columns just to the left and right of silicon in the periodic table. Elements to the right of silicon, such as phosphorous or arsenic, have an additional valence electron compared to silicon. Therefore, when one of these elements displaces a silicon atom in the crystal matrix, there is an extra electron that is easily excited or donated to the conduction band (see Figure 15.4b). These donated electrons in the conduction band serve as charge carriers and lead to *extrinsic conductivity*, or conductivity caused by doping. *Donors* such as

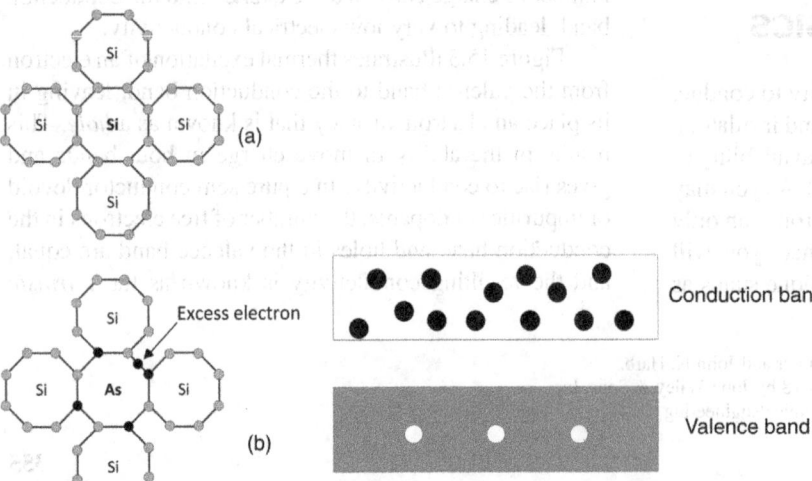

Figure 15.4 (a) Undoped Si. (b) An illustration of an *n*-type extrinsically doped Si with electrons as the majority carriers.

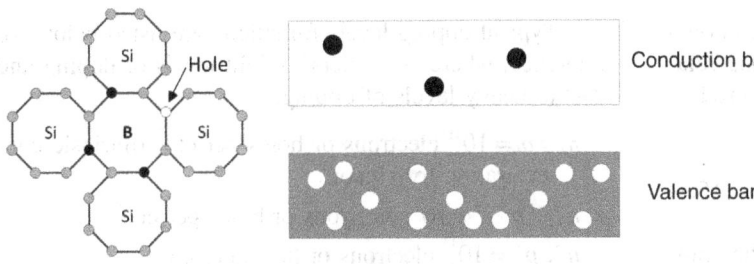

Figure 15.5 Illustration of a *p*-type extrinsically doped semiconductor with holes as the majority carriers.

phosphorous are called *n*-type dopants (*n* is for negative, since the resulting charge carriers are negative electrons).

In contrast, elements in the column to the left of silicon, such as boron, contain one less valence electron. When integrated into the silicon matrix, these elements tie up or "accept" an electron, giving the dopant atoms a net negative charge, as shown in Figure 15.5. Since that electron comes from the valence band, it generates a hole in the valence band, resulting in extrinsic conductivity or conductivity that is due to the presence of a dopant. Acceptors such as boron are called *p*-type dopants (*p* is for positive, since the resulting charge carriers are positive holes). As noted previously, electrical conduction can occur by the holes or the electrons. The more abundant carriers are called *majority carriers*, and the less abundant carriers are called *minority carriers*. In a *p*-type semiconductor, holes are the majority carriers while electrons are the minority carriers.

Figure 15.6 qualitatively illustrates the energy levels associated with the dopant atoms and their relation to the energy band of the semiconductor. Because of the proximity of the energy levels to the respective band, the thermal energy at room temperature is sufficient to excite essentially all of the electrons from the dopant atoms to the conduction band (*n*-type), or to permit all of the dopant atoms to capture an electron from the valence band (*p*-type). This condition is referred to as full ionization, and it is a good assumption for all temperatures at or above room temperature. The net result is a large concentration of the majority carriers, a smaller concentration of the minority carriers, and fixed charges associated with ionized dopant atoms that are incorporated into the crystal structure of the semiconductor. This situation is illustrated in Figure 15.7, where the charges that are circled represent the fixed charges. The net charge is zero as one would expect. The fixed charges associated with the dopant atoms play an important role, especially at the interface, in the behavior of semiconductors in electrochemical systems as will be seen later in this chapter.

Some of the important quantities associated with semiconductors are defined below. Note that the "number of" portion of the units if frequently left off and is implied. For example, *n* is frequently expressed in cm^{-3} rather than number of electrons cm^{-3}.

n = number of free electrons cm^{-3}

n_i = number of free electrons cm^{-3} in the intrinsic (undoped) semiconductor.

p = number of holes cm^{-3}

N_D = number of donor atoms cm^{-3}

N_A = number of acceptor atoms cm^{-3}

Assuming full ionization,

$n \sim N_D$ for *n*-type semiconductors,

$p \sim N_A$ for *p*-type semiconductors.

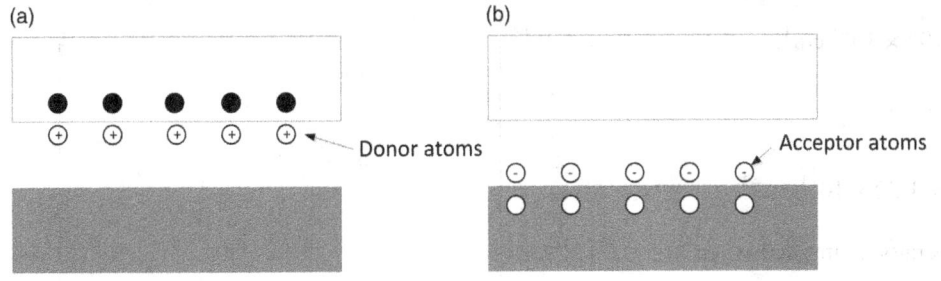

Figure 15.6 Illustration of energy levels and ionization for (a) *n*-type and (b) *p*-type dopants.

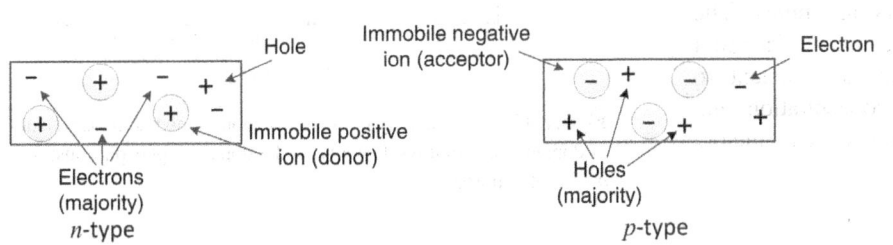

Figure 15.7 Carriers and immobile ions in extrinsic semiconductors.

If the concentration of the majority carriers is known (e.g., from the dopant concentration), the following relation, known as the law of mass action, can be used to calculate the concentration of the minority carriers:

$$np = n_i^2. \quad (15.1)$$

Illustration 15.1 demonstrates how to determine the number of carriers from the dopant level. The conductivity of intrinsic semiconductors is limited by the number of charge carriers available, and even low dopant concentrations can dramatically influence the number of charge carriers and the electrical conductivity.

ILLUSTRATION 15.1

Extrinsic Doping of Silicon
Silicon is doped with 1 ppb (by weight) of indium. What is the dopant concentration (cm^{-3}) at room temperature? Is the doping level sufficient to alter the properties of the silicon? Which is the majority carrier, electrons or holes? The density of silicon is 2.33 g·cm^{-3}; the molecular weight of indium is 114.82 g·mol^{-1}.

Solution:

Assume 1×10^{-9} g indium and 1 g silicon. First calculate the number of In atoms in 1×10^{-9} g.

$$(1 \times 10^{-9} \text{ g In}) \left(\frac{\text{mol}}{114.82 \text{ g}} \right) \left(\frac{6.02 \times 10^{23} \text{ atoms}}{\text{mol}} \right) = 5.243 \times 10^{12} \text{ atoms.}$$

Now calculate the volume from the silicon density:

$$\left(\frac{\text{cm}^3}{2.33 \text{ g}} \right) (1 \text{ g Si}) = 4.292 \times 10^{-1} \text{ cm}^3.$$

The dopant concentration is

$$N_A = \frac{5.243 \times 10^{12} \text{ atoms}}{4.292 \times 10^{-1} \text{ cm}^3} = 1.22 \times 10^{13} \text{ cm}^{-3}.$$

Note that "atoms" in the numerator is implied when you write cm^3.

Since In is a p-type dopant (three electrons in outer shell), the majority carriers are holes. The intrinsic carrier concentration for silicon is about 10^{10} cm^{-3}. Therefore, even 1 ppb is sufficient to significantly change the carrier concentration and hence the electronic properties of the semiconductor.

Typical doping levels for silicon are listed below for reference, where "−" refers to light levels of doping and "+" to heavy levels of doping.

$n_i = p_i = 10^{10}$ electrons or holes per cm^3 (intrinsic concentration for silicon)
$n^{--}, p^{--} < 10^{14}$ electrons or holes per cm^3
$n^{-}, p^{-} = 10^{15}$ electrons or holes per cm^3
$n, p = 10^{17}$ electrons or holes per cm^3
$n^{+}, p^{+} = 10^{19}$ electrons or holes per cm^3
$n^{++}, p^{++} > 10^{20}$ electrons or holes per cm^3

The resistivity of silicon as a function of the dopant concentration is shown in Figure 15.8 for both n-type and p-type dopants. Note that the resistivity of the p-type semiconductors is higher, owing to the lower mobility of the holes relative to the mobility of electrons. Perhaps more important, all of the values of resistivity shown in Figure 15.8 are significantly higher than those typically associated with metals. Why, then, are semiconductors useful? The answer to this question lies, at least in part, in the fact that we can locally modify semiconductor properties to create very small devices called transistors that form the basis for the microelectronics industry. Such devices are beyond the scope of this text, but we encourage students to explore this topic on their own! In the rest of this chapter, we will examine the behavior of semiconductors as part of electrochemical systems.

15.2 ENERGY SCALES

Before going any further, we need to relate the energy scale commonly used with semiconductors to the standard hydrogen electrode (SHE) scale familiar to electrochemists and electrochemical engineers. As you know, the SHE scale uses

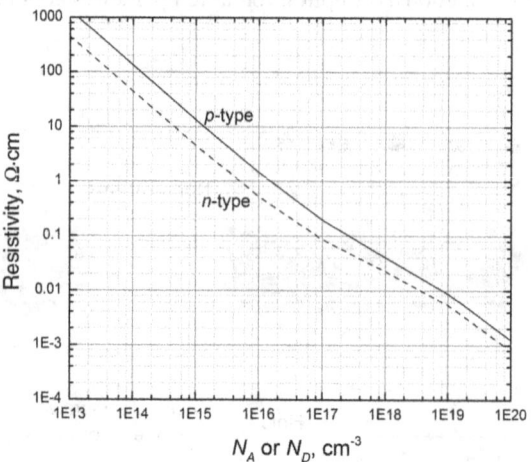

Figure 15.8 Resistivity of silicon at room temperature as function of dopant concentration. For n-type, the dopant is phosphorous; for p-type, it is boron.

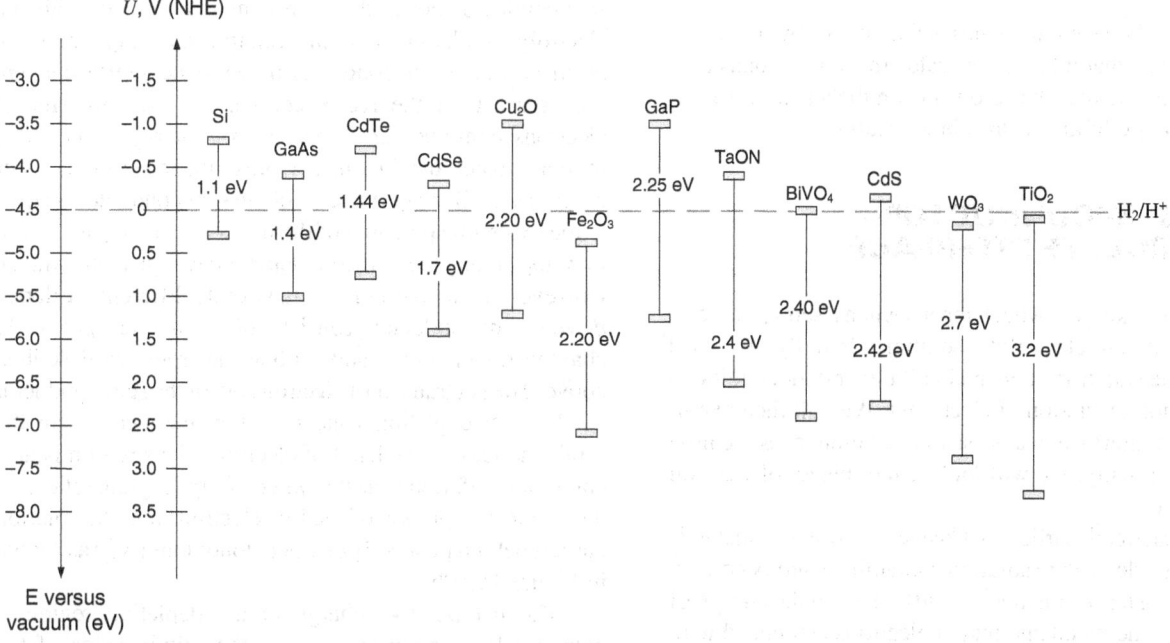

Figure 15.9 Band gap energies at 300 K as well as location of the band edges for various semiconductors. Both vacuum and hydrogen scales are shown. Solids with band gaps larger than about 3 eV are effectively insulators.

the hydrogen electrode under standard conditions as the reference potential for all electrochemical reactions. In contrast, the most common reference state used for semiconductors is an electron energy equal to zero in a complete vacuum ($E_{vacuum} = 0$). On this vacuum scale, the energy of an electron bound in a solid (expressed in electron volt [eV]) is typically negative since its energy is lower than that of a free electron in a vacuum. The SHE potential scale can be converted to an energy scale (eV) by multiplying by the charge on an electron. In doing so, one finds that the scale increment is the same for both scales (see Illustration 15.2) and that the hydrogen scale is simply shifted relative to the vacuum scale. Specifically,

$$E_{vacuum} \text{ [eV]} = -4.44 - qU^\theta. \qquad (15.2)$$

As we discussed earlier in Chapter 3, redox couples with positive potentials relative to hydrogen have lower electron energies. Therefore, positive values of U^θ yield energies more negative than -4.44 eV on the vacuum scale. Figure 15.9 shows the band gap energy and positions of the band edges (E_C and E_V) for several different semiconductors on (i) the vacuum scale and (ii) the SHE scale.

Energy levels are also important. The Fermi level, E_F, in a semiconductor is defined as the hypothetical energy level at which it is equally probable that the level is either occupied by an electron or vacant. In addition to this probabilistic definition, thermodynamically we can think of the Fermi level as the electrochemical potential of an electron in the solid. For an intrinsic (undoped) semiconductor, E_F lies in the middle of the band gap. In contrast, the presence of donor states in an n-doped semiconductor increases the probability that electrons exist at a higher energy state, resulting in an increase in E_F such that it lies slightly below E_c (about 0.1 eV for typical doping levels). Similarly, the presence of low-energy acceptor states in p-type semiconductors means that E_F is located just above E_V. The relative energies of a redox couple in solution and the Fermi level of the semiconductor determine what will happen when the semiconductor is placed in solution, as discussed in the next section.

ILLUSTRATION 15.2

Energy Units and Voltage
Although it may surprise you, an electron volt is actually an energy unit rather than a voltage (V or J·C^{-1}). Specifically, 1 eV is equal to 1.602×10^{-19} J and is, by definition, *the amount of energy gained (or lost) as the charge of a single electron is moved through an electric potential difference of 1 V.*

You may be wondering how we can put energy in eV on the same scale as potential (V versus SHE). To get the difference in energy [eV] that is comparable to a specified change in potential [V], we start with a voltage difference and then multiply by the charge on an electron. For a voltage difference of 1 V, it follows:

$$(1.0) \text{ J C}^{-1} (1.6 \times 10^{-19}) \text{ C} = 1.6 \times 10^{-19} \text{ J} = 1 \text{ eV}.$$

Therefore, a change in voltage of 1 V on the hydrogen scale is numerically equivalent to an energy change of 1 eV. In other words, the increment for both the vacuum scale [eV] and the potential scale [V] is the same. The difference between the scales is that the potential scale is shifted so that it has a zero value that corresponds to hydrogen, whereas the zero value on the vacuum scale is the energy of an electron in a vacuum.

15.3 SEMICONDUCTOR–ELECTROLYTE INTERFACE

We will now examine what happens when a semiconductor is placed in an electrolyte solution. Initially, we will consider the situation at open circuit under dark conditions (i.e., no photoexcitation of electrons). We will then examine how the interface that is established influences the flow of current. Finally, we will look at the impact of light on current flow.

As described earlier in Chapter 3, when a metal is placed in an electrolyte solution containing a redox couple, electron transfer occurs due to a difference in the energy of electrons in the metal and that of electrons associated with electroactive species in the electrolyte. If there is a net transfer of electrons from the metal to solution-containing species, the metal is left with a positive charge that is balanced by a negative charge on the solution side of the interface. The charge on the metal resides at the surface of the electrode. In contrast, the charge on the solution side is distributed throughout the double layer. For solutions with a significant concentration of ions, the thickness of the double layer is quite small, the diffuse portion of the double layer is not important, and nearly all of the charge resides in the Helmholtz layers.

The situation for a semiconductor is somewhat different. For illustration purposes, we consider an n-type semiconductor whose Fermi energy is higher than that of the redox couple in solution, as illustrated in Figure 15.10a, where all energy bands in the semiconductor are shown to be flat prior to contact with the electrolyte. When the n-type semiconductor electrode is brought into contact with the electrolyte under open-circuit conditions, the higher energy electrons in the semiconductor move to the interface where they react with the redox couple. This net transfer of electrons continues until the electron energy is the same in both materials; in other words, the electron transfer continues until the Fermi level of the semiconductor is equal to the reversible potential of the redox couple. Up to now, it may appear that the semiconductor behaves just like a metal. However, the situation is actually quite different. A doped n-type semiconductor consists of immobile, positively charged donor atoms and mobile electrons, as described above. The net transfer of electrons out of the semiconductor results in a depletion zone near the interface where the semiconductor is depleted of electrons. This region is also commonly referred to as the *space-charge* region, reflecting the fact that migration of the free electrons into the solution has left behind positively charged donor ions (N_D^+) as shown in Figure 15.10b.

The net positive charge in the depletion region is balanced by negative charge on the solution side of the interface (Figure 15.10b). Because the double layer is thin relative to the thickness of the depletion region, the negative charge on the solution side lies very close to the interface. As described by Poisson's equation, the local imbalance of charge results in an electric field and bending of the semiconductor bands within the depletion region, as shown in Figure 15.10c. The electric field and the associated band bending are greatest at the electrolyte–semiconductor interface, and the potential difference between the energy bands in the bulk and at the interface of the semiconductor is the *space-charge voltage* (V_{SC}).

In this chapter, we consider the ideal case where the entire drop in potential at the interface is across the semiconductor. This assumption provides a reasonable approximation of the behavior of the semiconductor–electrolyte interface in general, and a relatively good

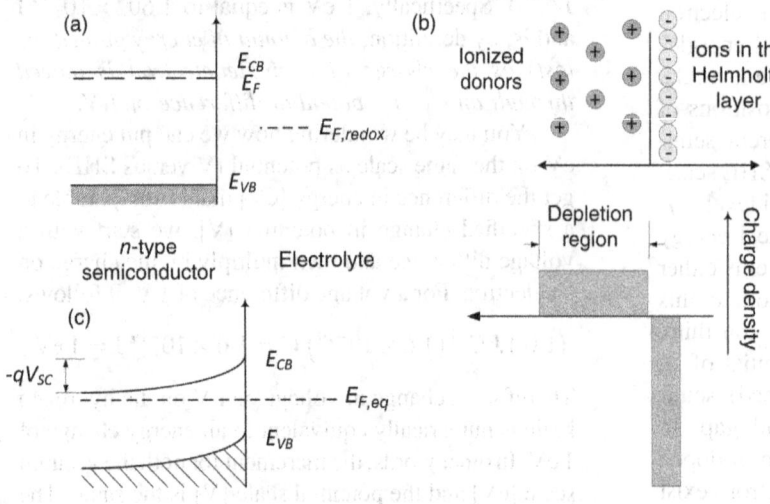

Figure 15.10 (a) Initial electron energy levels of semiconductor and redox couple, where the energy of the CB electron is higher than that of the redox couple. (b) Illustration of the physical distribution of charge after transfer of charge from n-type semiconductor (leaving a net positive charge on the left) to the electrolyte (on the right). (c) Band bending (different electron energy) at the interface after energy levels have equilibrated by the transfer of electrons between phases.

approximation for some systems. The physical situation it represents is very different than the metal–electrolyte interface that we have considered up to this point. In the metal–electrolyte system, the potential drop across the double layer was critical to our description of the interface and electrochemical reactions at the interface. Here, we assume that the electrolyte is sufficiently concentrated such that the diffuse part of the double layer is not important (see Chapters 3 and 11). We also assume that the potential drop across the Helmholtz portion of the double layer is sufficiently small, relative to the drop across the space-charge layer, that it can be neglected. This is, in general, a good assumption since the capacitance of the electrolyte double layer is typically much larger than that of the semiconductor; therefore, for a given amount of charge, the potential drop across the Helmholtz double layer will be much less than that across the depletion layer of the semiconductor. Finally, we ignore the thickness of the double layer and assume that the charge on the electrolyte side of the interface is located at the interface. This assumption is reasonable since the thickness of the Helmholtz portion of the double layer is ~0.5 nm compared to depletion layer thicknesses of 10–1000 nm in the semiconductor. Nevertheless, the physical situation described in this chapter is ideal, and additional factors (e.g., surface states in the band gap) will be important for some systems. A more comprehensive treatment can be found in sources such as those listed at the end of the chapter.

With these assumptions, we will now develop a description for the potential field and capacitance in the space-charge layer. The description is provided for an n-type semiconducting electrode, but analogous equations apply for a p-type semiconductor. This picture will allow us to determine V_{SC} and the thickness of the depletion layer as a function of the potential and the doping level. To do so, we will use Poisson's equation, which relates the electric field (gradient of the potential) to the charge density:

$$\frac{d^2\phi}{dx^2} = -\frac{\rho_e}{\varepsilon} \approx -\frac{qN_D}{\varepsilon}. \quad (15.3)$$

Here ρ_e is the charge density (charge per volume) and ε is the permittivity of the semiconductor, which is equal to the product of its dielectric constant (ε_r) and the permittivity of free space (ε_0). As a first approximation, we ignore any electrons in the depletion region and assume that all of the charge is due to the fixed donor atoms. Thus, the charge density, ρ_e, is a constant throughout the depletion region and is equal to qN_D. We define $x=0$ at the interface, and $x=W$ at the edge of the depletion region. Integration of Equation 15.3 yields

$$\frac{d\phi}{dx} = \frac{qN_D}{\varepsilon}(W-x), \quad \text{since} \quad \frac{d\phi}{dx} = 0 \quad \text{at} \quad x=W.$$

Integrating again :
$$\int_{\phi_{int}}^{\phi} d\phi = \int_0^W \frac{qN_D}{\varepsilon}(W-x)dx.$$

$$V_{SC} = \phi - \phi_{int} = V - V_{int} = \frac{qN_D W^2}{2\varepsilon}, \quad (15.4)$$

where the subscript *int* is used to designate the interface. Note that potential at the interface is negative relative to the bulk semiconductor (negative potential equates to higher electron energy versus the vacuum level) for the situation considered here (n-type semiconductor). For a p-type semiconductor, the bands will bend in the opposite manner, and the potential at the interface is more positive than it is in the bulk of the semiconductor.

The voltages V and V_{int} in Equation 15.4 are with respect to an unspecified reference electrode in solution. To avoid ambiguity, we incorporate the standard definition of overpotential into the equation:

$$\eta = V - U.$$

Subtracting the equilibrium value U from the two potentials, Equation 15.4 becomes

$$(V-U) - (V_{int} - U) = \eta + (U - V_{int}) = \frac{qN_D W^2}{2\varepsilon}.$$

As usual, $\eta = 0$ at the equilibrium potential. Significantly, when the overpotential is zero, there is still a potential difference between the reference electrode and the interface. This potential difference is $V_{SC,eq}$, and is an important characteristic of the semiconductor–electrolyte system. It is positive for an n-type semiconductor since the bulk potential is higher than that at the interface at equilibrium (Figure 15.11). To explore this potential further, consider what would happen if we start with the electrode at open circuit and then lower its

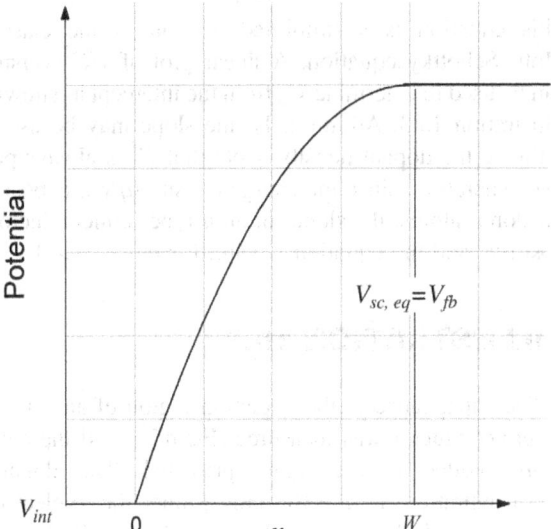

Figure 15.11 Variation in potential with distance from the interface when the overpotential is zero.

potential. As the potential of the electrode is made more negative (raising electron energy), the charge in the depletion region and its thickness will get smaller, and band bending will be reduced. We can continue reducing the potential until there is no charge ($Q=0$) in the depletion layer and no band bending at all—equivalent to the point of zero charge for metal electrodes. The potential difference across the space-charge layer is zero, and W is zero, but the overpotential is not zero. In fact, the overpotential for the condition where there is no band bending is simply $-V_{SC,eq}$. This quantity is called the *flat-band potential*, V_{fb}.

$$\eta + V_{SC, equiv} = \eta - V_{fb} = \frac{qN_D W^2}{2\varepsilon}. \quad (15.5)$$

We will use the flat-band potential as our reference point, since that is what is typically done. The width of the depletion layer can be expressed as

$$W = \sqrt{2\varepsilon(\eta - V_{fb})/qN_D}. \quad (15.6)$$

From Equation 15.6, we see that the width of the depletion layer, W, is a function of both the applied voltage (expressed as the overpotential) and the dopant concentration, N_D. It is valid for $\eta - V_{fb} > 0$.

Under the above assumptions, the charge in the depletion region of the semiconductor is

$$Q_{SC} = qN_D W = \sqrt{2q\varepsilon N_D(\eta - V_{fb})} \quad (15.7)$$
[charge per cross-sectional area]

Equation 15.7 can be differentiated to yield the capacitance per area for an *n*-type semiconductor:

$$\frac{1}{C^2} = \frac{2(\eta - V_{fb})}{\varepsilon A^2 q N_D}. \quad (15.8)$$

This equation is a simplified version of the classical Mott–Schottky equation. A linear plot of $1/C^2$ versus η can be used to determine V_{fb} from the intercept as shown in Illustration 15.3. Additionally, the slope may be used to estimate the dopant density. Note that C^2 is always positive. Therefore, since the charge density (qN_D) is positive for donor atoms, the slope for an *n*-type semiconductor is positive, and the equation is valid for $\eta - V_{fb} > 0$.

ILLUSTRATION 15.3

The capacitance of the depletion region of an *n*-type semiconductor was measured (F·cm^{-2}), and the data are plotted in the figure provided. The dopant concentration is 3×10^{15} cm^{-3}, and the dielectric constant of the semiconductor (ε_r) is 11.9. Please determine the voltage drop across and the width of the depletion region at open-circuit conditions.

Solution:

The potential of the semiconductor electrode at equilibrium will depend on the redox couple in solution, and V_{fb} cannot be extracted from a simple measurement of the potential of the semiconductor electrode relative to a reference electrode in solution. In order to estimate V_{fb}, we can measure the capacitance as a function of the overpotential. According to Equation 15.8, a plot of $1/C^2$ versus η will yield a straight line. This is called a Mott–Schottky plot. The *y*-intercept ($\eta = 0$) can be used to find V_{fb}. Using the data provided for the capacitance, a fit of $1/C^2$ versus η yields a line with a slope of 5.05×10^{15} and a *y*-intercept of 2.02×10^{15}.

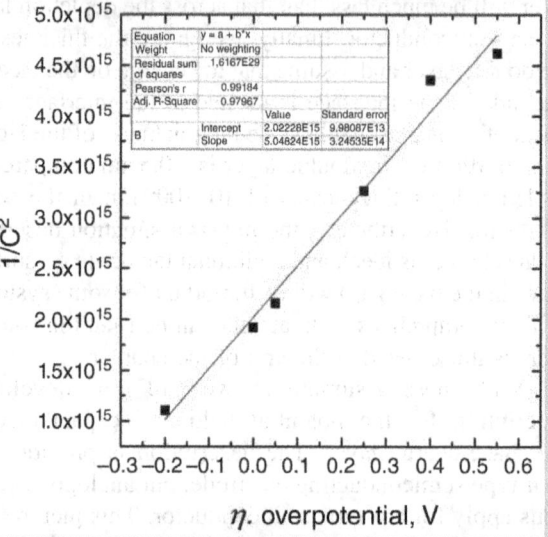

η. overpotential, V

The *y*-intercept is equal to $-mV_{fb}$, where m is the slope of the line $= 2/\varepsilon A^2 q N_D$ (see Equation 15.8. Therefore, $V_{fb} = -\frac{y\text{-intercept}}{m} = -0.40$ V.

We can now use V_{fb} to estimate W, the width of the depletion region. According to Equation 15.6, and noting that $\varepsilon = \varepsilon_r \varepsilon_o$,

$$W = \sqrt{\frac{2\varepsilon(\eta - V_{fb})}{qN_D}} = \sqrt{\frac{2\varepsilon_r\varepsilon_0(-V_{fb})}{qN_D}}$$

$$= \sqrt{\frac{2(11.9)(8.854 \times 10^{-14}\,\mathrm{C\cdot V^{-1}\cdot cm^{-1}})(0.40\,\mathrm{V})}{(1.602 \times 10^{-19}\,\mathrm{C})(3 \times 10^{15}\,\mathrm{cm^{-3}})}}$$

$$= 4.19 \times 10^{-5}\,\mathrm{cm}.$$

For an electrolyte with ion concentrations of 0.1 M, the electrochemical double layer is on the order of 1 nm thick, which is more than two orders of magnitude smaller than the depletion layer thickness given above.

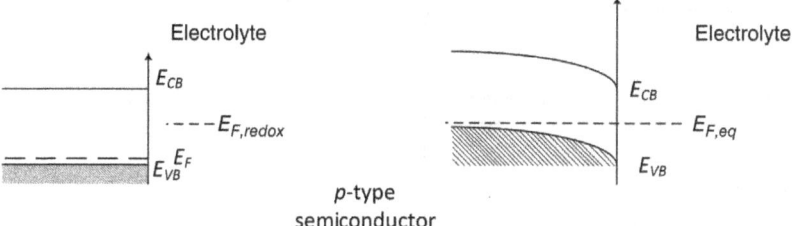

Figure 15.12 Energy diagrams for a *p*-type semiconductor before and after equilibration.

A similar discussion and derivation can be made for *p*-type semiconductors, where the band bending is in the opposite direction as shown in Figure 15.12. The Fermi level is just above the valence band. The charge density is negative for the acceptor atoms in the depletion region, and V_{fb} is positive. In addition, the slope of $1/C^2$ versus η line is negative for a *p*-type semiconductor, and Equations 15.7 and 15.8 apply for $\eta - V_{fb} < 0$.

15.4 CURRENT FLOW IN THE DARK

The space-charge layer introduced in the previous section plays a critical role in the flow of current for a semiconductor electrode. The band diagrams that correspond to open-circuit, positive overpotential, and negative overpotential for *n*-type semiconductors are shown in Figure 15.13. At open circuit ($\eta = 0$), the net flow of current is zero and the conduction band energy increases at the interface as reflected in the band bending illustrated in Figure 15.13a. For positive overpotentials, $\eta > 0$, the band bending is increased, and the width of the depletion region increases (Figure 15.13b). The band bending has a significant impact on current flow as will be discussed shortly. For negative overpotentials, Figure 15.13c, the band bending decreases or is reversed for *n*-type semiconductors.

For the ideal semiconductors that we are considering, the band edge positions at the interface are *pinned*, which means that the edge positions do not change with respect to the solution as the applied potential is varied. In contrast, the relative position of the Fermi level does change, as do the densities of electrons and holes at the surface of the semiconductor. As we'll see shortly, the density of electrons at the surface plays a key role in understanding the kinetic behavior of semiconductor electrodes. For an *n*-type semiconductor, the majority charge carriers are electrons in the conduction band, as shown in Figure 15.4, and the concentration of charge carriers in the bulk is n_b, which is approximately equal to N_D. However, the local concentration of electrons is affected by the formation of the depletion region and the associated potential field. Our earlier analysis assumed that all of the electrons were absent in the depletion region, leaving behind only positively charged dopant atoms at fixed sites. We now need to refine our physical model. Although the negative charges on the solution side of the interface make it energetically more difficult to have electrons on the surface of the semiconductor, thermal excitation must be considered. This balance between electrostatic repulsion and random thermal motion is expressed with a Boltzmann factor:

$$n(x) = n_b \exp\left\{\frac{-q(V - V(x))}{kT}\right\}. \quad (15.9)$$

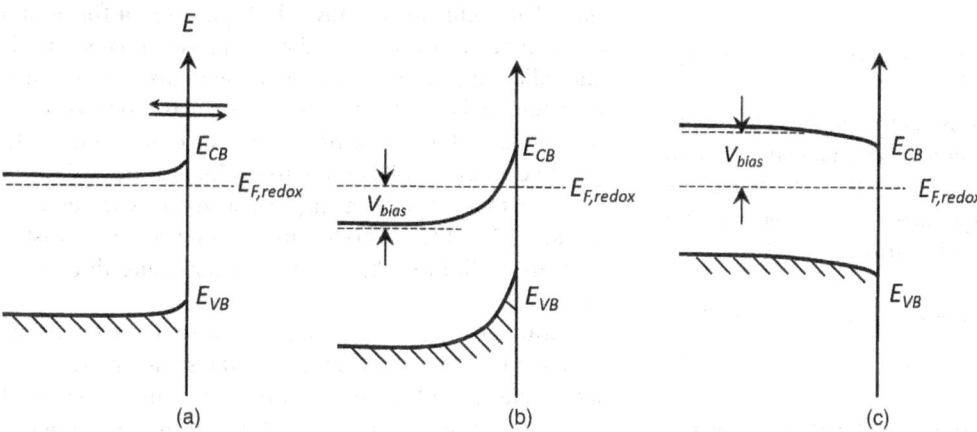

Figure 15.13 Energy diagrams for an *n*-type semiconductor at (a) open circuit where the net current is zero and the migration current is just balanced by the diffusion current as indicated by the arrows, (b) positive overpotential, which corresponds to reverse bias, and (c) negative overpotential, which corresponds to forward bias.

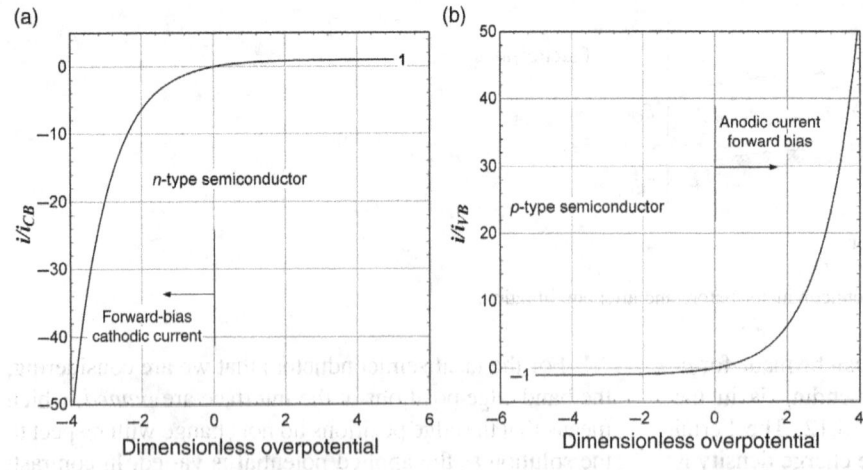

Figure 15.14 Current in the dark for n-type and p-type semiconductors exposed to a redox electrolyte.

V is the potential of the bulk semiconductor relative to a reference electrode in solution, and $V(x)$ is the potential in the depletion zone at a position x, relative to the same reference electrode. As we saw above for an n-type semiconductor, $V(x)$ is negative relative to the bulk (V). Therefore, the concentration of electrons decreases as you move from the bulk to the semiconductor–electrolyte interface. Altering the potential of the electrode relative to our reference causes the potential difference between the bulk and interface to vary, and thus the concentration of electrons at the interface can change. The surface concentration of electrons can be expressed as

$$n_s = n_b \exp\left\{\frac{-q(V - V_{int})}{kT}\right\} = n_b \exp\left\{\frac{-q(\eta + V_{SC,equiv})}{kT}\right\}$$
$$= n_b \exp\left\{\frac{-q(\eta - V_{fb})}{kT}\right\}. \tag{15.10}$$

The potential of the interface relative to the bulk reduces to $-V_{fb}$ when there is no applied potential ($\eta = 0$). We are now ready to examine the electrode kinetics. The heterogeneous electron-transfer reaction at the semiconductor electrode is written as

$$e^- + O \underset{k_a}{\overset{k_c}{\rightleftarrows}} v + R \tag{15.11}$$

where e^- represents electrons and v represents empty electron states in the conduction band. The concentration of empty states at the interface is high and v can be assumed constant and incorporated into k_a. The forward and reverse reaction rates can be written as follows:

$$\text{forward rate} = k_c n_s [O] \tag{15.12a}$$
$$\text{reverse rate} = k_a [R]. \tag{15.12b}$$

The forward (cathodic) reaction depends on the concentration of electrons at the surface, n_s, which depends on the potential. In contrast, with the anodic reaction, electrons are being injected into a nearly empty band. Thus, for n-type semiconductors, the reverse (anodic) reaction depends only on the relative position of the energy levels and, therefore, its rate remains constant with potential when $\eta > 0$. The potential change is absorbed by the semiconductor as the width of the depletion region changes while the current remains constant. Consequently, *the flow of current in the dark for an n-type semiconductor* is

$$i = i_{CB}^0 \left(1 - \exp\left\{\frac{-q\eta}{k_B T}\right\}\right) = i_{CB}^0 \left(1 - \exp\left\{\frac{-F\eta}{RT}\right\}\right), \tag{15.13}$$

where i_{CB}^0 is the saturation current, and CB refers to the conduction band. A similar expression applies to the valence band for p-type materials:

$$i = i_{VB}^0 \left(\exp\left\{\frac{F\eta}{RT}\right\} - 1\right). \tag{15.14}$$

From Equations 15.13 and 15.14, it is readily apparent that significant current is only able to flow in one direction. This behavior is shown in Figure 15.14 for both p- and n-type electrodes. For the n-type semiconductor, the cathodic current increases exponentially for negative overpotentials, similar to what we have observed for metal–electrolyte systems. In contrast, the current for positive overpotentials is relatively independent of potential and much lower in magnitude for n-type semiconductors. Note that cathodic current is negative, consistent with the definition that we have been using throughout the text.

Similar to the p–n junction of a semiconductor, the junction between an electrolyte and a semiconductor acts like a diode, with rectifying properties. In the terminology of the semiconductor field, when the magnitude of the current is large, the applied potential is called a *forward bias*. When the magnitude of the current is small, it is described as being under a *reverse bias*. We can use these terms for the junction between an electrolyte and a semiconductor. However, we must recognize that the behavior of a junction between an electrolyte and a semiconductor changes depending on whether it is *n*- or *p*-type. This difference is clearly shown in Figure 15.14. For an *n*-type semiconductor, large cathodic currents result under a forward bias, which corresponds to a negative overpotential. On the other hand, for a *p*-type semiconductor, large anodic currents result when a positive overpotential is applied; this also corresponds to a forward bias.

A Tafel plot can be used to determine i^0_{CB} or i^0_{VB} under forward bias if the equilibrium potential is known. Under the simplifying assumptions considered here, the equilibrium potential is determined by the redox couple in solution and the semiconductor equilibrates by passing charge to or from the solution.

ILLUSTRATION 15.4

An *n*-type semiconductor electrode operating in the dark has a current density under reverse bias of 6×10^{-5} A·cm^{-2}. The equilibrium potential is -1.20 V versus SCE. What are the potential and current density at a forward bias of 0.20 V with respect to the equilibrium potential? Assume 25 °C. Plot the current density versus potential.

Solution:

Forward bias for an *n*-type semiconductor results from application of a negative potential. Therefore, the voltage is -1.20 V $- 0.20$ V $= -1.40$ V, and $\eta = -0.20$ V. $i^0_{CB} = 6 \times 10^{-5}$ A·cm^{-2}. Substituting these values into Equation 15.13 yields

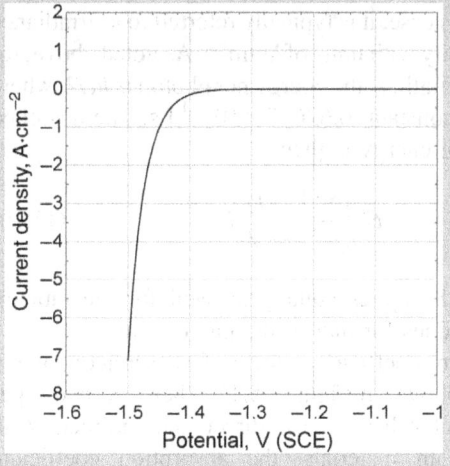

$$i = i^0_{CB}\left(1 - \exp\left\{\frac{-F\eta}{RT}\right\}\right)$$
$$= 6 \times 10^{-5}\left(1 - \exp\left\{\frac{(-96485)(-0.20)}{(8.314)(298)}\right\}\right)$$
$$= -0.145 \text{ A} \cdot \text{cm}^{-2},$$

where the negative sign indicates that the current is cathodic, consistent with the sign convention that we have been using in this text. The current density versus voltage curve (in the dark) is shown in the figure.

At negative overpotentials (potentials below -1.2 V), the current increases exponentially, similar to what we observed earlier for metal electrodes. In contrast, the current approaches a constant (small) value of i^0_{CB} at positive overpotentials for an *n*-type semiconductor. This curve illustrates the rectifying effect of semiconductors in the absence of illumination.

The language used in semiconductor photoelectrochemistry is a blend of expressions from solid-state physics, the semiconductor industry, and electrochemistry. You should be familiar with these terms.

Forward bias: Similar to *p–n* junctions of doped semiconductors, the interface between an electrolyte and a semiconductor acts like a diode. The lexicon from electrical engineering is often used to describe the interface between the electrolyte and a semiconductor. Namely, forward bias refers to the condition where the applied potential results in a current of large magnitude. However, bias and overpotential, η, are not the same. Most importantly, the behavior of a semiconductor–electrolyte interface is different depending on the type of doping.

n-type:	$\eta < 0$, forward bias, band bending decreases or is reversed, the thickness of the space-charge layer decreases, large cathodic current results
	$\eta > 0$, reverse bias, band bending increased, thickness of the space-charge layer increases, very small anodic current
p-type:	$\eta > 0$, forward bias, large anodic current
	$\eta < 0$, reverse bias, very small cathodic current

Drift current: What an electrochemist would call migration, many electrical engineers refer to as drift. The physics are the same, namely, the movement of charge particles (electrons or holes) in the presence of an electric field.

15.5 LIGHT ABSORPTION

Now that we have examined the electrochemical behavior of semiconductors in the dark, the next step is to include photoelectrochemical (PEC) effects. Toward that end, this section describes the absorption of light by semiconductors. The principal mechanism for the absorption of photon energy by a semiconductor is the excitation of an electron from the valence to the conduction band. This takes place when the energy of the light exceeds that of the band gap as shown in Figure 15.15. Consequently, there is little absorption of light whose energy is below that of the band gap, and the absorption that does occur is the result of additional energy states that are present owing to chemical impurities or physical defects.

Light is typically characterized by its wavelength (λ) or its frequency (ν). In order to determine if light of a particular wavelength will be absorbed, we need a relationship between the wavelength, frequency, and energy. Since the energy of a photon is $h\nu$ and $\nu = c/\lambda$,

$$E\text{ (eV)} = \frac{1.24}{\lambda\text{ (}\mu\text{m)}}. \tag{15.15}$$

The cutoff wavelength, λ_c, for absorption is the wavelength that corresponds to the band gap energy and is found as

$$\lambda_c\text{ (}\mu\text{m)} = \frac{1.24}{E_g\text{ (eV)}}. \tag{15.16}$$

Light absorption decreases precipitously for wavelengths greater than the cutoff value defined by Equation 15.16. In advanced photoelectrochemical systems, the photons that are not absorbed in one semiconductor are transmitted to a second semiconducting absorber layer having a lower band gap so that their energy is not wasted.

Light is absorbed gradually as it passes through a semiconductor, as illustrated in Figure 15.16. Therefore, the amount absorbed is a function of the thickness of the semiconductor through which the light is passing. The local rate of absorption is proportional to the flux at that

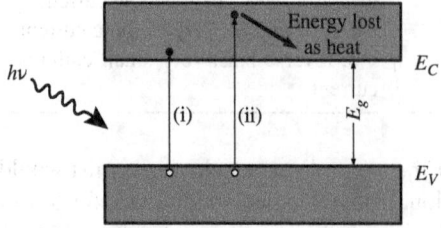

Figure 15.15 Optical absorption to excite electrons to the conduction band (i) where the photon energy is equal to the energy of the band gap, and (ii) where the photon energy is greater than the band gap energy.

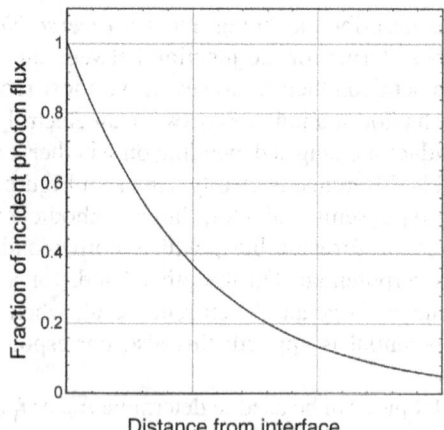

Figure 15.16 Optical absorption as a function of distance from the surface.

point, described mathematically as

$$\frac{dI''}{dx} = -\alpha I'', \tag{15.17}$$

where I'' is the photon flux per unit area, and the proportionality constant, α, is called the absorption coefficient.

The fraction absorbed as a function of distance from the surface can be obtained from Equation 15.17:

$$\frac{I''_0 - I''|_x}{I''_0} = 1 - \exp(-\alpha x), \tag{15.18}$$

where I''_0 is the entering flux (photons s^{-1} m^{-2}) and $I''|_x$ is the exiting flux at a distance x from the surface. The entering flux is equal to the incident flux in situations where reflection is not important. Otherwise, it is the difference between the incident flux and the fraction of that flux that is reflected. A larger α corresponds to higher absorption and, therefore, a shorter distance (thinner layer) to absorb a given amount of light. Note that the flux can also be expressed in terms of energy, rather than just photons. In this case, it is typically referred to as irradiance or energy density with units of W·m^{-2}. As noted above, for a given wavelength, λ, the energy per photon is hc/λ, where h is Planck's constant, 6.62607×10^{-34} J·s. The absorbed flux in units of energy is then

$$E''\left[\frac{W}{m^2}\right] = \frac{hc}{\lambda} I''_0. \tag{15.19}$$

Equation 15.18 applies equally as well for the ratio of energies as it does for that of the photon fluxes.

Absorption coefficients for a few semiconducting materials are shown in Figure 15.17. There are two types of semiconductor band gaps: direct and indirect. With direct band gap materials, the absorption coefficient increases rapidly for photons with an energy greater than E_g, which, for example, is 1.4 eV for GaAs. Light

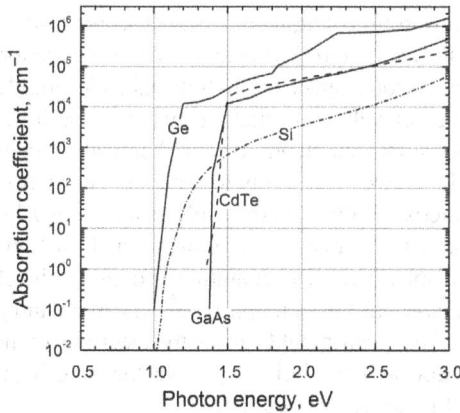

Figure 15.17 Absorption coefficients at 300 K for some common semiconducting materials. GaAs is direct and Si and Ge are indirect band gap materials.

absorption for indirect band gaps requires an additional transition involving the transfer of momentum from the electron to the crystal lattice; that is, in addition to the generation of a hole and electron, a phonon is created. For this reason, the increase in absorption coefficient with photon energy is more gradual for indirect band gap materials such as silicon.

The absorption coefficient for silicon is approximated by the following equation over the range of wavelengths from 400 to 1140 nm:

$$\ln \alpha = -5.7536 \times 10^{-8}\lambda^3 + 1.2221 \times 10^{-4}\lambda^2 \\ -9.3322 \times 10^{-2}\lambda + 32.699, \quad (15.20)$$

where λ is in nm and α is in cm^{-1}. Note that errors in the fit are magnified because of the logarithm, and measured data for the conditions of interest are always best.

What happens to photons with energy that are well above the band gap? Electrons are created with energies greater than the conduction band edge. It would be ideal if we could convert this to electrical work; however, most of this excess energy is quickly converted to heat. For some devices, this may not be an issue, but if we are trying to maximize the conversion of solar irradiance to electrical work, this energy is wasted and results in a loss of efficiency. This leads to an important trade-off that affects the overall efficiency of solar devices. Keep in mind that such devices are typically exposed to light with a broad range of wavelengths. A semiconductor with a small band gap can absorb a larger fraction of this light. However, much of the absorbed energy is lost since excess photon energy is dissipated as heat to the semiconductor crystal (see Figure 15.15). As the band gap increases, a smaller fraction of the available photons are absorbed, but the loss of energy as heat is also reduced. The net result is a maximum efficiency for the solar spectrum of about 34% at a band gap of 1.34 as shown in Figure 15.18.

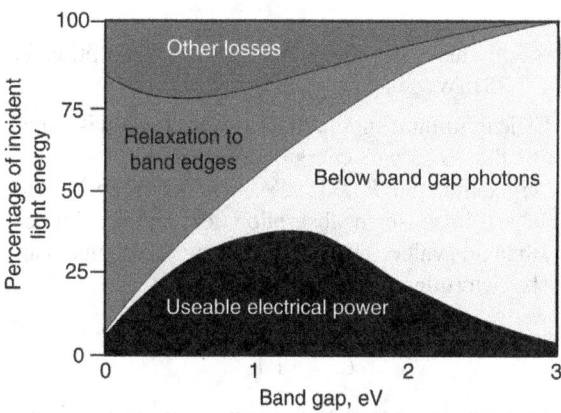

Figure 15.18 Useable electric power generated from light energy.

Importantly, silicon has an indirect band gap and light just below the critical wavelength tends to penetrate a significant distance before it is absorbed. Consequently, the thickness of silicon devices (~100 μm) is significantly greater than that required for direct band gap materials such as CdTe or CIGS (copper indium gallium (di)selenide) (~1 μm).

ILLUSTRATION 15.5

Light Absorption by Silicon

a. Single-crystal silicon is exposed to monochromatic light at a wavelength of 600 nm. The incident power due to the light is 10 mW. The thickness of the crystal (orthogonal to the incident light) is 0.5 μm. Please determine the total rate of energy absorption by the silicon. You may assume that the 10 mW is the energy that enters the silicon (you do not need to account for reflection at the surface).

b. What is the minimum energy of light that can be absorbed by silicon? To what wavelength does this correspond?

c. If light with energy in excess of the minimum energy is absorbed, what happens to the extra energy?

Solution:

(a) We need to determine the fraction of the light that enters the Si crystal that is absorbed in the 0.5 μm thickness of the crystal. The absorption coefficient for crystalline silicon at 600 nm is approximately 3.92×10^3 cm^{-1} (see Equation 15.20). From Equation 15.18, the fraction absorbed is

$$1 - \exp^{-\alpha W} = 1 - \exp\{-(3.92 \times 10^3 \text{ cm}^{-1})(0.5 \times 10^{-4} \text{ cm})\} \\ = 0.178.$$

Therefore, the rate of energy absorption is 1.78 mW.

(b) The minimum energy that can be absorbed is equal to the band gap energy. From Figure 15.17 for crystalline Si, $E_g \approx 1.1\,\text{eV}$, which is seen by the sharp increase in absorption for energies higher than this value. The corresponding wavelength can be determined from Equation 15.15:

$$\lambda(\mu\text{m}) = \frac{1.24}{E} = \frac{1.24}{1.1} = 1.13\ \mu\text{m}$$

(c) Energy in excess of the band gap goes to heat, which is dissipated into the crystal.

15.6 PHOTOELECTROCHEMICAL EFFECTS

To this point in the chapter, we have presented a brief introduction to semiconductors and have described the interface that develops when they are put into contact with an electrolyte. Current flow under dark conditions has also been described. Finally, we looked briefly at the light absorption characteristics of semiconductors. We are now ready to combine light and electrochemistry to explore photoelectrochemistry.

The absorption of light discussed in the previous section results in the excitation of an electron to the conduction band and the formation of an electron–hole pair. We would like to use the energetic electrons that are created to perform work by driving nonspontaneous electrochemical reactions. In order to do so, we need a way to separate the electrons and holes. Otherwise, the electron–hole pairs formed by light absorption will be lost to recombination. In a traditional solid-state solar cell, that separation is done by the electric field that exists at the junction between n-type and p-type materials. The electric field causes the electrons to move in one direction and the holes to move the other direction, effecting the desired separation.

As discussed above, an electric field is also present at a semiconductor–electrolyte interface. This field can be used to separate the electron–hole pairs, as illustrated in Figure 15.19a. To be separated, the electron–hole pairs must be generated in the space-charge layer (depletion region) where the field exists, or sufficiently close to that zone so that the charge carriers can diffuse to the zone before they are recombined (more on this shortly). The physical situation is illustrated in Figure 15.19b. The electron–hole pairs that are generated are separated by the field in the depletion zone. The strength of the electric field is enormous, and allows for effective collection of minority carriers at the surface of the semiconductor before they can undergo recombination. For an n-type semiconductor, the photogenerated minority-carrier holes migrate to the semiconductor–electrolyte surface, and

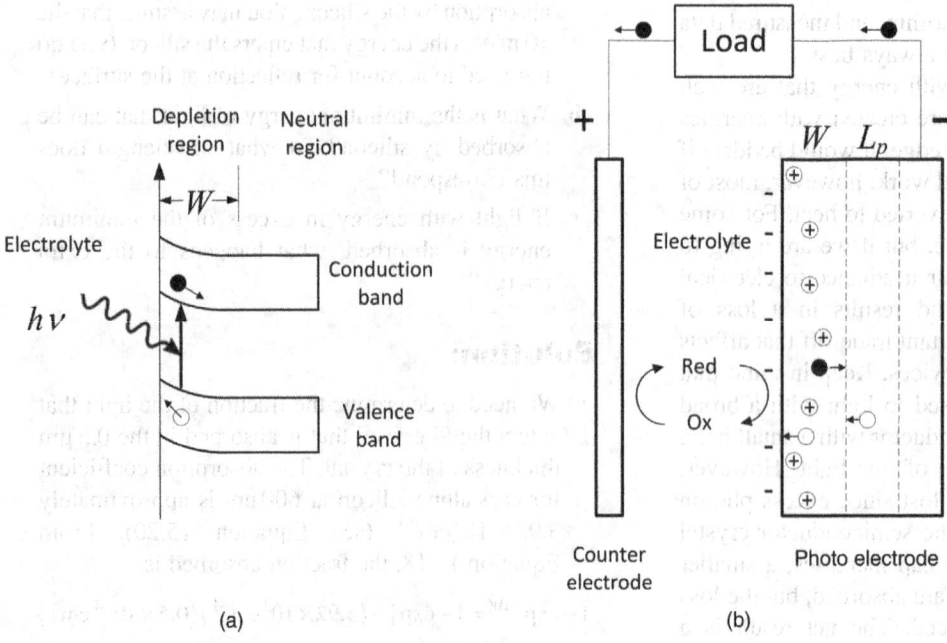

Figure 15.19 Generation and separation of electron–hole pairs through light absorption in the depletion region of an n-type semiconductor. (a) Band structure. (b) Physical situation.

the electrons migrate away from the surface to the bulk of the semiconductor and are collected at the back contact. The photoproduced current (*n*-type semiconductor) is anodic and associated with the movement of holes (the minority carrier) to the semiconductor–electrolyte interface. This anodic current is in contrast to the cathodic current produced by a forward bias in the dark. The photocurrent is often limited by the generation rate of electron–hole pairs and is thus directly connected to the rate of absorption of light in and near the depletion zone; "near" is interpreted as within a diffusion length, L_p, as defined below. The total current is simply the sum of the photocurrent and the potential-dependent majority current considered in a previous section. The relevant expression for an *n-type semiconductor* is

$$i = i_{ph} + i_{CB}^0 \left(1 - \exp\left\{\frac{-F\eta}{RT}\right\}\right). \quad (15.21)$$

The photogenerated anodic current (i_{ph}), which results from the movement of photogenerated holes (minority carrier), is positive and dominates when the interface is reverse biased. In contrast, the current is dominated by the flow of electrons (majority carrier) under conditions of forward bias. The shape of the current–voltage curve is shown in Figure 15.20 where the magnitude of i_{ph} depends on the level of light absorption as indicated in the figure. An analogous expression can be written for a *p*-type semiconductor, although the photocurrent would be negative (cathodic), and the forward bias would be at positive potentials and correspond to anodic current.

The Gärtner equation, written for an *n*-type semiconductor, provides a simplified expression for the photocurrent based on the absorption of light in the depletion region ($0 \leq x \leq W$) and within the diffusion region where holes are sufficiently close to the depletion zone to diffuse to the zone before recombining (see Figure 15.19).

$$|i_{ph}| = -qI_0'' \left[1 - \frac{\exp(-\alpha W)}{1 + \alpha L_p}\right], \quad (15.22)$$

where I_0'' is positive as it enters the semiconductor and the distance W is measured from the surface where the light enters. Remember that i_{ph} is positive (anodic) for *n*-type semiconductors and negative (cathodic) for *p*-type semiconductors. Comparison of Equation 15.22 with Equation 15.18 shows that the additional diffusion term increases the light absorption for a given distance W into the semiconductor. The diffusion region is characterized by the diffusion length, L_p, defined as

$$L_p = \sqrt{D_p \tau_p}, \quad (15.23)$$

where D_P is the diffusivity of the minority carriers and τ_P is the recombination time for the holes. The quantum yield, defined mathematically as

$$\Phi \equiv \frac{|i_{ph}|}{|qI_0''|} = \left[1 - \frac{\exp(-\alpha W)}{1 + \alpha L_p}\right], \quad (15.24)$$

can be readily obtained from Equation 15.22, where I_0'' is the entering light flux, as defined previously in conjunction with Equation 15.18. It turns out that for high-quality semiconductors such as silicon, the effective minority carrier diffusion length is much greater than the depletion region width, and the collection of photogenerated carriers is nearly independent of the applied potential and the depletion region width. For semiconductors with low lifetimes (small effective diffusion lengths), it is necessary to account for carrier collection that depends on the depletion region width and the potential.

15.7 OPEN-CIRCUIT VOLTAGE FOR ILLUMINATED ELECTRODES

What happens when the illuminated electrode is at open circuit? The answer to this question is perhaps most easily seen from Equation 15.21 for an *n*-type semiconductor, which includes both the photocurrent and the current due to the potential-dependent majority carrier current. At open circuit, the net current is equal to zero. In an *n*-type semiconductor, this happens when the negative charge in the bulk semiconductor (away from the interface) builds up due to electron–hole separation to the extent that the back flow of electrons toward the interface is equal to the anodic photocurrent due to holes. Under these conditions, Equation 15.21 can be solved for the open-circuit voltage of an *n*-type semiconductor to yield

$$\eta_{OC} = V_{OC} - U = -\frac{RT}{F} \ln\left(\frac{i_{ph}}{i_{CB}^0} + 1\right). \quad (15.25)$$

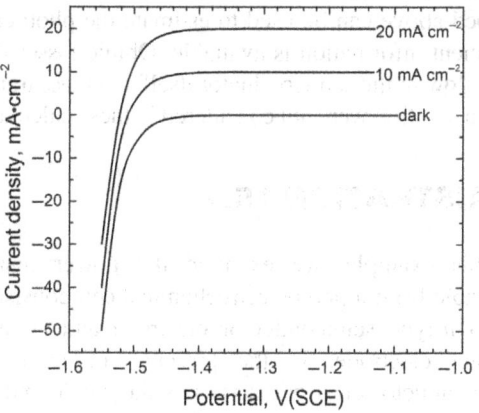

Figure 15.20 Current–voltage curves for an illuminated *n*-type semiconductor. The photocurrent is associated with minority carriers and depends directly on light absorption.

V_{OC} is the open-circuit voltage and U is the equilibrium voltage in the dark, which is determined by the equilibrium potential of the redox couple in solution for the simplifying assumptions that we have made. The open-circuit voltage is the maximum voltage (magnitude) that can be obtained from a solar cell. Its value depends on the intensity of the incident light and on the saturation current density, i_{CB}^0. Consequently, the absolute value of V_{OC} increases with increasing light intensity and with decreasing saturation current. Note that V_{OC} is lower (i.e., more negative) than the equilibrium voltage for n-type semiconductors.

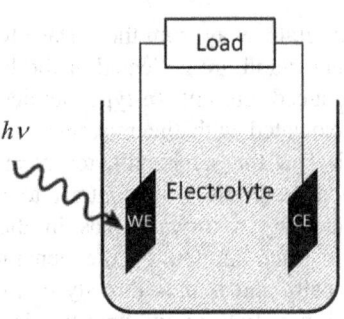

Figure 15.21 Schematic diagram of a photoelectrochemical cell.

ILLUSTRATION 15.6

For the n-type semiconductor from Illustration 15.4, calculate the open-circuit voltage (net current $= 0$), when the photocurrent is $25\,\text{mA}\cdot\text{cm}^{-2}$.

Solution:

The photocurrent for an n-type semiconductor, due to separation of electron–hole pairs at the semiconductor–electrolyte interface, is positive or anodic. From Equation 15.25, at open circuit with the parameter values from Illustration 15.4,

$$\eta_{OC} = V_{OC} - U = -\frac{RT}{F}\ln\left(\frac{i_{ph}}{i_{CB}^0}+1\right)$$

$$= -\frac{(8.314)(298.15)}{96485}\ln\left(\frac{25\times 10^{-3}}{6\times 10^{-5}}+1\right) = -0.155\,\text{V}$$

Therefore, $V_{OC} = -1.355\,\text{V}$.

η_{OC} is the overpotential at which the magnitude of the "dark current" (cathodic) equals that of the anodic photocurrent. The photocurrent is independent of the overpotential as long as there is a field of sufficient magnitude to separate the electron–hole pairs generated by light absorption. As shown in Figure 15.19, the photocurrent is the dominant current at reverse bias conditions.

15.8 PHOTOELECTROCHEMICAL CELLS

In this section, we consider a complete photoelectrochemical (PEC) cell such as that shown in Figure 15.21. The cell consists of a photoelectrode and a metal counter electrode that are immersed in a common electrolyte. For example purposes, we assume that the photoelectrode is an n-type semiconductor, which acts as the anode in the cell. The electrolyte contains both the oxidized and reduced forms of a redox couple whose equilibrium potential lies in between the conduction and valence band edges of the semiconductor (i.e., in the band gap).

During operation of the cell, electron–hole pairs are generated by illumination of the photoelectrode. The electron–hole pairs are separated due to the electric field in the space-charge region. The holes that are generated migrate to the surface of the semiconductor and participate in oxidation of the reduced form of the redox couple. The excited electrons move through the external circuit, including the load, to the counter electrode, where reduction of the oxidant species takes place. The doping level of the semiconductor influences the width of and voltage gradient in the space-charge layer. It also determines the Fermi level of the majority carriers (electrons for an n-type material), which is close to the conduction band for an n-type semiconductor. In this example with a single redox couple, there is no net reaction and no change in the electrolyte composition. The product of the PEC cell is therefore electrical power due to the work performed by the excited electrons as they move through the external circuit.

Illustration 15.7 provides a quantitative example of how such a PEC cell with an n-type photoanode works. Included in the calculations are surface overpotentials, ohmic losses in solution, and losses in the external circuit containing the load. The calculations are performed for a specified photocurrent. Alternatively, the methods described above can be used to estimate the photocurrent if sufficient information is available. Ohmic losses due to current flow in the semiconductor itself and mass transport losses in solution were not considered in these calculations.

ILLUSTRATION 15.7

In this example, we examine the power that is available from a photoelectrochemical cell consisting of an n-type semiconductor electrode and an inert counter electrode. In the presence of light, the semiconductor electrode serves as the anode and the counter electrode as the cathode. Energetic electrons from the anode pass through the load and then drive the cathodic reaction. We will consider ohmic losses in solution and through the load, as well as the surface overpotentials associated with the two reactions. The surface overpotentials and ohmic losses in solution all reduce the power that can be applied to the load. The current–voltage behavior of an n-type semiconductor electrode operating in the dark is described by the following expression:

$$i = i_{CB}^0 \left(1 - \exp\left\{\frac{-F\eta}{RT}\right\}\right).$$

i_{CB}^0 is 1×10^{-6} A·cm^{-2}. The photocurrent can be assumed to be constant and is equal to 25 mA·cm^{-2}. The aqueous electrolyte contains 0.1 M Fe^{2+}, 0.1 M Fe^{3+} in H$_2$SO$_4$. The conductivity of the electrolyte is approximately 0.06 S·cm^{-1}. The counter electrode is Pt, and the reversible iron reduction reaction on this electrode is characterized by an exchange current density of 0.005 A·cm^{-2} and a transfer coefficient of 0.5. The redox potential for the Fe^{2+}/Fe^{3+} electrode is 0.526 V with respect to a SCE reference electrode in this solution, and the equilibrium potential of the semiconductor electrode in the dark (versus SCE) can be assumed to have the same value. Mass-transfer effects may be neglected. Assume a load of 1 Ω. The distance between the two electrodes is 2 mm. The area of both the anode and cathode is 10 cm^2. Assume that only one side of each electrode is exposed to the electrolyte and that the current density is uniform. Please determine the following: (i) the voltage drop across the load, (ii) the cell current, (iii) the power available from the cell, (iv) the ohmic drop in solution, and (v) the surface overpotential associated with the counter electrode.

Solution:

Strategy: Write the appropriate expressions for the current in terms of the potential and set them equal to each other in order to solve for the values of potential in the cell. Once the potentials are known, the current is readily calculated. We will write expressions for each of the two electrodes, the ohmic drop in solution, and the voltage drop through the load.

Semiconductor electrode: As shown in Equation 15.21, the expression for an illuminated n-type semiconductor is

$$i = i_{ph} + i_{CB}^0 \left(1 - \exp\left\{\frac{-F\eta}{RT}\right\}\right),$$

where i_{ph}, the photocurrent, has been added to the expression for the current in the dark. For a PEC cell with an n-type electrode, the photocurrent is positive or anodic. Hence, i_{ph} is positive and equal to the oxidation that takes place as a result of holes created by light absorption. Neglecting resistive losses in the solid semiconductor electrode itself, $\eta_{anode} = (\phi_{semi} - \phi_{s,a}) - (\phi_{semi} - \phi_{s,a})_{eq}$ where the solution potentials are referred to SCE and $(\phi_{semi} - \phi_{s,a})_{eq} = U = 0.526$ V (the equilibrium potential of the semiconductor in solution in the dark). We arbitrarily set the absolute value of the potential of the semiconductor electrode $\phi_{semi} = 0$ as the reference point for the entire calculation (arbitrary), similar to what we did previously for a metal electrode. With this specification, everything in the above expression is known except the current density and the value of the potential in the solution at the anode.

Unknowns: i and $\phi_{s,a}$

Ohmic drop in solution: Assuming a 1D system,

$$i = -\kappa \frac{d\phi}{dx} = -\frac{\kappa}{\ell}(\phi_{s,c} - \phi_{s,a}) = \frac{\kappa}{\ell}(\phi_{s,a} - \phi_{s,c}).$$

Note that current flow is positive from the anode to the cathode, consistent with what we have done previously.

Unknowns: i and $\phi_{s,a}$, $\phi_{s,c}$.

Load: Electrons flow from the anode to the cathode through the load, with current flow in the opposite direction. Therefore,

$$iA = \frac{(\phi_m - \phi_{semi})}{R_\Omega} \quad \text{or} \quad i = \frac{\phi_m}{R_\Omega A},$$

where $iA = I =$ the total current that flows in the external circuit between the electrodes, and R_Ω is the resistance between the electrodes (i.e., the load) in ohms (specified as 1 Ω for this problem). ϕ_m is the potential of the metal at the cathode. Note that $\phi_{semi} = 0$, as specified above. The equation is just Ohm's law through the external circuit ($V = IR_\Omega$).

Unknowns: i and ϕ_m

Cathode:

$$i = -i_{cathode} = -i_0\left(\exp\left\{\frac{0.5F\eta}{RT}\right\} - \exp\left\{\frac{-0.5F\eta}{RT}\right\}\right),$$

where $\eta_{cath} = (\phi_m - \phi_{s,c}) - (\phi_m - \phi_{s,c})_{eq}$ and $(\phi_m - \phi_{s,c})_{eq} = V_{eq,c} = 0.526$ V. This is just a BV expression for the reduction of ferric ion to ferrous ion at the Pt cathode. Note that since cathodic current is defined as negative, $i = -i_{cathode}$.

Unknowns: i, ϕ_m and $\phi_{s,c}$.

Numerical Solution: The above expressions yield four equations and four unknowns: i, $\phi_{s,a}$, $\phi_{s,c}$, and ϕ_m.

We can eliminate i by setting the relationships equal to each other. This process gives us three equations for the three unknowns: $\phi_{s,a}$, $\phi_{s,c}$, and ϕ_m. This is similar to what we did in Section 3.4 for a Zn-Ni cell.

Simultaneous solution of these equations yields

$$\phi_{s,a} = -0.286 \text{ V}, \quad \phi_{s,c} = -0.332 \text{ V}, \quad \phi_m = 0.137 \text{ V}.$$

We can now answer the above questions:

1. The voltage drop across the load, which is the voltage of the PEC cell, $= \phi_m - \phi_{semi} = 0.137$ V, since $\phi_{semi} = 0$.
2. The cell current can be calculated from any of the equations for the current. The simplest of these is probably the one for the potential drop across the load:

$$i = \frac{\phi_m}{R_\Omega A} = \frac{0.137 \text{ V}}{1 \text{ }\Omega \times 10 \text{ cm}^2} = 0.0137 \frac{\text{A}}{\text{cm}^2}.$$

3. The power (IV) available from the cell is simply $iA\phi_m = 0.0188$ W.
4. The ohmic drop in solution is $\phi_{s,a} - \phi_{s,c} = 0.046$ V.
5. $\eta_{cath} = (\phi_m - \phi_{s,c}) - V_{eq,c} = -0.057$ V.

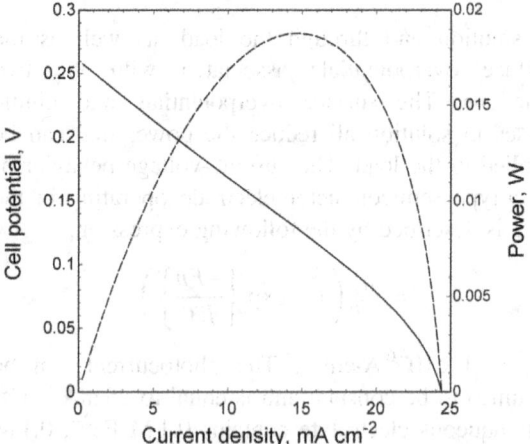

Figure 15.22 i–V curve, and on right ordinate, power versus cell potential for the cell from Illustration 15.7.

Maximum Power

The power available from a PEC varies with the potential of the cell (terminal to terminal) and the current. At one extreme, open circuit, the voltage difference is at its maximum value. However, the current and, therefore, the power output both are zero at open circuit. At the other extreme, the current is at its maximum when the cell is shorted, but the voltage difference between the terminals is zero, which again yields a power output of zero. The point on the i–V curve for the PEC cell where the product of i times V has the greatest value is the maximum power available from the cell. The i–V curve and the corresponding power curve for Illustration 15.7 both are shown in Figure 15.22. For the conditions given, the maximum power corresponds to a cell voltage of 0.13 V and a cell current density of 14.4 mA·cm^{-2}. This value depends on the potential losses in the cell (e.g., ohmic and kinetic losses), as well as on the magnitude of the external load.

Cell Efficiency

The energy conversion efficiency is expressed by

$$\text{Efficiency\%} = \frac{\text{Electrical power}}{\text{Incident solar power}} = \frac{iV}{P_{in}} \times 100. \tag{15.26}$$

Another measure of performance used in the solar community is called the *fill factor*, which is defined as follows:

$$\text{fill factor} = \frac{(iV)_{max}}{i_{SC}V_{OC}}. \tag{15.27}$$

The denominator is the product of the open-circuit voltage, which is the maximum voltage, and the short-circuit current, which is the maximum current; it represents a hypothetical power that is used as a convenient reference point.

How can we optimize efficiency? Part of this process is by now very familiar. We know that by reducing ohmic losses in the cell, for example, the maximum power $(iV)_{max}$ is increased, and this operating point occurs at a higher current density. For a photochemical cell, there are additional important aspects. A detailed analysis of efficiency is beyond our scope; but let's briefly consider a few points. First, it is critical to carefully align the load requirements with the conditions under which the power is maximized. As seen in Figure 15.21, the power can drop off quickly as conditions move away from the optimum. Depending on the application, this alignment will likely be achieved through a combination of cells in series and parallel. Also, the need to maximize light capture is frequently at odds with the reduction of electrochemical losses in the cell. For example, a clear pathway to light may require a circuitous path for current between the electrodes, increasing ohmic losses. Hence, the compact, bipolar stacks optimal from an electrochemical perspective stand in contrast to a structure that allows a clear path for radiation to enter and be absorbed. Additionally, the catalysts needed to reduce kinetic overpotential or the metal connections needed to reduce resistance losses may partially block access to light. Some incident radiation may be reflected or absorbed in the electrolyte, representing additional losses. The semiconductor must be thick enough to capture the light. On the other hand, a thicker electrode is more resistive and will increase ohmic losses. The magnitude of the saturation current is influenced by the difference between the potential of the redox reaction and that of the relevant valence or conduction band, as well as by the doping level of the semiconductor. Clearly, the design of PEC cells extends well beyond the factors typically considered in electrochemical cell design.

Types of PEC Cells

The type of PEC cell shown in Figure 15.21 is a *regenerative cell*, the product of which is electrical power. The cell is called a regenerative cell because the redox couple that reacts at the photoelectrode is regenerated at the counter electrode as illustrated in Figure 15.23a. No net chemical change takes place in this type of cell. In contrast, a PEC can also be used to generate chemical products (e.g., solar fuels) in a *photoelectrolytic cell* such as that shown in Figure 15.23b, which involves two separate redox reactions at different energies (potentials). Because they convert light into chemical energy, or fuels, photoelectrolytic cells are often referred to as "artificial photosynthesis." Note that a membrane is used to keep the desired chemical products from reacting at the other electrode.

The design requirements for an efficient photoelectrolytic cell are more stringent than those required for a regenerative cell. In both cases, the band gap must be suitable for light absorption. In addition, photoelectrolytic cells require alignment of both desired reactions with the appropriate band-edge energies. For example, for water to be oxidized at the photoanode, the valence band energy must be less than the energy that corresponds to the oxygen evolution reaction. On the SHE scale, this would require a potential greater than 1.229 V. The electron energy at the counter electrode must be greater than that of the hydrogen evolution reaction in order for hydrogen to be produced in the configuration shown in Figure 15.23. This means that the potential of the counter electrode would need to be more negative than the equilibrium potential of the hydrogen reaction. To handle both reactions, the band gap would need to be greater than 1.229 V (the difference between the reversible potentials for the O_2 and H_2 evolution reactions) and the band edges would need to be appropriately aligned to accommodate both reactions. The pH is also a consideration since the equilibrium potentials of both the O_2 and H_2 reactions, as well as the band-edge positions of the photoelectrodes, relative to a reference electrode in solution all change with pH.

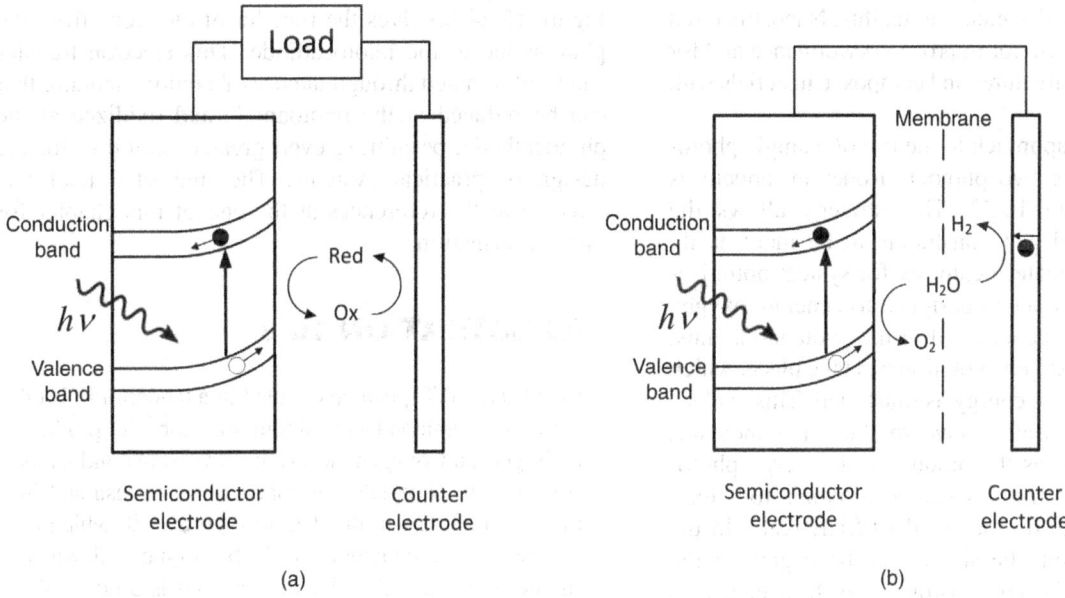

Figure 15.23 Energy diagrams for (a) a regenerative and (b) a photoelectrolytic cell. *Source:* Adapted from Grätzel 2001.

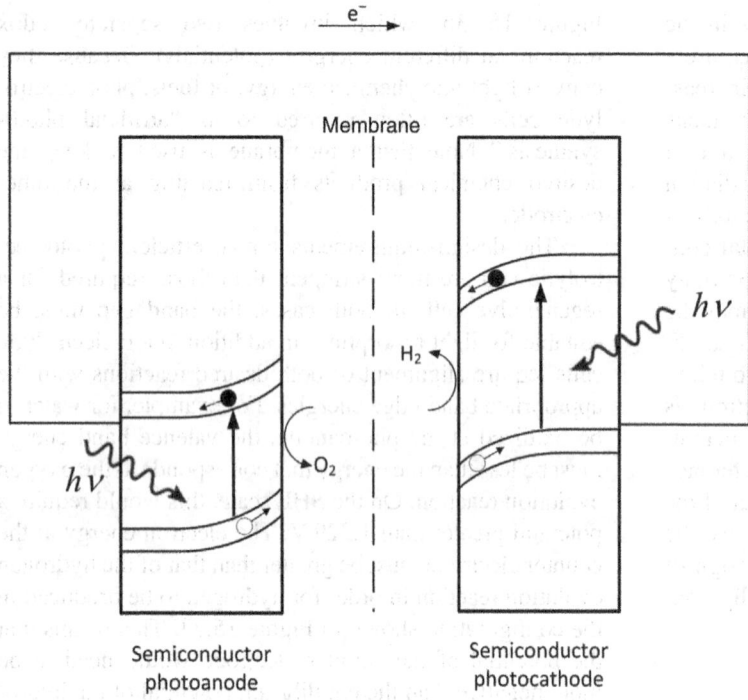

Figure 15.24 Photoelectrolytic cell with two photoelectrodes.

Satisfaction of the band gap and alignment constraints is, unfortunately, not sufficient. For example, CdS meets the band gap and alignment requirements, but is not suitable for use because it is not stable and undergoes photocorrosion. In addition, the hydrogen evolution reaction is very slow on CdS. Other factors that affect photoelectrode performance include the diffusion length for majority and minority carriers, carrier mobility, and rates of recombination. The use of cocatalysts to promote a desired electrochemical reaction (e.g., hydrogen or oxygen evolution) is an important strategy used to enhance performance. Protective layers are also being explored to enhance stability. Nanostructured materials show promise for improved performance and for the creation of new structures and composite materials with superior properties.

An alternative approach to the use of a single photoelectrode is to place two photoelectrodes in tandem as illustrated in Figure 15.23. This strategy allows the alignment and band gap constraints to be more easily met and provides greater flexibility for system optimization. For example, in a cell designed to generate oxygen and hydrogen, the valence band of the photoanode must be aligned so that oxygen evolution can take place, which requires a low electron energy as illustrated. This enables electrons from the water move to the semiconductor, creating oxygen gas as the product at the *n-type* photoanode. The low electron energy corresponds to a relatively high potential on the standard SHE scale. In the tandem configuration, the size of the band gap for the photoanode material is not constrained by the need to also accommodate the hydrogen reaction, and can be optimized for the system. On the other side of the cell, the electron energy of the conduction band of the photocathode must be high enough to reduce protons and form hydrogen gas (relatively low potential versus SHE). The band gap of this material is also not constrained. However, the energy of the conduction band of the photoanode must be greater than that of the valence band of the photocathode in order to transfer electrons between those two bands as shown. These concepts are shown in Illustration 15.8, which gives an example of a two photoelectrode system. As already noted, the strategy shown in Figure 15.24 involves the transfer of electrons from the photoanode to the photocathode. This electron transfer can be eliminated through the use of a redox mediator that can be reduced at the photoanode and oxidized at the photocathode, permitting even greater flexibility for the design of practical systems. The interested reader is directed to the references at the end of this chapter for more information.

ILLUSTRATION 15.8

Cu_2O and $BiVO_4$ can be coupled in a tandem PEC cell with two semiconductor–liquid junctions to produce hydrogen and oxygen, where the two semiconductors are wired to complete the connection as illustrated in Figure 15.23. Given the data in Figure 15.9, which of the two semiconductors should be *n-type* and which should be *p-type*? A ruthenium catalyst is used for the hydrogen evolution reaction (HER) and a cobalt catalyst for the oxygen evolution reaction (OER). With which electrode is each catalyst used? (Bornoz et al., *J. Phys. Chem. C*, **118**, 16959 (2014) provide further details.)

Solution:

From Figure 15.9, we approximate the following (hydrogen scale numbers calculated from the others):

Semiconductor	E_g	E_c	E_v	E_c H	E_v H
Cu_2O	2.2	−3.5	−5.7	−0.94	1.26
$BiVO_4$	2.4	−4.5	−6.9	0.06	2.46

From these energy data, Cu_2O better aligns with the hydrogen evolution reaction (cathodic), and $BiVO_4$ better aligns with the oxygen evolution reaction (anodic). This can be seen by considering each reaction separately. With $BiVO_4$, the valence band energy is quite low, permitting oxygen evolution. The corresponding voltage on the SHE scale is 2.46 V, well above the equilibrium potential for oxygen. The highest electron energy corresponds to the conduction band of Cu_2O. This corresponds to a potential on the SHE scale that is lower than the equilibrium potential for hydrogen and would therefore result in hydrogen evolution, the desired cathodic reaction. In both cases, the equilibrium potential (energy) for the preferred reaction lies near the middle of the band gap. Cu_2O is the photocathode and should be *p*-type. Similarly, $BiVO_4$ is the photoanode and should be *n*-type. The hydrogen catalyst should be integrated with the photocathode and the oxygen catalyst with the photoanode. Finally, the energy of the conduction band for $BiVO_4$ is higher than that of the valence band of Cu_2O, enabling the electron transfer needed to complete the circuit.

CLOSURE

We introduced the basic physics of semiconductors and described the interface that develops when they are put into contact with an electrolyte. Current flow under dark conditions has also been described. The most important feature of the semiconductor electrolyte system is that it is photoactive. Thus, light energy can be converted to electrical energy or to drive electrochemical synthesis. A key aspect of this chapter is the combination of light and electrochemistry to explore photoelectrochemistry.

FURTHER READING

Archer, M. D. and Nozik, A. J., eds. (2008) *Nanostructured and Photoelectrochemical Systems for Solar Photon Conversion*, Imperial College Press.

Chen, X., Shen, S., Guo, L., and Mao, S.S. (2010) Semiconductor-based photocatalytic hydrogen generation. *Chem. Rev.*, **110**, 6503–6570.

Memming, R. Semiconductor electrochemistry. Available at http://onlinelibrary.wiley.com/book/10.1002/9783527613069

Smith, W.A. Sharp, I.D. Strandwitz, N.C. and Bisquert, J. (2015) Interfacial band-edge energetics for solar fuels production. *Energy Environ. Sci.* **8**, 2851.

Tan, M. Laibinis, P.E. Nguyen, S.T. Kesselman, J.M. Stanton, C.E., and Lewis, N. (1994) Principles & Applications of Semiconductor Photoelectrochemistry, *Prog. Inorg. Chem.*, **41**, 20–144.

PROBLEMS

15.1 Sketch the interface between a metal and an electrolyte. Assume that the metal electrode has a small positive charge. Identify the inner and outer Helmholtz plane and the diffuse layer. What determines the thickness of the diffuse layer? Make a similar sketch for the semiconductor–electrolyte interface for an *n*-type semiconductor and identify the depletion region and the Helmholtz plane. What determines the thickness of the depletion region?

15.2 Thermal energy is represented by *kT*. Calculate the value of *kT* at room temperature and express the energy in terms of eV. Compare the value that you calculated with the band gap for Si. Explain why the intrinsic conductivity of semiconductors is low under normal conditions.

15.3 Determine the doping (in ppb) of phosphorous or arsenic that must be added to silicon to achieve a concentration of 10^{15} cm^{-3}. The density of silicon is 2329 kg·m^{-3}. Would this doping create an *n*- or a *p*-type semiconductor? What would be its resistivity?

15.4 Analogous to Figure 15.10, sketch the charge distribution and band bending for a *p*-type semiconductor brought into contact with an electrolyte. Assume that before equilibration the Fermi level of the redox couple is in the middle of the conduction and valence bands.

15.5 Data for the capacitance versus potential for an *n*-type Si semiconductor are provided in the table. The potential is measured with respect to a saturated calomel electrode, and the electrolyte is ammonium hydroxide. Create a Mott–Schottky plot (C^{-2} versus potential). From these data, determine the flat-band potential and the doping level in the semiconductor. Use a dielectric constant of 11.9. (Data are adapted from *J. Electrochem. Soc.*, **142**, 1705 (1995).)

Potential [V]	C [F·cm^{-2}]
−0.5	1.459E-08
0	1.101E-08
0.5	8.771E-09

(*continued*)

(Continued)

Potential [V]	C [F·cm^{-2}]
1	7.715E-09
1.5	7.036E-09

15.6 Repeat Problem 15.5, but use data that are provided for a *p*-type semiconductor. For a *p*-type semiconductor, Equation 15.8 is replaced with the following equation:

$$\frac{1}{C^2} = \frac{-2(\eta - V_{fb})}{\varepsilon A^2 q N_A}.$$

Potential [V]	C [F·cm^{-2}]
−1	2.108E-08
−1.5	1.571E-08
−2	1.265E-08
−2.5	1.125E-08
−3	1.031E-08

15.7 An intrinsic semiconductor has a 2 eV band gap. If the Fermi level of the electrode corresponds to 0.4 V relative to a SCE reference, sketch the energy levels and show the corresponding energies in eV of the band edges using a vacuum scale.

15.8 Calculate the strength of the electric field at the semiconductor–electrolyte interface. Use a dielectric constant of 20, and assume the thickness of the depletion layer is 100 nm, with a doping level of 10^{16} cm^{-3}.

15.9 For a 0.1 M 1:1 salt, calculate the concentration of charge carriers in solution. Compare this with typical values for doped semiconductors: 10^{15}–10^{18} cm^{-3}.

15.10 Using the semiconductor data from Illustration 15.3, determine at what electrolyte concentration is the Debye length of the same order of magnitude as the thickness of the depletion region for an *n*-type semiconductor doped to 10^{16} cm^{-3}. Assume a 1:1 electrolyte.

15.11 An *n*-type semiconductor is brought into contact with an electrolyte with a redox couple. Sketch the interface before and after contact with two different electrolyte solutions. The first where the redox potential is a little below the Fermi level of the semiconductor, and the second where the redox potential is a little above the valance band edge. Thus, four sketches are required. Be sure to show the excess charges in the semiconductor and the solution as well as the thickness of the depletion region and the Fermi levels.

15.12 Starting with Equation 15.17 (Beer–Lambert law), show that the penetration depth (distance at which the intensity of light is reduced by a factor of 1/*e*) is inversely proportional to the absorption coefficient. Using Equation 15.20, determine the penetration depth in crystalline silicon for light with wavelengths of 400, 600, and 1000 nm.

15.13 Describe three ways of (i) increasing the number of electrons in the conduction band of a semiconductor, and (ii) increasing the number of holes in the valence band of a semiconductor. In both cases, we are generating mobile charge carriers.

15.14 To effectively create a photocurrent, the semiconductor must absorb essentially all of the incident light. What's more, only light that is absorbed in or just outside the depletion layer results in charge separation. Discuss the implications of these restrictions when designing photoelectrodes for energy conversion. Specifically, compare the thicknesses and doping of Si and GaAs electrodes. Assume that the energies of the incident photons are 0.05 eV above their respective band gaps: Si 1.12 eV (indirect band gap) and GaAs 1.4 eV (direct band gap); and use the absorption coefficient from Equation 15.20 and Figure 15.17.

15.15 There are a number of parameters that arise is semiconductor–electrolyte systems: band gap, flat-band potential, doping levels, Fermi level, absorption coefficient, and width of the depletion layer. Describe each of these parameters. Specifically address whether these are intrinsic properties of the material, design variables that can be tuned, or one that will change depending on the operating conditions, such as light intensity or applied potential.

15.16 Given the spectral content of solar radiation, discuss the potential of the following semiconductors in (i) absorbing incident radiation and (ii) in efficient solar energy conversion: Si, GaAs, CdTe, and TiO$_2$. What about their use to electrolyze water?

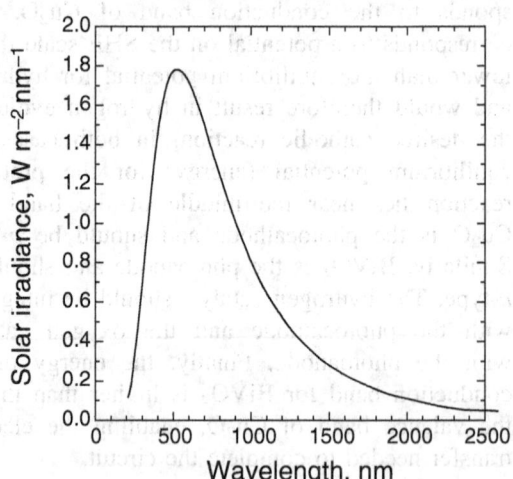

15.17 In Section 15.3, it is stated that "the potential drop across the Helmholtz double layer will be much less than that across the depletion layer of the semiconductor." Given how capacitors in series behave, justify this claim. Remember that the capacitance of the double layer is much higher than that of the depletion region in a semiconductor.

15.18 In Section 15.4, a simplified version of the Mott–Schottky equation was developed for an *n*-type semiconductor (see Equation 15.8). What is the analogous expression for a *p*-type semiconductor? For what range of overpotentials does it apply? Why?

15.19 Using the data from Illustration 15.3, calculate quantum yield at an overpotential of −0.1 V assuming that the diffusion length is 200 μm. How small could the diffusion length be and still have a quantum yield of 0.9 or more?

15.20 An approximate value for the photon flux in direct sunlight is 2×10^{21} photons/m^2·s. If the quantum yield were 1, what current density would this result from the photon flux? Name at least two reasons why this current density is not achieved in practice.

15.21 The open-circuit potential for the semiconductor electrode was determined in Section 15.7. Another important characteristic is the short-circuit current. Develop an expression for this current.

15.22 Using the data from Illustration 15.7, create a polarization curve (i versus V_{cell}) for photon currents of 50, 250, and 500 A·m^{-2}.

15.23 One method to determine V_{fb} is to measure the capacitance as was outlined in Illustration 15.3. Another approach is to measure the onset of photocurrent. The key to this method is to use monochromatic light of energy just slightly above the energy of the band gap. Under these conditions, α is tiny and only a small amount of the incident light is absorbed in the depletion region. Starting with Equation 15.22, show that the photocurrent is proportional to the width of the depletion region, W. Substitute this result into Equation 15.6 to obtain the following relationship:

$$\left(\frac{i_{ph}}{\alpha I_0''}\right)^2 = \frac{2\varepsilon}{qN_D}\left(\eta - V_{fb}\right).$$

15.24 Using the data provided in the table for TiO$_2$ and the result from the previous problem (Problem 15.23), determine the flat-band potential of the system.

η [V]	i_{ph} [mA·cm^{-2}]
0.704	0.909
0.557	0.845
0.441	0.778
0.321	0.706
0.200	0.612
0.129	0.553

15.25 For an n-type semiconductor with a flat-band potential of -0.4 V, what is the relative concentration of electrons at the surface of a semiconductor compared to the bulk: (i) when there is no applied potential ($\eta = 0$), and (ii) at an applied overpotential of 0.2 V. Comment on the implications of these results on the kinetics of the reaction and the current behavior.

15.26 In Section 15.3 the starting point for developing the description of the depletion layer was that the associated energy level of the redox couple was between the conduction and valence bands. What would change if the potential of the redox couple were either above the conduction band or below the valence band? Would it change with the doping (n or p)?

Ulick Richardson Evans

Ulick Richardson Evans was born on March 31, 1889 in Wimbledon. His father was a journalist and took Ulick to Switzerland where he introduced the young boy to mountaineering. Outdoor activities remained a passion throughout Evans' life. Ulick attended Marlborough College (1902–1907) and then Kings College, Cambridge (1907–1911). Subsequently, he studied electrochemistry in Wiesbaden and later at University College, London. These initial investigations of electrochemistry were ended by World War I when Evans served in the Army. He was stationed in the Middle East, returning to Cambridge in 1921.

When Evans began his career, there was an extremely poor understanding of corrosion. The attack of metals was all too familiar—There were lots of data and many empirical approaches to its prevention, but the scientific interpretation of corrosion was scant. It was accepted that corrosion associated with two dissimilar metals in contact was electrochemical in nature. However, when a metal is placed in an acid solution, the simultaneous dissolution of the metal and hydrogen evolution is observed. How could this homogeneous dissolution be anything more than a simple chemical replacement reaction? In what way could the corrosion of iron exposed to the atmosphere be driven by electrochemical processes? Why does the section of a sea pilings located away from the oxygen at the surface corrode faster than the section near the surface of the water?

Although W.R. Whitney presented the first convincing evidence of the electrochemical nature of corrosion in 1903, there was still much confusion at the start of the twentieth century. Evans quickly brushed away many of the inconsistencies and misunderstandings. His impact on the community is no doubt in part due to his ability to lecture in several languages and to his prolific writing. His first book, started before the war, was published in 1923, *Metals and Metallic Compounds*. He published a second book the next year, with three additional books (with multiple updates) and about 200 manuscripts over a long career. A second factor in his lasting influence can be attributed to his focus on both the theoretical and practical aspects of corrosion. His life-long avocation became addressing the lack of fundamental understanding of corrosion and the use of a scientific basis to quantify corrosion. Along with his picture is a figure from his 1929 paper of what is known today as an Evans diagram. This diagram shows the anodic and cathodic reactions for a corrosion system and illustrates the so-called corrosion potential. The Evans diagram has proved invaluable in the analysis of corroding systems.

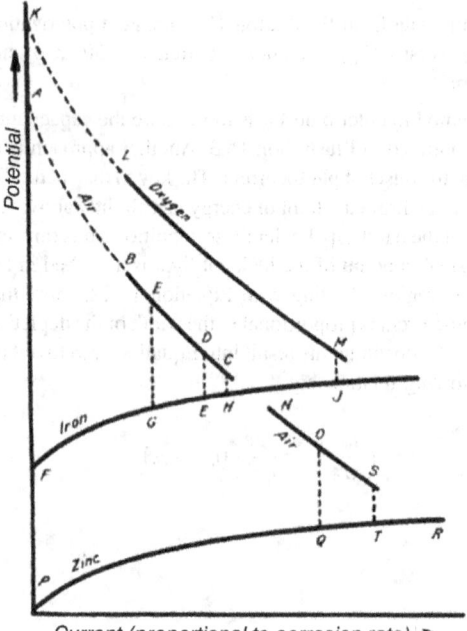

Source: From U.R. Evans (1929) The distribution and velocity of the corrosion of metals. *J. Franklin Inst.*, **208**, 45–58. Reprinted with permission from Elsevier.

Ulick Evans was elected Fellow of the Royal Society in 1949. He was awarded the Olin Palladium Medal from Electrochemical Society in 1955—the Society's highest technical award. The Institute of Materials, Minerals, and Mining now provides an annual award in honor of U.R. Evans for outstanding contributions to corrosion science. Two of his students, T.P. Hoar and R.B. Mears, established the basic principles for cathodic protection.

Evans died on April 3, 1980. Today, corrosion of metals is recognized as principally an electrochemical process. This was not the case at the start of the twentieth century. His legacy is aptly described as the father of the modern science of corrosion.

Image Source: Courtesy of National Portrait Gallery, London.

Chapter 16

Corrosion

Corrosion is the unwanted attack of metals by their environment. The many attractive properties of metals have led to their widespread use in industry and, frankly, in nearly every aspect of our lives. Metals, however, are susceptible to corrosion. It is therefore important that engineers understand the conditions under which corrosion is likely to occur. They should also be able to measure, predict (to the extent possible), and mitigate the negative impacts of corrosion. This chapter will help you to apply the principles of electrochemical engineering that you have learned to the analysis of corrosion systems. While the chapter addresses several important aspects of corrosion, it is by no means comprehensive. In fact, corrosion is an interdisciplinary topic, and a complete understanding requires the combined expertise of electrochemical engineers, material scientists, metallurgists, and mechanical engineers, among others.

We limit our discussion of corrosion to *electrochemical corrosion*, or corrosion that is the result of electrochemical processes. We also limit ourselves to aqueous corrosion, which is corrosion that involves an aqueous electrolyte. Electrochemical corrosion is the consequence of an electrochemical cell where the anode is the metal of interest. Such corrosion is, by definition, oxidation of the metal according to a reaction such as

$$Fe \rightarrow Fe^{2+} + 2e^- \qquad (16.1)$$

where iron has been used as an example. In other words, when we say that a metal is corroded, we are saying that a portion of the metal has been oxidized to form soluble products (e.g., Fe^{2+} ions in solution), insoluble products such as a salt or oxide layer (e.g., the rust layer on your old bike is a form of iron oxide), or a mixture of the two.

16.1 CORROSION FUNDAMENTALS

A complete electrochemical cell is required for electrochemical corrosion to occur. The components of that cell include the following:

- The metal (anode)
- A cathodic reactant with an equilibrium potential higher than that of the metal
- Electrical contact between the anode and cathode
- An electrolyte that permits current flow in solution between the anode and cathode

These constituents, of course, should not be a surprise by now. However, as shown in Figure 16.1, the form in which you find them can be quite different in corroding systems. None of these cell configurations looks like the typical electrochemical cell presented in Chapter 1. This lack of similarity partly explains why early researchers did not immediately connect corrosion to an electrochemical mechanism. Let's first examine galvanic corrosion where the link with a conventional electrochemical cell is the most direct. In galvanic corrosion, an active metal is electrically connected to a more noble metal (one that is less likely to corrode). This causes preferential corrosion of the more active metal as shown in Figure 16.1, and thus it is one type of localized corrosion. Recall the Daniel cell, Figure 1.1. There were two metals, Zn and Cu, separated by an electrolyte. Electrons flowed through an external circuit. Now imagine that in place of the resistor of Figure 1.1, the two metals are brought into direct contact. In fact, one important difference between conventional cells is that the anodes and cathodes in these corrosion cells are not only electrically connected but they are also short circuited. An electrolyte surrounds the electrodes, allowing

---

Electrochemical Engineering, First Edition. Thomas F. Fuller and John N. Harb.
© 2018 Thomas F. Fuller and John N. Harb. Published 2018 by John Wiley & Sons, Inc.
Companion Website: www.wiley.com/go/fuller/electrochemicalengineering

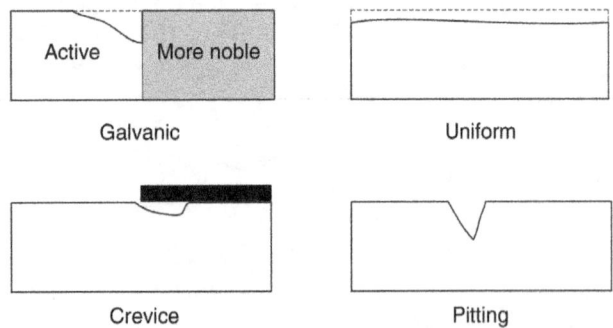

Figure 16.1 Examples of different types of corrosion.

ionic current to flow. As in the Daniell cell, the zinc (active metal) is oxidized. The cathodic reaction, however, will be different because copper ions are not present in solution. These cathodic or reduction reactions will be discussed shortly. You can see that the difference between the galvanic corrosion example and the typical electrochemical cell is small. We won't discuss all of the types shown in Figure 16.1 in detail now, but as you work through this chapter, try to bridge mentally between these corrosion cells and the prototypical electrochemical cell. A variety of geometries and electrolyte configurations are possible, but in the end the same electrochemical principles apply.

Pitting and crevice corrosion are also forms of localized corrosion, where the dissolution reaction takes place at discrete locations on the surface. In pitting corrosion, the local environment in the pit stabilizes the corrosion; the same is true in crevice corrosion for the local environment in the occluded area. There are many other forms of corrosion. We have chosen some of the simpler forms for our discussion in this chapter.

In the galvanic corrosion example, oxidation takes place on one metal and a reduction process occurs preferentially on the other. In contrast, for uniform corrosion, both the metal dissolution and cathodic reactions take place simultaneously over the entire surface of a single metal. This idea that both reactions can occur on the same surface seems counter to everything that we have learned and will require some careful thought (Section 16.3).

Knowledge of the cathodic reactions is critical to understanding corrosion. The most common cathodic reactions in aqueous systems are oxygen reduction and hydrogen evolution, shown below for both acidic and basic solutions:

$$O_2 + 4H^+ + 4e^- \rightarrow 2H_2O \qquad (16.2a)$$

$$O_2 + 2H_2O + 4e^- \rightarrow 4OH^- \qquad (16.2b)$$

$$2H^+ + 2e^- \rightarrow H_2 \qquad (16.3a)$$

$$H_2O + 2e^- \rightarrow H_2 + 2OH^- \qquad (16.3b)$$

Both of these reactions bring some interesting aspects to the corrosion problem. For example, oxygen gas (O_2) is only sparingly soluble in aqueous solutions. Consequently, its reaction rate is often limited by how fast it can get to the surface, which can vary widely with the thickness of the electrolyte layer. In the case of hydrogen evolution, the solvent is reacting. The equilibrium potentials of these reactions vary with pH. Also, both of these cathodic reactions can influence the local pH, which can also have a significant impact on corrosion as discussed below.

Other cathodic reactions are also possible, although less common. For example, metal ions in solution such as ferric ions may be reduced:

$$Fe^{3+} + e^- \rightarrow Fe^{2+} \qquad (16.4)$$

The reduction of oxidizing acids may also occur. In some situations, the plating of metals out of solution can contribute to the cathodic current.

16.2 THERMODYNAMICS OF CORROSION SYSTEMS

In this section, we examine the thermodynamics of corrosion cells. First, we should explain an apparent contradiction. As noted above, we described the corrosion system as being electrically shorted. How then can we apply thermodynamics and what can we learn about corrosion from an analysis of equilibrium? Clearly, we do not have an open-circuit voltage that is determined by the equilibrium of two half-cell reactions. *The key question that can be answered with thermodynamics is whether or not a corrosion reaction is favorable and may, therefore, occur spontaneously.* To do this, we imagine that the anodic and cathodic reactions take place on separate electrodes and consider whether there is a potential difference in the absence of current flow. *A thermodynamically favorable corrosion reaction is a reaction that has a negative free energy (positive cell potential) with the metal as the anode.* We follow the same procedure that you learned in Chapter 2, and will use the same simplified activity corrections in order to streamline our analysis. A more detailed analysis could be performed, but this reduced level of complexity is commonly used and is adequate for our purposes. Once again, we examine the driving force for corrosion by examining the potential change (Gibbs energy difference) associated with the overall reaction. Let's consider a few possibilities.

ILLUSTRATION 16.1

Please determine which of the following systems are thermodynamically favored to corrode: (i) iron/oxygen, (ii) iron/hydrogen, and (iii) copper/hydrogen. Assume a very low concentration of metal in the solution (10^{-6} M) and a pH of 4.

Solution:

$$\text{Fe (s)} \left| \begin{array}{c} \text{Aqueous} \\ \text{Electrolyte} \\ \text{pH} = 4 \end{array} \right| \begin{array}{c} \text{electrode(s),} \\ O_2(g) \end{array}$$

First we consider iron/oxygen.

Anodic reaction: $Fe \rightarrow Fe^{2+} + 2e^-$

Standard potential: -0.440 V

Cathodic reaction: $O_2 + 4H^+ + 4e^- \rightarrow 2H_2O$

Standard potential: 1.229 V

Overall reaction (4e$^-$): $O_2 + 4H^+ + 2Fe \rightarrow 2Fe^{2+} + 2H_2O$

Open-circuit potential (simplified activity corrections):

$$U = U^\theta - \frac{RT}{nF} \ln\left(\prod c_i^{s_i}\right)$$

$$= (1.23 - (-0.44)) - \frac{(8.314)(298)}{(4)(96485)} \ln\left(\frac{(10^{-6})^2}{(10^{-4})^4 (0.21)}\right)$$

$$= 1.60 \text{ V},$$

where a concentration (10^{-6} M) of ferrous ions was assumed to be present in solution, and a partial pressure of 21 kPa was assumed for oxygen. Clearly, the resulting potential is positive, and the reaction is favored thermodynamically. Similarly, for iron and hydrogen:

$$\text{Fe (s)} \left| \begin{array}{c} \text{Aqueous} \\ \text{Electrolyte} \\ \text{pH} = 4 \end{array} \right| \begin{array}{c} \text{electrode(s),} \\ H_2(g) \end{array}$$

Anodic reaction: $Fe \rightarrow Fe^{2+} + 2e^-$

Standard potential: -0.440 V

Cathodic reaction: $2H^+ + 2e^- \rightarrow H_2$

Standard potential: 0.0 V

Overall reaction (2e$^-$): $Fe + 2H^+ \rightarrow Fe^{2+} + H_2$

Open-circuit potential (via Nernst equation):

$$U = U^\theta - \frac{RT}{nF} \ln\left(\prod c_i^{s_i}\right)$$

$$= (0.0 - (-0.44)) - \frac{(8.314)(298)}{(2)(96485)} \ln\left(\frac{10^{-6}}{(10^{-4})^2}\right)$$

$$= 0.38 \text{ V}.$$

Here we have not included a pressure correction for hydrogen gas, since its partial pressure is not provided. The resulting potential is also positive, but less so than the iron/oxygen couple. Still, corrosion is thermodynamically favored. If we replace iron with copper under the same conditions:

$$\text{Cu (s)} \left| \begin{array}{c} \text{Aqueous} \\ \text{Electrolyte} \\ \text{pH} = 4 \end{array} \right| \begin{array}{c} \text{electrode(s),} \\ H_2(g) \end{array}$$

Anodic reaction: $Cu \rightarrow Cu^{2+} + 2e^-$

Standard potential: 0.337 V

Cathodic reaction: $2H^+ + 2e^- \rightarrow H_2$

Standard potential: 0.0 V

Overall reaction (2e$^-$): $Cu + 2H^+ \rightarrow Cu^{2+} + H_2$

Open-circuit potential:

$U = -0.399$ V.

In this case, the cell potential is negative, which means that copper will not react with water to evolve hydrogen. Therefore, copper is not susceptible to corrosion in deaerated (oxygen free) water at this pH. However, copper will corrode in the presence of dissolved oxygen.

If a corrosion reaction is not thermodynamically favorable, we can be confident that it will not take place. The term used to describe this condition in the corrosion field is *immunity*. What if the reaction is thermodynamically favorable? Does that mean that we need to worry about the metal dissolving away right before our eyes? Fortunately, no. A thermodynamic analysis does not provide information about the rate at which corrosion will happen (don't worry, we will talk about rate shortly). In fact, there are some metals like aluminum that are very stable and yet are thermodynamically favored to corrode.

If you look at the value for aluminum in Appendix A, you will see that it is very reactive, with standard potential of -1.66 V. Why, then, doesn't it readily corrode? Aluminum is stable because under many conditions it forms a dense oxide film on its surface that impedes the corrosion reaction. We call this film a *passive layer*, and refer to the metal as passivated. Metals are passive when they form stable, protective films upon oxidation. Many common metals are not stable thermodynamically, but are functionally stable because of passive films.

Thermodynamics can also help us determine when a metal is likely to be passive by considering the relative stability of different corrosion products. In electrochemical corrosion, we are interested in stability in electrolyte solutions of different composition. A critical variable in determining the stability of surface layers is the pH. The potential is also critical in determining corrosion behavior. We can use a Pourbaix diagram (Chapter 2) to identify

382 Electrochemical Engineering

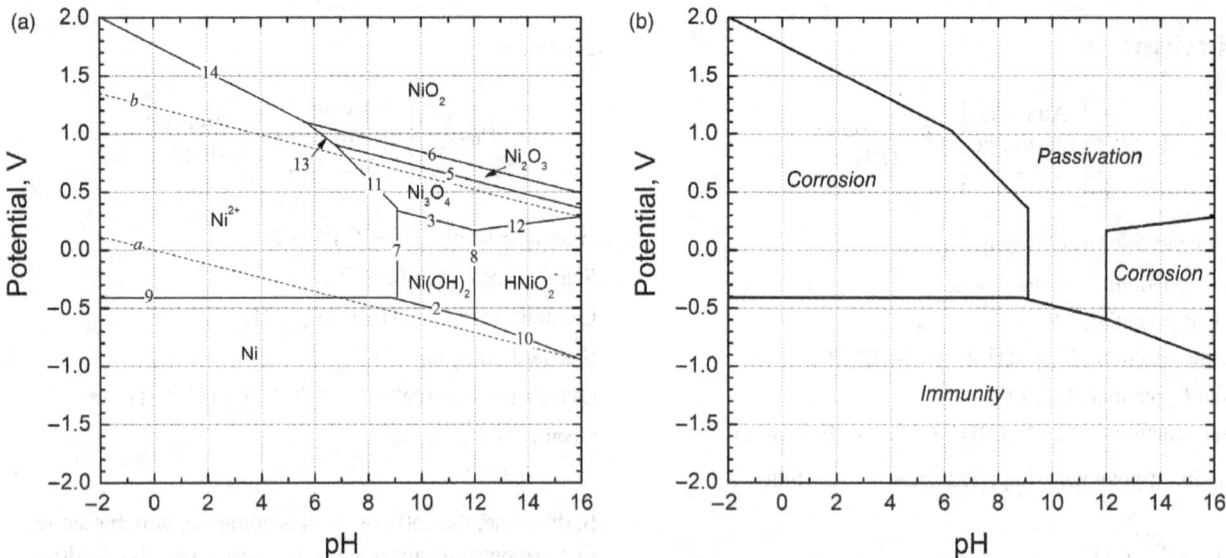

Figure 16.2 Pourbaix diagram for nickel and resulting corrosion map showing regions of immunity, corrosion, and passivation. The numbers and letters correspond to the equations below and in those used in Pourbaix's book.

regions according to the dominant form of the metal of interest as a function of potential (SHE) and pH.

A Pourbaix diagram for Ni is shown in Figure 16.2, recall that dashed lines a and b correspond to the hydrogen reaction and oxygen reaction, respectively. As you may remember, each solid line in the diagram represents equilibrium of an electrochemical or a chemical reaction. Therefore, the diagram is only as accurate as the reactions and species considered. For this diagram, the following reactions were included:

$$Ni + H_2O = Ni(OH)_2 + 2H^+ + 2e^- \quad (2)$$

$$3Ni(OH)_2 + H_2O = Ni_3O_4 + 2H^+ + 2e^- \quad (3)$$

$$2Ni_3O_4 + H_2O = 3Ni_2O_3 + 2H^+ + 2e^- \quad (5)$$

$$Ni_2O_3 + H_2O = 2NiO_2 + 2H^+ + 2e^- \quad (6)$$

$$Ni^{2+} + H_2O = Ni(OH)_2 + 2H^+ \quad (7)$$

$$Ni(OH)_2 + H_2O = HNiO_2^- + H^+ \quad (8)$$

$$Ni = Ni^{2+} + 2e^- \quad (9)$$

$$Ni + 2H_2O = HNiO_2^- + 3H^+ + 2e^- \quad (10)$$

$$3Ni^{2+} + 4H_2O = Ni_3O_4 + 8H^+ + 2e^- \quad (11)$$

$$3HNiO_2^- + H^+ = Ni_3O_4 + 2H_2O + 2e^- \quad (12)$$

$$2Ni^{2+} + 3H_2O = Ni_2O_3 + 6H^+ + 2e^- \quad (13)$$

$$Ni^{2+} + 2H_2O = NiO_2 + 4H^+ + 2e^- \quad (14)$$

You should be able to associate each of these reactions with the corresponding line in the diagram. For example, the first reaction describing the conversion of nickel to nickel hydroxide corresponds to line 2 in the diagram.

Each region in the diagram is labeled by the nickel-containing species that is thermodynamically favored for that region. Where needed, a concentration of 10^{-6} M was used to establish region boundaries. The region labeled Ni is where metallic nickel is favored. This region is labeled "immunity" in Figure 16.2b because there is no driving force for corrosion, and the metal is thermodynamically stable. In contrast, corrosion is favored in regions where a nickel-containing ion is the thermodynamically favored species. Stable ionic species include Ni^{2+} at neutral or acidic pH values, and $HNiO_2^-$ at high pH. Passivation is expected in regions where a solid nickel-containing species is stable; these species include $Ni(OH)_2$, Ni_3O_4, Ni_2O_3, and NiO_2, denoted as "passivation" in Figure 16.2b. An effective strategy for mitigating corrosion would move the metal into a region of immunity or passivity by, for example, changing the potential.

From thermodynamics, we have seen that we can determine if there is a driving force for corrosion and can gain insight into the products of corrosion reactions. We can also determine under what conditions those products are likely to lead to passivation of the surface. Such information is valuable for the assessment of corrosion. However, we must remember that our analysis is limited to the specific reactions considered and may not be accurate if important reactions have been left out. For example, complexation of metal ions with anions in solution may stabilize the metal ions and prevent the formation of solid, passivating products, significantly changing the corrosion characteristics of the system. Also, a thermodynamic analysis cannot provide information on the rate of corrosion or the integrity of surface layers that may be formed.

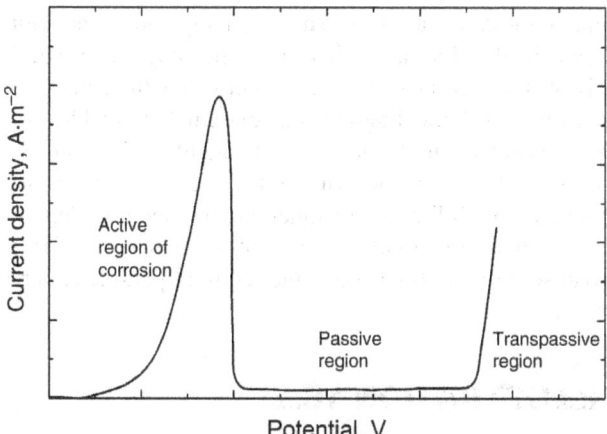

Figure 16.3 Current–voltage curve illustrating passivation.

In spite of these limitations, thermodynamics and Pourbaix diagrams provide valuable insight and are used frequently by corrosion engineers.

Another important aspect of metal behavior related to corrosion can be seen in Figure 16.2. If we follow a vertical line from the bottom to the top of the diagram, for example, at pH = 8, we see that, as the potential increases, we move from immunity, where metallic nickel is thermodynamically favored, to corrosion where Ni^{2+} is the stable species, and finally to passivation where nickel oxide is thermodynamically preferred. This type of passive behavior with increasing potential is a characteristic of several important metals, including iron, nickel, chromium, titanium, and their alloys. Figure 16.3 illustrates the I–V curve for a passivating metal. As you can see, the current drops abruptly at values of the potential where passive behavior is achieved. This phenomenon is the basis for anodic protection, a strategy that minimizes corrosion by increasing the potential so that the metal is in the passive region.

16.3 CORROSION RATE FOR UNIFORM CORROSION

As noted in the previous section, thermodynamics provides important insight into corrosion, but cannot offer information regarding the rate of corrosion. In this section, we examine a method for determining corrosion rate of a uniformly corroding surface.

In uniform corrosion, both the anodic (dissolution) and cathodic reactions take place evenly on the same surface. We've seen instances where multiple reactions occur on the same electrode. In fact, in Chapter 3 we introduced the current efficiency to describe such a situation. In those instances, the reactions were either both anodic or both cathodic. There is, however, no reason why oxidation and reduction reactions can't occur simultaneously on the same electrode. This situation is precisely what is occurring during uniform corrosion. One of the reactions that takes place is the oxidation of the metal, which is the corrosion reaction of interest.

For a uniformly corroding surface, the potential naturally moves to the *corrosion potential*, where the anodic and cathodic currents are equal. Why does this happen? To think about this, let's first consider the situation where the anodic current is higher than the cathodic current. Such a situation would lead to an accumulation of electrons in the metal, which would decrease its potential. As we learned previously, a decrease in potential would reduce the rate of the anodic reaction and increase that of the cathodic reaction. If the rate of the cathodic reaction were higher, there would be a net consumption of electrons and a corresponding rise in the potential of the metal. In this case, the increase in the potential would reduce the rate of the cathodic reaction and increase that of the anodic reaction. Both of these unbalanced situations move the system toward the stable condition where the anodic and cathodic currents are equal and there is no net buildup of charge.

Evans Diagrams

A diagram similar to the Tafel plot, Figure 3.7, can be used to illustrate and to help us analyze uniform corrosion. As shown in Figure 16.4, the current–voltage curves for both the anodic and cathodic reactions are drawn on the same plot. The potential is relative to a common reference

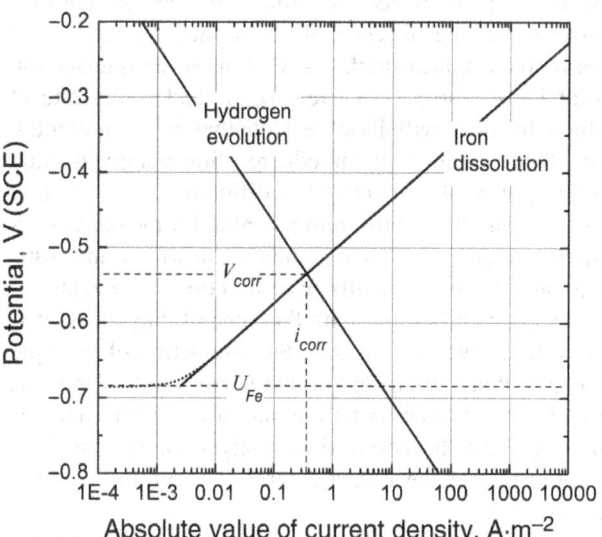

Figure 16.4 Evans diagram showing iron dissolution and hydrogen evolution (pH = 0.9, 1 N HCl, deaerated). *Source:* Adapted from R. J. Chin and K. Nobe 1972, J. Electrochem. Soc., 119, 1457.

electrode and is plotted as a function of the logarithm of the absolute value of the current for each reaction. Figure 16.4 shows the cathodic evolution of hydrogen and anodic dissolution of iron in deaerated solution, where Tafel expressions have been used for the kinetics. Tafel expressions are frequently used for corrosion systems and are applicable in situations where there is a significant difference in the equilibrium potentials for the two reactions. Typically, just one branch (anodic or cathodic) of a given reaction is plotted in the diagram, although this is not a requirement. In Figure 16.4, the cathodic branch has been plotted for the hydrogen reaction, and the magnitude of the cathodic current is seen to increase as the potential decreases, which corresponds to a higher cathodic overpotential. The slope of the resulting line is negative. In contrast, the anodic branch is plotted for iron, and the current increases with increasing potential, which corresponds to a higher anodic overpotential. The slope of the Tafel line for the anodic reaction is positive. The two curves are independent, and the current for a given reaction goes to zero at its equilibrium potential, shown in the diagram for iron (see dotted line asymptote on the log plot). The equilibrium potential for the hydrogen reaction is not shown.

The potential at the intersection point of the two curves is defined as the corrosion potential, which we will represent as V_{corr} to be consistent with the notation used in the rest of the text. We note, however, that E_{corr} is frequently used in the corrosion field. V_{corr} represents the potential at which the current from the cathodic reaction is equal to that from the anodic reaction. Hence, the net current at the corrosion potential is zero, since the rates of the two reactions are balanced. The corrosion potential is the potential at which the iron will undergo uniform corrosion in the absence of an applied potential, and can be readily measured. The current at the intersection point is the corrosion current (i_{corr}) and is the rate at which the metal will dissolve. Note that the overpotential for either the anodic or cathodic reaction at a given value of the potential is simply the difference between that potential and the equilibrium potential for the reaction of interest. Figure 16.4 is commonly known as an Evans diagram, and is frequently used by corrosion engineers. You should be familiar with this type of diagram as it is an effective way to illustrate the interaction of multiple reactions that participate to cause corrosion. Importantly, it includes information on the rate of corrosion that was missing from the equilibrium analysis of the previous section. Use of the Evans diagram is explored in Illustration 16.2.

As with any tool, Evans diagrams have their limitations. For example, both V_{corr} and i_{corr} are concentration dependent, reflecting the concentration dependence of the individual reactions. Changes with concentration and other important variables, such as temperature, are not shown in the diagram. Therefore, the diagram is valid only for the specific conditions under which the data were measured. Still, the diagram is an extremely valuable tool for conceptual understanding. If additional detail is required, it is possible with today's engineering tools to include the full compositional and temperature dependences in a numerical solution of the equations for a broad set of conditions, once the required parameters are known.

ILLUSTRATION 16.2

Use the Evans diagram in Figure 16.4 to answer the following questions:

1. What is the value of the corrosion potential, and what is the net current at this potential?
2. What is the current due to hydrogen evolution at the corrosion potential? What is the current due to iron dissolution at the corrosion potential?
3. What is the equilibrium potential for iron, and how does it compare to the corrosion potential?
4. What is the overpotential for the anodic (iron dissolution) reaction at the corrosion potential? Is use of the Tafel approximation justified?
5. What is the equilibrium potential for the hydrogen evolution reaction (not shown in the diagram)?

Solution:

1. The corrosion potential is the potential at which the anodic and cathodic currents are equal. The anodic and cathodic currents are equal at the intersection point in the graph. The potential at that point (see Figure 16.4), V_{corr}, is approximately -0.54 V versus SCE. The net current at the corrosion potential is zero since the anodic and cathodic currents are of equal magnitude and opposite sign.

2. Moving from the intersection point to the x-axis yields $\log|i| \approx -0.44$ Therefore, the hydrogen current at the corrosion potential is approximately -0.36 A·m^{-2}, and the iron dissolution current is approximately 0.36 A·m^{-2}. Note that the log scale is difficult to read accurately.

3. The equilibrium potential for iron is at the bottom of the diagram and is -0.684 V (SCE) for this problem. The potential would need to be lower than this value for immunity from corrosion.

> 4. The overpotential for iron at the corrosion potential is $-0.54 - (-0.684) = 0.144$ V. The Tafel approximation is valid at this large overpotential.
> 5. The equilibrium potential for the hydrogen reaction versus SHE at a pH of 0.9 = -0.053 V. Relative to a saturated calomel electrode, this is -0.297 V. Thus, even though the diagram shows the Tafel line only, the actual hydrogen current would taper off to zero at its equilibrium potential.

In situations where the exchange current densities and equilibrium potentials are known for a single anodic and cathodic reaction at the conditions of interest, V_{corr} and i_{corr} can be solved for analytically. Assuming that Tafel kinetics apply, where a refers to the anodic (metal) reaction and c refers to the cathodic reaction:

$$V_{corr} = \frac{\alpha_c U_c + \alpha_a U_a}{\alpha_a + \alpha_c} + \frac{RT}{F(\alpha_a + \alpha_c)} \ln\left(\frac{i_{o,c}}{i_{o,a}}\right)$$
$$= \frac{b_a U_c + |b_c| U_a}{b_a + |b_c|} + \frac{b_a |b_c|}{\ln(10)(b_a + |b_c|)} \log\left(\frac{i_{o,c}}{i_{o,a}}\right), \quad (16.5)$$

$$i_{corr} = i_{o,a}^{\left(\frac{\alpha_c}{\alpha_a+\alpha_c}\right)} i_{o,c}^{\left(\frac{\alpha_a}{\alpha_a+\alpha_c}\right)} \exp\left[\frac{F}{RT}\frac{\alpha_c \alpha_a}{\alpha_a + \alpha_c}(U_c - U_a)\right], \quad (16.6a)$$

$$= i_{o,a}^{\left(\frac{b_a}{b_a+|b_c|}\right)} i_{o,c}^{\left(\frac{|b_c|}{b_a+|b_c|}\right)} \exp\left[\frac{\ln(10)}{b_a + |b_c|}(U_c - U_a)\right], \quad (16.6b)$$

where b is the Tafel slope introduced in Chapter 3, which is negative for the cathodic reaction:

$$|b| = \frac{(\ln 10)RT}{\alpha F} = \frac{2.303 RT}{\alpha F}. \quad (3.25)$$

Note that, in contrast to the way these subscripts are used in the Butler–Volmer equation, the a and c refer to two different reactions: the anodic or corrosion reaction and the corresponding cathodic reaction. The concentration and temperature dependence of V_{corr} and i_{corr} are derived from the concentration and temperature dependence of both the equilibrium voltages and exchange-current densities.

Alternatively, V_{corr} and i_{corr} can be determined numerically for a variety of conditions if the parameters are known. The equations used in the numerical procedure are actually simpler, and numerical solution does not present a challenge with modern tools. Also, the numerical procedure can be used in situations where there are multiple cathodic reactions by simply adding terms for the additional reactions. Please see the illustration below in the section on "Multiple Reactions" for additional information.

Experimental Measurement of Corrosion

In practice, experimental measurement of the corrosion behavior is more common than prediction of that behavior from the fundamental kinetic properties of the system. The corrosion potential is the natural potential of the system and can be measured easily against a reference electrode. Common reference electrodes used for this purpose include Ag/AgCl, Cu/CuSO$_4$ (CSE), and to a lesser extent SCE and SHE. Placement of the reference electrode is not critical for measurement of the corrosion potential of uniformly corroding systems because the net current is zero and, therefore, there is no ohmic drop due to current flow. Note that the measured value of V_{corr} depends on environmental variables, such as composition and temperature, and should not be treated as a constant.

Measurement of the current is more problematic since the net current at the corrosion potential is zero, and it represents the sum of anodic and cathodic currents. The average corrosion rate can be determined directly for some samples by measuring the weight loss of the sample.

A corrosion-related polarization curve can also be measured by varying the potential of the sample both above and below the corrosion potential and measuring the current. In order to vary the potential and measure the current, the setup is modified as shown in the inset of Figure 16.5. The sample functions as the working electrode and a counter electrode is introduced. Appropriate ohmic corrections should be made to these measurements to account for the position of the reference electrode. The measured current represents the sum of both the cathodic

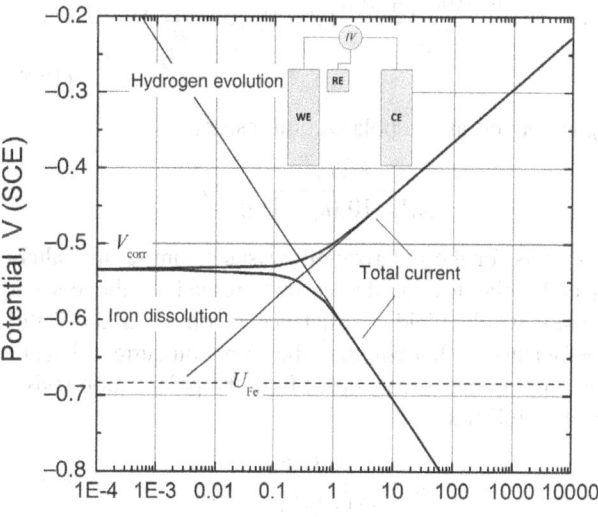

Figure 16.5 Relationship between the total current and the anodic and cathodic currents for a corrosion sample at different potentials.

and anodic currents as illustrated in Figure 16.5. Note that the combined curve approaches the anodic curve at potential values significantly above V_{corr} and the cathodic curve at potential values significantly below V_{corr}. Thus, it is possible to extract information about the individual reactions from these data. Care should be exercised, however. Movement away from the corrosion potential may introduce additional reactions that did not occur at an appreciable rate at the corrosion potential. Remember, although the polarization data in Figure 16.5 look like the data from a single reaction, V_{corr} does not represent an equilibrium potential, and the system does not approach equilibrium. If we assume Tafel kinetics and a single cathodic and anodic reaction, the following expression can be written for the combined current in terms of V_{corr} and i_{corr}. This equation is similar in form to the Butler–Volmer equation:

$$i = i_{corr}\left[\exp\frac{\alpha_a F}{RT}(V - V_{corr}) - \exp\frac{-\alpha_c F}{RT}(V - V_{corr})\right]. \quad (16.7a)$$

$$= i_{corr}\left[\exp\frac{\ln(10)}{b_a}(V - V_{corr}) - \exp\frac{\ln(10)}{b_c}(V - V_{corr})\right]. \quad (16.7b)$$

The limitations of Equation 16.7 should be considered when fitting data to this expression. Also, remember that this type of analysis is strictly valid only for systems undergoing uniform corrosion of the surface.

Equation 16.7 is the basis for a commonly used corrosion technique called *linear polarization resistance*. Linearization of Equation 16.7b about the corrosion potential yields

$$i = \frac{i_{corr}\ln(10)(b_a + |b_c|)}{b_a|b_c|}(V - V_{corr}) = \frac{1}{R_p}(V - V_{corr}). \quad (16.8)$$

R_p is defined as the polarization resistance:

$$\frac{b_a|b_c|}{i_{corr}\ln(10)(b_a + |b_c|)} \equiv R_p.$$

The units for the polarization resistance are $\Omega \cdot m^2$; alternatively, the use of the current instead of the current density would yield a resistance in ohms. If the Tafel coefficients are known, then the corrosion current density (or current) can be calculated from the polarization resistance as follows:

$$i_{corr} = \frac{b_a|b_c|}{\ln(10)(b_a + |b_c|)}\frac{1}{R_p}. \quad (16.9)$$

This calculation requires that the Tafel coefficients are known. Equation 16.9 is a form of the Stern–Geary equation.

Mass-Transfer Effects

Whereas Tafel kinetics is frequently well-suited to the analysis of corrosion rates, there are situations where a process other than kinetics controls the rate. One important example is that of oxygen reduction. Since the solubility of oxygen in water is quite low, the reduction of oxygen becomes limited by the mass transport of oxygen to the surface at high overpotentials as shown in Figure 16.6 for laminar flow over a flat plate. Note that the overpotential for the cathodic reduction of oxygen increases as the absolute value of the potential decreases. Once the limiting current is reached, oxygen reduction is no longer a function of potential (see Section 3.3). The limiting current, however, is a function of factors that influence the rate of mass transfer, such as the velocity of the fluid (Re) as illustrated in Figure 16.6.

In Chapter 3, we developed an expression for the current density that included mass transfer and was applicable for the entire range of potentials. This derivation was possible because both the mass-transfer expression and the kinetic expression were linearly dependent on a single surface concentration; that made it possible to eliminate the surface concentration and derive a single expression that does not explicitly include the concentration at the surface. Unfortunately, it is not always easy or possible to derive such an expression when the dependence of the kinetic expression on concentration(s) is nonlinear. This is the case for oxygen reduction. Since it is critical to account for both mass transfer and kinetics to accurately determine the corrosion potential, we need a procedure by which this can be done.

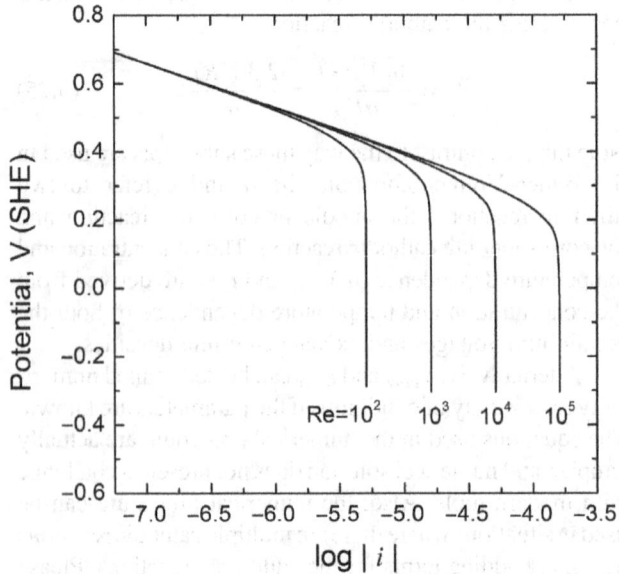

Figure 16.6 Oxygen reduction current at different flow rates for fully developed laminar flow over a flat plate.

Before we discuss the procedure, let's take a moment to examine the physical situation. The kinetic and the mass-transfer resistances are in series and, under steady conditions, the kinetic rate must be equal to the rate of mass transfer. The consumption rate of oxygen becomes constant once the mass-transfer limit is reached, even if we continue to change the potential. How is it possible for the kinetic rate and the rate of mass transfer to be equal if the potential continues to change even after the mass-transfer limit has been reached and remains constant? It turns out that an increase in the potential driving force for the reaction is offset by a reduction in the exchange-current density as the surface concentration decreases. After the surface concentration of oxygen decreases to 1% of the bulk value, the mass-transfer limited rate does not change appreciably. In contrast, the exchange-current density continues to decrease with decreasing surface concentration at a given potential until the kinetic rate is equal to the rate of mass transfer. From a practical perspective, it follows that we can use the mass-transfer expression alone to calculate the oxygen transport rate once the surface concentration is less than or equal to 1% of the bulk concentration, without introducing significant error. If the surface concentration is greater than that value, we should use the kinetic and mass-transfer expressions together to calculate the rate.

These physics are reflected in the following procedure:

1. Estimate the mass-transfer limited current for the flow conditions of interest. Often this can be done with a simple expression such as the mass-transfer portion of Equation 3.32. The required mass-transfer coefficient can be determined as illustrated below.

2. Approximate the voltage at which the limiting current is reached. This step can be done with the Tafel expression by assuming that the surface concentration is 1% of the bulk value, and that the current is equal to 99% the limiting current. We will refer to this potential as V_{99}.

3. Solve for the corrosion current and potential *assuming that the system is mass-transfer controlled*. In other words, find the potential at which the anodic current is equal to the limiting current of the cathodic reaction (or vice versa if the anodic curve is the one that is mass-transfer limited).

4. Evaluate the answer.
   (a) If the potential found in #3 is less than V_{99}, then the potential from #3 is the corrosion potential, and the corrosion current is equal to the mass-transfer limited current.
   (b) If the potential from #3 is greater than V_{99}, then redo the calculation using the kinetic expression

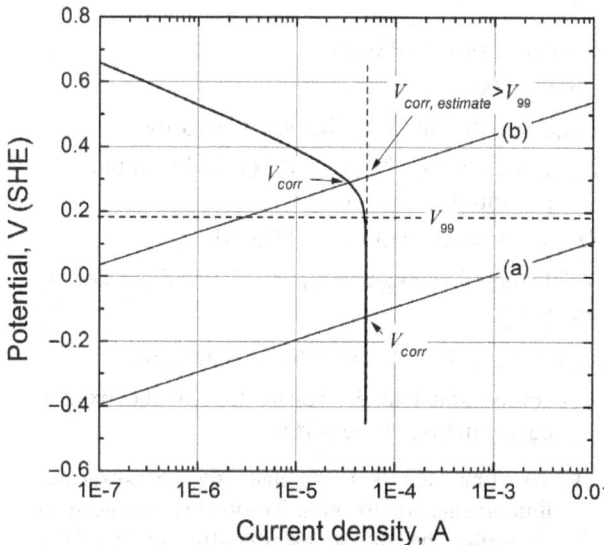

Figure 16.7 Illustration of corrosion potential determination for a cathodic reaction that is mass-transfer limited.

and the mass-transfer expression together, since the corrosion current will be less than the mass-transfer limited current. Note that both equations should be used since the surface concentration may be different than the bulk value because of mass transfer, and this will affect the calculated rate.

The two situations are illustrated graphically in Figure 16.7. The vertical portion of the cathodic curve is the mass-transfer limited region where the current is independent of the potential. Two anodic curves are shown: (a) one that intersects in the mass-transfer limited region and (b) one that intersects in the region where rate effects also contribute ($>V_{99}$). The procedure is provided in Illustration 16.3:

ILLUSTRATION 16.3

A metal corroding in an oxygen-containing environment has a Tafel slope of 60 mV per decade, an exchange-current density of $1 \times 10^{-6}\,\text{A}\cdot\text{cm}^{-2}$, and an equilibrium potential of $-0.4\,\text{V(SHE)}$. Water (pH = 6) flows over the metal, and the area of interest can be represented as fully developed laminar flow over a flat plate. The water is saturated with oxygen, and you may assume that the bulk concentration of oxygen remains essentially at saturation. Oxygen reduction is the only significant cathodic reaction that takes place on the surface, and you may assume that the Nernst equation can be used to estimate its equilibrium potential. Other parameters are as follows:

Length of plate = 100 cm
$Re_L = 10{,}000$
$\nu_{water} = 0.01 \text{ cm}^2 \cdot \text{s}^{-1}$, kinematic viscosity
$c_{O_2,sat} = 1.25 \times 10^{-6} \text{ mol} \cdot \text{cm}^{-3}$ (100 kPa partial pressure O_2, 25 °C)
$D_{O_2} = 2.0 \times 10^{-5} \text{ cm}^2 \cdot \text{s}^{-1}$, diffusivity
Tafel slope for oxygen = $b_{O_2} = -130$ mV per decade
$T = 25$ °C
$i_{0,ref} = 2.2 \times 10^{-9} \text{ A} \cdot \text{cm}^{-2}$ (O_2 at saturation)
i_{0,O_2} proportional to the square root of the oxygen concentration at the surface

1. The first step is to estimate the mass-transfer limited rate for oxygen. To do this, we need to determine the mass-transfer coefficient. For fully developed laminar flow over a flat plate, the relevant correlation is

$$Sh_{av} = 0.664 Re^{1/2} Sc^{1/3},$$

$$Sh_{av} = 0.664 Re^{1/2} \left(\frac{\nu}{D_{O_2}}\right)^{1/3} = 527,$$

$$k_c = \frac{Sh_{av} D_{O_2}}{L} = 1.05 \times 10^{-4} \text{ cm} \cdot \text{s}^{-1}.$$

We can now estimate the mass-transfer limiting current for oxygen from Equation 3.32:

$$i_L = nFk_c(c_{O_2,bulk} - c_{O_2,surface})$$
$$= nFk_c c_{O_2,sat} = 5.08 \times 10^{-5} \text{ A} \cdot \text{cm}^{-2}.$$

2. We can now determine V_{99}.

$$i_{0,O_2} = i_{o,ref}\left(\frac{c_{O_2,surface}}{c_{O_2,sat}}\right)^{1/2}$$
$$= 2.20 \times 10^{-9} \text{ A} \cdot \text{cm}^{-2}(0.01)^{1/2}$$
$$= 2.20 \times 10^{-10} \text{ A} \cdot \text{cm}^{-2},$$

$$i = 0.99 i_L = 5.03 \times 10^{-5} \text{ A} \cdot \text{cm}^{-2}.$$

For the oxygen reaction (acidic): $O_2 + 4H^+ + 4e^- \rightarrow 2H_2O$.

$$U_{O_2} = 1.23 - \frac{RT}{nF}\ln\left(\frac{1}{\left(\frac{c_{O_2,surface}}{c_{O_2,sat}}\right)(c_{H^+})^4}\right)$$

$$= 0.845 \text{ V (SHE)}.$$

$$V_{99} = U_{O_2} + b_{O_2}\left(\log(i/i_{0,O_2})\right) = 0.149 \text{ V}.$$

Note that we have assumed that the concentration of oxygen in the water is linearly proportional to its partial pressure (i.e., Henry's law behavior).

3. We next solve for V_{corr} assuming that oxygen is mass-transfer limited. This step is done by setting the anodic current equal to the mass-transfer limiting current of the oxygen. The only unknown is V_{corr}.

$$i = i_L = i_{0,a}\exp\left[\frac{\ln(10)}{b_a}(V_{corr} - U_a)\right];$$
$$V_{corr} = -0.298 \text{ V},$$

where $\ln(10) = 2.303$ and is needed because the Tafel slope is V per decade. In the corrosion field, it is more common to write

$$V_{corr} - U_a = b_a\bigl(\log(i/i_{0,a})\bigr),$$

which yields the same value for V_{corr}.

4. Finally, we evaluate this corrosion potential against V_{99}. In this case, the calculated corrosion potential is less than V_{99}. Therefore, our assumption that the oxygen reaction is mass-transport controlled is correct for this problem, since the rate of oxygen consumption is mass-transfer limited at values of the potential below (more negative than) V_{99}. The V_{corr} value that we calculated is the correct value, and the problem has been successfully completed. If the calculated value of V_{corr} had been greater than V_{99}, we would have needed to include a contribution from the Tafel equation for oxygen in our calculations, requiring simultaneous solution of the kinetic expression for oxygen, the mass-transfer relationship for oxygen, and the kinetic expression for anodic reaction, assuming that the anodic reaction is not mass-transfer limited. The resulting equations would have required numerical solution to yield the correct value for V_{corr}.

Multiple Reactions

In real systems undergoing uniform corrosion, it is possible to have multiple cathodic reactions. It is also possible to have multiple anodic reactions, but this involves different metals and an extra level of complexity as the assumption of uniform corrosion no longer applies. We will discuss multiple anodic reactions later in this chapter when we examine galvanic corrosion.

The situation with multiple cathodic reactions is shown in Figure 16.8. Here, the two cathodic reactions have been added to give a total cathodic curve. The total cathodic reaction rate from the sum of the two reactions must be equal to the anodic reaction rate. Therefore,

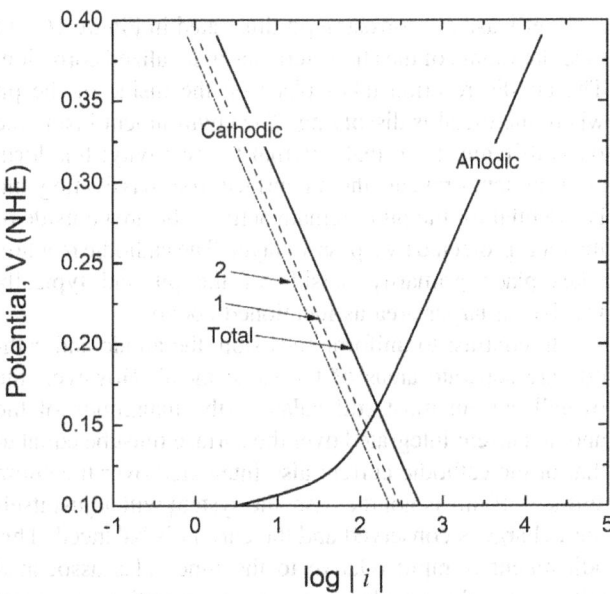

Figure 16.8 Multiple cathodic reactions for the corrosion of copper metal by copper(II) ions in acid.

V_{corr} is defined by the point at which the anodic curve intersects the total cathodic curve. Note that the logarithm of the total cathodic current at a given potential is not just the sum of the logarithms of each of the two cathodic reactions since $\log(i_1) + \log(i_2) \neq \log(i_1 + i_2)$. Also note that in the case illustrated, corrosion of copper metal by copper(II) ions, the anodic and cathodic currents intersect near the edge of the Tafel region for the anodic curve. Intersection outside of the region where the Tafel approximation is valid is possible when the overpotential for the cathodic reactions is much greater than that for the anodic reaction and is evident by a V_{corr} value that is close to the equilibrium value for the anodic reaction.

The procedure for calculating the corrosion potential for multiple cathodic reactions is illustrated below, and is just an extension of that used for the case of single reactions. Note that the problem is set up in terms of current since the currents must be added to give the total current.

ILLUSTRATION 16.4

The etching of copper takes place by corrosion in acid solution containing copper(II) ions (M. Georgiadou and R. Alkire (1993) *J. Electrochem. Soc.*, **140**, 1340). In this illustration, we study a solution of 0.5 M CuCl$_2$, 0.5 M KCl, and 0.5 M HCl. Copper(II) ions react homogeneously to form complexes with chloride ions in solution. For simplification, we will only consider one such complex, CuCl$^+$. Copper dissolution takes place by the following anodic reaction:

$$Cu + 3Cl^- \rightarrow CuCl_3^{2-} + e^-; \quad U^\theta = 0.233 \quad (1)$$

We consider two cathodic reactions:

$$Cu^{2+} + 3Cl^- + e^- \rightarrow CuCl_3^{2-}; \quad U^\theta = 0.441 \quad (2)$$

$$CuCl^+ + 2Cl^- + e^- \rightarrow CuCl_3^{2-}; \quad U^\theta = 0.419 \quad (3)$$

The anodic and both cathodic reactions produce the same Cu(II) ion complex. The ionic composition of the initial solution, which includes the complexes formed in solution, is as follows:

Cl$^-$	Cu^{2+}	CuCl$^+$	CuCl$_3^{2-}$	K$^+$	H$^+$
1.5510 M	0.1067 M	0.3822 M	0.0223 M	0.5 M	0.5 M

The kinetic expressions for each of the above reactions are as follows:

$$i_1 (A \cdot m^{-2}) = 200 \left(\frac{c_{Cl^-}}{1\,M}\right)^{1.5} \left(\frac{c_{CuCl_3^{2-}}}{1\,M}\right)^{0.5}$$
$$\left[\exp\left(\frac{0.5F}{RT}(V - U_1)\right) - \exp\left(\frac{-0.5F}{RT}(V - U_1)\right)\right].$$

$$i_2 (A \cdot m^{-2}) = 1.0 \left(\frac{c_{Cl^-}}{1\,M}\right)^{1.5} \left(\frac{c_{Cu^{2+}}}{1\,M}\right)^{0.5} \left(\frac{c_{CuCl_3^{2-}}}{1\,M}\right)^{0.5}$$
$$\left[\exp\left(\frac{0.5F}{RT}(V - U_2)\right) - \exp\left(\frac{-0.5F}{RT}(V - U_2)\right)\right].$$

$$i_3 (A \cdot m^{-2}) = 0.5 \left(\frac{c_{Cl^-}}{1\,M}\right)^{1.0} \left(\frac{c_{CuCl^+}}{1\,M}\right)^{0.5} \left(\frac{c_{CuCl_3^{2-}}}{1\,M}\right)^{0.5}$$
$$\left[\exp\left(\frac{0.5F}{RT}(V - U_3)\right) - \exp\left(\frac{-0.5F}{RT}(V - U_3)\right)\right].$$

The equilibrium potentials can be estimated from the Nernst equation:

$$U_1 = U_1^\theta - \frac{RT}{F} \ln\left(\frac{(c_{Cl^-})^3}{c_{CuCl_3^{2-}}}\right) = 0.1014 \text{ V}.$$

$$U_2 = U_2^\theta - \frac{RT}{F} \ln\left(\frac{c_{CuCl_3^{2-}}}{(c_{Cl^-})^3 c_{Cu^{2+}}}\right) = 0.5151 \text{ V}.$$

$$U_3 = U_3^\theta - \frac{RT}{F} \ln\left(\frac{c_{CuCl_3^{2-}}}{(c_{Cl^-})^2 c_{CuCl^+}}\right) = 0.5146 \text{ V}.$$

At V_{corr}, the total current is equal to zero. Therefore, we can substitute the above relationships into the following equation and solve for V, the only unknown:

$$\sum i_k = i_1 + i_2 + i_3 = 0.$$

Use of an equation solver yields $V_{corr} = 0.161$ V. Note that if we had used the Tafel expressions rather than the full BV expressions, we would have used the first term of reaction 1 (positive) and the second term of reactions 2 and 3 (both negative).

16.4 LOCALIZED CORROSION

In the previous section, we assumed that the corrosion occurred uniformly over the entire surface and that the surface area for the anodic reaction and cathodic reactions was the same and equal to the total surface area of the sample. Localized corrosion is very different. In localized corrosion, the anode and cathode are separated, and the anode is typically smaller than the cathode. Corrosion involving small anodes is important industrially because a large cathode-to-anode ratio can lead to fast dissolution of a small area of the metal, leading to leakage or rupture. Numerous different kinds of localized corrosion have been identified, with examples shown in Figure 16.1.

Many factors influence localized corrosion, and a complete treatment of all of these is beyond the scope of this text. For example, properties of the metal such as grain and grain-boundary composition and structure, the amount and properties of alloying components, and the presence of inclusions or defects influence its corrosion behavior. The dynamics of passive film formation and breakdown are also important in determining the initiation and stability of corrosion. Our objective in this section is to analyze localized corrosion using the fundamental relationships of electrochemical engineering. Specifically, we will examine how thermodynamics, kinetics, and transport influence the rate of corrosion, and how an understanding of the electrochemical behavior can lead to the development of engineering strategies to mitigate corrosion.

Let's use the corrosion pit illustrated in Figure 16.9 to elucidate some of the characteristics of localized corrosion. The anodic reaction takes place on the inside of the pit where the metal is dissolving. The environment inside the pit is different from that external to the cavity; this local environment prevents the pit surface from passivating and is essential for the pit to remain active. The area outside of the pit is protected by a passive layer. The cathodic reaction takes place primarily outside of the pit and typically involves a larger area as mentioned above.

In contrast to uniform corrosion, the anode and cathode are separate areas of the same metal. However, the overall current must still balance (the magnitude of the anodic current integrated over the surface must be equal to that of the cathodic current also integrated over the entire surface). If this is not the case, the system will adjust itself until charge is conserved and the current is balanced. This adjustment is rapid relative to the timescales associated with either the development of concentration gradients inside the pit or the growth (change in size) of the pit. In order for the corrosion process to proceed, there must be current flow between the anode and the cathode. Using our normal convention, current flows from the anode to the cathode in solution, and electrons flow from the anode to the cathode in the metal. Because the resistance of the solution is typically much higher than that of the metal, the potential loss due to current flow in the metal can be neglected and the metal is essentially at a constant potential.

You can prove the validity of assuming a constant metal potential to yourself by calculating the current density that would be required to produce an appreciable voltage drop in a metal. The potential drop in solution is an important aspect of localized corrosion; this is a voltage loss that was not present in uniform corrosion where the anodic and cathodic reactions occur simultaneously on the same surface. The current in solution is driven by a potential drop from the anode to the cathode. Therefore, the potential of the solution inside the pit is higher than that

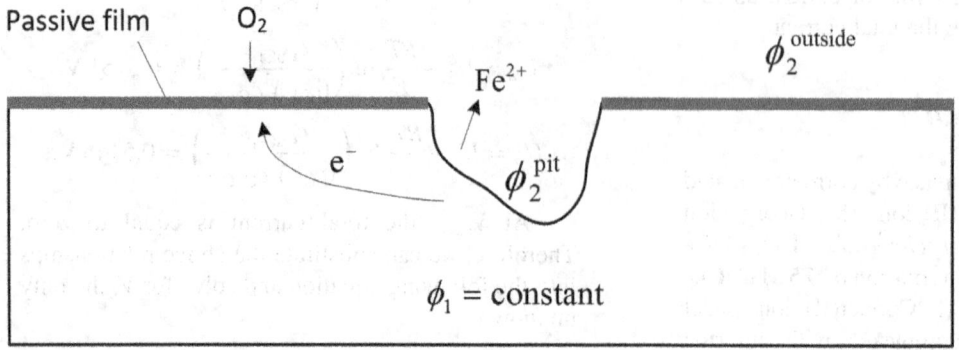

Figure 16.9 Schematic diagram of corrosion pit. Current flows from the inside of the pit to the outside, driven by a potential difference in solution. The potential is greater in the pit than outside the pit.

of the surrounding cathodic area. Most of the potential drop in solution occurs in and near the pit where current densities are highest. In conjunction with Figure 16.9, let's consider the following questions:

Is the potential measured inside the pit with a very small reference electrode greater or less than that which would be measured with the same reference electrode outside of the pit? The measured potential is $\phi_M - \phi_{ref}$. Since ϕ_M is the same at both locations and $\phi_{ref,pit} > \phi_{ref,outside}$, the measured potential inside the pit is less than that outside of the pit.

Doesn't the potential inside the pit favor the cathodic reaction? Why then does the cathodic reaction occur mostly outside of the pit? The potential inside the pit actually does favor the cathodic reaction relative to the potential outside of the pit. There are several reasons why the cathodic reaction occurs mostly outside of the pit. First, the cavity itself provides additional transport resistance, which can prevent cathodic reactants such as oxygen from reaching the surface. Also, the environment inside of the pit will influence the equilibrium potential, which may offset the effect of the potential difference. Perhaps most important, the kinetics of possible cathodic reactions are typically much slower than those of the anodic reaction inside the pit. Therefore, the large cathodic area available outside of the pit is needed to provide sufficient cathodic current.

What keeps a corrosion pit active? The driving force for pitting corrosion is the difference in the equilibrium potentials of the anodic and cathodic reactions. In environments where active dissolution of the metal is not favored, for example, due to stable passive layers, the local environment inside the pit is also critical for active dissolution since it is that local environment that prevents the re-passivation that would otherwise occur.

The local composition difference between the inside and outside of a corrosion pit is characteristic of localized corrosion where geometry typically facilitates its development. As mentioned above, this composition difference is critical to maintaining the stability of localized corrosion. In particular, oxygen depletion in restricted areas can promote corrosion because the surface is not passivated. For example, stainless steels, which rely on oxygen to maintain passivity, behave very differently in aerated and oxygen-free environments. Intuitively, you might expect that corrosion would be worse in oxygen compared to air. For pitting corrosion, the higher partial pressure of oxygen makes it more difficult for a break in the passivation layer to occur. As a result, pitting corrosion is worse in air. In addition, concentration differences contribute to the transfer of current from the anode to the cathode. Avoidance of geometries that significantly restrict transport and facilitate localized corrosion is important from a mitigation standpoint.

Galvanic Corrosion

Galvanic corrosion is a form of localized corrosion that can occur when two different metals are in contact and exposed to a common electrolyte. It is included under localized corrosion because the metals corrode at different rates and corrosion is favored near the location where the metals are in contact. Because of the high electrical conductivity of the metals, the two metals are at the same potential. Consider zinc and iron in electrical contact and exposed to the same electrolyte solution. The potential of the two metals equalizes by moving some of the higher energy electrons in the zinc to the iron. This situation is equivalent to applying a negative potential to the iron (a potential below its equilibrium potential) and a positive potential to the zinc (a potential higher than its equilibrium potential), resulting in an increase in the dissolution rate of zinc and a decrease in that of iron.

The rates of cathodic reactions also influence the final potential of the metals undergoing galvanic corrosion. The cathodic reaction(s) can occur on the surfaces of both metals, and the relative rate of reduction on the different metal surfaces will depend on effects of kinetics and mass transfer. As for the metal potential in our example, it will remain below the equilibrium potential of iron as long as the dissolution rate of zinc is sufficiently rapid to match the cathodic reaction rate. Where high rates of cathodic reaction are possible, both metals will corrode, but at different rates. Because the net current on the active metal is typically anodic, and the net current on the less active metal is cathodic, there must be current flow from one metal to the other and in solution between the two metals. The potential drop associated with current flow in the metals themselves is tiny because of high conductivity of the metal. In contrast, the potential drop in solution is significant and plays an important role in determining the local reaction rates of both the anodic and cathodic reactions, which will not be uniform. The direction of current flow in solution is from the zinc to the iron; hence, the potential of the solution is higher at the zinc electrode. As described previously for pitting corrosion, this means that the potential measured against a reference electrode will be lower at the zinc electrode than at the iron electrode.

To illustrate, we consider the situation where flat pieces of zinc and iron are in contact and exposed to a salt solution. Figure 16.10 shows a two-dimensional, semi-infinite representation: the zinc is 1 cm wide and the iron is 4 cm wide. To the right of the iron is an insulator. The left boundary can be thought of as an insulator or an axis of symmetry. As a first approximation, we neglect concentration effects and solve for the secondary current distribution; that is, Laplace's equation for the potential in solution is solved. This calculation requires expressions for the current density as a function of potential for the dissolution of iron and zinc, as well as kinetic expressions

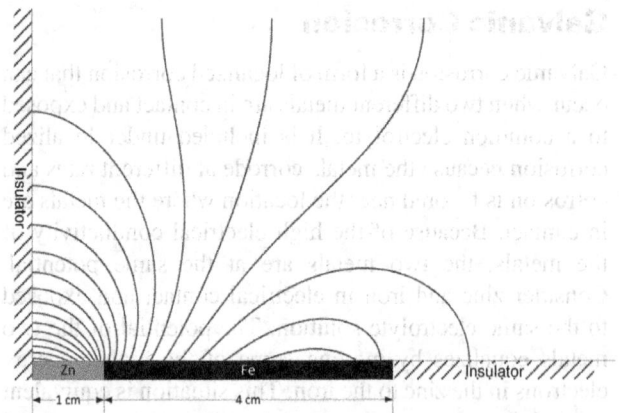

Figure 16.10 Potential field for galvanic coupling of Zn and Fe. *Source:* Data adapted from Abootalebi 2010.

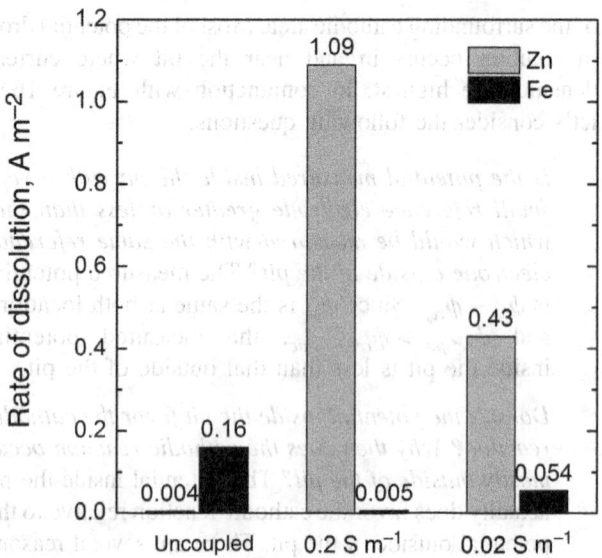

Figure 16.11 Average dissolution rates: (1) without galvanic coupling, (2) coupled with a solution conductivity of 0.2 S·m^{-1}, and (3) coupled with a solution conductivity of 0.02 S·m^{-1}.

for hydrogen evolution and oxygen reduction on each of the two metal surfaces. Tafel expressions suffice since we are well away from the equilibrium potentials for the reactions of interest except for iron. The reverse reaction or iron is not significant because of the negligible concentration of iron ions in solution. Since mass transfer in solution is not modeled explicitly, the rate expression(s) for oxygen reduction should include mass-transfer effects as oxygen is undoubtedly mass-transfer limited. Following our convention, the potential of the metal is ϕ_1, and the potential of the solution is ϕ_2. The metal potential is arbitrarily set to zero and the derivative of the solution potential is prescribed on the boundaries.

On the Fe electrode:

$$i_y = i_{Fe} + i_{O_2} + i_{H_2} = f(\phi_2) = -\kappa \frac{\partial \phi_2}{\partial y}.$$

Similarly, at the Zn electrode:

$$i_y = i_{Zn} + i_{O_2} + i_{H_2} = f(\phi_2) = -\kappa \frac{\partial \phi_2}{\partial y}.$$

The current, and therefore derivative of potential, is assumed to be zero at all other boundaries. The position of the outside boundaries is not critical as long as they are sufficiently far from the surface that they do not significantly distort the potential field. Solution of Laplace's equation with these boundary conditions yields the potential field shown in Figure 16.10. The closer the spacing of the contour lines are, the steeper the gradients in potential and the higher the current density. Note that the current density is highest at the intersection of the two metals. Figure 16.11 shows the average rates of dissolution of the zinc and iron before and after galvanic coupling of the two metals. These results show clearly that the dissolution rate of zinc increased significantly due to coupling with iron,

while that of the iron dropped to essentially zero. The impact of lower solution conductivity (0.02 S·m^{-1} rather than 0.2 S·m^{-1}) is also shown, where the increased resistance of the solution resulted in a significantly lower average current. The local current densities along both the zinc and iron electrodes are shown in Figure 16.12 for both values of the solution conductivity. The net current density on the zinc electrode is anodic, and that on the iron electrode is cathodic, and the current flow in solution is

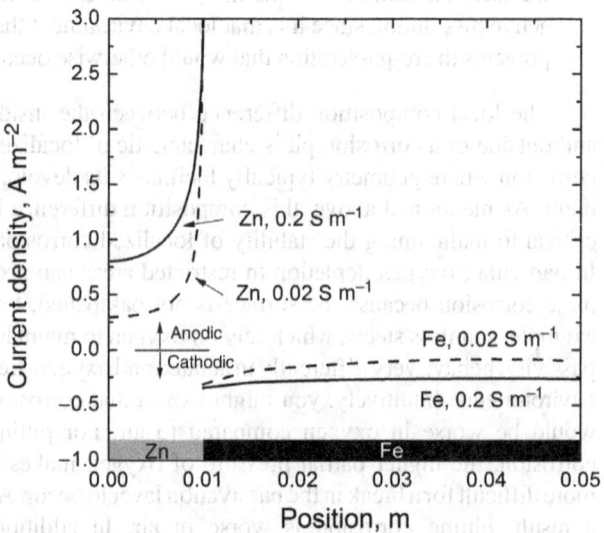

Figure 16.12 Local current density for galvanically coupled Zn and Fe at solution conductivities of 0.2 and 0.02 S·m^{-1}.

from the zinc electrode to the iron electrode. In both cases, the current is highest as the intersection of the metals is approached. In the case of the lower conductivity, the current distribution is more nonuniform and the average dissolution rate is lower. Importantly, the local dissolution rate of zinc at the junction of the two metals is actually higher; therefore, the lower average current density may actually be more problematic due to the high local dissolution rate of zinc.

You, as a student, should be able to qualitatively rationalize each of the results from this simulation. Illustration 16.5 is designed to help you do this.

ILLUSTRATION 16.5

Based on the material presented above and your understanding of electrochemical systems, please address the following questions related to galvanic corrosion:

Question 1: What is the total current (integrated across the surfaces of both metals)?
The total current integrated across both surfaces must be zero. The metal potential will adjust itself until this is the case, just as it did with uniform corrosion.

Question 2: Why is the corrosion rate of zinc higher when coupled with iron?
The addition of iron in contact with the zinc increases both the cathodic area and, since hydrogen kinetics are much faster on iron than on zinc, the rate at which the cathodic reaction can take place on the area available. The increased rate of the cathodic reaction is balanced by an increase in the corrosion rate of zinc.

Question 3: What role does the size of the electrodes play?
An increase in the electrode area increases the total current that can be driven by a given overpotential. Therefore, an increase in the surface area of the iron significantly impacts the magnitude of the cathodic current. This effect does not continue indefinitely and a size will be reached above which the corrosion rate does not continue to change, but is limited by the potential drop in solution. An increase in the zinc area would decrease the dissolution rate per area of the zinc, which results in a decrease in the zinc consumption rate (thickness per time).

Question 4. Are differences in the kinetics of the cathodic reactions on the two metal surfaces important? Would higher cathodic reaction rates on iron make the zinc corrosion better or worse? Why?

Yes, differences in the kinetics of the cathodic reaction(s) on the two metals can be essential, and these differences are important in the iron–zinc system. For example, hydrogen evolution is very slow on zinc, but much faster on iron. Therefore, electrical contact with iron provides not only additional area for the cathodic reaction but also area where the cathodic reaction is much faster.

Question 5: Why is the dissolution rate of iron lower when coupled with zinc?
The zinc dissolution rate is sufficiently rapid to provide the anodic current needed to balance the total cathodic current on the zinc and iron surfaces. Because of this, the potential at which the anodic and cathodic currents are balanced is near or below the equilibrium potential of iron, leading to little or no iron dissolution (corrosion).

Question 6: Why is the iron dissolution rate higher for the lower conductivity solution?
Because there is current flow from the zinc to the iron, the potential in the solution is lower adjacent to the iron than it is adjacent to the zinc. Also, the potential in solution at the surface of the iron is highest near the zinc–iron boundary, and lower as you move away from that boundary. Since the potential of the zinc metal and iron metal are the same, a lower solution potential is equivalent to a higher overpotential. If the overpotential exceeds the equilibrium potential of the iron, the iron will corrode. Iron corrosion is more likely as you move away from the zinc–iron interface. Eventually, you get far enough away from the zinc that its influence is no longer felt. This happens sooner for a low-conductivity solution.

Question 7: Why are the local corrosion rates nonuniform when the metals are coupled?
The rates are nonuniform because of the nonuniform potential in solution that drives the current flow from the zinc to the iron electrode. See the answer to the previous question for more information.

Question 8: Do cathodic reactions take place on the zinc electrode?
The overpotential values along the zinc electrode are actually more favorable for the cathodic reaction than those at the iron electrode. So yes, cathodic reactions can occur on the zinc electrode. Remember, however, that the kinetics for the cathodic reaction at the zinc electrode are less favorable. The zinc dissolution is sufficient to balance the cathodic reactions on the zinc and on the iron.

Question 9: What roles do the equilibrium potentials for iron and zinc play?
The difference between the equilibrium potentials of zinc and iron is actually quite important. A greater difference between these potentials means that we can have a higher driving force for zinc corrosion at potentials that remain below the equilibrium potential of iron, which reduces the probability of iron corrosion.

Question 10: Would the corrosion rate change if the solution were stirred? Why or why not?
Before answering this question, we note that none of the analysis in this illustration factored in the role of concentration gradients. In situations where the corrosion rates are sufficiently high such that concentration gradients can develop, stirring the solution may make a difference by changing the local conductivity. Stirring will also make a significant difference in situations where the cathodic reaction (e.g., oxygen reduction) is mass-transfer limited as described previously in this chapter. Mass transfer was accounted for in an approximate way in our simulation through use of a rate expression that included transport limitations for oxygen. While this expression does permit incorporation of mass-transfer effects to some degree, it is not sensitive to changing flow conditions (e.g., stirring).

As a closing comment, Evans diagrams can be used to gain some physical insight into galvanic coupling of metals, as traditionally presented in corrosion texts. However, the assumption of uniform corrosion rates on metal surfaces implicit in such an approach is clearly not correct, making it incomplete and frequently inadequate.

16.5 CORROSION PROTECTION

The topic of corrosion protection and mitigation is broad and includes many different approaches to a multitude of corrosion-related problems. For example, subtopics include materials design and selection, component design for corrosion avoidance, use of inhibitors to reduce corrosion rates, and a wide variety of surface treatments and coatings. Clearly, comprehensive treatment of this topic is beyond the scope of this book.

Electrochemical methods for addressing corrosion are more consistent with the objectives of the text and include the following: *cathodic protection, anodic protection, and e-coating*. The most important of these is cathodic protection, which is used extensively for protection of structures from corrosion. Electrophoretic coating or E-coating is widely used by the automotive industry and others for coating parts in order to provide, among other things, corrosion protection. In E-coating, it is the coating process itself that is electrochemical in nature. This important industrial process will not be treated in this chapter. Here, we provide a brief description of anodic protection and then spend the balance of the section on the fundamentals of cathodic protection.

Anodic Protection

Anodic protection is illustrated in Figure 16.13 and is applicable to metals that have an active to passive transition (Figure 16.3), such as iron, nickel, chromium, titanium, and their alloys. Simply stated, the strategy involves increasing the potential in the anodic direction in order to move the metal of interest from an active (a) to a passive state (b). This shift can be accomplished with use of a potentiostat to set the potential of the metal to a value in the passive region. For the example shown in the figure, the anodic corrosion current is reduced by almost two orders of magnitude by increasing the potential above the corrosion potential as shown. At the same time, the cathodic current is reduced even more. The potentiostat provides current to make up the difference between the current from the small cathodic reaction at the applied potential, i_c, and that required to balance the passive current of the metal, i_a, so that the magnitude of total cathodic current is equal to that of the anodic current. This applied current is much less than

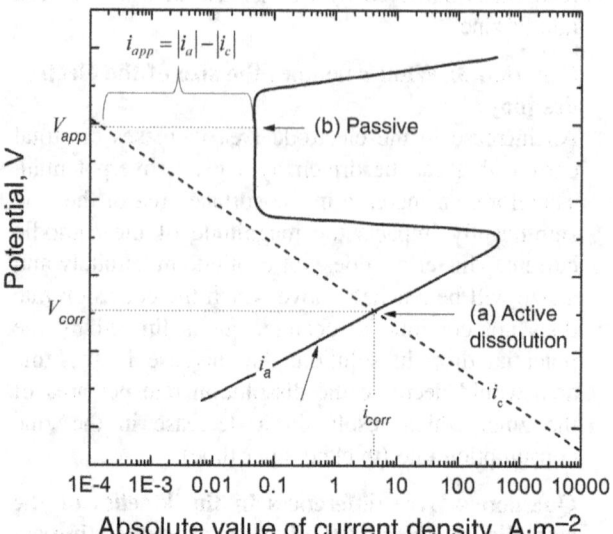

Figure 16.13 Illustration of using anodic protection to reduce the rate of corrosion for metals that passivate.

that required for cathodic protection where the potential is moved in the negative direction in order to protect the metal, and a relatively large external anodic current is required to balance the high cathodic current at the new potential. The lower required current and corresponding reduced operating costs represent the principal advantage of anodic protection. However, such protection requires a higher initial capital investment because the instrumentation needed to set the potential is more expensive than that required to apply a current for cathodic protection. As with galvanic corrosion, there is current flow between the anode and the cathode. The change in potential due to this current flow is not included in Figure 16.13. Still, the diagram represents the situation fairly accurately since the current density in the passive region is low; therefore, ohmic losses are not significant.

Another way to implement anodic protection is by changing the cathodic reaction so that the cathodic curve intersects the anodic curve in the passive region, rather than in the active region. This method is not always possible, but could avoid the need for the potentiostat. Let's take a minute to think about this curve and how it might be altered. The most common cathodic reaction is the reduction of oxygen. Given the sluggishness of the oxygen reduction reaction, Tafel kinetics apply. As seen in Figure 16.13, when the potential is plotted against the logarithm of cathodic current density, a straight line results.

Changing the exchange-current density for oxygen reduction is one way to shift the cathodic curve. Often, the exchange-current density is proportional to the concentration of oxygen; and therefore varying the amount of oxygen in the system shifts the cathodic curve. At first, this action seems counterintuitive. However, it is analogous to the example above where a current was applied to raise the potential in the anodic direction. In this instance, the oxygen concentration is raised to increase the cathodic current and to ensure that the metal is passivated. The approach is depicted in Figure 16.14. The cathodic curve labeled (a) is the original state, with a high corrosion current. The cathodic curve is raised by increasing the oxygen concentration (b), moving the potential into the passive region with a much lower corrosion current. Some care is needed when taking this approach—We could make the problem worse. Reflect that the rate of reduction of oxygen is often mass-transfer limited rather than controlled by kinetics. Given the low solubility of oxygen in water, this circumstance doesn't come as a surprise. The result is that the current associated with the cathodic curve will not increase indefinitely. The effect of mass transfer is shown with curve (c). Note that because of the limiting current, it is possible to be in a condition of high anodic (corrosion) current. Not only is it important to shift the curve up, but the mass-transfer limit must also be large enough to ensure that the metal remains passivated. It may be necessary to increase the mass-transfer limiting current as well as shown

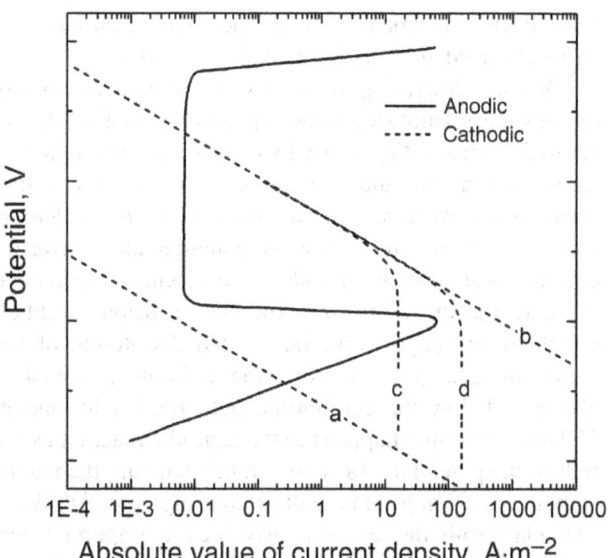

Figure 16.14 Importance of mass transfer on passivating the metal.

in curve (d). This increase could be accomplished by raising the flow rate or agitation of the fluid. A similar shift can be made by coupling the metal to be protected with metals that facilitate the cathodic reaction without adding a significant anodic component.

Anodic protection can be very effective in reducing corrosion in systems where it can be applied. It is particularly effective in aggressive environments where cathodic protection would require very large currents and would lead to breakdown of the solvent (i.e., water) to form hydrogen gas. However, relative to cathodic protection, which can be applied to nearly all metals, anodic protection is used much less commonly.

Cathodic Protection

Cathodic protection is used routinely to protect sea vessels, oil platforms, and other ocean structures, buried pipelines, and storage tanks. The concept is straightforward—lower the potential of the metal you are trying to protect in order to reduce or eliminate corrosion. Cathodic protection takes its name from the fact that the potential of the protected metal is moved in the cathodic direction in order to decrease its dissolution rate. Ideally, we would like to reduce the potential enough to put us into the immunity area of the Pourbaix diagram (see Figure 16.2). In order to reduce costs, we may settle for just lowering the corrosion rate rather than eliminating it completely.

Cathodic protection is implemented in one of two ways: (i) use of a sacrificial anode or (ii) use of an external power supply to apply a cathodic current to the metal to be protected. Both of these methods have associated operational expenses either to replace sacrificial anodes that have been consumed or to provide the electricity and hardware needed to apply the cathodic current.

Before describing these two strategies, we briefly discuss the relationship between cathodic protection and cathodic current. Figure 16.15 is an Evans diagram that shows the anodic and cathodic curves for a metal that undergoes corrosion. Without protection, the system is characterized by the corrosion potential and corrosion current as shown. With cathodic protection, we reduce the potential of the metal below the corrosion potential, which lowers (V_1) or eliminates (V_2) dissolution of the metal, depending on whether the cathodic potential is above or below the equilibrium potential for the metal, U. However, what happens to the cathodic reaction as we reduce the potential? As shown in the diagram, the rate of cathodic reaction goes up with decreasing potential. Also, secondary cathodic reactions may become important. For example, hydrogen evolution may have been insignificant at the corrosion potential, but may become important as the potential is lowered. Note that hydrogen evolution may contribute to hydrogen embrittlement on some metals such as high-strength steels.

Even though it is the potential that we are trying to drop, cathodic protection is often discussed in terms of the magnitude of the applied cathodic current, i_{app}. In essence, we need to provide sufficient current to drive the cathodic reactions at the desired protection potential. If this is not done, the potential will be higher and the metal will not be protected. Description of cathodic protection in terms of current rather than potential is not problematic since the two are related. Note that in contrast to anodic protection, the applied current can be large. Therefore, electrode spacing, electrical conductivity, transport conditions, and geometry may be important in relating current and potential. Hence, the relationship between the cathodic current and the potential of the metal varies from system to system and, where possible, direct measurement of the potential is used to verify the extent of protection.

Sacrificial Anodes

Use of a sacrificial anode is implemented by coupling the metal to be protected with a more active metal; that is, one with a lower equilibrium potential. It is an intentional application of galvanic corrosion where the active (sacrificial) metal is permitted to corrode in order to reduce the potential of the metal to be protected. Thus, all of the physics discussed above with respect to galvanic corrosion apply to sacrificial anodes. In particular, sacrificial anodes must be electrically connected to the metal they are protecting. The sacrificial anodes dissolve preferentially, and the cathodic reaction takes place primarily on the metal that is being protected. Therefore, there are separate anodic and cathodic regions, with current flowing between them.

Since the resistance to current flow in the metal is typically much smaller than the solution resistance, the effective area protected by a sacrificial anode is a function of the electrical conductivity of the electrolyte solution. A higher conductivity means lower losses in solution and greater protection. In contrast, a lower conductivity translates into a reduced area of protection and requires multiple sacrificial anodes on a single structure to be placed more closely together. The current distribution around a sacrificial anode is not uniform as we saw above for the coupling of zinc with iron. Therefore, it is possible to have on the same structure areas that are overprotected and areas that are underprotected. Typical metals used for sacrificial anodes include zinc, aluminum, and magnesium. These metals are frequently alloyed with other minor components to give the desired dissolution properties. Figure 16.16 illustrates the use of sacrificial anodes seen as light-colored blocks on the hull and rudder of the ship.

Design of a cathodic protection system that incorporates sacrificial anodes involves choosing the type, number, size, and spacing of the anodes. This process is frequently approached empirically. First, the current density needed to protect the metal of interest is chosen based on experience for the intended environmental conditions, and the total current required is determined by multiplying the protection current density by the total area of the metal to be protected. Next, the required anode surface area can be determined from knowledge of the dissolution rate of sacrificial anodes in the target environment. The final size and number of anodes reflect a balance between having a

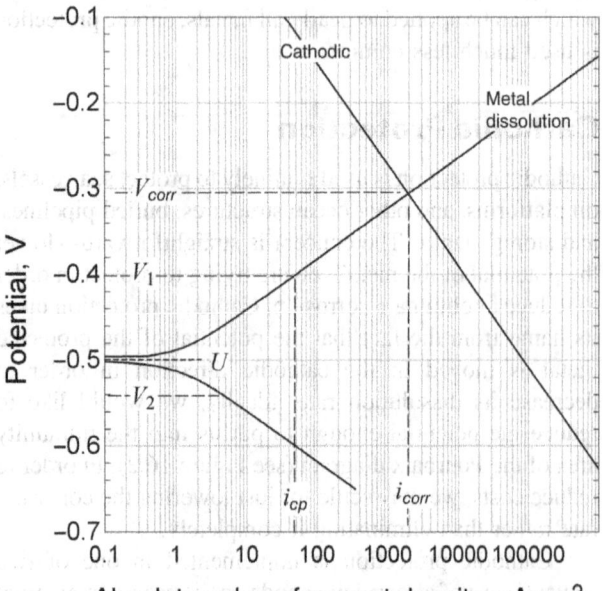

Figure 16.15 Evans diagram showing corrosion potential and the reduction of corrosion current with a reduction in potential.

Figure 16.16 Use of sacrificial anodes to protect against ship corrosion.

sufficient number of electrodes to provide adequate protection over the whole surface and the practical issues associated with installing and periodically replacing a large number of electrodes. Additional anodes can often be added to provide protection in areas when the initial system design was inadequate. Mathematical models and even physical models can be helpful in optimizing large, complex systems. This is particularly true for systems where the conductivity of the electrolyte is low and current densities are highly nonuniform.

Impressed Current Cathodic Protection (ICCP)

Cathodic protection can also be implemented with use of an external electrode and a DC power supply as illustrated in Figure 16.17, where the external electrode is the anode

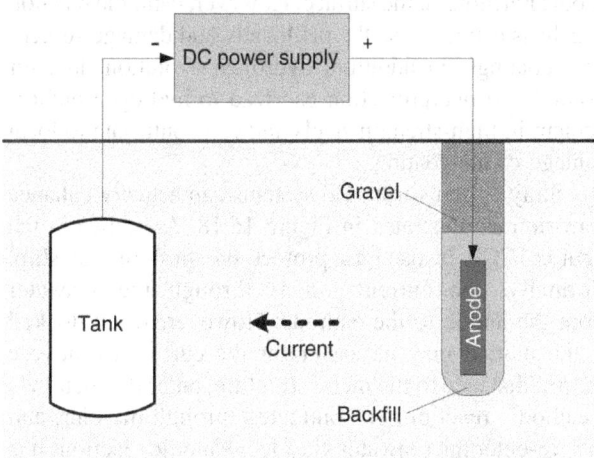

Figure 16.17 Impressed-current cathodic protection used to reduce corrosion of an underground tank.

and the structure to be protected is the cathode. The power supply drives the current necessary to protect the cathode. In situations where the anode is attached directly to the structure being protected (e.g., a ship), it is critical that it be electrically insulated. The desired reaction at the anode is oxygen evolution. In practice, voltages of 100 V and high current densities are possible. Operational costs can be substantial since continuous current flow is required to maintain protection.

Several different types of anode materials are available, including consumable anodes (e.g., scrap iron), semiconsumable anodes (graphite or high-silicon iron), or inert electrodes (e.g., lead alloys, platinized metals, or mixed metal oxide electrodes). Consumable electrodes, where the anodic reaction leads to dissolution of the electrode, are less expensive, but require regular replacement and typically cannot sustain high current densities. Inert electrodes offer the advantage of higher currents and longer lifetimes, but at a higher capital cost. Mixed metal oxide electrodes, based on technology borrowed from the chlor-alkali industry, operate at high current densities, are available in a variety of shapes and sizes, and cost less than platinized anodes, making them a popular electrode choice. The use of lead alloys (e.g., lead–silver) as a lower cost "inert" electrode has diminished due to concerns with the use of lead and improvements in alternative electrodes. Table 16.1 shows the characteristics of several anode materials for seawater applications.

Oxygen evolution is the most common anodic reaction at inert electrodes:

$$2H_2O \rightarrow O_2 + 4H^+ + 4e^- \qquad (16.10)$$

Chlorine may be produced at anodes in saline solution, presenting a potential problem in confined areas:

$$2Cl^- \rightarrow Cl_2 + 2e^- \qquad (16.11)$$

The design of an ICCP system involves choosing the number and location of external anodes. The data in

Table 16.1 Characteristics of ICCP Anodes for Seawater Applications

Material	Current density range (A·m^{-2})	Consumption rate kg·A^{-1}·yr^{-1}
Scrap steel	<1	9
Silicon iron	10–30	0.2–0.5
Graphite	10–20	0.3–0.5
Lead silver	200–300	0.1
Platinized anodes	250–2000	10^{-5}
Mixed metal oxide	250–600	0.5–4 × 10^{-6}

Table 16.1 can be used to provide an initial estimate of the anode surface area required as illustrated below.

ILLUSTRATION 16.6

A current density of approximately $0.1 \, \text{A} \cdot \text{m}^{-2}$ at the metal surface is required in order to protect steel in seawater applications. Estimate the anode surface area required to protect an undersea steel pipe with an outer diameter of 0.6 m and a length of 100 m.

Surface area of the pipe:
$\pi dL = \pi (0.6 \, \text{m})(100 \, \text{m}) = 188.5 \, \text{m}^2$

Total current required for protection:
$(0.1 \, \text{A} \cdot \text{m}^{-2})(188.5 \, \text{m}^2) = 18.85 \, \text{A}$

Choose a mixed-metal oxide anode operating at a current density of $500 \, \text{A} \cdot \text{m}^{-2}$

Required anode area:
$(18.85 \, \text{A})/(500 \, \text{A} \cdot \text{m}^{-2}) = 0.0377 \, \text{m}^2$

Because of the large difference in the current density at the anode and cathode, the difference in size is over three orders of magnitude. This size difference has some important implications for electrode placement as discussed briefly below.

Because the current density at the anode is so much larger than that at the cathode (structure to be protected), most of the potential drop in the system occurs near the small anode. The potential drop near the anode decreases with distance away from the anode as $1/r^n$, where r is a characteristic radius of anode and n is generally between 1 (cylindrical decay) and 2 (spherical decay), depending on the geometry of the system. Therefore, almost the entire potential drop occurs within 10–15 radii of the anode, which has important implications for anode placement. First, the current distribution at the cathode will be very nonuniform if the anode is closer than ~10 radii to the structure. Second, once the distance between the anode and cathode is at least 10–15 radii, further movement away from the cathode will not substantially increase the overall potential drop and power requirements of the system. However, placement of the anode further away can improve the level of protection by making the cathodic current density more uniform over the surface of the structure. There are practical factors, independent of the distribution of current, that limit the placement of anodes. Design of ICCP systems has traditionally been done empirically. However, mathematical models can be of great help in exploring options and overall system optimization and have found increased use in recent years.

Each of the two primary methods of cathodic protection has advantages and disadvantages. The following are the advantages of sacrificial anode systems:

1. Installation is simple.
2. Do not require a power supply.
3. Overprotection is easy to avoid.
4. Less prone to cause interaction with neighboring structures.
5. Moderately easy to obtain a uniform potential across the structure.

The most severe limitation associated with sacrificial anodes is the small potential driving force available, which limits use of this strategy to conductive environments or well-coated systems. The need to replace a large number of anodes on a regular cycle is also a disadvantage.

The following are the advantages of an ICCP system:

1. A large driving force to protect even large, uncoated structures in high resistivity environments.
2. Comparatively few anodes are needed.
3. A controllable system that can be adjusted to accommodate changes.

Disadvantages include the need for a large, specialized power supply, considerable variation of the potential over the surface, the possibility of considerable overprotection, the possibility of reversing electrical connections and the resulting enhancement of corrosion, and the significant costs associated with operation.

Cathodic protection can be used effectively in conjunction with coated surfaces to prevent localized corrosion at defects in the coating. The presence of a coating greatly reduces the current required to protect the target structure since the coating inhibits both cathodic and anodic reactions at the surface. However, cathodic reaction at defects can increase the pH locally and damage susceptible coatings. In addition, hydrogen evolution due, for example, to overprotection can lead to hydrogen embrittlement in high-strength steels and can contribute to local damage of the coating.

Stray currents from CP systems can actually enhance corrosion as illustrated in Figure 16.18. As shown in the figure, ICCP is used to protect the hull of the ship. Normally, the current moves through the seawater from the anode to the cathode. However, when docked at the pier, it may be easier for the current to move a shorter distance to the metal structure, enter the metal via a cathodic reaction, be conducted through the pier, and then re-enter the seawater via a local anodic reaction. It is the local anodic reaction that is problematic as this is a corrosion reaction that causes dissolution of the pier.

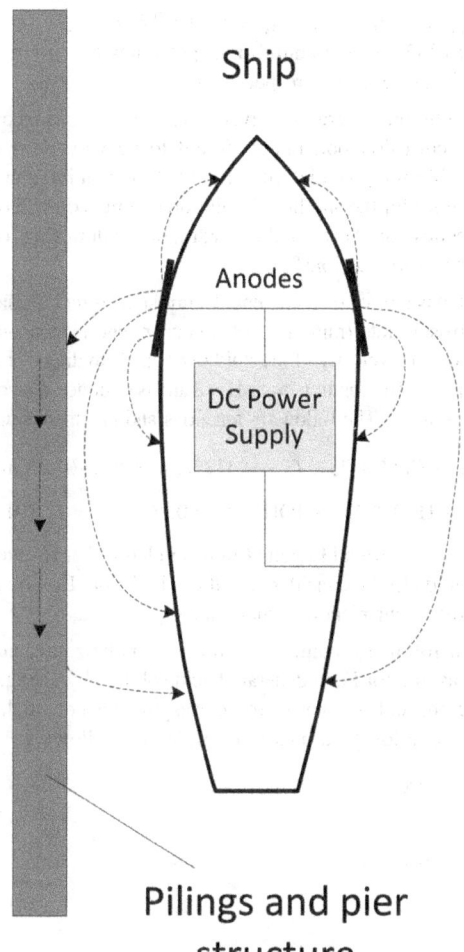

Figure 16.18 Stray currents lead to increased corrosion of pier.

Thus, the ICCP system protects the steel hull, but enhances corrosion in the unprotected pier. This, of course, is undesirable and problematic, and the impact of stray currents must be considered when designing cathodic protection systems. Difficulties are even more problematic for resistive environments, such an unprotected pipeline near the tank in Figure 16.17. Stray currents are most often associated with ICCP systems where electrode separation is typically greater and voltages are higher.

CLOSURE

Corrosion of metals and its prevention are important topics for electrochemical engineers. Corrosion is largely an electrochemical phenomenon and critical aspects of corrosion are described well with the fundamentals of thermodynamics, electrode kinetics, and transport. This chapter is focused on those aspects. The Pourbaix diagram is a useful starting point for analysis of corrosion. The practicing engineer should understand the regions of immunity, passivity, and corrosion for their application. The Evans diagram clearly shows the balance of anodic and cathodic reactions for uniform corrosion so that the corrosion potential and corrosion current can be identified. Anodic and cathodic protection schemes are available to reduce the rate of corrosion to tolerable levels. The level of corrosion that is acceptable is ultimately an economic question.

FURTHER READING

Fontana, M.G. and Green, N.D. (1978) *Corrosion Engineering*, McGraw-Hill Book Company, New York.

Kelly, R.G., Scully, J.R., Shoesmith, D.W., and Bucchheit, R.G. (2002) *Electrochemical Techniques in Corrosion Science and Engineering*, Marcel Dekker, New York.

McCafferty, E. (2010) *Introduction to Corrosion Science*, Springer, New York.

Roberge, P.R. (2008) *Corrosion Engineering: Principle and Practice*, McGraw-Hill.

Winston Revie, R. ed. (2011) *Uhlig's Corrosion Handbook*, John Wiley & Sons, Inc., Hoboken, NJ.

PROBLEMS

16.1 Please address the following qualitative questions:
  (a) What is corrosion and why does it occur?
  (b) What is the driving force for corrosion?
  (c) How would you expect temperature to affect the rate of corrosion for a structure in the ocean? Describe in detail the aspects of corrosion that would be affected by temperature and how temperature might affect those aspects.

16.2 An aqueous solution at pH = 5 contains 0.1 M ferric ion. From a thermodynamic perspective, is there a driving force for corrosion if this solution flows through nickel tubing? Support your answer quantitatively. Would you expect corrosion to occur?

16.3 Magnesium is being used as a structural material in cars, for example, because it is light and has good structural properties. However, there is concern regarding corrosion. Is that concern warranted? Please support your response quantitatively.

16.4 Using Gibbs energy values, determine the standard potential for the reaction represented by line 10 of Figure 16.2. Derive an expression for the equilibrium potential of the reaction represented by line 10 in Figure 16.2 as a function of pH. What assumption was made to get the values shown in the figure? What impact would a change in this assumption have on the equilibrium potential?

16.5 Using the Pourbaix diagram for nickel (Figure 16.2), is the corrosion of nickel in aqueous solutions more likely to be problematic in highly acidic or highly basic solutions?

Please justify your response. You should consider the stability of water.

16.6 There is a large driving force for the corrosion of zinc in deaerated aqueous solution, where the primary cathodic reaction would be hydrogen evolution. However, zinc is stable in such environments. The following kinetic parameters apply: for the zinc reaction, $\alpha_a = 1.5$ and $i_0 = 0.10$ A·cm^{-2}; for hydrogen evolution, $\alpha_c = 0.5$ and $i_0 = 10^{-9}$ A·cm^{-2}. Assume Tafel kinetics, and calculate the following:

(a) The corrosion potential

(b) The corrosion current for zinc

(c) The corrosion rate of zinc in mm·yr^{-1}

Why is the corrosion rate of zinc so low?

16.7 The following data are available for the corrosion of iron in acid solution where oxygen does not make a significant contribution to its rate of dissolution. The following parameters apply: $b_{Fe} = 118$ mV, $b_{hydrogen} = 120$ mV, $i_{0,Fe} = 0.002$ A·m^{-2}, and $i_{0,hydrogen} = 0.10$ A·m^{-2}. Use 0.0 and -0.44 V for the equilibrium potentials for the hydrogen and iron reactions, respectively.

(a) Calculate the corrosion potential and current

(b) Calculate the dissolution rate in mm·yr^{-1}

(c) Calculate the iron current, hydrogen current, and total current as a function of potential, and plot all three values on the same diagram.

16.8 The following data describe the corrosion of iron in seawater, where both oxygen and hydrogen contribute to the cathodic reaction. Assume Tafel kinetics for both the iron dissolution reaction and the hydrogen evolution reaction (HER). The expression for oxygen reaction includes an approximation of the mass-transfer limitations.

Reaction	U [V]	i_o [A·m^{-2}]	Expression
Fe → Fe^{2+} + 2e$^-$	-0.44 $\alpha_a = 0.5$	0.014	$i = i_0 \exp(\alpha_a F(V-U)/RT)$
2H$_2$O + 2e$^-$ → H$_2$ + 2OH$^-$	-0.42 $\alpha_c = 0.5$	1×10^{-6}	$i = -i_0 \exp(-\alpha_c F(V-U)/RT)$
O$_2$ + H$_2$O + 4e$^-$ → 4OH$^-$	0.815 $\alpha_c = 1$	1×10^{-16}	$i = -i_0(1 - i/i_{lim}) \exp(-\alpha_c F(V-U)/RT)$ $i_{lim} = 0.3$ A·m^{-2}

Please determine:

(a) The corrosion potential and corrosion current.

(b) The relative importance of the oxygen reaction (fraction of cathodic current from oxygen reduction).

(c) Plot the individual currents and the total current as a function of potential on the same diagram.

16.9 Find the corrosion potential and corrosion potential for lead in neutral solution. Under the conditions oxygen does not make a significant contribution to its rate of dissolution. The following parameters apply: $b_{Pb} = 60$ mV, $b_{H2} = 240$ mV, $i_{0,Pb} = 1$ A·m^{-2}, and $i_{0,H2} = 1 \times 10^{-6}$ A·m^{-2}. Use -0.41 and -0.126 V for the equilibrium potentials for the hydrogen and iron reactions, respectively.

16.10 In solutions where a passivation layer is not formed on zinc, the corrosion potential is found to be very close to the equilibrium potential for zinc. What can you infer about the kinetics for the anodic (dissolution of zinc) compared to the kinetics for the cathodic (hydrogen evolution or oxygen reduction) reactions?

16.11 What is the corrosion potential, corrosion current density, and corrosion rate (mm·yr^{-1}) of a copper pipe in laminar cross flow? The velocity of the fluid over the 3 cm diameter pipe is 2 m·s^{-1}. The solution is at pH = 8 and is saturated with oxygen (10 g·m^{-3}). The following reactions and parameters apply,

Cu → Cu^{2+} + 2e$^-$, $U = 0.337$ V, $i_0 = 5 \times 10^{-5}$ A m^{-2}

O$_2$ + H$_2$O + 4e$^-$ → 4OH$^-$, $U = 0.76$ V, $i_0 = 1 \times 10^{-7}$ A m^{-2}

Your friend Eugene Engineer claims that H$_2$ evolution should also be considered in the calculation. Do you agree? Justify your response quantitatively.

16.12 The following composite corrosion polarization curve is measured for Fe in deaerated acid solution by changing the current and measuring the corresponding potential. From the semi-log plot, please determine the following:

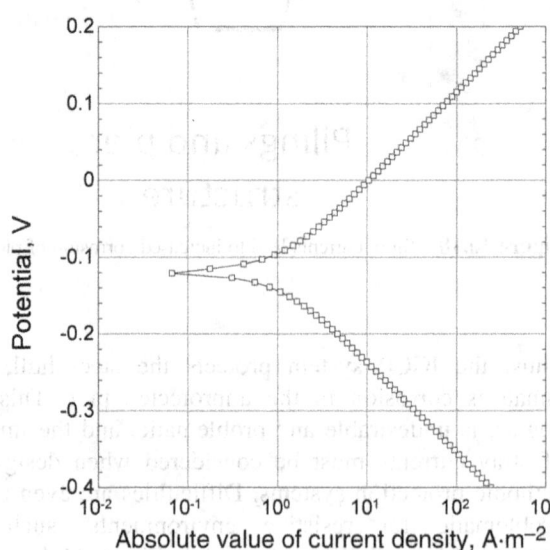

(a) The corrosion potential relative to SHE

(b) The corrosion current

(c) The Tafel slope of the anodic reaction

(d) The Tafel slope of the cathodic reaction

Make sure to state any assumptions that you make.

16.13 In Illustration 16.4, the dissolution of copper by copper(II) ions in acid chloride solution was explored. That example considered two cathodic reactions. It turns out that there are at least four possible cathodic reactions that involve additional species formed as a result of homogeneous reaction.

The additional reactions are as follows:

$$CuCl_2 + Cl^- + e^- \rightarrow CuCl_3^{2-}; \quad U^\theta = 0.431\partial \quad (4)$$

$$CuCl_3^- + e^- \rightarrow CuCl_3^{2-}; \quad U^\theta = 0.435 \quad (5)$$

Accounting for two additional equilibrium reactions, the concentrations are as follows:

Cl⁻	Cu²⁺	CuCl⁺	CuCl₃²⁻	CuCl₂	CuCl₃⁻	CuCl₄²⁻
1.23 M	0.067 M	0.1897 M	0.02 M	0.1517 M	0.071 M	0.0071 M

The kinetic expressions for the two additional reactions are as follows:

$$i_4 (A \cdot m^{-2}) = 0.5 \left(\frac{c_{Cl^-}}{1 M}\right)^{0.5} \left(\frac{c_{CuCl_2}}{1 M}\right)^{0.5} \left(\frac{c_{CuCl_3^{2-}}}{1 M}\right)^{0.5}$$
$$\left[\exp\left(\frac{0.5F}{RT}(V - U_4)\right) - \exp\left(\frac{-0.5F}{RT}(V - U_4)\right)\right]$$

$$i_5 (A \cdot m^{-2}) = 1.0 \left(\frac{c_{CuCl_3^{2-}}}{1 M}\right)^{0.5} \left(\frac{c_{CuCl_3^-}}{1 M}\right)^{0.5}$$
$$\left[\exp\left(\frac{0.5F}{RT}(V - U_5)\right) - \exp\left(\frac{-0.5F}{RT}(V - U_5)\right)\right].$$

(a) Plot the Evans diagram similar to the one in the illustration, but with all four cathodic reactions and a line for the total cathodic reaction. Use this plot to estimate (rough only) the corrosion potential of the system.

(b) Calculate the corrosion potential and the corrosion current density for the completed set of five reactions, one anodic reaction and four cathodic reactions using the full BV expressions.

(c) Repeat part (b) using Tafel expressions rather than the full BV expressions. Please comment on any differences between the two results. Is this what you expected? Why or why not?

16.14 A simplified cell, which results in a uniform current density during corrosion of the full surfaces, can be used to examine galvanic corrosion. If the two metals are not connected, each metal corrodes independent of the other. However, if a highly conductive wire is used to connect the two metals, they become galvanically coupled and their corrosion rates change dramatically. Examine corrosion of iron and zinc using the parameters from Figure 16.10.

The conductivity of the solution is constant and equal to $0.08 \, S \cdot m^{-1}$. The electrodes are infinite plates separated by a distance of 3 cm.

Please calculate the following:

(a) The uncoupled corrosion potential (SHE) and corrosion current density ($A \cdot m^{-2}$) for each of the two metals.

(b) The rate of corrosion for the coupled system.

You will need to account for the ohmic losses in solution, which will allow you to relate the potential of the solution at one electrode to that at the other. The sum of the cathodic and anodic currents on the two electrodes together must equal zero. Report the potential of each electrode versus a Ag/AgCl reference electrode located at the electrode surface. Also report the specific anodic and cathodic currents for each electrode.

16.15 Describe the difference between immunity and passivity. Explain the relationship between these conditions and anodic and cathodic protection strategies for corrosion mitigation.

16.16 Where does the cathodic reaction occur during pitting corrosion? Why does it occur there? What implications does this have for the growth of multiple corrosion pits on a surface?

16.17 Why does galvanic coupling increase the corrosion rate of the more active metal?

16.18 The corrosion rate of magnesium increases dramatically when it is coupled with iron, even if the area of the iron is just a fraction of that of the magnesium. Please explain why this might be so.

16.19 Explain why cathodic current is required in order to cathodically protect metal structures.

16.20 Does a sacrificial anode represent anodic protection or cathodic protection? Please explain.

16.21 You have been assigned to develop a system to protect a stationary sea oil drilling platform located in 400 m deep water. Based on your understanding of sacrificial anodes:

(a) What costs are associated with (i) installation of a sacrificial anode system, and (ii) operation of a sacrificial anodic protection system?

(b) Repeat part (a) for ICCP systems.

(c) Based on your answers to (a) and (b), which strategy do you expect to be more expensive? Which would be easier to implement?

16.22 You desire to use ICCP to protect a low-carbon steel surface from corrosion (area = $20 \, m^2$). Assuming a uniform current density on the steel surface, the pH of the solution is 8. For all potentials of interest, the reduction of oxygen is diffusion limited, $i_{lim} = 0.2 \, A \cdot m^{-2}$. For iron dissolution $\alpha_a = 0.5$, $i_{0,Fe} = 0.002 \, A \cdot m^{-2}$, $U_{Fe} = -0.64 \, V$, and for hydrogen $\alpha_c = 0.5$, and $i_{0,H2} = 0.05 \, A \cdot m^{-2}$.

(a) What value of the impressed current would be required to reduce the corrosion rate to $0.1 \, mm \cdot yr^{-1}$? The kinetics are known and given below.

(b) What is the corresponding potential of the electrode (versus SCE)?

(c) Often −850 mV versus SCE is specified as the potential required for protection of steel in seawater. Please comment on this value based on your results.

16.23 ICCP is to be used to protect a steel dock structure in seawater. The surface area of the structure is $15 \, m^2$. The anodes available are platinized Ti with an OD of 2.5 cm and a length of 1 m. Specific kinetic data and conductivity data for your system are not available.

(a) Please recommend an appropriate anode configuration.

(b) If electricity is $0.05/kWh, how much would it cost to operate this system for a year (assume continuous operation for the entire year and no losses in rectifying the electricity)? Assume a potential drop of 20 V.

(c) What factors might influence your placement of the ICCP anodes?

16.24 Data are shown for the passivation of iron in a phosphate solution at a pH = 9.7. (Adapted from *Corrosion Science*, **19**, 297 (1979). The solution is deaerated, hydrogen evolution is cathodic reaction. It is desired to protect 1 m^2 of iron to a corrosion rate of 1 mm·yr^{-1}. Compare the current required for anodic versus cathodic protection.

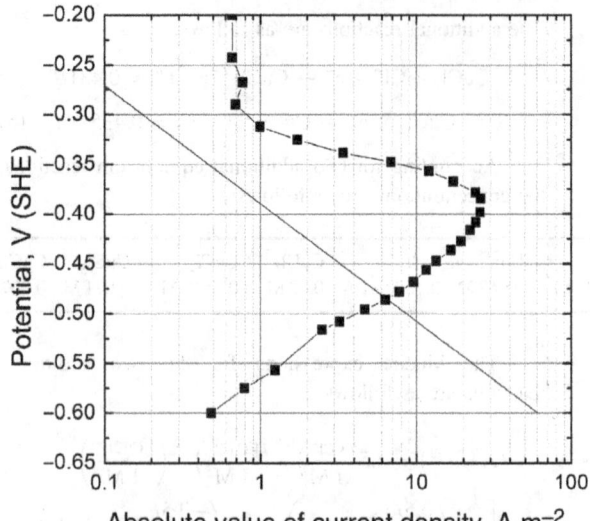

Appendix A

Electrochemical Reactions and Standard Potentials

	Electrochemical reaction	Standard potential, U^θ
1	$F_2 + 2e^- \rightarrow 2F^-$	2.87
2	$PbO_2 + SO_4^{2-} + 4H^+ + 2e^- \rightarrow PbSO_4 + 2H_2O$	1.685
3	$Cl_2 + 2e^- \rightarrow 2Cl^-$	1.3595
4	$O_2 + 4H^+ + 4e^- \rightarrow 2H_2O$	1.229
5	$Br_2(aq) + 2e^- \rightarrow 2Br^-$	1.078
6	$Ag^+ + e^- \rightarrow Ag$	0.7991
7	$Hg_2^{2+} + 2e^- \rightarrow 2Hg$	0.789
8	$Cu^+ + e^- \rightarrow Cu$	0.521
9	$O_2 + 2H_2O + 4e^- \rightarrow 4OH^-$	0.401
10	$Cu^{2+} + 2e^- \rightarrow Cu$	0.337
11	$Hg_2Cl_2 + 2e^- \rightarrow 2Hg + 2Cl^-$	0.2676
12	$AgCl + e^- \rightarrow Ag + Cl^-$	0.222
13	$Cu^{2+} + e^- \rightarrow Cu^+$	0.153
14	$HgO + H_2O + 2e^- \rightarrow Hg + 2OH^-$	0.098
15	$2H^+ + 2e^- \rightarrow H_2$	0
16	$Pb^{2+} + 2e^- \rightarrow Pb$	−0.126
17	$PbSO_4 + 2e^- \rightarrow Pb + SO_4^{2-}$	−0.356
18	$Fe^{2+} + 2e^- \rightarrow Fe$	−0.440
19	$Cr^{3+} + 3e^- \rightarrow Cr$	−0.74
20	$Zn^{2+} + 2e^- \rightarrow Zn$	−0.763
21	$2H_2O + 2e^- \rightarrow H_2 + 2OH^-$	−0.828
22	$Cr^{2+} + 2e^- \rightarrow Cr$	−0.91
23	$Mn^{2+} + 2e^- \rightarrow Mn$	−1.18
24	$Al^{3+} + 3e^- \rightarrow Al$	−1.66
25	$Mg^{2+} + 2e^- \rightarrow Mg$	−2.357
26	$Na^+ + e^- \rightarrow Na$	−2.714
27	$K^+ + e^- \rightarrow K$	−2.936
28	$Li^+ + e^- \rightarrow Li$	−3.045

Standard states, 25 °C, 100 kPa.: (a) gases, pure ideal gas, (b) liquids and solids, pure substance, (c) aqueous, hypothetical 1 molal solution.

Electrochemical Engineering, First Edition. Thomas F. Fuller and John N. Harb.
© 2018 Thomas F. Fuller and John N. Harb. Published 2018 by John Wiley & Sons, Inc.
Companion Website: www.wiley.com/go/fuller/electrochemicalengineering

Appendix B

Fundamental Constants

Nomenclature	Name	Value	Common units	Expressed in base SI units
R	Universal gas constant	8.3144621	$J \cdot mol^{-1} \cdot K^{-1}$	$m^2 \cdot kg\ mol^{-1} \cdot s^{-2} \cdot K^{-1}$
N_{AV}	Avogadro's number	6.022141×10^{23}	mol^{-1}	mol^{-1}
k	Boltzmann's constant	1.380649×10^{-23}	$J \cdot K^{-1}$	$m^2 \cdot kg \cdot s^{-2} \cdot K^{-1}$
q	Fundamental unit of charge	1.602177×10^{-19}	C	$A \cdot s$
F	Faraday's constant	96,485.34	C/equiv.	$A \cdot s$/equiv.
ϵ_0	Permittivity of vacuum	8.854188×10^{-12}	$F \cdot m^{-1}$	$m^{-3} \cdot kg^{-1} \cdot s^4 \cdot A^2$
c	Speed of light	2.99792×10^8	$m \cdot s^{-1}$	$m \cdot s^{-1}$
h	Planck's constant	6.626070×10^{-34}	$J \cdot s$	$m^2 \cdot kg \cdot s^{-1}$

Electrochemical Engineering, First Edition. Thomas F. Fuller and John N. Harb.
© 2018 Thomas F. Fuller and John N. Harb. Published 2018 by John Wiley & Sons, Inc.
Companion Website: www.wiley.com/go/fuller/electrochemicalengineering

Appendix C

Thermodynamic Data

Table C.1 Standard Enthalpy and Gibbs Energy of Formation from the Elements at 25 °C (298.15 K).

Chemical species	Formula	State	ΔG [kJ·mol^{-1}]	ΔH [kJ·mol^{-1}]
Acetic acid	CH_3COOH	Liquid	−389.9	−484.3
Acetic acid	CH_3COOH	Aqueous	−396.5	−486.1
Aluminum oxide	Al_2O_3	Solid, α	−1582.3	−1675.7
Ammonia	NH_3	Aqueous	−26.6	−80.3
Ammonia	NH_3	Gas	−16.4	−45.9
Bromine	Br_2	Gas	3.1	30.9
Carbon dioxide	CO_2	Gas	−394.359	−393.509
Carbon monoxide	CO	Gas	−137.2	−110.5
Cobalt(II) oxide	CoO	Solid	−214.2	−237.9
Formaldehyde	CH_2O	Gas	−102.5	−108.6
Formic acid	CH_2O_2	Liquid	−361.4	−425.0
Hydrogen bromide	HBr	Gas	−53.4	−36.3
Hydrogen chloride	HCl	Gas	−95.3	−92.3
Hydrogen peroxide	H_2O_2	Liquid	−120.4	−187.8
Hydrogen peroxide	H_2O_2	Gas	−105.6	−136.3
Lead(II) oxide	PbO	Solid	−187.9	−217.3
Lead(IV) oxide	PbO_2	Solid	−217.3	−277.4
Lead(II, IV) oxide	Pb_3O_4	Solid	−601.7	−718.8
Lead sulfate	$PbSO_4$	Solid	−813.302	−919.936
Lithia	Li_2O	Solid	−561.2	−597.9
Lithium hydroxide	$LiOH$	Aqueous	−451.1	−506.9
Lithium iodide	LiI	Solid	−270.3	−270.4
Lithium peroxide	Li_2O_2	Solid	−571.1	−632.6
Mercury(II) chloride	$HgCl_2$	Solid	−178.6	−224.3
Mercury(I) chloride	Hg_2Cl_2	Solid	−210.8	−265.2
Methanol	CH_3OH	Liquid	−166.6	−239.2
Methane	CH_4	Gas	−50.5	−74.6
Propane	C_3H_8	Gas	−24.3	−104.7
Silver oxide	Ag_2O	Solid	−11.21	−31.1
Silver sulfate	Ag_2SO_4	Solid	−618.4	−715.9
Sodium chloride	$NaCl$	Solid	−384.1	−411.2
Sodium chloride	$NaCl$	Aqueous	−393.1	−407.3

(*continued*)

Electrochemical Engineering, First Edition. Thomas F. Fuller and John N. Harb.
© 2018 Thomas F. Fuller and John N. Harb. Published 2018 by John Wiley & Sons, Inc.
Companion Website: www.wiley.com/go/fuller/electrochemicalengineering

Table C.1 (Continued)

Chemical species	Formula	State	ΔG [kJ·mol^{-1}]	ΔH [kJ·mol^{-1}]
Sodium hydroxide	NaOH	Aqueous	−419.2	−470.1
Sodium oxide	Na$_2$O	Solid	−375.5	−414.2
Sulfur dioxide	SO$_2$	Gas	−300.1	−296.8
Sulfuric acid	H$_2$SO$_4$	Aqueous	−744.530	−909.3
Water	H$_2$O	Liquid	−237.129	−285.830
Water	H$_2$O	Gas	−228.572	−241.572
Zinc oxide	ZnO	Solid	−320.48	−350.46

Ion	Formula	State	ΔG [kJ·mol^{-1}]	ΔH [kJ·mol^{-1}]
Hydrogen	H$^+$	Aqueous	0	0
Aluminum	Al^{3+}	Aqueous	−485.34	−531.37
Ammonium	NH$_4^+$	Aqueous	−79.37	−132.51
Calcium	Ca^{2+}	Aqueous	−553.54	−542.83
Copper(I)	Cu$^+$	Aqueous	50.00	71.67
Copper(II)	Cu^{2+}	Aqueous	65.52	64.77
Iron(II)	Fe^{2+}	Aqueous	−84.91	−89.12
Iron(III)	Fe^{3+}	Aqueous	−10.71	−48.53
Lead	Pb^{2+}	Aqueous	−24.39	−1.67
Lithium	Li$^+$	Aqueous	−293.3	−278.49
Magnesium	Mg^{2+}	Aqueous	−454.80	−466.85
Potassium	K$^+$	Aqueous	−283.26	−252.38
Silver	Ag$^+$	Aqueous	77.12	105.57
Sodium	Na$^+$	Aqueous	−261.66	−240.12
Zinc	Zn^{2+}	Aqueous	−147.03	−153.89
Bicarbonate	HCO$_3^-$	Aqueous	−586.85	−691.99
Bisulfate	HSO$_4^-$	Aqueous	−756.01	−887.34
Bisulfide	HS$^-$	Aqueous	12.6	−17.7
Bisulfite	HSO$_3^-$	Aqueous	−527.8	626.2
Bromide	Br$^-$	Aqueous	−103.97	−121.54
Carbonate	CO$_3^{2-}$	Aqueous	−527.89	−677.14
Chloride	Cl$^-$	Aqueous	−131.26	−167.16
Fluoride	F$^-$	Aqueous	−278.82	−332.63
Hydroxyl	OH$^-$	Aqueous	−157.29	−229.99
Iodide	I$^-$	Aqueous	−51.59	−55.19
Nitrate	NO$_3^-$	Aqueous	−111.34	−207.36
Perchlorate	ClO$_4^-$	Aqueous	−10.8	
Sulfate	SO$_4^{2-}$	Aqueous	−744.62	−909.3
Sulfide	S^{2-}	Aqueous	79.5	30.1
Sulfite	SO$_3^{2-}$	Aqueous	−486.6	−635.5

Standard states: (a) gases, pure ideal gas at 100 kPa, (b) liquids and solids, pure substance at 100 kPa, (c) aqueous, hypothetical 1 molal solution

N. de Nevers (2012) *Physical and Chemical Equilibrium for Chemical Engineers*, John Wiley & Sons, Inc..
D.R. Lide (2000) *CRC Handbook of Chemistry and Physics*, Boca Raton, FL, CRC Press.
M. Pourbaix (1974) *Atlas of Electrochemical Equilibria*, NACE, Houston.

Table C.2 Standard Molar Entropy of Substances at 25 °C (298.15 K) and 100 kPa

Species	Formula	State	Entropy [J·mol^{-1}·K^{-1}]
Hydrogen	H$_2$	Gas	130.5
Chlorine	Cl$_2$	Gas	223.1
Oxygen	O$_2$	Gas	205.03
Iron	Fe	Solid	27.28
Water	H$_2$O	Gas	188.7
Water	H$_2$O	Liquid	69.91
Fe(II)	Fe^{2+}	Aqueous	−137.654
Hydrogen	H$^+$	Aqueous	0
Chloride	Cl$^-$	Aqueous	56.48
Hydroxyl	OH$^-$	Aqueous	−10.75

Whereas the Gibbs energy and enthalpy are from an arbitrary datum, entropy values are absolute and therefore nonzero even for elements.
Dean, J.A. (1979) *Lange's Handbook of Chemistry*, 12th ed., McGraw-Hill: New York, NY, pp. 9-4–9-94.

Appendix D

Mechanics of Materials

D.1 STRESS AND STRAIN

A material subjected to a uniaxial load will deform. If compressed, the length of the material is reduced; if placed under tension, the length increases. The deformation or linear *strain* of a material of length L is

$$\epsilon = \frac{\delta}{L}, \tag{D.1}$$

where δ is the distance that the material deforms. Thus, this normal strain is a dimensionless quantity. Normal *stress* is simply the force of compression or tension in the material; and stress and strain are related with *Young's modulus of elasticity*:

$$E = \frac{\sigma}{\epsilon} = \frac{\text{stress}}{\text{strain}}. \tag{D.2}$$

This modulus is a property of the material and has units of pressure [Pa]. Figure D.1 shows the stress–strain relationship for a brittle and a ductile material. At low strain, the stress varies linearly, or said another way the modulus is constant. Furthermore, when the stress is relieved, the material returns to its original shape. Thus, Equation D.2 is equivalent to Hooke's law for a spring and describes *proportional* behavior. At some point the proportional limit is reached, above this strain the slope is no longer constant, but the behavior is still *elastic*, meaning that there is no permanent deformation. If the stress is excessive, the material can fail. For ductile materials, above the *elastic limit* the material continues to deform, but the deformation is permanent. This deformation is called *plastic*. Note that brittle materials fracture with little deformation. In contrast, a ductile material may deform significantly before failure. Failure of the material can depend on the application and what is acceptable for the design: elastic limit, yield point (where the material deforms without further increase in stress), or yield strength (where a fixed amount of permanent strain occurs, for example, 0.002).

Poisson's ratio, ν, is a second material property that needs to be introduced. The cube shown in Figure D.2 is put in tension along the x-axis. The length along the x-axis in tension increases, and simultaneously the width transverse to the axis in tension (y and z) decreases. Poisson's ratio is a means of describing how these two strains are related.

$$\nu = \frac{-\text{transverse strain}}{\text{axial strain}} = \frac{-\epsilon_{trans}}{\epsilon_{axial}}. \tag{D.3}$$

Typical values for metals are around 0.3. A variety of mechanical properties for different material relevant to this book are shown in Table D.1. As will be seen shortly, Poisson's ratio is used in calculations of spring behavior in the next section.

D.2 THERMAL EXPANSION

An applied force of compression or tension is not the only means of inducing strain in a material. If unrestrained, materials expand (with rare exceptions) when the temperature is raised. The amount of change depends on the material and is quantified by the coefficient of thermal expansion (CTE), which is commonly defined as

$$\alpha_L \equiv \frac{1}{L}\frac{dL}{dT}. \tag{D.4}$$

Here we restrict ourselves to expansion in one dimension or linear expansion. CTE is another material property, and typical values for α_L are listed in Table D.1. The stress-free thermal strain, ϵ_t, is just the change in length from a temperature change:

$$\epsilon_t = \alpha_L(\Delta T). \tag{D.5}$$

Electrochemical Engineering, First Edition. Thomas F. Fuller and John N. Harb.
© 2018 Thomas F. Fuller and John N. Harb. Published 2018 by John Wiley & Sons, Inc.
Companion Website: www.wiley.com/go/fuller/electrochemicalengineering

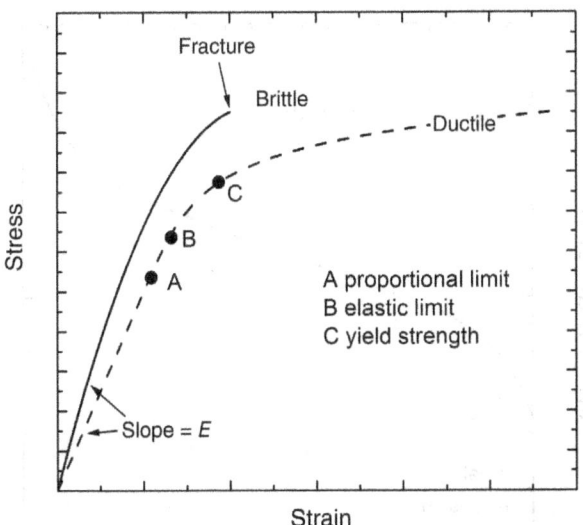

Figure D.1 Typical stress-strain behavior for brittle and ductile materials.

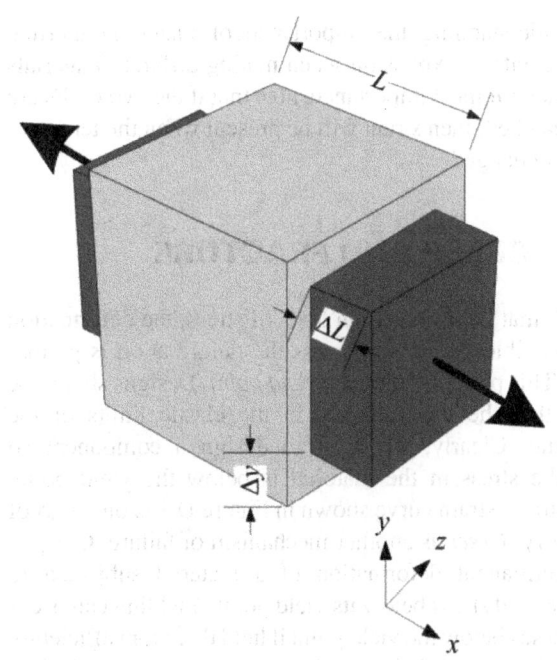

Figure D.2 Illustration of transverse and axial strain, quantified with Poisson's ratio.

To keep the material from expanding, a force would be required, which is expressed in terms of the stress in the material from Equations D.1 and D.2.

$$\sigma = E\alpha_L \Delta T. \qquad (D.6)$$

Thus, a stress is present in the material if it is constrained and the temperature changed from the stress-free state. In our rudimentary analysis, we will assume that one of the components is much stiffer than the other. Think of stiffness as the ability of the component to resist deformation. Stiffness is not simply a material property, but is affected by the thickness and shape of the component. This assumption means that one of the components, the support, completely resists deformation. The stress generated in the other component is related to the difference in CTE of the two materials:

$$\sigma_i = \frac{E_i}{1 - \nu_i}(\alpha_s - \alpha_i)(T - T_0). \qquad (D.7)$$

The subscript s refers to the support material, ν is Poisson's ratio, and T_0 is the stress-free temperature. This highly simplified analysis will give the engineer a starting point

Table D.1 Mechanical properties of select materials used in construction of fuel cell and springs

Material	α_L [K^{-1}] × 10^{-6}	Elastic modulus [GPa]	Poisson's ratio	Strength, stress at yield or failure [MPa]
Carbon steel	10.8 at (20 °C)	200	0.29	250 (yield)
Nickel	13 at 20 °C	177	0.31	150–480 (yield)
Stainless steel (304)	17.3	193–200	0.29	215 (yield)
Titanium	9.2 at 250 °C	80–125	0.31	880 (yield)
POCO graphite	8.4	11	0.27	69–207 (failure)
YSZ (8% doping)	10.5 (25–1000 °C)	220	0.22	416 (failure)
Porous LSM	12.4	130	0.36	–
Borosilicate glass	3.3–5.1	68–81	0.21	280 (ultimate)
Lanthanum chromite, calcium doped 15 %	9.8 (25–1200 °C)	150	–	150 (ultimate)
Teflon	100–140 (100 °C)	0.40–1.8	0.46	9–30

Use of these tabulated values is appropriate only for solving problems in this book. Other advanced texts should be consulted for detailed design work.

410   Appendix D: Mechanics of Materials

for understanding the importance of matching thermal coefficients of expansion when mating different materials together. Of most importance, note that if the two CTEs are not matched, then strain will be present when the temperature is changed.

D.3 CREEP AND FRACTURE

Recall that below a certain level of stress, the deformation is reversible; above this stress the deformation is permanent. This point is called the *yield point*. Designs should be such that the materials are in the elastic limits of the material. Clearly, we want to design a component so that the stress in the material is below the yield point. The stress–strain curve shown in Figure D.1 is only part of the story. *Creep* is another mechanism of failure. Creep is the permanent deformation of a material subjected to mechanical stress below its yield point. Yielding can occur at stresses below the yield point if held there for sufficiently long time or if held at elevated temperatures. Unlike *fracture*, discussed previously, creep is a slow process. Furthermore, some creep is generally acceptable as long as it was anticipated in the design.

In making assemblies of electrodes or cells, it is common to need a means of applying an axial load (normal to the electrode surface). The reasons for this load are first to hold the electrodes or cells together and second to reduce contact resistance between components. There are a number of methods to apply compressive loads that are used in electrochemical systems. The two idealized methods would be constant strain or constant stress. These provide a good framework for discussing creep and the stress–strain–time relationship. Figure D.3 displays typical creep behavior of a material under a constant tensile stress. The strain increases with time and is divided into three stages. The elastic strain occurs more or less instantly, this is the initial extension. The first stage of primary creep is one where the rate of creep decreases exponentially. This period is followed by a second stage that is linear, and of most relevance for long-term operation. In the third stage, the extension increases exponentially due to necking followed by fracture. We are often interested in compressive stress, where tertiary creep is less important. There isn't a single parameter or mechanical property that describes creep; in general, data like that shown in Figure D.3 are needed.

Constant strain is the other way to think about creep. This would be applicable if an electrochemical cell-sandwich were designed with a fixed thickness. Imagine that for a certain component that after a period of time, t_1, under constant stress as described above, the strain is now held constant. In this case the stress will decrease with time due to creep of the materials. This behavior is called

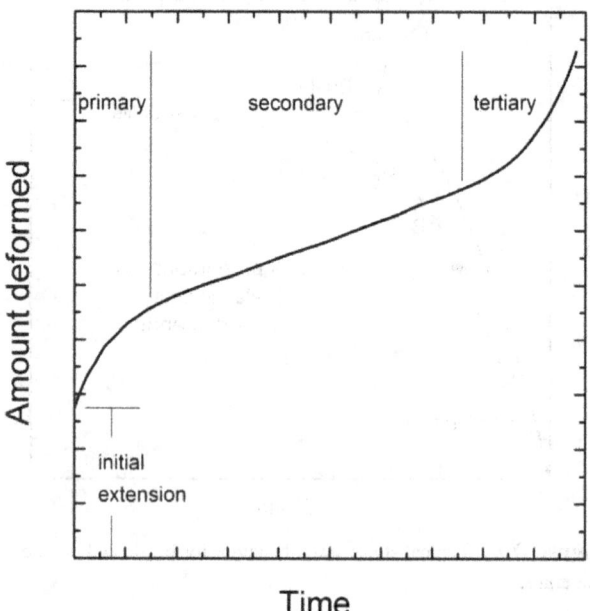

Figure D.3 Typical stages of creep under a constant tensile stress below elastic limit.

stress relaxation and is shown in Figure D.4. This relaxation could cause a large increase in contact resistance, for example.

The actual behavior in a well-designed system is unlikely to be solely constant stress or constant strain. The disadvantage of constant strain is readily apparent. As a result, springs are generally used to accommodate some creep but keep the stress more constant. Systems that do this are described as providing *load follow-up*. There are a multitude of approaches. In the next two sections, we just consider two types that are commonly used in batteries and fuel cells: the conical disk and wavy spring. The main

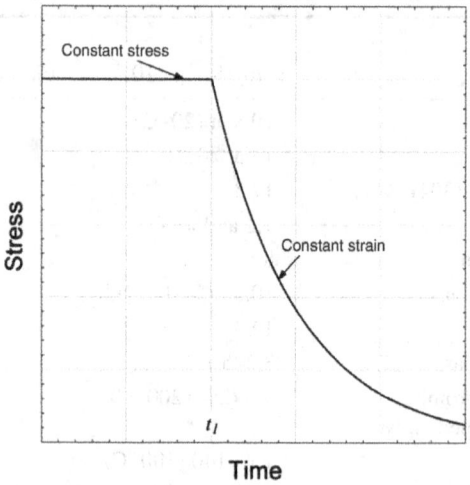

Figure D.4 Stress relaxation with constant strain.

advantage of these types of springs is that they can provide a high axial force, but are relatively compact. Although these two approaches are commonly used in industry and research laboratories, some of the details may not be readily accessible. Therefore, a few parts of the development are included here.

Belleville Washers

A common method of providing load follow-up is with Belleville washers. Figure D.5 shows the cross section that act as springs when added to a tie-rod (see Illustration 10.8, for instance). t is the thickness of the spring material. The cupped or cone-shaped washers are stacked together so that the desired spring properties are achieved. Because of the conical shape, a single washer has a known spring constant, analogous to Hooke's law. When washers are stacked in the same direction, the spring constants add in parallel, resulting in a stiffer assembly. When stacked in alternating directions, the spring constants add in series, resulting in an assembly with greater deflection for the same force. Thus, the spring constant can be tuned as needed. The downside is that extra space is needed for the load follow-up system, but this approach is relatively compact. The relationship between deflection, s, and force, F, is a bit more complicated than Hooke's law but can be expressed in terms of the geometry and material properties of the metal. Here friction is ignored.

$$F_x = \frac{4Et^4}{(1-\nu^2)\alpha D^2} \frac{s}{t} \left[\left(\frac{h_0}{t} - \frac{s}{t} \right) \left(\frac{h_0}{t} - \frac{s}{2t} \right) + 1 \right], \quad (D.8)$$

where $\quad \delta \equiv \dfrac{D}{d}, \quad h_0 \equiv H - t,$

and

$$\alpha \equiv \frac{1}{\pi} \times \frac{\left(\dfrac{\delta-1}{\delta}\right)^2}{\left(\dfrac{\delta+1}{\delta-1}\right) - \dfrac{2}{\ln \delta}}.$$

The relationship between force and deflection given by Equation D.8 is shown in Figure D.6. These washers are generally used in the linear range for low values of deflection where Hooke's law is closely approximated.

$$K = \frac{F}{s} \approx \frac{4Et^3}{(1-\nu^2)\alpha D^2} \left[\left(\frac{h_0}{t}\right)\left(\frac{h_0}{t}\right) + 1 \right]. \quad (D.9)$$

A stack or assembly of washers can be comprised of one or more groups. A *group* is where the washers are adjacent and oriented in the same direction; and each group contains n_i washers. Referring to Figure D.6, the first sequence is (1–1–1–1–1–1–1–1); and the second is (8). For an arrangement of g groups of washers, the total spring constant is

$$K_{\text{stack}} = \frac{K}{\sum_i^g \dfrac{1}{n_i}}. \quad (D.10)$$

n_i is the number of washers in the ith group. Thus, for Belleville washer in Figure D.6, Equation D.9 is used to find the spring constant for one washer. $K = 5755$ N/mm, which is the slope in Figure D.6. We can then compare this to K_{stack} for the two assemblies shown:

$$K_{\text{stack}} = \frac{K}{\sum_i^g \dfrac{1}{1}} = \frac{K}{8} = 719 \frac{\text{N}}{\text{mm}}.$$

Thus, for this arrangement, the stack is eight times less stiff than a single washer, and the stacking is referred to as in parallel. In contrast, for the other arrangement,

$$K_{\text{stack}} = \frac{K}{\dfrac{1}{8}} = 8K = 46 \frac{\text{kN}}{\text{mm}}.$$

This stacking is referred to as a series arrangement.

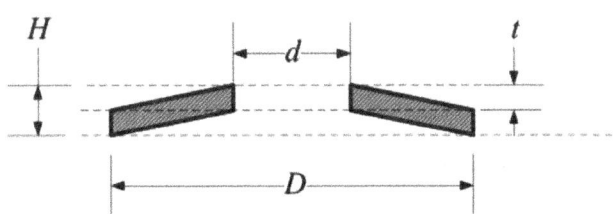

Figure D.5 Belleville washer in unloaded state.

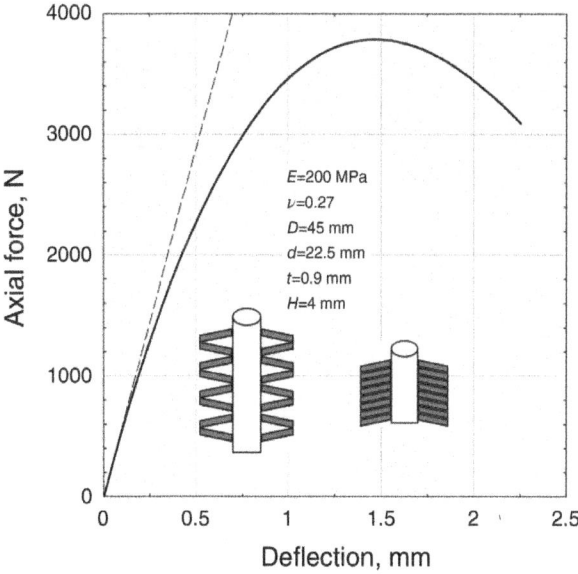

Figure D.6 Force deflection relationship for single Belleville washer. Alternate and same direction stacking are shown.

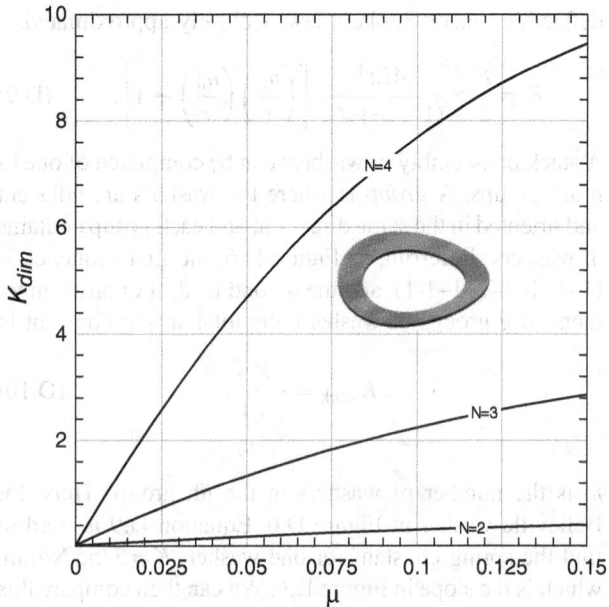

Figure D.7 Dimensionless spring constant for wave spring with $\lambda = 0.65$. Wave washer shown with N is two.

Wave Washers

Another method of applying axial load is with a wave spring washer shown in Figure D.7. A single wave or conical washer is commonly used in the assembly of coin cells for battery and electrochemical capacitor testing. For relatively small deflections, there is a constant of proportionality between force and deflection. This spring constant, K_{min}, depends on the dimensions of the washer: inner and outer diameters, thickness, and number of waves. The materials properties are also important. The relationship and analysis is complex, but for our purposes we can use the graph in Figure D.7 that provides a dimensionless spring constant.

$$K_{\dim} = \frac{K_{min} D_m^2}{E t^3}, \quad (D.11)$$

where D_m is the mean diameter,

$$D_m = \frac{D_0 + D_i}{2}, \quad (D.12)$$

and t is the thickness of the spring material (not the uncompressed height). The values plotted are for a fixed value of λ, the ratio of flexural to torsional rigidity, equal to 0.65. Whereas the modulus is a function of the material, rigidity depends on the structure of the object. The dimensionless spring constant is shown as a function of μ, which is

$$\mu = \frac{D_0 - D_i}{D_0 + D_i}. \quad (D.13)$$

The force needed to deflect the washer by distance δ is

$$F_x = K_{min} \delta. \quad (D.14)$$

FURTHER READING

Dragoni, E. (1988) A contribution to wave spring design. *J. Strain Anal.*, **23**, 145.

Wells, J.W. (1970) Wave springs. *Machine Des.*, **42** (20), 113.

Index

absorption
    and quantum yield, 369
    coefficient, 366, 367
    efficiency, 367, 369, 372
    losses, 367
    of photons, 366
acceptors, 357
activation energy, 44
activity, 23, 33
    coefficient, 33–34, 35–36
additives, for electrodeposition, 311
adion, 302
aluminum production, 4, 323
    energy balance, 342
    Hall-Herault process, 338
anode, 1
    battery, 151
    sacrificial, 396, 397
anodic protection, 394
anodic scan, 122
Archimedes number, 79
Arrhenius expression, 47

back of electrode, 108
balancing
    active, 188
    battery pack, 187
    passive, 188
band
    bending, 360, 363
    conduction, 356–357
    gap, 355–356, 359, 366–368, 373
    valence, 356–357
battery
    charging, 184
    chemistries, 154
    coin cell, 152, 193
    configuration, layout, 176
    cylindrical, 183, 184, 188, 191
    heat generation, 164, 165, 189
    module, 176
    primary, 152
    pulse power test, 185
    reserve, 153
    scaling, 178
    secondary, 152
    spirally wound, 183, 188
    state of charge (SOC), 157
    state of health (SOH), 186
    string, 175
    theoretical capacity, 156
    thermal management, 188
battery electric vehicle (BEV), 282
battery management system (BMS), 186
Belleville washer, 245, 411
binary electrolyte, 35, 67
bipolar
    design, 228
    plate, 230
Bode plot, 134
Boltzmann factor, 32
boundary layer, 76
brighteners, 312
Bruggemann equation, 98, 328
Butler-Volmer kinetics, 46

C-rate, 159–160
calomel electrode, 28
capacitance, 42, 251
    differential, 251
capacitor
    conventional, 252
    double layer, 251, 252
    electrolytic, 252

capacity
    battery, 156–164, 282, 283
    EDLC, 267, 269, 270, 286, 287
    fade, 167
    offset, 161
    turnover, 283
capillarity, 102
capillary pressure, 102
cathode, 1
    battery, 151
cathodic protection, 395–399
    impressed current (ICCP), 397–399
    sacrificial anodes, 396, 397
cathodic scan, 122
CCCV charging, 184
cell balancing
    active, 188
    passive, 188
cell potential, 10, 17
cell resistance, 12, 121, 161
cell stack assembly (CSA), 197
central ion, 31
cermet, 214
characteristic frequency, 263
charge coefficient, 184
charge density, 32, 270, 273, 361
charge depleting mode, 289, 290
charge number, 9
charge retention, for batteries, 167
charge separation, 42, 251, 254, 256, 273, 368
charge sustaining mode, 289
charge-transfer resistance, 83, 130, 317
charge-transfer coefficient, 46
charging time
    double layer, 118
    EDLC, 262

414 Index

charging, pulse, 184
chemical potential, 30–31, 208
chlor-alkali, 10–12, 21–23, 323–325
chronoamperometry, 116
coefficient of thermal expansion (CTE), 244
coin cell, 152, 183
cold cranking amps (CCA), 164, 182
complex variables, 131
concentration overpotential, 79
concentration profile, 64
conductance, 73
conduction band, 356–357
conductivity
    effective, 98, 328
    electrolyte, 11
    semiconductor, 356–358
    solid, 99
conservation of charge, 66
conservation of mass, 65
contact angle, 102
contact resistance, 241
control volume, 65
convection
    forced, 75–79
    natural, 76
conversion reaction (reconstruction), 153
corrosion
    galvanic, 391–394
    localized, 390
    pitting, 390
    types, 380
    uniform, 383
corrosion potential, 383
    influence of mass transfer on, 386
Cottrell equation, 71, 116
coulomb, 6
coulombic efficiency, 166
counter electrode, 113
creep, 245
critical cluster size
    for electrodeposition, 305
critical overpotential
    for electrodeposition, 305
current, 6
    efficiency, 58
    interruption, 120
    limiting, 55
    sign convention, 44–45
    superficial, 45, 57
current collector, 94, 108
    resistance, 180
current density, 9, 66
current distribution, 81–85
    impact on electrodeposition, 312–314
    porous electrode, 98
    primary, 84
    resistive substrates, 316–318
    secondary, 85
    tertiary, 85
cutoff frequency, 264
cutoff potential, 160
    and capacity fade, 168
    and cell balancing, 188
    and maximum power, 283
cycle life, 168–169, 283
cyclic voltammetry, 122
cylindrical cell, 184

Daniel cell, 2, 151–152
Debye Huckel
    estimation of activity coefficients, 35
    limiting law, 35
    theory, 31
Debye, length, 31–32, 42, 255
deflection, 411
degree of hybridization, 289
depletion layer, region, zone, 360–362
deposit morphology, 310
depth of discharge, 157–158
derived units, 5–6
diaphragm cell, 324–328
dielectric constant, 252
diffuse layer, 42
diffusion, 63
    layer, 42
    length, 117
    time constant, 120
dilute solution
    activity coefficients, 35
    theory, 64
dimensionally stable anode, 329
dimensionless number, 75, 82, 317
direct band gap, 366
discharge, 151
discharge time, 175
disk electrode, 136–139, 142–143
displacement reaction, 153
dissolution-precipitation mechanism, 153
divergence, 65
donor or donor atom, 356
dopant
    levels, 358
    types, 356–357
doping
    acceptors or acceptor atoms, 357–358
    and extrinsic conductivity, 356
    donors or donor atoms, 356–358
    of semiconductors, 356–358
double layer, 41–43
    capacitance, 42, 119–123, 130, 140, 143, 255–258
    capacitor, 251
    charging time, 118
drift current, 365

driving schedule, 279
dryout, 209

effective transport properties, 98
effectiveness factor, 105
efficiency
    coulombic, 166
    energy, 167
    faradaic, 9
    fuel, 225
        processing, 235
    mechanical, 226
    photoelectrochemical cell, 372
    power conditioning, 226
    system, 224
        thermal, 224
    thermal voltage, 225
    voltage, 166
    fuel cell, 225
elastic limit, 408
elasticity, modulus of, 408
electric field, 42, 63, 66
    and binary electrolyte, 68
    and excess supporting electrolyte, 69
electrical conductance, 65
electrochemical
    cell, 1
    impedance spectrospcopy (EIS), 129
    potential, 30–31
    reaction, 16
    reactor design, 331–337
    surface area (ECSA), 128
    system, 1
electrochemical double layer
    capacitor (EDLC), 254
    activation controlled leakage model, 271
    energy and power, 267
    equivalent resistance model, 264, 266
    for hybrid vehicles, 286–287
    impedance of, 263–266
    porous electrode model, 261–262
    zero order leakage model, 271
electrode, 1
    back, 108
    counter, 113
    flow by, 105
    flow through, 105
    front, 108
    inert, 3
    negative, 151
    positive, 151
    reference, 27, 113
    working, 113
electrolysis
    electrorefining, 339
    electrowinning, 338
    to produce inorganic chemicals, 337

to produce organic chemicals, 339
    water, 343–345
electrolyte
    binary, 34, 67
    density, 76
    excess supporting, 69
electrolytic cell, 11, 24
electron volt, 359
electroneutrality, 66
electroorganic synthesis, 339–340
    direct, 339
    indirect, 339
electroosmotic drag, 207
electroplating, 299
electropolishing, 90
electrorefining, 68, 339
electrowinning, 338
endplate, 230
energy
    bands, 355–357
    density, 163
    efficiency, 167
    gap, 355, 366
    scales, 359
engine map, 288
enthalpy of formation, 21, 224–225
equilibrium
    constant, 25
    dynamic, 15
equivalent, 6
    circuit, 118
    diameter, 89
    ionic conductance, 72–73
equivalent distributed resistance
    (EDR), 264
equivalent series resistance (ESR),
    266
Euler's formula, 130
Evans diagram, 383–384
excess supporting electrolyte, 69
exchange current density, 45
    concentration dependence, 47
    temperature dependence, 47

fanning friction factor, 236
faradaic
    current, 254
    efficiency, 8
    reaction, 8
Faraday law, 6, 7
Faraday's constant, 6
Fermi energy level, 359–360
ferricyanide-ferrocyanide, 47–49, 126
Fick's law, 63
fill factor, 372
flat-band potential, 362
flooded agglomerate model, 103–104,
    205–206
flooding, 209

flow-by electrode, 105
flow-through electrode, 105–107
flowfield, 235
flux
    molar, 9
foil gain, 252
forced convection, 75
formation reaction, 153
forward bias, 365
fracture, 244
free convection, 77
front of electrode, 108
fuel
    processing, 223
    starvation, 236
fuel cell
    direct methanol, 195
    flowfield design, 235–237
    hybrid vehicle, 291–293
    PEM, 206
    polarization curve, 198
    solid oxide, 211–215
fugacity coefficient, 33
full hybrid, 288
full-cell reactions, 3

gallery space, 156
galvanic cell, 11, 24
galvanic intermittent titration technique
    (GITT), 146
galvanostatic, 11, 113
Gartner equation, 369
gas diffusion layer (GDL), 230
gas effect parameter, 329
gas evolution, 77–79, 328–330
gas holdup, 79
Gauss's law, 32
Gibbs energy
    and cell potential, 17–18
    and maximum work, 17
    and standard potential, 19–20
Gibbs energy of formation, 20, 224
Grashof number, 76
Grotthuss mechanism, 207
growth of nuclei
    during electrodeposition, 308–310

half-cell, 1
half-cell reactions, 1
Hall-Heroult process, 338, 342
Haring-Blum cell, 313
heat conduction, 188–189
heat generation, 165
Helmholtz inner plane, 41
Helmholtz outer plane, 41
hemispherical electrode, 141
Henderson equation, 27
Henry's law, 87
$HgSO_4$ reference electrode, 39

higher heating value (HHV), 224
hole, 355
homogeneous reaction, 65
Hooke's law, 408
Hull cell, 90, 314
hydraulic diameter, 76
hydrogen electrode, 28

ideally polarizable electrode, 254
impedance
    definition, 131
    mass transfer, 134
    of simplified Randles circuit, 131
Impressed current cathodic protection
    ICCP, 397–399
impurities, 167, 271, 299
indirect band gap, 367
industrial electrolysis, 323
infinite dilution, 73
inner Helmholtz plane (IHP), 41
inner-sphere reaction, 49
insertion reaction, 155
insulator, 85, 251, 355–356
intercalation, 155
interconnect, 215
interlayer, 214
ionic strength, 35
IR compensation, 139, 141
IR correction, 141
IR drop, 141
irreversible
    losses, 165
    reactions, 125–126

Joule heating, 165
junction, 365

kinetics
    linear, 52
    Tafel, 49
Koutecky Levich plot, 138

Laplace's equation, 66–67
lead-acid battery, 24, 154, 155, 159,
    164, 166, 168, 287
leakage current, 270
levelers, 311
Levich plot, 138
light absorption, 366
limiting current, 55
linear kinetics, 52
linear sweep voltammetry, 122
liquid junction, 27
lithium-ion cell and battery, 73–74, 154,
    156, 159, 161, 162–163, 168,
    190
lower heating value (LHV), 224
Luggin capillary, 140

majority carrier, 357
Marcus theory, 49
mass activity of catalyst, 210
mass transfer, 54, 63
   and concentration overpotential, 80
   and supporting electrolyte,
      69–70
   battery electrodes, 74, 155
   coefficient, 75–79
   correlation, 75–79
   fuel-cell electrodes, 201
   limited porous electrodes, 105–106
membrane electrode assembly, 243
membrane transport, 86
mercuric oxide electrode, 29
microelectrodes, 141
migration, 63
minority carrier, 357
mixed ionic electronic conductor
      (MIEC), 205
mixed potential, 199
   corrosion potential as example,
      383–385
mobility, 71
module, 176
modulus of elasticity, 244–245, 408
monopolar design, 228
Mott-Schottky
   equation, 362
   plot, 362

n-type dopant, 357
Nafion, 206
negative electrode, 151
Nernst Einstein relation, 71
Nernst equation, 22
Nernst-Planck equation, 63
nominal voltage, 160
nonfaradaic current, 119, 254
normal hydrogen electrode (NHE), 28
nucleation, 303
   2D, 304
   3D, 303
   instantaneous, 308
   progressive, 308
nuclei growth, 308
Nyquist plot, 132

Ohm's law, 9
ohmic
   correction, 57, 140–141
   drop, 12, 56, 115
   resistance, 65
open-circuit, 2, 15
   potential, 12, 15
osmotic coefficient, 33
outer Helmholtz plane (OHP), 41
outer-sphere reaction, 49

overpotential, 46, 80
   concentration or mass transfer, 79
   kinetic or surface, 46
oxidation, 1
oxygen electrode, 200
oxygen vacancy conduction, 212

p-type dopant, 357
parallel hybrid, 279
partition coefficient, 87
passive region, 383, 394–395
penetration depth, 100–101
   EDLC, 262
permeability, 87
permittivity, 32, 251
Peukert equation, 162
pH, 26
photoelectrochemical cell, 370–375
   efficiency, 372
   fill factor, 372
   photoelectrolytic, 373
   regenerative, 373
   tandem cell, 374
planform area, 229
plug-in hybrid electric vehicle
      (PHEV), 291
point of zero charge, 255
Poisson's equation, 32
Poisson's ratio, 244, 409
polarization, 12, 80
polarization curve, 12, 210
pore size distribution, 97
porosity, 96
porous electrode, 94
   current distribution, 98
   resistance, 101
   Tafel kinetics, 109
   with fluid flow, 105
potential
   drop, 141
   effect of temperature, 22
   Galvani, 30
   standard, 18
   thermodynamic, 16
potentiostat, 115
potentiostatic, 12, 113
Pourbaix diagram, 25, 382
power density, 163
primary current distribution, 83
prismatic cell, 183
proton exchange membrane
      (PEM), 206
pseudocapacitance, 271
pulse charging, 185

quantum yield, 369
quasi-reversible
   reaction, 125–126

Ragone plot, 163–164
Randles equivalent circuit, 118, 163
reaction
   irreversible, 126
   quasi-reversible, 126
   reversible, 126
reaction order, 45
reaction zone, 205
rechargeable energy storage system
      (RESS), 277
reconstruction reaction, 153
recycle ratio, 234
recycle streams, 234
redox reaction, 44, 122
redox-flow battery, 348–349
reduction, 1
reference electrode, 27, 113
   Ag/AgCl, 29
   Ag/Ag$_2$SO$_4$, 29
   calomel, 28
   Hg/HgO, 29
   hydrogen, 28
   use of, 139–141
reference state
   activity and fugacity, 33–34
   infinite dilution, 34
reformation, 195, 223
regenerative braking, 281
relative permittivity, 252
reorganization energy, 49
reserve battery, 153, 154, 167
resistance
   cell, 121
   internal, 161–162, 176–178
   kinetic, charge transfer, 82
   ohmic, 65
resistive substrate, 316
resistivity, doped semiconductors, 358
reversible
   reactions, 16, 27, 126
   system, 16
   work, 16, 35
Reynolds number, 76
rotating disk electrode (RDE), 136–139
rotating ring disk electrode, 136
run-time, 163, 283

sacrificial anode, 396
salt bridge, 151
Sand equation, 117
saturation level, 103
scale up, 182, 282, 337
Schmidt number, 75
sealing, 242
secondary current distribution, 83
self-discharge, 153, 167
semiconductor, 355–356
separator plate, 230

series hybrid, 285
Shepherd equation, 162
Sherwood number, 75–79
shuttle mechanism
  for self discharge, 167
side reaction, 8
silver sulfate reference electrode, 29
silver zinc battery, 152, 154
silver/silver chloride electrode, 29
solid oxide fuel cell (SOFC), 211–215
solubility product, 25
space-charge voltage, 360
specific capacitance, 270
specific energy, 163
specific fuel consumption (SFC), 288
specific interfacial area, 96, 108
specific power, 163
spiral growth, 302
spirally wound cell, 183
standard hydrogen electrode, 18
standard potential, 18
  activity correction, 33
  effect of temperature, 21
  simplified concentration effect, 22
standard state, 16, 34
start-stop hybrid, 284, 285
starting lighting ignition (SLI), 181
state of charge (SOC), 157
state of discharge (SOD), 157–158
state of health, 186
stoichiometric coefficient, 7
strain, 408
stress, 408
  relaxation, 245
stripping analysis, 127

supercapacitor, 258
superficial area, 108
supporting electrolyte, 69
surface overpotential, 46
surface roughness, 46
surface tension, 102
switching potential, 122
symmetric electrolyte, 255
symmetry factor, 45
system of units, 5

tabs, current collection, 180
Tafel
  approximation, 50
  kinetics, 50–51
  plot, 50, 200
  slope, 50, 200
tertiary current distribution, 83
theoretical capacity, battery, 157
thermal conductivity, 188
thermal expansion, 244
thermal management, battery, 188–189
thermoneutral potential, 165
three-electrode setup, 114
three-phase boundary, 103, 205
throwing power, 313
tie rod, 243
time constant
  diffusion, 74–75, 120
  double layer charging, 118, 120
tortuosity, 97
transfer coefficient, 46
transference number, 72
transition time, 118
transpassive region, 383

transport problem types, 66
transport properties, 71–72
  effective, 98
triple-phase boundary, 205
tubular SOFC design, 215

ultracapacitor, 258
utilization, 231
  fuel, 233
    system, 234
  oxidant, 231
  sweep, 232

vacancy, 212
vacancy diffusion, 212
valence band, 356–357
vehicle dynamics, 295
vehicle mechanism of conduction, 207
vehicle model, 279
void volume fraction, 96
Volmer-Tafel mechanism, 49
voltage efficiency, 167

Wagner number, 83, 313, 317
Warburg impedance, 134
waste water treatment, 106
water balance, fuel cell, 238
water electrolysis, 343–345
wave washers, 412
working electrode, 113

yield point, 408
Young's modulus, 408
yttria stabilized zirconia (YSZ), 211